GLYCOCHEMICAL SYNTHESIS

GLYCOCHEMICAL SYNTHESIS

Strategies and Applications

Edited by

SHANG-CHENG HUNG
MEDEL MANUEL L. ZULUETA

WILEY

Published by John Wiley & Sons, Inc., Hoboken, New Jersey
Published simultaneously in Canada

Library of Congress Cataloguing-in-Publication Data:

Names: Hung, Shang-Cheng, editor. | Zulueta, Medel Manuel L., editor.
Title: Glycochemical synthesis : strategies and applications / edited by Shang-Cheng Hung, Medel Manuel L. Zulueta.
Description: Hoboken, New Jersey : John Wiley & Sons, Inc., [2016] | Includes bibliographical references and index.
Identifiers: LCCN 2016023028 | ISBN 9781118299845 (cloth)
Subjects: | MESH: Glycoconjugates–chemical synthesis | Glycosylation | Glycomics | Drug Discovery
Classification: LCC QP702.G577 | NLM QU 75 | DDC 572/.567–dc23
LC record available at https://lccn.loc.gov/2016023028

Set in 10/12pt Times by SPi Global, Pondicherry, India

10 9 8 7 6 5 4 3 2 1

CONTENTS

CONTRIBUTORS

Hajjaj H. M. Abdu-Allah, Department of Pharmaceutical Organic Chemistry, Faculty of Pharmacy, Assiut University, Assiut, Egypt

Avijit Kumar Adak, Department of Chemistry, National Tsing Hua University, Hsinchu, Taiwan

Thomas Jan Boltje, Cluster for Molecular Chemistry, Institute for Molecules and Materials, Radboud University Nijmegen, Nijmegen, The Netherlands

Geert-Jan Boons, Complex Carbohydrate Research Center, University of Georgia, Athens, GA, USA

Kasemsiri Chandarajoti, Department of Pharmaceutical Chemistry, Faculty of Pharmaceutical Sciences, Prince of Songkla University, Hat Yai, Thailand

Shih-Huang Chang, Genomics Research Center, Academia Sinica, Taipei, Taiwan

David Crich, Department of Chemistry, Wayne State University, Detroit, MI, USA

Benjamin G. Davis, Department of Chemistry, Chemistry Research Laboratory, University of Oxford, Oxford, UK

Yukari Fujimoto, Graduate School of Science and Technology, Keio University, Kanagawa, Japan

Koichi Fukase, Department of Chemistry, Graduate School of Science, Osaka University, Osaka, Japan

Jacquelyn Gervay-Hague, Department of Chemistry, University of California, Davis, Davis, CA, USA

Zhongwu Guo, Department of Chemistry, Wayne State University, Detroit, MI, USA

Yu-Peng Hu, Genomics Research Center, Academia Sinica, Taipei, Taiwan

Xuefei Huang, Department of Chemistry, Michigan State University, East Lansing, MI, USA

Shang-Cheng Hung, Genomics Research Center, Academia Sinica, Taipei, Taiwan

Mattan Hurevich, Institute of Chemistry, The Hebrew University of Jerusalem, Jerusalem, Israel

Hideharu Ishida, Department of Applied Bioorganic Chemistry, Gifu University, Gifu, Japan

Akihiro Ishiwata, Synthetic Cellular Chemistry Laboratory, RIKEN, Saitama, Japan

Yukishige Ito, Synthetic Cellular Chemistry Laboratory, RIKEN, Saitama, Japan

Charles Johnson, Department of Chemistry, Wayne State University, Detroit, MI, USA

Yasuhiro Kajihara, Department of Chemistry, Graduate School of Science, Osaka University, Osaka, Japan

Jeyakumar Kandasamy, Department of Chemistry, Indian Institute of Technology (BHU)–Varanasi, Varanasi, India

Makoto Kiso, Department of Applied Bioorganic Chemistry, Gifu University, Gifu, Japan

Suvarn S. Kulkarni, Department of Chemistry, Indian Institute of Technology–Bombay, Mumbai, India

Chun-Cheng Lin, Department of Chemistry, National Tsing Hua University, Hsinchu, Taiwan

Shu-Yi Lin, Genomics Research Center, Academia Sinica, Taipei, Taiwan

Jian Liu, Division of Chemical Biology and Medicinal Chemistry, Eshelman School of Pharmacy, University of North Carolina at Chapel Hill, Chapel Hill, NC, USA

Lin Liu, Complex Carbohydrate Research Center, University of Georgia, Athens, GA, USA

Weigang Lu, State Key Laboratory of Natural and Biomimetic Drugs, School of Pharmaceutical Sciences, Peking University, Beijing, China and Complex Carbohydrate Research Center, University of Georgia, Athens, GA, USA

Yoshiyuki Manabe, Department of Chemistry, Graduate School of Science, Osaka University, Osaka, Japan

Masumi Murakami, Department of Chemistry, Graduate School of Science, Osaka University, Osaka, Japan

Chandrasekhar Navuluri, Department of Chemistry, Wayne State University, Detroit, MI, USA

Mauro Pascolutti, Institute for Glycomics, Griffith University, Gold Coast, Queensland, Australia

Macarena Sánchez-Navarro, Institute for Research in Biomedicine, The Barcelona Institute of Science and Technology, Barcelona, Spain and Department of Chemistry, Chemistry Research Laboratory, University of Oxford, Oxford, UK

Matthew Schombs, Department of Chemistry, University of California, Davis, Davis, CA, USA

Peter H. Seeberger, Department of Biomolecular Systems, Max Planck Institute of Colloids and Interfaces, Potsdam, Germany

Atsushi Shimoyama, Department of Chemistry, Graduate School of Science, Osaka University, Osaka, Japan

Katsunori Tanaka, Biofunctional Synthetic Chemistry Laboratory, RIKEN, Saitama, Japan

Carlo Unverzagt, Bioorganische Chemie, Universität Bayreuth, Bayreuth, Germany

Mark von Itzstein, Institute for Glycomics, Griffith University, Gold Coast, Queensland, Australia

Cheng-Chung Wang, Institute of Chemistry, Academia Sinica, Academia Road, Nangang, Taipei, Taiwan

Chung-Yi Wu, Genomics Research Center, Academia Sinica, Taipei, Taiwan

Bo Yang, Department of Chemistry, Michigan State University, East Lansing, MI, USA

Xin-Shan Ye, State Key Laboratory of Natural and Biomimetic Drugs, School of Pharmaceutical Sciences, Peking University, Beijing, China

Keisuke Yoshida, Department of Chemistry, Michigan State University, East Lansing, MI, USA

Ching-Ching Yu, Department of Chemistry, National Tsing Hua University, Hsinchu, Taiwan

Medel Manuel L. Zulueta, Genomics Research Center, Academia Sinica, Taipei, Taiwan

FOREWORD

The importance of carbohydrates in biological systems and, particularly, in cell–cell interaction has been witnessed by a great explosion of new knowledge that received attention and recognition. Carbohydrates found in nature are generally constituents of glycosides, oligosaccharides, and polysaccharides or of quite complex glycoconjugates. The sugar units are attached to one another or to aglycones commonly through O- or N-glycosidic linkages. As most of these compounds are not readily accessible by isolation from natural sources, tremendous efforts have been undertaken to develop efficient procedures for their chemical synthesis, whereby the main focus was and still is on the generation of glycosidic linkages. Hence, a compilation of recent developments and future directions of this field is of great significance.

Chemical glycosylation usually involves the coupling of a fully protected glycosyl donor with a leaving group at the anomeric center and a suitably protected glycosyl acceptor generally containing only one free hydroxy group. Hence, protecting group strategies are essential in the design and success of glycosidic bond formation. Additionally, the selection of the leaving group and its activation in the glycosylation step through promoters are equally essential aspects. As different leaving group types have been profiled, a general discussion of their power and eventual weaknesses is needed. Together with solvent effects, these factors are generally decisive for efficient intermolecular glycosylation reactions.

A dream in the synthesis of complex glycans is the self-controlled chain extension of the individual building blocks. This requires, in addition to the efficiency in each glycosylation step, subtle differences in the reactivities. Hence, the building block design has to provide, through variation of the protecting group array, leaving groups, and manner of activation, reactivity differences that permit successful one-pot assembly of oligosaccharides. Alternatively, automated oligosaccharide synthesis on solid phase could be employed. As glycosylation efficiencies are often inferior to those reached in peptide and nucleotide synthesis, tremendous efforts went into this ambitious project. Moreover, sophisticated approaches as microfluidic techniques were probed. In this way, the desirable optimization of the reaction conditions for efficient glycosidic bond formation is more readily possible. Because many factors influencing carbohydrate synthesis are optimized by nature through enzyme-mediated transformations, it is tempting to combine chemistry and biology to perform sugar acquisition by chemoenzymatic means. Though access and stability of the enzymes are still limited, a documentation of the increasing importance of the chemoenzymatic approach is highly desirable.

The successful synthesis of complex glycoconjugates found in biological systems poses often specific questions. Among them, the glycosaminoglycans display a large variety of vital biological functions often associated with their variably polyanionic chain. Interest in their biological and therapeutic applications stimulated the synthesis of structurally defined partial structures. Posttranslational glycosylation of proteins, forming glycoproteins, impacts the physicochemical properties of the biomolecule, and the attached glycans are often involved in the cellular recognition phenomena, influencing many biological events including pathogenic disorders. Hence, the successful construction of these complex structures is not only an intellectual satisfaction but also an important issue. Like glycoproteins, glycosphingolipids affect various cellular processes due to the diversity in their oligosaccharides. The synthetic design considers not only the amphiphilic character of these molecules but also the general presence of double bonds and the linkage to the lipophilic ceramide aglycone. Similarly, glycosylphosphatidylinositol (GPI) anchors are found in eukaryotes and protozoan parasites where they usually ligate plasma-membrane proteins to the cell surface. The synthetic strategies for GPIs require a combination of lipid, phosphate, glycan, and peptide/protein chemistries. As bacterial envelope glycans often contain unique monosaccharide constituents, they are generally more complex than those of higher organisms. Additionally, the envelope glycans are often immunogenic; thus, the synthesis of bacterial cell envelope components has gained great interest for the study and cure of bacterial infections and for vaccination studies.

The multitude of carbohydrate epitope–receptor recognition events, chemically based on the high functionality and the stereochemical bias of these compounds, requires fast methods in order to decipher the information stored in the carbohydrate structures. This can be achieved by carbohydrate arrays that consider such huge structural diversity. Hence, a discussion of the state of the art on discoveries and applications of glycan arrays is valuable. Many physiological and pathological processes are mediated by multivalent protein–carbohydrate and carbohydrate–carbohydrate interactions. A study of multivalency in this context is an important task that can be based on the generation of glyconanoparticles, glycodendrimers, and glycoclusters of defined structures. Thus, information, for instance, on geometrical positioning of the receptors as well as on the strength of individual and multimeric epitope–receptor binding is provided.

The importance of small molecule carbohydrates and carbohydrate mimetics as drugs has been questioned by pharmaceutical chemists. With the successful development of very efficient neuraminic acid-derived anti-influenza agents that are recognized by influenza virus lectins, such statements were proven wrong. Recently, also a lot of attention was directed toward sialic acid-binding immunoglobulin-like lectins (siglecs). Their occurrence, important biological role, and ligands were discovered, thus enabling the design and synthesis of ligands and antagonists of siglecs as immune response modifiers. But not only small carbohydrate-based molecules but also glycoproteins have become of great importance as therapeutics. As they are not readily accessible containing the human glycan pattern, sugar–protein hybrids are a valuable alternative. Hence, nonnatural but highly efficient methods were employed to link sugars to proteins, providing glycoprotein mimetics with minor structural modifications for biomedical studies.

This book, written by highly reputed experts in the field, will be a great source of inspiration for many scientists. Hence, it will greatly advance the further development of this exciting research area that attracted my attention for many years.

Richard R. Schmidt
Emeritus Professor
University of Konstanz, Germany

PREFACE

For over a century, much progress in glycochemistry accumulated over the foundations laid upon by that scientific edifice, Emil Fischer. Since then, our understanding has certainly grown, and our techniques have steadily advanced in great proportions. The realization that glycans are important players in biological systems beyond their traditional roles as energy sources and structural support molecules instilled an astonishing amount of interest in defining their functions. Glycans are now known to impart widespread influence over important biological processes from fertilization to development to cancer progression and from bacterial and viral infection to immune response. Not surprisingly, they have become inspirations for the design of several pharmaceuticals and vaccines.

That sugars carry dense structural information, by virtue of the multiple chiral centers and the numerous ways the glycosidic bond can be formed, has been deemed critical in many recognition, adhesion, and signal transduction events. Another aspect has, however, kept studies at bay: glycan formation is not directly template driven, leading to an abundance of glycoforms that made isolation of homogeneous materials such an arduous procedure. Thus, the synthesis of sugars in their many forms and sizes became a necessary prerequisite of structure–activity relationship evaluations as well as in the pursuit of pharmaceutical and industrial applications.

The rapid explosion of techniques, especially in the last few decades, provided carbohydrate chemists of today with many tools in their disposal to form their desired glycan structures. Still, the main challenge remains to be the efficient assembly of glycans in a regioselective and stereoselective manner. As described in this book, the reliance on conformational effects, anomeric effect, solvent effect, neighboring group participation, and more recent techniques such as intramolecular aglycone delivery have proven valuable in effective glycosidic bond formation. Within this framework, novel synthetic designs were developed to form the sugar backbones in solution in a single vessel, by automation in solid phase, or in microfluidic environments and with enzymatic transformations as an aid to assembly and backbone modifications. These strategies have been applied to address the complex combinations of the nine common monosaccharide building blocks found in mammalian cells apart from the unusual ones displayed by other organisms. The available tools that pushed glycochemistry to remarkable heights enabled the preparation of glycoconjugates, hybrid carbohydrate–protein vaccines, multivalent sugar assemblies, glycan arrays, and many others.

This book covers the fundamentals and the exciting developments in the fields around glycochemical synthesis. It could serve as a concise resource material about the methods and approaches toward creating glycan structures and its conjugates. Glycochemical synthesis is loosely divided into strategies and applications in this book. Starting from the basic aspects of glycochemistry, the volume expands into specific solutions to challenges unique to oligosaccharide assembly, particularly in protecting group selection and regioselective and stereoselective glycosylations. Clever strategies to the synthesis of sugars, techniques in the preparation of (neo) glycoconjugates, and the contributions of carbohydrate synthesis to understanding structure–activity relationships and drug development are highlighted. All 19 chapters are written by experts and are meant to convey the knowledge and expertise that have made glycochemistry quite an exciting field of study. Selected illustrative examples are described all throughout the book to guide the readers in understanding the concepts and strategies. Citations to primary literature are provided. While we acknowledge that this book may not include every notable strategy and application related to glycochemical synthesis, we offer with confidence a very useful reference for advanced undergraduate and graduate students with keen interests in glycochemistry and for academic and industrial scientists focused on carbohydrate research.

Shang-Cheng Hung
Medel Manuel L. Zulueta

1

GLYCOCHEMISTRY: OVERVIEW AND PROGRESS

MATTHEW SCHOMBS AND JACQUELYN GERVAY-HAGUE

Department of Chemistry, University of California, Davis, Davis, CA, USA

1.1 INTRODUCTION

Officially, the International Union of Pure and Applied Chemistry (IUPAC) defines glycan as "synonymous with polysaccharides," meaning compounds consisting of a large number of monosaccharides linked to each other through glycosidic bonds [1]. Practically however, the term glycan is all encompassing and often used to describe the carbohydrate portion of glycoconjugates such as glycoproteins and glycolipids. Carbohydrates are the most abundant organic molecules on Earth and are the main products through which the energy of the sun is harnessed and stored. Glucose polysaccharides, such as starch in plants and glycogen in bacteria and animals, serve as a source of energy for essentially all organisms. However, the complex roles of carbohydrates are not limited to simply that of biological fuel stocks or biosynthetic starting materials. DNA and RNA, which transmit and store genetic information, have sugar backbones. Other carbohydrate polymers are essential structural and protective components of the cell walls of plants as cellulose, bacteria as peptidoglycan, and the exoskeletons of arthropods as chitin. They are important constituents of secreted and cell-surface proteins, membrane components in the form of glycolipids and gangliosides, as well as various types of extracellular matrix molecules [2]. The significance of the carbohydrate domains of glycoproteins and glycolipids is further exhibited in their roles as cell-surface recognition elements and as determinants in blood-group typing [3, 4]. Carbohydrates are also appended to various natural products including antibiotics [5]. As such, glycans mediate a wide range of biological processes from

Glycochemical Synthesis: Strategies and Applications, First Edition.
Edited by Shang-Cheng Hung and Medel Manuel L. Zulueta.

embryonic development to differentiation, signaling, host–pathogen interactions, metastasis, intracellular trafficking, and localization [6].

The many hydroxy groups that adorn the carbon backbone of glycans give rise to multiple stereoisomers, a fact that has been capitalized on for use as chiral synthons in organic synthesis [7]. The nine common monosaccharides found in mammalian cells can be linked in an astonishing number of ways, resulting in much higher complexity than is possible from amino acid or nucleotide building blocks. Unlike proteins and DNA, glycans encode immense biological information without being template driven or encoded by the genome. The first draft of the human genome revealed a relatively small number of genes associated with the human species—approximately 32,000—as compared to less complex organisms such as fly or worm, which encompasses roughly 13,000 or 18,000 genes, respectively [8–10]. While the origin of biological complexity remains a largely debated topic, one hypothesis accounting for this paradox is the posttranslational modifications of proteins.

Glycosylation is one of the most ubiquitous forms of posttranslational modification and is widely recognized as a modulator of protein structure, localization, and function. Because glycosylation is not under tight genetic control, often complex and unpredictable mixtures of glycoforms with varying properties are produced [11, 12]. Therefore, access to homogeneous glycolipids, glycopeptides, and glycoproteins is an essential step toward furthering our understanding of these important molecules. Over the past century, significant developments have occurred, from the establishment of a carbohydrate nomenclature to discovering the simple building blocks that make up oligosaccharides and how they combine to create unique structures. These advances have enabled studies that reveal the multifaceted roles of glycans.

1.2 NOMENCLATURE, STRUCTURES, AND PROPERTIES OF SUGARS

Most simple sugars have the general formula $C_n(H_2O)_n$, where n is between three and nine. Early nineteenth-century French chemists generically defined carbohydrates as "hydrates de carbone" because they were thought to consist solely of carbon and water in a 1:1 ratio. However, the term is used today in a much broader sense. Saccharides can be roughly split into two categories: monosaccharides and complex saccharides such as oligosaccharides and polysaccharides. Depending on their size, oligosaccharides and polysaccharides tend to exhibit different chemical and physical properties as compared to monosaccharides. Polysaccharides can form stable secondary and tertiary structures and are hydrolyzed into smaller subunits upon treatment with aqueous acid, while monosaccharides can be found in a variety of forms including linear and cyclic structures. Monosaccharides are the building blocks from which oligosaccharides and polysaccharides are constructed. They include polyhydroxyaldehydes (aldoses) and polyhydroxyketones (ketoses) as well as the resulting compounds derived thereof by either the reaction of the carbonyl group, via oxidation to form carboxylic acids, or by replacing one or more hydroxy groups with hydrogen, amino, acetamide, thiol, or other functional groups (Fig. 1.1).

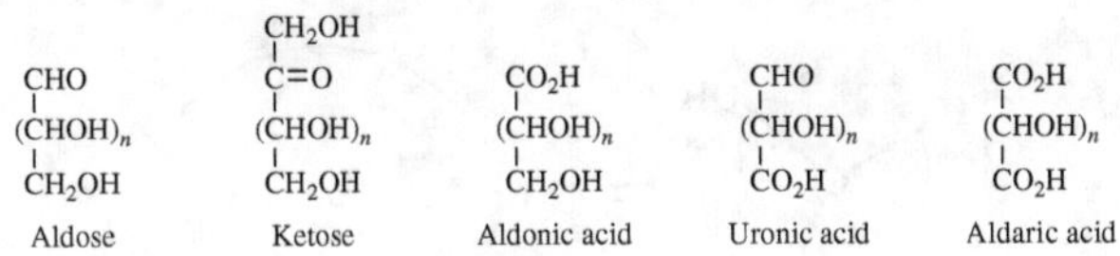

FIGURE 1.1 Common carbohydrate oxidation levels.

Monosaccharides are classified according to the number of carbons in their skeleton per IUPAC recommendations [13]. The suffix -ose is used to indicate an aldose, while -ulose denotes a ketose. Accordingly, the common aldoses in ascending order would be trioses, tetroses, pentoses, hexoses, etc. Prior to their structures being known and the subsequent systematization developed by Emil Fischer, carbohydrates were named based on either their sources (fructose for fruit sugar, glucose for grape sugar, lactose for milk sugar, and sucrose for cane sugar) or physical properties (dextrose for glucose because it rotates plane-polarized light in a clockwise manner (dextrorotation) and levulose for fructose because of its levorotatory nature). Note that each secondary carbon of the sugar alcohols is sp^3 hybridized and represents a stereogenic or chiral center. A uniform method to visualize this tetrahedral geometry in two dimensions came in the form of the Fischer projection. While the Fischer proof is discussed later, this work largely eliminated inconsistencies in the representation and naming of sugars.

1.2.1 Fischer Projection

The Fischer projection is a convenient way of showing the configurations of the linear forms of monosaccharides. This convention depicts the concepts of stereochemistry established by Jacobus Henricus van 't Hoff and Joseph Achille Le Bel in a simplified form. While these abbreviated structural formulas are simple to write and easy to visualize, there are some guidelines that should be taken into account when converting a three-dimensional structure into a Fischer projection and in its manipulation (Fig. 1.2):

1. Orient the molecule in such a way that the chiral center is in the plane of the paper with its vertical bonds toward the back and the horizontal bonds coming out in front.
2. Position the carbon atoms of the chain on the vertical plane with the carbonyl group on top and the primary alcohol at the bottom. The hydrogen and hydroxy moieties should be oriented horizontally. Numbering of the carbon atoms begins with the carbonyl group in the case of aldoses or the terminal carbon closest to the carbonyl group in the case of ketoses.
3. Flatten the resulting model by "pulling" the vertical bonds toward the plane of the paper and "pushing" the horizontal bonds into the plane of the paper. The stereogenic carbon and the attached hydrogen atoms can then be omitted for clarity.

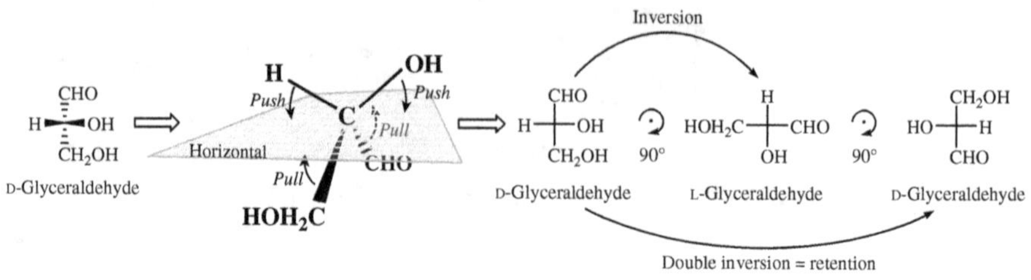

FIGURE 1.2 Fischer projection of glyceraldehyde and its manipulation.

4. Define the most distant stereogenic center from the carbonyl group as the reference atom. The stereoisomer in which the highest priority substituent of the reference atom is pointed to the right is assigned the prefix "D" and to the left is assigned the prefix "L." This prefix designates the absolute configuration at the center of reference. A trivial name, on the other hand, is used to indicate the relative configuration of all other chiral centers in relation to the reference stereogenic center (Fig. 1.3). Consequently, the D- and L-isomers of a given trivial name are mirror images of each other.
5. Fischer projections must only be rotated in increments of 180° as a 90° rotation represents an inversion of configuration at the stereogenic center.

1.2.2 Linear Forms of Monosaccharides

Emil Fischer deduced the stereochemical relationship between monosaccharides using D-glyceraldehyde as the reference molecule. Ultimately, Fischer applied his proof to create the D-aldose family tree (Fig. 1.3), which is still in use to this day. The abbreviated names for aldopentoses and aldohexoses consist of the first three letters of their trivial names except only for "Glc," which is used for glucose ("Glu" had already been assigned to glutamic acid). The "D" (or "L") prefix in the abbreviated names may be omitted when referring to the more abundant isomer. Epimers are carbohydrates that differ only in the configuration at one stereocenter, a relationship that is readily apparent by comparing their Fischer projections. For example, glucose is the C2 epimer of mannose. The trivial names of aldoses may form configurational prefixes, such as *glycero*, *erythro*, *arabino*, *xylo*, *galacto*, *manno*, and *gluco*, in combination with the "D" or "L" notation to describe other sugars. These prefixes point to analogous, but not necessarily contiguous, sequences of chiral centers present in the molecule and may be combined to reflect the stereochemistries embedded in monosaccharides larger than hexoses [13]. Figure 1.4 shows the structures and the trivial and derived names for the D-ketoses as their Fischer projections.

While both the configuration and naming can also be assigned using the Cahn–Ingold–Prelog system, it is typically only used to describe attached chiral substituents as opposed to the stereochemistry of the sugars themselves. The main disadvantages to the application of the Cahn–Ingold–Prelog convention are the lengthy and complicated names that result and the fact that replacement of the terminal carbon may result in prefix changes to unchanged centers of analogues with the same configuration.

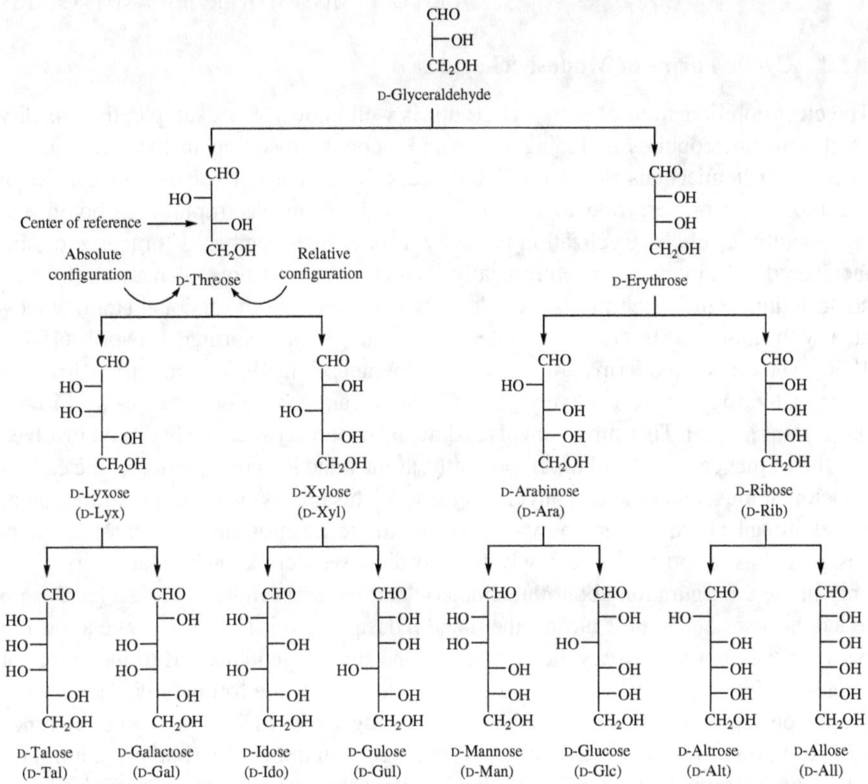

FIGURE 1.3 The family tree of D-aldoses with the trivial and abbreviated names.

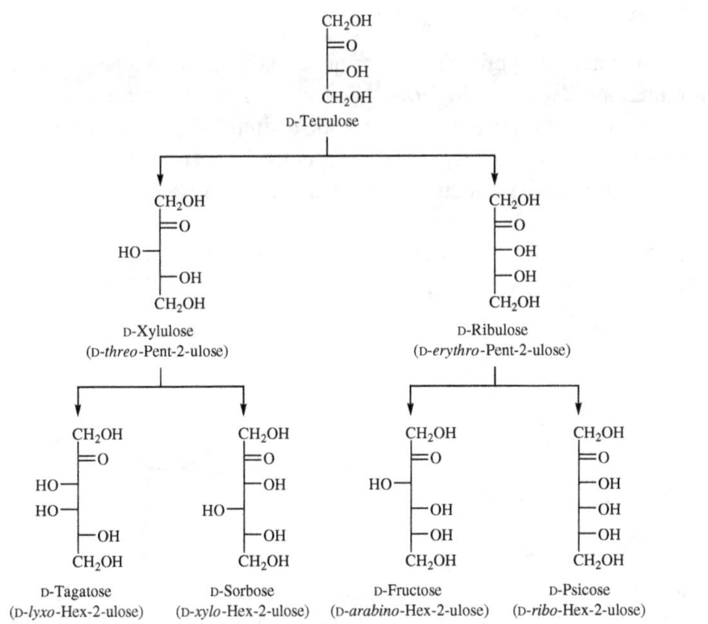

FIGURE 1.4 The family tree of D-ketoses with the trivial and derived names.

1.2.3 Cyclic Forms of Monosaccharides

The electrophilic nature of carbonyl groups is well known. For example, they readily react with nucleophiles including water and alcohols, resulting in the formation of hydrates or hemiacetals in a reversible process. In the case of sugars, this can be an intramolecular process due to the presence of both an electrophilic carbonyl and nucleophilic alcohols. Cyclization typically affords five-membered furanoses or six-membered pyranoses and is entropically favored over intermolecular attack. Further contributing to their stability is the relatively small amount of torsional strain associated with these constructs. As an interesting aside, it was Norman Haworth (1883–1950) who coined the terms "furanose" and "pyranose" in 1927 from tetrahydrofuran and tetrahydropyran, respectively [14]. Cyclic sugars can be depicted using Fischer projections as well. This simply involves drawing a loop between the hydroxyl involved and the former carbonyl carbon. Importantly, as the bond is a continuation of the carbon skeleton, it must enter the carbonyl from the top. This process leads to the formation of an additional chiral center known as the anomeric position and two diastereomers designated as α and β. These newly formed diastereomers known as anomers differ only in the configuration about the anomeric carbon. For simple monosaccharides up to aldohexoses and hept-2-uloses, the α and β designation is based on the relationship between the anomeric exocyclic substituent and the oxygen attached to the center of reference (Fig. 1.5). The anomer is α if these substituents are formally *cis* in a Fischer projection and β if they are formally *trans*. In larger monosaccharides, the reference atom used for the anomeric assignment is the highest-numbered carbon in a configurational prefix formed by the group of chiral centers closest to the anomeric carbon.

1.2.4 Haworth and Mills Projections

A major drawback of cyclic Fischer projections is the unrealistic manner in which the structures are depicted. In 1929, Haworth designed a representation to address this deficiency. Haworth projections provide a simple way to represent cyclic monosaccharides with a three-dimensional perspective. The following process allows the conversion of a Fischer projection into a Haworth representation:

1. Identify the hydroxy group that will be reacting with the carbonyl carbon. In the aldopyranose form, this is 5-OH.

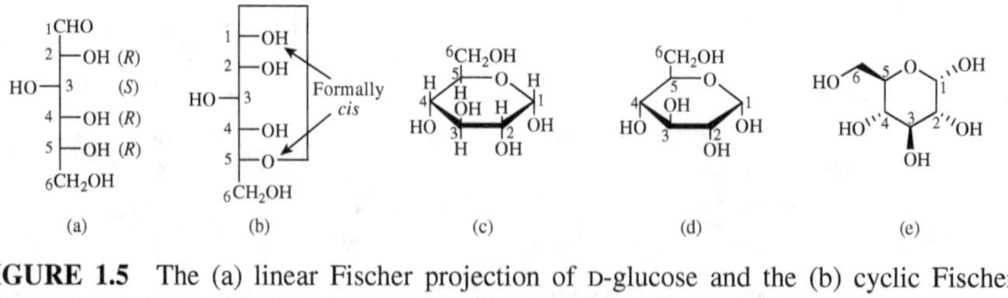

FIGURE 1.5 The (a) linear Fischer projection of D-glucose and the (b) cyclic Fischer, (c) Haworth, (d) simplified Haworth, and (e) Mills projections of α-D-glucopyranose.

2. Manipulate the Fischer projection such that this hydroxyl is at the bottom after exchange with the terminal (C6) functional group.
3. Draw the carbohydrate skeleton such that the ring is drawn on its side. The face closest to the viewer is drawn at the lower side and with a thicker line than the more distant upper side. The ring oxygen is located in the upper right-hand corner for pyranoses and at the top for furanoses.
4. Populate the ring substituents such that those on the right side of the Fischer projection are on the bottom face of the Haworth projection and those on the left side are on the top. Hydrogen atoms are typically omitted for clarity.

William Mills described a similar convention to depict the structures of monosaccharides. While the ring atoms of the Haworth projections are oriented perpendicular to the paper, Mills chose to depict the carbon skeleton in the plane of the paper (Fig. 1.5). Although Fischer, Haworth, and Mills projections are useful tools for depicting the structures of carbohydrates, the planar nature of these representations does not provide an accurate picture of the actual geometry of the molecules. In order to understand carbohydrate function and reactivity, recognition of each distinct conformation and the properties associated with it is required [15].

1.2.5 Reeves Projection

In 1949, Richard Reeves remodeled the Haworth projection by applying the ring conformations of cyclohexane to describe the structures of pyranoses in solution [16]. The Reeves convention is based on the similarity between the geometry of pyranoses to that of the model. Importantly, the assumption that the ring oxygen only introduced a slight conformational perturbation from that of cyclohexane was later confirmed by nuclear magnetic spectroscopy (NMR) spectroscopy. A major advantage of this convention is that it closely resembles the actual shape of the molecule, thereby allowing one to predict the distances and dihedral angles between the substituents. These values can be compared to those determined using the Karplus equation [17, 18] and applied to the interpretation of the NMR spectra. Taken together, it is a relatively trivial task to resolve the predominant averaged conformations of a monosaccharide. Figure 1.6 shows the Reeves projections for all furanose and pyranose structures associated with the α-D-pentoses and α-D-hexoses.

1.2.6 Conformational Analysis

Most molecules tend to favor one conformer over the others based on the stereochemistry of the particular monosaccharide and the steric bulk of the groups that are appended to it. For example, most aldohexoses prefer the chair conformation that places the bulky C5 hydroxymethyl group in the equatorial position. Having said that, the energy barrier between the two possible chair conformations is

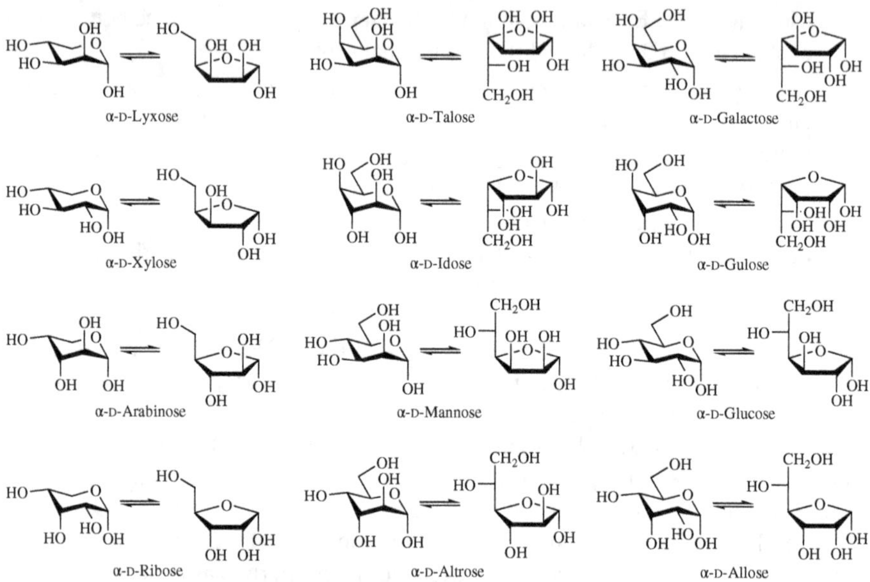

FIGURE 1.6 Reeves projections for α-D-pentoses and α-D-hexoses.

generally low enough to allow conformational flexibility and equilibrium to be established. Interconversion between the two chair conformations involves the rotation of ring atoms and bonds. This process requires the molecule to adopt several distinct conformations with respect to energy and the position of the ring atoms. The current convention for describing a particular conformation begins with assigning reference points. When naming conformations, the reference point above the plane is denoted as a superscript preceding the conformational descriptor and is followed by the one below the plane expressed as a subscript. The names and descriptors of the main conformations of pyranoses are boat (*B*), chair (*C*), envelope (*E*), half-chair (*H*), and skew (*S*). There are 2 discrete chair, 6 boat, 6 skew, 12 half-chair, and 12 envelope conformations (Fig. 1.7) [15]. The reference plane of the boat conformation consists of the two parallel sides of the boat. Of the remaining two out-of-plane atoms, one must be the lowest-numbered ring carbon. The same two parallel sides define the chair conformations as long as the lowest-numbered ring carbon resides above or below the plane. In the half-chair conformation, the reference plane is determined by four adjacent coplanar atoms, leaving the remaining two atoms on opposite sides of the plane. The reference plane of the envelope conformation includes the five adjacent coplanar atoms. The skew conformation contains two exoplanar atoms, one of which must be the lowest-numbered carbon atom. The reference plane is defined as the three adjacent atoms and the remaining coplanar nonadjacent one.

Furanose rings also exhibit a degree of conformational mobility, albeit to a lesser extent. The two predominant conformations adopted by these five-membered rings are envelope (*E*) and twist (*T*). There are 10 individual envelope

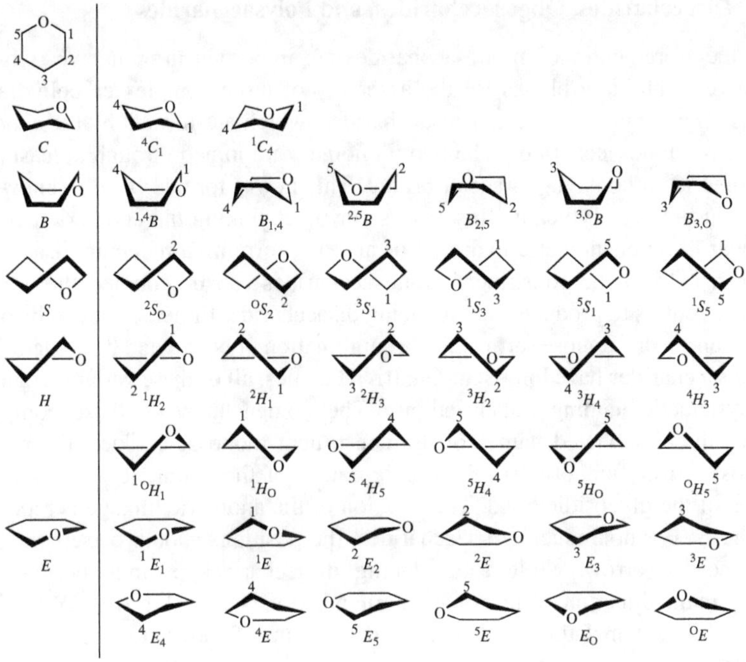

FIGURE 1.7 Pyranose ring nomenclature and conformations.

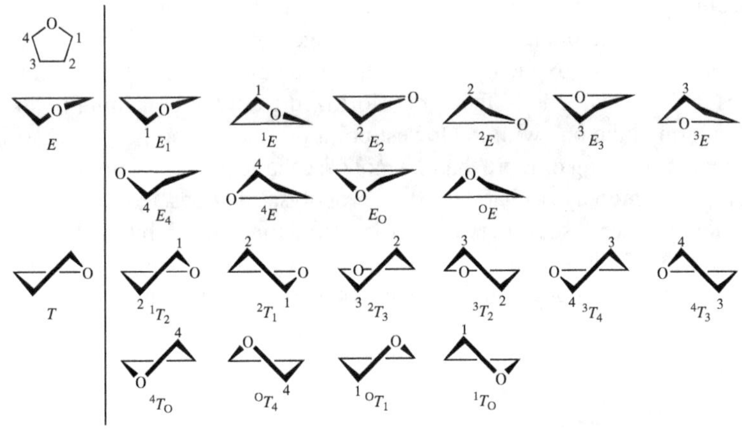

FIGURE 1.8 Furanose ring nomenclature and conformations.

and 10 twist conformations (Fig. 1.8) [15]. The reference plane of the envelope conformation is defined by the four adjacent coplanar atoms, with the remaining one either above or below this plane. For the twist, the reference plane is defined by three contiguous coplanar atoms, with the remaining two atoms placed on opposite sides of the plane.

1.2.7 Disaccharides, Oligosaccharides, and Polysaccharides

While the aforementioned monosaccharides are important in their own right, they also serve as building blocks for the assembly of more complex carbohydrates or glycans. The simplest of these is disaccharides, which are formed by the condensation of two monosaccharides. The two monomers are joined through at least one of the anomeric carbons via an acetal bridge. This newly formed bond is known as a glycosidic linkage. Typically, this process involves linking the anomeric carbon of one sugar to a nonanomeric hydroxyl of another, forming a reducing disaccharide. However, both of the constituent monosaccharides can be coupled through their anomeric centers to produce a nonreducing disaccharide. Either sugar can be present in its pyranose or furanose form and in a combination of both α and β anomers. While many disaccharides have long-standing trivial names, all of these variables can make their systematic naming rather tedious. The nomenclature of these compounds includes the abbreviated names of the constituent monosaccharides, the ring size (pyranose (*p*) or furanose (*f*)), the configuration of the anomeric centers, and the location of the glycosidic bond. The position of the anomeric linkage is reported in parenthesis. For nonreducing disaccharides, the positions should be separated by a double-headed arrow, while for reducing disaccharides a single-headed arrow pointing in the direction of the nonanomeric position is used (Fig. 1.9). A condensed form of this nomenclature is also allowed wherein the descriptors referring to the more abundant form of the sugar residue are omitted.

Even though the term "oligosaccharide" is not rigorously defined, it is generally used to describe complex carbohydrates composed of between 2 and 10 monosaccharide units. Oligosaccharides can be linear or branched in nature but are of distinct structure. If there are branches present, the longer arm is designated as the parent, and all connections are explicitly stated within square brackets. If two or more of the branches are of equal length, they are prioritized from the branching point. Thus, the parent arm is the one with the lowest point of attachment. Polysaccharides are biopolymers consisting of more than 10 monosaccharides. These complex structures are found as homopolysaccharides or heteropolysaccharides. Polysaccharides can assume highly ordered secondary and tertiary structures or exhibit random behavior. Their physical properties are determined largely by chain conformation, intra- and intermolecular interactions, and the solvent they are dissolved in.

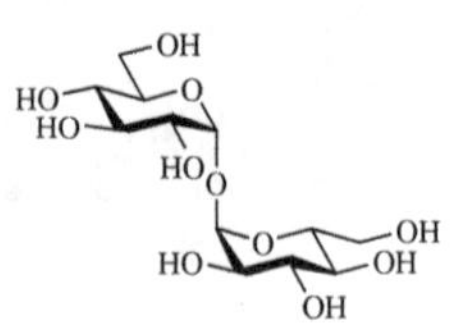

Trehalose

Extended form: α-D-Glc*p*-(1↔1)-α-D-Glc*p*

Condensed form: Glc(α1↔α1)Glc

Nonreducing disaccharide

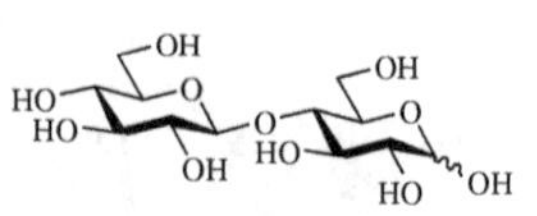

Cellobiose

Extended form: β-D-Glc*p*-(1→4)-D-Glc

Condensed form: Glc(β1→4)Glc

Reducing disaccharide

FIGURE 1.9 Nomenclatures and structures of selected disaccharides.

1.2.8 Anomeric Effect

It is widely known that equatorial substituents of six-membered cyclic hydrocarbons are less sterically demanding and energetically preferred as compared to their axial counterparts. This is largely due to unfavorable 1,3-diaxial interactions associated with axial substituents. However, in contrast to cyclohexane, pyranose rings contain an endocyclic oxygen atom (O5 in aldohexoses) adjacent to the anomeric center (C1). In 1955, John Edward identified an axial bias for the anomeric substituents of sugars [19]. Raymond Lemieux later defined this phenomenon as the "anomeric effect" [20]. Although there are a number of theories, this effect is largely attributed to a combination of hyperconjugative and electrostatic effects [21]. In both cases, the nonbonding electron pairs of O5 play a major role. This is especially apparent when there is an electronegative moiety at the anomeric position. The combination of the adjacent ring oxygen and an electronegative substituent X having nonbonding electrons (where X is defined as O, S, N, F, Cl, Br, or I) at the anomeric position renders C1 particularly electron deficient. When X is in the axial position, a lone pair of electrons from O5 is positioned antiperiplanar to the C1—X antibonding orbital. Delocalization of these electrons can stabilize the electron-deficient anomeric center through hyperconjugation. This stabilizing n–σ^* interaction is not possible when X is in an equatorial position (Fig. 1.10a) and also explains why the anomeric effect becomes more dominant as the electronegativity of X increases. Electrostatics also plays a role in the observed axial preference of electronegative anomeric substituents. While hyperconjugation is a stabilizing interaction, some electronic effects can be described in terms of alleviating unfavorable dipole–dipole interactions. When X is in an equatorial orientation, its exocyclic lone pairs exhibit a strong repulsive

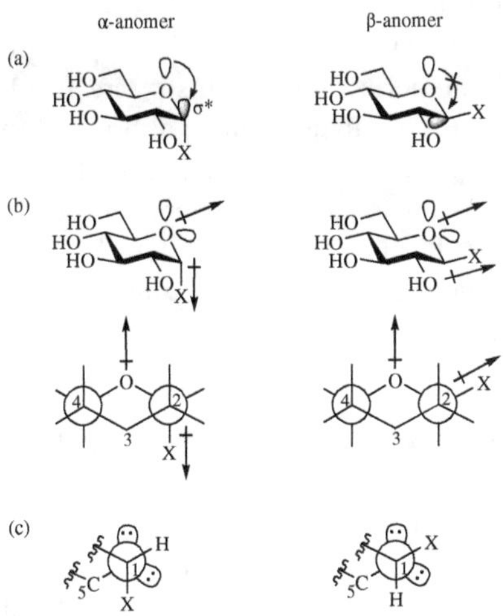

FIGURE 1.10 The anomeric effect. (a) The n–σ^* interaction stabilizes the α anomer. (b) The β anomer experiences unfavorable dipole–dipole interaction that is reduced in the α anomer. (c) Greater electrostatic repulsion between the lone-pair electrons of the endocyclic oxygen and the electronegative anomeric substituent in the β anomer.

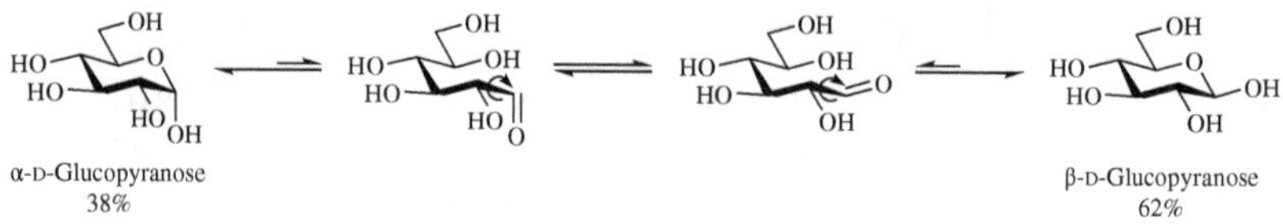

FIGURE 1.11 Mutarotation of D-glucose in water at pH 7.

electrostatic interaction with the O5 lone-pair electrons. These destabilizing interactions are drastically reduced when X is in the axial position (Fig. 1.10b). Another feature attributed to the anomeric effect is the preference for synclinal (*gauche*) over antiperiplanar (*anti*)conformations such as in the system C5—O5—C1—X (Fig. 1.10c). Looking along the C1—O5 bond, one can see that, in the *anti*-conformer, the electronegative heteroatom is placed between two lone pairs, resulting in greater electrostatic repulsion.

There is a balance between the stabilizing anomeric effect (which favors the α anomer) and other factors that contribute to the anomeric preference of a particular sugar, such as solvent effects and sterics (which can favor the β anomer). A good example to illustrate this duality is the mutarotation process.

1.2.9 Mutarotation

Mutarotation [22] is defined as a change in optical rotation due to epimerization. When a crystalline sample of pure α-D-glucose is dissolved in water at neutral pH, its initial optical rotation value is +112°; but after approximately 3 h at 20°C, this value decreases until an equilibrium value of +52.7° is reached. The same equilibrium value is observed if one starts with a sample of pure β-D-glucose. In the crystalline form, the ring size and anomeric stereochemistry are fixed. However, in solution, ring opening and hydrolysis contribute to an equilibrium, resulting in a combination of both anomers (Fig. 1.11) and to a lesser degree a mixture of pyranose and furanose forms. This process results in a change in the optical rotation of the solution as the equilibrium is established. If anomeric or steric effects were solely responsible for the configuration at the anomeric position, the equilibrium would lie heavily to one side, which is not the case.

1.3 HISTORICAL OVERVIEW OF CARBOHYDRATE RESEARCH

Sugars, such as fructose and glucose from honey, have been harvested and processed by humans since the Stone Age [23]. The use of sucrose as a sweetener dates back to the eighth century BC and could only be afforded by royalty and the very wealthy [24]. More recently, these natural products become critical in a variety of industries focused on the production of paper, pulp, textiles, and pharmaceuticals. As often is the case, industrial applications ultimately provided the economic impetus for investigations into carbohydrate synthesis, purification, and characterization in the late nineteenth century.

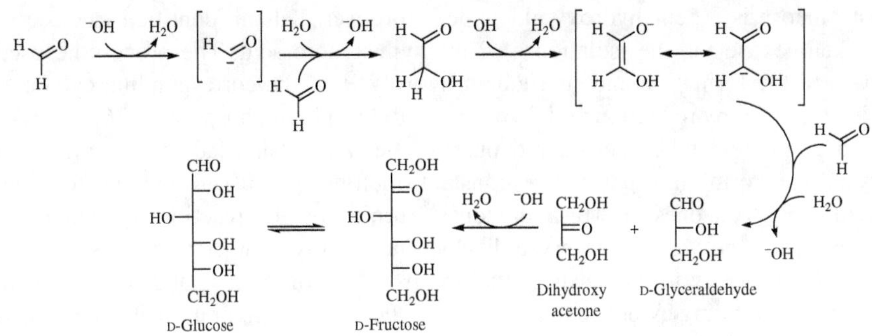

SCHEME 1.1 The formose reaction.

Any discussion of the beginning of carbohydrate chemistry should include Alexander Butlerov's discovery of the formose reaction in 1861 [25]. Subsequent experiments identified that out of all the possible formose products, glucose was present in the highest concentration [26]. Indeed, glucose is the most abundant sugar found in nature. This finding is of historical significance as the formose reaction provides a plausible route to ribose and other sugars from simple formaldehyde building blocks (Scheme 1.1).

1.3.1 Emil Fischer (1852–1919): The Father of Carbohydrate Chemistry

The structures and properties described in the previous sections would not be possible without the knowledge and insight provided by the "father of carbohydrate chemistry," Emil Fischer. Until the work of Fischer [27, 28], progress was made largely by the empirical observations of alchemists. In 1891, his pioneering work and that of his students provided the structural characterization and relative configurations of monosaccharides via a combination of analytical (polarimetry) and chemical techniques [29, 30]. It is a combination of the complexity and limited chemical and analytical tools available that made the Fischer proof so profound. While this skillfully planned and beautifully executed work has been reported in detail elsewhere [27, 28, 31–34], its significance warrants highlighting.

In 1870, only two aldohexoses (glucose and galactose) and one ketose (fructose) were known. Three disaccharides (lactose, maltose, and sucrose) had also been identified. Perhaps the first milestone on a path that ultimately led Fisher to the Nobel Prize in Chemistry (1902) was the serendipitous discovery of the reagent phenylhydrazine in 1875 [35]. Although he found it reacted rapidly with aldehydes, resulting in the formation of the corresponding phenylhydrazones, it would be 9 years until he applied this tool to the characterization of carbonyl compounds and an additional 7 years before his structural assignment of the sugars was achieved. At that time, it was known that glucose was composed of 6 carbons, 6 oxygens, and 12 hydrogens and that it reduced Tollens' reagent. Heinrich Kiliani also described the conversion of glucose and galactose to *n*-heptanoic acid, thereby confirming that they are aldohexoses. Taken together, these experiments supported the hypothesis that glucose is a pentahydroxy aldehyde. Moreover, Kiliani identified fructose as a 2-ketohexose via the isolation of 2-methylhexanoic acid. The process he used involved the formation and subsequent hydrolysis of the corresponding cyanohydrins followed by reduction with hydrogen iodide and red phosphorus [36]. In order for a compound to be considered pure at the time, it needed to be isolated in crystalline form and possess a constant melting point and optical rotation. Fortunately, osazones (1,2-bishydrazones) formed by the reaction of sugars and phenylhydrazine were often crystallizable and readily characterized. Fischer utilized this property to demonstrate that the osazones of glucose and fructose were identical, providing evidence that they share the same configuration at C3, C4, and C5. In a subsequent paper, Fischer reported that the isolation and identification of phenylhydrazone intermediates were possible when the reaction was conducted at a reduced temperature. It was this discovery that ultimately led to the conclusion that glucose and mannose are C2 epimers, as they yielded the same osazone but different hydrazones (Scheme 1.2).

Although the postulate of Le Bel and van 't Hoff was based solely on theoretical considerations, it provided an explanation for the occurrence of the numerous isomers that were inexplicable on the basis of the structural formulas of the time [37, 38]. Fischer applied this theory as the foundation of his stereochemical deductions, ultimately resulting in the assignment of a tetrahedral geometry for carbon atoms. The next major breakthrough came in 1889 when Fischer discovered that sodium amalgam could be used to reduce the lactones of sugar acids to their corresponding aldoses. For example, mannonic acid lactone was reduced to mannose in this manner [39]. The combination of this reaction with the known cyanohydrin procedure led to the conclusion that the D-enantiomers of arabinose, glucose, mannose, and fructose all share the same configuration at the three highest-numbered chiral centers. Polarimetry was critical for establishing the final piece of evidence needed to determine the configuration of the aldopentoses. Of the 1,5-dicarboxylic acids derived from the nitric acid oxidation of pentoses, D-arabinose was optically active, while those of D-ribose and D-xylose were not. Therefore, they were identified as *meso*-compounds, allowing correlation between the configuration and optical activity of aldopentoses.

Finally, oxidation experiments on D-gluconic and D-gulonic acid produced the head-to-tail enantiomers D/L-glucaric acid. Thus, it was apparent that these acids could only be derived from D-glucose and D-gulose. In a seminal report published in 1891, Fischer

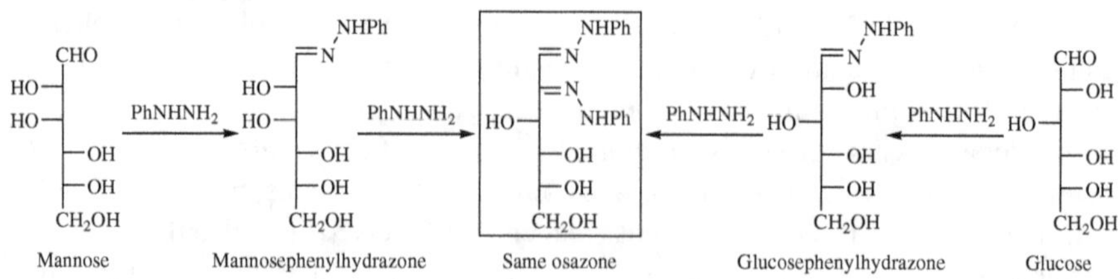

SCHEME 1.2 Convergence of glucose and mannose to the same osazone.

described the configuration of glucose, mannose, and fructose in accordance with the van 't Hoff convention [29]. Just over 2 months later, Fischer replaced the van 't Hoff notation with his own projections to eliminate the confusion associated with the depictions [30]. Fischer projections have since become the universally accepted way to represent the linear form of sugars in two dimensions [40]. As an interesting aside, Fischer arbitrarily decided to place the hydroxy substituent of the lowest chiral center of D-glucose (C5) to the right, a choice that was later confirmed by X-ray crystallography some 60 years later [41]. It should be emphasized that Fischer set out to accomplish this monumental task with limited knowledge of carbohydrate chemistry, without an understanding of the concept of stereochemistry, having crystallization as the primary method of purification, and limited access to reference compounds. In the end, it was a mixture of brilliance, determination, and luck that resulted in the Fischer solution. He not only provided a strong foundation upon which the disciplines of organic chemistry and biochemistry were built, but his inspiration was transferred to over 300 doctoral students and postdoctoral researchers, stimulating the next generation of scientists.

1.3.2 Koenigs–Knorr Reaction

With limited information about the structure and function of carbohydrates, the work of a few brilliant scientists during this early period provided the foundation for the advances that were just around the corner. For example, Arthur Michael reported the first chemical glycosylation in 1879. The reaction involved the nucleophilic displacement of an anomeric halide by the potassium salts of various phenols (Scheme 1.3a) [42, 43]. The beginning of the twentieth century was marked by the discovery of perhaps the most commonly used glycosylation method, the Koenigs–Knorr reaction (Scheme 1.3b) [44, 45]. Initial reaction conditions involved the displacement of an anomeric halide with excess Ag_2CO_3 in methanol to afford the corresponding methyl glycoside. Since then, the reaction has been successfully applied to the synthesis of a wide range of alkyl and aryl O-glycosides as well as O-linked oligosaccharides. In fact, the procedure was utilized for the first stereoselective formation of an α-linked glycoside [46]. However, low reaction efficiencies with unreactive acceptors, stoichiometric amounts of toxic heavy metals, halophilic promoters, and facile donor 1,2-elimination provided motivation for a significant number of modifications and refinements [47–51].

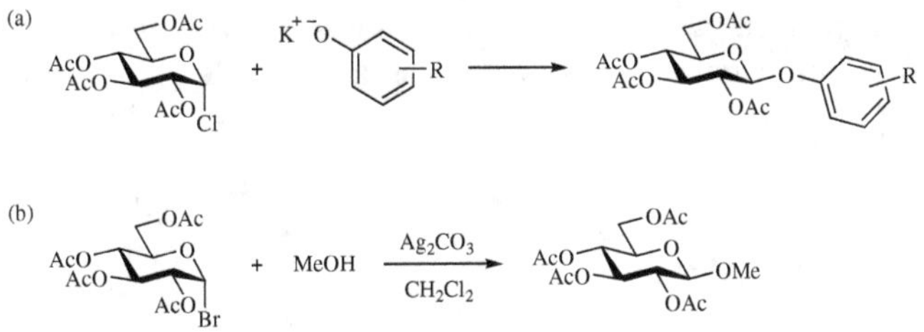

SCHEME 1.3 Early examples of chemical glycosylations. (a) The first reported chemical glycosylation. (b) A Koenigs–Knorr reaction. Ac, acetyl.

1.3.3 Karl Freudenberg (1886–1983)

Karl Freudenberg [52] conducted his doctoral and postdoctoral studies under the guidance of Emil Fischer. As such, it is not surprising that a major portion of his research was dedicated to investigations on the absolute configuration of "sterically related compounds." His observations led to the "optical shift rule," which has been frequently invoked to assign the absolute configuration of molecules with one or two chiral centers [53–56]. Freudenberg is perhaps better known for his contributions to the field of carbohydrate chemistry. His investigations of sugar cyclic acetals provided the structures of di-*O*-isopropylidene derivatives of glucose, galactose, mannose, and xylose [56]. Some of these compounds served as regioselectively functionalized substrates for pioneering studies with tosyl esters. Freudenburg was the first to report the synthesis of 3-*O*-tosyl-D-glucose and 6-*O*-tosyl-D-galactose, which have since become valuable intermediates for the construction of complex glycosides as well as modified sugars such as deoxy, deoxyamino, and deoxyhalogeno [57–62]. The work of Freudenberg et al. also contributed to our understanding of the structure and functions of glycans such as amylose, cellulose, and cyclodextrins. In 1939, he postulated that hydrogen bonding would cause macromolecules such as starch and proteins to assume a helical structure [63]. His work on the natural product witch hazel tannin resulted in the identification of the first naturally occurring branched-chain sugar, hamamelose (Fig. 1.12).

1.3.4 Burckhardt Helferich (1887–1982)

Concurrent with Freudenberg, another Fischer alumnus was at the center of a highly contested debate over the most stable conformations assumed by carbohydrates. Burckhardt Helferich [64] began his studies into the cyclization of hydroxyaldehydes in 1919. He used these simple substrates as model systems to analyze their cyclic and linear characteristics. Until that time, it was largely accepted that cyclic sugars assumed a 1,4-furanosyl structure. Helferich's investigations pointed toward a 1,5-pyranoid structure, which was ultimately shown to be correct [65]. Arguably, the central theme of his research was the discovery and development of methods to assemble complex carbohydrates. In order to facilitate the chemical construction of these molecules, he sought to develop methods to obtain regioselectively and orthogonally protected carbohydrate building blocks and to efficiently couple them. In one such study, Helferich determined that the yield of the Koenigs–Knorr reaction could be improved by replacing Ag_2CO_3 or Ag_2O with mercury(II) salts and by conducting the reaction in more polar solvents such as acetonitrile or nitromethane [66].

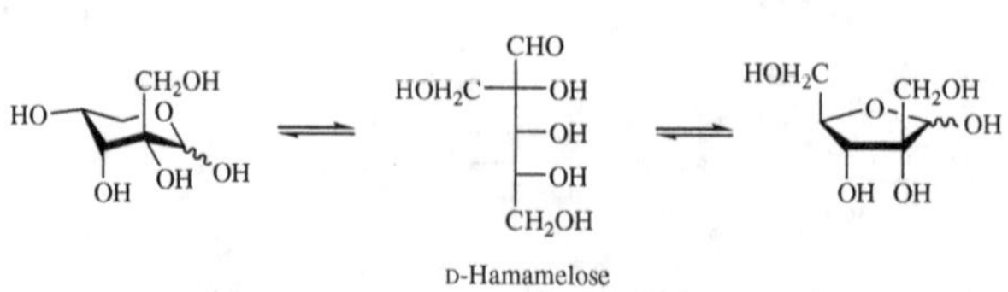

FIGURE 1.12 The first identified branched-chain sugar, hamamelose.

An attractive feature of these conditions was that no water was formed as the reaction progressed, alleviating the need to add drying agents as required under more classical conditions. These optimized conditions were later termed the Helferich modification [67, 68]. In 1923, his discovery of the trityl group revolutionized the way in which chemists approached the regioselective protection of organic molecules [69]. Helferich later employed the trityl moiety for the chemical synthesis of several di-, tri-, and tetrasaccharides. He also explored the synthetic utility of glycals in glycosylation reactions [70]. Like Freudenberg, he saw the usefulness of sulfonyl esters for organic synthesis. Indeed, it was his laboratory that introduced the methanesulfonyl (mesyl) group.

1.3.5 Hermann Fischer (1888–1960)

Hermann Fischer [56] conducted his postdoctoral research in the laboratory of his father, Emil Fischer, before leaving for World War I in 1914. Shortly after returning from 4 years of service, his father passed away in July of 1919. Hermann started his independent career focusing primarily on the difficult chemistry of triose phosphates. His synthesis of D-glyceraldehyde and D-glyceraldehyde-3-phosphate from 1,2,5,6-di-*O*-isopropylidene-D-mannitol is worthy of note as these molecules are chiral synthons for a wide range of biological and industrial processes. His other main research interests involved the natural products quinic acid and shikimic acid. These unique natural products are formed biosynthetically from the phosphate precursors described earlier. Hermann Fisher's studies of the structure and absolute configuration of quinic acid ultimately resulted in its correct assignment [71].

In 1948, Hermann Fischer joined the faculty of the University of California at Berkeley, where he continued his work on the synthesis of amino sugars and phospho sugars among other compounds of biological importance. During the beginning of the twentieth century, there was a significant power shift taking place in the field of carbohydrate chemistry. The discipline, which had largely been dominated by German scientists, was transitioning to American leadership, and Hermann Fischer was one of several exceptional scientists that contributed to this change.

1.3.6 Claude Hudson (1881–1952)

Claude Hudson [72] was awarded a Ph.D. (*magna cum laude*) from Princeton in 1907. He spent the better part of the next decade moving between institutions, serving in a variety of capacities, before settling into a long-term relationship with US government laboratories. One such position was as a visiting researcher in the laboratory of van 't Hoff in Berlin. Hudson's research interests were primarily concerned with the stereochemistry of the reducing or anomeric carbon of sugars. Indeed, he conducted extensive kinetic studies of the mutarotation and oxidation of lactose and glucose with the assistance of his colleague Horace Isbell [73–76]. He also extended van 't Hoff's ideas on optical superposition to a wide range of optically active substrates (sugars), which laid the groundwork for Hudson's *isorotation rule* [77].

As with any rule, there are exceptions. In particular, some inconsistencies were noted between his studies and those of Haworth's methylation analysis [78]. However, they were both indispensable methods to assign the anomeric configuration and structure of carbohydrates for several decades until they were replaced with physical methods such as NMR and X-ray diffraction. In addition to the large amount of information gained from the analysis of these compounds, Hudson published several papers concerned with the isolation, preparation, and purification of the numerous sugars that were required in high purity for the aforementioned studies. In search of a more fitting substitute for Fischer's phenylosazones, Hudson found phenylosotriazole derivatives, which he readily obtained through oxidation of the corresponding phenylosazone with $CuSO_4$ [79]. Another interest of Hudson was the way in which enzymes act on carbohydrates. For example, his studies on the hydrolysis of sucrose with invertase revealed that the reaction was irreversible and that α-D-glucose was liberated as a product of the hydrolysis [80–83].

1.3.7 Horace Isbell (1898–1992)

In 1926, Horace Isbell [84] earned his Ph.D. degree at the University of Maryland where his studies focused on the research of organogold compounds. In 1927, he obtained a position at the National Bureau of Standards (NBS) in Washington, DC, where he met Claude Hudson and remained there for more than 40 years. During his tenure at the NBS and later at American University, he allocated the bulk of his research to carbohydrate chemistry, resulting in several notable discoveries. Indeed, in a seminal report, Isbell identified the important roles neighboring groups play during the course of reactions [85]. He also developed the current system for describing the conformation of pyranoid sugars [86], which built on the pioneering work of Haworth [87, 88], and investigated the effects these conformations have on reactivity. Perhaps the most significant contribution of Isbell was his development of the first practical methods to synthesize ^{14}C- and ^{3}H-radiolabeled sugars and their derivatives [89]. This work revolutionized the way in which complex biological processes were probed and visualized.

1.3.8 Melville Wolfrom (1900–1969)

Melville Wolfrom [90] earned a Ph.D. in 1927 from Northwestern University. He then began postdoctoral studies under the mentorship of Claude Hudson at NBS. From there, Wolfrom moved to the Rockefeller Institute for Medical Research, where he worked in the laboratory of Phoebus Levene. His research focused on the structural elucidation of biologically relevant carbohydrates. In 1929, Wolfrom accepted a position at Ohio State University, where he remained for the rest of his career. His research interests can be described by the broad heading of carbohydrate structure and reactivity. He developed methods to obtain acetylated straight-chain sugars, including their dithioacetals, and demonstrated their use as reactive sugar intermediates [90]. Some of these acyclic sugars (*keto*-acetates) were utilized for the synthesis of branched carbohydrates. Extending his dithioacetal work, Wolfrom developed a method for their reductive desulfurization [91]. One of his exceptional postdoctoral researchers, Raymond Lemieux, later employed this reaction to correlate the stereochemistry of amino acids and sugars through the transformation of 2-amino-2-deoxy-D-glucose to an L-alanine derivative.

1.3.9 "Sugar" Raymond Lemieux (1920–2000)

One of the most prolific scientists of the second half of the twentieth century was Raymond Lemieux. His discoveries spanned a wide range of scientific disciplines including carbohydrate chemistry, organic synthesis, NMR, stereochemistry, and their resulting biological implications. While a postdoctoral researcher under Melville Wolfrom, his research primarily dealt with the structural elucidation of streptomycin [92]. In 1947, Lemieux started his independent research career at the University of Saskatchewan. It was during this time that he began investigating the chemical and physical properties of carbohydrates. The results of these studies provided the foundation for the first chemical synthesis of sucrose [93]. In 1954, Lemieux accepted the position of professor and chair of the Department of Chemistry at the University of Ottawa. It was there that he collaborated with Harold Bernstein and William Schneider of the National Research Council to study sugars using NMR. This formative work not only showed a correlation between chemical shift and the local environment of the protons, but also it demonstrated the utility of ^{1}H—^{1}H couplings for the determination of the preferred conformation of per-O-acetylated sugars in solution for the first time. These studies were perhaps the most significant development in the field of carbohydrate chemistry since the Fischer proof as they experimentally validated the Karplus equation before it was even published [94]. About this time, Lemieux also published a seminal report detailing the anomeric effect. Moreover, Lemieux's studies of the conformation of glycosides led to his identification of the reverse anomeric [95] and *exo*-anomeric effects [96–98]. He believed the *exo*-anomeric effect was the reason for the orientation of the sugars in higher-order structures such as polysaccharides and oligosaccharides in solution, a theory that was ultimately confirmed by NMR. It was the combination of theoretical results and extensive mechanistic studies that paved the way for the development of novel glycosylation methods, which was a central theme of his research interests.

A long-standing problem facing carbohydrate chemists is the formation of 1,2-*cis*-glycosides. Lemieux's answer to this challenge was the halide ion-catalyzed glycosylation reaction. This extraordinary work permitted, for the first time, the efficient and reproducible synthesis of 1,2-*cis*-glycosides in a completely stereoselective manner. At the heart of this approach was a rapid equilibrium between the relatively stable α-halide and its far more reactive β anomer [99]. This process is known as Lemieux-type *in situ* anomerization and is still one of only a few methods to obtain α-linked glycosides stereoselectively. Some recent methods that have been particularly effective in achieving stereochemical control during the formation of 1,2-*cis*-glycosides include the use of stereospecific activators, novel participating groups, and intramolecular aglycone delivery systems [12].

Lemieux sought to answer biological questions through chemistry. The discovery of the multifaceted roles of glycans in key biological processes not only inspired him but also marked the dawn of glycobiology and chemical biology. Indeed, his group synthesized numerous biologically relevant natural products including six human blood-group antigens, which he then utilized for immunization and animal studies. The resulting monoclonal antibodies against these synthetic sugars were harvested and purified, and their binding affinities were quantified [100]. For the first time, a picture of the complex interactions between glycans and their protein receptors (lectins) emerged on a molecular level. Lemieux attracted top-tier students and postdoctoral fellows from around the world. Together they helped cement the way in which chemistry was applied to answer biological questions.

1.3.10 Ascent of De Novo Sugar Synthesis

During the time of Lemieux, the field of organic synthesis was rapidly expanding. In laboratories across the globe, complex natural products were submitting to total synthesis. Occasionally, total synthesis endeavors would include carbohydrates, but for the most part, this was left to those skilled in the art. The high density of largely equivalent functional groups, poor solubility in organic solvents, and nontrivial purifications associated with carbohydrate synthesis posed considerable challenge. However, as the biological relevance of glycans became increasingly evident, carbohydrate chemistry began to garner the attention of scientists from a wide range of backgrounds. The synthesis of carbohydrate-based molecules became a target for noncarbohydrate chemists who expanded the chemical toolbox to include carbenes [101], carbanions [102], carbocations [43], organometallics [103], and radicals [104–106] for the synthesis of sugars and sugar derivatives. Although these reactions often afforded product mixtures, novel purification and characterization methodologies were also introduced. As a result, a wide range of elongated, branched, carbocyclic, and C-linked glycosides and nucleosides were created.

A particularly elegant example is the application of Danishefsky's diene [107] to the total synthesis of carbohydrates and carbohydrate derivatives. While it was known that activated aldehydes undergo cycloaddition with electron-rich dienes, the process was not efficient with typical aldehydes under thermal conditions. A major breakthrough was realized [108] with the development of the Lewis acid-catalyzed diene–aldehyde cyclocondensation (LACDAC) reaction, which provided a new strategy for the synthesis of carbohydrates and other polyoxygenated natural products (Scheme 1.4) [109].

Initially, the de novo synthesis of enantiomerically pure carbohydrates [110] and glycolipids [111] using transition metal complexes and chiral auxiliaries afforded only modest success. Ultimately, it was the use of enantiomerically pure aldehydes, such as the *R* and *S* enantiomers of 2-(phenylseleno)propionaldehyde, to convey facial selectivity upon the LACDAC reaction that enabled the synthesis of optically pure glycals. Syntheses of several complex monosaccharides such as the main sialic acid-type *N*-acetylneuraminic acid (Neu5Ac) and *rac*-3-deoxy-*manno*-2-octulosonic acid (KDO) were accomplished with this technology [112, 113]. The LACDAC

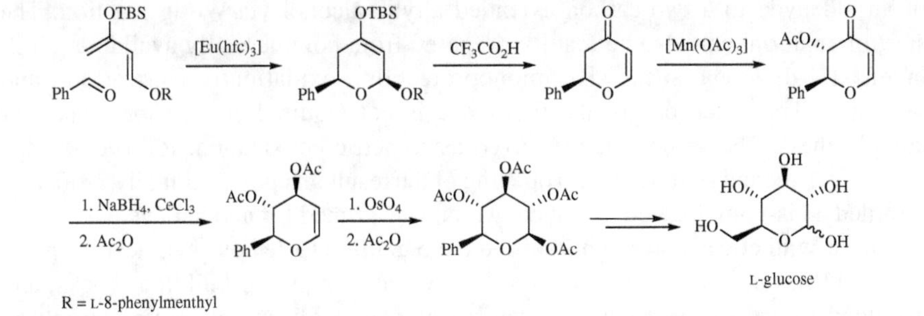

SCHEME 1.4 Enantioselective synthesis of L-glucose via the LACDAC reaction. hfc, 3-(heptafluoropropylhydroxymethylene)-D-camphorato; TBS, *tert*-butyldimethylsilyl.

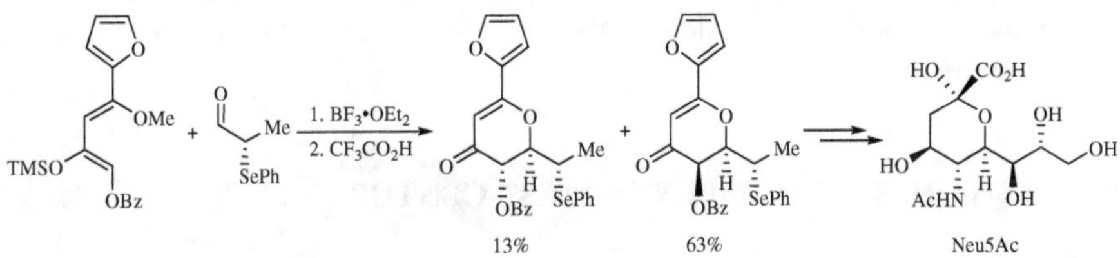

SCHEME 1.5 LACDAC reaction in the total synthesis of Neu5Ac. Bz, benzoyl; TMS, trimethylsilyl.

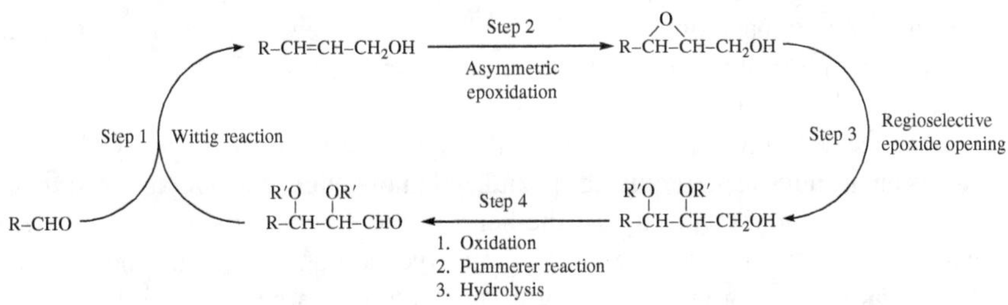

SCHEME 1.6 General scheme for a reagent-controlled approach to the total synthesis of all eight L-hexoses.

reaction that eventually led to the total synthesis of Neu5Ac is shown in Scheme 1.5. A noteworthy critical element of the synthesis is that a furan ring was employed as the carboxylic acid surrogate.

While Samuel Danishefsky and coworkers were optimizing the LACDAC reaction, many other groups were also developing methods for the total synthesis of natural and nonnatural sugars. For example, William Roush et al. reported the stereoselective synthesis of several dideoxyhexoses from allylic alcohol precursors [114–116]. Moreover, in a seminal report by Saturo Masamune, K. Barry Sharpless, and coworkers, a reagent-controlled approach to the total synthesis of all eight L-hexoses was achieved via a reiterative two-carbon extension cycle consisting of four key transformations (Scheme 1.6) [117]. This cycle began with the conversion of an aldehyde to a two-carbon extended allylic alcohol via Wittig reaction. The starting material can also be readily prepared from commercially available (*Z*)-2-butene-1,4-diol via successive monoprotection, oxidation/isomerization, and reduction. Thus, step one of the first cycle is not required but is shown here for completeness. The second step involved asymmetric epoxidation, followed by the regioselective (and stereospecific) opening of the resultant epoxide. Finally, oxidation afforded a bis-homologated aldehyde, which was primed for another cycle.

These synthetic achievements shifted the paradigm to de novo synthesis of pyranoses [118], rather than begin with naturally occurring sugar building blocks, and provided the necessary tools to assemble glycans with unprecedented structural complexity. In fact, the origin of glycomics can be traced back to this period. Numerous novel methods for the manipulation of sugars were discovered at a rapid pace. Application of the methodologies afforded reasonable quantities of both natural and unnatural analogues that were available for biological evaluation for the first time.

1.4 ONWARD TO THE TWENTY-FIRST CENTURY

The structural complexity and numerous isoforms found in naturally occurring glycans pose significant challenges in isolating pure and homogeneous samples of glycolipids (variations in carbohydrates, linkages, and lipids), glycoconjugates (differences in sugars and connectivity), and glycoproteins (existing as a diverse collection of posttranslational modifications). Because even slight impurities can intensely affect bioactivity, access to structurally and compositionally defined samples is essential for biological evaluation. Much of the responsibility for providing these samples rests on the shoulders of the synthetic chemists. The synthesis of glycans is an arduous task, requiring specialized knowledge, considerable resources, and, most importantly, creativity. At the heart of the matter is the regio- and stereoselective assembly of oligosaccharides and glycoconjugates. While the diversity of orthogonal protecting groups currently available has largely solved the issue of regioselectivity, achieving stereoselective formation of glycosidic linkages in a controlled manner remains one of the central challenges of modern synthetic chemistry. Fortunately, the chemists of today have many more tools at their disposal than those of only a few decades ago.

1.4.1 Glycosyl Donors and Glycosylation Systems

Perhaps the one area that has yielded the highest dividends during the latter part of the twentieth century is the design and development of new glycosylation methods. These processes traditionally involve two components: the glycosyl donor and the glycosyl acceptor. A glycosyl donor is the species that contributes the anomeric center to the resulting glycoside and is typically electrophilic in nature. The acceptor, in majority of glycosylation reactions, provides the nucleophile. For O-glycosides, the glycosylation pathway generally begins with donor activation upon addition of a

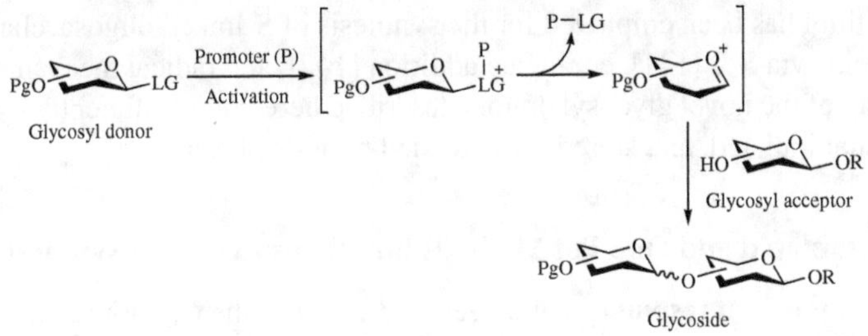

SCHEME 1.7 General glycosylation pathway.

promoter. Once activated, the donor reacts with a nucleophilic hydroxy group of the glycosyl acceptor or aglycone (Scheme 1.7). There are many factors that can have a profound effect on the reactivity and selectivity of a glycosylation event and a wide range of mechanisms by which the reaction can proceed [119]. This section provides a few examples of key developments in glycosylation techniques and applications from a historical perspective. More detailed analyses can be found in subsequent chapters of this book.

Some donors such as glycosyl bromides and chlorides were employed in the early chemical glycosylations and are still in use today, albeit under optimized conditions. Within the last two decades, glycosyl iodides have been increasingly employed due to our ability to tame their reactivity through careful choice of protecting groups. Although stable glycosyl iodides protected with acetates had been discovered by Emil Fischer in the first part of the twentieth century [120], ether-protected analogues were typically found to be too reactive to be useful glycosyl donors [121–123]. Thanks, in part, to the work of Conrad Schuerch [124], Joachim Thiem and Bernd Meyer [125], and others [126], these valuable donors enjoyed increasing popularity since the 1970s. This is especially true for the synthesis of 1,2-*cis*-glycosides [127–132]. While glycosyl fluorides had been known since 1923, it was not until Teruaki Mukaiyama introduced them as glycosyl donors in 1981 that their popularity increased [133]. Another glycosyl donor introduced at about this time is the novel O-imidate leaving group by Pierre Sinaÿ in 1977 [134]. Further refinements by Richard Schmidt led to the trichloroacetimidate donors 3 years later [135]. Trichloroacetimidates enjoyed widespread application due to their stability, efficiency, and the relatively mild conditions required for activation. Possibly the most versatile family of glycosyl donors to date is the thioglycosides. First reported in 1909 by Emil Fischer [136], the anomeric thiol moiety is stable toward a wide range of reaction conditions including those typically required for the manipulation of protecting groups. Thioglycosides can be activated under relatively mild conditions that are often orthogonal to those required for other donors [137–139]. This property made them particularly useful for the synthesis of oligosaccharides in one-pot and iterative couplings [140]. As workhorses of carbohydrate chemistry, thioglycosides are readily converted into other glycosyl donors such as hemiacetals, imidates, halides, and sulfoxides. When used in this manner, one can consider the thioacetal moiety as a transient anomeric protecting group. The nucleophilicity of the anomeric thiol has been employed for the synthesis of S-linked oligosaccharides and glycoproteins via S_N2 [141], conjugate addition [142], and radical mechanisms [143, 144]. Many of the novel glycosyl donors described herein are sufficiently stable to be purified, manipulated, and stored for extended periods of time.

1.4.2 Automated and One-Pot Methods for Oligosaccharide Synthesis

Recently, significant resources have been dedicated to the development of one-pot strategies for the synthesis of oligosaccharides. While many variations were reported, they generally utilize chemoselective, orthogonal, or preactivation strategies [145]. Two examples of particular significance are Chi-Huey Wong's automated one-pot synthesis of oligosaccharides and Shang Cheng Hung's regioselective one-pot protection method. In 1999, Wong and coworkers developed a custom computer program for the automated one-pot synthesis of oligosaccharides that they named OptiMer. To accomplish this, they derived and tabulated the relative reactivity values (RRVs) of a library of thioglycoside building blocks. Their studies showed that selective activation of the anomeric leaving group could be achieved through careful choice of the protecting groups. A database containing the reactivity profiles and the target oligosaccharide sequence was loaded into Optimer. The software predicted the optimal set of these building blocks and the order in which they should be added to accomplish the synthesis [146]. A drawback of this technology is the need to synthesize large libraries of orthogonally protected donors offering a wide range of RRVs. In 2007, Hung et al. addressed this shortcoming with a trimethylsilyl triflate-catalyzed one-pot approach for the direct and efficient preparation of hundreds of thioglucoside building blocks (Scheme 1.8) [147]. This technology represents a paradigm shift for the way in which regioselectively protected monosaccharides are obtained.

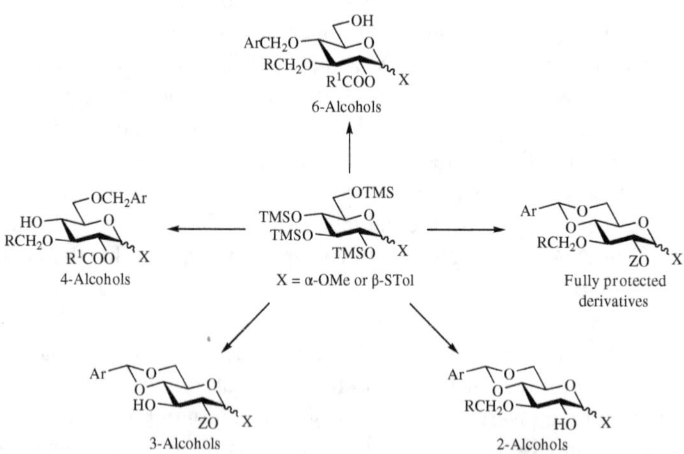

SCHEME 1.8 Regioselective one-pot protection of carbohydrates by Hung et al.

1.4.3 Solid-Phase Oligosaccharide Synthesis

Inspired by the success of Robert Bruce Merrifield's solid-phase peptide synthesis (SPPS) [148], Jean Frechet and Conrad Schuerch disclosed the first synthesis of oligosaccharides on solid support in 1971 [149]. Although the utility of the process was readily apparent, the technology went largely unused for over 20 years due to limitations associated with reaction monitoring, the large excesses required of costly reagents, and the lack of automation. Interest in solid-phase oligosaccharide synthesis was rejuvenated in 1993 with the advent of new glycosyl donors and linkers [150]. In 2001, Peter Seeberger modified a peptide synthesizer to produce the first automated solid-phase carbohydrate synthesizer [150–153]. This technology resulted in the generation of large oligosaccharide libraries, which could prove to be particularly useful for high-throughput screening assays [154]. In general, most solid-phase oligosaccharide strategies may be categorized as either donor-bound, acceptor-bound or bidirectional, referring to the reactant component that is attached to the solid support. In a seminal report in which the donor-bound strategy was utilized, Danishefsky employed silicon to tether glycal donors to the resin. The glycosylations were performed with an excess of the solution-based acceptor. Following iterative coupling reactions, the oligosaccharide was released upon addition of tetrabutylammonium fluoride and acetic acid. One striking advantage of this process over solution-based chemistries is that the excess acceptor and promoter can be removed by rinsing after each coupling reaction [150]. The results for acceptor-bound [149, 155] and bidirectional [153] strategies are equally promising. While progress has been made, solid-phase oligosaccharide synthesis remains an area of intense investigation with the goal of simplifying the process to the extent that nonchemists would be able to perform the synthesis of complex glycosides in an automated and programmable fashion.

1.4.4 Natural Product Synthesis

The aforementioned synthetic tools and other discoveries enabled the assembly of carbohydrates with unprecedented structural complexity, such as that of the potent adjuvant QS-21A (Fig. 1.13). Investigations into the dynamic functions of these

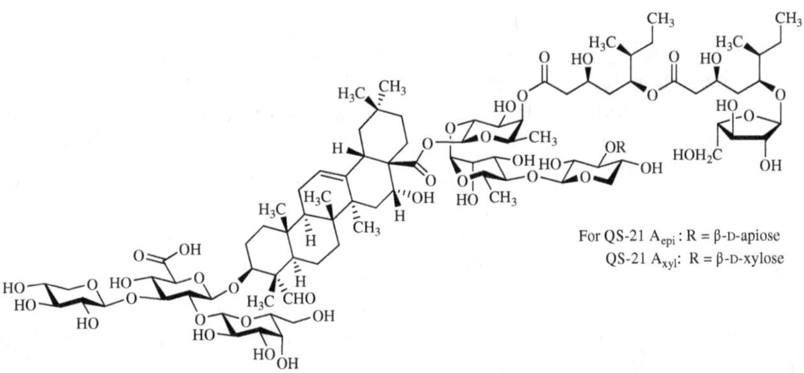

FIGURE 1.13 The immunological adjuvant QS-21A.

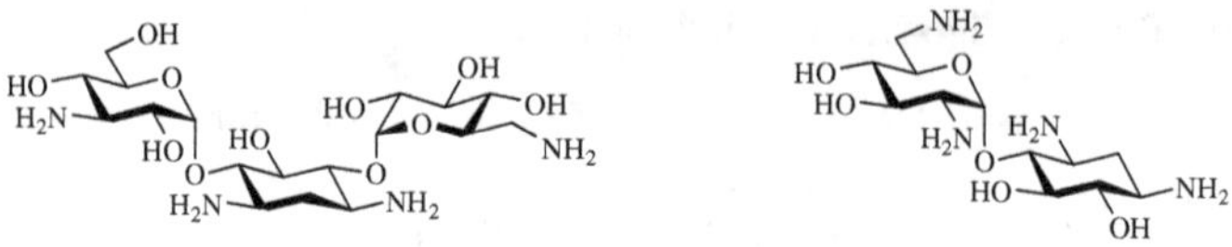

FIGURE 1.14 Structures of the aminoglycoside antibiotics kanamycin and neamine.

compounds often begin with their isolation from natural sources. This complex triterpene–oligosaccharide–normonoterpene conjugate is a heavily glycosylated saponin originally isolated from *Quillaja saponaria* Molina as both apiose and xylose forms [156].

Many of the same chromatographic and spectroscopic techniques were applied to the isolation and identification of natural products such as carbohydrates and glycolipids. The discovery, isolation, computational analysis, synthesis, and structure–activity relationship studies of antibiotics such as the aminoglycosides kanamycin and neamine are of particular note (Fig. 1.14) [157, 158].

1.4.5 Carbohydrate-Based Therapeutics

Carbohydrates in the form of glycoproteins, GPI anchors, or glycoconjugates such as glycolipids and gangliosides (Fig. 1.15) are ubiquitous cell-surface components of animal as well as some plant cells [159, 160]. They are also found on the surface of virions and bacterial cells. Gangliosides, such as GM1, are found on the cell surface in lipid rafts. They are believed to modulate signal transduction and are considered as possible therapeutics for neurodegenerative disorders. Members of this class of biologically relevant molecules facilitate a myriad of biological and pathological processes including cell–cell communication, growth, fertility, recognition, adhesion, fusion, replication, metastasis, and immune system evasion [161, 162].

In 1993, a Kirin Pharmaceuticals research team isolated the first reported α-linked galactosylceramides (α-GalCer) from an extract of the marine sponge *Agelas mauritianus* off the coast of Okinawa, Japan [163]. These glycolipids, also known as agelasphins, were found to possess antitumor activity. The carbohydrate moiety of these compounds varied from mono- to tetrasaccharides, and related compounds have also been isolated from the pathogenic microbe *Borrelia burgdorferi* [164, 165]. Subsequent studies into the mode of action of these unique compounds revealed that their therapeutic effects stemmed from activation of invariant natural killer T cells in a CD1d-dependent manner. The formation of the T-cell receptor/α-GalCer/CD1d triplex results in the initiation of a cascade of immunological events. The progression primarily involves the secretion of the proinflammatory T helper 1 cytokine interferon-γ as well as the immunoregulatory T helper 2 cytokine interleukin-4. Research has shown that the immunological response can be tuned by altering the structure of the glycolipid. As such, research into the identification of analogues that elicit biased cytokine production is of high interest for the treatment of a wide range of maladies ranging from viral and bacterial infections and tumor

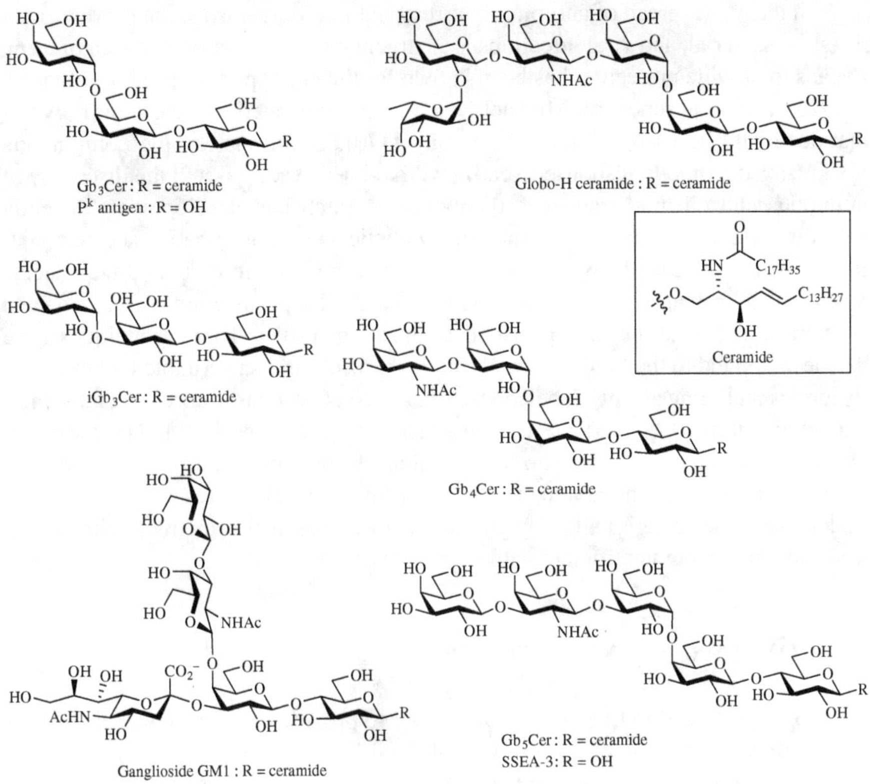

FIGURE 1.15 Structures of immunological glycolipids and gangliosides.

growth inhibition (tumor immunotherapy) to certain autoimmune diseases such as type 1 diabetes and multiple sclerosis. The dense surface distribution and characteristic glycan composition presented by a wide range of pathogens and malignant cells render them attractive targets for vaccines [166]. Indeed, glycoconjugates such as trehalose glycolipids and glycosylceramides are being investigated for use as adjuvants and key vaccine components [167–169]. The structure and properties of glycolipids, including their interactions with proteins, are not fully understood due to several reasons including the lack of synthetic methods for their efficient preparation; their properties that are closely correlated to their local environment such as microdomains or lipid rafts, which are hard to mimic; and their existence as heterogeneous mixtures with a high degree of structural flexibility. The combination of spectroscopic techniques, molecular dynamic simulations, biomimetic membrane chemistry, and carbohydrate chemistry has recently begun to shed some light on the multifaceted roles of these multifunctional compounds. As a result, interest in the synthesis and application of glycolipids and gangliosides has been increasing over the last two decades. A wide range of industries are interested in capitalizing on their highly amphiphilic character for use as environmentally friendly detergents, surfactants, and emulsifiers.

With the rapid spread of antimicrobial-resistant microorganisms, the prevention of parasitic, bacterial, and viral infections is an urgent global necessity. The idea of using glycans to provide protection has been known for the better part of a century. Indeed, in 1923, a seminal report by Michael Heidelberger indicated that capsular polysaccharides could be used to induce immunity [170]. Unfortunately, the combination of short-lived antibody response to carbohydrate-based vaccines and the discovery of antibiotics and chemotherapeutics dampened research and development. Recently however, rational vaccine design, modern synthetic and semisynthetic vaccine conjugates, and the advent of glycomics brought attention back to carbohydrate-vaccine development. Advances in glycan analysis, synthesis, purification, screening, and structural determination have provided astonishing results. These techniques have also been applied to the field of nanotechnology, resulting in a veritable tool chest for glycomics including affinity-labeled species, neoglycoproteins, fluorescent tags, multivalent quantum dots, and targeted magnetic nanoparticles [171]. Nanoparticles bearing carbohydrates revolutionized the diagnosis, imaging, and treatment of a wide range of biological phenomenon. The fruits of this translational research allowed the modern scientist to rationally design carbohydrate-based therapeutics with higher efficacies and in a more efficient manner than ever before.

1.5 CONCLUSION AND OUTLOOK

The topics presented in this chapter are meant to provide an overview of the evolution of glycan chemistry from the time of Emil Fischer to the burgeoning field of glycomics. Worthy of note is the way in which pioneering chemists contributed to the understanding of a broad range of sciences such as organic stereochemistry, carbohydrate chemistry, computational science, biology, and immunology. Although we are gaining a better understanding of the biological roles played by these polyols, glycobiology and chemical glycobiology are scientific disciplines still in their infancy. While carbohydrate chemistry has enjoyed remarkable progress, the search for alternative glycosylation strategies for the synthesis of biologically important compounds is an ongoing endeavor. Oligosaccharide synthesis is in no way routine. Some of the challenges that remain include the stereoselective formation of 1,2-*cis*-glycosides and the incompatibility associated with matched–mismatched donor–acceptor pairs. Advances in solid-supported and one-pot protocols are being successfully applied to the synthesis of oligosaccharides, which should help to streamline the synthesis and purification processes, thereby increasing the overall efficiency and will undoubtedly play a major role in the expansion of the field to nonchemists.

In nature, most oligosaccharides are covalently linked to peptides, proteins, or lipids. The advent of protein and peptide conjugation methods including SPPS, site-selective protein modification [172], native chemical ligation [173], and expressed protein ligation [174] has provided extraordinary access to the corresponding glycopeptides and glycoproteins. The efforts of chemists and biologists combined with recent advances in computational and spectroscopic techniques will undoubtedly yield answers to many of the questions surrounding the structure and function of glycolipids, gangliosides, and the microdomains in which they reside.

The diverse repertoire of glycoconjugates available today also expedited the identification of glycolipid- and glycoprotein-based ligands and inhibitors. This could not have happened at a better time as the emergence of antibiotic resistance has become a worldwide crisis. One solution may be the rational design of potent inhibitors to block entry, propagation, or other enzymatic processes such as glycosylation or hydrolysis. Recent developments in the multivalent presentation of carbohydrate-based high-affinity ligands resulted in some of the most potent inhibitors to date [175]. Another weapon in our arsenal is the development of synthetic and semisynthetic glycan-based vaccines and adjuvants. In combination with high-throughput screening methods such as carbohydrate microarrays, the future of carbohydrate-based vaccines appears bright. Indeed, the intrinsic diversity and complex relationship between nucleic acids, proteins, glycolipids, and carbohydrates will certainly put the technologies described herein to the test. However, if the past is any indication of the future, these challenges will continue to inspire researchers to invent even more ingenious solutions.

REFERENCES

[1] McNaught, A. D.; Wilkinson, A. *IUPAC. Compendium of Chemical Terminology*, 2nd ed.; Blackwell Scientific: Oxford, **1997**.

[2] Moremen, K. W.; Tiemeyer, M.; Nairn, A. V. *Nat. Rev. Mol. Cell Biol.* **2012**, *13*, 448–462.

[3] Lowe, J. B. In *The Molecular Basis of Blood Diseases*, 3rd ed.; Stamatoyannopoulos, G., Nienhuis, A. W., Majerus, P. W., Varmus, H., Eds.; Saunders: Philadelphia, **1987**, p 293.

[4] Wang, Z.-G.; Zhang, X.; Visser, M.; Live, D.; Zatorski, A.; Iserloh, U.; Lloyd, K. O.; Danishefsky, S. J. *Angew. Chem. Int. Ed.* **2001**, *40*, 1728–1732.

[5] Mallams, A. K. In *Carbohydrate Chemistry*; Kennedy, J. F., Ed.; Oxford Science: Oxford, **1988**, pp 73–170.

[6] Ghazarian, H.; Idoni, B.; Oppenheimer, S. B. *Acta Histochem.* **2011**, *113*, 236–247.

[7] Hannessian, S. *The Total Synthesis of Natural Products: The Chiron Approach*; Pergamon: Oxford, **1983**.

[8] The *C. elegans* Sequencing Consortium. *Science* **1998**, *282*, 2012–2018.

[9] Venter, J. C. et al. *Science* **2001**, *291*, 1304–1351.

[10] Grogan, M. J.; Pratt, M. R.; Marcaurelle, L. A.; Bertozzi, C. R. *Annu. Rev. Biochem.* **2002**, *71*, 593–634.

[11] Davis, B. G. *Pure Appl. Chem.* **2009**, *81*, 285–298.

[12] Bertozzi, C. R.; Kiessling, L. L. *Science* **2001**, *291*, 2357–2364.

[13] McNaught, A. D. *Pure Appl. Chem.* **1996**, *68*, 1919–2008.

[14] Goodyear, E. H.; Haworth, W. N. *J. Chem. Soc.* **1927**, 3136.

[15] Satoh, H.; Manabe, S. *Chem. Soc. Rev.* **2013**, *42*, 4297–4309.

[16] Reeves, R. E. *J. Am. Chem. Soc.* **1949**, *71*, 215–217.
[17] Karplus, M. *J. Am. Chem. Soc.* **1963**, *85*, 2870–2871.
[18] Karplus, M. *J. Chem. Phys.* **1959**, *30*, 11–15.
[19] Edward, J. T. *Chem. Ind.* **1955**, 1102–1104.
[20] Lemieux, R. U.; Chu, P. In *133rd National Meeting of the American Chemical Society*; American Chemical Society: Washington, DC, **1958**, p 31N.
[21] Juaristi, E.; Cuevas, G. *The Anomeric Effect*; CRC Press: Boca Raton, FL, **1995**.
[22] Capon, B. *Chem. Rev.* **1969**, *69*, 407–498.
[23] Levy, D. E.; Fügedi, P. *The Organic Chemistry of Sugars*; CRC Press/Taylor & Francis: New York, **2006**.
[24] Rolph, G. M. *Something about Sugar: Its History. Growth, Manufacture and Distribution*; John J. Newbegin: San Francisco, **1917**.
[25] Boutlerow, A. *Liebigs Ann. Chem.* **1861**, *120*, 295–298.
[26] Melendez-Hevia, E.; Montero-Gomez, N.; Montero, F. *J. Theor. Biol.* **2008**, *252*, 505–519.
[27] Freudenberg, K. *Adv. Carbohydr. Chem.* **1966**, *21*, 1–38.
[28] Lichtenthaler, F. W. *Eur. J. Org. Chem.* **2002**, 4095–4122.
[29] Fischer, E. *Ber. Dtsch. Chem. Ges.* **1891**, *24*, 1836–1845.
[30] Fischer, E. *Ber. Dtsch. Chem. Ges.* **1891**, *24*, 2683–2687.
[31] Lichtenthaler, F. W. *Angew. Chem. Int. Ed. Engl.* **1992**, *31*, 1541–1556.
[32] Hudson, C. S. *J. Chem. Educ.* **1941**, *18*, 353–357.
[33] Hudson, C. S. *Adv. Carbohydr. Chem.* **1951**, *3*, 1–22.
[34] Kunz, H. *Angew. Chem. Int. Ed.* **2002**, *41*, 4439–4451.
[35] Fischer, E. *Ber. Dtsch. Chem. Ges.* **1876**, *9*, 880–891.
[36] Kiliani, H. *Ber. Dtsch. Chem. Ges.* **1885**, *18*, 3066–3072.
[37] Le Bel, J. A. *Bull. Soc. Chim. Fr.* **1874**, *22*, 337–347.
[38] van 't Hoff, J. H. *Die Lagerung der Atome im Raume*; Vieweg: Braunschweig, **1877**.
[39] Fischer, E. *Ber. Dtsch. Chem. Ges.* **1889**, *22*, 2204–2205.
[40] Maehr, H. *Tetrahedron: Asymmetry* **1992**, *3*, 735–748.
[41] Bijvoet, J. M.; Peerdeman, A. F.; Van Bommel, A. J. *Nature* **1951**, *168*, 271–272.
[42] Michael, A. *Am. Chem. J.* **1879**, *1*, 305–312.
[43] Mydock, L. K.; Demchenko, A. V. *Org. Biomol. Chem.* **2010**, *8*, 497–510.
[44] Flowers, H. M. *Carbohydr. Res.* **1971**, *18*, 211–218.
[45] Koenigs, W.; Knorr, E. *Chem. Ber.* **1901**, *34*, 957–981.
[46] Brigl, P.; Keppler, H. *Ber. Dtsch. Chem. Ges.* **1926**, *59*, 1588–1591.
[47] Zemplen, G.; Gerecs, A. *Chem. Ber.* **1930**, *63*, 2720–2729.
[48] Garegg, P. J.; Konradsson, P.; Kvarnstrom, I.; Norberg, T.; Svensson, S. C. T.; Wigilius, B. *Acta Chem. Scand., Ser. B* **1985**, *39*, 569–577.
[49] Igarashi, K. In *Advances in Carbohydrate Chemistry and Biochemistry*; Tipson, R. S., Horton, D., Eds.; Academic: New York, **1977**; Vol. *34*, pp 243–283.
[50] Paulsen, H. *Angew. Chem. Int. Ed. Engl.* **1982**, *21*, 155–173.
[51] Kurti, L.; Czako, B. *Strategic Applications of Named Reactions in Organic Synthesis*; Academic: Burlington, MA, **2005**.

[52] Oesper, R. E. *J. Chem. Educ.* **1951**, *28*, 426–427.
[53] Freudenberg, K.; Kuhn, W. *Ber. Dtsch. Chem. Ges.* **1931**, *64*, 703–734.
[54] Freudenberg, K.; Kuhn, W.; Bumann, I. *Ber. Dtsch. Chem. Ges.* **1930**, *63*, 2381–2390.
[55] Freudenberg, K. *Stereochemie*; F. Deuticke Verlag: Leipzig, **1933**.
[56] Lichtenthaler, F. W. *Carbohydr. Res.* **1987**, *164*, 1–22.
[57] Freudenberg, K.; Brauns, F. *Ber. Dtsch. Chem. Ges.* **1922**, *55*, 3233–3238.
[58] Freudenberg, K.; Doser, A. *Ber. Dtsch. Chem. Ges.* **1923**, *56*, 1243–1247.
[59] Freudenberg, K.; Ivers, O. *Ber. Dtsch. Chem. Ges.* **1922**, *55*, 929–941.
[60] Freudenberg, K.; Raschig, K. *Ber. Dtsch. Chem. Ges.* **1927**, *60*, 1633–1636.
[61] Freudenberg, K.; Raschig, K. *Ber. Dtsch. Chem. Ges.* **1929**, *62*, 373–383.
[62] Tipson, R. S. *Adv. Carbohydr. Chem.* **1953**, *8*, 107–215.
[63] Freudenberg, K.; Schaaf, E.; Dumpert, G.; Ploetz, T. *Naturwissenschaften* **1939**, *27*, 850–853.
[64] Oesper, R. E. *J. Chem. Educ.* **1952**, *29*, 459.
[65] Helferich, B.; Malkomes, T. *Ber. Dtsch. Chem. Ges.* **1922**, *55*, 702–708.
[66] Helferich, B.; Wedemeyer, K. F. *Liebigs Ann.* **1949**, *563*, 139–145.
[67] Helferich, B.; Weis, K. *Chem. Ber.* **1956**, *89*, 314–321.
[68] Helferich, B.; Zirner, J. *Ber. Dtsch. Chem. Ges.* **1962**, *95*, 2604–2611.
[69] Helferich, B.; Speidel, P. E.; Toeldte, W. *Ber. Dtsch. Chem. Ges.* **1923**, *56*, 766–770.
[70] Helferich, B.; Schmitz-Hillebrecht, E. *Chem. Ber.* **1933**, *66B*, 378–383.
[71] Fischer, H. O. L.; Dangschat, G. *Ber. Dtsch. Chem. Ges.* **1932**, *65*, 1009–1031.
[72] Small, L. F.; Wolfrom, M. L. *Claude Silbert Hudson (1881–1952)*; The National Academies: Washington, DC, **1958**; Vol. *32*.
[73] Hudson, C. S. *Z. Physik. Chem.* **1903**, *44*, 487–494.
[74] Hudson, C. S. *J. Am. Chem. Soc.* **1907**, *29*, 1571–1576.
[75] Isbell, H. S.; Hudson, C. S. *BS J. Res.* **1932**, *8*, 327–338.
[76] Isbell, H. S. *BS J. Res.* **1932**, *8*, 615–624.
[77] Hudson, C. S. *J. Am. Chem. Soc.* **1909**, *31*, 66–86.
[78] Wang, Z. In *Comprehensive Organic Name Reactions and Reagents*; Wiley: Hoboken, NJ, **2009**.
[79] Hudson, C. S.; Hann, R. M. *J. Am. Chem. Soc.* **1944**, *66*, 735–738.
[80] Hudson, C. S. *J. Am. Chem. Soc.* **1908**, *30*, 1160–1166.
[81] Hudson, C. S. *J. Am. Chem. Soc.* **1908**, *30*, 1564–1583.
[82] Hudson, C. S. *J. Am. Chem. Soc.* **1909**, *31*, 655–664.
[83] Hudson, C. S.; Paine, H. S. *J. Am. Chem. Soc.* **1910**, *32*, 774–779.
[84] El Khadem, H. S. *Adv. Carbohydr. Chem. Biochem.* **1995**, *51*, 1–13.
[85] Frush, H. L.; Isbell, H. S. *J. Res. Natl. Bur. Stand.* **1941**, *27*, 413–428.
[86] Isbell, H. S.; Tipson, R. S. *J. Res. Natl. Bur. Stand.* **1959**, *64A*, 171–176.
[87] Drew, H. D. K.; Haworth, W. N. *J. Chem. Soc.* **1926**, *129*, 2303–2310.
[88] Haworth, W. N. *The Constitution of Sugars*; Edward Arnold & Co.: London, **1929**.
[89] Whelan, W. J. *Annu. Rev. Biochem.* **1960**, *29*, 105–130.
[90] Horton, D.; Hassid, W. Z. *Melville Lawrence Wolfrom (1900–1969)*; The National Academies: Washington, DC, **1975**; Vol. *67*.

[91] Horton, D. *Adv. Carbohydr. Chem. Biochem.* **1971**, *26*, 1–47.

[92] Lemieux, R. U.; Wolfrom, M. L. *Adv. Carbohydr. Chem. Biochem.* **1948**, *3*, 337–384.

[93] Lemieux, R. U.; Huber, G. A. *J. Am. Chem. Soc.* **1953**, *75*, 4118–4119.

[94] Lemieux, R. U.; Kullnig, R. K.; Bernstein, H. J.; Schneider, W. G. *J. Am. Chem. Soc.* **1958**, *80*, 6098–6105.

[95] Lemieux, R. U.; Morgan, A. R. *Can. J. Chem.* **1965**, *43*, 2205–2213.

[96] Lemieux, R. U.; Koto, S. *Tetrahedron* **1974**, *30*, 1933–1944.

[97] Lemieux, R. U. *Abstract of Papers*; American Chemical Society: Washington, DC, **1959**; Vol. *135*, p 5E.

[98] Lemieux, R. U.; Pavia, A. A.; Martin, J. C.; Watanabe, K. A. *Can. J. Chem.* **1969**, *47*, 4427–4439.

[99] Lemieux, R. U.; Hendriks, K. B.; Stick, R. V.; James, K. *J. Am. Chem. Soc.* **1975**, *97*, 4056–4062.

[100] Lemieux, R. U.; Venot, A. P.; Spohr, U.; Bird, P.; Mandal, G.; Morishima, N.; Hindsgaul, O.; Bundle, D. R. *Can. J. Chem.* **1985**, *63*, 2664–2668.

[101] Dötz, K. H.; Ehlenz, R. *Chem. Eur. J.* **1997**, *3*, 1751–1756.

[102] Hanessian, S.; Pernet, A. G. *Can. J. Chem.* **1974**, *52*, 1266–1279.

[103] Soengas, R. G.; Estevez, A. M. *Curr. Org. Synth.* **2013**, *10*, 183–209.

[104] RajanBabu, T. V. *J. Am. Chem. Soc.* **1987**, *109*, 609–611.

[105] RajanBabu, T. V. *J. Org. Chem.* **1988**, *53*, 4522–4530.

[106] RajanBabu, T. V.; Fukunaga, T.; Reddy, G. S. *J. Am. Chem. Soc.* **1989**, *111*, 1759–1769.

[107] Danishefsky, S.; Kitahara, T. *J. Am. Chem. Soc.* **1974**, *96*, 7807–7808.

[108] Danishefsky, S.; Kerwin, J. F.; Kobayashi, S. *J. Am. Chem. Soc.* **1982**, *104*, 358–360.

[109] Danishefsky, S. J.; Bilodeau, M. T. *Angew. Chem. Int. Ed. Engl.* **1996**, *35*, 1380–1419.

[110] Bednarski, M.; Danishefsky, S. *J. Am. Chem. Soc.* **1986**, *108*, 7060–7067.

[111] Bednarski, M.; Danishefsky, S. *J. Am. Chem. Soc.* **1983**, *105*, 6968–6969.

[112] DeNinno, M. P.; Danishefsky, S. J.; Schulte, G. *J. Am. Chem. Soc.* **1988**, *110*, 3925–3929.

[113] Danishefsky, S. J.; DeNinno, M. P.; Chen, S. H. *J. Am. Chem. Soc.* **1988**, *110*, 3929–3940.

[114] Roush, W. R.; Brown, R. J. *J. Org. Chem.* **1982**, *47*, 1371–1373.

[115] Roush, W. R.; Brown, R. J. *J. Org. Chem.* **1983**, *48*, 5093–5101.

[116] Roush, W. R.; Straub, J. A.; Brown, R. J. *J. Org. Chem.* **1987**, *52*, 5127–5136.

[117] Ko, S. Y.; Lee, A. W. M.; Masamune, S.; Reed, L. A., III; Sharpless, K. B.; Walker, F. J. *Tetrahedron* **1990**, *46*, 245–264.

[118] Schmidt, R. R. *Pure Appl. Chem.* **1987**, *59*, 415–424.

[119] Demchenko, A. V. *Handbook of Chemical Glycosylation*; Wiley: Weinheim, **2008**.

[120] Fisher, E.; Fischer, H. *Chem. Ber.* **1910**, *43*, 2521–2536.

[121] Gervay, J. In *Organic Synthesis: Theory and Applications*; Hudlicky, T., Ed.; JAI: Greenwich, **1998**; Vol. *4*, pp 121–153.

[122] Meloncelli, P. J.; Martin, A. D.; Lowary, T. L. *Carbohydr. Res.* **2009**, *344*, 1110–1122.

[123] Helferich, B.; Gootz, R. *Chem. Ber.* **1929**, *63*, 2788–2792.

[124] Kronzer, F. J.; Schuerch, C. *Carbohydr. Res.* **1974**, *34*, 71–78.

[125] Thiem, J.; Meyer, B. *Chem. Ber.* **1980**, *113*, 3075–3085.

[126] Gervay-Hague, J. *Acc. Chem. Res.* **2016**, *49*, 35–47.

[127] Du, W.; Gervay-Hague, J. *Org. Lett.* **2005**, *7*, 2063–2065.

[128] Hadd, M. J.; Gervay-Hague, J. *Carbohydr. Res.* **1999**, *320*, 61–69.

[129] Jensen, K. J. *J. Chem. Soc., Perkin Trans. 1* **2002**, 2219–2233.

[130] Lam, S. N.; Gervay-Hague, J. *Carbohydr. Res.* **2002**, *337*, 1953–1965.

[131] Schombs, M.; Park, F. E.; Du, W.; Kulkarni, S. S.; Gervay-Hague, J. *J. Org. Chem.* **2010**, *75*, 4891–4898.

[132] van Well, R. M.; Kartha, K. P. R.; Field, R. A. *J. Carbohydr. Chem.* **2005**, *24*, 463–474.

[133] Mukaiyama, T.; Murai, Y.; Shoda, S.-I. *Chem. Lett.* **1981**, 431–432.

[134] Pougny, J. R.; Jacquinet, J. C.; Nassr, M.; Duchet, D.; Milat, M. L.; Sinay, P. *J. Am. Chem. Soc.* **1977**, *99*, 6762–6763.

[135] Schmidt, R. R.; Michel, J. *Angew. Chem. Int. Ed. Engl.* **1980**, *19*, 731–732.

[136] Fischer, E.; Delbrück, K. *Ber. Dtsch. Chem. Ges.* **1909**, *42*, 1476–1482.

[137] Ferrier, R. J.; Hay, R. W.; Vethaviyasar, N. *Carbohydr. Res.* **1973**, *27*, 55–61.

[138] Garegg, P. J.; Henrichson, C.; Norberg, T. *Carbohydr. Res.* **1983**, *116*, 162–165.

[139] Nicolaou, K. C.; Seitz, S. P.; Papahatjis, D. P. *J. Am. Chem. Soc.* **1983**, *105*, 2430–2434.

[140] Codee, J. D. C.; Litjens, R. E. J. N.; van den Bos, L. J.; Overkleeft, H. S.; van der Marel, G. A. *Chem. Soc. Rev.* **2005**, *34*, 769–782.

[141] Hashimoto, H.; Shimada, K.; Horito, S. *Tetrahedron Lett.* **1993**, *34*, 4953–4956.

[142] Bernardes, G. J. L.; Chalker, J. M.; Errey, J. C.; Davis, B. G. *J. Am. Chem. Soc.* **2008**, *130*, 5052–5053.

[143] Floyd, N.; Vijayakrishnan, B.; Koeppe, J. R.; Davis, B. G. *Angew. Chem. Int. Ed.* **2009**, *48*, 7798–7802.

[144] Conte, M. L.; Staderini, S.; Marra, A.; Sanchez-Navarro, M.; Davis, B. G.; Dondoni, A. *Chem. Commun.* **2011**, *47*, 11086–11088.

[145] Boltje, T. J.; Buskas, T.; Boons, G. J. *Nat. Chem.* **2009**, *1*, 611–622.

[146] Zhang, Z.; Ollmann, I. R.; Ye, X. S.; Wischnat, R.; Baasov, T.; Wong, C. H. *J. Am. Chem. Soc.* **1999**, *121*, 734–753.

[147] Wang, C.-C.; Lee, J.-C.; Luo, S.-Y.; Kulkarni, S. S.; Huang, Y.-W.; Lee, C.-C.; Chang, K.-L.; Hung, S.-C. *Nature* **2007**, *446*, 896–899.

[148] Merrifield, R. B. *J. Am. Chem. Soc.* **1963**, *85*, 2149–2154.

[149] Schuerch, C.; Frechet, J. M. *J. Am. Chem. Soc.* **1971**, *93*, 492–496.

[150] Danishefsky, S.; McClure, K.; Randolph, J.; Ruggeri, R. *Science* **1993**, *260*, 1307–1309.

[151] Kahne, D. *Curr. Opin. Chem. Biol.* **1997**, *1*, 130–135.

[152] Plante, O. J.; Palmacci, E. R.; Seeberger, P. H. *Science* **2001**, *291*, 1523–1527.

[153] Seeberger, P. H.; Haase, W.-C. *Chem. Rev.* **2000**, *100*, 4349–4394.

[154] Carrel, F. R.; Seeberger, P. H. *J. Org. Chem.* **2007**, *73*, 2058–2065.

[155] Frechet, J. M.; Schuerch, C. *J. Am. Chem. Soc.* **1972**, *94*, 604–609.

[156] Galonic, D. P.; Gin, D. Y. *Nature* **2007**, *446*, 1000–1007.

[157] Vacas, T.; Corzana, F.; Jiménez-Osés, G.; González, C.; Gómez, A. M.; Bastida, A.; Revuelta, J.; Asensio, J. L. *J. Am. Chem. Soc.* **2010**, *132*, 12074–12090.

[158] Zhang, W.; Chen, Y.; Liang, Q.; Li, H.; Jin, H.; Zhang, L.; Meng, X.; Li, Z. *J. Org. Chem.* **2012**, *78*, 400–409.

[159] Kiessling, L. L.; Splain, R. A. *Annu. Rev. Biochem.* **2010**, *79*, 619–653.

[160] Peri, F. *Chem. Soc. Rev.* **2013**, *42*, 4543–4556.

[161] Hakomori, S. *Acta Anat.* **1998**, *161*, 79–90.

[162] DeMarco, M. L. *Biochemistry* **2012**, *51*, 5725–5732.

[163] Natori, T.; Koezuka, Y.; Higa, T. *Tetrahedron Lett.* **1993**, *34*, 5591–5592.

[164] Du, W.; Kulkarni, S. S.; Gervay-Hague, J. *Chem. Commun.* **2007**, 2336–2338.

[165] Kulkarni, S. S.; Gervay-Hague, J. *Org. Lett.* **2006**, *8*, 5765–5768.

[166] Seeberger, P. H.; Werz, D. B. *Nat. Rev. Drug Discov.* **2005**, *4*, 751–763.

[167] Khan, A. A.; Stocker, B. L.; Timmer, M. S. *Carbohydr. Res.* **2012**, *356*, 25–36.

[168] Sarpe, V. A.; Kulkarni, S. S. *J. Org. Chem.* **2011**, *76*, 6866–6870.

[169] Jensen, H. H.; Bols, M. *Acc. Chem. Res.* **2006**, *39*, 259–265.

[170] Heidelberger, M.; Avery, O. T. *J. Exp. Med.* **1923**, *38*, 73–79.

[171] Ratner, D. M.; Adams, E. W.; Disney, M. D.; Seeberger, P. H. *ChemBioChem* **2004**, *5*, 1375–1383.

[172] Chalker, J. M.; Bernardes, G. J. L.; Davis, B. G. *Acc. Chem. Res.* **2011**, *44*, 730–741.

[173] Dawson, P. E.; Muir, T. W.; Clark-Lewis, I.; Kent, S. B. *Science* **1994**, *266*, 776–779.

[174] Muir, T. W.; Sondhi, D.; Cole, P. A. *Proc. Natl. Acad. Sci. U. S. A.* **1998**, *95*, 6705–6710.

[175] Kitov, P. I.; Shimizu, H.; Homans, S. W.; Bundle, D. R. *J. Am. Chem. Soc.* **2003**, *125*, 3284–3294.

2

PROTECTING GROUP STRATEGIES IN CARBOHYDRATE SYNTHESIS

SHANG-CHENG HUNG[1] AND CHENG-CHUNG WANG[2]

[1]*Genomics Research Center, Academia Sinica, Taipei, Taiwan*
[2]*Institute of Chemistry, Academia Sinica, Taipei, Taiwan*

2.1 INTRODUCTION

Protecting groups are widely used in organic synthesis to prevent a particular functionality in a molecule from being acted upon by reagents of a reaction [1–4]. After the group has served its purpose, further treatment with another set of reagents should allow for its cleavage and the exposure of the original functionality. In the chemical synthesis of carbohydrates, protecting groups play more important and diverse roles [5–7]. Compared to proteins and nucleic acids, the chemical assembly of sugars is much more complicated and challenging. The linear or branching possibilities and the regiochemical and stereochemical considerations for glycosidic bond formation are largely absent in the sequential construction of peptide bonds and phosphodiester linkages.

A carbohydrate molecule carries multiple hydroxy groups with nearly similar reactivities and may hold numerous glycosidic bonds. Additionally, amino functionalities are, at times, engrained in the core structure. Protecting groups are generally utilized to mask the hydroxy and amino groups in a sugar and to control the regioselectivity and stereoselectivity of the glycosylation step [8–16]. Some protecting groups have to be removed selectively during the course of the synthesis to either elongate the oligosaccharide skeletons or to modify functional groups according to the target molecules. The protecting group pattern of each building block, thus, has

Glycochemical Synthesis: Strategies and Applications, First Edition.
Edited by Shang-Cheng Hung and Medel Manuel L. Zulueta.

to fit the logic of orthogonality—the protecting group relationship in which certain groups are able to tolerate the reaction conditions used to remove another—among all the present protecting groups. In other words, prior to the assembly of the desired oligosaccharide skeleton, all of these protecting groups should be judiciously selected and placed properly on each building block. Protecting groups not only prevent the hydroxy groups from reacting, but their inductive effects also influence the reactivity of the building blocks. Apart from these roles, they could also enhance solubility in organic solvents and benefit compound purification and structure elucidation. Owing to their importance in carbohydrate chemistry [17–22], new protecting groups and novel strategies to regioselectively install and orthogonally manipulate different protecting groups are continuously being developed and reported in the literature.

2.2 GENERAL CONSIDERATIONS FOR PROTECTING GROUP SELECTION

2.2.1 Retrosynthesis

Prior to the synthesis of oligosaccharides, the regioselectivity and stereoselectivity issues as well as the future functional group modification of the target molecules should be taken into account. As applied to the synthesis of the model trisaccharide **1** (Scheme 2.1), protecting groups are typically classified as permanent, temporary, and stereodirecting. Trisaccharide **1** can be reasonably synthesized through precursors **2** and **3**, which are derived from building blocks **4–6** and the conjugation molecule **7**. In precursor **2**, the hydroxyls destined to be free in the target molecule **1** need to be masked with permanent protecting groups (*p*P). These groups should withstand most of the planned protection, deprotection, and functionalization conditions as they are

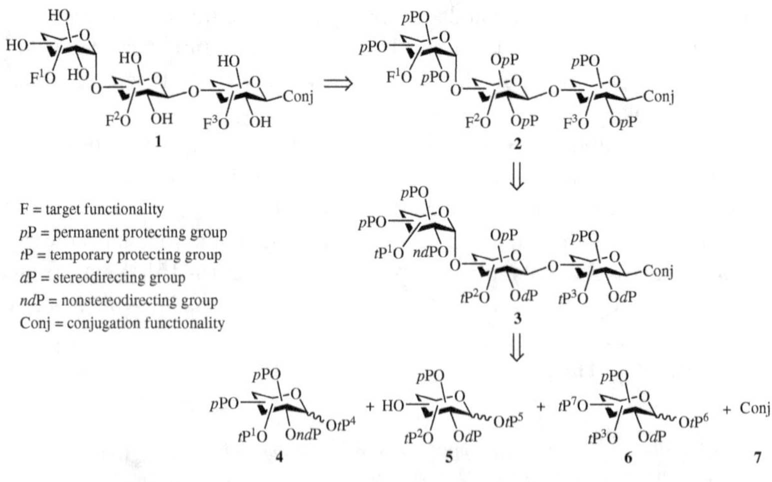

SCHEME 2.1 A retrosynthetic analysis of the model trisaccharide **1**.

usually not removed until the final step of the synthesis. Moreover, their removal should be sufficiently smooth without decomposing the assembled molecule.

To install functional groups (F^1–F^3) at specific positions, temporary protecting groups (*t*P) are placed at the corresponding positions in precursor **3**. Temporary protecting groups are cleaved at intermediate stages during the course of the synthesis, well before the removal of permanent protecting groups. However, they should be able to withstand various reagent treatments until such time that they have to be removed. Several temporary protecting groups may be needed depending on the diversity of functionalities that are to be established. These functionalities could be acetate, sulfate, phosphate, another sugar unit, and others. Consequently, the orthogonality of the multiple types of required protecting groups should be considered in the selection process. One common practice before each glycosylation reaction is the conversion of the anomeric hydroxyl of the glycosyl donor into a labile leaving group that may not survive other reaction conditions. In such case, a temporary protecting group is needed at the anomeric position of each building block (e.g., tP^4 to tP^6 in **4–6**). The hydroxy group designated for the chain elongation in the acceptor also needs a temporary protecting group when it is not currently in use to avoid interference with other reactions. For example, tP^7 in building block **6** is necessary to prevent unwanted side products during the downstream glycosylation between **6** and **7**. After the necessary adduct is generated, tP^7 should be cleaved to permit further coupling with the nonreducing end disaccharide donor produced from building blocks **4** and **5**.

2.2.2 Neighboring Group Participation

The substituent of the carbon adjacent to the anomeric center of each glycosyl donor affects the stereoselectivity of the glycosylation reaction. Protecting groups located at C2 positions of building blocks **4–6** are, therefore, critical for the proper construction of the two 1,2-*trans*- and one 1,2-*cis*-glycosidic bonds in trisaccharide **1**. Generally, a carbonyl-containing functionality, such as ester, amide, and carbonate, installed at the C2 position of the donor directs the formation of the glycosidic bond in nearly exclusive 1,2-*trans* stereoselectivity. The groups attached to form these functionalities are, therefore, defined as stereodirecting groups (*d*P) [11, 23]. The mechanistic aspect of this type of neighboring group participation is discussed in Section 4.2.1. The use of the adjacent stereodirecting group to form a 1,2-*trans*-glycosidic bond is usually very reliable, although a considerable amount of 1,2-*cis* product may be isolated in some rare cases brought about by unreactive acceptors or the poorly nucleophilic participating functionality of the donor. Obviously, such stereodirecting groups cannot be used if a 1,2-*cis*-glycosidic bond is desired. These instances call for nonstereodirecting groups (*nd*P), such as those that form ethers and azide. However, the glycosylation reactions, which now depend on other less effective factors such as the anomeric effect [24, 25], may likely result in mixtures of α and β anomers. In consideration of the glycosidic bonds of compound **1**, a non-stereodirecting protecting group should be placed at O2 of building block **4**, and acyl-type protecting groups are needed at O2 of both building blocks **5** and **6**. Moreover, despite the lack of direct evidence, experimental data [26–30] support the long-range participating effect of the 6-*O*-acyl or carbonate group, directing the preferential formation of, for instance, α-glycosides in D-hexoses by shielding the β face of the anomeric center [31–36].

2.2.3 Inductive Effect

The electron-donating or electron-withdrawing properties of protecting groups have strong influence on the reactivity of both the glycosyl donors and acceptors in a glycosylation reaction. For glycosyl donors of the same backbone bearing the same leaving group, those equipped with more or stronger electron-donating groups, such as ethers, are more reactive than those with more or stronger electron-withdrawing groups, such as esters. Subsequently, reactive donors can be activated by mild promoters. This effect has similar consequences when comparing acceptors of similar type and position of the free hydroxy group. In fact, the chemoselective activation of a donor with more electron-donating groups ("armed") over a donor with more electron-withdrawing groups ("disarmed") proved feasible and has been utilized in the "armed–disarmed" one-pot sequential glycosylation [37–52]. Furthermore, the inductive effect brought about by protecting groups can already be quantified [53, 54].

2.3 COMMON PROTECTING GROUPS IN CARBOHYDRATE SYNTHESIS

The protecting groups used in carbohydrate chemistry are no different from those used in normal organic synthesis, but several types are often needed for a single monosaccharide building block. Carbohydrate chemists employ numerous protecting groups, which are introduced into the molecules in different stages and under diverse conditions [1, 2]. Presented here are protecting groups commonly encountered in carbohydrate synthesis and the typical methods employed for their installation and eventual cleavage.

2.3.1 Benzyl Ethers

Benzyl (Bn) and substituted Bn ethers are generally electron-donating and nonstereodirecting groups encompassing several members. They are widely used in the synthesis of complex oligosaccharides and glycoconjugates, with each member capable of being selectively removed under appropriate conditions. Different approaches are available for their formation (Scheme 2.2), although the Williamson ether synthesis using an arylmethyl halide and NaH is one of the most convenient and frequently used [1, 2]. In spite of the longer reaction time, a Ag_2O-mediated benzylation is particularly useful when a base-sensitive moiety or protecting group is already present [55, 56]. The same purpose can also be achieved by using Bn imidate as the benzylation reagent under acid catalysis [57–60]. Et_3SiH-reductive etherification of a trimethylsilyl (TMS) ether with benzaldehyde or substituted benzaldehyde promoted by trimethylsilyl triflate (TMSOTf) offers another alternative for O-benzylation under acidic condition [61].

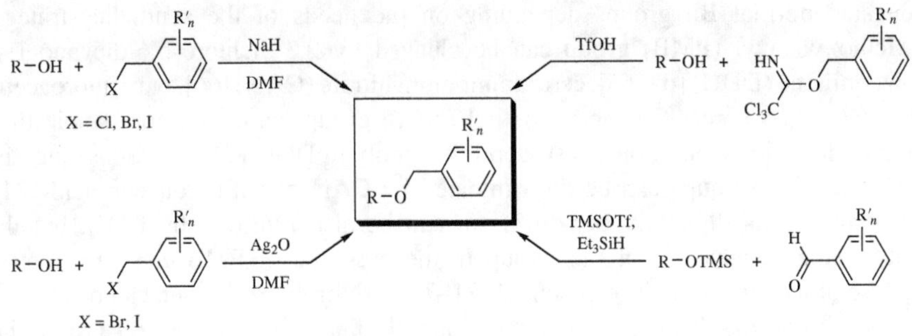

SCHEME 2.2 Bn and substituted Bn ether formations. DMF, dimethylformamide.

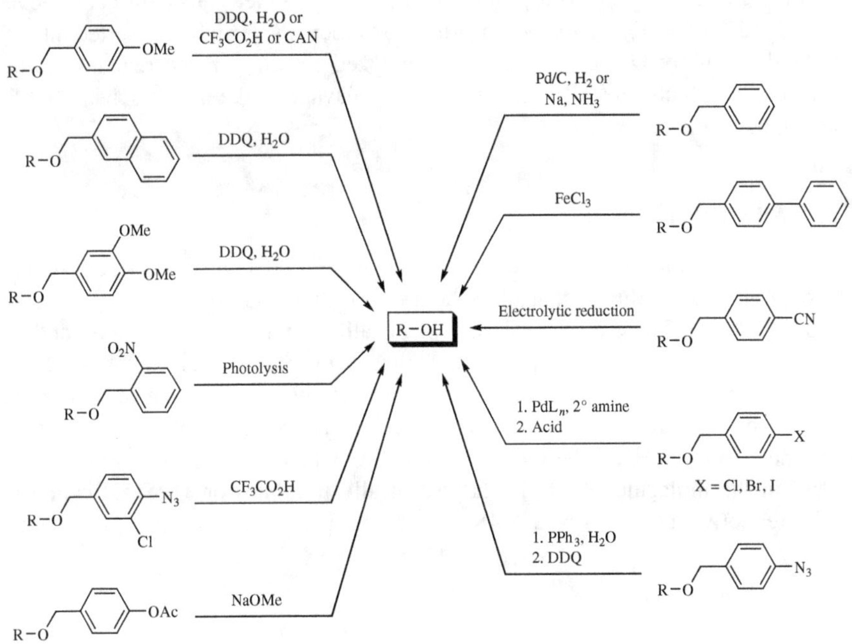

SCHEME 2.3 Cleavage conditions of Bn and substituted Bn ethers.

Different substituents allow Bn-type ethers to be cleaved at different conditions and therefore provide good orthogonality among themselves (Scheme 2.3). The Bn group is resistant to both strong acids and bases and usually serves as permanent protecting group. It can be removed by Birch reduction or hydrogenolysis in the final step of the synthesis to release free hydroxy groups. However, if the target molecules contain groups that are sensitive to hydrogenolysis, other protecting groups need to be considered in place of Bn. With unique reagent combinations that can be applied for the cleavage of their ethers [1–4], substituted Bn moieties can act as temporary protecting groups. They could also be removed by reduction or hydrogenolysis alongside normal Bn groups depending on the needs of the synthetic strategy. *p*-Methoxybenzyl (PMB) group can be cleaved by 2,3-dichloro-5,6-dicyano-1,4-benzoquinone (DDQ) [62–65], ceric ammonium nitrate (CAN) [66], or trifluoroacetic acid [67–70], whereas 2-naphthylmethyl (NAP) group, which is more stable than PMB under acidic conditions, is susceptible mainly to DDQ [71–73]. Consequently, PMB and NAP groups can be discriminated by CAN or trifluoroacetic acid [74]. Furthermore, 3,4-dimethoxybenzyl group can be also removed by DDQ, but the DDQ-mediated cleavage of this group in the presence of PMB can be achieved because of its lower oxidation potential [71–73]. *o*-Nitrobenzyl group can be cleaved by photolysis [75–77]. Electrolytic reduction eliminates *p*-cyanobenzyl group [78, 79], $FeCl_3$ is used to remove *p*-phenylbenzyl group [80], and neat trifluoroacetic acid cleaves 4-azido-3-chlorobenzyl group [81]. *p*-Azidobenzyl group can undergo Staudinger reaction to afford the *p*-aminobenzyl group, which can be further removed by DDQ [82]. Similarly, halogen-substituted Bn ethers [83] can be converted into the acid-labile aminobenzyl ethers by a Pd-catalyzed reaction in the reactivity order I>Br>Cl>F, which allows for the orthogonal cleavage of the halide-substituted Bn ethers in the same order.

2.3.2 Allyl Ethers

Allyl-type protecting groups are easy to form, have moderate stability, and require mild deprotection conditions that allow them to be orthogonal to many other protecting groups [1, 2, 84]. There are many ways of installing and cleaving the allyl and the substituted allyl groups such as prenyl and cinnamyl groups (Schemes 2.4 and 2.5).

Allylation conditions are similar to those used for benzylation to a certain extent. Williamson's etherification using allyl bromides and NaH smoothly produces the allyl ethers [85]. NaH can be replaced with BaO when base-sensitive moieties are present in the molecule [86, 87]. The use of allyl imidates or TMSOTf-catalyzed Et_3SiH-reductive allylation of a TMS ether using allyl aldehydes also provides the

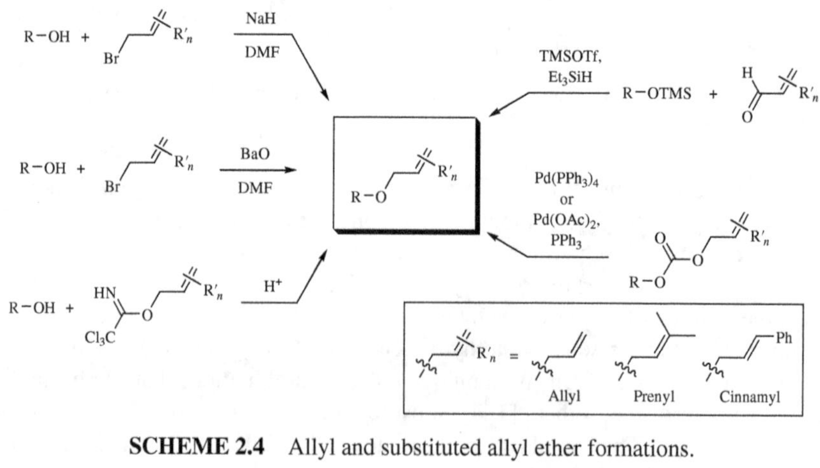

SCHEME 2.4 Allyl and substituted allyl ether formations.

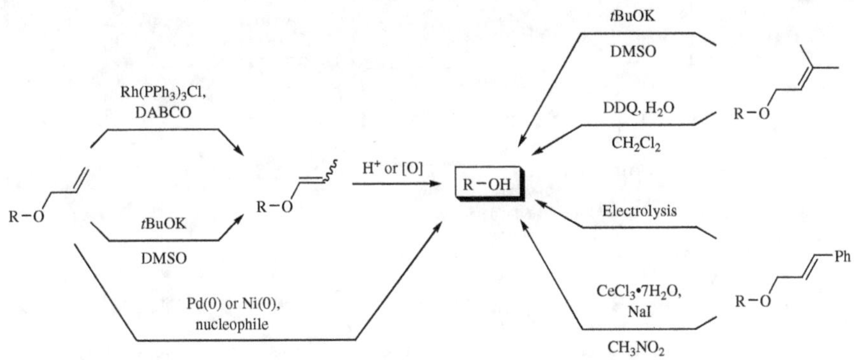

SCHEME 2.5 Cleavage conditions of allyl and substituted allyl ethers.

acidic conditions for allyl etherification [57, 88, 89]. The conversion of allyl carbonates into the corresponding allyl ethers by $Pd(PPh)_4$ or by $Pd(OAc)_2$ and PPh_3 allows allylation in a nearly neutral environment, although an extra synthetic step for the allyl carbonate formation is needed [90].

To cleave allyl ethers, a frequently used method is a two-step procedure that includes the isomerization of the allyl to a vinyl group by treatment with potassium *tert*-butoxide (*t*BuOK) in dimethyl sulfoxide (DMSO) or with Wilkinson's catalyst [$Rh(PPh_3)_3Cl$] and 1,4-diazabicyclo[2.2.2]octane (DABCO) [91, 92], followed by acid hydrolysis [57, 88, 93] or oxidation [94–98]. On the other hand, Pd(0) or Ni(0) in combination with K_2CO_3 in MeOH was found to cleave the allyl group in one step [99–106]. Prenyl groups can be removed in one step by *t*BuOK in DMSO through a γ-elimination mechanism rather than isomerization [92]. DDQ in wet CH_2Cl_2 also readily cleaves the prenyl group [107, 108]. The cleavage of cinnamyl group can be achieved by electrolysis [109, 110] or the combination of $CeCl_3$ and NaI and thus provide orthogonality with other protecting groups [111].

2.3.3 Silyl Ethers

Silyl ethers, such as TMS, triethylsilyl (TES), triisopropylsilyl (TIPS), *tert*-butyldimethylsilyl (TBS), and *tert*-butyldiphenylsilyl (TBDPS) ethers (Scheme 2.6), are frequently used in carbohydrate chemistry due to their unique reactivity and steric effect (see Section 2.4.1) [1, 2]. These groups are typically introduced onto the molecule by treatment with the corresponding silyl halide or triflate (OTf) and a base, such as Et_3N, pyridine, 2,6-lutidine, or imidazole.

The stabilities of silyl ethers are closely related to the electronic and steric effect of the substituents on the silicon atom and are generally proportional to the steric hindrance provided by the substituents. Moreover, electron-withdrawing substituents on the silicon atom increase the stability of the silyl groups toward acid but decrease their stability toward base. Consequently, their stability in acid follows the order: TMS < TES < TBS < TIPS < TBDPS, whereas in base TMS < TES < TBS ~ TBDPS < TIPS [1, 112, 113]. Selective deprotection among these groups can thus be achieved

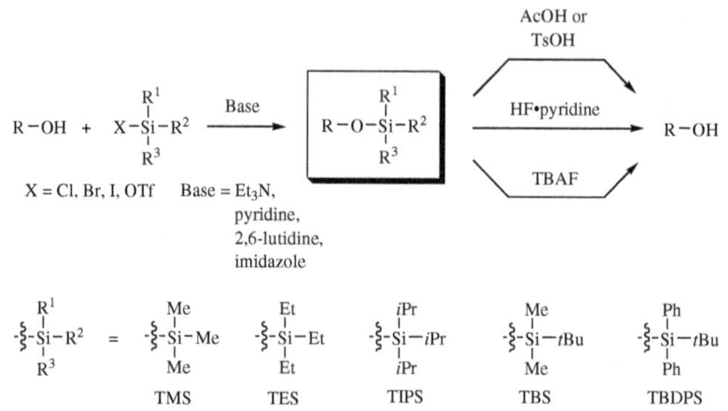

SCHEME 2.6 Typical methods for silyl ether formation and cleavage.

in a well-controlled reaction condition [114, 115]; however, migration of the silyl group can sometimes be observed [116–118]. The steric bulkiness of TBDPS, TIPS, and TBS groups is usually utilized to selectively protect the primary alcohols in the presence of secondary alcohols.

Silyl ethers can be cleaved by acid hydrolysis (Scheme 2.6), but for the more acid-resistant TIPS and TBDPS groups, strong acids, which may also cleave glycosidic bonds, are necessary. One of the features that makes silyl ethers unique among other protecting groups is that they can be readily cleaved by fluoride ion because of the strong Si—F bond (590 kJ·mol^{-1}) [114]. The fluoride ion sources commonly used for such purposes are HF·pyridine complex and tetrabutylammonium fluoride (TBAF).

2.3.4 Esters

Ester groups are base sensitive, which makes acyl moieties as suitable choices as temporary protecting groups for sugar hydroxyls. Their popularity can also be attributed to the ease of installation and removal, as well as the stereodirecting effect by neighboring group participation toward the anomeric center. As mentioned previously, esters reduce the reactivity of both glycosyl donors and acceptors due to their electron-withdrawing property. The commonly used esters in carbohydrate chemistry include acetate, chloroacetate, dichloroacetate, trichloroacetate, benzoate, levulinate, and pivaloate (Scheme 2.7) [1–6]. They are generally introduced by treatment with the corresponding acid chloride or acid anhydride in the presence of pyridine or Et_3N with 4-dimethylaminopyridine (DMAP) often serving as catalyst. Acid anhydride treatment with TMSOTf as catalyst provides an acidic condition for esterification [119, 120]. The Steglich esterification allows for the direct coupling of an acid and an alcohol under mild conditions by using *N,N′*-dicyclohexylcarbodiimide (DCC) and DMAP [121]. The steric bulkiness of the Piv group allows it to selectively acylate more reactive sugar hydroxyls [122].

The removal of acyl groups to form alcohols can be carried out under basic conditions (Scheme 2.8). The simplest method uses catalytic NaOMe in MeOH. Other bases such as NaOH and ammonia gas can also cleave esters. The base sensitivity of acetate, chloroacetate, dichloroacetate, and trichloroacetate increases with the number of the

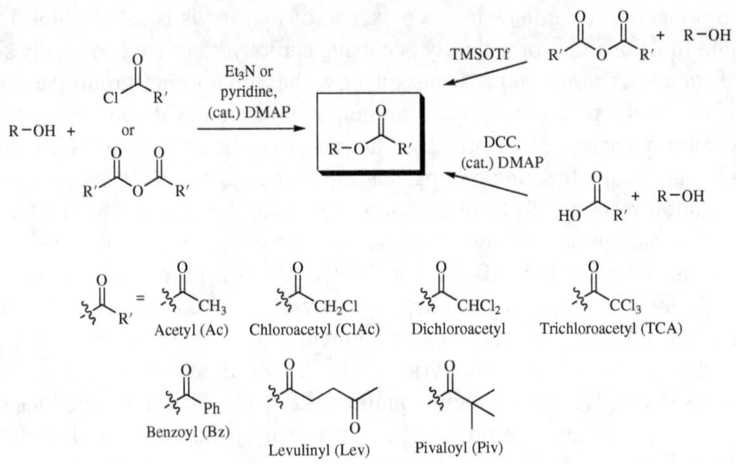

SCHEME 2.7 Commonly used methods of ester formation.

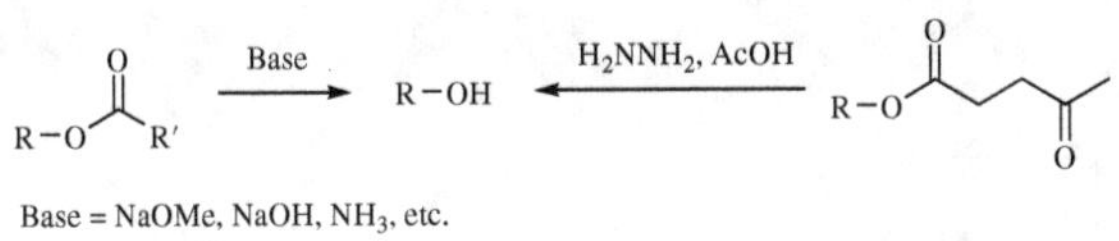

SCHEME 2.8 Commonly used conditions to cleave esters into alcohols.

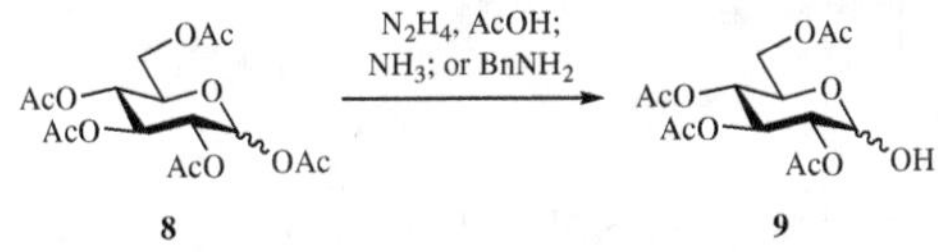

SCHEME 2.9 Selective cleavage of the anomeric acetate.

electron-withdrawing chlorine atoms [123]. Benzoate and pivaloate groups, which are relatively more stable with bases, need longer reaction times or harsher conditions to cleave [1, 2, 122, 124]. They are also less prone to migration unlike the simpler acetates. Moreover, the levulinate group can be selectively removed under a nearly neutral condition by hydrazinium acetate [125]. Furthermore, the selective cleavage of an anomeric ester, such as that in the peracetylated glucose **8**, can be easily carried out by using hydrazinium acetate, ammonia, or benzylamine as the base to afford the hemiacetal **9** (Scheme 2.9) [125, 126]. The exposed anomeric hydroxyl of **9** can be readily converted into an anomeric leaving group, such as imidate.

2.3.5 Amine Protecting Groups

Amino-type groups are also present in carbohydrate molecules although the number is usually much fewer than that of the hydroxy groups. Selective protections among the amino groups are therefore less frequent. Because of their stronger nucleophilicity, selective protection of amines in the presence of hydroxyls is usually not difficult. The amino functionalities of naturally occurring carbohydrates are commonly situated at C2 of monosaccharide units. Consequently, the neighboring group participating effect of the amine protecting group can help determine the stereoselectivity of the glycosylation reactions. Most protected amine groups such as amides, imides, and carbamates are stereodirecting groups, leading to the 1,2-*trans*-glycosidic linkage in a glycosylation reaction. To form a 1,2-*cis*-glycosidic bond, masking the C2 amine as an azide is one of the very few choices. The conversion of an amine into an azide is usually carried out using triflic azide (TfN_3), which is prepared by mixing NaN_3 and triflic anhydride (Tf_2O) with $CuSO_4$ as catalyst (Scheme 2.10) [127]. Amides and carbamates are typically generated upon treatment of the amine with the corresponding acid chloride or chloroformate in pyridine [1, 2, 128]. Phthaloyl (Phth)-type groups also possess strong 1,2-*trans* directing nature [128, 129]. Their introductions require two steps, which include opening of the phthalic anhydride to form the amide and cyclization by treatment with Ac_2O in pyridine. The variants of the Phth group that are also in use include dichlorophthaloyl (DCPhth) [130] and tetrachlorophthaloyl (TCP) protecting groups [128].

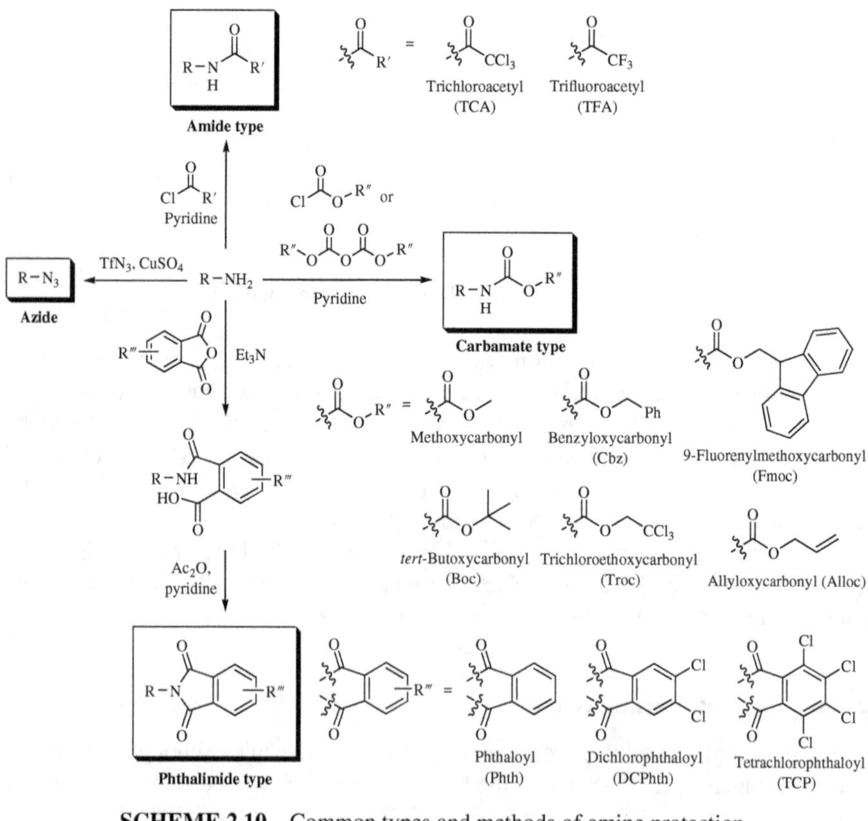

SCHEME 2.10 Common types and methods of amine protection.

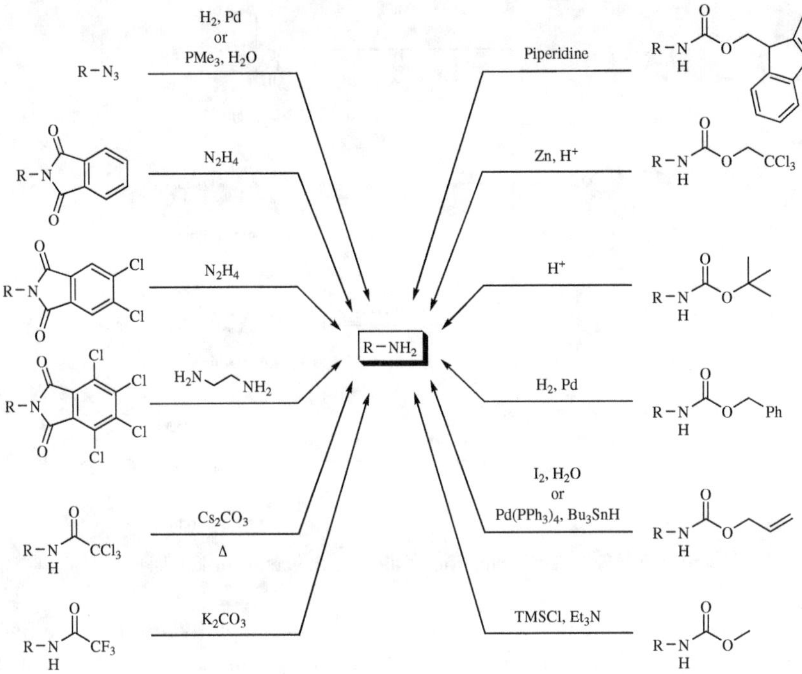

SCHEME 2.11 Cleavage of amine protecting groups.

The azido group is both base and acid resistant but is transformed to the amine by reduction methods, such as the Staudinger reaction [131] and hydrogenolysis [1, 2] (Scheme 2.11). On the other hand, the phthalimido-type groups are all susceptible to hydrazine, but the sensitivity increases with the number of the chlorine atoms. Therefore, the removal of DCPhth or TCP group can be achieved under a milder condition as compared to that used for Phth group [128, 130]. Carbamates can be cleaved under different conditions. Fmoc group, which is widely used in peptide synthesis, is acid stable but base sensitive and is usually cleaved by using piperidine [132]. Troc group can be cleaved by zinc reduction under acidic conditions [133]. Boc group is acid sensitive and therefore can be removed under a variety of acidic conditions [1]. Cbz group is removed by hydrogenolysis [134], whereas Alloc group can be acted upon by iodine in the presence of water or by Pd(0) [135, 136]. Methyl carbamate group is base sensitive and can be cleaved by using TMSCl and Et_3N [137]. The amide bond formed by TCA and TFA groups can be cleaved by Cs_2CO_3 [138] and K_2CO_3 [139, 140], respectively. However, the TCA group is more often converted directly into the naturally occurring Ac group by Bu_3SnH reduction or zinc reduction in acetic acid [141, 142].

2.3.6 Diol Protection with Acetals and Ketals

Acetals and ketals protect two hydroxyls at a time by forming either a 1,3-dioxane or a 1,3-dioxolane ring (Scheme 2.12). The most common ketal and acetal protecting groups are isopropylidene or acetonide and the benzylidene-type groups with

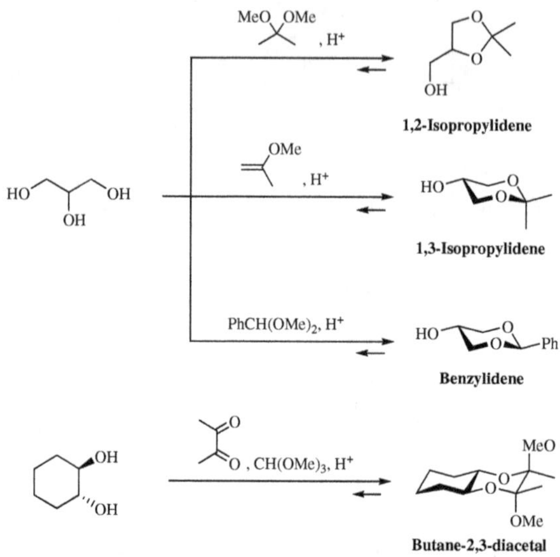

SCHEME 2.12 Protection of diols with acetals or ketals.

typical aryl moieties as phenyl, *p*-methoxyphenyl, and 2-naphthyl groups. The five-membered isopropylidene ring formation is thermodynamically favored. Hence, treatment with 2,2-dimethoxypropane under acidic condition should supply the 1,2-*O*-isopropylidene derivative as the major product [143]. On the other hand, 2-methoxypropene prefers the kinetically controlled six-membered ketal ring and therefore is used to afford the 1,3-*O*-isopropylidene product [144, 145]. Benzylidene-type groups also favor the formation of the six-membered acetal ring with the aryl group in the equatorial orientation [146]. Conversely, protection of *trans*-diols with the aforementioned acetal or ketal is difficult because of the ring strain of the product. A recently developed butane-2,3-diacetal (BDA) protection by using butane-2,3-dione as the reagent exclusively favors the *trans*-diol due to the chair form of the product [147]. Ketals and acetals are removed completely by acid treatment. The utility and manipulations of these diol protecting groups are further discussed in Section 2.4.2.

2.4 REGIOSELECTIVE PROTECTION OF MONOSACCHARIDES

The differentiation of the hydroxy (and amino) groups of monosaccharides to form building blocks is the primary task encountered in carbohydrate synthesis. The anomeric position is usually the first to be functionalized. Because of its proximity to the ring oxygen, the anomeric hydroxyl is the most acidic and also the most readily eliminated in an acid-catalyzed reaction to be replaced by alcohols, thiols, or selenols to afford a glycoside. However, the differentiation and transformation of the remaining hydroxyls are generally difficult and inefficient. Multistep protection–deprotection sequences are frequently needed for the preparation of each building block.

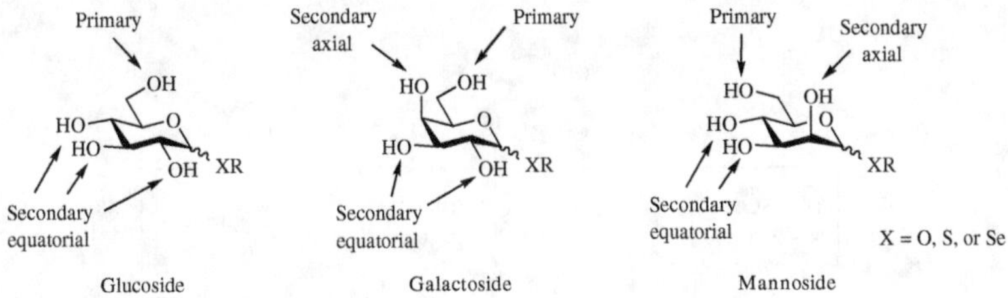

FIGURE 2.1 The hydroxy groups of glucosides, galactosides, and mannosides.

Because of the diversity of monosaccharides, the protection of the most naturally abundant glucose, galactose, and mannose are discussed here in more detail. Other than the primary 6-hydroxyl, the 2-, 3-, and 4-hydroxyl of glucosides are all secondary and equatorial, but galactoside and mannoside contain an axial secondary hydroxyl at C4 and C2, respectively (Fig. 2.1). Carbohydrate chemists utilize the reactivity order: primary hydroxyl > secondary equatorial hydroxyl > secondary axial hydroxyl as guide for regioselective protection. However, this can be a very tricky task and hence require experience and, sometimes, trial and error. Once the anomeric center is secured, the manipulation of these hexoses typically shifts toward either of the two most commonly used methodologies: (i) treatment with bulky reagents to generate the 6-O-protected products and (ii) use of acetal protecting groups to mask O4 and O6 or other vicinal *cis*-diols [1–7].

2.4.1 Through Differentiation of Primary Hydroxyls Using Bulky Protecting Groups

Primary hydroxyls can be selectively protected in the presence of their secondary counterparts with bulky protecting groups, such as TBDPS, TBS, and triphenylmethyl (trityl, Tr) groups, under basic conditions with good yields and high regioselectivity. Accordingly, the 6-O-protected monosaccharides **11–13** could be generated from the starting compound **10** with the proper application of the corresponding chloride reagents (Scheme 2.13). Aside from TBDPS and TBS groups being selectively removed by fluoride sources, the trityl group is susceptible to mild acids such as acetic acid. After protection at O2, O3, and O4 with Bn, allyl, Ac, or Bz groups to produce the fully protected derivatives **14–16**, the removal of bulky groups could effectively provide the 6-alcohol **17** [148–153].

After 6-O-protection, the outright partial alkylation or acylation of the remaining three secondary hydroxyls usually results in product mixtures and poor yields (Scheme 2.14). For example, monoacetylation of glucoside **18** under the typical basic procedure gave the 3-acetate **20** as the major product in 48% yield, but the 2-acetate **19** and the 4-acetate **21** were also produced, necessitating the cumbersome separation of these regioisomers [154]. Regarding the random acetylation of galactoside **22** [155], the axial 4-hydroxyl was still observed as the most unreactive, but the difference is not sufficient to provide good selectivity. Fortunately, for sugars containing *cis*-diols, one can utilize the dialkylstannylene acetal or trialkylstannyl ether generated

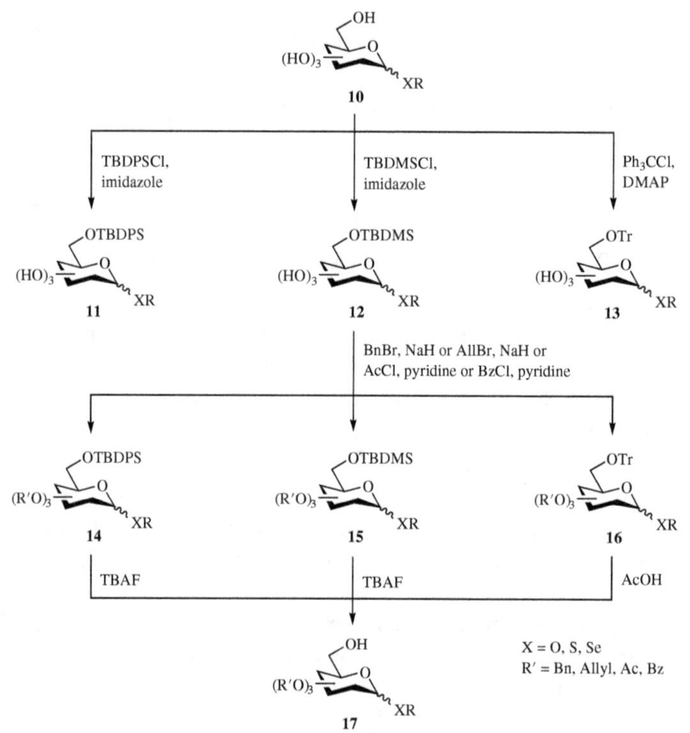

SCHEME 2.13 Selective protection of the primary hydroxyl using bulky groups.

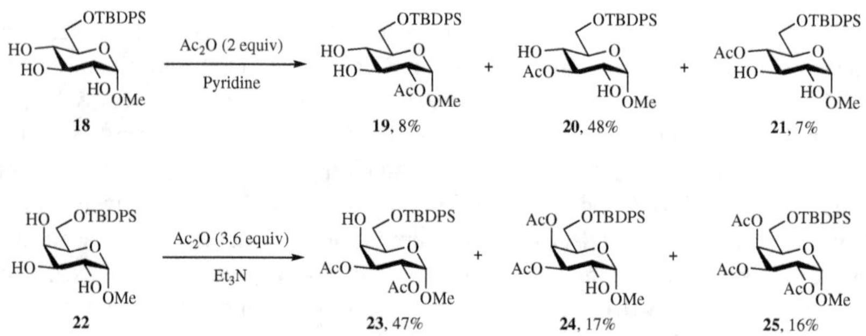

SCHEME 2.14 Acetylation of the 6-*O*-protected glucoside **18** and galactoside **22**.

in situ by treatment with a dialkyltin oxide or a bis(trialkyltin) oxide reagent, respectively, to perform the highly regioselective acylation or alkylation at the equatorial hydroxyl [156–161]. For example, after the stannyl acetal formation using Bu_2SnO on both the *cis*-diols of galactoside **26** and mannoside **28**, the Bn and allyl groups can be regioselectively installed at the equatorial 3-hydroxyl of both compounds (Scheme 2.15) [162, 163]. The precise mechanism and the exact complex structure remain unclear, but the nucleophilicity enhancement of the equatorial hydroxyl by the tin complexes is

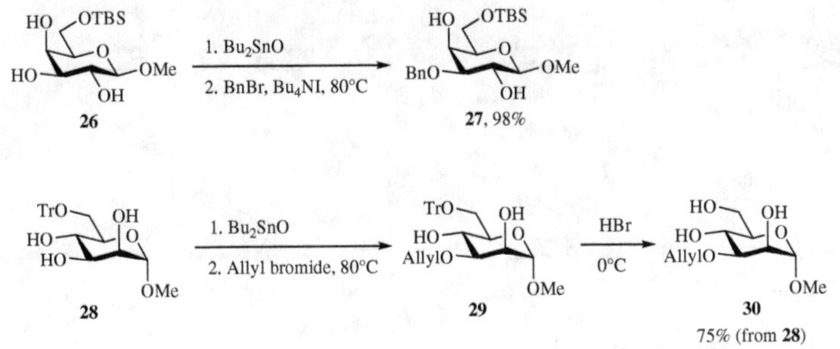

SCHEME 2.15 Regioselective tin-activated etherification.

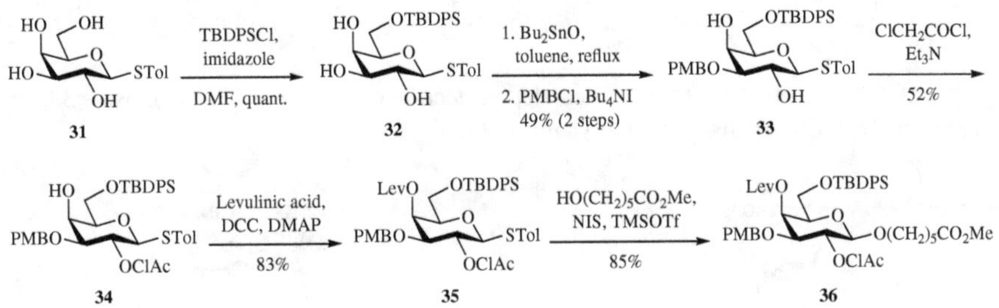

SCHEME 2.16 Preparation of the orthogonally protected galactoside **36**. NIS, *N*-iodosuccinimide; Tol, 4-tolyl.

likely responsible for the high regioselectivity [156–161]. While it was once thought that more than an equimolar amount of the tin reagent was needed, catalytic amounts of Bu_2SnO or Me_2SnCl_2 was found to be sufficient for regioselective acylation and sulfonylation [164–166]. Recently, a borinic acid-catalyzed methodology that also results in the acylation of the equatorial hydroxyl of *cis*-diols was developed [167].

A galactoside building block preparation involving a combination of 6-*O*-TBDPS protection and stannylene acetal-mediated etherification was reported by Chi-Huey Wong and coworkers (Scheme 2.16) [168]. Silylation of thiogalactopyranoside **31** with TBDPSCl and imidazole quantitatively furnished the 2,3,4-triol **32**, which was regioselectively *p*-methoxybenzylated at O3 by using the tin-mediated benzylation methodology to produce the 2,4-diol **33**. As expected, the ensuing chloroacetylation of **33** favored the equatorial over the axial hydroxyl to give the 4-alcohol **34**. Further protection with the Lev group gave the fully protected glycosyl donor **35**, which was coupled to methyl 6-hydroxyhexanoate to afford the β-linked derivative **36**. The protecting groups in **36** are orthogonal to each other and, under different reaction conditions, could supply different glycosyl acceptors for oligosaccharide library synthesis.

Interestingly, Osamu Onomura and coworkers reported that the Me_2SnCl_2-catalyzed regioselective benzoylation of methyl α-D-glucopyranoside (**37**) furnished the 2-benzoate **40** (Scheme 2.17) [166]. Meanwhile, exposure of galactoside **38** and mannoside **39** to the same condition gave the equatorial benzoates **41** and **42**,

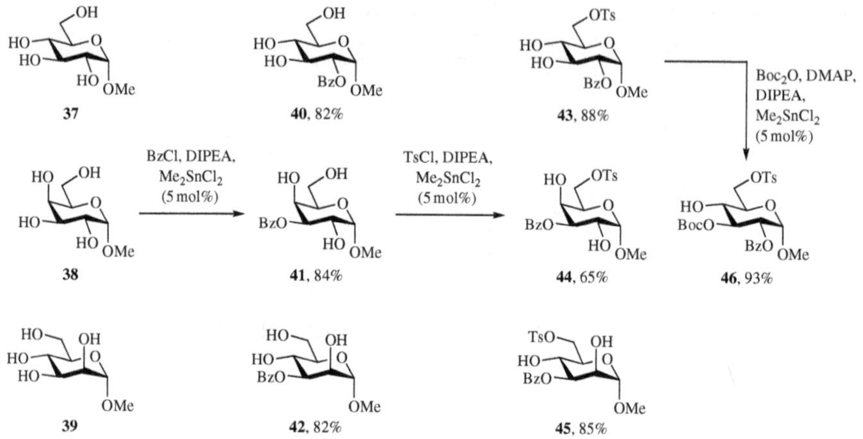

SCHEME 2.17 Regioselective tin-mediated protection of glucoside **37**, galactoside **38**, and mannoside **39**. DIPEA, diisopropylethylamine; Ts, tosyl.

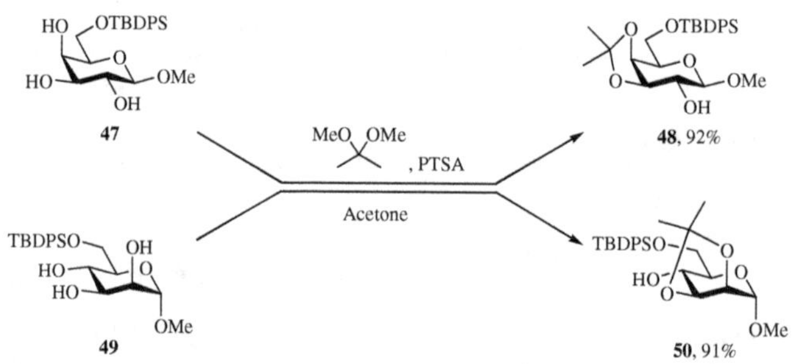

SCHEME 2.18 Regioselective isopropylidenation among the three secondary hydroxyls. PTSA, *p*-toluenesulfonic acid.

respectively. Using the same catalyst, further tosylation of the aforementioned triols occurred at the 6-O-position and afforded the diols **43–45**. With the 3,4-diol **43**, Boc group installation in the presence of Me_2SnCl_2 favored O3 to give the 4-alcohol **46**.

Another approach that led to differentiation is the combination of bulky group and isopropylidene ketal. With the ketal strongly favoring formation onto vicinal *cis*-diols, the 6-O-protected galactoside **47** [169, 170] and mannoside **49** [171] provided the 2-alcohol **48** and 4-alcohol **50** in high yields, respectively (Scheme 2.18).

2.4.2 Through Protection of Diols with Acetals or Ketals

The orientations of the available hydroxy groups in a sugar help determine the regioselectivity as well as the ring size of the acetal and ketal formation. There are no vicinal *cis*-diols in a glucopyranoside, and therefore, regardless of the reagent used, the isopropylidene- and benzylidene-type groups get installed mainly at O4 and O6 to form 1,3-dioxane rings. Also, as shown in Scheme 2.18, treatment with

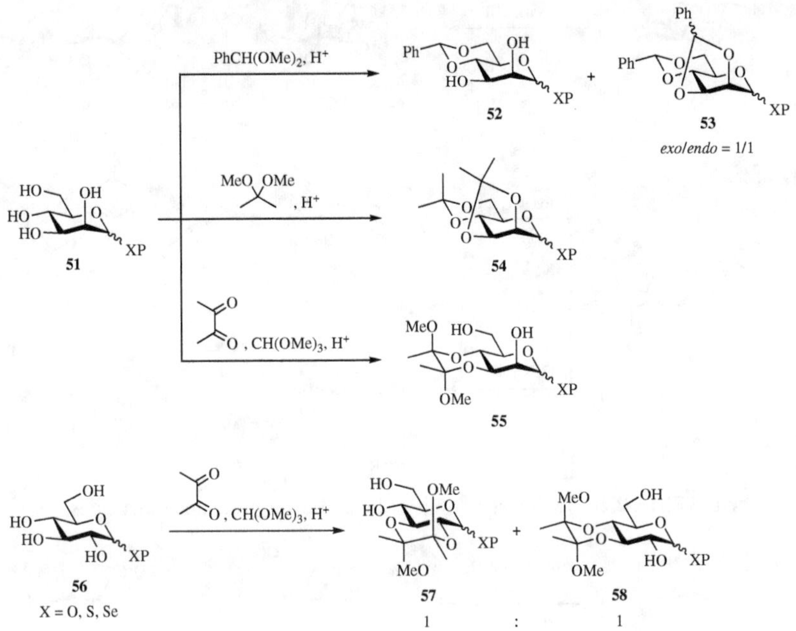

SCHEME 2.19 Notable formations of acetals and ketals on pyranosides.

2,2-dimethoxypropane readily protects vicinal *cis*-diols to generate an isopropylidene ketal. However, there is poor regioselectivity for both benzylidene and isopropylidene formation between the 2,3-*cis*-diol and the 4,6-diol in an unprotected mannopyranoside (Scheme 2.19). The benzylidenation of mannoside **51** leads to the mixture of the 4,6-*O*-benzylidene product **52** as the minor and the 2,3:4,6-di-*O*-benzylidene mannoside **53** as the major product. Unfortunately, selective hydrolysis among these two benzylidenes is not available so far [1–6]. While the phenyl moiety is exclusively equatorial in the 4,6-*O*-benzylidene group, the 2,3-*O*-benzylidene of **53** shows no stereoselectivity and forms the *exo*- and *endo*-isomer usually in a 1:1 ratio. Although there is no stereochemistry involved in the isopropylidene formation, the regioselectivity is also poor and gives the fully protected compound **54** as the major product. BDA protection, fortunately, has a good regioselectivity toward the vicinal 3,4-*trans*-diol of **51** and gives compound **55**, leaving the axial 2-hydroxyl and the primary 6-hydroxyl for easy differentiation [147]. Conversely, the same method for BDA formation shows no selectivity between the two *trans*-diol sets of glucopyranoside **56**, resulting in a 1:1 mixture of the 4,6-diol **57** and the 2,6-diol **58**.

After the simultaneous protection at O4 and O6, the differentiation of the remaining 2-hydroxyl and 3-hydroxyl is puzzling for glucosides because they are both secondary and equatorial. Conditions reported in the literature usually result in mixtures of regioisomers [6]. For example, the regioselective protection of diol **59** via a copper complex intermediate gave the 3-O-acylated derivatives **61** and **63** as the major products in moderate yields (Scheme 2.20) [172]. Nonetheless, 2-O-benzoylation

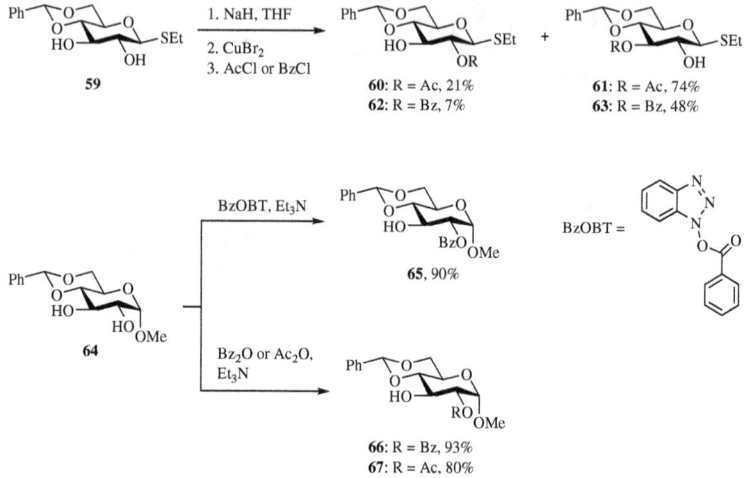

SCHEME 2.20 Monoacylations of benzylidene-protected glucosides.

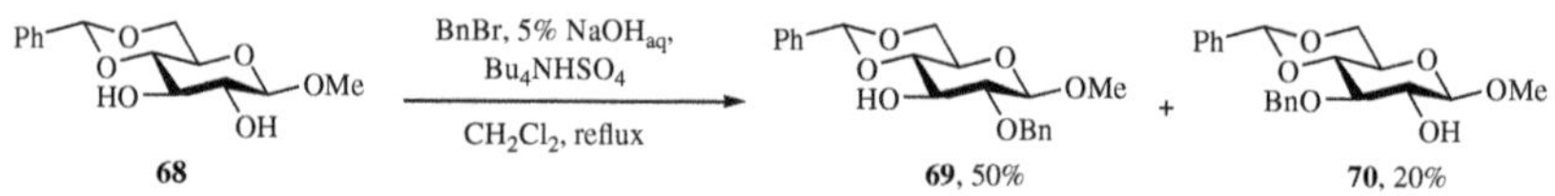

SCHEME 2.21 Regioselective benzylation of the diol **68**.

and acetylation of diol **64** can be achieved by using benzoyloxybenzotriazole (BzOBT) or the simple Bz_2O and Ac_2O with Et_3N as a base in high yield and regioselectivity [173]. Such regioselectivity could be attributed to the inductive effect of the two oxygens attached to the anomeric center, causing the adjacent 2-hydroxyl to be more acidic than the 3-hydroxyl.

Similarly, regioisomers are usually obtained upon alkylation of the 2,3-diol. For instance, a mild reaction between diol **68** and BnBr with NaOH as a base under biphasic condition gave the 2-O- and 3-O-benzylated products **69** and **70** in 50 and 20% yield, respectively (Scheme 2.21) [174]. Conversely, tin-activated acylation or alkylation using Bu_2SnO or $Bu_2Sn(OMe)_2$ usually shows high regioselectivity toward the 3-hydroxyl of 4,6-O-benzylidenated glucosides. Thus, the corresponding tin-mediated benzylation and benzoylation of diols **59** and **64** afforded the 3-O-functionalized products **71–73** in moderate to good yields (Scheme 2.22) [156, 161, 175].

Despite the good regioselectivity, the drawbacks of using organotin reagents are their high toxicity and sometimes the difficulty in separating the desired products from these reagents. Alternatively, a TMSOTf-catalyzed Et_3SiH-reductive benzylation of the 2,3-bis-TMS ether **74** with benzaldehyde at −78°C was found to form the Bn ether in highly regioselectivity at C3 (Scheme 2.23) [89]. Changing the catalyst to $Cu(OTf)_2$ did not affect the regioselectivity, although the yield is slightly lower due to the hydrolysis of the bis-TMS ether [176].

With orientations similar to those in glucopyranosides, the remaining hydroxyls of 4,6-O-protected galactopyranosides are difficult to selectively protect. Taking

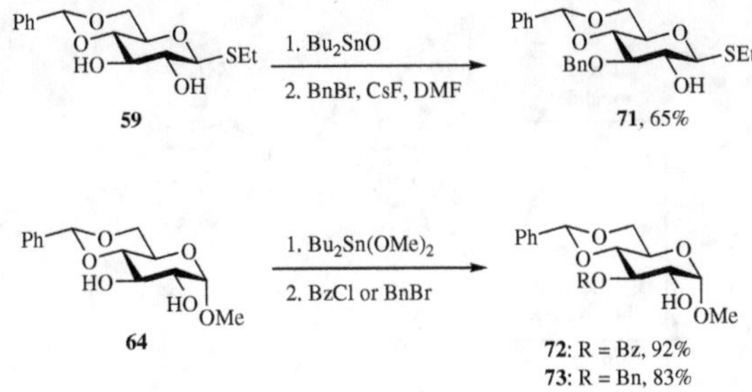

SCHEME 2.22 Tin-mediated benzylation or benzoylation of diols **59** and **64**.

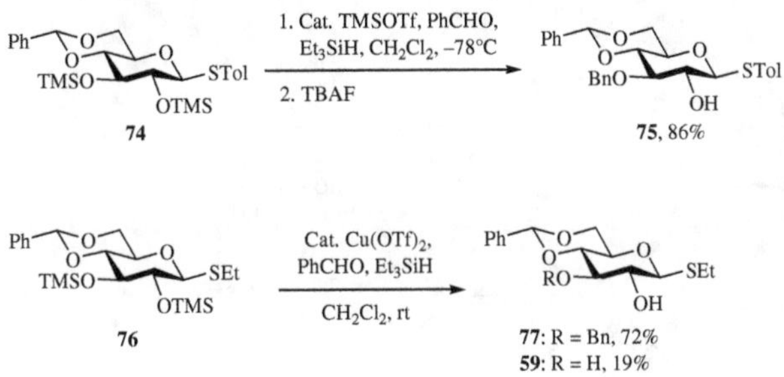

SCHEME 2.23 TMSOTf- or $Cu(OTf)_2$-catalyzed Et_3SiH-reductive 3-O-benzylation of 4,6-*O*-benzylidenated glucopyranosides.

the benzoylation of galactopyranoside **78** as an example, both the yield and regioselectivity were poor under the typical BzCl in pyridine (Scheme 2.24a) [177]. Fortunately, some methods that provide good selectivity have been reported. In the presence of aqueous NaOH and tetrabutylammonium chloride as the phase-transfer catalyst, the regioselectivity of benzoylation of **78** with BzCl was found dependent on the composition of the solvent. The reaction in benzene gave the 2-benzoate **79** in 69% yield, whereas the mixture of benzene and hexamethylphosphoric triamide (HMPT) produced the 3-benzoate **80** in 62% yield [178].

Interestingly, the configuration at the anomeric center was also found influential to the acylation of 4,6-O-benzylidenated galactopyranoside. The Ag_2O-mediated acylation of the α-galactoside **78** with AcCl or BzCl in the presence of potassium iodide gave the 2-O-acylated compounds **81** and **79** in excellent yields. On the other hand, the acylation of the β-galactoside **82** led to the 3-O-acylated products **83** and **84** also in excellent yields (Scheme 2.24b) [179]. Additionally, tin-mediated benzylation of the same starting galactoside installed the Bn group at O3, similar to the results with glucosides [180].

With substantial reactivity differences, the differentiation of the 2,6-diol or the 4,6-diol of galactosides is relatively easier (Schemes 2.25 and 2.26). The acetylation of the 2,6-diol **85** by using AcCl and Et_3N gave the 6-acetate **86** and the 2,6-diacetate **87** in 62 and 24% yields, respectively [181]. Replacement of Et_3N with the more hindered *sym*-collidine as base and lowering the temperature to −78°C resulted in a better selectivity and yield for **86** (88%). Treatment of diol **88** with BnBr and NaH

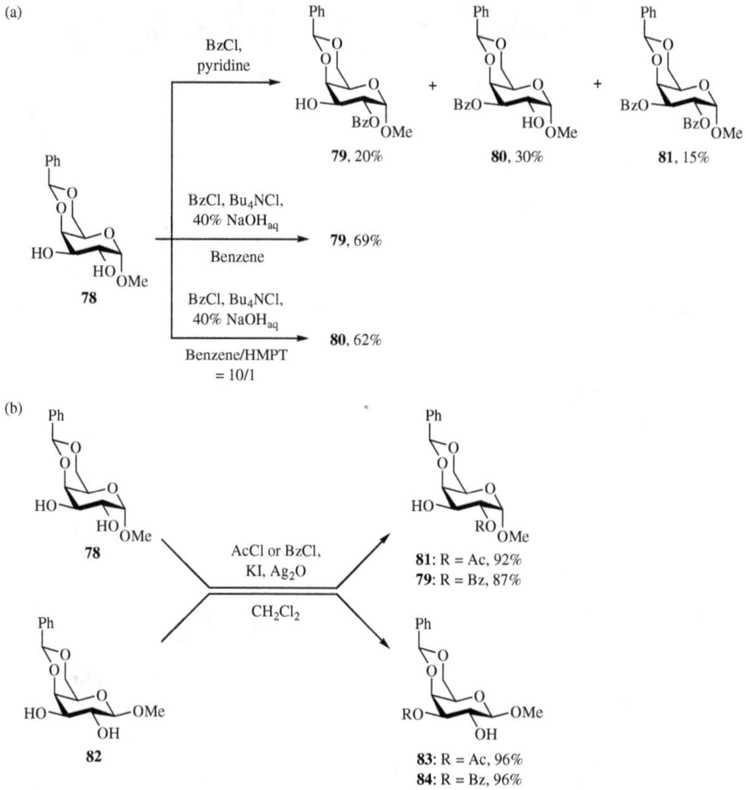

SCHEME 2.24 Dependence of the regioselective acylation of 4,6-*O*-benzylidene-D-galactopyranosides with (a) solvent and (b) anomeric configuration.

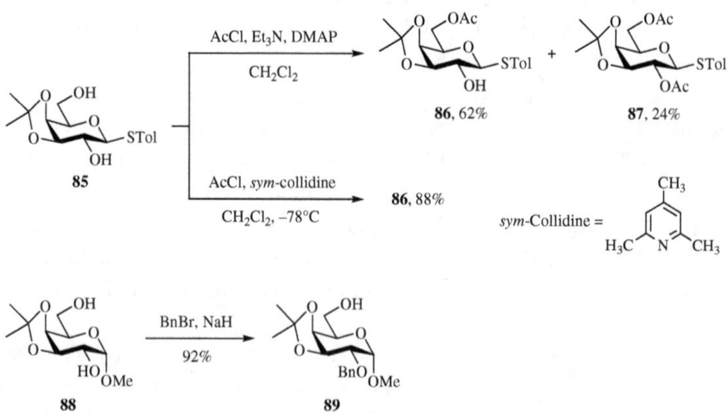

SCHEME 2.25 Differentiation of the 2,6-diol in galactopyranosides.

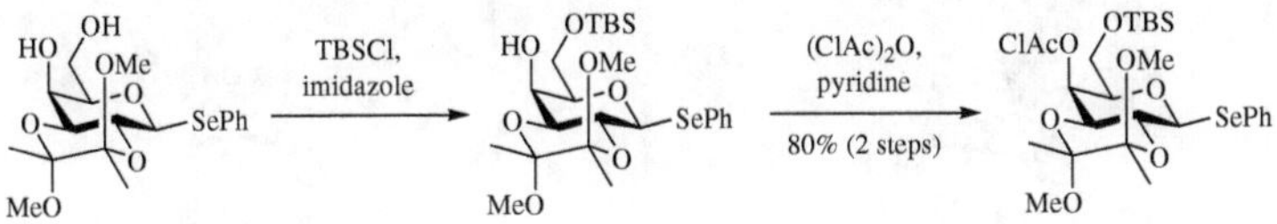

SCHEME 2.26 Differentiation of 4,6-diol in galactopyranosides.

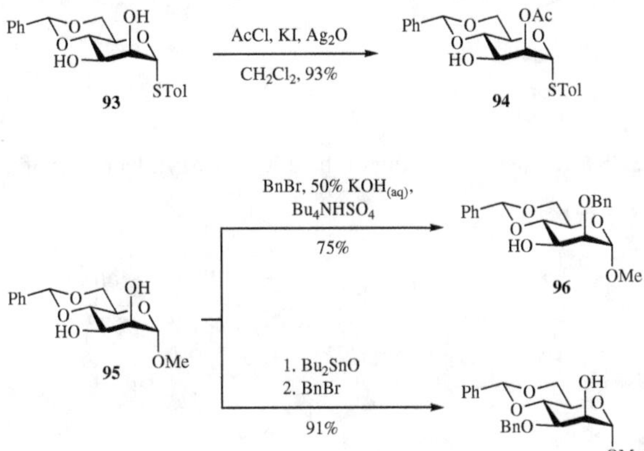

SCHEME 2.27 Differentiation of the 2,3-diol in mannopyranosides.

led to the 2-O-benzylated product **89** in 92% yield probably owing to the aforementioned inductive effect of the nearby oxygen atoms [182]. The 4-hydroxyl and 6-hydroxyl of the 2,3-O-protected galactoside **90** were easily distinguished by the bulky TBS group to afford the 4-alcohol **91**. Further chloroacetylation supplied the fully protected compound **92** [183].

Despite complications for the generation of 4,6-*O*-benzylidene-D-mannopyranosides (**52**), strategies that regioselectively protect the remaining 2,3-*cis*-diol have been reported (Scheme 2.27). The 2-acetate **94** was obtained in a Ag_2O-mediated acetylation of mannopyranoside **93** with AcCl and potassium iodide [179]. Benzylation of mannoside **95** by using aqueous potassium hydroxide and BnBr under a two-phase system gave the 2-O-benzylated product **96** [184], whereas the Bu_2SnO-mediated benzylation furnished the 3-O-benzylated product **97** [185].

With dibenzylidenation of mannosides as an easy transformation to execute, selective reductive opening of the 2,3-*O*-benzylidene ring is quite helpful for differentiation. Diisopropyl aluminum hydride (DIBAL-H) or HCl/$NaBH_3CN$ gave the monobenzylation product with regioselectivity that is highly dependent on the configuration of the phenyl moiety in the benzylidene group. Accordingly, while DIBAL-H treatment of the *exo*-isomer **98** led to the 2-alcohol **97** (78%), the *endo*-isomer **99** gave the 3-alcohol **96** (85%) (Scheme 2.28) [186, 187].

BDA protection of mannosides leaves the secondary 2-hydroxyl and the primary 6-hydroxyl unprotected. The tin-activated chloroacetylation of diol **100** gave the 6-chloroacetate **101** in excellent yield (Scheme 2.29) [147, 183]. Notably, benzylation

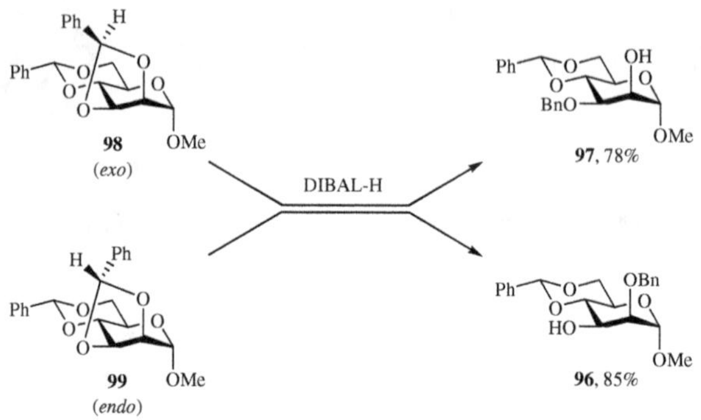

SCHEME 2.28 Reductive opening of the 2,3-*O*-benzylidene ring of mannosides.

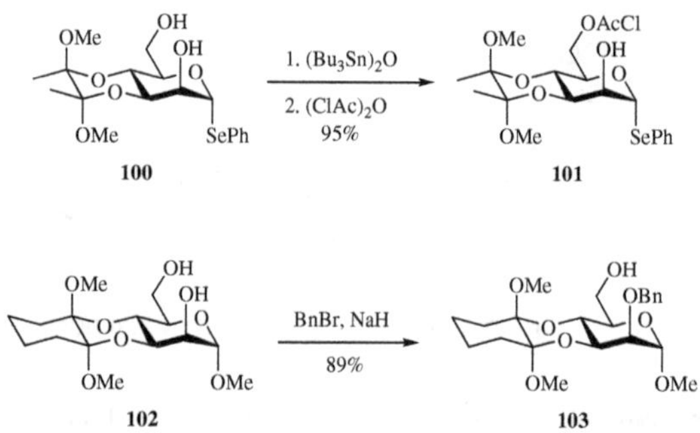

SCHEME 2.29 Regioselective protection of the 2,6-diol in mannopyranosides.

of diol **102** under the common BnBr and NaH condition occurred at O2 rather than O6 as a likely result of inductive effect [147, 188].

The importance of the 4,6-*O*-benzylidene acetal and its derivatives has been broadened by their regioselective ring openings, which release either the primary 6-hydroxyl or the secondary 4-hydroxyl by using different hydride and acid tandems (Table 2.1). The Et_3SiH/trifluoroacetic acid [189], $Bu_3SnH/BF_3{\cdot}OEt_2$ [190], and HCl/$NaBH_3CN$ [191, 192] reagent systems open the arylmethylidene acetals in **104** at O4 and give the 6-arylmethyl ether products **105** in good yields. On the other hand, $LiAlH_4/AlCl_3$ [193, 194], DIBAL-H [186], and $BH_3{\cdot}THF/Bu_2BOTf$ or TMSOTf [195] open the arylmethylidene acetal ring at O6 to deliver the 4-arylmethyl ether **106**. $Cu(OTf)_2$ also catalyze the ring opening at either O4 or O6 with high yield and regioselectivity according to the reactivity of the reducing agents [196, 197]. Consequently, the $Cu(OTf)_2$-catalyzed ring opening with Me_2EtSiH as the reducing agent resulted in opening at O4, whereas opening at O6 occurred when $BH_3{\cdot}THF$ was used as the hydride source.

The usefulness of acetal or ketal groups can be exhibited in the following example (Scheme 2.30): Chien-Tien Chen et al. carried out a $VO(OTf)_2$-catalyzed 4,6-*O*-benzylidene acetal protection of tetraol **107** with benzaldehyde to give the

TABLE 2.1 Benzylidene Acetal Opening Conditions

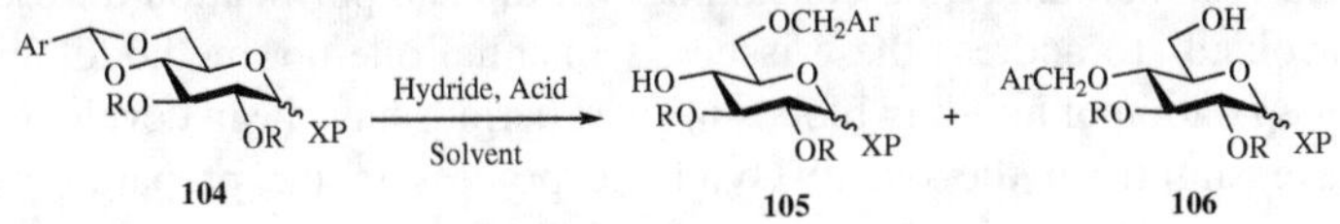

Entry	Ar	R	Hydride	Acid	Solvent	Yield (%) 105	Yield (%) 106
1	Ph	Bn	Et_3SiH	CF_3CO_2H	CH_2Cl_2	81	—
2	Ph	Ac	Bu_3SnH	$BF_3{\cdot}OEt_2$	CH_2Cl_2	63	—
3	Ph	Bn	$NaBH_3CN$	$HCl_{(g)}$	THF	81	—
4	Ph	Bn	$LiAlH_4$	$AlCl_3$	CH_2Cl_2/Et_2O	—	90
5	Ph	Bn	DIBAL-H	—	CH_2Cl_2	4	88
6	Ph	Bn	$BH_3{\cdot}THF$	Bu_2BOTf or TMSOTf	CH_2Cl_2	—	87
7	Ph	Bn	Me_2EtSiH	$Cu(OTf)_2$	CH_3CN	84	—
8	Ph	Bn	$BH_3{\cdot}THF$	$Cu(OTf)_2$	CH_2Cl_2	—	94

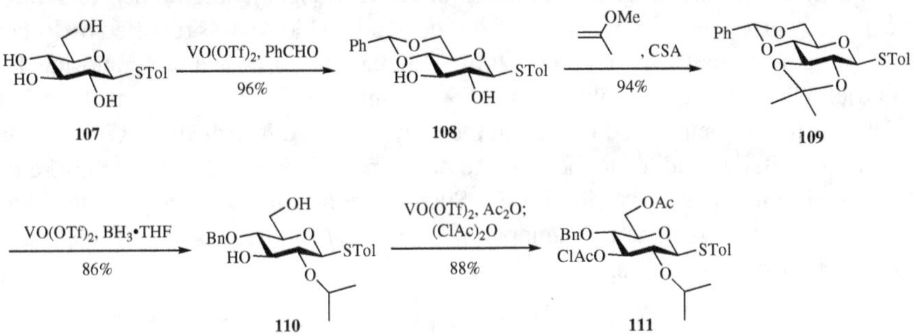

SCHEME 2.30 Orthogonal protection of the thioglucoside **107** via acetal and ketal ring openings.

2,3-diol **108** in excellent yield. Subsequent ketalization with 2-methoxypropene in the presence of camphorsulfonic acid (CSA) provided compound **109**. Then, a $VO(OTf)_2$-catalyzed regioselective reductive ring opening of both the benzylidene acetal and the isopropylidene ketal furnished the 3,6-diol **110**. Finally, regioselective 6-O-acetylation and 4-O-chloroacetylation in one pot afforded the fully protected building block **111**. Thus, isopropyl, chloroacetyl, Bn, and Ac groups were introduced at various positions in just four steps [198].

2.5 ONE-POT PROTECTION METHODS

Despite the numerous methodologies that have been used to differentiate the multiple hydroxyls in sugars, the preparations of fully protected building blocks and building blocks with free hydroxyl(s) at defined position(s) still need many steps and usually require protection–deprotection sequences without a common strategy. In such cases, lengthy processes, including time-consuming workup and purification for each step, cannot be avoided. To address these issues, sequential one-pot methodologies that enable the preparation of building blocks in short periods have been developed. Such methods have simplified the overall synthetic process in the preparation of the building blocks for sugar assembly.

A one-pot tritylation–silylation–acylation of glycosides was developed by Fanzuo Kong et al. [199]. The DMAP-catalyzed 6-O-tritylation of the methyl α-glycosides **37–39** with trityl chloride followed by TBS installation and subsequent diacylation led to the fully protected derivatives **112–115** in a one-pot manner (Scheme 2.31a). The key regioselective silylation took place at O2 of glucoside **37** and at O3 of mannoside **39** but showed poor selectivity between O2 and O3 of galactoside **38**. However, simply changing the α-methoxy into the β-thio group resulted in a different regioselectivity (Scheme 2.31b). After the 6-O-tritylation, the silylation of both thioglucoside **116** and thiogalactoside **117** occurred at O3 in a highly regioselective manner. Further treatment with Ac_2O in one pot produced the 2,4-diacetates **118** and **119** in 75 and 83% yields, respectively.

An orthoesterification–benzylation–orthoester rearrangement approach to synthesize partially protected mono- and disaccharide building blocks was reported by Robert Field and coworkers (Scheme 2.32) [200]. The reaction of L-rhamnopyranoside **120** with triethyl orthoacetate catalyzed by PTSA formed the 4-alcohol intermediate **121**, which was further benzylated to furnish the fully protected intermediate **122**. Without workup, 1 N HCl was added to the same vessel to rearrange the orthoester to afford the 3-alcohol **123** in a one-pot yield of 78%. Such protocol can also be applied to sugars containing *cis*-diol, such as D-mannopyranoside, D-galactopyranoside, L-fucopyranoside, and lactosyl derivatives.

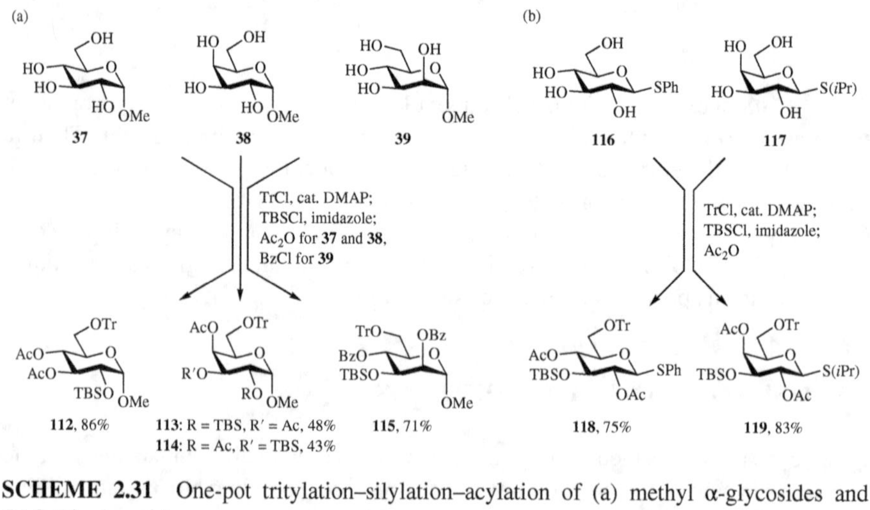

SCHEME 2.31 One-pot tritylation–silylation–acylation of (a) methyl α-glycosides and (b) β-thioglycosides.

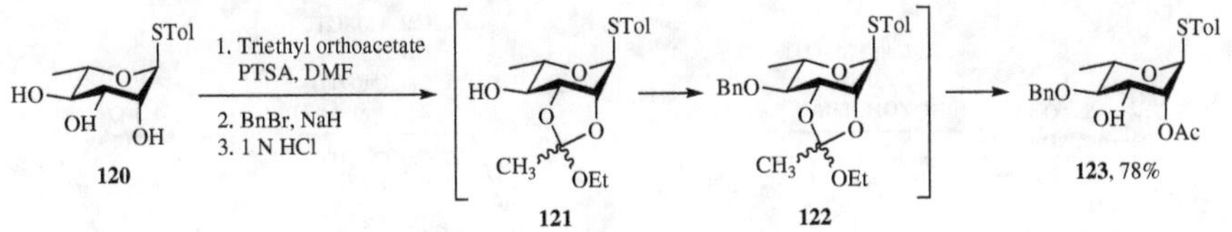

SCHEME 2.32 One-pot orthoesterification–benzylation–orthoester rearrangement methodology for the protection of the L-rhamnopyranoside **120**.

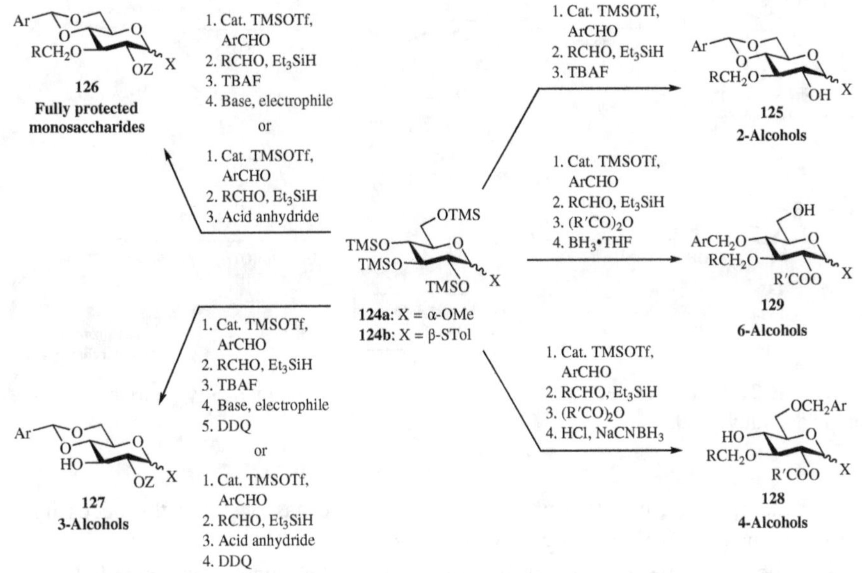

SCHEME 2.33 TMSOTf-catalyzed regioselective one-pot protection of glucopyranosides.

A one-pot methodology that allows the preparation of fully protected hexopyranosides or building blocks with free hydroxyl at C2, C3, C4, or C6 was developed by Shang-Cheng Hung and coworkers using the per-O-trimethylsilylated glucosides **124a** and **124b** as starting materials (Scheme 2.33) [201, 202]. These one-pot protocols are (i) a TMSOTf-catalyzed arylmethylidene formation at O4 and O6 [203] followed by a regioselective reductive arylmethylation at O3 to get the 2-alcohols **125**; (ii) a TMSOTf-catalyzed 4,6-O-arylmethylidenation, a reductive 3-O-arylmethylation, and a base-promoted etherification or a TMSOTf-catalyzed acylation at O2 to get the fully protected building blocks **126**; (iii) a TMSOTf-catalyzed 4,6-O-arylmethylidenation, a temporary PMB or NAP protection at O3, an etherification or acylation at O2, and finally the removal of the PMB or NAP group to get the 3-alcohols **127**; and (iv) a 4,6-O-arylmethylidenation, a 3-O-arylmethylation, a 2-O-acylation, and the regioselective opening of the arylmethylidene acetals at either O4 or O6 to get the 4-alcohols **128** or the 6-alcohols **129**, respectively. Consequently, compounds **124a** and **124b** were efficiently transformed into hundreds of building blocks bearing different protecting groups. It was shown recently that a TMSOTf-catalyzed silylation using hexamethyldisilazane

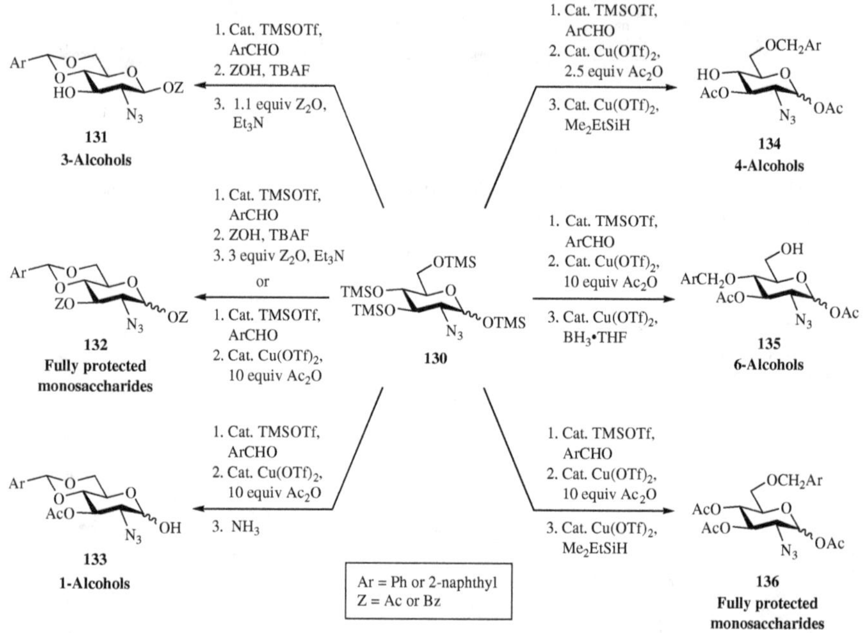

SCHEME 2.34 Regioselective one-pot protection of the anomerically nonfixed D-glucosamine derivative **130**.

can be incorporated into this protocol and therefore has obviated the necessity of preparing the per-O-silylated pyranosides in advance [204].

The regioselective one-pot protection strategy has also been extended to the anomerically nonfixed D-glucosamine derivative **130** (Scheme 2.34) [205]. Numerous compounds, including fully protected monosaccharides and 1-, 3-, 4-, and 6-alcohols, could be synthesized in a series of one-pot reactions. A TMSOTf-catalyzed 4,6-O-benzylidenation of the tetrakis-TMS ether **130** followed by removal of the remaining TMS groups with TBAF and a regioselective 1-O-acylation [173, 206] with acid anhydrides using Et_3N as a base gave the 3-alcohols **131**. By increasing the amount of acid anhydride, the 1,3-di-O-acylated products **132** was readily obtained. After benzylidene formation and without cleaving the TMS groups, the same fully protected monosaccharides were obtained through direct 1,3-di-O-acetylation using excess Ac_2O mediated by catalytic amounts of $Cu(OTf)_2$. Subsequent treatment with ammonia in the same vessel allowed the selective cleavage of the 1-acetate to afford the hemiacetals **133**. After full acetylation, the Me_2EtSiH- and BH_3·THF-reductive ring opening of the benzylidene acetal yielded the 4-alcohols **134** and 6-alcohols **135**, respectively. A sequential benzylidene acetal formation, $Cu(OTf)_2$-catalyzed esterification using excess Ac_2O, and acetal opening at O4 afforded the 1,3,4-triacetates **136** in a one-pot manner.

The one-pot protection of 4-tolyl 2,3,4,6-tetra-*O*-trimethylsilyl-1-thio-α-D-mannopyranoside (**137**) was also accomplished by Hung et al. (Scheme 2.35) [207].

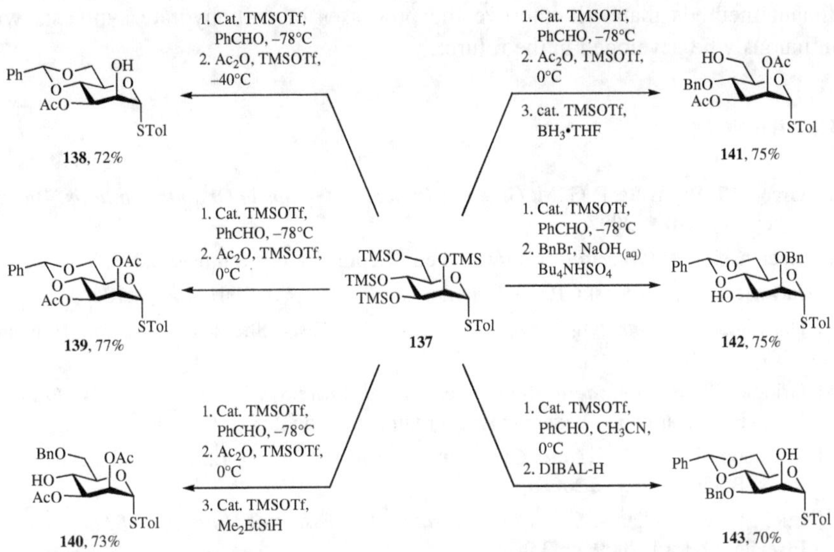

SCHEME 2.35 TMSOTf-catalyzed regioselective one-pot protection of mannopyranoside **137**.

Although the benzylidene acetal formations in mannosides can result in a mixture of **52** and **53** as mentioned earlier, the TMSOTf-catalyzed benzylidene formation with benzaldehyde at −78°C afforded only the 4,6-*O*-benzylidene group without generating 2,3-*O*-benzylidene. By this discovery, the 4,6-*O*-benzylidene formation at −78°C followed by acetylation at −40°C gave the 2-alcohol **138** in good regioselectivity and yield. On the other hand, the acetylation conducted at 0°C delivered the fully protected mannoside **139**. After getting **139** in the reaction vessel, subsequent benzylidene opening at O4 or O6 resulted in the 4-alcohol **140** and 6-alcohol **141**. By using the aforementioned biphasic benzylation using aqueous NaOH as base (Scheme 2.27), regioselective 2-O-benzylation was achieved, affording the 3-alcohol **142**. Moreover, the TMSOTf-catalyzed benzylidene formation at 0°C in acetonitrile produced the 2,3;4,6-di-*O*-benzylidene derivative, but only the *exo*-isomer of the 2,3-*O*-benzylidene was generated. Such excellent stereoselectivity enabled the formation of the 2-alcohol **143** through further regioselective DIBAL-H-mediated benzylidene ring opening at O3.

2.6 CONCLUSION

This chapter introduced protecting groups that are commonly and currently utilized in carbohydrate synthesis, the logic of selecting protecting groups, and the concept of orthogonality. Approaches for the installation and removal of protecting groups have been summarized. The methodologies used to differentiate the hydroxyls in sugars and the preparation of building blocks in either a stepwise or in a one-pot manner were also described. It is expected that better protecting groups and more efficient methods that will improve the processes of carbohydrate synthesis will continuously be developed in the future.

REFERENCES

[1] Greene, T. W.; Wuts, P. G. M. *Greene's Protective Groups in Organic Synthesis*, 4th ed.; Wiley: New York, **2007**.

[2] Kocienski, P. J. *Protecting Groups*, 3rd ed.; Georg Thieme: Stuttgart, **2005**.

[3] Jarowicki, K.; Kocienski, P. *J. Chem. Soc. Perkin Trans. 1*, **2001**, 2109–2135.

[4] Hanson, J. R. *Protecting Groups in Organic Synthesis*. Sheffield Academic: England, **1999**.

[5] Grindley, T. B. In *Modern Methods in Carbohydrate Synthesis*; Khan, S. H., O'Niell, R. A., Eds.; Harwood Academic: Amsterdam, **1996**, pp 225–250.

[6] Oscarson, S. In *The Organic Chemistry of Sugars*; Levy, D. E.; Fügedi, P., Eds.; CRC Press: Boca Raton, FL, **2006**, pp 53–88.

[7] Stick, R. V.; Williams, S. J. *Carbohydrates: The Essential Molecules of Life*, 2nd ed.; Elsevier: Oxford, **2009**, pp 35–74.

[8] Paulsen, H. *Angew. Chem. Int. Ed.* **1982**, *21*, 155–173.

[9] Schmidt, R. R. *Angew. Chem. Int. Ed. Engl.* **1986**, *25*, 212–235.

[10] Toshima, K.; Tatsuta, K. *Chem. Rev.* **1993**, *93*, 1503–1531.

[11] Nukada, T.; Berces, A.; Zgierski, M. Z.; Whitfield, D. M. *J. Am. Chem. Soc.* **1998**, *120*, 13291–13295.

[12] Kim, J.-H.; Yang, H.; Park, J.; Boons, G.-J. *J. Am. Chem. Soc.* **2005**, *127*, 12090–12097.

[13] Flitsch, S. L. *Nature* **2005**, *437*, 201–202.

[14] Demchenko, A. V. *Synlett* **2003**, 1225–1240.

[15] Pellissier, H. *Tetrahedron* **2005**, *61*, 2947–2993.

[16] Baressi, F.; Hindsgaul, O. In *Modern Synthetic Methods*; Ernst, B., Leumann C., Eds.; Wiley: Switzerland, **1995**, pp 281–330.

[17] Bertozzi, C. R.; Kiessling, L. L. *Science* **2001**, *291*, 2357–2364.

[18] Ernst, B.; Hart, G. W.; Sinaÿ, P. *Carbohydrates in Chemistry and Biology*; Wiley: Weinheim, **2000**; Vol. *1*.

[19] Wang, P. G.; Bertozzi, C. R. *Glycochemistry*; Marcel Dekker: New York, **2001**.

[20] Schifiela, L.; Hewitt, M. C.; Evans, K.; Siomos, M.-A.; Seeberger, P. H. *Nature* **2002**, *418*, 785–789.

[21] Holemann, A.; Seeberger, P. H. *Curr. Opin. Biotechol.* **2004**, *15*, 615–622.

[22] Prescher, J. A.; Dube, D. H.; Bertozzi, C. R. *Nature* **2004**, *430*, 873–877.

[23] Fife, T. H.; Bembi, R.; Natarajan, R. *J. Am. Chem. Soc.* **1996**, *118*, 12956–12963.

[24] Lemieux, R. U. *Pure Appl. Chem.* **1971**, *25*, 527–548.

[25] Demchenko, A. V. In *Handbook of Chemical Glycosylation*; Demchenko, A. V., Ed.; Wiley: Weinheim, **2008**, pp 1–27.

[26] Ishikawa, T.; Fletcher, H. G. *J. Org. Chem.* **1969**, *34*, 563–571.

[27] Eby, R.; Schuerch, C. *Carbohydr. Res.* **1974**, *34*, 79–90.

[28] Koto, S.; Yago, K.; Zen, S.; Tomonaga, F.; Shimada, S. *Bull. Chem. Soc. Jpn.* **1986**, *59*, 411–414.

[29] Dasgupta, F.; Garegg, P. J. *Carbohydr. Res.* **1990**, *202*, 225–238.

[30] Mukaiyama, T.; Suenaga, M.; Chiba, H.; Jona, H. *Chem. Lett.* **2002**, 56–57.

[31] van Boeckel, C. A. A.; Beetz, T.; van Aelst, S. F. *Tetrahedron* **1984**, *40*, 4097–4107.

[32] Fei, C. P.; Chan, T. H. *Tetrahedron Lett.* **1987**, *28*, 849–852.

[33] Houdier, S.; Vottero, P. J. A. *Carbohydr. Res.* **1992**, *232*, 349–352.

[34] Fukase, K.; Kinoshita, I.; Kanoh, T.; Nakai, Y.; Hasuoka, A.; Kusumoto, S. *Tetrahedron* **1996**, *52*, 3897–3904.

[35] Fukase, K.; Kakai, Y.; Kanoh, T.; Kusumoto, S. *Synlett* **1998**, 84–86.

[36] Fukase, K.; Nakai, Y.; Egusa, K.; Porco, J. A.; Kusumoto, S. *Synlett* **1999**, 1074–1078.

[37] Mootoo, D. R.; Konradsson, P.; Udodong, U.; Fraser-Reid, B. *J. Am. Chem. Soc.* **1988**, *110*, 5583–5584.

[38] Konradsson, P.; Mootoo, D. R.; McDevitt, R. E.; Fraser-Reid, B. *J. Chem. Soc. Chem. Commun.* **1990**, 270–272.

[39] Fraser-Reid, B.; Udodong, U. E.; Wu, Z.; Ottosson, H.; Merritt, J. R.; Rao, C. S.; Roberts, C. *Synlett* **1992**, 927–942.

[40] Raghavan, S.; Kahne, D. *J. Am. Chem. Soc.* **1993**, *115*, 1580–1581.

[41] Yamada, H.; Harada, T.; Miyazaki, H.; Takahashi, T. *Tetrahedron Lett.* **1994**, *35*, 3979–3982.

[42] Veeneman, G. H.; Van Boom, J. H. *Tetrahedron Lett.* **1990**, *31*, 275–278.

[43] Veeneman, G. H.; Van Leeuwen, S. H.; Van Boom, J. H. *Tetrahedron Lett.* **1990**, *31*, 1331–1334.

[44] Friesen, R. W.; Danishefsky, S. J. *J. Am. Chem. Soc.* **1989**, *111*, 6656–6660.

[45] Barrena, M. I.; Echarri, R.; Castillon, S. *Synlett* **1996**, 675–676.

[46] Cheung, M. K.; Douglas, N. L.; Hinzen, B.; Ley, S. V.; Pannecoucke, X. *Synlett* **1997**, 257–260.

[47] Nguyen, H. M.; Poole, J. L.; Gin, D. Y. *Angew. Chem. Int. Ed.* **2001**, *40*, 414–417.

[48] Hashimoto, S.; Sakamoto, H.; Honda, T.; Ikegami, S. *Tetrahedron Lett.* **1997**, *38*, 5181–5184.

[49] Hashimoto, S.; Sakamoto, H.; Honda, T.; Abe, H.; Nakamura, S.; Ikegami, S. *Tetrahedron Lett.* **1997**, *38*, 8969–8972.

[50] Chiba, H.; Funasaka, S.; Kiyota, K.; Mukaiyama, T. *Chem. Lett.* **2002**, *31*, 746–747.

[51] Chiba, H.; Mukaiyama, T. *Chem. Lett.* **2003**, *32*, 172–173.

[52] Chiba, H.; Funasaka, S.; Mukaiyama, T. *Bull. Chem. Soc. Jpn.* **2003**, *76*, 1629–1644.

[53] Zhang, Z.; Ollmann, I. R.; Ye, X.-S.; Wischnat, R.; Baasov, T.; Wong, C.-H. *J. Am. Chem. Soc.* **1999**, *121*, 734–753.

[54] Ye, X.-S.; Wong, C.-H. *J. Org. Chem.* **2000**, *65*, 2410–2431.

[55] Kuhn, R.; Löw, I.; Trischmann, H. *Chem. Ber.* **1957**, *90*, 203–218.

[56] Van Hijfte, L.; Little, R. D. *J. Org. Chem.* **1985**, *50*, 3940–3942.

[57] Iversen, T.; Bundle, D. R. *J. Chem. Soc. Chem. Commun.* **1981**, 1240–1241.

[58] White, J. D.; Reddy, G. N.; Spessard, G. O. *J. Am. Chem. Soc.* **1988**, *110*, 1624–1626.

[59] Widmer, U. *Synthesis* **1987**, 568–569.

[60] Eckenberg, P.; Groth, U.; Huhn, T.; Richter, N.; Schmeck, C. *Tetrahedron* **1993**, *49*, 1619–1624.

[61] Hatakeyama, S.; Mori, H.; Kitano, K.; Yamada, H.; Nishizawa, M. *Tetrahedron Lett.* **1994**, *35*, 4367–4370.

[62] Oikawa, Y.; Yoshioka, T.; Yonemitsu, O. *Tetrahedron Lett.* **1982**, *23*, 885–888.

[63] Oikawa, Y.; Tanaka, T.; Horita, K.; Yonemitsu, O. *Tetrahedron Lett.* **1984**, *25*, 5397–5400.

[64] Horita, K.; Yoshioka, T.; Tanaka, T.; Oikawa, Y.; Yonemitsu, O. *Tetrahedron* **1986**, *42*, 3021–3028.

[65] Tanaka, T.; Oikawa, Y.; Hamada, T.; Yonemitsu, O. *Tetrahedron* **1986**, *42*, 3651–3659.

[66] Johansson, R.; Samuelsson, B. *J. Chem. Soc. Perkin Trans. 1* **1984**, 2371–2374.

[67] Yan, L.; Kahne, D. *Synlett* **1995**, 523–524.

[68] Bouzide, A.; Sauve, G. *Tetrahedron Lett.* **1999**, *40*, 2883–2886.

[69] De Medeiros, E. F.; Herbert, J. M.; Taylor, R. J. K. *J. Chem. Soc. Perkin Trans. 1* **1991**, 2725–2730.

[70] Li, D. R.; Sun, C. Y.; Su, C.; Lin, G.-Q.; Zhoum, W.-S. *Org. Lett.* **2004**, *6*, 4261–4264.

[71] Xia, J.; Piskorz, C. F.; Alderfer, J. L.; Locke, R. D.; Matta, K. L. *Tetrahedron Lett.* **2000**, *41*, 2773–2776.

[72] Xia, J.; Alderfer, J. L.; Piskorz, C. F.; Matta, K. L. *Chem. Eur. J.* **2000**, *6*, 3442–3451.

[73] Fuwa, H.; Fujikawa, S.; Tachibana, K.; Takakura, H.; Sasaki, M. *Tetrahedron Lett.* **2004**, *45*, 4795–4799.

[74] Wright, J. A.; Yu, J.; Spencer, J. B. *Tetrahedron Lett.* **2001**, *42*, 4033–4036.

[75] Zehavi, U.; Ami, B.; Patchornik, A. *J. Org. Chem.* **1972**, *37*, 2281–2285.

[76] Zehavi, U.; Patchornik, A. *J. Org. Chem.* **1972**, *37*, 2285–2288.

[77] Ohtsuka, E.; Tanaka, S.; Ikehara, M. *J. Am. Chem. Soc.* **1978**, *100*, 8210–8213.

[78] van der Stouwe, C.; Schäfer, H. J. *Tetrahedron Lett.* **1979**, *20*, 2643–2646.

[79] van der Stouwe, C.; Schäfer, H. J. *Chem. Ber.* **1981**, *114*, 946–958.

[80] Park, M. H.; Takeda, R.; Nakanishi, K. *Tetrahedron Lett.* **1987**, *28*, 3823–3824.

[81] Egusa, K.; Fukase, K.; Kusumoto, S. *Synlett* **1997**, 675–676.

[82] Fukase, K.; Hashida, M.; Kusumoto, S. *Tetrahedron Lett.* **1991**, *32*, 3557–3558.

[83] Plante, O. J.; Buchwald, S. L.; Seeberger, P. H. *J. Am. Chem. Soc.* **2000**, *122*, 7148–7149.

[84] Guibe, F. *Tetrahedron* **1997**, *40*, 13509–13556.

[85] Corey, E. J.; Suggs, W. J. *J. Org. Chem.* **1973**, *38*, 3224.

[86] Jacquinet, J.-C.; Sinaÿ, P. *J. Org. Chem.* **1977**, *42*, 720–724.

[87] Hindsgaul, O.; Norberg, T.; Le Pendu, J.; Lemieux, R. U. *Carbohydr. Res.* **1982**, *109*, 109–142.

[88] Iversen, T.; Bundle, D. R. *J. Chem. Soc. Perkin Trans. 1* **1985**, 2247–2250.

[89] Wang, C.-C.; Lee, J.-C.; Luo, S.-Y.; Fan, H.-F.; Pai, C.-L.; Yang, W.-C.; Lu, L.-D.; Hung, S.-C. *Angew. Chem. Int. Ed.* **2002**, *41*, 2360–2362.

[90] Guibe, F.; Saint M'Leux, Y. *Tetrahedron Lett.* **1981**, *22*, 3591–3594.

[91] Gent, P. A.; Gigg, R. *J. Chem. Soc. Chem. Commun.* **1974**, 277–278.

[92] Gigg, R. *J. Chem. Soc. Perkin Trans. 1* **1980**, 738–740.

[93] Nicolaou, K. C.; Caulfield, T. J.; Kataoka, H.; Stylianides, N. A. *J. Am. Chem. Chem.* **1990**, *112*, 3693–3697.

[94] Cunningham, J.; Gigg, R.; Warren, D. C. *Tetrahedron Lett.* **1964**, *5*, 1191–1196.

[95] Smith, A. B., III; Rivero, R. A.; Hale, K. J.; Vaccaro, H. A. *J. Am. Chem. Soc.* **1991**, *113*, 2092–2112.

[96] Yamada, H.; Harada, T.; Takahashi, T. *J. Am. Chem. Soc.* **1994**, *116*, 7919–7920.

[97] Lamberth, C.; Bednarski, M. D. *Tetrahedron Lett.* **1991**, *32*, 7369–7372.

[98] Barili, P. L.; Berti, G.; Bertozzi, D.; Gatelani, G.; Colonna, F.; Corsetti, T.; D'Andrea, F. *Tetrahedron Lett.* **1990**, *46*, 5365–5368.

[99] Nonda, M.; Morita, H.; Nagakura, I. *J. Org. Chem.* **1997**, *62*, 8932–8936.

[100] Taniguchi, T.; Ogasawara, K. *Angew. Chem. Int. Ed. Engl.* **1998**, *37*, 1136–1137.

[101] Roush, W. R.; Sciotti, R. J. *J. Am. Chem. Soc.* **1998**, *120*, 7411–7419.

[102] Opatz, T.; Kunz, H. *Tetrahedron Lett.* **2000**, *41*, 10185–10188.

[103] Chandrasekhar, S.; Reddy, C. R.; Rao, R. J. *Tetrahedron* **2001**, *57*, 3435–3438.

[104] Tsukamoto, H.; Kondo, Y. *Synlett* **2003**, 1061–1063.

[105] Vutukuri, D. R.; Bharathi, P.; Yu, Z.; Rajasekaran, K.; Tran, M.-H.; Thayumanavan, S. *J. Org. Chem.* **2003**, *68*, 1146–1149.

[106] Murakami, H.; Minami, T.; Ozawa, F. *J. Org. Chem.* **2004**, *69*, 4482–4486.

[107] Vatele, J.-M. *Synlett* **2002**, 507–509.

[108] Vatele, J.-M. *Tetrahedron* **2002**, *58*, 5689–5698.

[109] Solis-Oba, A.; Hudlicky, T.; Koroniak, L.; Frey, D. *Tetrahedron Lett.* **2001**, *42*, 1241–1243.

[110] Hansen, J.; Freeman, S.; Hudlicky, T. *Tetrahedron Lett.* **2003**, *44*, 1575–1578.

[111] Bartoli, G.; Gupone, G.; Dalpozzo, R.; De Nino, A.; Maiuolo, L.; Marcantoni, E.; Procopio, A. *Synlett* **2001**, 1897–1900.

[112] Shimizu, N.; Takesue, N.; Yamamoto, A.; Tsutsumi, T.; Yasuhara, S.; Tsuno, Y. *Chem. Lett.* **1992**, *21*, 1236–1266.

[113] Shimizu, N.; Takesue, N.; Yasuhara, S.; Inazu, T. *Chem. Lett.* **1993**, *22*, 1807–1810.

[114] Nelson, T. D.; Crouch, R. D. *Synthesis* **1996**, 1031–1069.

[115] Crouch, D. R. *Tetrahedron* **2004**, *60*, 5833–5871.

[116] Wuts, B. G. M.; Bigelow, S. S. *J. Org. Chem.* **1988**, *53*, 5023–5024.

[117] Mulzer, J.; Schöllhorn, B. *Angew. Chem. Int. Ed. Engl.* **1990**, *29*, 431–432.

[118] Marshall, J. A.; Bourbeau, M. P. *J. Org. Chem.* **2002**, *67*, 2751–2754.

[119] Procopiou, P. A.; Baugh, S. P. D.; Flack, S. S.; Inglis, G. G. A. *Chem. Commun.* **1996**, 2625–2626.

[120] Procopiou, P. A.; Baugh, S. P. D.; Flack, S. S.; Inglis, G. G. A. *J. Org. Chem.* **1998**, *63*, 2342–2347.

[121] Neises, B.; Steglich, W. *Angew. Chem. Int. Ed. Engl.* **1978**, *17*, 522–524.

[122] Jiang, L.; Chan, T.-H. *J. Org. Chem.* **1998**, *63*, 6035–6038.

[123] Reese, C. B.; Stewart, J. C. M.; van Boom, J. H.; de Leeuw, H. P. M.; Nagel, J.; de Rooy, J. F. M. *J. Chem. Soc. Perkin Trans. 1*, **1975**, 934–942.

[124] Byramova, N. E.; Ovchinnikov, M. V.; Backinowsky, L. V.; Kochetkov, N. K. *Carbohydr. Res.* **1983**, *124*, C8–C11.

[125] Best, W. M.; Dunlop, R. W.; Stick, R. V.; White, S. T. *Aust. J. Chem.* **1994**, *47*, 433–440.

[126] Mikamo, M. *Carbohydr. Res.* **1989**, *191*, 150–153.

[127] Alper, P. B.; Hung, S.-C.; Wong, C.-H. *Tetrahedron Lett.* **1996**, *37*, 6029–6032.

[128] Debenham, J. R.; Rodebaugh, R.; Fraser-Reid, B. *Liebigs Ann.* **1997**, 791–802.

[129] Wang, Z.-G.; Zhang, X.; Visser, M.; Live, D.; Zatorski, A.; Iserloh, U.; Lloyd, K. O.; Danishefsky, S. J. *Angew. Chem. Int. Ed.* **2001**, *40*, 1728–1732.

[130] Lergenmüller, M.; Ito, Y.; Ogawa, T. *Tetrahedron* **1998**, *54*, 1381–1394.

[131] Staudinger, H.; Meyer, J. *Helv. Chim. Acta* **1919**, *2*, 635–646.

[132] Atherton, E.; Sheppard, R. C. In *The Peptides*; Udenfriend, S.; Meienhofer, J., Eds.; Academic: Orlando, **1987**; Vol. *9*, pp 1–38.

[133] Just, G.; Grozinger, K. *Synthesis* **1976**, 457–458.

[134] Bergmann, M.; Zervas, L. *Ber. Dtsch. Chem. Ges.* **1932**, *65*, 1192–1201.

[135] Szumigala, R. H.; Onofiok, E.; Karady, S.; Armstrong, J. D., III; Miller, R. A.; *Tetrahedron Lett.* **2005**, *46*, 4403–4405.

[136] Guibé, F. *Tetrahedron* **1998**, *54*, 2967–3042.

[137] Yeung, B. K. S.; Adamski-Werner, S. L.; Bernard, J. B.; Poulenat, G.; Petillo, P. A. *Org. Lett.* **2000**, *2*, 3135–3138.

[138] Urabe, D.; Sugino, K.; Nishikawa, T.; Isobe, M. *Tetrahedron Lett.* **2004**, *45*, 9405–9407.

[139] Bergeron, R. J.; Garlich, J. R.; Stolowich, N. J. *J. Org. Chem.* **1984**, *49*, 2997–3001.

[140] Bergeron, R. J.; McManis, J. J. *J. Org. Chem.* **1988**, *53*, 3108–3111.

[141] Blatter, G.; Beau, J.-M.; Jacquinet, J.-C. *Carbohydr. Res.* **1994**, *260*, 189–202.

[142] Vibert, M.; Lopin-Bon, C.; Jacquinet, J. C. *Tetrahedron Lett.* **2010**, *51*, 1867–1869.

[143] Evans, M. E.; Parrish, F. W.; Long, L., Jr. *Carbohydr. Res.* **1967**, *3*, 453–462.

[144] Corey, E. J.; Kim, S.; Yoo, S.; Nicolaou, K. C.; Melvin, L. S., Jr.; Brunelle, D. J.; Falck, J. R.; Trybulski, E. J.; Lett, R.; Sheldrake, P. W. *J. Am. Chem. Soc.* **1978**, *100*, 4620–4622.

[145] Fanton, E.; Gelas, J.; Horton, D. *J. Chem. Soc. Chem. Commun.* **1980**, 21–22.

[146] Richtmyer, N. K. *Methods Carbohydr. Chem.* **1962**, *1*, 107–113.

[147] Ley, S. V.; Baeschlin, D. K.; Dixon, D. J.; Foster, A. C.; Ince, S. J.; Priepke, H. W. M.; Reynolds, D. J. *Chem. Rev.* **2001**, *101*, 53–80.

[148] Yoshimura, J.; Horito, S.; Tamura, J.-I.; Hashimoto, H. *Chem. Lett.* **1985**, 1335–1338.

[149] Ottosson, H. *Carbohydr. Res.* **1990**, *197*, 101–107.

[150] Ziegler, T.; Sutoris, H.; Glaudemans, C. P. J. *Carbohydr. Res.* **1992**, *229*, 271–291.

[151] Zuurmond, H. M.; van der Klein, P. A. M.; van der Marel, G. A.; van Boom, J. H. *Tetrahedron* **1993**, *49*, 6501–6514.

[152] Das, S. K.; Roy, N. *J. Carbohydr. Chem.* **1995**, *14*, 417–428.

[153] Kramer, S.; Nolting, B.; Ott, A.-J.; Vogel, C. *J. Carbohydr. Chem.* **2000**, *19*, 891–921.

[154] Hanessian, S.; Katotani, M. *Carbohydr. Res.* **1990**, *202*, 67–79.

[155] Moradei, O.; du Mortier, C.; Cirelli, F.; Thiem, J. *J. Carbohydr. Chem.* **1995**, *14*, 525–532.

[156] Hanessian, S.; David, S. *Tetrahedron* **1985**, *41*, 643–663.

[157] Ricci, A.; Roelens, S.; Vannucchi, A. *J. Chem. Soc. Chem. Commun.* **1985**, 1457–1458.

[158] Roelens, S. *J. Chem. Soc. Perkin Trans. 2* **1988**, 1617–1625.

[159] Reginato, G.; Ricci, A.; Roelens, S.; Scapecchi, S. *J. Org. Chem.* **1990**, *55*, 5132–5139.

[160] Roelens, S. *J. Org. Chem.* **1996**, *61*, 5257–5263.

[161] Grindley, T. B. *Adv. Carbohydr. Chem. Biochem.* **1998**, *53*, 17–142.

[162] Egbertson, M.; Danishefsky, S. J. *J. Org. Chem.* **1989**, *54*, 12–15.

[163] Varma, A. J.; Schuerch, C. *J. Org. Chem.* **1981**, *46*, 799–803.

[164] Iwasaki, F.; Maki, T.; Onomura, O.; Nakashima, W.; Matsumura, Y. *J. Org. Chem.* **2000**, *65*, 996–1002.

[165] Martinelli, M. J.; Vaidyanathan, R.; Pawlak, J. M.; Nayyar, N. K.; Dhokte, U. P.; Doecke, C. W.; Zollars, L. M. H.; Moher, E. D.; Van Khau, V.; Kosmrjl, B. *J. Am. Chem. Soc.* **2002**, *124*, 3578–3585.

[166] Demizu, Y.; Kubo, Y.; Miyoshi, H.; Maki, T.; Matsumura, Y.; Moriyama, N.; Onomura, O. *Org. Lett.* **2008**, *10*, 5075–5077.

[167] Lee, D.; Taylor, M. S. *J. Am. Chem. Soc.* **2011**, *133*, 3724–3727.

[168] Wong, C.-H.; Ye, X.-S.; Zhang, Z. *J. Am. Chem. Soc.* **1998**, *120*, 7137–7138.

[169] Betaneli, V. I.; Ott, A. Y.; Brukhanova, O. V.; Kochetkov, N. K. *Carbohydr. Res.* **1988**, *179*, 37–50.

[170] Dubey, R.; Reynolds, D.; Abbas, S. A.; Matta, K. L. *Carbohydr. Res.* **1988**, *183*, 155–162.

[171] Khan, S. H.; Jain, R. K.; Matta, K. L. *Carbohydr. Res.* **1990**, *207*, 57–69.

[172] Osborn, H. M. I.; Brome, V. A.; Harwood, L. M.; Suthers, W. G. *Carbohydr. Res.* **2001**, *332*, 157–166.

[173] Lu, X.-A.; Chou, C.-H.; Wang, C.-C.; Hung, S.-C. *Synlett* **2003**, *9*, 1364–1366.

[174] Garegg, P. J.; Iversen, T.; Oscarson, S. *Carbohydr. Res.* **1976**, *53*, C5–C7.

[175] Boons, G.-J.; Castle, G. H.; Clase, J. A.; Grice, P.; Ley, S. V.; Pinel, C. *Synlett* **1993**, 913–914.

[176] Yang, W.-C.; Lu, X.-A.; Kulkarnia, S. S.; Hung, S.-C. *Tetrahedron Lett.* **2003**, *44*, 7837–7840.

[177] Dang, N.; Munasinghe, V. R. N.; Overend, W. G. *J. Chem. Soc. Perkin Trans. 1* **1983**, 257–270.

[178] Szeja, W. *Synthesis* **1979**, 821–822.

[179] Wang, H.; She, J.; Zhang, L.-H.; Ye, X.-S. *J. Org. Chem.* **2004**, *69*, 5774–5777.

[180] Falguni, D.; Garegg, P. J. *Synthesis* **1994**, 1121–1123.

[181] Khiar, N.; Martin-Lomas, M. *J. Org. Chem.* **1995**, *60*, 7017–7021.

[182] Fraser-Reid, B.; Nair, A. G.; Nair, L. G.; Radhakrishnan, K. V.; Cristobal, L. J.; Gomez, A.; Uriel, C. *Aust. J. Chem.* **2002**, *55*, 123–130.

[183] Baeschlin, D. K.; Chaperon, A. R.; Charbonneau, V.; Green, L. G.; Ley, S. V.; Lucking, U.; Walther, E. *Angew. Chem. Int. Ed.* **1998**, *37*, 3423–3428.

[184] Crich, D.; Li, W.; Li, H. *J. Am. Chem. Soc.* **2004**, *126*, 15081–15086.

[185] Boger, D. L.; Honda, T. *J. Am. Chem. Soc.* **1994**, *116*, 5647–5656.

[186] Mikami, T.; Asano, H.; Mitsunobu, O. *Chem. Lett.* **1987**, 2033–2036.

[187] Ennis, S. C.; Cumpstey, I.; Fairbanks, A. J.; Butters, T. D.; Mackeen, M.; Wormald, M. R. *Tetrahedron* **2002**, *58*, 9403–9411.

[188] Grice, P.; Ley, S. V.; Pietruszka, J.; Priepke, H. W. M.; Warriner, S. L. *J. Chem. Soc. Perkin Trans. 1* **1997**, 351–363.

[189] Denino, M. P.; Etienne, J. B.; Duplantier, K. C. *Tetrahedron Lett.* **1995**, *36*, 669–672.

[190] Zheng, B. Z.; Yamauchi, M.; Dei, H.; Kusaka, S. I.; Matsui, K.; Yonemitsu, O. *Tetrahedron Lett.* **2000**, *41*, 6441–6445.

[191] Garegg, P. J.; Hultberg, H. *Carbohydr. Res.* **1981**, *93*, C10–C11.

[192] Garegg, P. J.; Hultberg, H.; Wallin, S., *Carbohydr. Res.* **1982**, *108*, 97–101.

[193] Liptak, A.; Jodal, I.; Nasai, P. *Carbohydr. Res.* **1975**, *44*, 1–11.

[194] Liptak, A.; Jodal, I.; Nasai, P. *Carbohydr. Res.* **1976**, *52*, 17–22.

[195] Jiang, L.; Chan, T. H. *Tetrahedron Lett.* **1998**, *39*, 355–358.

[196] Wang, C.-C.; Luo, S.-Y.; Shie, C.-R.; Hung, S.-C. *Org. Lett.* **2002**, *4*, 847–849.

[197] Shie, C.-R.; Tzeng, Z.-H.; Kulkarni, S. S.; Uang, B.-J.; Hsu, C.-Y.; Hung, S.-C. *Angew. Chem. Int. Ed.* **2005**, *44*, 1665–1668.

[198] Chen, C.-T.; Weng, S.-S.; Kao, J.-Q.; Lin, C.-C.; Jan, M.-D. *Org. Lett.* **2005**, *7*, 3343–3346.

[199] Du, Y.; Zhang, M.; Kong, F. *Org. Lett.* **2000**, *2*, 3797–3800.

[200] Mukhopadhyay, B.; Field, R. A. *Carbohydr. Res.* **2003**, *338*, 2149–2152.

[201] Wang, C.-C.; Lee, J.-C.; Luo, S.-Y.; Kulkarni, S. S.; Huang, Y.-W.; Lee, C.-C.; Chang, K.-L.; Hung, S.-C. *Nature* **2007**, *446*, 896–899.

[202] Wang, C.-C.; Kulkarni, S. S.; Lee, J.-C.; Luo, S.-Y.; Hung, S.-C. *Nat. Protoc.* **2008**, *3*, 97–113.

[203] Tsunoda, T.; Suzuki, M.; Noyori, R. *Tetrahedron Lett.* **1980**, *21*, 1357–1358.

[204] Joseph, A. A.; Verma, V. P.; Liu, X.-Y.; Wu, C.-H.; Dhurandhare, V.; Wang, C.-C. *Eur. J. Org. Chem.* **2012**, 744–753.

[205] Chang, K.-L.; Zulueta, M. M. L.; Lu, X.-A.; Zhong, Y.-Q.; Hung, S.-C. *J. Org. Chem.* **2010**, *75*, 7424–7427.

[206] Luo, S.-Y.; Kulkarni, S. S.; Chou, C.-H.; Liao, W.-M.; Hung, S.-C. *J. Org. Chem.* **2006**, *71*, 1226–1229.

[207] Patil, P. S.; Lee, C.-C.; Huang, Y.-W.; Zulueta, M. M. L.; Hung, S.-C. *Org. Biomol. Chem.* **2013**, *11*, 2605–2612.

3

GENERAL ASPECTS IN O-GLYCOSIDIC BOND FORMATION

XIN-SHAN YE[1] AND WEIGANG LU[1,2]

[1]*State Key Laboratory of Natural and Biomimetic Drugs, School of Pharmaceutical Sciences, Peking University, Beijing, China*
[2]*Complex Carbohydrate Research Center, University of Georgia, Athens, GA, USA*

3.1 INTRODUCTION

The efficient assembly of naturally occurring complex oligosaccharides or their analogues has long been pursued by chemists for a broader understanding of these "sweet molecules." Homogeneous oligosaccharides, which are very difficult to gain from nature, would enable scientists to investigate their three-dimensional structures and conformations to establish the structure–activity relationships of these biomolecules, to reveal the molecular mechanism of carbohydrate–protein interactions, and to discover leading structures of potential therapeutic value.

3.2 SOME BASIC CONCEPTS

Glycosidic bond formation is at the center of oligosaccharide synthesis. Generally, a glycoside is a cyclic acetal (a glycone) that bonds another glycone or an aglycone at the anomeric center (Scheme 3.1). Because O-glycoside is the most abundant type of glycoside in nature, without specifying, the term glycoside refers to O-glycoside in

Glycochemical Synthesis: Strategies and Applications, First Edition.
Edited by Shang-Cheng Hung and Medel Manuel L. Zulueta.

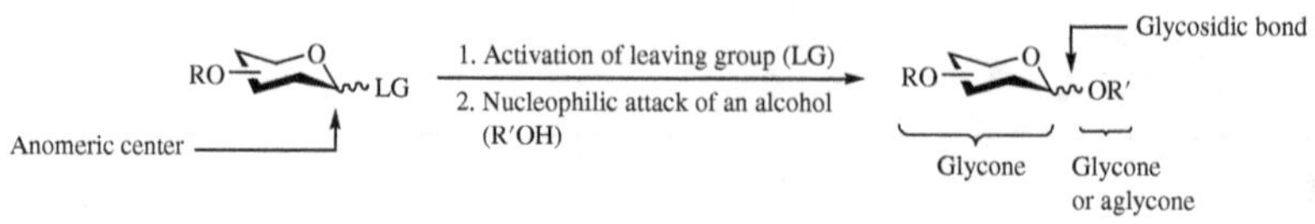

SCHEME 3.1 The glycosidic bond formation.

this chapter. Moreover, unless otherwise stated, the general discussions on monosaccharides refer to aldohexopyranoses. The glycosidic bond formation, also known as glycosylation, involves the transfer of the sugar moiety (as an electrophile) from a glycosyl donor to a glycosyl acceptor (as a nucleophile). Usually, from a chemical synthesis point of view, the glycosyl donor is fully protected with a leaving group at the anomeric center, whereas the glycosyl acceptor is suitably protected, leaving a free hydroxy group, which is glycosylated in the glycosylation reaction.

3.2.1 Mechanism of Glycosylation

The exact mechanism of glycosylation is not very clear yet. It is mostly interpreted according to the nucleophilic displacement mechanism. Because nucleophilic displacement occurs at the anomeric secondary carbon atom with a weak nucleophile, an S_N1-type mechanism is more commonly accepted, although an S_N2-type attack may occur in many cases. Thus, glycosylation seems to involve (i) the departure of the leaving group in the glycosyl donor assisted by a catalyst or promoter to form a reactive intermediate and (ii) the coupling of the glycosyl intermediate and the glycosyl acceptor to provide the glycoside product (Scheme 3.1).

3.2.2 Stereoselectivity

As the nucleophilic alcohol (glycosyl acceptor) can attack the anomeric carbon from either the top or bottom of the sugar ring, the coupling product may be a mixture of α-linked and β-linked glycosides. Therefore, specific care should be taken with regard to stereoselectivity. Various factors, such as protecting groups, leaving groups, promoters, additives, solvents, temperature, concentrations, substrates, and even the sequence of addition of reactants, may influence the stereochemical outcome of the glycosylation reaction. To reach satisfactory stereoselectivity and acceptable yield, the glycosylation reaction needs to be performed in a highly controlled manner to keep a delicate balance between reactivity and selectivity.

3.2.3 Anomeric Effect

The anomeric effect or Edward–Lemieux effect is a stereoelectronic effect that describes the tendency of heteroatomic substituents adjacent to a heteroatom within a six-membered ring to prefer the axial orientation instead of the less-hindered equatorial orientation that would be expected from steric considerations. For sugars, substituents, such as halide, at the anomeric center of pyranosides prefer to adopt the axial orientation, which would be exclusive if the equilibrium could be reached. As described in detail in Section 1.2.8, the main driving forces behind this phenomenon include the dipole–dipole repulsion between the heteroatoms bonded to C1 and the stabilizing hyperconjugation between the unshared electron pair of the endocyclic oxygen (O5) and the σ* orbital of the axial C1–X bond. Consequently, the anomeric effect is partially responsible for the stereochemical outcome of a glycosylation reaction.

It was also observed that an electropositive group, particularly nitrogen-containing heterocycles, at the anomeric center tends to be oriented equatorially. This phenomenon is referred to as "reverse anomeric effect." However, it is accepted that the reverse anomeric effect is not generalized [1, 2].

3.2.4 Participation by Functional Groups in the Glycosyl Donor

The involvement of ester groups located at the C2 position of glycosyl donors in a glycosylation reaction leads to an acyloxonium intermediate, which is formed from the initially produced oxocarbenium ion (see Chapter 4). In this neighboring group participation, nucleophilic attack of the glycosyl acceptor can only occur from the *trans* side of anomeric center, resulting to the 1,2-*trans*-glycoside. If the nucleophilic attack occurs at the dioxolane ring carbon of the oxocarbenium ion, the orthoester will form and it can be further isomerized to the corresponding glycoside in the presence of a Lewis acid. Thus, 1,2-*trans*-glycosides are readily produced when C2-participating ester groups are involved. On the other hand, a nonparticipating group at C2 usually leads to a mixture of both 1,2-*trans*- and 1,2-*cis*-glycosides. The synthesis of 1,2-*cis*-glycosides, such as α-glucosides and α-galactosides, is relatively difficult, and especially, the synthesis of β-mannosides is a tricky exercise.

The concept of neighboring group participation is not confined to ester groups, as exemplified by Geert-Jan Boons' work on the introduction of chiral auxiliaries at C2 for efficient stereoselective glycosylations [3]. As shown in Scheme 3.2, the nucleophilic phenylsulfanyl moiety at the C2 position of compound **1** participates after the formation of the oxocarbenium ion. Guided by favorable steric and electronic factors, the resulting intermediate sulfonium ion assumes a *trans* decalin-type orientation. Displacement of the equatorial anomeric sulfonium ion by a glycosyl acceptor, thus, leads to the formation of the 1,2-*cis*-glycoside **2**.

Although remote substituents at positions other than C2 are of less importance, in some cases, especially at C6, they are strong enough to serve as participating groups

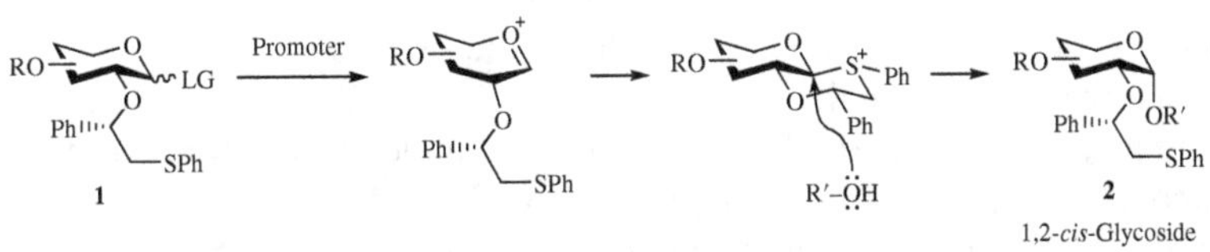

SCHEME 3.2 Stereoselective glycosylation by using chiral auxiliaries.

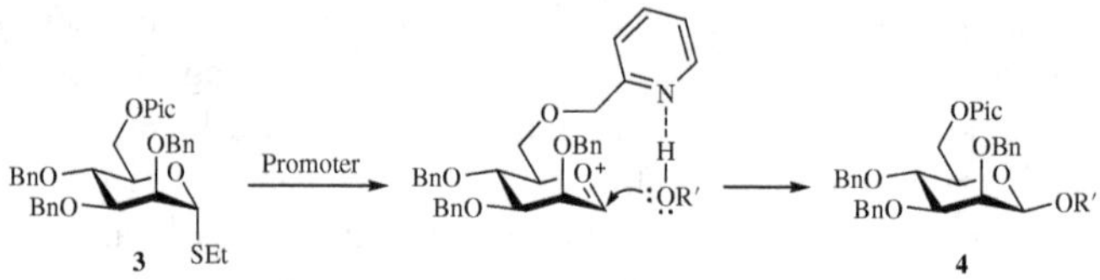

SCHEME 3.3 Synthesis of β-mannosides via remote participation. Bn, benzyl; Pic, picolinyl.

via either hydrogen bond formation or stereoelectronic effect. The recent progress by Alexei Demchenko and coworkers gives a good example of this concept [4]. *O*-Picolinyl at remote positions (C3, C4, or C6) can mediate glycosylation reactions by providing high or even complete facial selectivity for the attack of the glycosyl acceptor. Evidence supports the intermolecular H-bond tethering between the glycosyl donor and the glycosyl acceptor counterpart, leading the acceptor to attack the anomeric center from only one side. This strategy has been utilized to synthesize challenging glycosidic linkages like β-mannosides (Scheme 3.3).

The various ways in which functionalities in both the glycosyl donors and glycosyl acceptors contribute to the stereoselectivity of the glycosylation reaction are further described in Chapter 4.

3.2.5 The Armed–Disarmed Concept

The "armed–disarmed" concept, put forward by Bert Fraser-Reid [5], refers to the observations that, in general, acyl protecting groups reduce the reactivity at the anomeric center of the sugar donor, while ether groups increase it. The influence of substitution pattern on the reactivity of the glycosyl donor was noticed by Hans Paulsen in the late 1970s [6]. The rationale behind this phenomenon may be that electron-donating groups like ethers favor the development of positive charge in the transition state of the sugar ring during glycosylation, while electron-withdrawing groups like esters discourage such positive charge. As a consequence, the armed (more reactive) sugar donor can react with the disarmed (less reactive) acceptor (with the same anomeric protecting group as the leaving group of the donor) while avoiding the unwanted self-coupling of the glycosyl acceptors. In principle, with further manipulations of the protecting groups, the resulting disaccharide can be armed for the next cycle of the glycosylation reaction. Iterative glycosylation is, thus, made possible.

3.2.6 Additives

Additives, in some cases, are used to modulate the stereochemistry of glycosylation by directly participating in the reaction process. Raymond Lemieux and coworkers, for example, demonstrated that α-pyranosyl bromides would transform into the β-pyranosyl bromides *in situ* in the presence of tetraalkylammonium bromide [7]. The more reactive β-pyranosyl bromides would react with glycosyl acceptors in an S_N2-type manner to give the kinetically controlled α-linked glycosides.

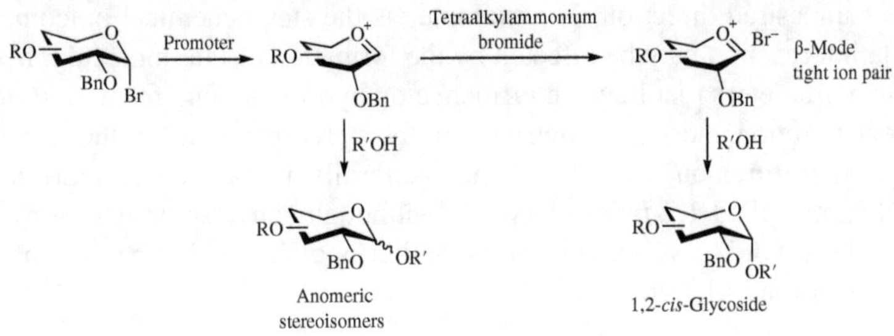

SCHEME 3.4 Synthesis of 1,2-*cis*-glycosides using halide additives.

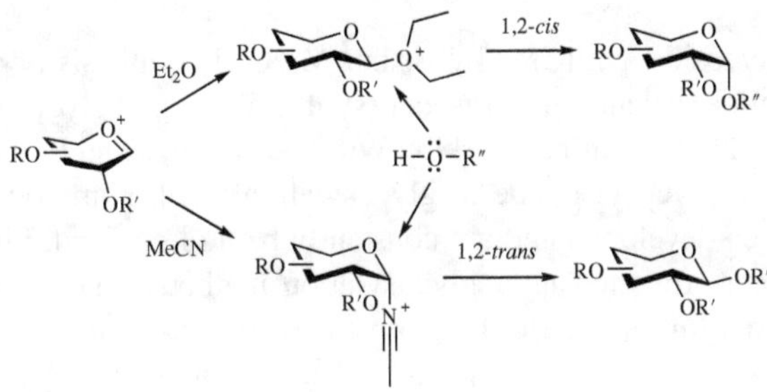

SCHEME 3.5 Stereoselective glycosylation via solvent participation.

The so-called "*in situ* anomerization" by halide catalysis is one of the important approaches for the synthesis of 1,2-*cis*-glycosides (Scheme 3.4).

3.2.7 Solvent Effect

Solvent is an important factor that contributes to the rate, yield, and stereoselectivity of glycosylation reactions. It affects the stereochemical outcome of the reaction not only through solvation but also by directly participating in the reaction process, similar to the role of additives. Preferential formation of 1,2-*cis*- or 1,2-*trans*-glycosides may occur if different participating solvents are used. One well-known example of the so-called "solvent effect" is the stereoselective formation of glycosidic bonds using either acetonitrile or diethyl ether as the solvent when a C2-participating group is absent [8]. In acetonitrile, the *in situ*-generated nitrilium moiety adopts the axial orientation, while in diethyl ether, the equatorial intermediate is preferentially formed. These behaviors allow for the stereoselective formation of equatorially substituted glycosides in acetonitrile and axially substituted glycosides in diethyl ether because the nucleophilic substitution proceeds in an S_N2-like manner (Scheme 3.5).

3.2.8 Effects of Other Factors on Glycosylation

In addition to the aforementioned factors, any other factors or conditions may unexpectedly influence the reaction transition state and steer the reaction course to a desired or undesired direction. For example, as the stereochemical outcome of the glycosylation reaction may be affected by the formation of intermolecular hydrogen bond, any parameters that have an influence on hydrogen bond formation (temperature, concentration, etc.) may contribute to the stereoselectivity of the reaction [4]. Lower temperature, on the other hand, generally favors the formation of the kinetically controlled 1,2-*trans*-glycosides, although converse observations are not excluded. In a word, a systematic investigation is often needed before an optimal reaction condition is determined.

3.3 METHODS FOR GLYCOSIDIC BOND FORMATION

Since the first glycoside syntheses by Arthur Michael (synthesis of aryl glycosides from glycosyl halide) [9] and Emil Fischer (synthesis of alkyl glycosides from hemiacetals), followed by the seminal work by Wilhelm Koenigs and Eduard Knorr (general application of glycosyl halide to glycosylation) [10], numerous glycosylation methods have been developed and are constantly being improved. Figure 3.1 shows the glycosyl donors for the major glycosylation methods together with the usual activating reagents for each case. In some cases, for example, in the synthesis of 2-deoxy glycosides, glycals are of special value. Although each glycosyl donor has its unique application and is sometimes irreplaceable for certain glycoside synthesis, it seems that the three most widely used types, namely, glycosyl halides, glycosyl trichloroacetimidates, and thioglycosides, are sufficient for most oligosaccharide syntheses.

3.3.1 Glycosyl Halides

The application of glycosyl halides in glycosylation, as developed by Koenigs and Knorr [10], had been, for a long time, the only available method for the stereospecific synthesis of 1,2-*trans*-glycosides. Today, they are still widely used in the syntheses

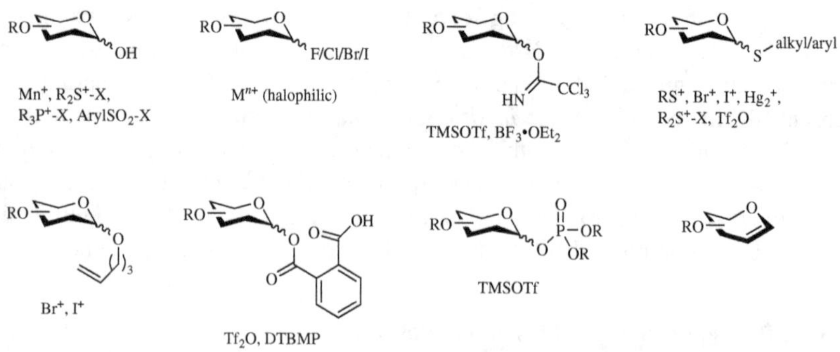

FIGURE 3.1 Typical glycosyl donors and their corresponding promoters or activating reagents. DTBMP, 2,6-di-*tert*-butyl-4-methylpyridine; TMS, trimethylsilyl; Tf, triflyl.

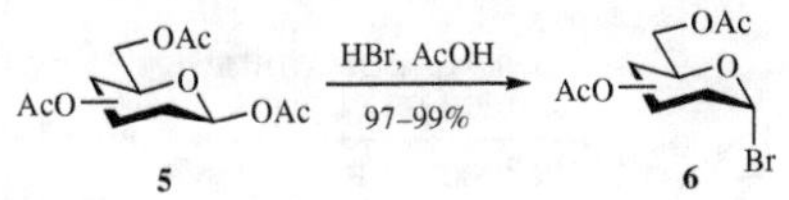

SCHEME 3.6 Synthesis of glycosyl bromides. Ac, acetyl.

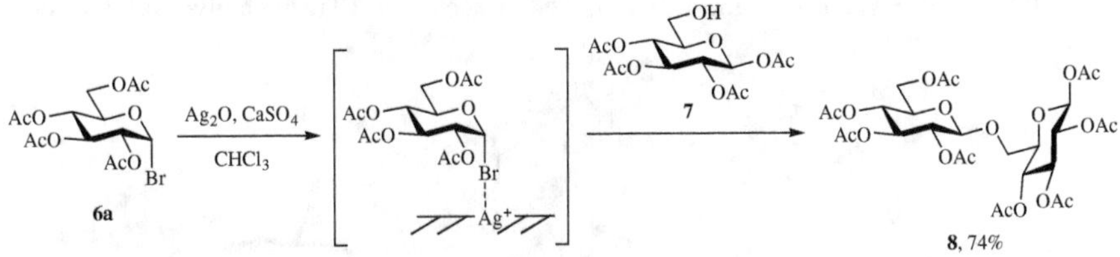

SCHEME 3.7 Koenigs–Knorr glycosylation promoted by insoluble silver salts.

of numerous glycosides. One advantage of this method is the easy availability of glycosyl donors. Acylated glycosyl bromides can be prepared by treatment of per-O-acetylated sugars with HBr in acetic acid (Scheme 3.6). Glycosyl chlorides are obtained from the same starting material using various chlorinating agents such as $SnCl_4$, $TiCl_4$, or $AlCl_3$ in high yields. Due to the anomeric effect, the product is exclusively the α anomer in usual cases.

The classic glycosyl donors of the Koenigs–Knorr glycosylation are glycosyl bromides and glycosyl chlorides. Glycosyl bromides react faster than the corresponding chlorides and are therefore preferred for most reactions. Insoluble salts such as Ag_2O and Ag_2CO_3 as well as those soluble ones such as AgOTf and $AgClO_4$ serve as promoters. Mercury salts such as $HgBr_2$ and $Hg(CN)_2$, which was introduced by Burckhardt Helferich [11], are also employed. In classic Koenigs–Knorr reactions, neighboring participating groups at C2 normally lead to 1,2-*trans*-glycosides, especially when insoluble silver salts are used as promoters. In this heterogeneous reaction system, the reaction is thought to proceed on the surface of the silver salts, assisting the formation of the *trans* glycoside by shielding the α face of the glycosyl donor (Scheme 3.7) [12]. Alternatively, *in situ* anomerization by treatment of tetraalkylammonium halide can be used for selective synthesis of 1,2-*cis*-glycosides. Other factors such as solvent participation and temperature may also play a role in the stereocontrol of the glycosylation reaction.

Despite its wide application in glycoside synthesis, the Koenigs–Knorr reaction suffers from several disadvantages: (i) the glycosyl halides are unstable, (ii) excess toxic heavy metals are needed to activate the donor, and (iii) a desiccant (to absorb any liberated water) and an acid acceptor (to absorb the liberated hydrogen halide and can be a promoter itself) are often needed to increase the yield and suppress side reactions.

Once thought too inert to be of synthetic value, glycosyl fluorides have emerged as versatile glycosyl donors for the stereoselective formation of glycosidic bonds. They can be obtained by treatment of the free sugar building block with *N,N*-diethylaminosulfur trifluoride (DAST) [13] or the per-O-acetylated sugar with HF·pyridine [14]. Compared

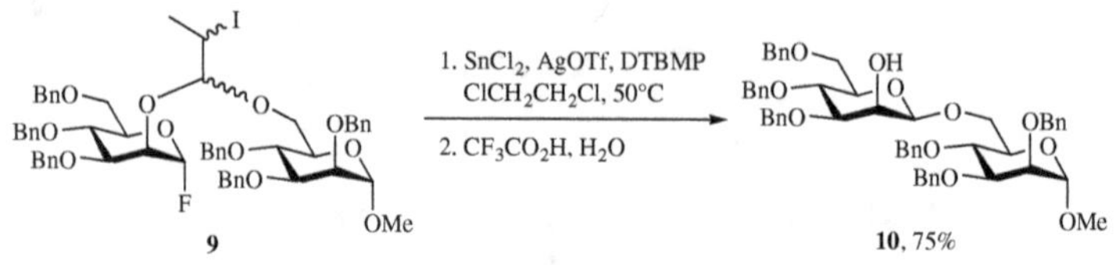

SCHEME 3.8 Synthesis of the 1,2-*cis*-linked disaccharide **10** using a glycosyl fluoride.

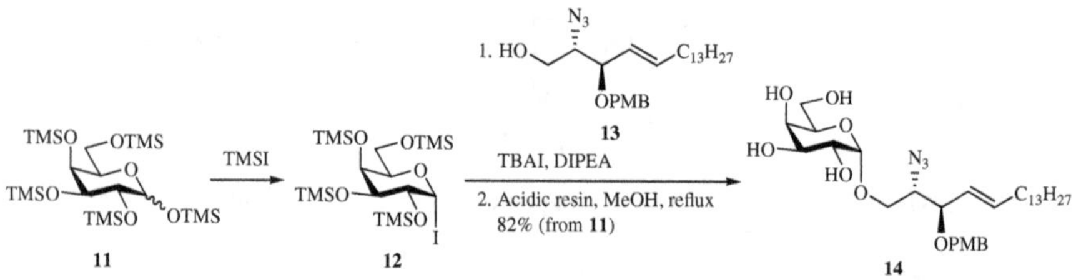

SCHEME 3.9 Synthesis of the α-glycolipid **14** using a glycosyl iodide. DIPEA, diisopropylethylamine; PMB, *p*-methoxybenzyl.

to other relatively labile glycosyl halides, glycosyl fluorides display remarkable stability toward chromatography on silica gel and can resist hydrolysis in basic conditions. Also, β-glycosyl fluorides can be obtained in a kinetically controlled manner. They can be orthogonally activated with Lewis acids under mild conditions over other glycosyl donors such as thioglycosides. Nowadays, they are frequently used for the synthesis of various natural glycosides. For example, the challenging 1,2-*cis*-linked disaccharide **10** was synthesized using a mannosyl fluoride (Scheme 3.8) [15].

Glycosyl iodides, generally considered too reactive to be of synthetic utility, have also been demonstrated to offer advantageous properties in certain glycosylation reactions in terms of activity, efficiency, and stereoselectivity. The most reliable and commonly used method to prepare glycosyl iodides involves the treatment of anomeric acetates with TMSI [16]. In general, glycosyl iodides are activated under basic conditions to give β-glycosides with high stereoselectivity. The utility of glycosyl iodides as donors for stereoselective synthesis of glycosides was demonstrated by the successful construction of the backbone of a biologically active α-glycolipid in the presence of tetrabutylammonium iodide (TBAI) (Scheme 3.9) [17].

3.3.2 Glycosyl Trichloroacetimidates

Among all the synthetic methods developed for glycosidic bond formation, the trichloroacetimidate method developed by Richard Schmidt and coworkers [18] is probably the most popular and widely used for the synthesis of complex oligosaccharides. Glycosyl trichloroacetimidates can be readily prepared by a base-catalyzed addition of the anomeric hydroxy group to trichloroacetonitrile. The stereochemistry can be influenced by the addition of different bases; the use of K_2CO_3 favors formation of the β anomer, whereas NaH or 1,8-diazabicyclo[5.4.0]undec-7-ene (DBU) favors the α anomer (Scheme 3.10).

$BF_3 \cdot Et_2O$ and catalytic amount of TMSOTf are the most commonly used promoters, although other Lewis acids such as $Yb(OTf)_3$ [19] and $Sm(OTf)_3$ [20] are also applied on occasion. In the trichloroacetimidate method, neighboring group participation of 2-*O*/*N*-acyl groups normally leads to 1,2-*trans*-glycosides. For donors having a non-participating group at C2, treatment with relatively weak Lewis acid catalysts ($BF_3 \cdot Et_2O$) generally results in the formation of an inverted glycoside, probably by an S_N2-type process, whereas treatment with stronger Lewis acid (TMSOTf) supports the formation of the thermodynamically more stable products. Additionally, factors such as solvent effect and protecting group patterns can be utilized to manipulate the stereochemistry of the reactions. The trichloroacetimidate method has been used for the construction of numerous glycoconjugates and oligosaccharides. Scheme 3.11 shows two examples of its successful application [21, 22].

Glycosyl trichloroacetimidates are activated under relatively mild conditions, and the promoter is used in catalytic amounts, making it possible to prepare various trichloroacetimidates from not only monosaccharides but also oligosaccharides

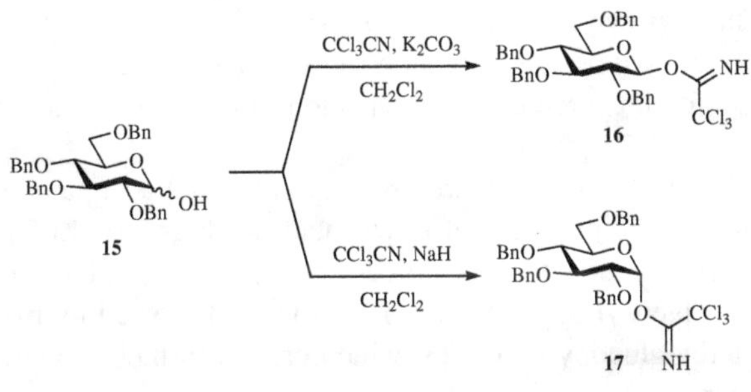

SCHEME 3.10 Preparation of glycosyl trichloroacetimidates.

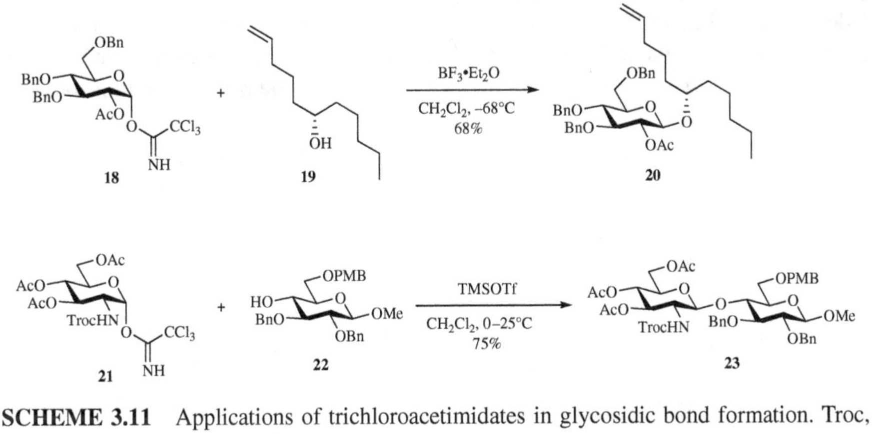

SCHEME 3.11 Applications of trichloroacetimidates in glycosidic bond formation. Troc, 2,2,2-trichloroethoxycarbonyl.

carrying a variety of protecting groups. These donors, especially the β anomers, are also more stable than the corresponding glycosyl halides and can be stored at low temperature for months. Moreover, no toxic heavy metals are needed in the reaction process. However, the high reactivity of glycosyl trichloroacetimidates sometimes leads to side reactions or even decomposition of the glycosyl donors. When the glycosyl acceptor is very unreactive, the donor may rearrange to the glycosyl trichloroacetamide, which cannot serve as glycosyl donor. This could be avoided by the so-called "inverse procedure" in which the glycosyl acceptor and the catalyst are mixed together before the glycosyl donor is added [23].

In attempts at finding new glycosyl donors, a series of N-substituted glycosyl trihaloacetimidate derivatives were also prepared. Initial experiments revealed that glycosylation with trifluoroacetimidates were generally less efficient than those with trichloroacetimidates in terms of product yield. Later, Biao Yu and coworkers explored the application of glycosyl *N*-phenyltrifluoroacetimidates and reported particularly good reactivity for some specific glycosylation reactions [24].

3.3.3 Thioglycosides

First evaluated as glycosyl donors by Robert Ferrier and coworkers in the 1970s [25], thioglycosides have grown to become an important donor of choice for the synthesis of oligosaccharides. Numerous protocols have been reported and are constantly being developed for the preparation, activation, and glycosylation application of thioglycosides.

Thioglycosides are usually prepared by treating per-O-acetylated sugar building blocks with thiols in the presence of Lewis acids such as $BF_3 \cdot Et_2O$, providing the predominant 1,2-*trans* products in the process (Scheme 3.12). Alternatively, treatment of the acetylated glycosyl bromide **6a** with thiourea, followed by hydrolysis, gives the acetylated 1-thioglucopyranose **25**, which can be further alkylated to form the desired thioglycosides **26**.

Thioglycosides can be activated by a variety of thiophilic promoter systems, often comprising a thiophile source in equimolar or excess amounts in tandem with a catalyst. *N*-Iodosuccinimide (NIS), together with catalytic TfOH [26] or AgOTf [27], is often used. In addition to the iodonium system, organosulfur promoters such as Ph_2SO with Tf_2O [28], 1-benzenesulfinylpiperidine with Tf_2O [29], benzenesulfinylmorpholine

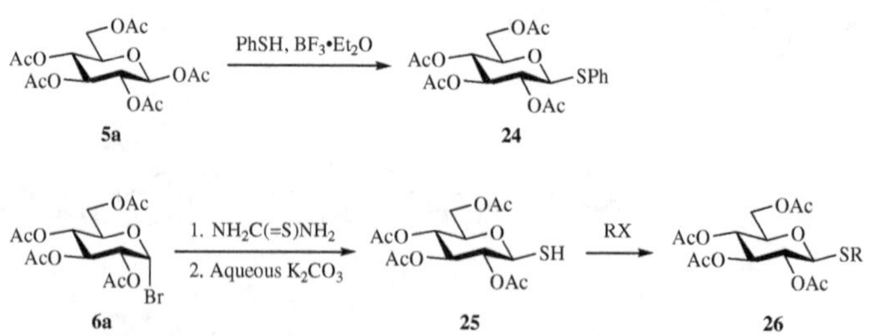

SCHEME 3.12 Preparation of thioglycosides.

with Tf_2O [30], and dimethyl(methylthio)sulfonium triflate (DMTST) (Scheme 3.13) [31] have been developed. Thiophilic heavy metals such as mercury [25, 32–34] and silver salts [35] are also used as promoters in some cases. The stereochemical outcome with thioglycosides follows the general rules as discussed in other glycosylation methods. A participating ester group at C2 usually leads to the formation of 1,2-*trans*-glycosides, while a nonparticipating group at C2 is necessary but often not sufficient to ensure 1,2-*cis* glycosidic bond formation.

The advantage of thioglycosides lies in their great stability under a wide range of conditions for protecting group manipulation and orthogonal activation over other glycosyl donors. In addition, they can be smoothly converted to other glycosyl donors directly or by a two-step procedure. These features, combined with the tunable reactivity of thioglycosides, are often exploited for the efficient syntheses of complex oligosaccharides, as demonstrated in the one-pot assembly of the linear trisaccharide **30** (Scheme 3.14) [36].

A variety of thioglycoside derivatives has also been investigated as glycosyl donors. Glycosyl sulfoxides or sulfinyl glycosides, readily available by the oxidation of thioglycosides with *m*-chloroperbenzoic acid, can be promoted by Tf_2O combined with hindered bases such as DTBMP [37]. Other promoting systems have also been suggested. The synthesis of disaccharide **33** using glycosyl sulfoxide, which can be selectively activated over thioglycosides, is illustrated in Scheme 3.15 [38].

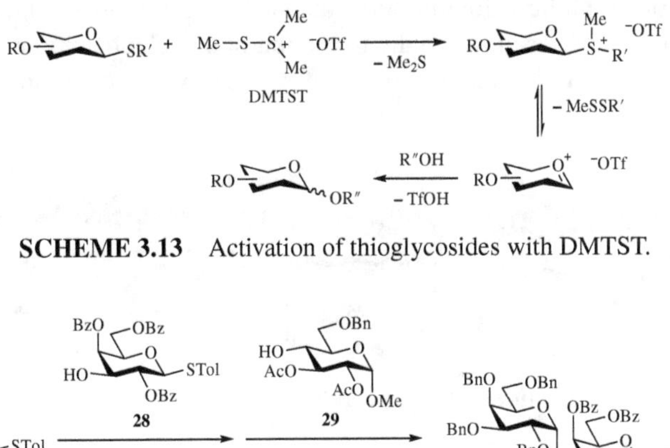

SCHEME 3.13 Activation of thioglycosides with DMTST.

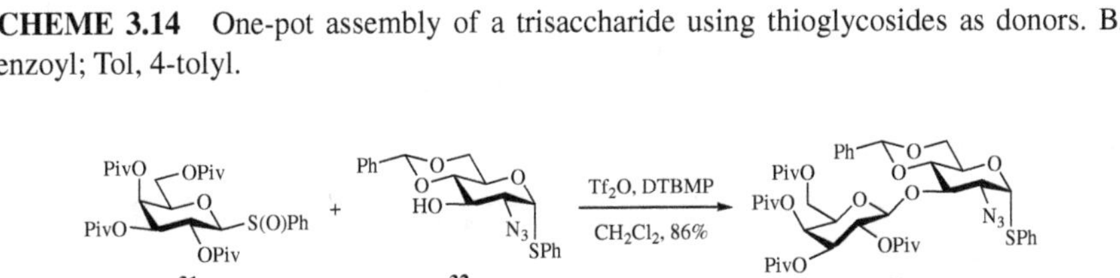

SCHEME 3.14 One-pot assembly of a trisaccharide using thioglycosides as donors. Bz, benzoyl; Tol, 4-tolyl.

SCHEME 3.15 Synthesis of disaccharide **33** using glycosyl sulfoxide. Piv, pivaloyl.

Glycosyl thioimidates, containing a $SCR^1=NR^2$ aglycone, was first introduced by Robert Woodward and coworkers as glycosyl donors in the total synthesis of erythromycin [39]. They can be easily prepared by Lewis acid-promoted displacement of anomeric acetoxy groups with thio aglycones or substitution of anomeric halogens with thiolate anions [40]. As thioimidates possess the properties of both glycosyl imidates and thioglycosides, they can be activated not only by thiophilic promoters such as NIS/TMSOTf but also with conventional promoters for glycosyl imidates such as $BF_3 \cdot Et_2O$. One advantage of glycosyl thioimidates is their high stability, which enables a wide range of protecting group manipulations. For example, they can withstand rather harsh reaction conditions such as acetylation, benzylation, and deacetylation. In addition, the possibility of orthogonal activation of glycosyl thioimidates over thioglycosides and *n*-pentenyl glycosides further extends their synthetic utility in oligosaccharide assembly. Scheme 3.16 demonstrates the utility of glycosyl thioimidates in the form of *S*-benzoxazolyl thioglycoside in the synthesis of disaccharide **36** [41].

3.3.4 *n*-Pentenyl Glycosides

Fraser-Reid introduced the *n*-pentenyl glycosides as viable glycosyl donors in the late 1980s [42]. *n*-Pentenyl glycosides can be prepared by Fischer glycosylation of the aldose of interest, by a Koenigs–Knorr procedure (Scheme 3.17), or from *n*-pentenyl orthoesters. Electrophilic halonium ions liberated from promoters interact with the pentenyl double bond, resulting to the formation of a cyclic halonium intermediate. The cyclic halonium intermediate then rearranges to an oxonium intermediate

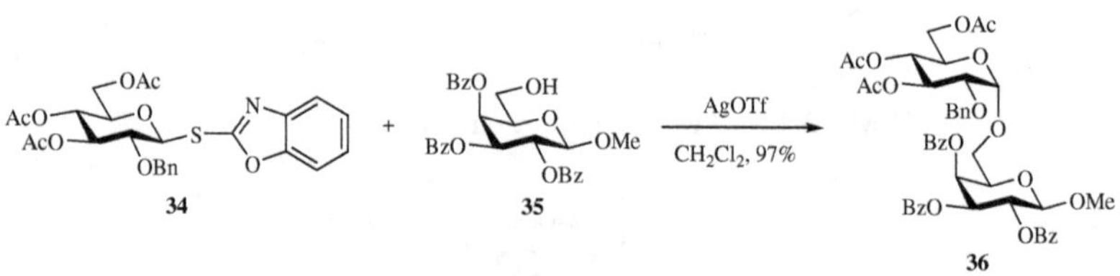

SCHEME 3.16 Synthesis of disaccharide **36** using a thioimidate glycosyl donor.

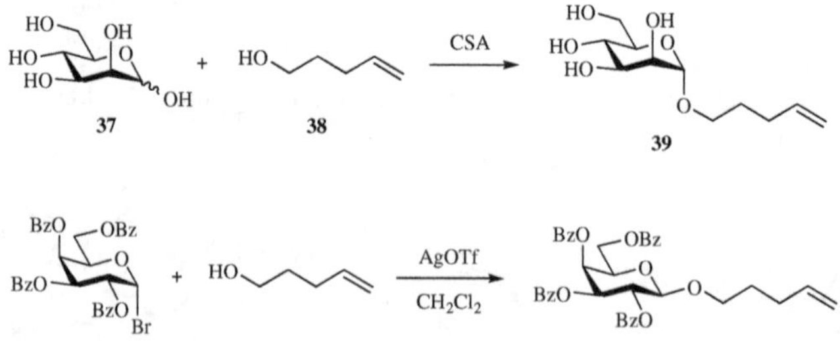

SCHEME 3.17 Synthesis of *n*-pentenyl glycosides.

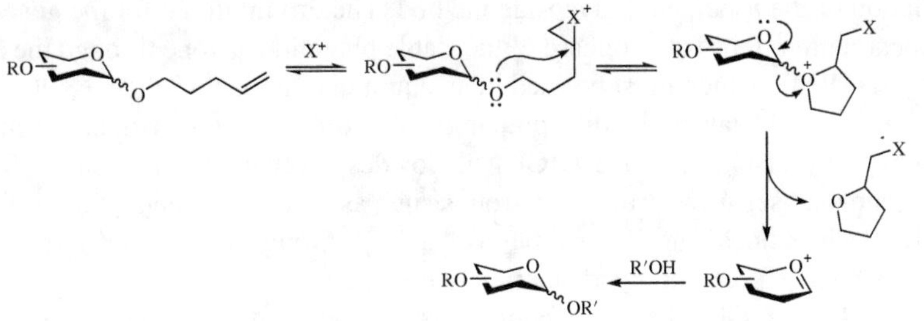

SCHEME 3.18 Mechanism of glycosylation involving *n*-pentenyl glycosides.

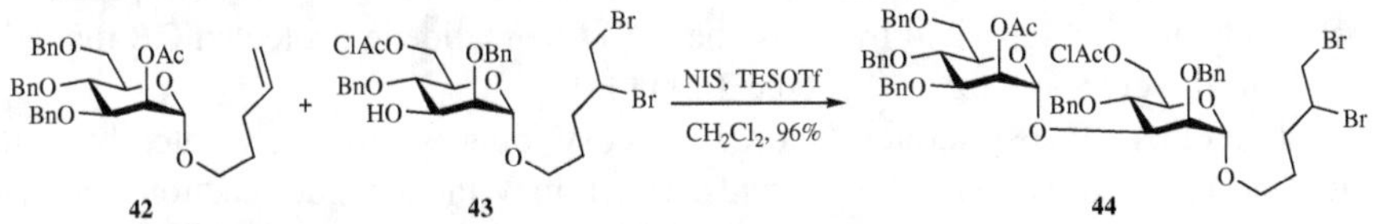

SCHEME 3.19 Glycosylation using *n*-pentenyl glycoside. ClAc, chloroacetyl.

containing the leaving group 2-halomethyltetrahydrofuran. Elimination of this leaving group leads to the oxocarbenium ion, which then reacts with a glycosyl acceptor to form the glycoside (Scheme 3.18).

N-Bromosuccinimide (NBS) and NIS are the most widely used electrophiles to activate *n*-pentenyl glycosides. Catalytic amounts of acids such as TfOH or triethylsilyl triflate (TESOTf) are introduced to speed up the generation of bromonium or iodonium ions [43]. The stereochemical outcome depends, as usual, on whether a participating group exists at C2. An alternative promoter is iodonium dicollidine perchlorate (IDCP), whose lower reactivity allows it to selectively activate certain "disarmed" *n*-pentenyl glycosides [44].

The *n*-pentenyl glycoside has gained popularity because of its stability under most conditions (except reductive hydrogenation), easy preparation and promotion, and tunable reactivity. In addition, the reactivity of the *n*-pentenyl glycosides can be switched off by bromination of the double bond [5], which can later be restored using Zn dust, samarium iodide, or sodium iodide. The "switch" function enables a wide range of synthetic options such as the consecutive assembly of oligosaccharides with the same building block. An application of *n*-pentenyl glycoside as the glycosyl donor for disaccharide construction is depicted in Scheme 3.19 [45].

Anomeric 4-pentenoate esters have also been shown to behave as glycosyl donors, liberating 4-iodomethyl-γ-lactone as the by-product [46].

3.3.5 Carboxybenzyl Glycosides

In 2001, Kwan Soo Kim and coworkers reported the 2-carboxybenzyl (CB) glycosides as a new type of glycosyl donor, which undergo glycosylation in high yield and with high stereoselectivity [47]. Conceptually, this type of glycosylation is a variation of the *n*-pentenyl glycoside method. The driving force for the generation of oxocarbenium ions is the release of the stable phthalide lactone through the action of Tf_2O/DTBMP, which must be used in an equimolar amount (Scheme 3.20).

CB glycosides can be readily prepared by hydrogenolysis of their precursors, 2′-(benzyloxycarbonyl)benzyl (BCB) glycosides, even in the presence of other hydrogenolysis-sensitive protecting groups such as benzyl and benzylidene groups. The BCB glycosides can, in turn, be synthesized through Koenigs–Knorr method. They can also be prepared by anomeric O-alkylation.

The utility of CB glycosyl donors was demonstrated in the construction of numerous glycosides, including the challenging β-mannosides. Generally, the stereochemical outcome is highly dependent on the protecting groups of CB glycosides and not affected by the anomeric configuration. The β-mannoside **47** was produced exclusively and in high yield by using the 4,6-*O*-benzylidene-protected CB mannoside **45** as the glycosyl donor (Scheme 3.21) [47].

Unlike other glycosyl donors, the CB glycosyl donors can be activated by both acids and bases; thus, many Lewis acids, which may induce side reactions, can be avoided by using bases as activators. In addition, the precursor BCB glycosides can be seen as latent glycosyl donors, the donor activity of which can be gained by hydrogenolysis. This property, similar to the "switch" function of *n*-pentenyl glycoside donors, further extends the utility of the CB glycoside method.

Seemingly a combination of *n*-pentenyl glycoside method and CB glycoside method, an efficient glycosylation protocol with glycosyl *ortho* alkynylbenzoates as the donors under $Ph_3PAuOTf$ catalysis was developed by Yu and coworkers in 2008 [48]. Activation of the benzylic triple bond in glycosyl hexynylbenzoate with Au(I) invites nucleophilic attack of the proximal carbonyl oxygen, resulting in the

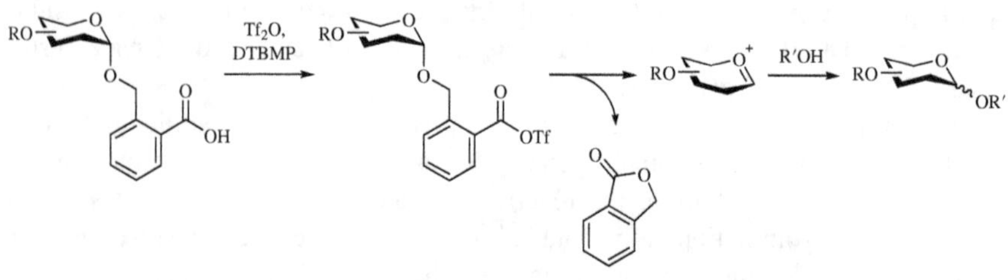

SCHEME 3.20 Mechanism of glycosylation using CB glycosides.

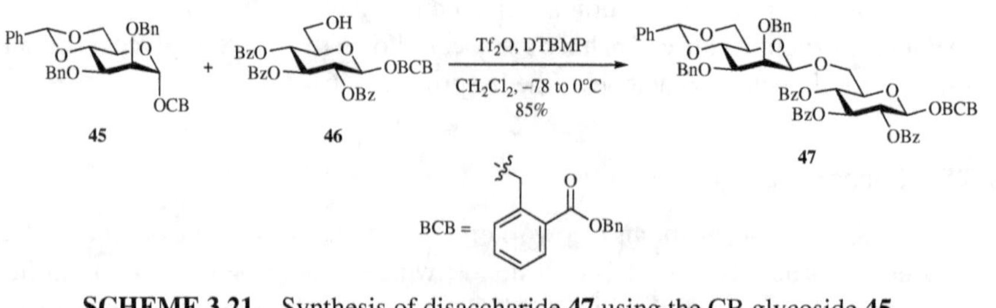

SCHEME 3.21 Synthesis of disaccharide **47** using the CB glycoside **45**.

subsequent cleavage of the glycosidic bond to give isocoumarin and the sugar oxo-carbenium ion (Scheme 3.22). This method is quite attractive for catalytic, neutral, and orthogonal promotion systems and has been used for the efficient synthesis of complex glycosides.

3.3.6 Glycosyl Phosphates/Phosphites

Due to their biological importance, the synthesis of glycosyl phosphates has attracted much attention since the early 1980s [49]. However, it was not until 2001 that an extensive investigation of their reactivity as glycosyl donors by Peter Seeberger and coworkers established the role of glycosyl phosphates in chemical glycosylation [50]. These glycosyl donors can be readily prepared from glycal precursors, glycosyl trichloroacetimidates, or *n*-pentenyl glycosides.

Various Lewis acids were screened as promoters, but only equimolar silyl triflate reagents such as TMSOTf and *tert*-butyldimethylsilyl triflate (TBSOTf) could ensure high-yielding glycosylation reactions. An interesting feature of glycosyl phosphate donors is the formation of 1,2-*trans*-glycosides at low temperature, even with non-participating groups at C2 [51, 52]. The anomeric isomers of glycosyl phosphates also have different reactivity, enabling orthogonal activation of one over the other; the β isomers could be activated at −78°C, whereas the α isomers usually require a higher temperature for activation (Scheme 3.23) [53].

A phosphite forms the phosphonium salt by interaction with a Lewis acid owing to their basic character. When a glycosyl phosphite is transformed to the phosphonium salt, the resulting phosphonooxy group acts as a strong leaving group in a nucleophilic substitution at the anomeric carbon. Since the early reports on their use

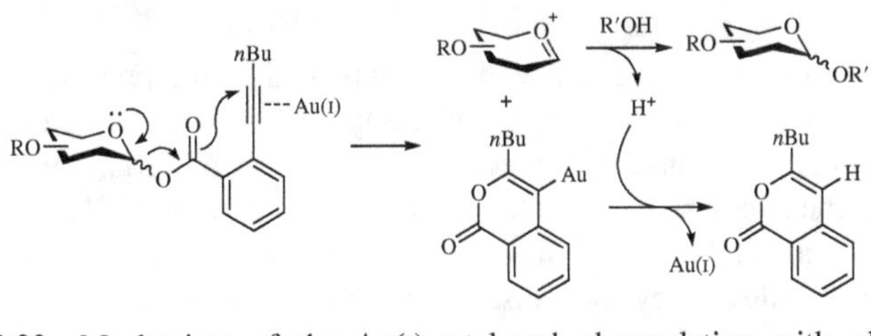

SCHEME 3.22 Mechanism of the Au(I)-catalyzed glycosylation with glycosyl *ortho* alkynylbenzoates.

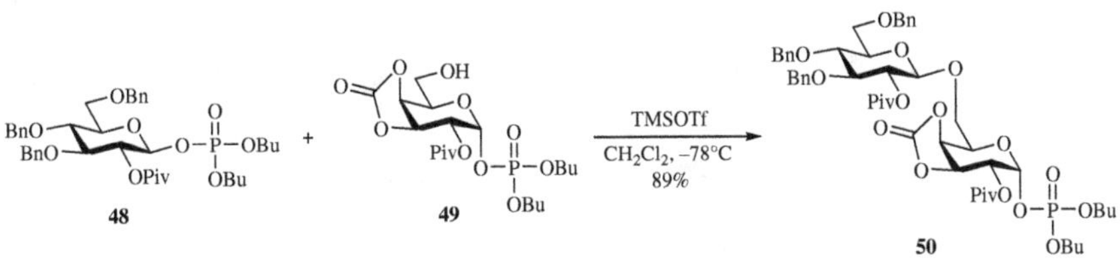

SCHEME 3.23 Orthogonal activation of isomeric glycosyl phosphates.

as glycosyl donors [54, 55], various promoters, typically strong Lewis acid such as TMSOTf [54, 55] as well as some mild ones including 2,6-*tert*-butylpyridinium iodide [56] and TfOH salts of tertiary amines [57], have been used for the activation of glycosyl phosphites. They are one of the most reactive glycosyl donors developed so far and have been applied to the synthesis of a wide variety of oligosaccharides.

3.3.7 Dehydrative Glycosylation

Apart from the common practice of using glycosyl donors equipped with leaving groups at the anomeric position, relatively less attention has been paid to glycosidic bond formation through dehydrative glycosylation, which uses 1-hydroxy sugar directly as the glycosyl donor. A number of conditions have been applied to dehydrative glycosylation from the classic Fischer glycosylation to the elegant works of David Gin and coworkers. The Fischer glycosylation—an acid-catalyzed reaction for the direct replacement of the anomeric oxygen atom—has been applied successfully to the synthesis of many glycosylation substrates. However, the reversibility of this reaction limits its utility in the synthesis of complex oligosaccharides and glycoconjugates. The pioneering works by Gin and coworkers with Tf_2O/Ph_2SO as the promoter system represent the modern version of dehydrative glycosylation. The promoter system permits an irreversible *in situ* activation in the presence of a base (Scheme 3.24) [58, 59].

Other conditions for dehydrative glycosylation such as the combination of phosphines and alkyl halides (PR_3/CX_4) [60, 61], silyl halides [62, 63], and bis(triphenyl) oxophosphonium triflate (Hendrickson reagent) [64] have also been investigated. However, the majority of these promoter systems require stoichiometric or even an excess amount, and only a few applications in glycoside synthesis have been reported.

3.3.8 Glycals

The treatment of a suitable glycal with a Lewis acid in the presence of a reactive acceptor can give rise to a glycoside via electrophilic addition, cycloaddition, nucleophilic addition, or rearrangement reactions. Glycals are important glycosyl donors, especially for the synthesis of 2-deoxyglycosides and 2-amino-2-deoxyglycosides.

A reliable approach for the synthesis of 2-deoxyglycosides is through haloalkoxylation of glycals followed by the subsequent reductive removal of the halide at C2.

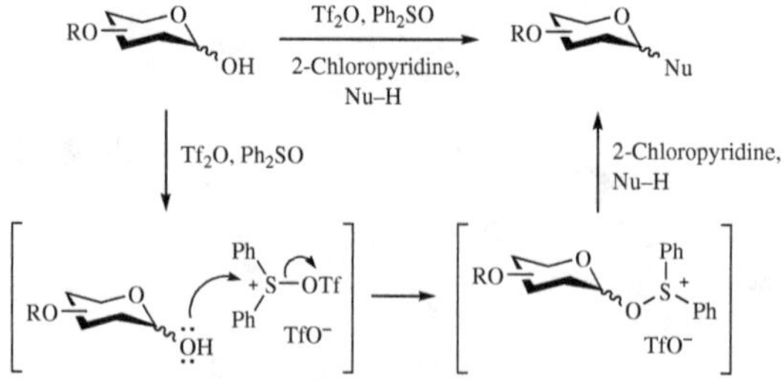

SCHEME 3.24 Dehydrative glycosylation promoted by Tf_2O/Ph_2SO.

IDCP [65] and NIS [66] are frequently used as promoters, leading to the formation of cyclic iodonium intermediate in which, due to stabilization by the reverse anomeric effect, the β face of the sugar ring is sterically shielded. The nucleophilic attack from the α face by a glycosyl acceptor then gives the 1,2-*trans* 2-deoxy-2-iodo-α-glycoside. The iodo functional group is reduced by either catalytic hydrogenation or radical reduction, as exemplified in the synthesis of the disaccharide **54** (Scheme 3.25) [66]:

2-Amino-2-deoxyglycosides can be obtained from glycals through a number of approaches, for example, through an azidonitration reaction by treatment of galactals or glucals with sodium azide and ceric ammonium nitrate (CAN). Hydrolysis of the generated 2-azido-2-deoxyglycosyl nitrates gives the corresponding 2-azido-2-deoxypyranoses, which can be further converted into other 2-deoxy glycosyl donors for glycosylation reactions and functional group manipulations (Scheme 3.26) [67, 68]. Another way is through the "azaglycosylation" approach pioneered and developed by Samuel Danishefsky and coworkers [69]. The reaction process involves the *trans* addition of iodonium and benzenesulfonamide, leading to 2-iodo-1-benzenesulfonamidoglycosides. The latter can be coupled with a glycosyl acceptor in the presence of lithium

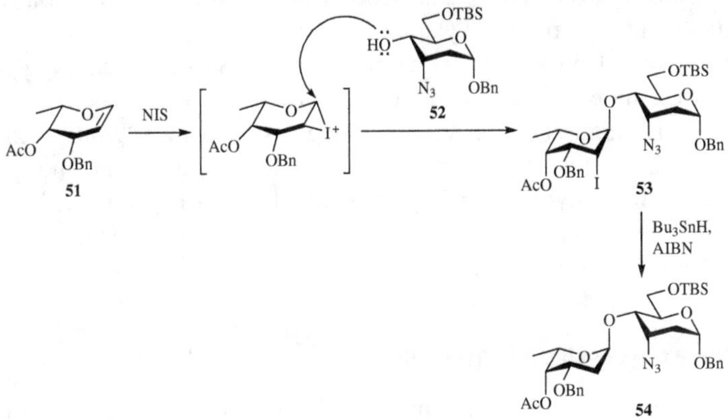

SCHEME 3.25 Synthesis of a 2-deoxyglycoside using the glycal approach. AIBN, azobisisobutyronitrile.

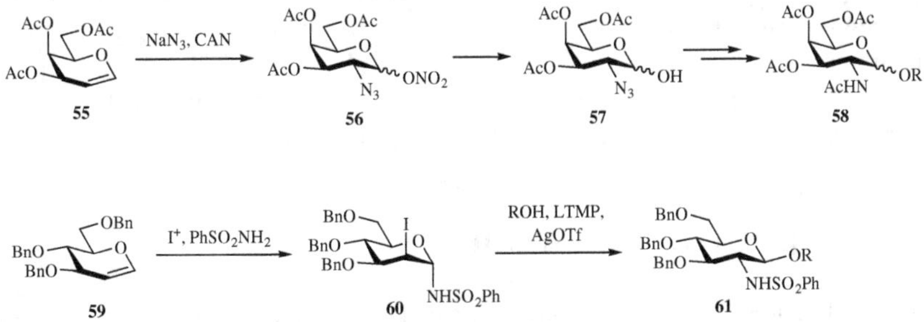

SCHEME 3.26 Approaches utilized for the synthesis of 2-amino-2-deoxyglycosides from glycals.

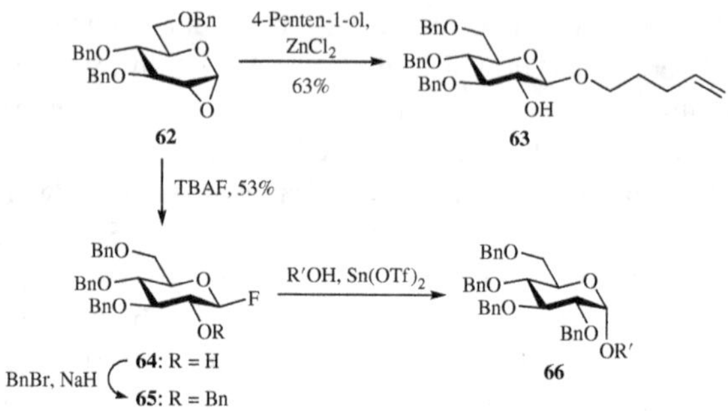

SCHEME 3.27 Stereoselective control of glycosylation by using glycosyl epoxide. TBAF, tetrabutylammonium fluoride.

tetramethylpiperidide (LTMP) and AgOTf. This approach has been employed for the preparation of complex oligosaccharides.

The conversion of glycals by treatment with dimethyldioxirane (DMDO) to form glycosyl epoxides developed by Danishefsky and coworkers substantially extends the potential of glycals in glycosylation reactions [70–72]. Firstly, it is possible to control the stereoselectivity of the anomeric center. Secondly, an oxygen atom is reinstated at C2 stereoselectively in the glycosylation process. The epoxide is usually formed with high stereoselectivity, being installed in a "*trans*" sense to the substituent at C3. Thus, in the presence of a Lewis acid, the alcohol attacks an α-epoxide to form a β-glycoside and a β-epoxide to form an α-glycoside. Alternatively, the anomeric epoxide can be transformed into other glycosyl donors to manipulate the stereochemistry, as demonstrated in Scheme 3.27 [73].

3.3.9 Other Glycosylation Protocols

In addition to the aforementioned methods for glycosidic bond formation, many other glycosyl donors such as seleno- [74, 75]/telluroglycosides [76], 1,2-orthoesters of aldoses [77], glycosyl carbonates [78], various heteroaryl glycosides [79–82], various N-substituted glycosyl carbamates [83–87], glycosyl disulfides [88, 89], N-glycosyl amides [90, 91], and glycosyl phthalates [92–94] to name a few have also been reported. Furthermore, new promoter systems for activating the currently used glycosyl donors are constantly emerging. However, not a single method is universally applicable, and the construction of complex oligosaccharides and glycoconjugates still poses a great challenge for chemists.

3.4 GLYCOSYLATION STRATEGIES

Substantial factors have to be taken into consideration in oligosaccharide synthesis. The glycosyl donor and promoter system need to be chosen. Extensive protecting group manipulations have to be performed in order to differentiate multiple hydroxy groups, especially for the glycosyl acceptor of which usually only one hydroxyl is exposed to be glycosylated. The choice of protecting groups also influences the reactivity of glycosyl building blocks as well as the stereochemical outcome of the reaction. Besides the problem of regioselectivity and stereoselectivity, economy has to be taken into account, particularly for the synthesis of a complex target. As a general principle, a convergent synthesis has advantages over a stepwise or linear synthesis in these cases. The traditional stepwise assembly remains as the most common method of generating oligosaccharides. Apart from this, specific methodologies have emerged, which enabled more rapid access to sugar backbones. Although many advances have been made in the area, new strategies for more efficient and convenient oligosaccharide synthesis are still in great demand.

3.4.1 One-Pot Glycosylation

One-pot glycosylation method refers to one in which several glycosyl donors are allowed to react sequentially in the same flask, resulting in a single main oligosaccharide product. Three major strategies have been employed most frequently in one-pot synthesis, namely, anomeric reactivity-based glycosylation, orthogonal glycosylation, and preactivation strategy. Brief descriptions of these strategies are presented herein. The readers are referred to Chapter 6 for more elaborate discussions.

The anomeric reactivity-based glycosylation is a chemoselective strategy that takes advantage of the different reactivities of glycosyl donors and acceptors on the basis of the armed–disarmed concept. In this type of one-pot glycosylation, the most reactive donor is condensed with the less reactive bifunctional donor–acceptor to provide a new glycoside, which can subsequently glycosylate the least reactive sugar building block.

The orthogonal glycosylation strategy is based on selective activation of one leaving group over another. In this approach, different anomeric groups (e.g., thioglycoside and glycosyl trichloroacetimidate) are exploited either as an anomeric protecting group or as a leaving group because they are mutually stable to the conditions used to activate the other anomeric functionality. A clear advantage of the orthogonal strategy is that it allows the condensation of glycosyl building blocks independent of their reactivities.

The preactivation strategy [95, 96] employs the activation of a glycosyl donor in the absence of a glycosyl acceptor to provide a reactive intermediate, which is immediately treated with a second building block (glycosyl acceptor) to yield a coupling saccharide with an identical leaving group at the reducing end. This process can be repeated in the same flask to extend the sugar sequence. This strategy combines the advantages of both the anomeric reactivity-based glycosylation (activation under a single set of glycosylation conditions) and the orthogonal strategy (independent of relative reactivity). Because it does not require the tuning of the building blocks in an armed–disarmed fashion, excessive synthetic operation on building blocks is avoided.

One-pot procedure integrates several glycosylation steps into one synthetic operation to furnish the target oligosaccharide in a short period of time without

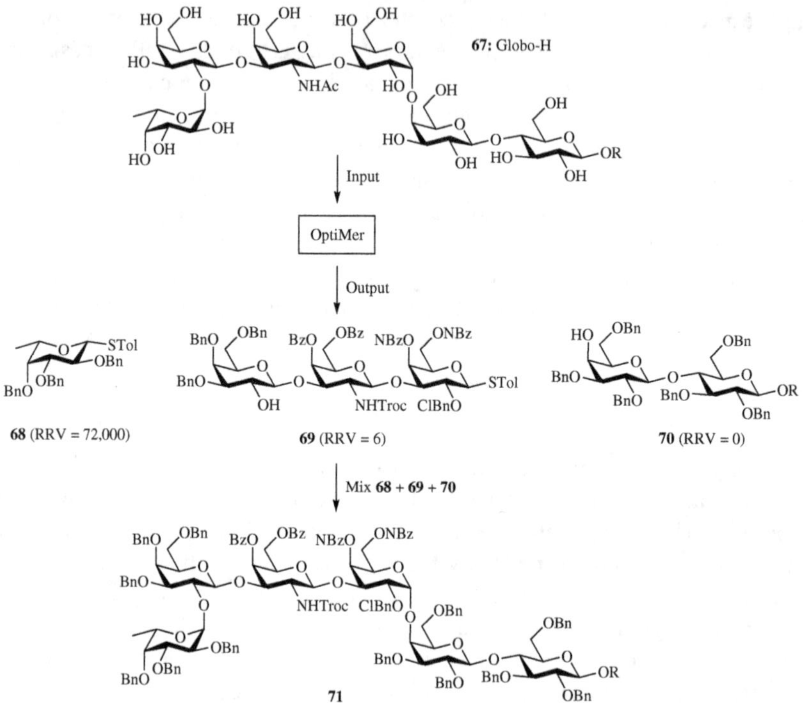

SCHEME 3.28 Programmable one-pot synthesis of Globo-H. ClBn, *o*-chlorobenzyl; NBz, *p*-nitrobenzoyl; RRV, relative reactivity value.

the need for protecting group manipulation and intermediate isolation. It is, in principle, much more efficient than traditional synthesis of oligosaccharides, which requires tedious multiple protection and deprotection steps and time-consuming purification processes. For example, programmable one-pot glycosylation, a highly promising protocol based on the chemoselective strategy, has been devised by Chi-Huey Wong and coworkers [36]. With a thioglycoside reactivity database determined by HPLC analysis at hand, they developed the "OptiMer" computer program for use as a database search tool and guide for the selection of suitable building blocks for the one-pot assembly of a desired oligosaccharide. The synthesis of the Globo-H (**67**) was carried out by this manner (Scheme 3.28, also see Section 12.4.1) [97].

3.4.2 Solid-Phase Oligosaccharide Synthesis

Initial attempts at solid-phase oligosaccharide synthesis (SPOS) were met with little success due to the limited range of glycosylation methods [98]. It was not until much later when new methods emerged that SPOS was reexplored and has achieved remarkable success in the construction of complex oligosaccharides.

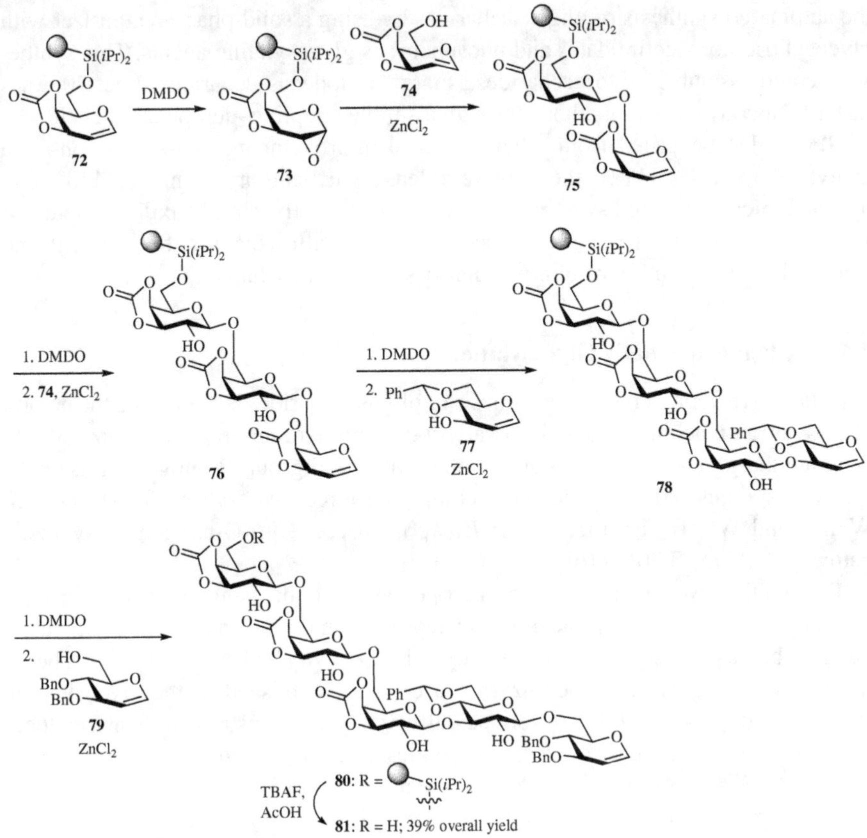

SCHEME 3.29 Solid-phase synthesis of a pentasaccharide with glycals.

In the early 1990s, Danishefsky and coworkers successfully applied the glycal method to SPOS, whereby the sugar chain was elongated from the nonreducing end to the reducing end [99–101]. They linked the first glycal unit to Merrifield resin through a disilane linkage that could be readily cleaved after the completion of the synthesis by treatment with fluoride. The glycal was, then, oxidized to glycosyl epoxide to couple with a glycal acceptor. Repeating the oxidation/coupling steps led to the desired oligosaccharide (Scheme 3.29). However, this sort of donor-bound strategy could not simply be extended to other donors as most side reactions during glycosylation reactions involve the donor, thus, preventing further the elongation of the sugar chain in such cases.

Since then, the most common glycosyl donors such as glycosyl sulfoxides [102, 103], glycosyl trichloroacetimidates [104, 105], thioglycosides [106, 107], *n*-pentenyl glycosides [108], and glycosyl phosphates [109, 110] have been investigated as glycosylating agents in acceptor-bound approach for SPOS.

A breakthrough in carbohydrate chemistry was the appearance of the first automated oligosaccharide synthesizer [111]. Seeberger and coworkers carried out the automated synthesis of oligosaccharides by using a solid-phase synthesizer with glycosyl trichloroacetimidates and phosphates as glycosylating agents. This synthesizer could assemble oligosaccharides as large as dodecasaccharides about 20 times faster than conventional methods through a simple coupling–deprotection cycle.

SPOS also benefited from other technical improvements such as the on-resin analytical methods [112], the capture–release purification technique [113], and hydrophobically assisted switching phase [114]. Recently, chiral auxiliary-mediated 1,2-*cis*-glycosylations were introduced to SPOS, allowing for the efficient and stereoselective assembly of oligosaccharides [115] (see Chapter 7).

3.4.3 Chemoenzymatic Glycosylation

An alternative approach for the synthesis of oligosaccharides is exploiting the natural glycosylation machinery, namely, a variety of carbohydrate-processing enzymes, to generate the glycosidic bond. For example, the blood group B antigen tetrasaccharide was synthesized by sequentially employing three glycosyltransferases (WbnJ, WbnK, and WbnJ) identified in the *Escherichia coli* O86 O-antigen biosynthetic pathway (Scheme 3.30) [116].

Enzymatic glycosylation can be carried out without using protecting groups. Thus, no multiple steps of protection and deprotection for the construction of suitable building blocks are required. Meanwhile, regioselectivity and stereoselectivity, being inherent characteristics of the enzymes used, are established in the glycosylation step. Consequently, such characteristics also imply the constraints of this method, such as restricted access to specific enzymes and the required cofactors and the tedious aqueous work-up procedures for the removal of buffer salts.

The combined use of enzymatic and chemical methods under the so-called chemoenzymatic synthesis offers the flexibility of chemical synthesis and the high regioselectivity and stereoselectivity of enzyme-catalyzed reactions to achieve highly efficient synthesis of complex carbohydrates. Chemoenzymatic glycosylation is

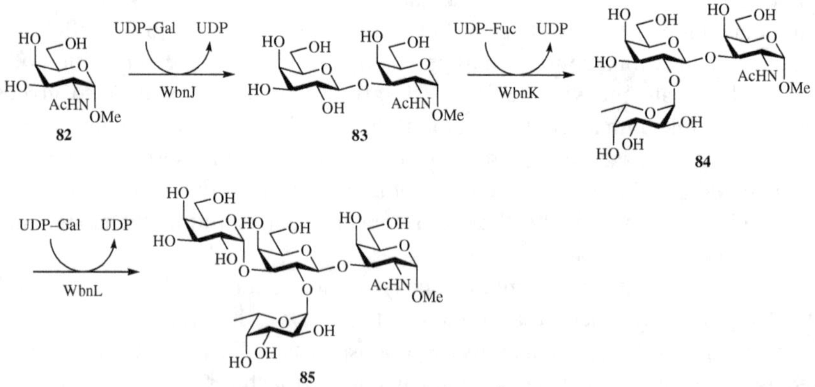

SCHEME 3.30 Enzymatic synthesis of the blood group B antigen tetrasaccharide. UDP, uridine diphosphate.

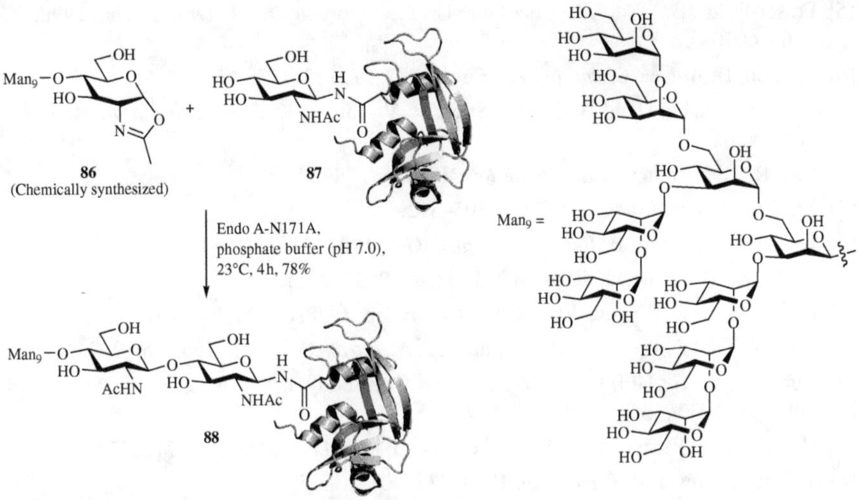

SCHEME 3.31 Chemoenzymatic synthesis of high-mannose type RNase B.

particularly valuable in obtaining challenging targets including structurally diverse homogeneous glycopeptides/glycoproteins, glycolipids, glycosylated natural products, sialosides, bacterial polysaccharides, heparan sulfate or heparin, and other complex oligosaccharides and glycoconjugates. For example, a highly efficient chemoenzymatic synthesis of N-glycoproteins carrying intact natural N-glycans was accomplished by Lai-Xi Wang and coworkers (Scheme 3.31) [117].

3.5 CONCLUSION

The assembly of complex oligosaccharides is still a great challenge for chemists. Significant efforts are constantly being devoted to the development of efficient protocols for oligosaccharide synthesis. For decades, carbohydrate chemists dreamed of assembling complex oligosaccharides rapidly, economically, and automatically, just as what have been done with the synthesis of peptides and oligonucleotides. Obviously, breakthroughs in methodology development for the glycosidic bond formation are needed to make this dream a reality.

REFERENCES

[1] Perrin, C. L. *Tetrahedron* **1995**, *51*, 11901–11935.

[2] Randell, K. D.; Johnston, B. D.; Green, D. F.; Pinto, B. M. *J. Org. Chem.* **2000**, *65*, 220–226.

[3] Kim, J.-H.; Yang, H.; Park, J.; Boons, G.-J. *J. Am. Chem. Soc.* **2005**, *127*, 12090–12097.

[4] Yasomanee, J. P.; Demchenko, A. V. *J. Am. Chem. Soc.* **2012**, *134*, 20097–20102.

[5] Fraser-Reid, B.; Wu, Z.; Udodong, U. E.; Ottosson, H. *J. Org. Chem.* **1990**, *55*, 6068–6070.

[6] Paulsen, H. *Angew. Chem. Int. Ed. Engl.* **1982**, *21*, 155–173.

[7] Lemieux, R. U.; Hendriks, K. B.; Stick, R. V.; James, K. *J. Am. Chem. Soc.* **1975**, *97*, 4056–4062.

[8] Eby, R.; Schuerch, C. *Carbohydr. Res.* **1974**, *34*, 78–90.

[9] Michael, A. *Am. Chem. J.* **1879**, *1*, 305–312.

[10] Koenigs, W.; Knorr, E. *Ber. Dtsch. Chem. Ges.* **1901**, *34*, 957–981.

[11] Helferich, B.; Zirner, J. *Chem. Ber.* **1962**, *95*, 2604–2611.

[12] Reynolds, D. D.; Evans, W. L. *J. Am. Chem. Soc.* **1938**, *60*, 2559–2561.

[13] Rosenbrook W., Jr.; Riley, D. A.; Lartey, P. A. *Tetrahedron Lett.* **1985**, *26*, 3–4.

[14] Buchanan, J. G.; Hill, D. G.; Wightman, R. H.; Boddy, I. K.; Hewitt, B. D. *Tetrahedron* **1995**, *51*, 6033–6050.

[15] Cumpstey, I.; Fairbanks, A. J.; Redgrave, A. J. *Org. Lett.* **2001**, *3*, 2371–2374.

[16] Thiem, J.; Meyer, B. *Chem. Ber.* **1980**, *113*, 3075–3085.

[17] Du, W.; Gervay-Hague, J. *Org. Lett.* **2005**, *7*, 2063–2065.

[18] Schmidt, R. R. *Angew. Chem. Int. Ed. Engl.* **1986**, *25*, 212–235.

[19] Adinolfi, M.; Barone, G.; Iadonisi, A.; Mangoni, L.; Schiattarella, M. *Tetrahedron Lett.* **2001**, *42*, 5967–5969.

[20] Adinolfi, M.; Barone, G.; Guariniello, L.; Iadonisi, A. *Tetrahedron Lett.* **2000**, *41*, 9005–9008.

[21] Fursstner, A.; Jeanjean, F.; Razon, P. *Angew. Chem. Int. Ed. Engl.* **2002**, *41*, 2097–2101.

[22] Yeung, B.; Hill, D. C.; Janicka, M.; Petillo, P. A. *Org. Lett.* **2000**, *2*, 1279–1282.

[23] Schmidt, R. R.; Toepfer, A. *Tetrahedron Lett.* **1991**, *32*, 3353–3356.

[24] Yu, B.; Tao, H. *Tetrahedron Lett.* **2001**, *42*, 2405–2407.

[25] Ferrier, R. J.; Hay, R. W.; Vethaviyasar, N. *Carbohydr. Res.* **1973**, *27*, 55–61.

[26] Veeneman, G. H.; van Leeuwen, S. H.; van Boom, J. H. *Tetrahedron Lett.* **1990**, *31*, 1331–1334.

[27] Konradsson, P.; Udodong, U. E.; Fraser-Reid, B. *Tetrahedron Lett.* **1990**, *31*, 4313–4316.

[28] Codee, J. D. C.; Litjens, R.; den Heeten, R.; Overkleeft, H. S.; van Boom, J. H.; van der Marel, G. A. *Org. Lett.* **2003**, *5*, 1519–1522.

[29] Crich, D.; Smith, M. *J. Am. Chem. Soc.* **2001**, *123*, 9015–9020.

[30] Wang, G. N.; Wang, H. S.; Huang, X. F.; Zhang, L. H.; Ye, X.-S. *Synlett.* **2006**, 2846–2850.

[31] Fugedi, P.; Garegg, P. J. *Carbohydr. Res.* **1986**, *149*, 9–12.

[32] Tsai, T. Y. R.; Jin, H.; Wiesner, K. *Can. J. Chem.* **1984**, *62*, 1403–1405.

[33] Van Gleve, J. W. *Carbohydr. Res.* **1979**, *70*, 161–164.

[34] Hanessian, S.; Bacquet, G.; Lehong, N. *Carbohydr. Res.* **1980**, *80*, G17–G22.

[35] Lear, M. J.; Yoshimura, F.; Hirama, M. *Angew. Chem. Int. Ed.* **2001**, *40*, 946–949.

[36] Zhang, Z.; Ollumann, I. R.; Ye, X.-S.; Wischnat, R.; Baasov, T.; Wong, C.-H. *J. Am. Chem. Soc.* **1999**, *121*, 734–753.

[37] Crich, D.; Sun, S. *J. Am. Chem. Soc.* **1997**, *119*, 11217–11223.

[38] Yan, L.; Kahne, D. *J. Am. Chem. Soc.* **1996**, *118*, 9239–9248.

[39] Woodward, R. B. et al. *J. Am. Chem. Soc.* **1981**, *103*, 3215–3217.

[40] El Ashry, E. S. H.; Award, L. F.; Atta, A. I. *Tetrahedron* **2006**, *62*, 2943–2998.

[41] Demchenko, A. V.; Malysheva, N. N.; De Meo, C. *Org. Lett.* **2003**, *5*, 455–458.

[42] Mootoo, D. R.; Date, V.; Fraser-Reid, B. *J. Am. Chem. Soc.* **1988**, *110*, 2662–2663.

[43] Konradsson, P.; Mootoo, D. R.; McDevitt, R. E.; Fraser-Reid, B. *J. Chem. Soc. Chem. Commun.* **1990**, 270–272.

[44] Mootoo, D. R.; Konradsson, P.; Udodong, U.; Fraser-Reid, B. *J. Am. Chem. Soc.* **1988**, *110*, 5583–5584.

[45] Merritt, J. R.; Fraser-Reid, B. *J. Am. Chem. Soc.* **1992**, *114*, 8334–8336.

[46] Kunz, H.; Wernig, P.; Schultz, M. *Synlett* **1990**, 631–632.

[47] Kim, K. S.; Kim, J. H.; Lee, Y. J.; Lee, Y. J.; Park, J. *J. Am. Chem. Soc.* **2001**, *123*, 8477–8481.

[48] Li, Y.; Yang, Y.; Yu, B. *Tetrahedron Lett.* **2008**, *49*, 3604–3608.

[49] Schmidt, R. R.; Stumpp, M.; Michel, J. *Tetrahedron Lett.* **1982**, *23*, 405–408.

[50] Plante, O. J.; Palmacci, E. R.; Andrade, R. B.; Seeberger, P. H. *J. Am. Chem. Soc.* **2001**, *123*, 9545–9554.

[51] Hariprasad, V.; Singh, G.; Tranoy, I. *Chem. Commun.* **1998**, 2129–2130.

[52] Hariprasad, V.; Singh, G.; Tranoy, I. *Tetrahedron: Asymmetry* **2001**, *12*, 1373–1381.

[53] Bernardes, G. J. L.; Gamblin, D. P.; Davis, B. G. *Angew. Chem. Int. Ed.* **2006**, *45*, 4007–4011.

[54] Martin, T. J.; Schmidt, R. R. *Tetrahedron Lett.* **1992**, *33*, 6123–6126.

[55] Kondo, H.; Ichikawa, Y.; Wong, C.-H. *J. Am. Chem. Soc.* **1992**, *114*, 8748–8750.

[56] Tanaka, H.; Sakamoto, H.; Sano, A.; Nakamura, S.; Nakamura, M.; Nakajima, M.; Hashimoto, S. *Chem. Commun.* **1999**, 1259–1260.

[57] Matsumura, F.; Tatsumi, S.; Oka, N.; Wada, T. *Carbohydr. Res.* **2010**, *345*, 1211–1215.

[58] Garcia, B. A.; Poole, J. L.; Gin, D. Y. *J. Am. Chem. Soc.* **1997**, *119*, 7597–7598.

[59] Garcia, B. A.; Gin, D. Y. *J. Am. Chem. Soc.* **2000**, *122*, 4269–4279.

[60] Chretien, F.; Chapleur, Y.; Castro, B.; Gross, B. *J. Chem. Soc. Perkin Trans. 1*, **1980**, 381–384.

[61] Shingu, Y.; Nishida, Y.; Dohi, Y.; Matsuda, K.; Kobayashi, K. *J. Carbohydr. Chem.* **2002**, *21*, 605–611.

[62] Koto, S.; Morishima, N.; Zen, S. *Chem. Lett.* **1976**, *5*, 61–64.

[63] Koto, S.; Morishima, N.; Zen, S. *Bull. Chem. Soc. Jpn.* **1979**, *52*, 784–788.

[64] Mossotti, M.; Panza, L. *J. Org. Chem.* **2011**, *76*, 9122–9126.

[65] Tatsuta, K.; Fujimoto, K.; Kinoshita, M. *Carbohydr. Res.* **1977**, *54*, 85–104.

[66] Thiem, J.; Karl, H.; Schwentner, J. *Synthesis*, **1978**, 696–698.

[67] Lemieux, R. U.; Ratcliffe, R. M. *Can. J. Chem.* **1979**, *57*, 1244–1251.

[68] Paulsen, H.; Lorentzen, J. P. *Carbohydr. Res.* **1990**, *133*, G1–G4.

[69] Griffith, D. A.; Danishefsky, S. J. *J. Am. Chem. Soc.* **1990**, *112*, 5811–5819.

[70] Halcomb, R. L.; Danishefsky, S. J. *J. Am. Chem. Soc.* **1989**, *111*, 6661–6666.

[71] Gervay, J.; Danishefsky, S. J. *J. Org. Chem.* **1991**, *56*, 5448–5451.

[72] Liu, K. K. C.; Danishefsky, S. J. *J. Org. Chem.* **1994**, *59*, 1895–1897.

[73] Gordon, D.; Danishefsky, S. J. *Carbohydr. Res.* **1990**, *206*, 361–366.

[74] Mehta, S.; Pinto, B. M. *Tetrahedron Lett.*, **1991**, *32*, 4435–4438.

[75] Mehta, S.; Pinto, B. M. *J. Org. Chem.* **1993**, *58*, 3269–3276.

[76] Yamago, S.; Kokubo, K.; Murakami, H.; Mino, Y.; Hara, O.; Yoshida, J.-I. *Tetrahedron Lett.* **1998**, *39*, 7905–7908.

[77] Kong, F. *Carbohydr. Res.* **2007**, *342*, 345–373.

[78] Mukaiyama, T.; Miyazaki, K.; Uchiro, H. *Chem. Lett.* **1998**, 635–636.

[79] Huchel, U.; Schmidt, C.; Schmidt, R. R. *Tetrahedron Lett.* **1995**, *36*, 9457–9460.

[80] Hannessian, S.; Conde, J. J.; Lion, B. *Tetrahedron Lett.* **1995**, *36*, 5865–5868.

[81] *Preparative Carbohydrate Chemistry*; Hanessian, S., Ed.; Marcel Dekker: New York, **1997**.

[82] Yasukochi, T.; Fukase, K.; Kusumoto, S. *Tetrahedron Lett.* **1999**, *40*, 6591–6593.

[83] Herzner, H.; Berling, J.; Schultz, M.; Zimmer, J.; Kunz, H. *J. Carbohydr. Chem.* **1998**, *17*, 759–776.

[84] Hinklin, R. J.; Kiessling, L. L. *J. Am. Chem. Soc.* **2001**, *123*, 3379–3380.

[85] Knoben, H. P.; Schlüter, U.; Redlich, H. *Carbohydr. Res.* **2004**, *339*, 2821–2833.

[86] Jayakanthan, K.; Vankar, Y. D. *Carbohydr. Res.* **2005**, *340*, 2688–2692.

[87] Matsuo, J. I.; Shirahata, T.; Omura, S. *Tetrahedron Lett.* **2006**, *47*, 267–271.

[88] Davis, B. G.; Ward, S. J.; Rendle, P. M. *Chem. Commun.* **2001**, 189–190.

[89] Grayson, E. J.; Ward, S. J.; Hall, A. L.; Rendle, P. M.; Gamblin, D. P.; Batsanov, A. S.; Davis, B. G. *J. Org. Chem.* **2005**, *70*, 9740–9754.

[90] Pleuss, N.; Kunz, H. *Angew. Chem. Int. Ed.* **2004**, *42*, 3174–3176.

[91] Pleuss, N.; Kunz, H. *Synthesis* **2005**, 122–130.

[92] Kim, K. S.; Lee, Y. J.; Kim, H. Y.; Kang, S. S.; Kwon, S. Y. *Org. Biomol. Chem.* **2004**, *2*, 2408–2410.

[93] Meppen, M.; Wang, Y.; Cheon, H. S.; Kishi, Y. *J. Org. Chem.* **2007**, *72*, 1941–1950.

[94] Hsu, M. C.; Lee, J.; Kishi, Y. *J. Org. Chem.* **2007**, *72*, 1931–1940.

[95] Huang, X.; Huang, L.; Wang, H.; Ye, X.-S. *Angew. Chem. Int. Ed.* **2004**, *43*, 5221–5224.

[96] Crich, D.; Sun, S. *J. Org. Chem.* **1996**, *61*, 4506–4507.

[97] Burkhart, F.; Zhang, Z.; Wacowich-Sgarbi, S.; Wong, C.-H. *Angew. Chem. Int. Ed.* **2001**, *40*, 1274–1277.

[98] Fréchet, J. M. J.; Schuerch, C. *J. Am. Chem. Soc.* **1971**, *93*, 492–496.

[99] Danishefsky, S. J.; McClure, K. F.; Randolph, J. T.; Ruggeri, R. B. *Science* **1993**, *260*, 1307–1309.

[100] Randolph, J. T.; Danishefsky, S. J. *Angew. Chem. Int. Ed.* **1994**, *33*, 1470–1473.

[101] Randolph, J. T.; McClure, K. F.; Dannishefsky, S. J. *J. Am. Chem. Soc.* **1995**, *117*, 5712–5719.

[102] Yan, L.; Taylor, C. M.; Goodnow, R.; Kahne, D. *J. Am. Chem. Soc.* **1994**, *116*, 6953–6954.

[103] Liang, R.; Yan, L.; Loebach, J.; Ge, M.; Uozumi, Y.; Sekanina, K.; Horan, N.; Gildersleeve, J.; Thompson, C.; Smith, A.; Biswas, K.; Still, W. C.; Kahne, D. *Science* **1996**, *274*, 1520–1522.

[104] Rademann, J.; Schmidt, R. R. *Tetrahedron Lett.* **1996**, *37*, 3989–3990.

[105] Hunt, J. A.; Roush, W. R. *J. Am. Chem. Soc.* **1996**, *118*, 9998–9999.

[106] Nicolaou, K. C.; Winssinger, N.; Pastor, J.; DeRoose, F. *J. Am. Chem. Soc.* **1997**, *119*, 449–450.

[107] Nicolaou, K. C.; Watanabe, N.; Li, J.; Pastor, J.; Winssinger, N. *Angew. Chem. Int. Ed.* **1998**, *37*, 1559–1561.

[108] Rodebaugh, R.; Joshi, S.; Fraser-Reid, B.; Geysen, H. M. *J. Org. Chem.* **1997**, *62*, 5660–5661.

[109] Andrade, R. B.; Plante, O. J.; Melean, L. G.; Seeberger, P. H. *Org. Lett.* **1999**, *1*, 1811–1814.

[110] Melean, L. G.; Love, K. R.; Seeberger, P. H. *Carbohydr. Res.*, **2002**, *337*, 1893–1916.

[111] Plante, O. J.; Palmacci, E. R.; Seeberger, P. H. *Science* **2001**, *291*, 1523–1527.

[112] Kanemitsu, T.; Wong, C.-H; Kanie, O. *J. Am. Chem. Soc.* **2002**, *124*, 3591–3599.

[113] Egusa, K.; Kusumoto, S.; Fukase, K. *Eur. J. Org. Chem.* **2003**, 3435–3445.

[114] Bauer, J.; Rademann, J. *J. Am. Chem. Soc.* **2005**, *127*, 7296–7297.

[115] Boltje, T. J.; Kim, J.; Park, J.; Boons, G.-J. *Nat. Chem.* **2010**, *2*, 552–557.

[116] Yi, W.; Shao, J.; Zhu, L.; Li, M.; Singh, M.; Lu, Y.; Lin, S.; Li, H.; Ryu, K.; Shen, J.; Guo, H.; Yao, Q.; Bush, C. A.; Wang, P. G. *J. Am. Chem. Soc.* **2005**, *127*, 2040–2041.

[117] Huang, W.; Li, C.; Li, B.; Umekawa, M.; Yamamoto, K.; Zhang, X.; Wang, L.-X. *J. Am. Chem. Soc.* **2009**, *131*, 2214–2223.

4

CONTROLLING ANOMERIC SELECTIVITY, REACTIVITY, AND REGIOSELECTIVITY IN GLYCOSYLATIONS USING PROTECTING GROUPS

THOMAS JAN BOLTJE[1], LIN LIU[2], AND GEERT-JAN BOONS[2]

[1] *Cluster for Molecular Chemistry, Institute for Molecules and Materials, Radboud University Nijmegen, Nijmegen, The Netherlands*
[2] *Complex Carbohydrate Research Center, University of Georgia, Athens, GA, USA*

4.1 INTRODUCTION

Synthetic oligosaccharides and glycoconjugates are increasingly used as probes for biological research and as lead compounds for drug and vaccine discovery. Unlike oligonucleotide and peptide synthesis, there are no general protocols for the preparation of complex carbohydrates of biological importance. As a result, the chemical synthesis of each target glycan or glycoconjugate is a research project, which may take many months and sometimes years to complete. In recent years, considerable progress has been made in method development for complex oligosaccharide synthesis. In this chapter, recently developed methods for controlling the anomeric selectivity of glycosylations are discussed, and particular attention is focused on protecting group effects. Furthermore, convergent strategies exploiting differences in glycosyl donor and acceptor reactivities are described. Finally, the influences of protecting groups on the regioselective outcome of glycosylations are outlined.

Glycochemical Synthesis: Strategies and Applications, First Edition.
Edited by Shang-Cheng Hung and Medel Manuel L. Zulueta.

4.2 PROTECTING GROUP AND CONTROL OF ANOMERIC SELECTIVITY OF GLYCOSYLATIONS

Chemical glycosylation—the condensation of a glycosyl donor and a glycosyl acceptor—is the most important reaction in carbohydrate synthesis. According to the relative and absolute stereochemistry at C1 (the typical anomeric center) and C2, the resulting glycosidic linkage can be classified as 1,2-*cis*- and 1,2-*trans*-2-D-*glycero* series (allo-, gluco-, gulo-, and galactopyranosides) and 1,2-*cis*- and 1,2-*trans*-2-L-*glycero* series (altro-, manno-, ido-, and talopyranosides) (Fig. 4.1). Apart from these types of glycosides, a number of miscellaneous glycosidic linkages can be identified including 2-deoxyglycosides and 3-deoxy-2-keto-ulo(pyranosylic) acids.

Anomeric selectivity in a glycosylation is substantially influenced by protecting groups installed in the glycosyl donor. In the case when the protecting groups are nonstereodirecting, the reaction conditions (e.g., solvent, temperature, and promoter) may help promote anomeric selectivity. This section highlights the state-of-the-art methods for stereoselective glycosylation that do not employ reaction parameters such as solvent and temperature. Instead, the use of intrinsic properties of certain carbohydrates and protecting groups as well as the purposely designed auxiliaries for stereoselective glycosylation are discussed.

4.2.1 Neighboring Group Participation of C2 Esters to Afford 1,2-*trans*-Glycosides

The nature of the protecting group at C2 of a glycosyl donor is a major determinant of the anomeric selectivity of a glycosylation [1]. After the departure of the anomeric leaving group, neighboring group participation of a 2-*O*-acyl functionality can stabilize the intermediate oxocarbenium ion **2** to give the intermediate five-membered dioxolenium ion **3** (Scheme 4.1). Due to ring strain, the latter intermediate can only be formed as a 1,2-*cis*-fused ring system. The dioxolenium ion has been detected by NMR spectroscopy [2], and its existence has also been demonstrated by X-ray crystallography [3]. Nucleophilic attack at the anomeric center can only take place from the β face leading to the stereoselective formation of a 1,2-*trans*-glycoside (**5**). Alternatively, nucleophilic attack at the dioxolenium ion leads to the formation of orthoester **4**, which is not stable under acidic conditions and can rearrange to the desired 1,2-*trans*-glycoside. Numerous 2-*O*-acyl protecting groups were developed to reduce orthoester formation, including pivaloyl and 4-acetoxy-2,2-dimethylbutanoyl groups. The geminal dimethyl substitution in the latter [4] suppresses orthoester formation while promoting cyclization by the *gem*-dialkyl or

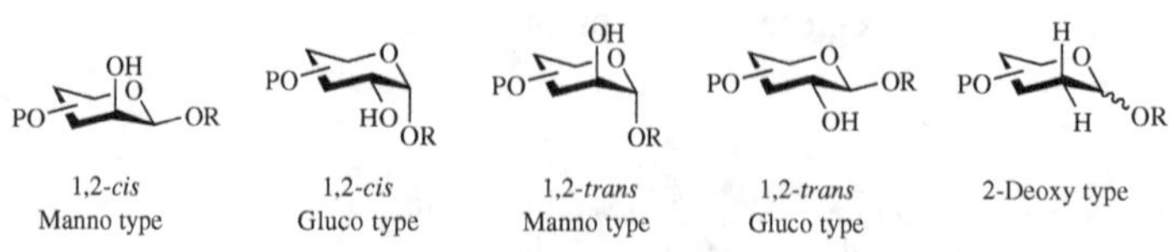

FIGURE 4.1 Type of glycosides based on the orientations of substituents at C1 and C2.

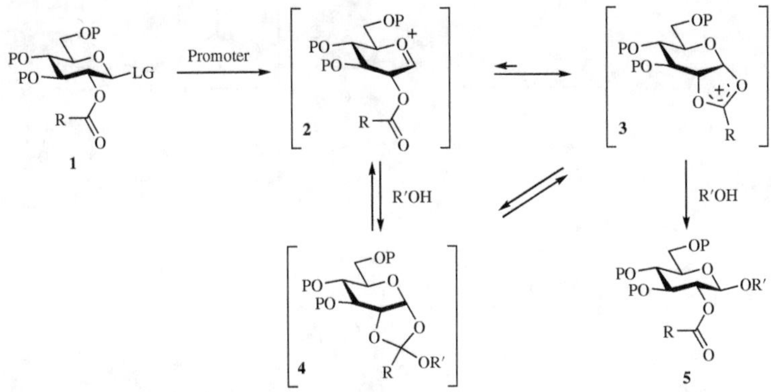

SCHEME 4.1 Neighboring group participation of a 2-*O*-acyl functionality to provide 1,2-*trans*-glycosides.

Thorpe–Ingold effect [5–7]. Aside from C2-ester groups, C2-amino protecting groups, including phthaloyl (Phth), tetrachlorophthaloyl (TCP), *N,N*-dimethylmaleoyl, trichloroacetyl, allyloxycarbonyl, and trichloroethoxycarbonyl (Troc) groups, can also perform neighboring group participation. Such functionalities allow the formation of 2-deoxy-2-aminoglycosides that are common constituents of glycoproteins, glycolipids, and proteoglycans.

4.2.2 Remote Neighboring Group Participation

Neighboring group participations by esters in positions other than C2 have been implicated in the stereochemical control of glycosylations [8–11]. However, this mode of anomeric control is controversial and evidence for its existence rarely extends beyond stereochemical arguments [12–14]. Attempts were made to trap the dioxolenium ion arising from remote participation of the *tert*-butoxycarbonyl (Boc) group (Scheme 4.2) [15]. It was envisioned that upon participation of the Boc group, an intermediate dioxolenium ion would subsequently lose isobutylene, leading to the formation of a stable cyclic carbonate. Indeed, treatment of thioglycoside **6** with *N*-iodosuccinimide (NIS) and AgOTf resulted in the formation of the 1,2-carbonate **8** in high yield. Consequently, a series of thioglycoside donors carrying a Boc group at either equatorial or axial C3, C4, and C6 positions were prepared. However, only allose donor **9**, which has an axial 3-*O*-Boc group, led to the formation of the corresponding 1,3-carbonate **10**, thereby supporting the possibility of anomeric control by remote participation. The other thioglycosides did not form bicyclic carbonates but instead hydrolyzed or formed glycosides when an alcohol was added, leaving the Boc group intact. Hence, it was concluded that no other positions could be involved in remote participation.

A subsequent report by Kwan Soo Kim and coworkers indicated that remote participation from an equatorial C3 position is indeed possible [16]. In this case, a trichloroacetimidate was used as a nucleophilic trap. Upon low-temperature activation of thioglycoside **11** using 1-benzenesulfinylpiperidine (BSP) and Tf_2O, bicyclic

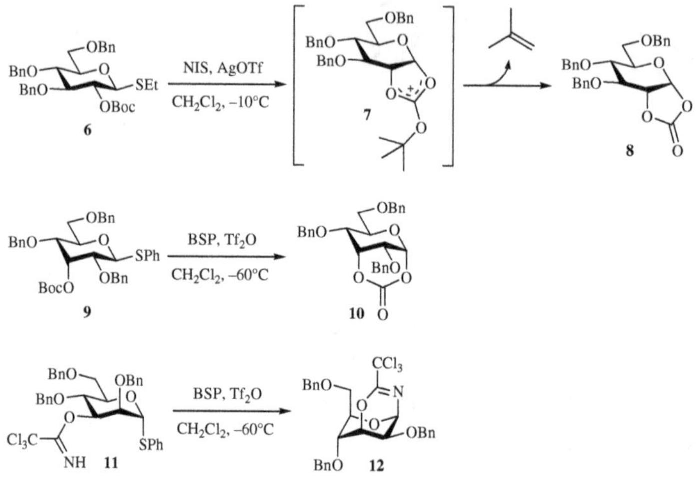

SCHEME 4.2 Nucleophilic traps to explore neighboring group participation from nonvicinal positions. Bn, benzyl; Tf, triflyl.

trichlorooxazine product **12** was formed in 85% yield. However, the 4-*O*- and 6-*O*-trichloroacetimidoyl groups did not form bicyclic trichlorooxazine products. Thus, remote participation can play an important role, although it seems to depend on the nature of the sugar configuration and conformation as well as the structure of the participating group.

4.2.3 Neighboring Group Participation by Other Functional Groups

Electron-withdrawing ester groups reduce the reactivity of a glycosyl donor by destabilization of the intermediate oxocarbenium ion [17]. This reduced reactivity usually leads to incomplete activation and lower glycosylation yield. Considerable efforts have been directed at resolving this issue by either modification of the participating group [18] or by the use of reactivity-enhancing groups [19]. Alexei Demchenko and coworkers reported the use of modified ether groups that can perform neighboring group participation to ensure stereoselectivity while not adversely affecting the reactivity of the glycosyl donor. The most promising group for this goal is the 2-pyridylmethyl (or picolyl) ether, which was found suitable for selective 1,2-*trans*-glycosylation (Scheme 4.3).

Although the picolyl ether is less electron-withdrawing and therefore less deactivating than an ester group, reaction rates were slow, and heating was needed to drive glycosylations to completion [20]. Activation of the thioglycoside donor **13** with $Cu(OTf)_2$ was completed within 1 h, but the ensuing glycosylation was found to be sluggish. NMR studies showed that a mixture of two stable pyridinium ions (**15** and **16**) were generated upon activation, with the 1,2-*cis*-pyridinium ion **16** formed predominantly. The pyridinium ions are remarkably stable and could be isolated by silica gel chromatography. After purification, the 1,2-*cis*-pyridinium ion **16** reacted smoothly with methanol or sodium methoxide to afford the 1,2-*trans*-glycoside **17**, whereas the

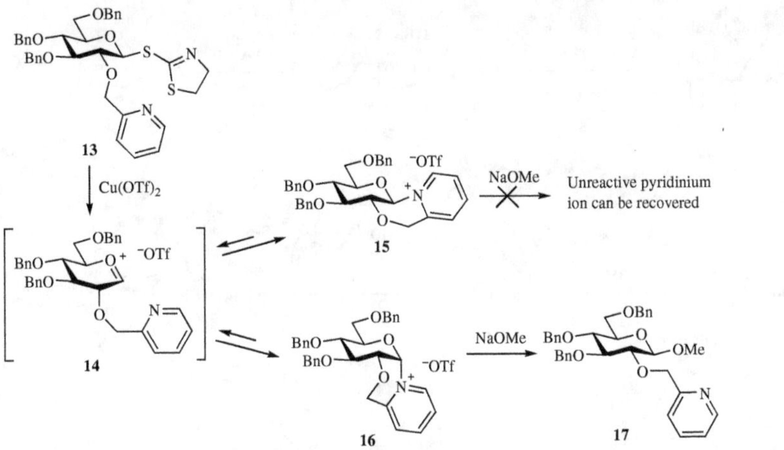

SCHEME 4.3 Use of the picolyl ether for the stereoselective synthesis of 1,2-*trans*-glycosides.

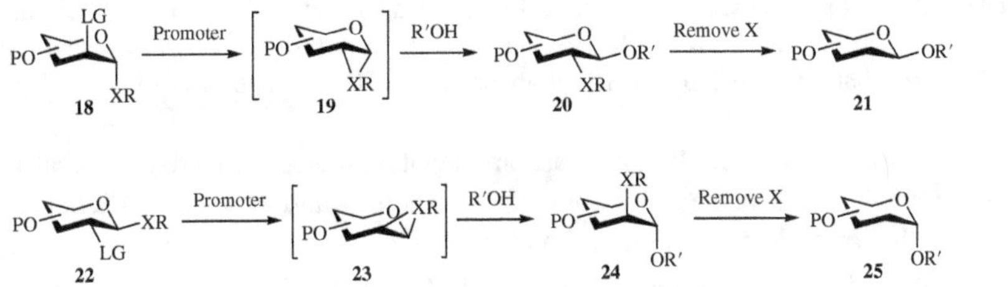

SCHEME 4.4 1,2-Migration of an anomeric participating group to afford α- or β-deoxyglycosides.

1,2-*trans*-pyridinium ion **15** was inert under these conditions and was recovered. Although the activation of glycosyl donors proceeded readily, the unreactive nature of the intermediate pyridinium ions hampered broad applicability of the methodology.

Neighboring group participation followed by migration is particularly useful in the stereoselective synthesis of 2-deoxyglycosides (Scheme 4.4) [21]. Typically, the leaving group is placed at C2, while the participating functionality (XR) is at the anomeric center in a 1,2-*trans* configuration (**18**, **22**). The participating group of the glycosyl donor should be α-oriented if a β-anomeric product is desired and vice versa. Promoter-assisted departure of the leaving group and participation of the C2 functional group results in the formation of a three-membered ring intermediate (**19**, **23**), which, after displacement in an S_N2 fashion by the glycosyl acceptor, leads to stereoselective formation of disaccharides. The participating group is usually a thio- or seleno-ether, which can be reductively removed using Raney nickel to afford 2-deoxyglycosides (**21**, **25**). In addition, three-membered halonium intermediates can also be used, which afford haloglycosides that can be reduced to 2-deoxyglycosides or used to incorporate additional functionalities.

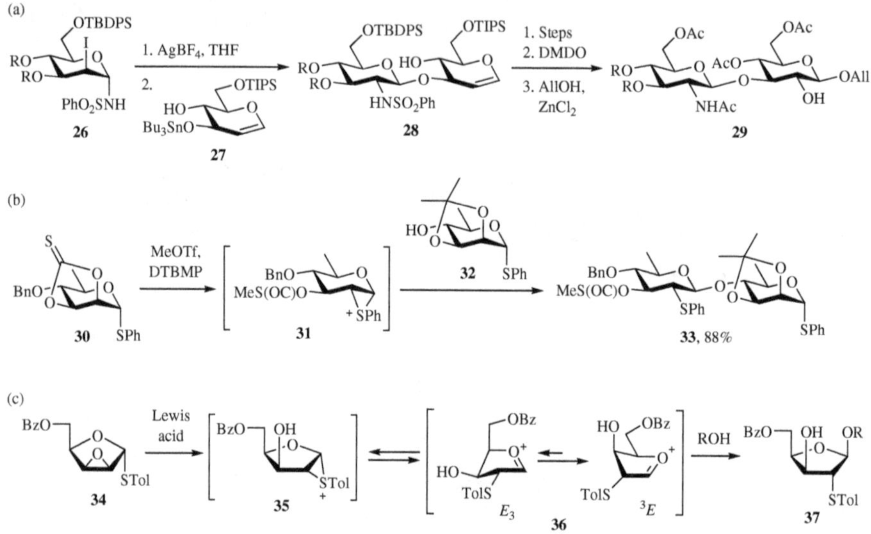

SCHEME 4.5 1,2-Migration in the stereoselective synthesis of 1,2-*trans* (amino)-glycosides and 2-deoxyglycosides as reported by (a) Danishefsky, (b) Yu, and (c) Lowary. Ac, acetyl; All, allyl; Bn, benzyl; Bz, benzoyl; DTBMP, 2,6-di-*tert*-butyl-4-methylpyridine; TBDPS, *tert*-butyldiphenylsilyl; THF, tetrahydrofuran; TIPS, triisopropylsilyl; Tol, 4-tolyl.

An early example of the 1,2-participation–migration strategy termed the "glycal assembly method" was developed by Samuel Danishefsky and coworkers (Scheme 4.5a) [22]. In this approach, the double bond of a glycal is converted into an epoxide or 1,2-*trans*-diaxial iodosulfonamide. Opening of the epoxide with an aglycone affords 1,2-*trans*-glycosides. In the presence of a strong base, the 1,2-*trans*-diaxial iodosulfonamide gives a 1,2-*N*-sulfonylaziridine, which can react with an appropriate alcohol to afford 1,2-*trans*-aminoglycosides. This method was applied for the synthesis of Le^y and Le^b neoglycoconjugates [23]. Key steps involved the preparation of a glycal intermediate that was transformed by iodonium di-*sym*-collidine perchlorate (IDCP) and benzenesulfonamide into the iodosulfonamide **26** in a stereoselective manner [24]. Activation of **26** with $AgBF_4$ led to the intermediate 1,2-*N*-sulfonylaziridine, which reacted with the tributyltin ether glycal **27** to afford the dimer **28**. The second glycal was activated with dimethyldioxirane (DMDO) to supply the intermediate epoxide that was treated with $ZnCl_2$ and allyl alcohol to afford compound **29**.

Recently, Biao Yu and Zunyi Yang reported 2,3-*O*-thionocarbonyl-1-thioglycosides as viable glycosyl donors for the construction of 2-deoxy-β-glycosides (Scheme 4.5b) [25]. This method was applied to the synthesis of the landomycin A hexasaccharide, which contains two β-D-olivose-(1→4)-β-D-olivose disaccharides [26]. The 2,3-*O*-thionocarbonate of the olivosyl donor **30** was activated by MeOTf, leading to the 1,2-migration of the thiophenyl moiety to provide the corresponding episulfonium ion **31**. The latter reaction intermediate was displaced by the olivose acceptor **32** to give the β-linked disaccharide **33** with absolute control of anomeric selectivity.

Todd Lowary and coworkers have shown that activation of 2,3-anhydro-thioglycosides with $Cu(OTf)_2$ results in the formation of 1,2-*trans*-glycosides (Scheme 4.5c).

Initially, it was thought that the glycosylation proceeded through the episulfonium intermediate **35**. However, the measured kinetic isotope effect (KIE) of 1.17–1.20 suggests an oxocarbenium ion intermediate rather than an episulfonium ion [27]. Furthermore, computational studies also indicated that the ^{3}E oxocarbenium ion **36** is considerably more stable than the strained episulfonium ion **35**. The observed stereoselectivity was explained by an inside attack on the ^{3}E oxocarbenium ion.

4.2.4 Neighboring Group Participation Using Chiral Auxiliaries to Obtain 1,2-*cis*-Glycosides

Until recently, 1,2-*cis*-glycosides could only be obtained by employing glycosyl donors having a nonassisting functionality at C2. In these glycosylations, reaction parameters such as solvent, temperature, and promoter, as well as the constitution of the glycosyl donor and acceptor (e.g., type of saccharide, leaving group at the anomeric center, protection, and substitution pattern), control anomeric selectivity. The introduction of 1,2-*cis*-glycosides often leads to mixtures of anomers despite extensive efforts to optimize the reaction conditions.

An approach toward 1,2-*cis*-glycosides using neighboring group participation by a chiral auxiliary at C2 of the glycosyl donor was recently reported [28]. Here, a chiral auxiliary was designed to trap an oxocarbenium ion from the β face, and displacement of this moiety should result to 1,2-*cis*-glycosides. The nucleophilic group is placed in such a way to form a six-membered ring upon participation, resulting in either a *cis*- (**40, 45**) or *trans*-decalin (**41, 47**) system (Scheme 4.6). It was envisioned that a chiral substituent could be used to influence the formation of the *cis*- or *trans*-decalin system leading to 1,2-*trans*- or 1,2-*cis*-glycosides, respectively. In the case of

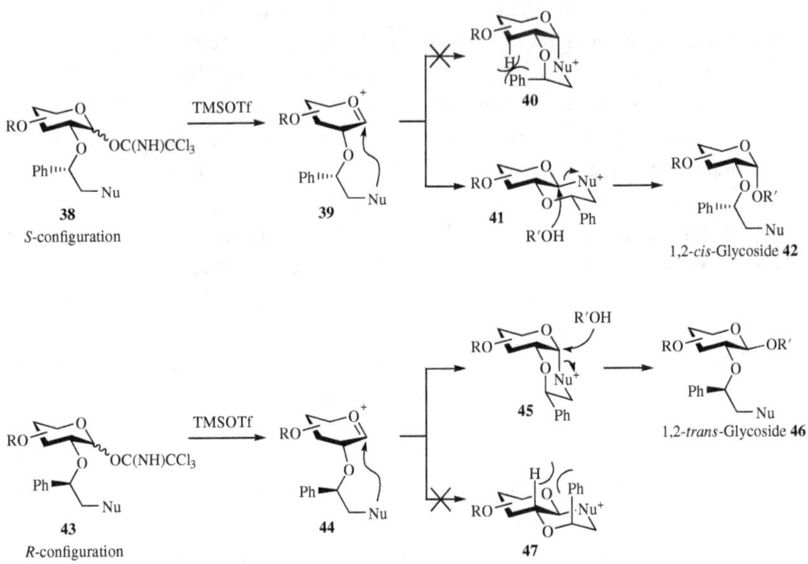

SCHEME 4.6 Chiral auxiliaries designed to form *cis*- or *trans*-decalin oxonium intermediates to afford 1,2-*trans*- or 1,2-*cis*-glycosides, respectively. Nu, nucleophile; TMS, trimethylsilyl.

TABLE 4.1 C2 Chiral Auxiliaries in Stereoselective Glycosylation

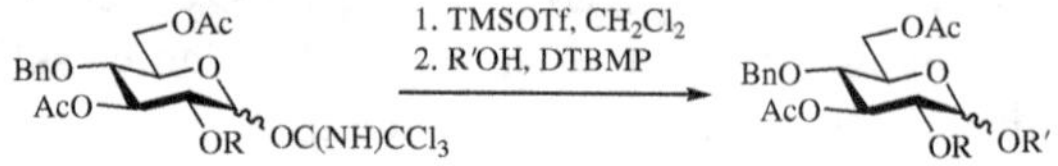

Entry	R	α/β	Yield (%)
1	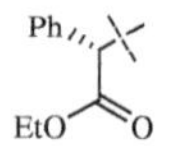	20/1–4/1	92–94
2	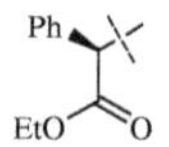	1/8–1/1	88–94
3	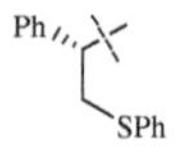	α only	95
4	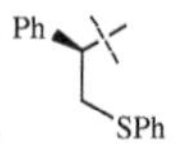	1/1	88
5	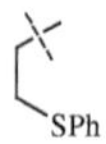	8/1	86
6	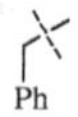	3/1	94

an *S*-configured auxiliary, formation of the *cis*-decalin intermediate **40** results in axial placement of the chiral substituent, which leads to an unfavorable 1,4-diaxial steric interaction. In contrast, formation of the *trans*-decalin intermediate **41** leads to a more favorable equatorial substitution. In addition, the *trans*-decalin system experiences fewer unfavorable gauche interactions than the *cis*-decalin system. Therefore, the *S* configuration is expected to favor the 1,2-*cis*-glycoside by selectively forming the *trans*-decalin intermediate. Similarly, the *R*-configured auxiliary is expected to favor the 1,2-*trans*-glycoside formation because the 1,2-*cis*-decalin intermediate **45** leads to equatorial placement of the phenyl substituent and prevents the 1,3-diaxial steric interaction that the *trans*-decalin **47** experiences.

The first generation of chiral auxiliaries employed an (*S*)-ethoxycarbonylbenzyl moiety as the nucleophilic head group [28, 29]. As expected, the *S*-configured auxiliary led to the selective formation of 1,2-*cis*-glycosides with α/β selectivities ranging from 20/1 to 4/1 depending on the nature of the glycosyl acceptor (Table 4.1, entry 1). On the other hand, the use of the *R*-configured auxiliary formed the 1,2-*trans*-glycosides (entry 2), albeit with lower selectivity (α/β ratios range between 1/8 and 1/1). These results clearly demonstrated that the chirality of the auxiliary controls the anomeric selectivity, albeit with modest selectivity in some cases. The lack of absolute anomeric stereoselectivity was attributed to the sp^2 character of the (*S*)-ethoxycarbonylbenzyl moiety that does not allow the formation of a perfect *trans*-decalin system.

The second generation of chiral auxiliary employed a more nucleophilic thiophenyl ether [30]. This modification significantly improved 1,2-*cis* stereoselectivities

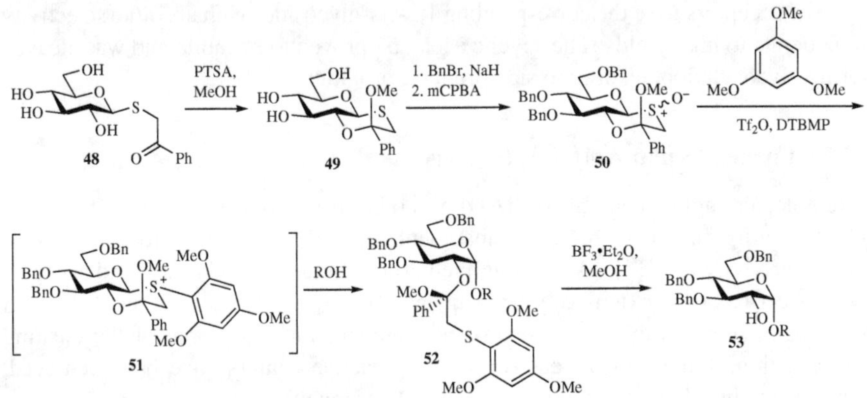

SCHEME 4.7 Generation of sulfonium ions using electrophilic aromatic substitution of oxathiane glycosyl donors. PTSA, *p*-toluenesulfonic acid.

(entry 3). In general, glycosylations using the (*S*)-(phenylthiomethyl)benzyl moiety were rather slow, indicating the presence of a slow-reacting intermediate. Low-temperature NMR experiments demonstrated the selective formation of a β-linked sulfonium ion intermediate in a decalin system [30]. An auxiliary lacking a chiral substituent was less selective than the (*S*)-(phenylthiomethyl)benzyl moiety (entry 5). The loss of 1,2-*cis* selectivity compared to the (*S*)-(phenylthiomethyl)benzyl moiety can be explained by the fact that a more substituted ring system is more stable and that additional substituents also promote cyclization (Thorpe–Ingold effect). Finally, a glycosyl donor having a C2 benzyl ether gave poor anomeric selectivity (entry 6).

The (*S*)-(phenylthiomethyl)benzyl moiety could be introduced by reaction of the sugar alcohol with (*S*)-(phenylthiomethyl)benzyl acetate in the presence of $BF_3{\cdot}OEt_2$. The removal of this group with concurrent exchange to an acetyl group was achieved by treatment with $BF_3{\cdot}OEt_2$ in Ac_2O. Both reactions proceed through an intermediate episulfonium ion with overall retention of configuration.

Recently, W. Bruce Turnbull and coworkers reported an elegant modification of the (*S*)-(phenylthiomethyl)benzyl auxiliary that involves prearranged oxathiane glycosyl donors (Scheme 4.7) [31, 32]. The intramolecular reaction of glycosyl ketone **48** in acidic methanol afforded the corresponding oxathiane derivative **49** in high yield and absolute stereoselectivity. The methoxy group orientation in **49** is stabilized by the anomeric effect and the equatorial placement of the larger substituent. A major benefit of this approach is that no chiral starting material other than glucose is needed and that the O2 position is selectively modified without the use of protecting groups. Protection of the remaining hydroxy groups and oxidation of the thioether using *m*-chloroperoxybenzoic acid (mCPBA) afforded the sulfoxide **50**. Activation of the oxathiane donor **50** was performed using Tf_2O and 1,3,5-trimethoxybenzene in the presence of 2,6-di-*tert*-butyl-4-methylpyridine (DTBMP). The triflation of the sulfoxide led to a sulfonium ion, which reacted with 1,3,5-trimethoxybenzene by an electrophilic aromatic substitution to afford the intermediate **51**. Addition of various glycosyl acceptors gave the corresponding 1,2-*cis*-glycosides with absolute selectivity in moderate to high yields. The acyclic acetal **52** proved to be labile and was cleaved during glycosylation, leading to side product formation.

4.2.5 Intramolecular Aglycone Delivery

Intramolecular aglycone delivery (IAD) is another innovative approach that exploits relative configuration of (remote) substituents of a glycosyl donor to control anomeric selectivity [33, 34]. IAD through tethering of the glycosyl donor and acceptor by a spacer has been extensively investigated. The tether prearranges the donor and acceptor in such a way that it directs the stereochemical outcome of the ensuing glycosylation. Tetherings at reactive and nonreactive centers have been achieved, although the installation and removal of the tether involve additional steps.

Anomeric tethers have been successfully implemented; however, the anomeric selectivity is difficult to control because the tether needs to be detached for glycosylation to take place (Scheme 4.8a). This approach leads to competition between intra- and intermolecular reaction pathways, resulting in a loss of stereoselectivity. Linking monosaccharides prior to glycosylation through a nonreactive center was found more suitable to ensure stereoselective glycosylation. The ability of an enzyme to prearrange a glycosyl donor and acceptor has inspired the use of rigid spacers (Scheme 4.8b). Use of such spacers on nonreacting centers, however, requires additional steps for introduction and removal, and the stereoselectivity can be hard to predict.

Arguably the most successful version of IAD to date involves a tether between the O2 of the glycosyl donor and a hydroxyl of an acceptor (Schemes 4.8c and 4.9). This type of IAD is distinct from the prearranging strategy because the acceptor alcohol is directly connected to the tether, usually resulting in a five-membered transition state (TS) and the formation of a 1,2-*cis*-glycoside. The tether is designed to stabilize the positive charge and fragment upon glycosylation. Mixed acetals and the silyl acetal

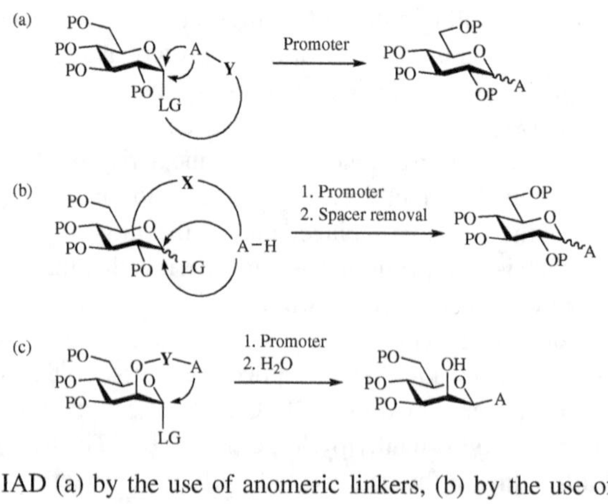

SCHEME 4.8 IAD (a) by the use of anomeric linkers, (b) by the use of rigid spacers at nonreacting centers, and (c) by tethering from C2 to form 1,2-*cis*-glycosides.

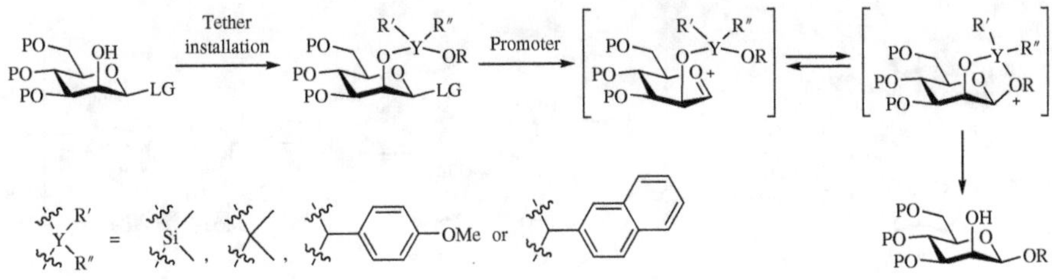

SCHEME 4.9 IAD by acetal tethering to afford 1,2-*cis*-glycosides.

tether, which can be installed using Me_2SiCl_2, are suitable for this purpose. The mixed isopropylidene acetal can be readily obtained under mild acidic conditions from an isopropenyl functionalized sugar and a glycosyl acceptor. The isopropenyl functionality can be obtained from an acetyl ester precursor upon treatment with Tebbe's reagent. The latter step somewhat limits this technology because no other ester should be present.

Another approach to prepare a mixed acetal exploits the oxidation of an electron-rich aryl ether in the presence of an alcohol acceptor. The mixed acetal is usually prepared by 2,3-dichloro-5,6-dicyano-1,4-benzoquinone (DDQ) oxidation of a glycosyl donor carrying a 2-naphthylmethyl (NAP), *p*-methoxybenzyl (PMB), or 3,4-dimethoxybenzyl (DMB) ether at C2 under dry conditions in the presence of a glycosyl acceptor. The position of the substituted aryl ether can also be switched from the acceptor to the donor, an approach referred to as inverse tethering, which can improve the yield of mixed acetal formation in some cases. Upon activation of the glycosyl donor, the mixed aryl acetal ensures that the acceptor oxygen is delivered at the same face as the O2 substituent (1,2-*cis*) of the glycosyl donor. Usually, the resulting product is isolated as a free 2′-alcohol, which can directly serve as glycosyl acceptor. However, the oxocarbenium intermediate arising from the substituted aryl ether can also be trapped using Et_3SiH, succinimide (formed when using NIS as promoter), or a 3-*O*-TMS ether affording a 2-*O*-aryl ether, 2-*O*-succinyl mixed acetal, or 2,3-*O*-aryl acetal, respectively.

The benefits of trapping the intermediate oxocarbenium ion by using neighboring protecting group was demonstrated by Yukishige Ito and coworkers in an elegant synthesis of β-L-rhamnosides (Scheme 4.10) [35]. The mixed acetal **56** was prepared by using the NAP-protected rhamnosyl donor **54** and the glucosyl acceptor **55** upon oxidation with DDQ. Activation of the thiomethyl leaving group of **56** with MeOTf led to the formation of an oxonium ion intermediate **57** that was trapped by O3′, resulting in the formation of the naphthylmethylidene acetal **58**. Selective reductive opening of the acetal using diisobutylaluminum hydride (DIBAL-H) furnished the 3′-alcohol **59**, which could act as glycosyl acceptor for the ensuing glycosylation. IAD has been used to prepare other 1,2-*cis*-pyranosides and furanosides such as α-D-glucoside, β-D-mannoside, β-L-rhamnopyranoside, β-D-arabinofuranoside, β-D-fructofuranoside, and β-D-fucofuranoside.

IAD proved very useful to the preparation of α,α-trehaloses such as sulfolipid-1 (SL-1) (Scheme 4.11). The key step in the synthesis of SL-1 analogues by Carolyn Bertozzi

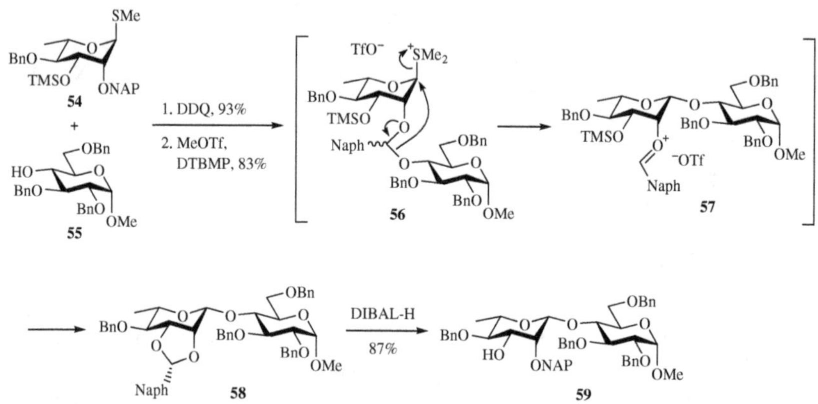

SCHEME 4.10 Trapping of an intermediate oxocarbenium ion using a 3-*O*-TMS ether in the synthesis of β-L-rhamnosides. Naph, 2-naphthyl.

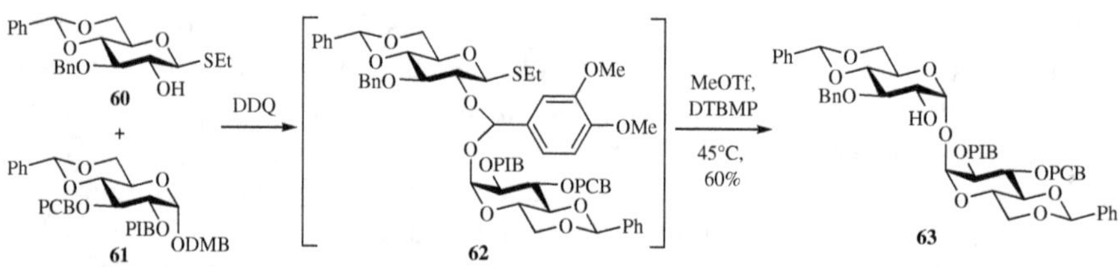

SCHEME 4.11 Dual anomeric control using IAD in the synthesis of α,α-trehalose. PCB, *p*-chlorobenzyl; PIB, *p*-iodobenzyl.

and Clifton Leigh involved a DMB mixed acetal [36]. The synthesis of α,α-trehaloses is made difficult by the need to control two anomeric centers. Also, the anomeric hydroxyl cannot be used as acceptor because it can mutarotate and convert to the unwanted β anomer. IAD lends itself well because the anomeric hydroxyl can be masked as a stable α-anomeric ether and directly transferred to the donor anomeric center through inverse tethering. Thus, DDQ oxidation of the glucosyl acceptor **61** carrying an α-anomeric DMB ether in the presence of the thioglycoside donor **60** afforded the mixed acetal **62**. The ensuing glycosylation using MeOTf as promoter afforded only the α,α-trehalose product **63** in good yield.

4.2.6 Anomeric Control by Electronic and Steric Effects

David Crich and coworkers found that preactivation of 4,6-*O*-benzylidene-protected mannosyl sulfoxide donors at low temperature, followed by addition of a glycosyl acceptor, leads to the stereoselective formation of β-mannosides (Scheme 4.12). The necessity for preactivation at low temperature to obtain good β selectivity led to the notion that a stable intermediate is first formed, which then reacts in a stereoselective manner with the nucleophile to generate β-mannosides. In addition to preactivation, the 4,6-*O*-benzylidene acetal was found to be indispensable for high

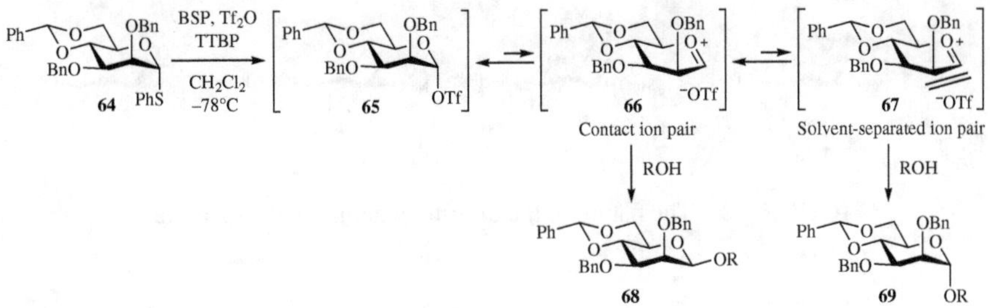

SCHEME 4.12 Use of the 4,6-*O*-benzylidene acetal in mannosyl donors to favor the α-triflate intermediate. TTBP, 2,4,6-tri-*tert*-butylpyrimidine.

β selectivity. Crich and coworkers proposed that, under preactivation conditions, the oxocarbenium ion is trapped by a triflate anion to form the more stable α-triflate **65**. After addition of the acceptor, the α-triflate intermediate can then be displaced in an S_N2-like manner to afford a β-mannoside product (**68**). The formation of α-glycosyl triflates was confirmed by ^1H, ^{13}C, and ^{19}F NMR analyses of the activated mannosyl donor recorded at low temperature [37]. The experimentally determined KIE is approximately 1.12, which is consistent with an oxocarbenium-like TS [38]. It was hypothesized that the α-triflate converts into the contact ion pair **66** in which the triflate anion remains at the α face or that an exploded TS is formed where the nucleophile is loosely associated with the oxocarbenium ion as the triflate departs [39, 40]. The α product **69** can be explained by the formation of the solvent-separated ion pair **67** where the counterion is solvated and facial selectivity is lost.

The influence of the 4,6-*O*-benzylidene acetal in β-mannosylations was believed to be due to its torsional disarming effect. In an earlier work, Bert Fraser-Reid and coworkers found that a 4,6-*O*-benzylidene acetal locks the pyranose ring in a stable chair conformation [41]. Therefore, it disfavors the formation of an intermediate oxocarbenium ion, which requires rehybridization and flattening of the sugar ring, usually leading to a half-chair conformation. This torsional effect, combined with the strong *endo*-anomeric effect in mannose, favors the α-triflate intermediate.

Mikael Bols and coworkers studied the electronic and torsional effects of 4,6-cyclic acetals on the hydrolysis rate of anomeric dinitrophenyl (DNP) glycosides (Fig. 4.2) [42]. It was established that the donor can be deactivated by torsional effects. In addition, the electronic influence the 4,6-*O*-acetal has on the O6 substituent was found to be significant and is caused by the fixed arrangement of the O6 substituent. Compounds **71–73** are all torsionally deactivated as opposed to compound **70** because they all have a fused bicyclic ring system. With variable orientation of the O6 substituent, a marked difference in the hydrolysis rate was observed and can be explained by the direction of the O6 dipole. A 4,6-acetal forces the O6 substituent to adopt the *tg* (**73**) conformation, which places its dipole approximately antiparallel to the electron-deficient center in the TS, thereby destabilizing it. In the *gg* (**71**) and *gt* (**72**) conformations, the dipoles are more or less perpendicular to the developing positive charge and thus less destabilizing. This electronic disarming effect was found to be approximately equal to the torsional disarming effect of a 4,6-acetal.

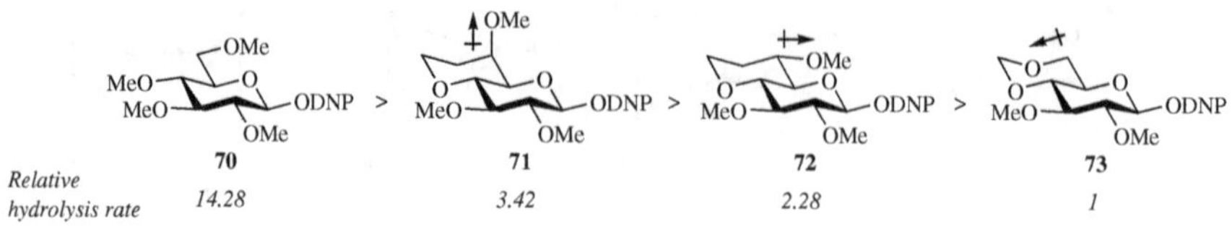

FIGURE 4.2 The nature of the disarming ability of 4,6-*O*-acetals.

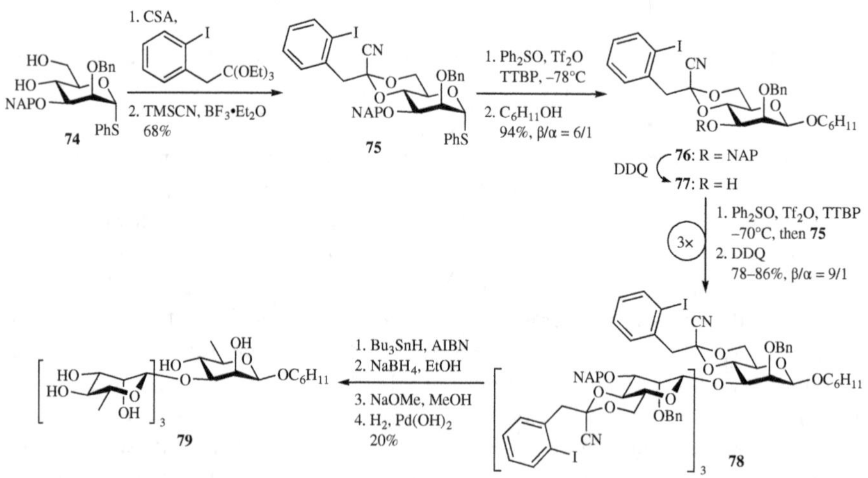

SCHEME 4.13 Application of the 1-cyano-2-(2-iodophenyl)-ethylidene acetal for the synthesis of a β(1→3)-D-rhamnotetraose. CSA, 10-camphorsulfonic acid.

The use of a 4,6-acetal to ensure 1,2-*cis*-glycosylation has been extended to the syntheses of L-*glycero*-β-D-*manno*-heptopyranoses [43] and β-D-rhamnose (β-D-6-deoxy-mannose) derivatives [44]. To this end, the 1-cyano-2-(2-iodophenyl)-ethylidene acetal was developed [45] for stereoselective β-mannosylation as well as to perform radical fragmentation/reduction to finally afford a β(1→3)-D-rhamnotetraose (Scheme 4.13). The 1-cyano-2-(2-iodophenyl)-ethylidene acetal **75** was prepared through the 4,6-ethoxy-orthoester formation followed by Lewis acid-promoted cyanation [46]. Ensuing cycles of Ph_2SO/Tf_2O promoted glycosylation and oxidative NAP group removal gave the β(1→3)-D-rhamnotetraose **78** in high yield and stereoselectivity. Treatment with Bu_3SnH and azobisisobutyronitrile (AIBN) led to the radical fragmentation [47, 48] of the 1-cyano-2-(2-iodophenyl)-ethylidene acetal and C6 deoxygenation of the tetrasaccharide. $NaBH_4$ was used to remove any tin residues, followed by transesterification and hydrogenation to afford the deprotected β(1→3)-D-rhamnotetraose **79**.

In addition to the disarming effect of the 4,6-*O*-benzylidene acetal, highly electron-withdrawing protecting groups were also shown to improve 1,2-*cis* stereoselectivity [49–53]. It was noted that a highly electron-withdrawing group at C2 of a mannosyl donor led to the selective formation of β-mannosides. Recently, Kim and coworkers reported a systematic study on the influence of highly electron-withdrawing

TABLE 4.2 Use of Strongly Electron-Withdrawing Groups to Ensure β-Mannosylation through α-Triflate Formation

Entry	LG	R¹	R²	R³	Promoter	α/β ratio
1	$OC(NH)CCl_3$	SO_2Bn	Bn	Bn	TMSOTf	1/15.9
2	$OC(NH)CCl_3$	Bn	SO_2Bn	Bn	TMSOTf	1/10.7
3	$OC(NH)CCl_3$	Bn	Bn	SO_2Bn	TMSOTf	1/13.8
4	Br	SO_2Bn	Bn	Bn	AgOTf	1/3.9
5	Br	SO_2Bn	Bn	Bn	$AgClO_4$	2.1/1

protecting groups at positions other than C2 on β-mannosylation [16]. A clear correlation between the electron-withdrawing capacity and β selectivity was observed regardless of the position at the sugar ring (Table 4.2). The highest selectivity was observed when a benzylsulfonate ester was placed at C3 (entry 1). Low-temperature NMR experiments identified an α-triflate intermediate from the mannosyl trichloroacetimidate donors. The β selectivity, together with the NMR experiments, excludes remote participation of the electron-withdrawing protecting groups, which should lead to α-glycosylation if positioned at C3. These observations led to the conclusion that the S_N2-like displacement of an α-triflate intermediate is responsible for the high β selectivity. The α-triflate intermediate is expected to be more stable than an oxocarbenium ion that is electronically destabilized by the electron-withdrawing protecting group. To further demonstrate the importance of the counterion, activation of mannosyl bromides with AgOTf selectively formed β-mannosides (entry 4), whereas $AgClO_4$, which has a nonparticipating counterion, gave mainly α-mannosides (entry 5).

Uronic acids are an important class of glycans, which are notoriously difficult to glycosylate due to the electron-withdrawing 5-carboxy group [54]. Thioglycoside derivatives of uronic acids (thiouronic acid donors) can be prepared through the use of 2,2,6,6-tetramethylpiperidinyloxy (TEMPO) and [bis(acetoxy)iodo]benzene (BAIB) in a highly selective oxidation [55]. This reagent combination selectively oxidizes primary sugar alcohols without affecting the anomeric thioether. The thiouronic acid donors could be activated by using the highly thiophilic Ph_2SO/Tf_2O promoter system (Scheme 4.14). Interestingly, thiomannuronic acid donors exhibit high β selectivity, although a 3-*O*-acetyl moiety seemed to diminish the β selectivity presumably through participation [56]. This methodology was later extended to the stereoselective synthesis of alginates composed of β-D-mannuronic acid and its C5-epimer α-L-guluronic acid [57]. The unexpected high 1,2-*cis* selectivity was attributed to a unique combination of factors [58]. Firstly, the orientation of the 5-carboxylate ester in the presumed half-chair oxocarbenium ion TS (**85**, **86**) was proposed to be a dominating factor in influencing the stereoselectivity [59, 60]. It was suggested that the $^3H_4$ conformer **85** is favored because it places the methyl ester in an axial orientation and can donate electron density to the anomeric oxocarbenium ion. In

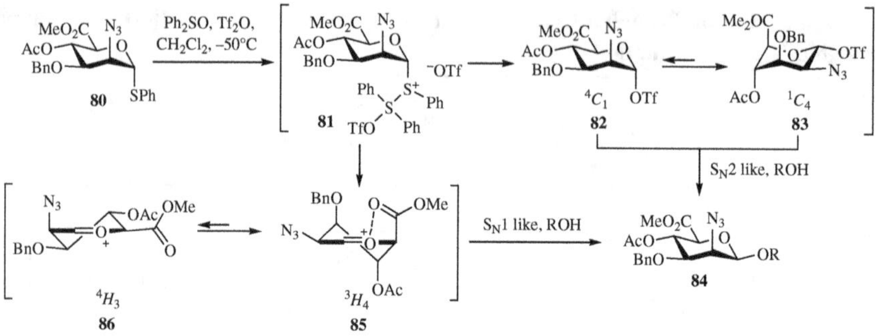

SCHEME 4.14 β selectivity in the glycosylation of mannuronate esters.

addition, the axial orientation at C5 places its dipole antiperiplanar to the oxocarbenium ion, minimizing its destabilizing effect. Computational studies indicated that the ${}^3H_4$ conformer **85** is more stable by 3.4 kcal/mol compared to the ${}^4H_3$ conformer **86**. The nucleophile attacks this stabilized oxocarbenium ion with facial selectivity to form the lower energy chair β product as opposed to the twist-boat α adduct. L-Gulose, the C5-epimer of D-mannose, prefers the ${}^4H_3$ conformation due to the axial orientation of the 5-carboxylate ester leading to the stereoselective formation of α-L-gulose products [57]. Secondly, low-temperature NMR experiments revealed the presence of two anomeric α-triflate species **82** and **83** as a mixture of ${}^4C_1$ and ${}^1C_4$ conformers [61]. The electronic deactivating effects of the 2-azide and 5-methylester and the strong *endo*-anomeric effect in mannose are thought to be responsible for the preferential formation of these triflate intermediates. Thus, the nucleophilicity of the glycosyl acceptor probably determines the reaction path. Highly nucleophilic acceptors are expected to promote an S_N2-like mechanism, whereas acceptors of reduced nucleophilicity will react in an S_N1-like manner, both leading to 1,2-*cis*-glycosides.

4.2.7 Conformational Selection Using a 3,5-*O*-Di-*tert*-Butylsilylidene Protecting Group

Small differences between anomeric effects of α and β anomers and inherent ring flexibilities have complicated the development of a general method for stereoselective introduction of 1,2-*cis*-furanosides. By examining possible conformers of the arabinofuranosyl oxocarbenium ion, factors that favor attack from the β face (a 1,2-*cis* attack) were identified. Oxocarbenium ions of furanosides can adopt ${}^3E$ or E_3 as two possible low-energy conformations (Scheme 4.15). The Newman projection of the E_3 conformer of D-arabinofuranose showed that nucleophilic attack from the α face is disfavored because an eclipsed 2-H will be encountered. On the other hand, an approach from the β face is expected to be more favorable because it experiences only staggered substituents.

Computational studies indicated that the arabinofuranosyl oxocarbenium ion can be locked in the E_3 conformation by employing a 3,5-*O*-di-*tert*-butylsilylidene (3,5-*O*-DTBS) protecting group, which places O3 and O5 in a pseudoequatorial

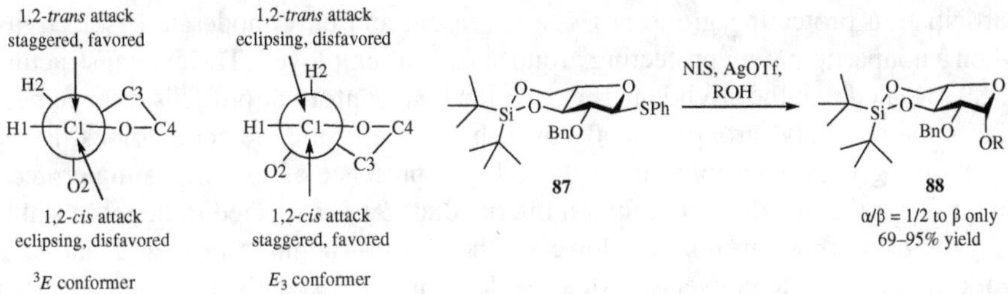

SCHEME 4.15 Securing of the E_3 conformer using the 3,5-*O*-DTBS protecting group to ensure 1,2-*cis*-glycosylation.

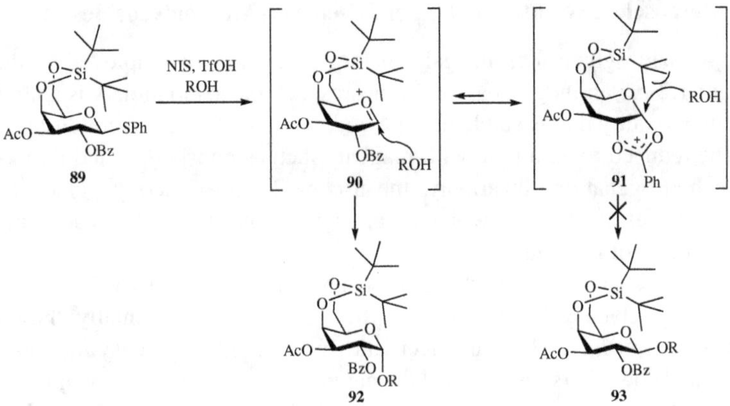

SCHEME 4.16 Use of the DTBS group in α-galactosylation.

orientation, resulting in a perfect chair conformation of the protecting group [62]. Indeed, the DTBS-protected arabinofuranosyl donor **87** gave excellent β selectivities in glycosylations with acceptors having primary and secondary alcohols. In addition to the 3,5-*O*-DTBS protecting group, selection of the promoter system proved to be important to obtain good 1,2-*cis* selectivity. This method was successfully employed in the synthesis of an arabinogalactan fragment derived from plant cell walls and the arabinogalactan and lipoarabinomannan domains of mycobacteria. Lowary and coworkers prepared an octadecasaccharide arabinan domain found in *Mycobacterium tuberculosis* containing four 1,2-*cis*-linked arabinofuranosides using this method [63].

Crich and coworkers showed that preactivation of the DTBS-protected arabinofuranosyl donor followed by addition of the glycosyl acceptor led to nonselective reactions [64]. NMR experiments confirmed that these reactions did not proceed by S_N1-like attack of the oxocarbenium ion but through other intermediates such as glycosyl triflates. Furthermore, glycosylations with galactosyl donor **89**, which is protected with a 4,6-*O*-DTBS group, gave the corresponding α-galactosides (**92**) with very high α-anomeric selectivities even in the presence of a 2-*O*-benzoyl group that can perform neighboring group participation (Scheme 4.16) [65]. In stark contrast, the 4,6-*O*-benzylidene counterpart of **89** showed only β-glycosylation when a participating protecting group at C2 was present and only moderate α selectivity when a nonparticipating protecting group at C2 was employed. The crystal structure of **89** revealed that the silylidene ring is in the boat conformation. This arrangement places the *tert*-butyl groups directly over the β face of the molecule, blocking the attack of a glycosyl acceptor at this face. In the presence of a participating protecting group at C2, the dioxolenium ion intermediate **91** is expected to be more stable than the oxocarbenium ion **90**. However, the Curtin–Hammett principle can be at work when the interconversion of dioxolenium and oxocarbenium ion is faster than glycosylation, and thus, α-selective glycosylation of the oxocarbenium ion can predominate.

4.2.8 Stereoselective Introduction of 2-Deoxy-2-Aminoglycosides

2-Deoxy-2-azido-glucosides and galactosides are commonly employed for the introduction of 1,2-*cis*-aminoglycosides. In this respect, the azido moiety is a nonparticipating functionality that is stable under a wide variety of reaction conditions but can readily be reduced to an amine with reagents such as phosphines and thiols and by catalytic hydrogenation. Invariably, the use of 2-azido-2-deoxy glycosyl donors leads to mixtures of anomers, which requires time-consuming purification protocols resulting in loss of material.

2,3-*N*,*O*-Oxazolidinone-protected glucosamine derivatives were developed in an effort to improve the glycosyl-accepting properties [66–68]. Coincidentally, these derivatives also exhibited very high α selectivities when used as glycosyl donors. Stefan Oscarson and coworkers investigated the nature of this selectivity by using *N*-benzyl- and *N*-acetyl-2,3-*N*,*O*-oxazolidinone-protected glycosyl donors (Scheme 4.17). The *N*-benzyl derivative **94** provided the α product, whereas the *N*-acetyl derivative **95** produced the β anomer when activated in the presence of a glycosyl acceptor [66]. However, when prolonged reaction times or higher concentrations of AgOTf were employed, the β anomer was converted into the α anomer **98**. NMR studies showed that the β anomer was stable for approximately 30 min at a low concentration of AgOTf (0.1 equiv.) and was completely anomerized in 4 h. In the presence of excess AgOTf, the anomerization was complete within 1 h.

The anomerization proceeds by an intramolecular reaction involving endocyclic C–O bond cleavage to give mainly or exclusively an α-glucoside. Evidence for

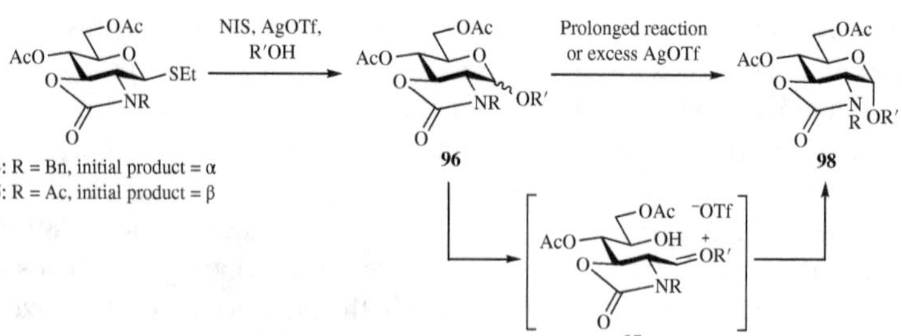

94: R = Bn, initial product = α
95: R = Ac, initial product = β
96
97
98

SCHEME 4.17 Anomerization of 2,3-*N*,*O*-oxazolidinone-protected glycosides.

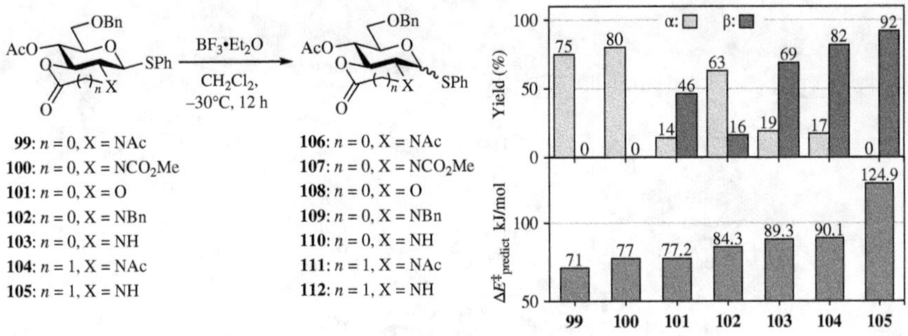

FIGURE 4.3 Correlation of the calculated TS energies from inner strain caused by the two fused rings and the experimental results. Source: Satoh et al. [70]. Reproduced with permission of American Chemical Society.

endocyclic cleavage was provided by intra- and intermolecular Friedel–Crafts reactions, chloride addition, and reduction of the intermediate **97** [69]. It is important to note that pyranosides are locked in chair conformation for endocyclic opening, and furanosides with distorted conformations do not undergo this transformation.

Ito and coworkers explored the driving force for the C–O bond cleavage using computational and experimental methods [70]. They found that the 2,3-*trans*-fused protecting group induces ring strain that lowers the TS energy of endocyclic bond cleavage. The trend between the calculated TS energy and degree of anomerization (α/β ratio) is in agreement (Fig. 4.3). To promote anomerization, the thioglycosides **99–105** were treated with $BF_3{\cdot}Et_2O$ at −30°C. The compounds with fused five-membered rings (**99–103**) all showed anomerization as predicted by the calculated activation energies. The less strained six-membered ring derivatives **104** and **105** only anomerized to a small extent consistent with the ring strain hypothesis. Furthermore, the TS energy was found to correlate directly to the ring strain. It takes much less computing time to calculate the ring strain than to calculate the TS energy, and therefore, ring strain can be used to rapidly predict the degree of anomerization in a virtual screening.

4.3 USE OF PROTECTING GROUPS FOR CHEMOSELECTIVE GLYCOSYLATIONS

Aside from affecting the stereoselectivity of glycosylations, protecting groups can also modulate the reactivity of glycosyl donors [71–74]. Hans Paulsen and coworkers first noted that the rate of hydrolysis of glycosyl halides was influenced by the electronic nature of the protecting groups [75]. They found that glycosyl halides bearing benzoyl esters hydrolyzed at a much slower rate than their benzyl ether counterparts. This observation was rationalized by the electron-withdrawing properties of the ester-forming protecting groups that destabilize an oxocarbenium ion-like TS. Subsequently, Fraser-Reid and coworkers showed that differences in anomeric reactivity can be exploited for chemoselective glycosylations and coined this approach

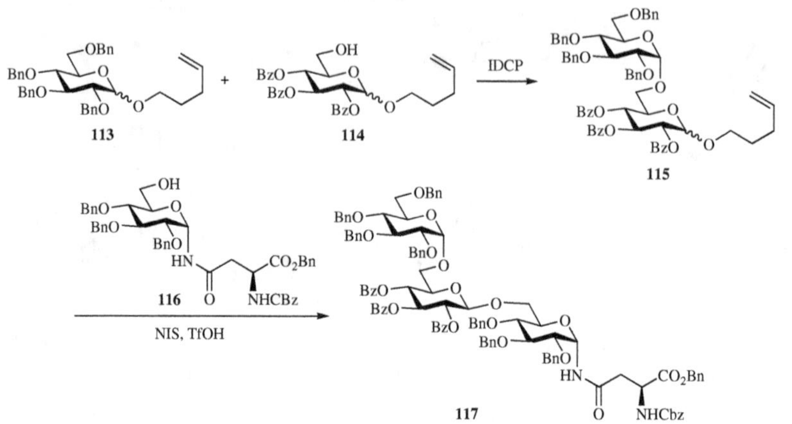

SCHEME 4.18 Selective glycosylation using the armed–disarmed glycosylation strategy. Cbz, benzyloxycarbonyl.

"armed–disarmed glycosylations" [76, 77]. They demonstrated that *n*-pentenyl glycosides protected by benzyl ethers (armed) can selectively be activated in the presence of an *n*-pentenyl glycosyl acceptor that is protected with electron-withdrawing esters (disarmed) (Scheme 4.18). No self-coupling of **114** was observed, supporting the hypothesis that the electron-withdrawing nature of the esters prevented activation of the anomeric center. The addition of one equivalent of activator to a mixture of armed and disarmed glycosyl donors forms two intermediates that are in equilibrium favoring the armed cation [78]. However, the disarmed donor does not readily form an oxocarbenium ion, and hence product formation proceeds much faster in case of the armed donor. By using a more reactive promoter system (NIS and TfOH), the disarmed disaccharide **115** could be activated and coupled to the amino acid building block **116** to afford saccharide **117** [79]. An obvious advantage of this strategy is that the glycan can be assembled from the nonreducing end, thereby minimizing protecting group manipulations.

The nature of the torsional effects imposed by cyclic protecting groups such as acetals can profoundly influence glycosyl donor reactivity. These effects have also been exploited in chemoselective glycosylations [41, 42, 80]. In this respect, Steven Ley and coworkers developed novel acetals such as the dispiroketal (dispoke) and cyclohexane-1,2-diacetal (CDA) to selectively protect 1,2-diequatorial diols [81]. These acetals reduce anomeric reactivity by exerting torsional strain, thereby resisting the flattening of the chair conformation to accommodate the sp^2-hybridized oxocarbenium ion (see Section 4.2.6). The disarming effect of the CDA acetal was used to assemble oligosaccharides in one pot (see Section 6.2.1) [82].

Ley and coworkers also showed that the degree of arming/disarming is a continuum and proposed that identification of donors with several different levels of reactivity would make it possible to prepare rather complex oligosaccharides by a one-pot procedure without the need for intermediate purifications. Subsequently, Chi-Huey Wong and coworkers reported a comprehensive study determining relative reactivities of

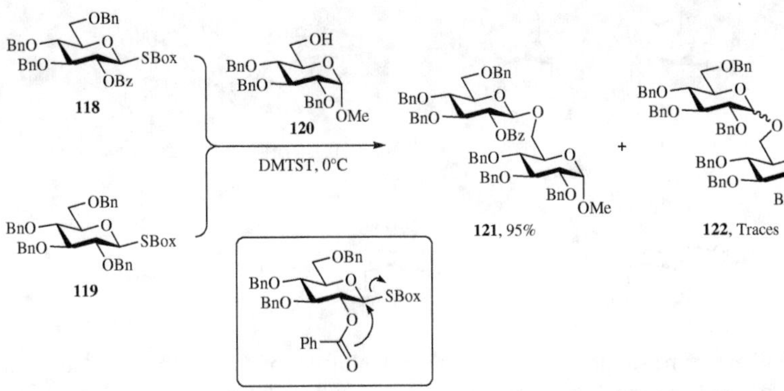

SCHEME 4.19 Competition experiment showing the super-arming effect of a participating group at C2 in *trans* orientation with respect to the leaving group. Box, benzoxazolyl; DMTST, dimethyl(methylthio)sulfonium triflate.

different thioglycoside donors [17]. Useful insights in the relative reactivities of glycosyl donors with respect to protecting groups, type of sugar (glucose, galactose, mannose, etc.), and number of free hydroxy groups were identified. The results of the evaluation and the successful applications are described in Section 6.2.1. Regarding the influence of the substituent on the relative reactivities, the deactivating power was found being N_3>OAcCl>NPhth>OBz>OBn.

The scope of anomeric reactivities was further extended by the introduction of super-armed donors, which are even more reactive than the armed per-O-benzylated donors. Super-armed donors can be obtained by either increasing the propensity of a leaving group to depart by participation or by conformational activation. For example, Demchenko's group showed that glycosyl donors carrying a C2 participating group are more reactive than the corresponding per-*O*-benzyl derivative even though the C2 group is electron-withdrawing [83, 84]. The activation of a glycosyl donor involves the initial activation of the leaving group followed by its departure, which proceeds through an oxocarbenium ion-like TS. The latter step is rate-determining. Even though the electron-withdrawing group slows down the first step, the second and rate-determining step is drastically accelerated by the participation of the C2 ester. It is, however, critical that the leaving group is positioned *trans* to the C2 participating functionality to achieve super-arming. It is only in this case where a direct displacement of the leaving group by the C2 ester is possible. As shown in Scheme 4.19, a competition experiment with the super-armed donor **118** and the armed donor **119** led to the formation of disaccharide **121** in 95% yield, while only trace amounts of disaccharide **122**, arising from glycosylation by **119**, was detected [84].

Steric effect also dramatically increases anomeric reactivity by altering the sugar ring conformation [85, 86]. As depicted in Scheme 4.20a, a glucoside in the $^4C_1$ conformation hydrolyzes much slower than the corresponding $^1C_4$ glucoside [86]. An axially oriented substituent carrying a polar group is deemed less destabilizing than an equatorial group due to the orientation of the dipole. Moreover, a galactosyl donor is approximately five times more reactive than a similar glucosyl donor, which

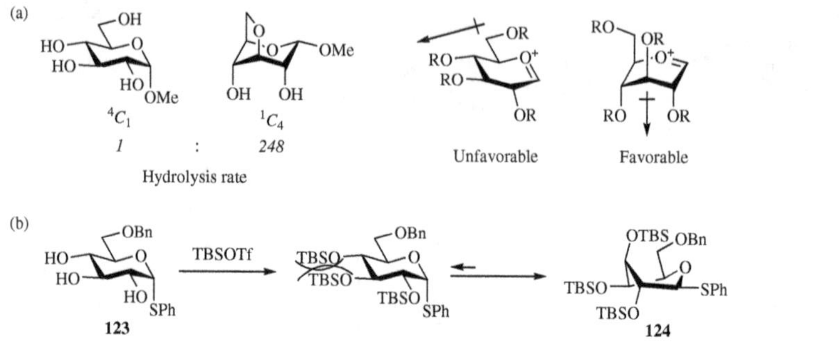

SCHEME 4.20 Super-arming by steric effect. (a) Influence of conformation on the reactivity of glycosides. (b) Conformational change induced by bulky silyl ethers.

has one more equatorial substituent [86, 87]. In the $^4C_1$ conformation, the dipoles of the O3 and O4 substituents of the glucoside are equatorial and therefore antiperiplanar to the oxocarbenium ion. For the $^1C_4$ conformation, the O3 and O4 dipoles are approximately perpendicular to the positive charge and therefore less destabilizing, which increases the reactivity. Bols and coworkers showed that bulky silyl ethers, such as *tert*-butyldimethylsilyl (TBS) ether, can change the conformation of glucosides from $^4C_1$ to $^1C_4$ by steric repulsion (Scheme 4.20b) [85]. However, in many cases, mixtures of conformers were obtained, and for glucose, a twist-boat conformation was observed. Upon activation of a per-O-silylated derivative, only a 1,6-anhydro product was obtained. To prevent this unwanted reaction, O6 was protected as a benzyl ether to give compounds such as **124**. These super-armed donors were readily activated at low temperature and were more reactive than per-O-benzylated glycosides [88]. The super-arming effect of silyl ethers is strictly due to conformational and not electronic effects.

4.4 PROTECTING GROUPS IN REGIOSELECTIVE GLYCOSYLATIONS

Regioselective glycosylations are attractive for oligosaccharide synthesis because it requires, in general, fewer steps for the preparation of glycosyl acceptors. The regioselectivity of a glycosylation is affected by many parameters. In particular, the nature of protecting groups of the glycosyl donor and acceptor influences the outcome of the reaction. Generally, primary hydroxyls exhibit higher reactivities than secondary hydroxyls. This difference in reactivity is often exploited for regioselective glycosylations. Interestingly, by employing trityl ethers, the selectivity of a primary over a secondary hydroxyl can be reversed. In this respect, glycosylations of a glycosyl acceptor protected at a primary and secondary hydroxyl with trityl ethers occur preferentially at the tritylated secondary hydroxyl [89, 90]. The trityl-based strategy has been successfully employed for a broad range of acceptors, including 2,6-, 3,6-, and 4,6-di-*O*-trityl ethers of glucosides, mannosides, galactosides, and

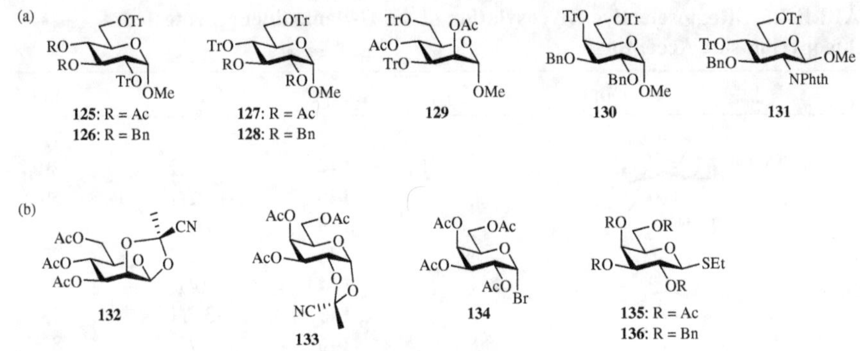

FIGURE 4.4 Examples of (a) glycosyl acceptors and (b) glycosyl donors used in trityl ether-assisted regioselective glycosylation. Tr, trityl.

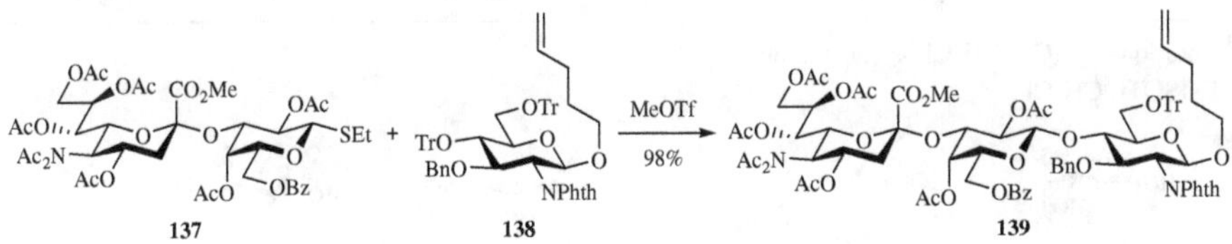

SCHEME 4.21 Trityl ether-assisted regioselective glycosylation.

glucosamines (Fig. 4.4). The selective glycosylation of a secondary trityl ether was elegantly employed for the preparation of a complex oligosaccharide derived from the capsular polysaccharide of *Streptococcus* group B type III. Using the sialic acid-containing disaccharide **137** as donor, the 4,6-di-O-tritylated acceptor **138** was selectively glycosylated at the O4 position in 98% yield (Scheme 4.21) [91].

Achieving regioselectivity between secondary hydroxy groups is challenging. Several examples are described below to highlight factors that affect such selectivities. In particular, it is shown that seemingly bystander protecting groups can affect the regioselective outcome of a glycosylation.

Andrea Vasella and coworkers demonstrated that the highly reactive per-O-benzylated diazirine donor **140** (Table 4.3, entries 1–4) exhibits low regioselectivity to four different glycosyl acceptors (**143–146**) bearing different anomeric protecting groups [92]. However, in a related work by Fanzuo Kong, the use of the per-O-benzoylated glucosyl trichloroacetimidate donor **141** resulted in modest to good regioselectivities (entries 5–8) [93]. Interestingly, the use of the methyl α-glucoside **143** as acceptor (entry 5) resulted in modest preference for glycosylation of the 3-hydroxyl, whereas the allyl α-glucoside **144** (entry 6) gave only the 3-O-glycosylated product. On the other hand, the methyl β-glucoside **145** (entry 7) was selectively glycosylated at C2, whereas no regioselectivity was observed for the allyl β-glucoside **146** (entry 8). Although it is difficult to rationalize these observations, the results clearly demonstrated that both the nature and configuration of the anomeric protecting group of the glycosyl acceptor affect the regioselective outcome of glycosylation. The nature of

TABLE 4.3 Regioselective Glycosylation of 4,6-*O*-Benzylidene-Protected Glucopyranoside Acceptors

Entry	Donor	Acceptor	O2/O3	Yield (%)
1	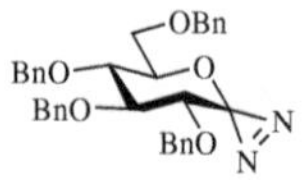 140	**143**	1/1	73[a]
2		**144**	1/1.2	89[a]
3		**145**	1.2/1	59[a]
4		**146**	1/1.1	94[a]
5	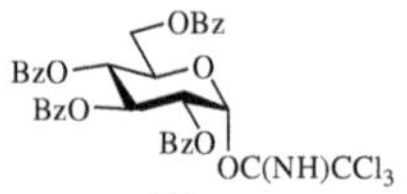 141	**143**	1/2.2	77[b]
6		**144**	0/1	75[b]
7		**145**	3.2/1	83[b]
8		**146**	1/1	82[b]
9	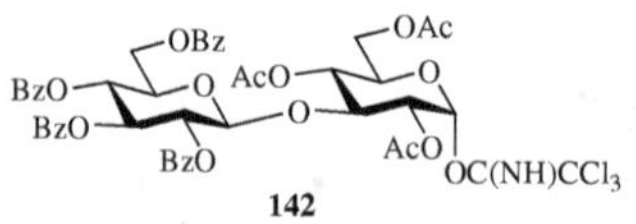 142	**143**	1/0	64[b]
10		**144**	1/0	59[b]
11		**145**	1/0	70[b]
12		**146**	1/0	59[b]

Conditions: [a] $ClCH_2CH_2Cl$ or 1,4-dioxane.
[b] TMSOTf, CH_2Cl_2.

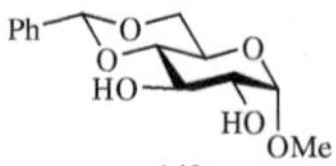
143

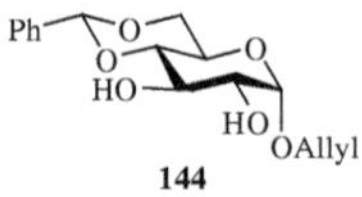
144

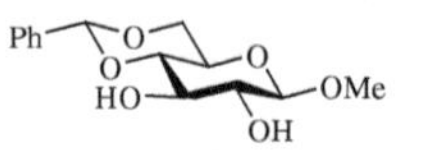
145

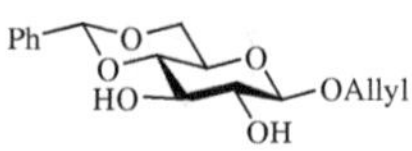
146

the glycosyl donor also influences the regioselectivity, and it appears that less reactive donors give higher selectivities. In particular, the use of the disaccharide donor **142** (entries 9–12) [93] resulted exclusively in C2 glycosylation of the four different diols. In this case, the greater steric hindrance exerted by the disaccharide donor was the likely determinant of the outcome of the glycosylations regardless of the differences in reactivity of the hydroxyls of the acceptors.

While studying different acceptors using diazirine donors, Vasella and coworkers uncovered the importance of hydrogen bonding on the regioselective outcome of glycosylations of diols [92, 94]. A hydrogen bond-accepting hydroxyl possesses an increased acidity and a lowered nucleophilicity, while a hydrogen bond-donating hydroxy group exhibits reduced acidity and increased nucleophilicity [95, 96]. For the diazirine-derived glycosylidene carbene-based donors, the increased acidity leads to an increased reaction rate, and therefore, these glycosylations favor a hydrogen bond-accepting hydroxyl. However, for glycosylations with classical glycosyl donors, the increased nucleophilicity determines the preferred site of reaction. The factors affecting the strength of intramolecular hydrogen bonding includes the O–H distances and the O–H–O angle. IR and NMR spectroscopy were used to probe possible hydrogen bonding in the acceptors [92, 94, 97]. However, a recent computational study indicated that small shifts in IR or NMR may not be sufficient to identify hydrogen bonds [98]. It is clear that additional calculations need to be performed to uncover their existence.

The hydrogen bonding model for rationalizing regioselectivities was further studied by Manuel Martín-Lomas and coworkers. Density functional theory (DFT) calculations

TABLE 4.4 Regioselective Glycosylations on *trans*-Diols[a]

Entry	Donor	Acceptor	O2/O3	Yield (%)
1	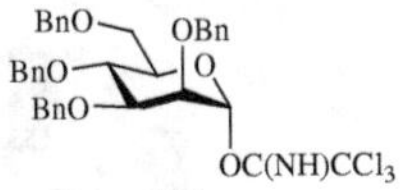 **147**	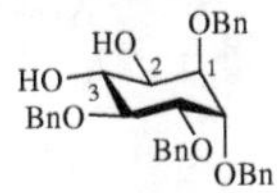 **149**	1.5/1	49[b]
2		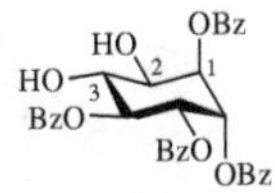 **150**	2.5/1	60
3	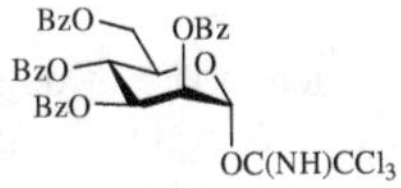 **148**	**149**	8/1	46
4		**150**	3.8/1	47

Conditions: [a]TMSOTf, Et_2O.
[b]Along with 11% β and 16% trisaccharide.

were used to examine the presence of hydrogen bonds in diol acceptors [99]. Based on experimental and theoretical methods, it was concluded that the formation of a hydrogen bond between a given hydroxyl and a neighboring protecting group could significantly increase the nucleophilicity of the hydroxyl, thereby influencing the regioselective outcome of a glycosylation. In cases where both hydroxy groups have hydrogen bond connections, the strength of the hydrogen bonds on a certain hydroxy group determines the regioselectivity trend.

trans-Diol acceptors protected by either benzyl ethers (**149**) or benzoyl esters (**150**) (Table 4.4) were employed by Martín-Lomas' group. The per-O-benzylated and per-O-benzoylated mannopyranosyl trichloroacetimidate donors **147** and **148** preferentially glycosylated O2 over the O3 positions, albeit in different ratios. This regioselectivity was rationalized by an intramolecular hydrogen bond in the acceptor. The hydrogen bond patterns for **149** and **150** are shown in Figure 4.5a. It was postulated that the hydrogen bond between the O2–H and the adjacent axial substituent at C1 is stronger than the hydrogen bond between the O3–H and its adjacent equatorial C4 substituent. The stronger hydrogen bonds lead to weaker O2–H bond and, consequently, result in an increase in nucleophilicity at O2. A DFT study involving simplified model compounds supported both the existence and differences in strength of the hydrogen bonds. Based on the comparison between the per-O-benzylated and per-O-benzoylated mannopyranosyl trichloroacetamide donors, it was concluded that higher regioselectivities are achieved by using less reactive donors.

The strength of hydrogen bonds to predict the regioselectivity of diols works well for acceptors having 1,2-*cis*-diaxial, 1,2-*cis*-diequatorial, and 1,2-*trans*-diaxial diols (Fig. 4.5). However, more studies are needed to explain regioselectivities observed for 1,2-*trans*-diequatorial diols as in 4,6-*O*-benzylidene-protected glucopyranosides.

Fraser-Reid and coworkers studied the influence of different O2 protecting groups of glycosyl donors on the regioselective outcome of glycosylations (Table 4.5). Both

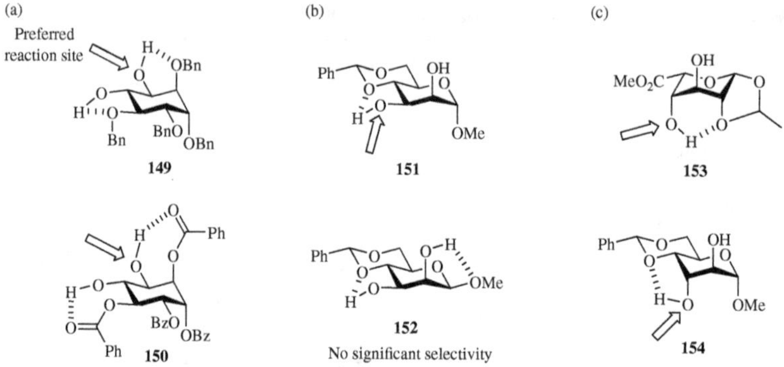

FIGURE 4.5 Martín-Lomas' rationale for the preferred reaction site of diols. (a) 1,2-*cis*-Diequatorial diols with Bn or Bz protecting groups. (b) 1,2-*cis*-Axial-equatorial diols in D-mannosides. (c) 1,2-*trans*-Diaxial diols in L-iduronic acid and D-altrose.

TABLE 4.5 Effect of O2 Protecting Groups on Regioselective Glycosylation[a]

Entry	Acceptor	Donor	O2/O3	Yield (%)
1	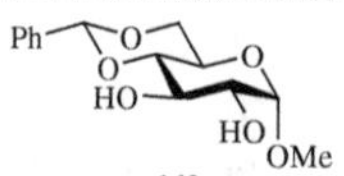 143	**113β**	1.7/1	53
2		**155**	3.0/1	42
3		**156**	1/3.0	36
4		**157**	2.4/1	62
5	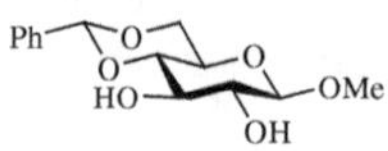 145	**113β**	1.2/1	51
6		**155**	1.8/1	11
7		**156**	1/2.9	35
8		**157**	2.3/1	46

Conditions: [a]NIS, TBSOTf, CH_2Cl_2.

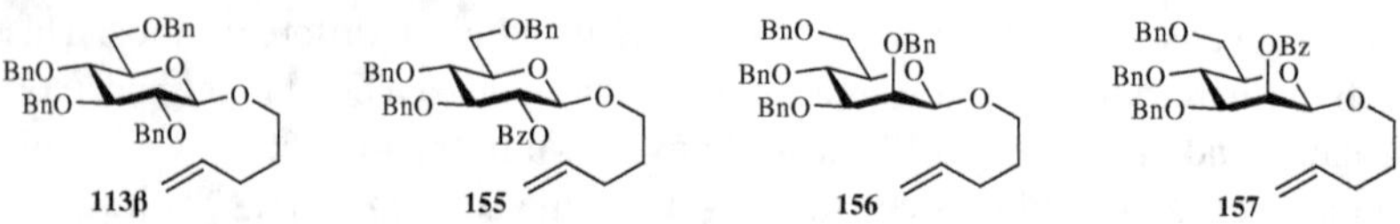

2-O-benzylated and 2-O-benzoylated *n*-pentenyl donors were used for glycosylations of the 4,6-*O*-benzylidene-protected glucopyranosyl acceptors **143** and **145** [100]. In contrast to the results reported by Kong and coworkers, the anomeric configuration of the acceptor did not exert much influence on the regioselective outcome of the glycosylations. The glucopyranosyl donors **113β** and **155** both preferred glycosylation at O2 over O3 (entries 1, 2, 5, and 6). The regioselectivity ratios were slightly higher when a 2-O-benzoylated donor was used. However, when mannopyranosyl donors were employed, the glycosylations gave opposite regioselectivities between the 2-O-benzylated and 2-O-benzoylated donor. The 2-O-benzylated donor **156** preferred 3-O-glycosylation (entries 3 and 7), while the 2-O-benzoylated donor **157** preferred 2-O-glycosylation (entries 4 and 8).

TABLE 4.6 Donor-Dependent Regioselective Glycosylations[a]

Entry	Donor	Acceptor	Regioselectivity	Yield (%)
1	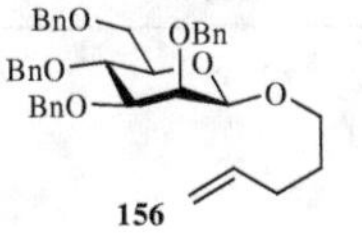 **156**	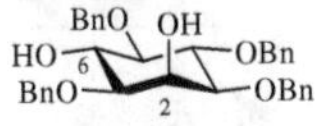 **159**	O2/O6=0/1	73
2		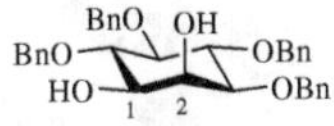 **160**	O1/O2=2/1	71
3	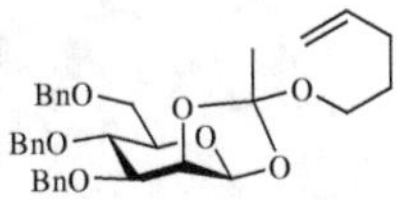 **158**	**159**	O2/O6=3.3/1	64
4		**160**	O1/O2=6/1	73

Conditions: [a]NIS, TBSOTf, CH_2Cl_2.

Based on the observation that the O2 protecting group of the donor can influence the outcome of the regioselective glycosylations, Fraser-Reid extended Paulsen's "match–mismatch" concept to the regioselectivity of glycosylations [101]. The concept was termed "reciprocal donor–acceptor selectivity" (RDAS) [102, 103] and underscores the unique preference exhibited by each hydroxy group of a diol for different donors. The reactivity of a given hydroxyl has to match the reactivity of the donor to achieve good regioselectivity. Table 4.6 shows examples demonstrating this principle [102, 104]. For the inositol-based 2,6-diol acceptor **159**, the 4-pentenyl mannopyranosyl donor **156** glycosylated exclusively at the O6 equatorial hydroxyl (entry 1), while the 4-pentenyl orthoester donor **158** glycosylated preferentially at the O2 axial hydroxyl in a 3.3/1 ratio (entry 3). The inositol-based 1,2-diol acceptor **160** did not exhibit such a prominent change in regioselectivity. In this case, donor **156** gave a 2/1 ratio (entry 2), while donor **158** favored the O1 position in a 6/1 ratio (entry 4). RDAS shows that the reactivity difference in the acceptor is only one factor for the regioselective outcome of a glycosylation; the donor also plays a key role in determining the regioselectivity.

The regioselectivity of glycosylations of the 3,4-diol of various N-protected glucosamine was studied by several groups owing to its importance in the synthesis of biologically important compounds such as SLe^x- and Le^x-containing oligosaccharides (Table 4.7). It was found that the 2-amino protecting group plays a crucial role. When azide was used as an amino masking group, the preferred glycosylation site is O3 as observed for the case of the per-O-acetylated galactopyranosyl trichloroacetimidate donor **161** (entry 1) [105]. When the amino protecting group is changed, the regioselectivity changes as well. For example, the glycosylation of the acetamido-containing acceptor **163** with **161** gave complete O4 regioselectivity (entry 2) [106]. The configuration of the anomeric protecting group also plays a crucial role. For example, the Troc-protected acceptor **164** having an anomeric allyl group in the α configuration led to the exclusive glycosylation at O3 (entry 3) in contrast to the result with the β anomer **163** [107].

TABLE 4.7 Regioselective Glycosylations of Glucosamine Acceptors Having Different N-Protecting Groups

Entry	Donor	Acceptor	O3/O4	Yield (%)
1	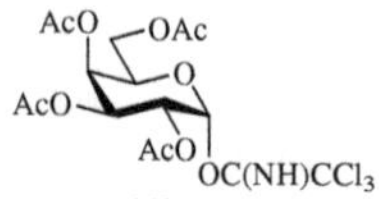 **161**	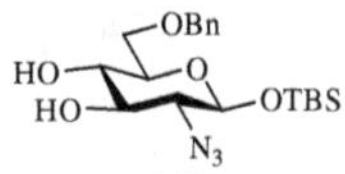 **162**	3/1	86[a]
2	**161**	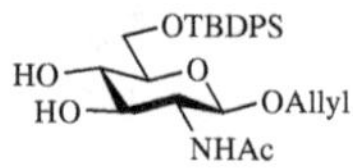 **163**	0/1	68[b]
3	**161**	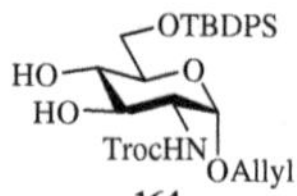 **164**	1/0	59[c]
4	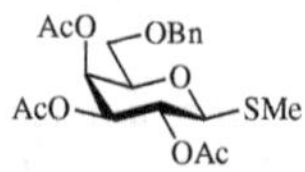 **165**	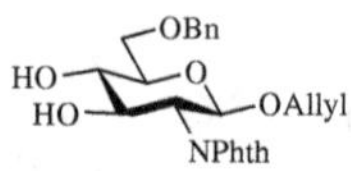 **166**	1/3.6	77[d]
5	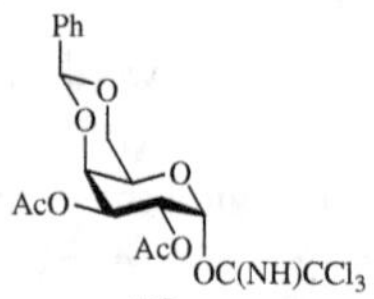 **167**	**166**	0/1	81[e]
6	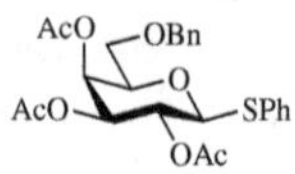 **168**	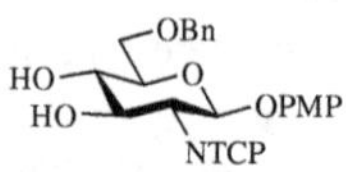 **169**	1/84	85[f]
7	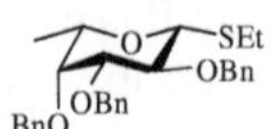 **170**	**169**	3.6/1	84[g]
8	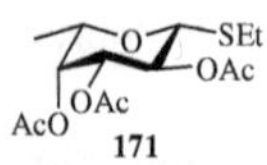 **171**	**169**	0/1	58[f]

PMP, *p*-methoxyphenyl.
Conditions: [a] $BF_3 \cdot Et_2O$, CH_2Cl_2/*n*-hexane, −25°C.
[b] $BF_3 \cdot Et_2O$, CH_2Cl_2/*n*-hexane, −45°C.
[c] $HClO_4$/silica, $ClCH_2CH_2Cl$, −10°C.
[d] MeOTf, $PhCH_3$, −15°C.
[e] TMSOTf, CH_2Cl_2, −25°C.
[f] AgOTf, CH_3CN, −78°C; then MeSBr, $ClCH_2CH_2Cl$, −70°C.
[g] AgOTf, CH_3CN, −45°C; then MeSBr, $ClCH_2CH_2Cl$, −70°C.

Glycosyl acceptors having an *N*-phthaloyl-type protecting group and a β-anomeric substituent (**166** and **169**) gave good selectivity toward glycosylation at O4 when galactopyranosyl donors (**165**, **167**, and **168**) are employed (entries 4 [108], 5 [109], and 6 [110]). For the synthesis of Le^x antigens, it was reported that the NTCP group displayed greater influence on selectivity than the NPhth group [110]. Furthermore, a dramatic change in regioselectivity occurred when a fucosyl instead of a galactosyl donor was used for coupling with a TCP-protected acceptor (entries 7 and 8) [110]. This is an example of Fraser-Reid's RDAS concept and emphasizes the importance of matching the donors and acceptors.

4.5 CONCLUSION

The examples described here show that protecting groups of glycosyl donors play key roles in the stereoselective outcome of glycosylations. Furthermore, glycosyl donor reactivity is controlled by the nature and site of protecting groups and differences in reactivity can be exploited for chemoselective glycosylations. Protecting groups of glycosyl donors and acceptors also play critical roles in the regioselective outcome of glycosylations. Dramatic changes in regioselectivity when seemingly small changes are made in the protecting group pattern highlight the importance of the "match–mismatch" concept. Hydrogen bonds between hydroxyls and neighboring protecting groups provide an attractive explanation for observed regioselectivities; however, care must be taken not to overuse the rules as only a limited number of examples have been reported. The frequently cited statement made by Paulsen in 1982 [75] still holds true: "Although we have now learned to synthesize oligosaccharides, it should be emphasized that each oligosaccharide synthesis remains an independent problem, whose resolution requires considerable systematic research and a good deal of know-how. There are no universal reaction conditions for oligosaccharide syntheses."

REFERENCES

[1] Capon, B.; McManus, S. P. *Neighboring Group Participation*; Plenum: New York, **1976**.

[2] Crich, D.; Dai, Z. M.; Gastaldi, S. *J. Org. Chem.* **1999**, *64*, 5224–5229.

[3] Paulsen, H.; Herold, C.-P. *Chem. Ber.* **1970**, *103*, 2450–2462.

[4] Yu, H.; Williams, D. L.; Ensley, H. E. *Tetrahedron Lett.* **2005**, *46*, 3417–3421.

[5] Hammond, G. S. In *Steric Effects in Organic Chemistry*; Newman, M. S., Ed.; Wiley: New York, **1956**, pp. 425–470.

[6] Allinger, N. L.; Zalkow, V. *J. Org. Chem.* **1960**, *25*, 701–704.

[7] Jager, J.; Graafland, T.; Schenk, H.; Kirby, A. J.; Engberts, J. B. F. N. *J. Am. Chem. Soc.* **1984**, *106*, 139–143.

[8] Demchenko, A. V.; Rousson, E.; Boons, G. J. *Tetrahedron Lett.* **1999**, *40*, 6523–6526.

[9] Mukaiyama, T.; Suenaga, M.; Chiba, H.; Jona, H. *Chem. Lett.* **2002**, *31*, 56–57.

[10] Cheng, Y. P.; Chen, H. T.; Lin, C. C. *Tetrahedron Lett.* **2002**, *43*, 7721–7723.

[11] De Meo, C.; Kamat, M. N.; Demchenko, A. V. *Eur. J. Org. Chem.* **2005**, 706–711.

[12] van Boeckel, C. A. A.; Beetz, T.; van Aelst, S. F. *Tetrahedron* **1984**, *40*, 4097–4107.

[13] Smid, P.; Deruiter, G. A.; van der Marel, G. A.; Rombouts, F. M.; van Boom, J. H. *J. Carbohydr. Chem.* **1991**, *10*, 833–849.

[14] Ustyuzhanina, N.; Komarova, B.; Zlotina, N.; Krylov, V.; Gerbst, A.; Tsvetkov, Y.; Nifantiev, N. *Synlett* **2006**, 921–923.

[15] Crich, D.; Hu, T. S.; Cai, F. *J. Org. Chem.* **2008**, *73*, 8942–8953.

[16] Baek, J. Y.; Lee, B. Y.; Jo, M. G.; Kim, K. S. *J. Am. Chem. Soc.* **2009**, *131*, 17705–17713.

[17] Zhang, Z.; Ollmann, I. R.; Ye, X.-S.; Wischnat, R.; Baasov, T.; Wong, C.-H. *J. Am. Chem. Soc.* **1999**, *121*, 734–753.

[18] Jiao, H. L.; Hindsgaul, O. *Angew. Chem. Int. Ed.* **1999**, *38*, 346–348.

[19] Crich, D.; Dudkin, V. *J. Am. Chem. Soc.* **2001**, *123*, 6819–6825.

[20] Smoot, J. T.; Pornsuriyasak, P.; Demchenko, A. V. *Angew. Chem. Int. Ed.* **2005**, *44*, 7123–7126.

[21] Hou, D. J.; Lowary, T. L. *Carbohydr. Res.* **2009**, *344*, 1911–1940.

[22] Seeberger, P. H.; Bilodeau, M. T.; Danishefsky, S. J. *Aldrichim. Acta* **1997**, *30*, 75–92.

[23] Danishefsky, S. J.; Behar, V.; Randolph, J. T.; Lloyd, K. O. *J. Am. Chem. Soc.* **1995**, *117*, 5701–5711.

[24] Griffith, D. A.; Danishefsky, S. J. *J. Am. Chem. Soc.* **1990**, *112*, 5811–5819.

[25] Yu, B.; Yang, Z. *Org. Lett.* **2001**, *3*, 377–379.

[26] Yu, B.; Wang, P. *Org. Lett.* **2002**, *4*, 1919–1922.

[27] Hou, D. J.; Taha, H. A.; Lowary, T. L. *J. Am. Chem. Soc.* **2009**, *131*, 12937–12948.

[28] Kim, J. H.; Yang, H.; Boons, G.-J. *Angew. Chem. Int. Ed.* **2005**, *44*, 947–949.

[29] Kim, J. H.; Yang, H.; Khot, V.; Whitfield, D.; Boons, G.-J. *Eur. J. Org. Chem.* **2006**, 5007–5028.

[30] Kim, J. H.; Yang, H.; Park, J.; Boons, G.-J. *J. Am. Chem. Soc.* **2005**, *127*, 12090–12097.

[31] Fascione, M. A.; Turnbull, W. B. *Beilstein J. Org. Chem.* **2010**, *6*, 19.

[32] Fascione, M. A.; Adshead, S. J.; Stalford, S. A.; Kilner, C. A.; Leach, A. G.; Turnbull, W. B. *Chem. Commun.* **2009**, 5841–5843.

[33] Jung, K. H.; Muller, M.; Schmidt, R. R. *Chem. Rev.* **2000**, *100*, 4423–4442.

[34] Cumpstey, I. *Carbohydr. Res.* **2008**, *343*, 1553–1573.

[35] Lee, Y. J.; Ishiwata, A.; Ito, Y. *J. Am. Chem. Soc.* **2008**, *130*, 6330–6331.

[36] Leigh, C. D.; Bertozzi, C. R. *J. Org. Chem.* **2008**, *73*, 1008–1017.

[37] Crich, D.; Sun, S. X. *J. Am. Chem. Soc.* **1997**, *119*, 11217–11223.

[38] Crich, D.; Chandrasekera, N. S. *Angew. Chem. Int. Ed.* **2004**, *43*, 5386–5389.

[39] Crich, D. *Acc. Chem. Res.* **2010**, *43*, 1144–1153.

[40] Huang, M.; Garrett, G. E.; Birlirakis, N.; Bohé, L.; Pratt, D. A.; Crich, D. *Nat. Chem.* **2012**, *4*, 663–667.

[41] Fraser-Reid, B.; Wu, Z. F.; Andrews, C. W.; Skowronski, E.; Bowen, J. P. *J. Am. Chem. Soc.* **1991**, *113*, 1434–1435.

[42] Jensen, H. H.; Nordstrom, L. U.; Bols, M. *J. Am. Chem. Soc.* **2004**, *126*, 9205–9213.

[43] Crich, D.; Banerjee, A. *J. Am. Chem. Soc.* **2006**, *128*, 8078–8086.

[44] Crich, D.; Bowers, A. A. *Org. Lett.* **2006**, *8*, 4327–4330.

[45] Crich, D.; Bowers, A. A. *J. Org. Chem.* **2006**, *71*, 3452–3463.

[46] Utimoto, K.; Wakabayashi, Y.; Horiie, T.; Inoue, M.; Shishiyama, Y.; Obayashi, M.; Nozaki, H. *Tetrahedron* **1983**, *39*, 967–973.

[47] Beckwith, A. L. J.; Easton, C. J. *J. Am. Chem. Soc.* **1981**, *103*, 615–619.

[48] Rychnovsky, S. D.; Swenson, S. S. *Tetrahedron* **1997**, *53*, 16489–16502.

[49] Srivastava, V. K.; Schuerch, C. *Carbohydr. Res.* **1980**, *79*, C13–C16.

[50] Srivastava, V. K.; Schuerch, C. *J. Org. Chem.* **1981**, *46*, 1121–1126.

[51] Awad, L. F.; Elashry, E. S. H.; Schuerch, C. *Bull. Chem. Soc. Jpn.* **1986**, *59*, 1587–1592.

[52] Abdel-Rahman, A. A. H.; Jonke, S.; El Ashry, E. S. H.; Schmidt, R. R. *Angew. Chem. Int. Ed.* **2002**, *41*, 2972–2974.

[53] Crich, D.; Hutton, T. K.; Banerjee, A.; Jayalath, P.; Picione, J. *Tetrahedron: Asymmetry* **2005**, *16*, 105–119.

[54] van den Bos, L. J.; Codee, J. D. C.; Litjens, R. E. J. N.; Dinkelaar, J.; Overkleeft, H. S.; van der Marel, G. A. *Eur. J. Org. Chem.* **2007**, 3963–3976.

[55] van den Bos, L. J.; Codee, J. D. C.; van der Toorn, J. C.; Boltje, T. J.; van Boom, J. H.; Overkleeft, H. S.; van der Marel, G. A. *Org. Lett.* **2004**, *6*, 2165–2168.

[56] van den Bos, L. J.; Dinkelaar, J.; Overkleeft, H. S.; van der Marel, G. A. *J. Am. Chem. Soc.* **2006**, *128*, 13066–13067.

[57] Dinkelaar, J.; van den Bos, L. J.; Hogendorf, W. F. J.; Lodder, G.; Overkleeft, H. S.; Codee, J. D. C.; van der Marel, G. A. *Chem. Eur. J.* **2008**, *14*, 9400–9411.

[58] Walvoort, M. T. C.; Dinkelaar, J.; van den Bos, L. J.; Lodder, G.; Overkleeft, H. S.; Codee, J. D. C.; van der Marel, G. A. *Carbohydr. Res.* **2010**, *345*, 1252–1263.

[59] Codee, J. D. C.; van den Bos, L. J.; de Jong, A. R.; Dinkelaar, J.; Lodder, G.; Overkleeft, H. S.; van der Marel, G. A. *J. Org. Chem.* **2009**, *74*, 38–47.

[60] Dinkelaar, J.; de Jong, A. R.; van Meer, R.; Somers, M.; Lodder, G.; Overkleeft, H. S.; Codee, J. D. C.; van der Marel, G. A. *J. Org. Chem.* **2009**, *74*, 4982–4991.

[61] Walvoort, M. T. C.; Lodder, G.; Mazurek, J.; Overkleeft, H. S.; Codée, J. D. C.; van der Marel, G. A. *J. Am. Chem. Soc.* **2009**, *131*, 12080–12081.

[62] Zhu, X. M.; Kawatkar, S.; Rao, Y.; Boons, G. J. *J. Am. Chem. Soc.* **2006**, *128*, 11948–11957.

[63] Joe, M.; Bai, Y.; Nacario, R. C.; Lowary, T. L. *J. Am. Chem. Soc.* **2007**, *129*, 9885–9901.

[64] Crich, D.; Pedersen, C. M.; Bowers, A. A.; Wink, D. J. *J. Org. Chem.* **2007**, *72*, 1553–1565.

[65] Imamura, A.; Ando, H.; Korogi, S.; Tanabe, G.; Muraoka, O.; Ishida, H.; Kiso, M. *Tetrahedron Lett.* **2003**, *44*, 6725–6728.

[66] Olsson, J. D.; Eriksson, L.; Lahmann, M.; Oscarson, S. *J. Org. Chem.* **2008**, *73*, 7181–7188.

[67] Benakli, K.; Zha, C. X.; Kerns, R. J. *J. Am. Chem. Soc.* **2001**, *123*, 9461–9462.

[68] Manabe, S.; Ishii, K.; Ito, Y. *J. Am. Chem. Soc.* **2006**, *128*, 10666–10667.

[69] Manabe, S.; Ishii, K.; Hashizume, D.; Koshino, H.; Ito, Y. *Chem. Eur. J.* **2009**, *15*, 6894–6901.

[70] Satoh, H.; Manabe, S.; Ito, Y.; Lüthi, H. P.; Laino, T.; Hutter, J. *J. Am. Chem. Soc.* **2011**, *133*, 5610–5619.

[71] Boons, G.-J. *Tetrahedron* **1996**, *52*, 1095–1121.

[72] Codée, J. D. C.; Litjens, R.; van den Bos, L. J.; Overkleeft, H. S.; van der Marel, G. A. *Chem. Soc. Rev.* **2005**, *34*, 769–782.

[73] Premathilake, H. D.; Demchenko, A. V. In *Reactivity Tuning in Oligosaccharide Assembly*; Fraser-Reid, B., Lopez, J. C., Eds.; Topics in Current Chemistry *301*; Springer: Berlin, **2011**, pp. 189–221.

[74] Koeller, K. M.; Wong, C.-H. *Chem. Rev.* **2000**, *100*, 4465–4494.

[75] Paulsen, H. *Angew. Chem. Int. Ed. Engl.* **1982**, *21*, 155–173.

[76] Fraserreid, B.; Udodong, U. E.; Wu, Z. F.; Ottosson, H.; Merritt, J. R.; Rao, C. S.; Roberts, C.; Madsen, R. *Synlett* **1992**, 927–942.

[77] Mootoo, D. R.; Konradsson, P.; Udodong, U.; Fraser-Reid, B. *J. Am. Chem. Soc.* **1988**, *110*, 5583–5584.

[78] Fraser-Reid, B.; Wu, Z.; Udodong, U. E.; Ottosson, H. *J. Org. Chem.* **1990**, *55*, 6068–6070.

[79] Ratcliffe, A. J.; Konradsson, P.; Fraser-Reid, B. *J. Am. Chem. Soc.* **1990**, *112*, 5665–5667.

[80] Douglas, N. L.; Ley, S. V.; Lucking, U.; Warriner, S. L. *J. Chem. Soc. Perkin Trans. 1* **1998**, 51–65.

[81] Ley, S. V.; Priepke, H. W. M.; Warrine, S. L. *Angew. Chem. Int. Ed. Engl.* **1994**, *33*, 2290–2292.

[82] Grice, P.; Ley, S. V.; Pietruszka, J.; Osborn, H. M. I.; Priepke, H. W. M.; Warriner, S. L. *Chem. Eur. J.* **1997**, *3*, 431–440.

[83] Premathilake, H. D.; Mydock, L. K.; Demchenko, A. V. *J. Org. Chem.* **2010**, *75*, 1095–1100.

[84] Mydock, L. K.; Demchenko, A. V. *Org. Lett.* **2008**, *10*, 2107–2110.

[85] Pedersen, C. M.; Nordstrøm, L. U.; Bols, M. *J. Am. Chem. Soc.* **2007**, *129*, 9222–9235.

[86] McDonnell, C.; López, O.; Murphy, P.; Fernández Bolaños, J. G.; Hazell, R.; Bols, M. *J. Am. Chem. Soc.* **2004**, *126*, 12374–12385.

[87] Wang, Y.; Ye, X.-S.; Zhang, L.-H. *Org. Biomol. Chem.* **2007**, *5*, 2189–2200.

[88] Jensen, H. H.; Pedersen, C. M.; Bols, M. *Chem. Eur. J.* **2007**, *13*, 7576–7582.

[89] Tsvetkov, Y. E.; Kitov, P. I.; Backinowsky, L. V.; Kochetkov, N. K. *Tetrahedron Lett.* **1993**, *34*, 7977–7980.

[90] Tsvetkov, Y. E.; Kitov, P. I.; Backinowsky, L. V.; Kochetkov, N. K. *J. Carbohydr. Chem.* **1996**, *15*, 1027–1050.

[91] Demchenko, A. V.; Boons, G.-J. *Tetrahedron Lett.* **1998**, *39*, 3065–3068.

[92] Muddasani, P. R.; Bozó, E.; Bernet, B.; Vasella, A. *Helv. Chim. Acta* **1994**, *77*, 257–290.

[93] Zeng, Y.; Kong, F. *Carbohydr. Res.* **2003**, *338*, 843–849.

[94] Muddasani, P. R.; Bernet, B.; Vasella, A. *Helv. Chim. Acta* **1994**, *77*, 334–350.

[95] Uhlmann, P.; Vasella, A. *Helv. Chim. Acta* **1992**, *75*, 1979–1994.

[96] Bozó, E.; Vasella, A. *Helv. Chim. Acta* **1994**, *77*, 745–753.

[97] Bernet, B.; Vasella, A. *Helv. Chim. Acta* **2000**, *83*, 995–1021.

[98] Klein, R. A. *J. Comput. Chem.* **2003**, *24*, 1120–1131.

[99] Belen Cid, M.; Alfonso, F.; Alonso, I.; Martín-Lomas, M. *Org. Biomol. Chem.* **2009**, *7*, 1471–1481.

[100] Fraser-Reid, B.; Lopez, J. C.; Radhakrishnan, K. V.; Mach, M.; Schlueter, U.; Gomez, A. M.; Uriel, C. *J. Am. Chem. Soc.* **2002**, *124*, 3198–3199.

[101] Fraser-Reid, B.; Lopez, J. C.; Radhakrishnan, K. V.; Nandakumar, M. V.; Gomez, A. M.; Uriel, C. *Chem. Commun.* **2002**, 2104–2105.

[102] Fraser-Reid, B.; López, J. C.; Gómez, A. M.; Uriel, C. *Eur. J. Org. Chem.* **2004**, *2004*, 1387–1395.

[103] Fraser-Reid, B.; Lopez, J. C.; Radhakrishnan, K. V.; Mach, M.; Schlueter, U.; Gomez, A.; Uriel, C. *Can. J. Chem.* **2002**, *80*, 1075–1087.

[104] Uriel, C.; Gómez, A. M.; López, J. C.; Fraser-Reid, B. *Eur. J. Org. Chem.* **2009**, *2009*, 403–411.

[105] Toepfer, A.; Schmidt, R. R. *J. Carbohydr. Chem.* **1993**, *12*, 809–822.

[106] Gan, Z.; Cao, S.; Wu, Q.; Roy, R. *J. Carbohydr. Chem.* **1999**, *18*, 755–773.

[107] Mukhopadhyay, B.; Maurer, S. V.; Rudolph, N.; van Well, R. M.; Russell, D. A.; Field, R. A. *J. Org. Chem.* **2005**, *70*, 9059–9062.

[108] Numomura, S.; Iida, M.; Numata, M.; Sugimoto, M.; Ogawa, T. *Carbohydr. Res.* **1994**, *263*, C1–C6.

[109] Figueroa-Pérez, S.; Verez-Bencomo, V. *Tetrahedron Lett.* **1998**, *39*, 9143–9146.

[110] Ellervik, U.; Magnusson, G. *J. Org. Chem.* **1998**, *63*, 9314–9322.

5

STEREOCONTROLLED SYNTHESIS OF SIALOSIDES

Chandrasekhar Navuluri and David Crich
Department of Chemistry, Wayne State University, Detroit, MI, USA

5.1 INTRODUCTION

Sialic acids (Fig. 5.1) are the most structurally diverse group among the vast family of carbohydrates. They are involved in a wide array of biological phenomena whose complexities are rapidly being unraveled [1], leading to their incorporation as key elements in the design of glycosidase inhibitors and of vaccines. Of the 50 known sialic acids, *N*-acetylneuraminic acid (Neu5Ac, **1**) is the most widely occurring, is present at the outermost positions of cells in vertebrates and invertebrates, and is the most well studied of the series [2–4]. The second most important sialic acid derivative, *N*-glycolylneuraminic acid (Neu5Gc, **2**), is far less abundant, is known to occur in minute quantities in humans, and has been found abnormally expressed on human cancer cells. The other prominent member of this family is the 5-deamino sialic acid (3-deoxy-D-*glycero*-D-*galacto*-non-2-ulosonic acid, KDN, **3**) [5], which occurs in various human tissues albeit in far lower concentration than neuraminic acid. All types of glycosidic linkage known to neuraminic acid are also found in the KDN series [6]. The important posttranslational modifications on the sialic acid skeleton include acetylation, sulfonation, methylation, and phosphorylation [7]. The wide range of biological functions attributed to the sialic acid glycosides coupled with the difficulties encountered in the isolation of all but minute quantities of homogeneous glycoforms necessitates the development of synthetic methods for sialic acid glycosides, some of which are described in this chapter.

Glycochemical Synthesis: Strategies and Applications, First Edition.
Edited by Shang-Cheng Hung and Medel Manuel L. Zulueta.

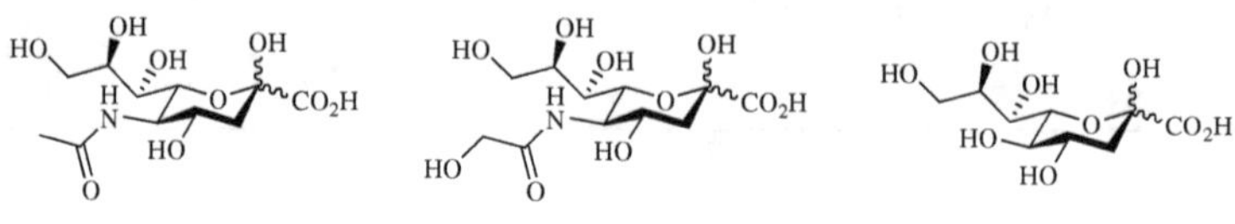

FIGURE 5.1 Commonly occurring sialic acids.

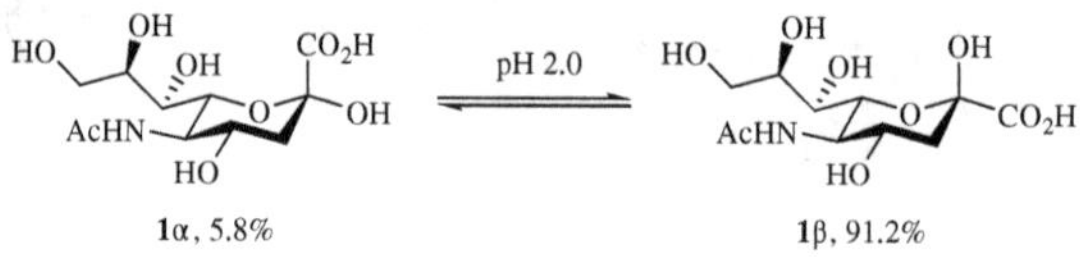

SCHEME 5.1 The anomeric equilibrium in Neu5Ac. Ac, acetyl.

The efficient stereocontrolled synthesis of the sialic acid glycosides, particularly the equatorial or α anomers, has traditionally been viewed as one of the more formidable challenges in the glycochemistry arena. It is also one of the more important challenges in view of the widespread occurrence of the sialic acid glycosides in biologically relevant oligosaccharides. Accordingly, the subject has been widely investigated and has been reviewed numerous times [3, 8–15].

Under equilibrating conditions, Neu5Ac exists as a >90/10 β/α mixture in water over a wide range of pH [16, 17] (Scheme 5.1). Methyl sialosides display a similarly high preference for the axial site [18, 19]. Therefore, kinetic conditions are mandated for the preparation of α-sialosides.

Numerous factors combine to render the α-selective formation of sialyl glycosides difficult, the most obvious being the fully substituted nature of the anomeric carbon and the steric consequences pertaining to it, the electron-withdrawing nature of the carboxylate ester group, and the unsubstituted nature of C3. The latter factor precludes neighboring group participation and results in a strong tendency toward the formation of glycal by-products. In view of these problems, most early solutions centered on the use of preintroduced artificial participating groups at the C3 position. Nonetheless, such methods have largely been abandoned because of the extra steps they engender for the introduction and eventual removal of the auxiliary [3, 8–14].

5.2 CONFORMATIONAL ANALYSIS OF SIALYL OXOCARBENIUM IONS

If one subscribes to the oxocarbenium-centric theory of glycosidic bond formation [20, 21], it is apparent from Scheme 5.2 that the grounds for stereoselective bond formation are slim. This is because neither of the two possible interconverting half-chair conformers of the oxocarbenium ion ($^4H_5$ and $^5H_4$) appear to exhibit any overwhelming steric preference for one face of the system over the other [22].

An analysis of related boat conformations for the oxocarbenium ion leads to a similar conclusion. It is clear that other factors have to be introduced into the equation if stereoselectivity is to be achieved.

5.3 ADDITIVES IN SIALYLATIONS

Since its introduction to the sialic acid field by Akira Hasegawa and coworkers at the end of the 1980s, acetonitrile and acetonitrile-based mixtures as the glycosylation solvent are frequently used to improve α selectivities [23, 24]. The effect on stereoselectivity is attributed to participation by the solvent and the generation of the β-glycosyl nitrilium ions (Fig. 5.2) that are subsequently displaced by the incoming acceptor alcohol. More recently, evidence has been presented for the formation of other adducts with the sialyl oxocarbenium ion such as those with diaryl sulfoxides that can lead to enhanced stereoselectivity in solvents such as CH_2Cl_2 [25]. The use of alternative participating solvents has been explored by Cristina De Meo and coworkers [26].

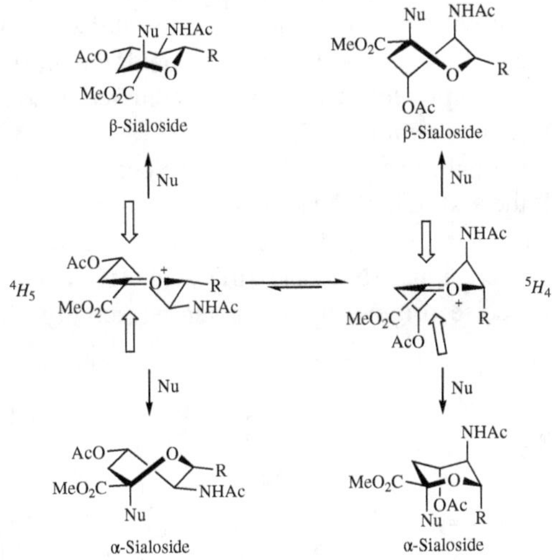

SCHEME 5.2 Stereoelectronic considerations for the reaction of interconverting conformers of a sialyl oxocarbenium ion.

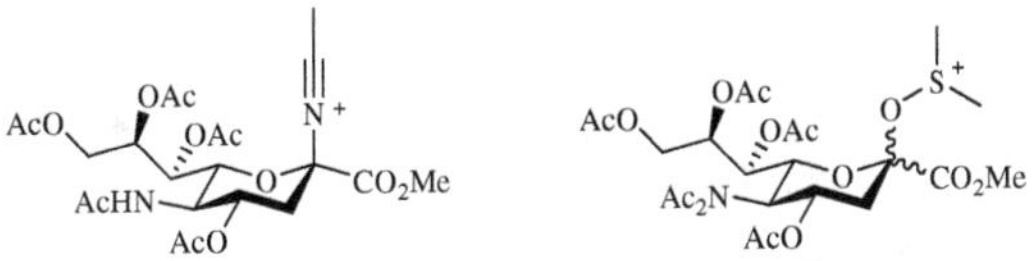

4 5

FIGURE 5.2 Nitrile and sulfoxide adducts.

5.4 LEAVING GROUPS IN SIALYLATIONS

Almost all types of leaving groups have been explored in sialyl donors with, until recently, two notable exceptions: the venerable imidates and the sulfoxides. While the preparation of stable, isolable sialyl trichloroacetimidates has still to be demonstrated, the recent innovation of the Biao Yu group—the *N*-phenyltrifluoroacetimidates—has proven equal to the task and has provided access to new potent class of sialyl donors [27]. Stable sulfoxides of sialyl thioglycosides have also been described recently and employed as glycosyl donors on activation with triflic anhydride (Tf_2O) [28]. The *S*-benzoxazolyl (*S*-Box) thiosialosides constitute another recent addition to the armory of sialyl donors [29, 30].

5.5 INFLUENCE OF THE N5 PROTECTING GROUP ON REACTIVITY AND SELECTIVITY

The most important advances in stereoselective sialylation in the last 15 years, beginning with the work of Geert-Jan Boons and Alexei Demchenko in 1998 [31], have centered around the discovery that modification of the N5 protecting group influences both the reactivity of sialyl donors and the stereoselectivity of their coupling reactions. In view of their importance to the field, the most important features of modifications of the N5 protecting groups are discussed in some detail here even if they have been reviewed previously [32, 33].

In their initial report on the use of the 5-*N*-acetyl Neu5Ac donors, Demchenko and Boons observed that the addition of the second acetyl group to the standard donor **6**, resulting in the imide **7**, increased the reactivity of the donor considerably on activation with *N*-iodosuccinimide (NIS) and triflic acid (TfOH) at –40°C in acetonitrile (Scheme 5.3). The precise origins of this increased activity were not identified,

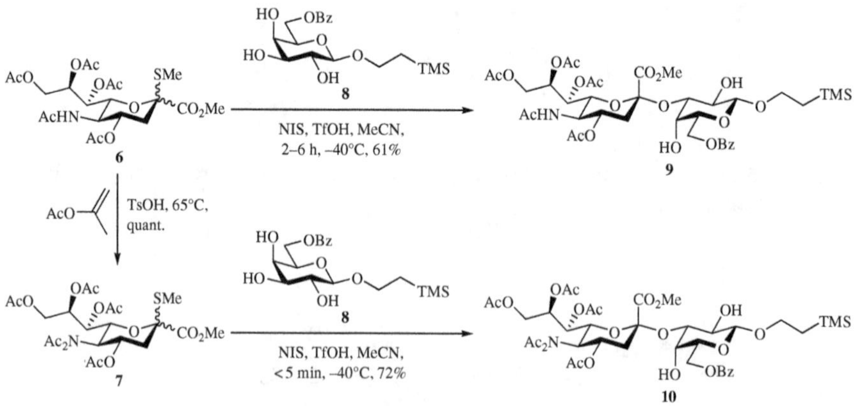

SCHEME 5.3 Effect of *N,N*-diacetylation of a thioglycoside donor on sialylation. Bz, benzoyl; TMS, trimethylsilyl; Ts, toluenesulfonyl.

although it was suggested that they are related to the reduced nucleophilicity of the amide group [31]. Subsequent work from France-Isabelle Auzanneau's laboratory [34, 35], demonstrating the capture of activated glycosyl donors by the *N*-acetylglucosamine derivatives in the form of glycosyl imidates, tends to support this hypothesis, but an alternative explanation has been advanced by Leonid Kononov and coworkers based on the disruption of intermolecular hydrogen bonding networks (see Section 5.9) [36].

Demchenko and Boons later revealed the beneficial effect of the *N,N*-diacetyl moiety on the reactivity of sialic acid acceptors, particularly the 8-alcohols (Scheme 5.4) [37]. Thus, coupling of donor **7** with the *N,N*-diacetyl-protected acceptor **13** gave a 50% yield for disaccharide **14**, whereas the use of the corresponding simple *N*-acetyl acceptor **11** gave only a 16% yield for disaccharide **12** under the same conditions. As noted by Demchenko and Boons and consistent with the reactivity of the imide functionality, the extraneous *N*-acetyl group is readily removed from the coupled products under simple Zemplén conditions leaving the native acetamide group in place [31, 37].

Other laboratories subsequently prepared and commented on the advantageous reactivity of the different classes of sialyl donor carrying the *N,N*-diacetyl protecting group system. For example, Kononov and coworkers described the chloride **15** and its reactivity with simple anionic nucleophiles, including thiolates and phosphates, in the absence of a promoter (Scheme 5.5) [38]. As these substitutions proceeded with high degrees of inversion even in polar solvents, an S_N2-type mechanism was considered to be the likely pathway. With simple alcohols as nucleophile, however, anomeric mixtures were obtained [38]. The *N,N*-diacetyl system has been used in conjunction with a stereodirecting phenylthio group installed at the C3 position, although no significant advantage appears to accrue from this combination in view of the more complex synthesis required for its preparation [39, 40]. Alternative variations on the general theme

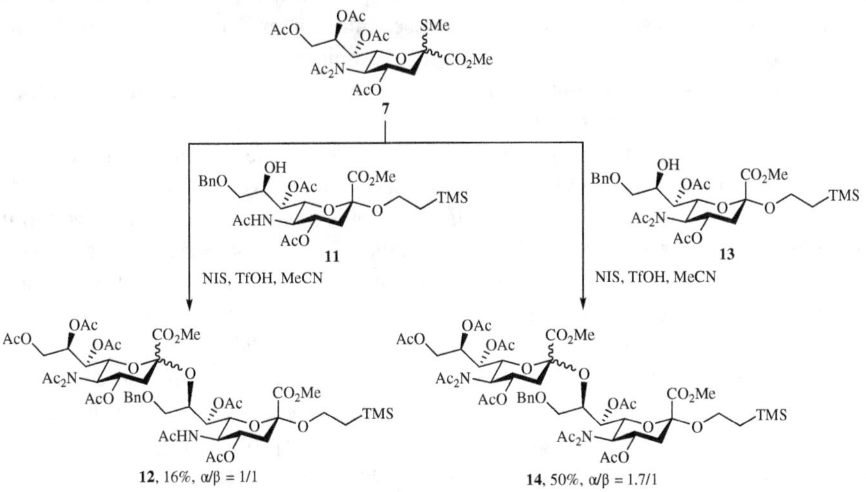

SCHEME 5.4 Beneficial effect of the imide function on acceptor reactivity. Bn, benzyl.

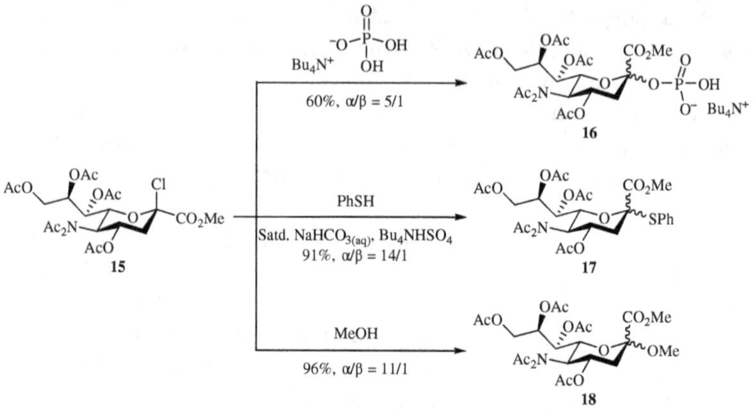

SCHEME 5.5 Selectivity in the reactions of an *N,N*-diacetylsialyl chloride.

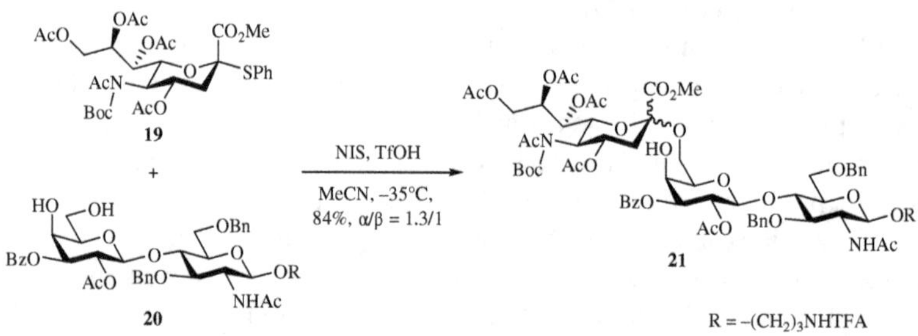

SCHEME 5.6 Sialylation with an *N*-acetyl-*N*-Boc-protected donor. TFA, trifluoroacetyl.

include the *N-tert*-butyloxycarbonyl (*N*-Boc) acetamide **19** (Scheme 5.6), which, in addition to increased reactivity as a donor in comparison to the simple acetamide, has the advantage of facilitating cleavage of the acetamide under Zemplén-type conditions and so allowing the ready installation of alternative amides such as the glycolamide [41]. In contrast, an *N*-benzoyl-*N*-acetylthiosialoside protected with benzoate esters was found to be β- rather than α-selective [42]. It was subsequently suggested, however, that this inversion of selectivity is more a function of the use of benzoate rather than acetate esters, particularly at O7, than of the mixed imide group [43].

The use of *N*-phthaloyl-protected sialic acid donors was first reported by Mamoru Sugimoto's group en route to a GM2 ganglioside derivative [44], but the real advantages of this protecting group in synthesis were realized by Koichi Fukase and coworkers in conjunction with the *N*-phenyltrifluoroacetimidate class of donors. Thus, it was reported that activation of donor **22** with trimethylsilyl triflate (TMSOTf) at −78°C in propionitrile in the presence of diol **23** gave the disaccharide **24** in 77% yield and 97/3 α/β selectivity (Scheme 5.7) [45]. Comparable yields and selectivities were obtained with other five-membered cyclic imide type protecting groups including the tetrachlorophthalimido and the dimethylmaleimido system, but significantly worse results were obtained with the acyclic *N,N*-diacetyl group. When propionitrile was replaced by CH_2Cl_2 as solvent, the stereoselectivity of the system was inverted.

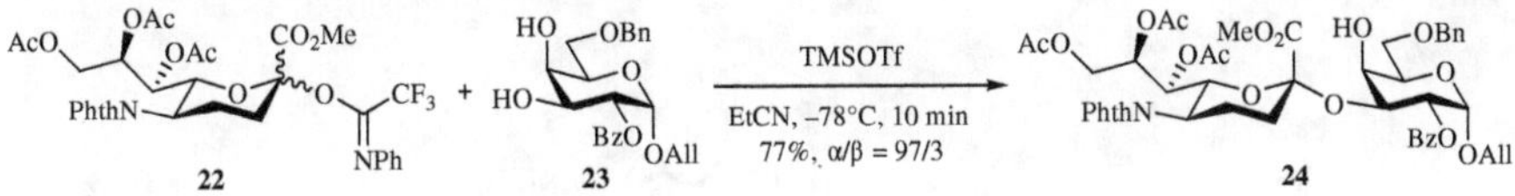

SCHEME 5.7 Use of a phthalimide-protected sialyl donor. All, allyl; Phth, phthaloyl.

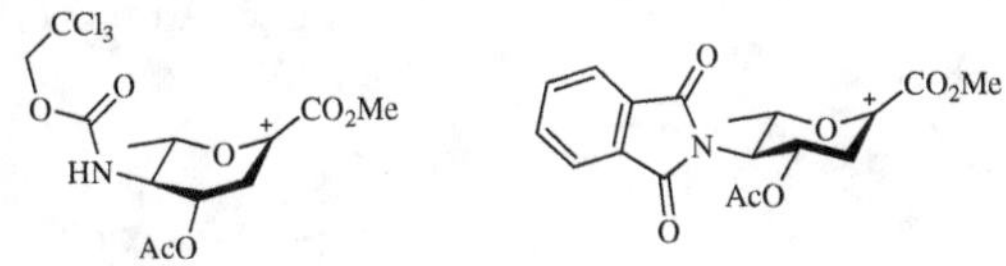

FIGURE 5.3 Computationally optimized structures of oxocarbenium ions **25** and **26**.

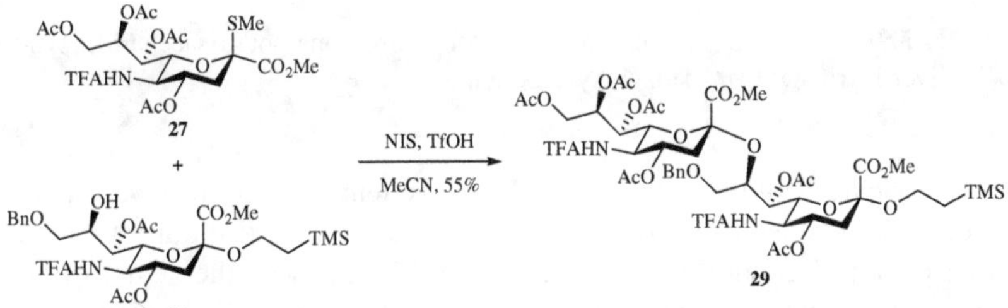

SCHEME 5.8 Use of an *N*-TFA-protected donor.

Seeking to explain the strong α-directing properties of the phthalimido group, computational work was conducted, which led to the suggestion that an *N*-phthaloyl-protected sialyl oxocarbenium ion would adopt a chair-like conformation (**26**) in which the positive charge at the anomeric center is stabilized by the fixed dipole of the phthalimido group resulting in preferential formation of the β-nitrilium ion and then, by S_N2-like attack, of the α-glycosides (Fig. 5.3). The well-known preferential adoption of the *trans* conformation by the acetamide moiety, which has the effect of orienting the overall group dipole in the opposite direction to that of the necessarily *cis*-configured phthalimide, was advanced to explain the lower selectivity of the *N*,*N*-diacetyl-protected donors [45].

The postulate by Demchenko and Boons that the *N*,*N*-diacetyl system enhances the reactivity by reducing the nucleophilicity of the amide group led to the development of the *N*-TFA protecting group for sialic acid donors [46–49]. In particular, it was demonstrated that such donors are generally more reactive than their *N*-acetyl counterparts and can be effective for the highly stereoselective coupling to the unreactive 8-OH of sialic acid (Scheme 5.8) [50]. It was also demonstrated in the course of a synthesis of an oligo-(2→9)-linked sialoside using an *N*-TFA-protected sialyl phosphite donor that the trifluoroacetamido system is an effective protecting group for sialyl 9-OH acceptors [49].

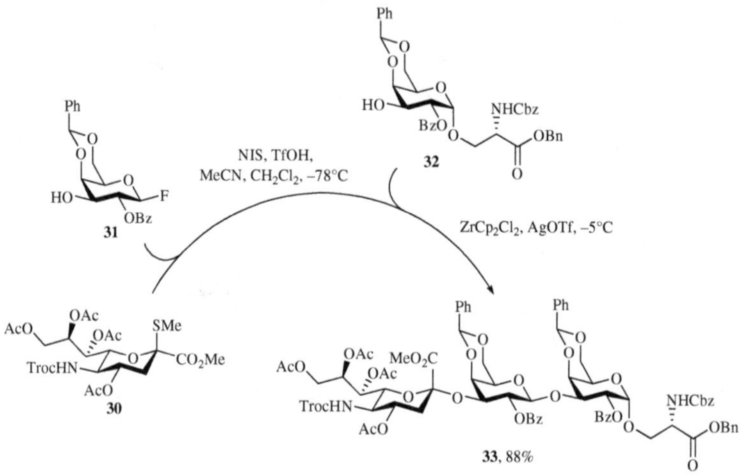

SCHEME 5.9 Use of an *N*-Troc-protected donor in a one-pot trisaccharide synthesis. AgOTf, silver(I) triflate; Cbz, benzyloxycarbonyl.

The *N*-trichloroethoxycarbonyl (*N*-Troc) system has been investigated by numerous groups since its introduction to the field of sialic acid donors by the laboratories of Shih-Hsiung Wu and Makoto Kiso [51, 52] with the demonstration of greater reactivity than the native acetamide-based donors [52]. By combining an *N*-Troc-protected thiosialoside donor with a partially protected glycosyl fluoride-based acceptor, the Takashi Takahashi group achieved the one-pot syntheses of sialic acid-containing trisaccharides by this method, as illustrated in Scheme 5.9. Both the Fukase [45] and the Takahashi [53] groups have successfully employed the *N*-Troc protecting system in conjunction with the *N*-phenyl trifluoroacetimidate class of donors, but the Fukase group noted that lower stereoselectivities were obtained than with the corresponding *N*-phthaloyl-protected donors. In a similar vein, Chun-Cheng Lin and coworkers, while finding an *N*-Troc-protected sialyl phosphite to be highly reactive, noted it to be less stereoselective than the corresponding *N*-TFA system [49].

The high reactivity and good α selectivity of the *N*-Troc-protected donors have been rationalized on the basis of computational work by the Fukase group. Thus, it was considered that the intermediate sialyl oxocarbenium ion (**25**) adopts a $B_{O,4}$ conformation (Fig. 5.3). In such form, the *N*-Troc group is better able to stabilize the oxocarbenium ion through space electrostatically than the simple amide group, leading to preferential attack by the incoming acceptor on the opposite face [45].

Various other types of amine protection have been investigated as replacements for the native *N*-acetyl group in sialyl donors including Boc [54], allyloxycarbonyl [55], 9-fluorenylmethylcarbonyl [55], trichloroacetyl [55], and Cbz [55, 56] groups, but no particular advantages of these systems are apparent at present.

Following a brief initial report by Carlo Unverzagt [57], 5-azido-based sialyl donors were prepared by various groups and their reactivity investigated [58–61].

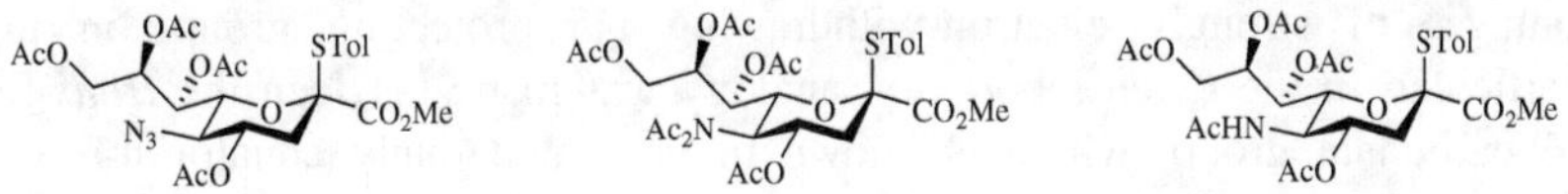

34 (RRV = 0.692) **35** (RRV = 3.45) **36** (RRV = 2.55)

FIGURE 5.4 RRVs of various sialyl donors relative to 4-tolyl peracetyl-α-D-thiomannopyranoside.

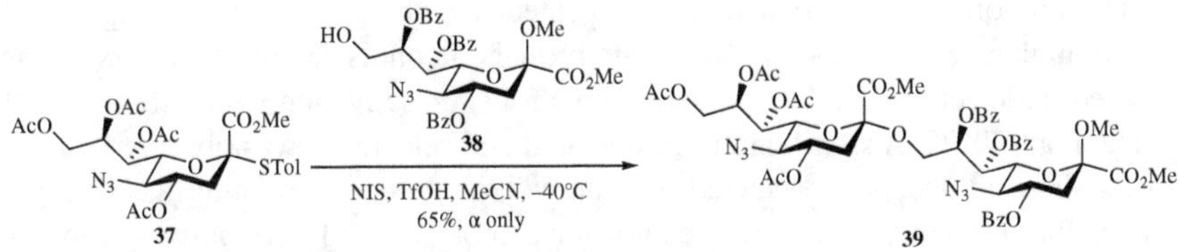

37 **39**

SCHEME 5.10 Use of an azide-protected sialyl donor.

Unlike the other N5 modifications described previously, a 5-azido thioglycoside was found to be less reactive than the simple amide as is clear from the measured relative reactivity values (RRVs) (Fig. 5.4) [59]. However, it was subsequently reported by Teruaki Mukaiyama and coworkers that a 5-azido sialyl fluoride was more reactive than both the corresponding acetamide and the *N*,*N*-diacetyl derivatives toward activation with the combination of $SnCl_4$ and $AgB(C_6F_5)_4$, although this may be a consequence of the differing anomeric stereochemistries employed [60].

All examples reported for sialylations with 5-azido-protected donors, even with the more reactive perbenzyl-protected system, involve coupling to primary alcohols [59–61], and attention has been drawn to the loss of yield and selectivity with the more hindered primary alcohols and secondary alcohols [61]. Nevertheless, the sialic acid 9-OH group was found to be a good acceptor toward azide-protected donors with moderate yield and complete α selectivity (Scheme 5.10), in contrast to the lower selectivity observed for the same acceptor on coupling to the corresponding 5-acetamido system and the *N*,*N*-diacetyl system [59]. The high selectivity observed for some primary acceptors on coupling to azide-protected donors has been ascribed to the low steric bulk of the azide moiety and its ability to stabilize the intermediate glycosyl nitrilium ions, thereby promoting S_N2-like displacement reactions [59].

5.6 4-*O*,5-*N*-OXAZOLIDINONE GROUP AND ITS STEREODIRECTING INFLUENCE ON SIALYLATIONS

The most significant advance in sialylation chemistry in recent years is the introduction of the 4-*O*,5-*N*-oxazolidinone protecting group introduced by the Takahashi [62] and De Meo [63] groups and its *N*-acetyl variant by the David Crich [64, 65] laboratory. Cyclic conformation-restricting protecting groups are known to have beneficial effects on the stereochemical outcome of some types of glycosylation reaction [66–68] (see Section 4.2), and the oxazolidinone system was conceived with this and the advantages of strongly electron-withdrawing N5 protecting groups in mind. In particular, in the Crich laboratory, analogy was made between the *trans*-fused 3,4-*O*-carbonate group, which is known to give moderately equatorial-selective coupling reactions in the rhamnopyranose series [69] and the topologically equivalent 4-*O*,5-*N*-oxazolidinone in the sialic acids; the *N*-acetyl group was included in the design to electronically mimic the *N*,*N*-diacetyl systems described previously.

In their initial publication on the use of the oxazolidinone system, Takahashi and coworkers reported that an oxazolidinone-protected β-thioglycoside gave high yields and very high α selectivities on coupling to primary alcohols and to an oxazolidinone-protected sialic acid 8-OH derivative (Scheme 5.11) on activation with NIS and TfOH in CH_2Cl_2 at −78°C. A significantly lower yield and selectivity was observed with an 8-OH sialic acid acceptor protected with an *N*-Troc group, illustrating the ability of the oxazolidinone to activate the acceptor for glycosylation. Importantly and in contrast to almost all other sialyl donors, the formation of the 2,3-glycal by-product was not noted in these reactions, thereby enabling the high yields observed [62]. Analogous high yields, alcohol-dependent high selectivity, and absence of glycal formation were

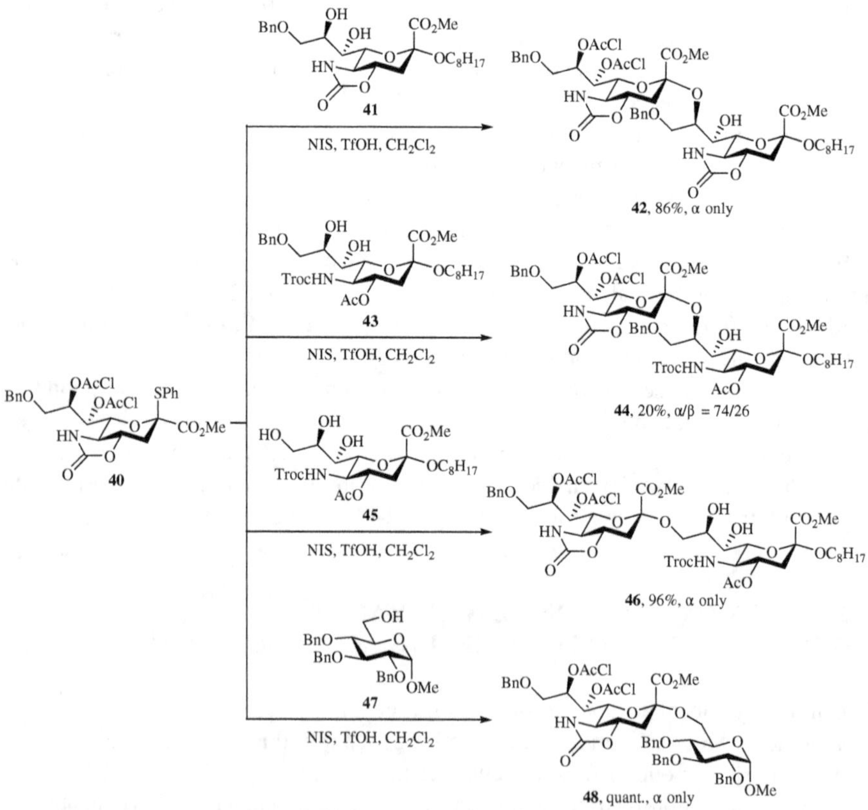

SCHEME 5.11 Examples of sialylations with a 4-*O*,5-*N*-oxazolidinone-protected sialyl thioglycoside. AcCl, chloroacetyl.

reported shortly thereafter for an analogous α-thioglycoside protected with the oxazolidinone group on activation with NIS and TfOH in acetonitrile at –40°C [63].

The relatively robust nature of the oxazolidinone protecting group is reflected in the conditions reported for its removal. LiOH in aqueous ethanol at 80°C was used in one laboratory [62] and $Ba(OH)_2$ in ethanol at 65°C in another [63]. Both cases were followed by reinstallation of the native acetamide group.

Using this methodology and the ability of the oxazolidinone to control glycosylation stereoselectivity and activate sialic acid-based hydroxy groups toward glycosylation, Takahashi and coworkers were able to synthesize oligomeric α(2→9)- and α(2→8)-linked sialic acid polymers by iterative methods as exemplified in Scheme 5.12 [62, 70, 71]. An impressive convergent synthesis of the GP1c ganglioside epitope that featured multiple formation of α(2→8) linkages between neuraminic acid residues was also completed by the same laboratory using the same methodology [72].

In the Crich laboratory, moderate to excellent α-selective sialylations were conducted with the *N*-acetyloxazolidone-protected phenyl thioglycosides on activation with NIS and TfOH in mixtures of CH_2Cl_2 and acetonitrile at −40°C [64], but better results were obtained at −78°C with the more readily activated 1-adamantanyl thioglycosides (Scheme 5.13) [65]. An analogous *N*-acetyloxazolidinone-protected *p*-tolyl thioglycoside described by Guo-wen Xing and coworkers [73] was employed in the synthesis of the tumor-associated ganglioside GM3 [74]. The advantage of the *N*-acetyloxazolidinone protecting system over the simple oxazolidinone arises from

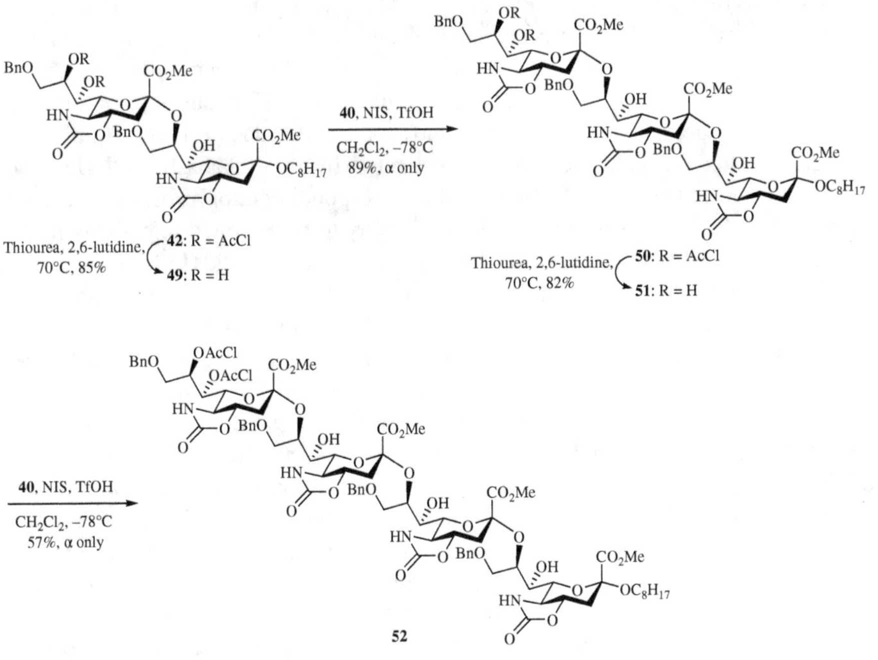

SCHEME 5.12 Convergent synthesis of the GP1c ganglioside epitope assisted by the 4-*O*,5-*N*-oxazolidinone group.

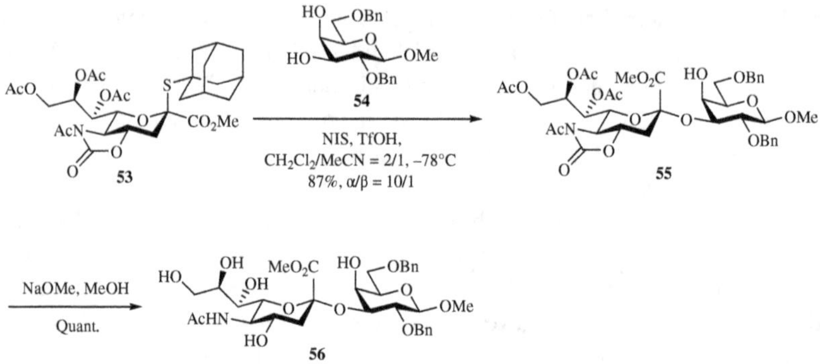

SCHEME 5.13 Sialylation with an *N*-acetyl-4-*O*,5-*N*-oxazolidinone-protected donor and subsequent removal of the oxazolidinone group.

the very mild conditions used for deprotection—NaOMe in methanol at room temperature, which selectively remove the oxazolidinone while leaving the necessary acetamide group in place [64].

N-Acetyloxazolidinone-protected *p*-tolyl thioglycosides have been activated for glycosylation at −70°C in CH_2Cl_2 by a combination of diphenyl sulfoxide and Tf_2O [75] first introduced to carbohydrate chemistry by David Gin and coworkers [76], applied to thioglycosides by the Jacques van Boom group [77], and originally applied in sialic acids by David Crich and Wenju Li [25]. Similarly, it has been demonstrated that the *N*-acetyloxazolidinone-protected thiosialosides may be employed in a preactivation protocol employing a combination of AgOTf and *p*-toluenesulfenyl chloride [78], much as previously described for the β-mannosides [79]; subsequent addition of a range of acceptors gave excellent yields for α-sialosides [78]. Oxazolidinone-protected *S*-Box thiosialosides were investigated by the Takahashi and De Meo groups and showed similar selectivity to the corresponding thioglycosides; the main advantage of the *S*-Box systems lies in their ability to be activated selectively in the presence of other thioglycosides with AgOTf and other Lewis acids [30, 71]. An interesting development concerns the oxidation of an *N*-acetyloxazolidinone-protected thiosialoside to give the corresponding isolable glycosyl sulfoxide in the form of an undefined mixture of isomers at sulfur, which could be stored for months at −20°C [28]. Preactivation of this sulfoxide with Tf_2O at −70°C in CH_2Cl_2 followed by addition of a simple primary alcohol gave low yield for the anticipated α-glycoside along with the 2,3-glycal as the major product. However, preactivation with a combination of Tf_2O and ditolyl sulfoxide resulted in reduced formation of the glycal and significantly improved yields of the glycoside, albeit with a loss of stereoselectivity as the amount of ditolyl sulfoxide employed increased. The optimum conditions were found to involve the use of 2 equiv of ditolyl sulfoxide and 1.2 equiv of Tf_2O, providing good to excellent yields with excellent selectivity for a range of glycosides [28]. Building on earlier results for the activation of thiosialosides with the same combination of reagents [75], the excess diaryl sulfoxide is considered to trap and thereby stabilize any intermediate oxocarbenium ion in the form of a pair of anomeric glycosyl sulfoxonium ions (**5**, Fig. 5.2), such that displacement of the β anomer by the acceptor yields the α-glycoside.

True one-pot syntheses of oligosaccharides (see Chapter 6) carrying sialic acid residues at the nonreducing end have long been targeted by many groups but one that has been complicated by the low reactivity of typical thiosialosides from which such a synthesis must begin and because of the low stereoselectivities of most sialylation methods. For example, the use of simple thiosialosides in the programmable one-pot synthesis developed by Chi-Huey Wong's group is thwarted by the inability to activate thiosialosides selectively [80]. Such problems can, in principle, be overcome by the use of the so-called active–latent strategies in which an electron-rich thiosialoside is activated in the presence of an electron-deficient thioglycoside-based acceptor as in the work of René Roy and coworkers, but the subsequent activation of the latent thioglycoside necessitates additional steps [81, 82]. Alternative approaches use pairs of orthogonal donors such as thiosialosides and glycosyl fluorides (Scheme 5.9) [55] or *S*-Box thiosialosides and simple thioglycosides, but stereoselectivity remains a problem, and the need for two sets of activating conditions complicates matters [29, 30]. The introduction of the reactive adamantanyl thiosialosides in conjunction with use of the *N*-acetyloxazolidinone protecting system, however, has recently provided a solution to this problem. Thus, it was demonstrated that the *N*-acyloxazolidinone-protected adamantanyl thiosialoside **57** could be activated by NIS and TfOH at −78°C in a CH_2Cl_2/acetonitrile mixture in the presence of the 4-chlorophenyl thioglycoside **58**, leading to the formation of the disaccharide **59** (Scheme 5.14). Further addition of a second acceptor, for example, **60**, followed by NIS and TfOH and raising the temperature to 0°C resulted in the formation of the trisaccharide **61** in the 51% overall yield for the one-pot sequence (Scheme 5.15) [83]. The use of the *N*-(acetoxyacetyl) oxazolidinone protecting system in this study, which was intended to lead directly to the Neu5Gc-based oligosaccharides, revealed the exquisite balance of reactivity in the *N*-acyloxazolidinones needed for selective deprotection. Thus, as noted previously, the *N*-acetyl system (Scheme 5.13) undergoes selective removal of the oxazolidinone ring conveniently leaving in place the acetamide, whereas, in the case of the *N*-(acetoxyacetyl) oxazolidinone system, the greater electrophilicity of the acetoxyacetyl moiety did not allow for such selectivity and necessitated recourse to more vigorous conditions for complete deprotection followed by reinstallation of the glycolyl unit (Scheme 5.15) [83].

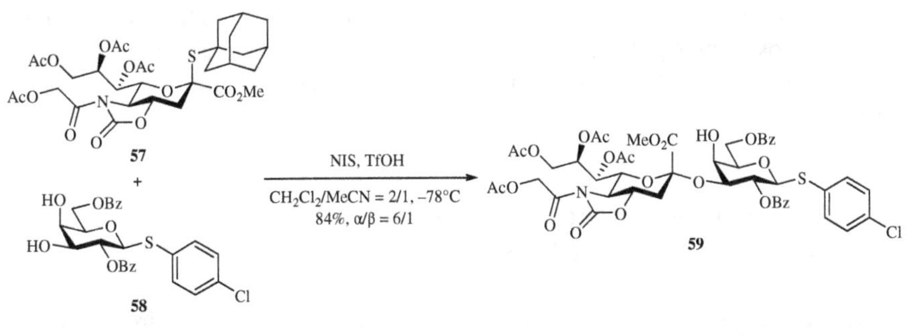

SCHEME 5.14 Synthesis of an α-Neu5Gc-based disaccharide.

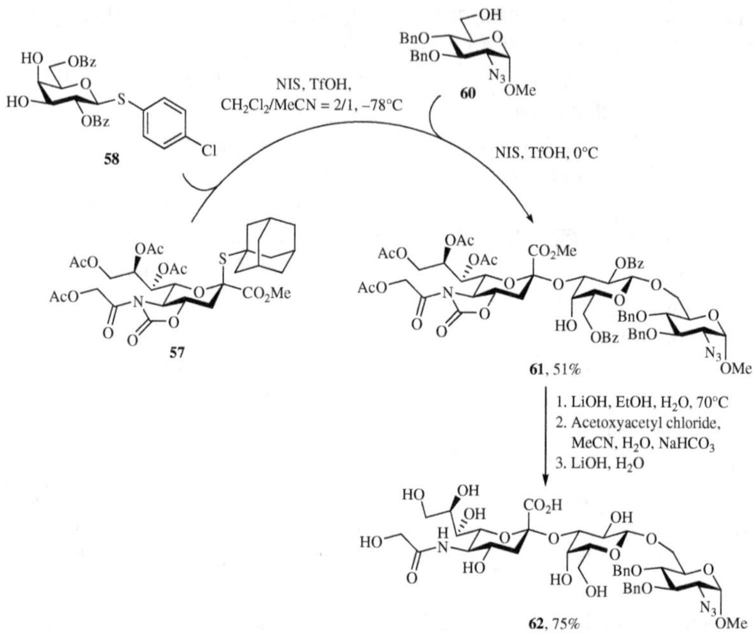

SCHEME 5.15 One-pot synthesis and subsequent deprotection of an Neu5Gc-based trisaccharide.

Wong and coworkers refined the oxazolidinone-directed sialylation method by converting the thiosialoside donors to the corresponding dibutylphosphates by activation with NIS/TfOH in the presence of dibutylphosphoric acid. The phosphates, which are stable at room temperature and whose anomers can be separated by silica gel chromatography, serve as excellent α-selective sialyl donors at low temperature in CH_2Cl_2 (−78°C for the α-configured donor and 50°C for the β anomer) on activation with TMSOTf. As in all other oxazolidinone-directed sialylations, little or no glycal formation is observed in these reactions [84]. It was demonstrated that activation of the phosphate could be achieved cleanly in the presence of standard thioglycosides, thereby enabling initiation of programmable one-pot oligosaccharide syntheses with the sialyl donor as exemplified in Scheme 5.16 [84]. Variations on this general theme enabled the high yielding and highly stereoselective synthesis of the tumor-associated stage-specific embryonic antigen 4 [84] and of the α(2→9)-linked oligosialosides containing as many as 12 sialic acid units [85].

5.7 4,5-*O*-CARBONATE PROTECTING GROUP IN α-SELECTIVE KDN DONORS

KDN, a rare sialic acid that is characterized by the presence of a hydroxy group at the 5-position rather than the more customary amido substituent, is undergoing a resurgence of interest as a possible cancer marker in humans, having been found on

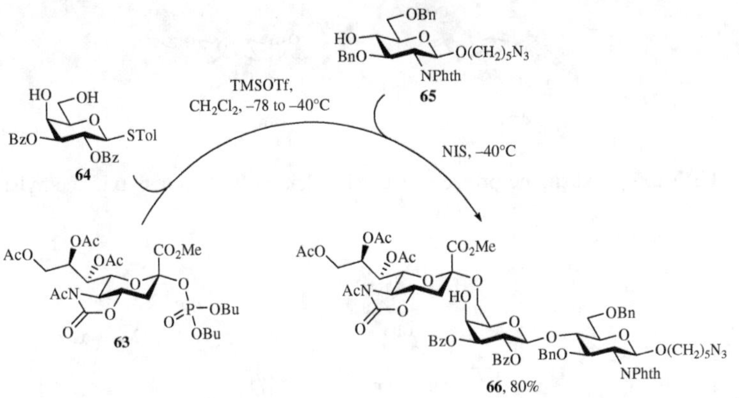

SCHEME 5.16 Use of an *N*-acetyl-4-*O*,5-*N*-oxazolidinone-protected sialyl phosphate in a one-pot trisaccharide synthesis.

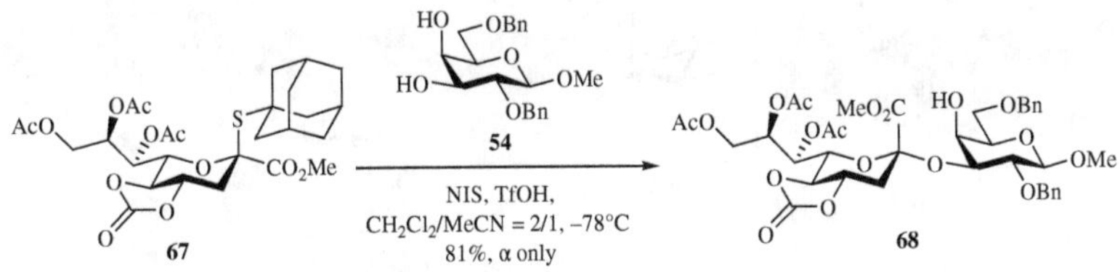

SCHEME 5.17 Use of a 4,5-*O*-carbonate-protected KDN donor.

ovarian cancer and human red blood cells among others [6]. The chemical synthesis of its α-glycosides suffers from the same problems as that of the neuraminic acid glycosides, namely, poor selectivity and elimination. Building on the success of the oxazolidinones in the neuraminic acid series, David Crich and Chandrasekhar Navuluri synthesized the 4,5-*O*-carbonate-protected adamantanyl thioglycosides and showed them to be excellent α-selective donors in acetonitrile and CH_2Cl_2 at −78°C on activation with the usual NIS/TfOH combination (Scheme 5.17) [86]. No glycal formation was observed in the course of these reactions in contrast to the analogous glycosylations conducted with the peracetylated donor; the carbonate is easily removed under Zemplén conditions [86]. In subsequent work from the Crich laboratory, conversion of the carbonate-protected thioglycoside to the corresponding phosphates was demonstrated, but this has yet to be applied in glycoside synthesis [22].

5.8 OTHER CYCLIC AND BICYCLIC PROTECTING SYSTEMS FOR SIALYL DONORS

Seeking to emulate the benzylidene effect in β-mannosylation [68] (see Section 4.2.6), David Crich and Baolin Wu [87] and Hiromune Ando and coworkers [88] prepared 5-*N*,7-*O*-oxazinone-protected thiosialosides **69** and **70** (Fig. 5.5). Unfortunately, both groups found this class of donors to be moderately β-selective under a variety of activating conditions and even in the presence of acetonitrile.

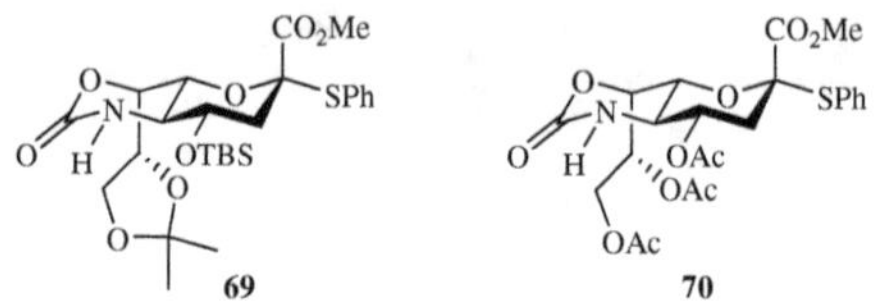

69 70

FIGURE 5.5 Oxazinone-protected sialoside donors. TBS, *tert*-butyldimethylsilyl.

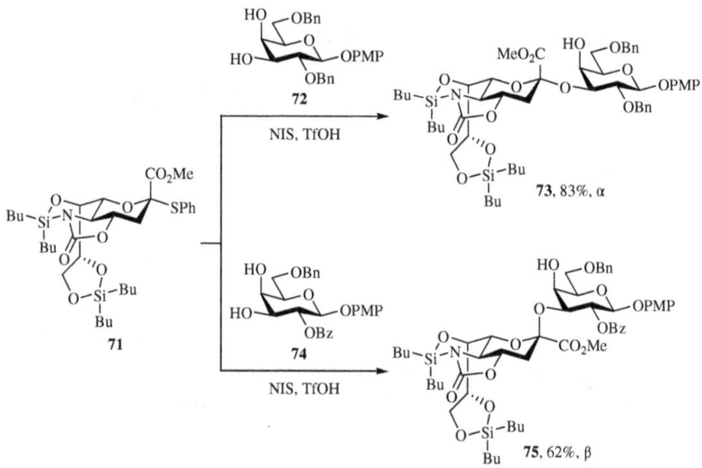

SCHEME 5.18 Use of a tricyclic sialyl donor and the influence of the acceptor protecting system on stereoselectivity. PMP, *p*-methoxyphenyl.

Shinya Hanashima and coworkers prepared two tricyclic sialyl donors in which the 4-*O*,5-*N*-position was blocked by an oxazolidinone ring and the 5-*N*,7-*O*-position by a di-*tert*-butylsilylidene acetal, with the 8,9-diol carrying either a second di-*tert*-butylsilylidene acetal or a tetraisopropyldisiloxane group [89]. These donors were found to be highly α-selective toward primary carbohydrate acceptors and to galactosyl 3,4-diols bearing benzyl ethers at the C2 position (Scheme 5.18) on activation with NIS/TfOH at −40°C. Curiously, however, high β selectivity was observed with galactosyl 3,4-diols protected by esters at the C2 position [89].

5.9 MECHANISTIC ASPECTS OF SIALYLATION WITH CYCLICALLY PROTECTED SIALYL DONORS

Crich and coworkers used a mass spectral method to investigate the influence of the oxazolidinone ring and its *N*-acetyl counterpart on the activation of sialyl donors [22]. They found that a minimum cone voltage of 78 V is required to cause fragmentation of the glycosyl phosphate **76** under electrospray ionization conditions (Fig. 5.6). The comparable oxazolidinone phosphate (**77**) did not undergo mass spectral fragmentation

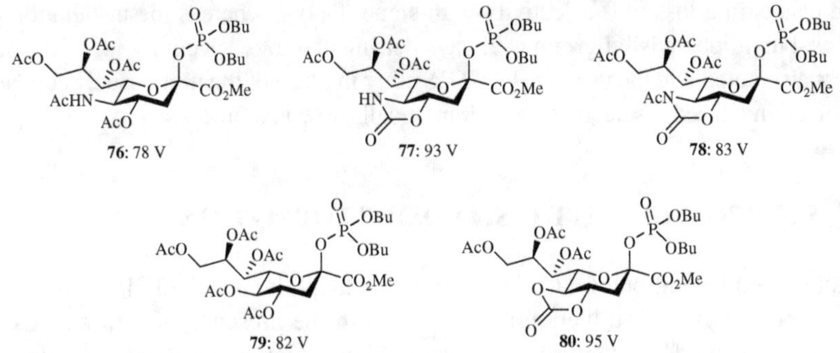

76: 78 V **77**: 93 V **78**: 83 V **79**: 82 V **80**: 95 V

FIGURE 5.6 Threshold cone voltages for the fragmentation of sialyl phosphates.

below 93V whereas the *N*-acetyl oxazolidinone system (**78**) required only 83V, enabling the conclusion that the oxazolidinone-protected systems destabilize the glycosyl oxocarbenium ions and thereby favor S_N2-like mechanisms for glycosidic bond formation. The origin of the strong electron-withdrawing effect of the oxazolidinone protecting system is considered to be the strong dipole moment of the protecting system, which aligns with the C4—O4 and C5—N5 bond dipoles in the mean plane of the pyranose ring and reinforces their inherent electron-withdrawing properties. The influence of the 4,5-*O*-carbonate in the KDN series is analogously explained.

An alternative hypothesis for the influence of N5 protecting groups, in general, on sialyl donor reactivity and selectivity was advanced by Kononov and coworkers, according to which both phenomena are concentration dependent and related to the degree of supramolecular aggregation through intermolecular hydrogen bonding [36, 90, 91]. In particular, it has been noted that the selectivity of a sialylation reaction changes with the extent of conversion, and it has been suggested that this effect is related to the changing degree of aggregation in the course of the reaction, owing to the changing composition of the reaction mixture [36, 90, 91]. While this explanation is plausible, similar variations in selectivity during the course of a reaction would also arise when one anomer is formed by an S_N1-type process and the other by an S_N2-like mechanism as has recently been demonstrated to be the case for the 4,6-*O*-benzylidene-directed mannopyranosylation [92, 93].

5.10 INFLUENCE OF HYDROXY PROTECTING GROUPS ON SIALYL DONOR REACTIVITY AND SELECTIVITY

As is generally the case for glycosyl donors with certain exceptions [94, 95], the use of ether protecting groups in sialyl donors, while not widely exploited, leads to more armed systems than that of the more common ester blocking systems [59]. The effect of hydroxy protecting groups on the stereoselectivity of sialyl donors has only recently begun to be investigated in a systematic manner. Thus, following on from an initial report by Xin-Shan Ye [42] using a series of perbenzoylated donors, De Meo and coworkers have determined that the replacement of a 7-*O*-acetyl group by a 7-*O*-benzoyl group results in a loss of α selectivity with some donors, whereas the installation of a 7-*O*-*tert*-butyldiphenylsilyl group is always detrimental to α selectivity [43]. The same authors also noted that the use of a 4-*O*-TBS ether in place of the more usual acetyl ester leads to an increased α selectivity and dramatically so with some acceptors [43].

5.11 STEREOSELECTIVE C-SIALOSIDE FORMATION

The attempted formation of a C-sialoside by nucleophilic attack of allyltributylstannane on peracetyl *N*-acetylneuraminyl chloride in the presence of Lewis acids was reported early by Hans Paulsen and Peter Matschulat to result in the exclusive formation of the 2,3-glycal [96]. Susumu Satoh and coworkers found that the elimination reaction could be minimized by using sialyl acetate as donor and electron-rich arenes as acceptor on activation with a combination of silver trifluoroacetate and $SnCl_4$, but selectivities and yields remained modest in most cases [97, 98]. In view of these somewhat unpromising results, most efforts toward C-glycosylation of sialic acids have involved sequences of the opposite polarity in which sialyl enolates [99, 100] or their equivalents, generated by the action of reducing metals on suitable derivatives such as glycosyl sulfones, halides, and thioglycosides, are either alkylated or quenched with suitable electrophiles (Scheme 5.19) [101–104]. A variant on this theme, developed by the Mikiko Sodeoka group, revolves around the use of the Ireland–Claisen rearrangement (Scheme 5.20) [105].

The ability of the 4-*O*,5-*N*-oxazolidinone group to limit glycal formation prompted Crich and coworkers to revisit the use of electrophilic sialyl donors for the formation of C-sialosides. Their effort resulted in the discovery that the *N*-acetyloxazolidone-protected dibutylphosphates, on activation with TMSOTf at −78°C in the presence of

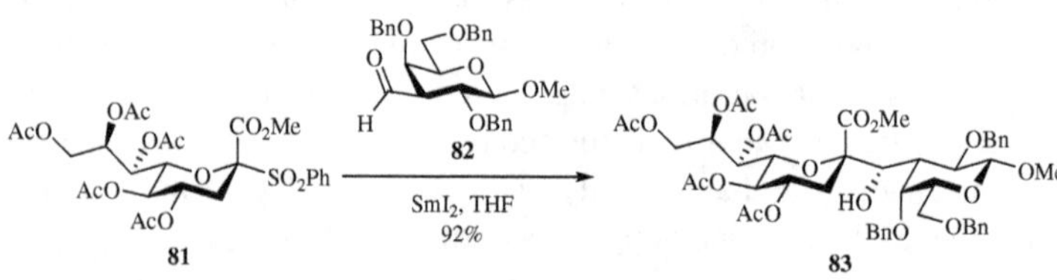

SCHEME 5.19 Nucleophilic C-sialylation. THF, tetrahydrofuran.

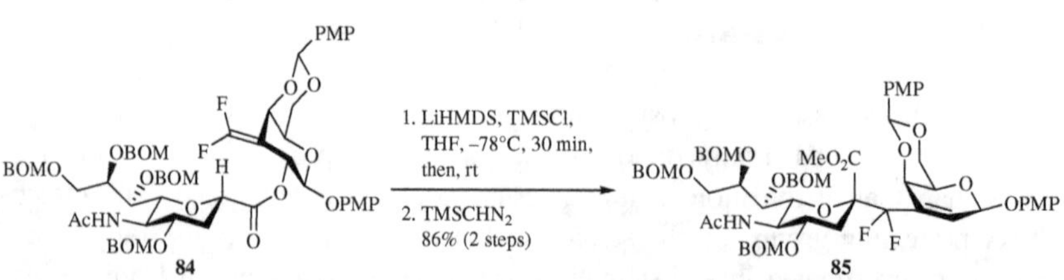

SCHEME 5.20 Use of the Ireland–Claisen rearrangement in C-sialylation. BOM, benzyloxymethyl; LiHMDS, lithium bis(trimethylsilyl)amide; TMS, trimethylsilyl.

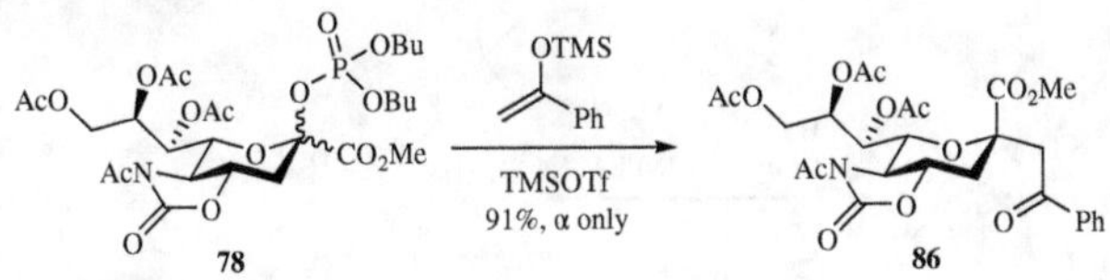

SCHEME 5.21 Electrophilic C-sialidation with an *N*-acetyloxazolidinone-protected sialyl donor.

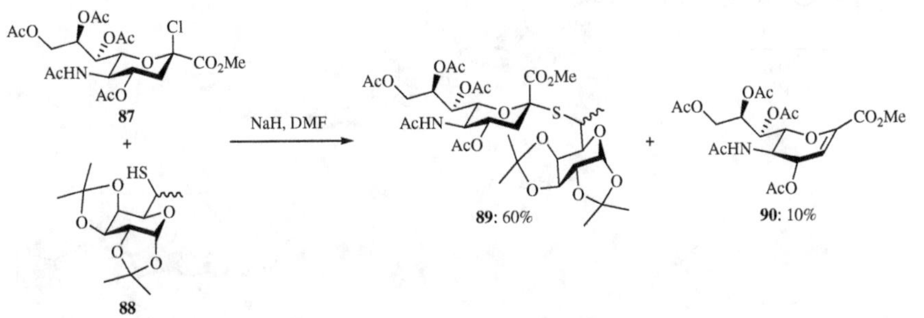

SCHEME 5.22 Stereoselective S-sialoside formation by reaction of a β-sialyl chloride with a thiolate. DMF, *N,N*-dimethylformamide.

potent C-nucleophiles such as allyltributylstannane and silyl enol ethers, afford C-glycosides with excellent yield and α selectivity (Scheme 5.21) [106]. Interestingly, under free radical conditions, no selectivity is observed in the formation of C-glycosides by the trapping of anomeric sialyl radicals either in the presence or absence of the oxazolidinone system [22, 96, 107].

5.12 STEREOSELECTIVE S-SIALOSIDE FORMATION

The formation of S-sialosides has been approached by two avenues, either the reaction of β-configured electrophilic sialyl donors with thiolate anions or the alkylation of anomeric sialyl thiolates. The former approach suffers from the need to prepare anomerically pure sialyl chlorides but otherwise functions well for simple thiolates and even for some more biologically relevant ones (Scheme 5.22) [108–111], albeit that elimination to give the 2,3-glycal has been reported to be a major problem in some instances [112]. The anomeric thiolate alkylation approach [109, 113, 114], which is also reported to suffer from elimination of the electrophilic partner [112], has been improved recently through the use of anomeric *tert*-butyl α-sialyl disulfides, which allow clean entry into the thiolate under mild conditions, enabling the synthesis of S-linked α(2→8) oligosialic acids among others (Scheme 5.23) [115, 116].

The *N*-acetyloxazolidinone-protected sialyl phosphates have recently made possible the formation of complex S-sialosides under classical glycosylation

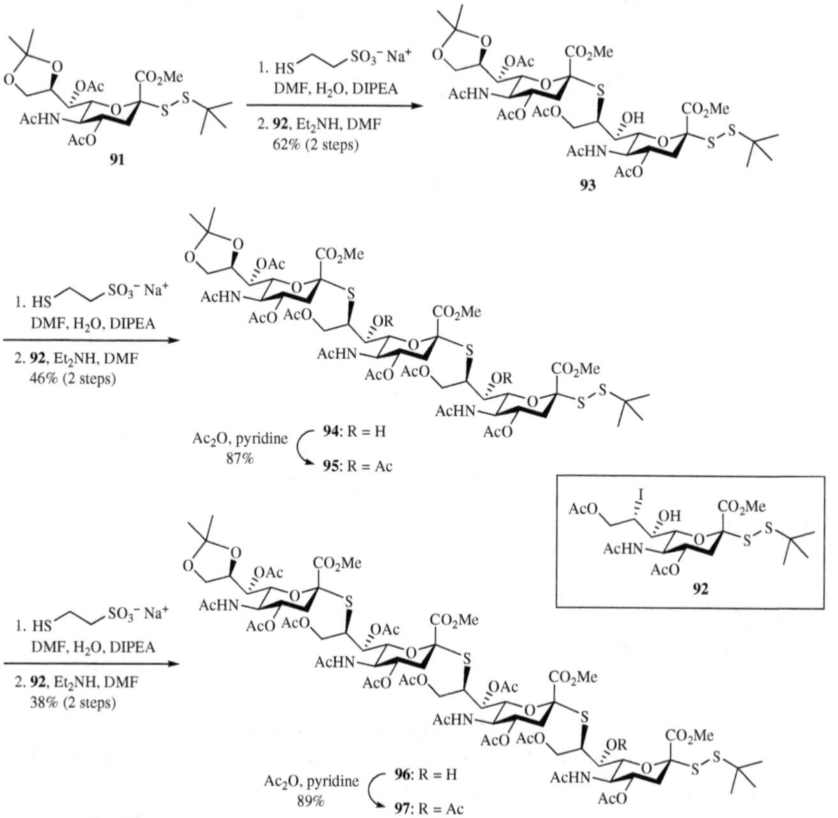

SCHEME 5.23 Oligothiosialoside synthesis by iterative alkylation of a sialic acid-based anomeric thiolate. DIPEA, diisopropylethylamine.

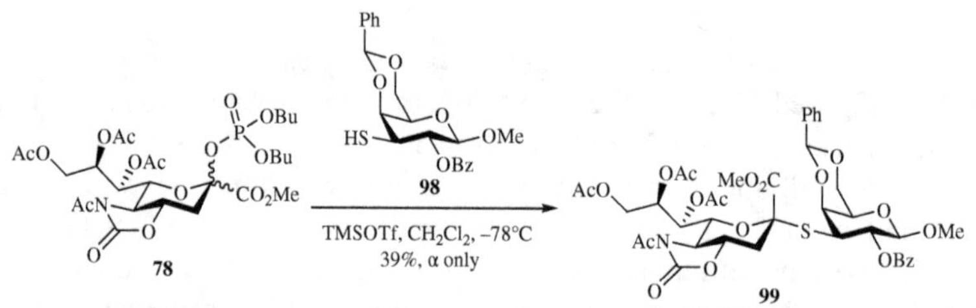

SCHEME 5.24 Sialylation of a thiol under Lewis acid-mediated conditions by an *N*-acetyl oxazolidinone-protected sialyl phosphate.

conditions, that is, on activation of the donors with a Lewis acid in the presence of an acceptor thiol. Moderate to good yields and α selectivities were observed under these conditions (Scheme 5.24) [117].

5.13 CONCLUSION

Considerable advances have been made in the stereoselective formation of the sialic acid glycosides and their C- and S-analogues in the last 12 years, stemming from the recognition of the importance of the N5 protecting group by Boons and Demchenko and the subsequent introduction of the *O*-4,*N*-5-oxazolidione system by three laboratories. This latter system functions with a number of different classes of donor, of which Wong's phosphates are currently the optimum, and is especially powerful for the synthesis of a wide range of O-, C-, and S-sialosides with excellent yield and selectivity.

REFERENCES

[1] Varki, A.; Gagneux, P. *Ann. N. Y. Acad. Sci.* **2012**, *1253*, 16–36.

[2] Angata, T.; Varki, A. *Chem. Rev.* **2002**, *102*, 439–469.

[3] Boons, G.-J.; Demchenko, A. V. In *Carbohydrate-Based Drug Discovery*; Wong, C.-H., Ed.; John Wiley & Sons, Inc.: Weinheim, **2003**; Vol. *1*, pp. 55–102.

[4] Chen, X.; Varki, A. P. *ACS Chem. Biol.* **2010**, *5*, 163–176.

[5] Inoue, S.; Kitajima, K. *Glycoconj. J.* **2006**, *23*, 277–290.

[6] Inoue, S.; Kitajima, K.; Sato, C.; Go, S. *Adv. Exp. Med. Biol.* **2011**, *705*, 669–678.

[7] Padler-Karavani, V.; Yu, H.; Cao, H.; Chokhawala, H.; Karp, F.; Varki, N.; Chen, X.; Varki, A. *Glycobiology* **2008**, *18*, 818–830.

[8] Hasegawa, A. In *Modern Methods in Carbohydrate Synthesis*; Khan, S. H., O'Niell, R. A., Eds.; Harwood Academic: Amsterdam, **1996**, pp. 277–300.

[9] Hasegawa, A.; Kiso, M. In *Preparative Carbohydrate Chemistry*; Hanessian, S., Ed.; Marcel Dekker: New York, **1997**, pp. 357–379.

[10] Boons, G.-J.; Demchenko, A. V. *Chem. Rev.* **2000**, *100*, 4539–4565.

[11] Magnusson, G.; Nilsson, U. J. In *Glycoscience: Chemistry and Chemical Biology*; Fraser-Reid, B., Kuniaki, T., Thiem, J., Eds.; Springer: Berlin, **2001**; Vol. *2*, pp. 1543–1587.

[12] Kiefel, M. J.; von Itzstein, M. *Chem. Rev.* **2002**, *102*, 471–490.

[13] Kiso, M.; Ishida, H.; Ito, H. In *Carbohydrates in Chemistry and Biology: Chemistry of Saccharides*; Ernst, B., Hart, G. W., Sinaÿ, P., Eds.; John Wiley & Sons, Inc.: Weinheim, **2000**; Vol. *1*, pp. 345–365.

[14] Ando, T.; Ando, H.; Kiso, M. *Trends Glycosci. Glycobiol.* **2001**, *13*, 573–586.

[15] De Meo, C.; Boons, G. J.; Demchenko, A. V. In *Comprehensive Glycoscience*; Kamerling, J., Ed.; Elsevier: Amsterdam, **2007**; Vol. *1*, pp. 583–604.

[16] Wilson, J. C.; Angus, D. I.; von Itzstein, M. *J. Am. Chem. Soc.* **1995**, *117*, 4214–4217.

[17] Klepach, T.; Carmichael, I.; Serianni, A. S. *J. Am. Chem. Soc.* **2008**, *130*, 11892–11900.

[18] Kuhn, R.; Lutz, P.; Macdonald, D. L. *Chem. Ber.* **1966**, *99*, 611–617.

[19] Yu, P. K.; Ledeen, R. *J. Biol. Chem.* **1969**, *244*, 1306–1313.

[20] Walvoort, M. T. C.; Dinkelar, J.; van den Bos, L. J.; Lodder, G.; Overkleeft, H. S.; Codée, J. D. C.; van der Marel, G. A. *Carbohydr. Res.* **2010**, *345*, 1252–1263.

[21] Bohé, L.; Crich, D. *C. R. Chim.* **2011**, *14*, 3–16.

[22] Kancharla, P. K.; Navuluri, C.; Crich, D. *Angew. Chem. Int. Ed.* **2012**, *51*, 11105–11109.

[23] Murase, T.; Kameyama, A.; Kartha, K. P. R.; Ishida, H.; Kiso, M.; Hasegawa, A. *J. Carbohydr. Chem.* **1989**, *8*, 265–283.

[24] Schmidt, R. R.; Behrendt, M.; Toepfer, A. *Synlett* **1990**, 694–696.

[25] Crich, D.; Li, W. *Org. Lett.* **2006**, *8*, 959–962.

[26] De Meo, C.; Farris, M.; Ginder, N.; Gulley, B.; Priyadarshani, U.; Woods, M. *Eur. J. Org. Chem.* **2008**, 3673–3677.

[27] Cai, S.; Yu, B. *Org. Lett.* **2003**, *5*, 3827–2830.

[28] Gu, Z. Y.; Zhang, J. X.; Xing, G. W. *Chem. Asian J.* **2012**, *7*, 1524–1528.

[29] Wang, C.-C.; Lee, J.-C.; Luo, S.-Y.; Kulkarni, S. S.; Huang, Y.-W.; Lee, C.-C.; Chang, K.-L.; Hung, S.-C. *Nature* **2007**, *446*, 896–899.

[30] Harris, B. N.; Patel, P. P.; Gobble, C. P.; Stark, M. J.; De Meo C. *Eur. J. Org. Chem.* **2011**, 4023–4027.

[31] Demchenko, A. V.; Boons, G.-J. *Tetrahedron Lett.* **1998**, *39*, 3065–3068.

[32] De Meo, C.; Priyadarshani, U. *Carbohydr. Res.* **2008**, *343*, 1540–1552.

[33] De Meo, C. In *Frontiers in Modern Carbohydrate Chemistry*; Demchenko, A. V., Ed.; ACS Symposium Series 960; American Chemical Society: Washington, DC, **2007**, pp. 118–131.

[34] Cheng, A.; Hendel, J. L.; Colangelo, K.; Bonin, M.; Auzanneau, F.-A. *J. Org. Chem.* **2008**, *73*, 7574–7579.

[35] Liao, L.; Auzanneau, F.-I. *J. Org. Chem.* **2005**, *70*, 6265–6273.

[36] Kononov, L. O.; Malysheva, N. N.; Kononova, E. G.; Orlova, A. V. *Eur. J. Org. Chem.* **2008**, 3251–3254.

[37] Demchenko, A. V.; Boons, G.-J. *Chem. Eur. J.* **1999**, *5*, 1278–1283.

[38] Orlova, A. V.; Shpirt, A. M.; Kulikova, N. Y.; Kononov, L. O. *Carbohydr. Res.* **2010**, *345*, 721–730.

[39] Hossain, N.; Magnusson, G. *Tetrahedron Lett.* **1999**, *40*, 2217–2220.

[40] Hossain, N.; Zapate, A.; Wilstermann, M.; Nilsson, U. J.; Magnusson, G. *Carbohydr. Res.* **2002**, *337*, 569–580.

[41] Sherman, A. A.; Yudina, O. N.; Shashkov, A. S.; Menshov, V. M.; Nifant'ev, N. E. *Carbohydr. Res.* **2001**, *330*, 445–458.

[42] Wang, Y.; Ye, X.-S. *Tetrahedron Lett.* **2009**, *50*, 3823–3826.

[43] Premathilake, H. D.; Gobble, C. P.; Pornsuriyasak, P.; Hardimon, T.; Demchenko, A. V.; De Meo, C. *Org. Lett.* **2012**, *14*, 1126–1129.

[44] Fujita, S.; Numata, M.; Sugimoto, M.; Tomita, K.; Ogawa, T. *Carbohydr. Res.* **1994**, *263*, 181–196.

[45] Tanaka, K.; Goi, T.; Fukase, K. *Synlett* **2005**, 2958–2962.

[46] De Meo, C.; Demchenko, A. V.; Boons, G. J. *Aust. J. Chem.* **2002**, *55*, 131–134.

[47] Meijer, A.; Ellervik, U. *J. Org. Chem.* **2004**, *69*, 6249–6256.

[48] Tsvetkov, Y. E.; Nifantiev, N. E. *Synlett* **2005**, 1375–1380.

[49] Lin, C.-C.; Huang, K.-T.; Lin, C.-C. *Org. Lett.* **2005**, *7*, 4169–4172.

[50] De Meo, C.; Demchenko, A. V.; Boons, G. J. *J. Org. Chem.* **2001**, *66*, 5490–5497.

[51] Ren, C.-T.; Chen, C.-S.; Wu, S.-H. *J. Org. Chem.* **2002**, *67*, 1376–1379.

[52] Ando, H.; Koike, Y.; Ishida, H.; Kiso, M. *Tetrahedron Lett.* **2003**, *44*, 6883–6886.

[53] Tanaka, H.; Nishiura, Y.; Adachi, M.; Takahashi, T. *Heterocycles* **2006**, *67*, 107–112.

[54] Ito, Y.; Nunomura, S.; Shibayama, S.; Ogawa, T. *J. Org. Chem.* **1992**, *57*, 1821–1831.

[55] Tanaka, H.; Adachi, M.; Takahashi, T. *Chem. Eur. J.* **2005**, *11*, 849–862.

[56] Sparks, M. A.; Williams, K. W.; Lukacs, C.; Schrell, A.; Priebe, G.; Spaltenstein, A.; Whitesides, G. M. *Tetrahedron* **1993**, *49*, 1–12.

[57] Kuznik, G.; Hörsch, B.; Kretzschmar, G.; Unverzagt, C. *Bioorg. Med. Chem. Lett.* **1997**, *7*, 577–580.

[58] Schneider, R.; Freyhardt, C. C.; Schmidt, R. R. *Eur. J. Org. Chem.* **2001**, 1655–1661.

[59] Yu, C.-S.; Niikura, K.; Lin, C.-C.; Wong, C.-H. *Angew. Chem. Int. Ed.* **2001**, *40*, 2900–2903.

[60] Mukaiyama, T.; Mandai, H.; Jona, H. *Chem. Lett.* **2002**, *31*, 1182–1183.

[61] Lu, K.-C.; Tseng, S.-Y.; Lin, C.-C. *Carbohydr. Res.* **2002**, *337*, 755–760.

[62] Tanaka, H.; Nishiura, Y.; Takahashi, T. *J. Am. Chem. Soc.* **2006**, *128*, 7124–7125.

[63] Farris, M. D.; De Meo, C. *Tetrahedron Lett.* **2007**, *48*, 1225–1227.

[64] Crich, D.; Li, W. *J. Org. Chem.* **2007**, *72*, 2387–2391.

[65] Crich, D.; Li, W. *J. Org. Chem.* **2007**, *72*, 7794–7797.

[66] Codée, J. D. C.; Ali, A.; Overkleeft, H. S.; van der Marel, G. A. *C. R. Chim.* **2011**, *14*, 178–193.

[67] Litjens, R. E. J. N.; van den Bos, L. J.; Codée, J. D. C.; Overkleeft, H. S.; van der Marel, G. *Carbohydr. Res.* **2007**, *342*, 419–429.

[68] Crich, D. *Acc. Chem. Res.* **2010**, *43*, 1144–1153.

[69] Crich, D.; Vinod, A. U.; Picione, J. *J. Org. Chem.* **2003**, *68*, 8453–8458.

[70] Tanaka, H.; Nishiura, Y.; Takahashi, T. *J. Org. Chem.* **2009**, *74*, 4383–4386.

[71] Tanaka, H.; Tateno, Y.; Nishiura, Y.; Takahashi, T. *Org. Lett.* **2008**, *10*, 5597–5600.

[72] Tanaka, H.; Nishiura, Y.; Takahashi, T. *J. Am. Chem. Soc.* **2008**, *130*, 17244–17245.

[73] Liang, F. F.; Chen, L.; Xing, G. W. *Synlett* **2009**, 425–428.

[74] Xing, G. W.; Chen, L.; Liang, F. F. *Eur. J. Org. Chem.* **2009**, 5963–5970.

[75] Wang, Y. J.; Jia, J.; Gu, Z. Y.; Liang, F. F.; Li, R. C.; Huang, M. H.; Xu, C. S.; Zhang, J. X.; Men, Y.; Xing, G. W. *Carbohydr. Res.* **2011**, *346*, 1271–1276.

[76] Garcia, B. A.; Poole, J. L.; Gin, D. Y. *J. Am. Chem. Soc.* **1997**, *119*, 7597–7598.

[77] Codée, J. D. C.; van den Bos, L. J.; Litjens, R. E. J. N.; Overkleeft, H. S.; van Boeckel, C. A. A.; van Boom, J. H.; van der Marel, G. A. *Tetrahedron* **2004**, *60*, 1057–1064.

[78] Sun, B.; Jiang, H. *Tetrahedron Lett.* **2011**, *52*, 6035–6038.

[79] Crich, D.; Sun, S. *J. Am. Chem. Soc.* **1998**, *120*, 435–436.

[80] Zhang, Z.; Niikura, K.; Huang, X.-F.; Wong, C.-H. *Can. J. Chem.* **2002**, *80*, 1051–1054.

[81] Roy, R.; Andersson, F. O.; Letellier, M. *Tetrahedron Lett.* **1992**, *33*, 6053–6056.

[82] Shiao, T. C.; Roy, R. *Top. Curr. Chem.* **2011**, *301*, 69–108.

[83] Crich, D.; Wu, B. *Org. Lett.* **2008**, *10*, 4033–4035.

[84] Hsu, C.-H.; Chu, K.-C.; Lin, Y.-S.; Han, J.-L.; Peng, Y.-S.; Ren, C.-T.; Wu, C.-Y.; Wong, C.-H. *Chem. Eur. J.* **2010**, *16*, 1754–1760.

[85] Chu, K.-C.; Ren, C.-T.; Lu, C.-P.; Hsu, C.-H.; Sun, T.-H.; Han, J.-L.; Pal, B.; Chao, T.-A.; Lin, Y.-F.; Wu, S.-H.; Wong, C.-H.; Wu, C.-Y. *Angew. Chem. Int. Ed.* **2011**, *50*, 9391–9395.

[86] Crich, D.; Navuluri, C. *Angew. Chem. Int. Ed.* **2010**, *49*, 3049–3052.

[87] Crich, D.; Wu, B. *Tetrahedron* **2008**, *64*, 2042–2047.

[88] Tanaka, H.; Ando, H.; Ishihara, H.; Koketsu, M. *Carbohydr. Res.* **2008**, *343*, 1585–1593.

[89] Hanashima, S.; Sato, K.-I.; Ito, Y.; Yamaguchi, Y. *Eur. J. Org. Chem.* **2009**, 4215–4220.

[90] Kononov, L. O.; Malysheva, N. N.; Orlova, A. V. *Eur. J. Org. Chem.* **2009**, 611–616.

[91] Kononov, L. O.; Malysheva, N. N.; Orlova, A. V.; Zinin, A. I.; Laptinskaya, T. V.; Kononova, E. G.; Kolotyrkina, N. G. *Eur. J. Org. Chem.* **2012**, 1926–1934.

[92] Huang, M.; Garrett, G. E.; Birlirakis, N.; Bohé, L.; Pratt, D. A.; Crich, D. *Nat. Chem.* **2012**, *4*, 663–667.

[93] Huang, M.; Retailleau, P.; Bohé, L.; Crich, D. *J. Am. Chem. Soc.* **2012**, *134*, 14746–14749.

[94] Fraser-Reid, B.; Lopez, C. *Top. Curr. Chem.* **2011**, *301*, 1–9.

[95] Crich, D.; Li, M. *Org. Lett.* **2007**, *9*, 4115–4118.

[96] Paulsen, H.; Matschulat, P. *Liebigs Ann. Chem.* **1991**, 487–495.

[97] Kuribayashi, T.; Ohkawa, N.; Satoh, S. *Tetrahedron Lett.* **1998**, *39*, 4537–4540.

[98] Kuribayashi, T.; Mizuno, Y.; Gohya, S.; Satoh, S. J. *Carbohydr. Chem.* **1999**, *18*, 371–382.

[99] Wallimann, K.; Vasella, A. *Helv. Chim. Acta* **1991**, *74*, 1520–1532.

[100] Somsak, L. *Chem. Rev.* **2001**, *101*, 81–135.

[101] Vlahov, I. R.; Vlahova, P. I.; Linhardt, R. J. *J. Am. Chem. Soc.* **1997**, *119*, 1480–1481.

[102] Abdallah, Z.; Doisneau, G.; Beau, J.-M. *Angew. Chem. Int. Ed.* **2003**, *42*, 5209–5212.

[103] Papin, C.; Doisneau, G.; Beau, J.-M. *Chem. Eur. J.* **2009**, *15*, 53–57.

[104] Malapelle, A.; Abdallah, Z.; Doisneau, G.; Beau, J.-M. *Heterocycles* **2009**, *77*, 1417–1424.

[105] Watanabe, T.; Hirai, G.; Kato, M.; Hashizume, D.; Miyagi, T.; Sodeoka, M. *Org. Lett.* **2008**, *10*, 4167–4170.

[106] Noel, A.; Delpech, B.; Crich, D. *Org. Lett.* **2012**, *14*, 1342–1345.

[107] Nagy, J. O.; Bednarski, M. D. *Tetrahedron Lett.* **1991**, *32*, 3953–3956.

[108] Lubineau, A.; LeGallic, J. *J. Carbohydr. Chem.* **1991**, *10*, 263–268.

[109] Rich, J. R.; Wakarchuk, W. W.; Bundle, D. R. *Chem. Eur. J.* **2006**, *12*, 845–858.

[110] Eisele, T.; Schmidt, R. R. *Liebigs Ann. Recl.* **1997**, 1303–1313.

[111] Sabesan, S.; Neira, S.; Davidson, F.; Duus, J. O.; Bock, K. *J. Am. Chem. Soc.* **1994**, *116*, 1616–1634.

[112] Turnbull, W. B.; Field, R. A. *J. Chem. Soc. Perkin Trans. 1* **2000**, 1859–1866.

[113] Bennett, S.; von Itzstein, M.; Kiefel, M. J. *Carbohydr. Res.* **1994**, *259*, 293–299.

[114] Kiefel, M. J.; Beisner, B.; Bennett, S.; Holmes, I. D.; von Itzstein, M. *J. Med. Chem.* **1996**, *39*, 1314–1320.

[115] Liang, C.-F.; Yan, M.-C.; Chang, T.-C.; Lin, C.-C. *J. Am. Chem. Soc.* **2009**, *131*, 3138–3139.

[116] Liang, C. F.; Kuan, T.-C.; Chang, T.-C.; Lin, C.-C. *J. Am. Chem. Soc.* **2012**, *134*, 16074–16079.

[117] Noel, A.; Delpech, B.; Crich, D. *Org. Lett.* **2012**, *14*, 4138–4141.

6

STRATEGIES FOR ONE-POT SYNTHESIS OF OLIGOSACCHARIDES

Bo Yang, Keisuke Yoshida, and Xuefei Huang

Department of Chemistry, Michigan State University, East Lansing, MI, USA

6.1 INTRODUCTION

Traditionally, oligosaccharide synthesis is carried out in a stepwise fashion. Upon the successful formation of a glycosidic linkage, further elongation of the glycan chain requires that the newly formed oligosaccharide is either deprotected to generate a new glycosyl acceptor or transformed into a new glycosyl donor by modifying its aglycone. Although many complex oligosaccharides have been assembled in this manner, the stepwise approach is tedious and time consuming due to the need for multiple protecting group adjustment and aglycone leaving group manipulation on oligosaccharide intermediates. In order to improve the overall synthetic efficiency, one-pot strategies have been designed.

One-pot oligosaccharide synthesis refers to approaches by which one glycosyl building block is subjected to successive chemical reactions in the same flask without the need to purify the intermediates. It can potentially improve the speed of the overall synthetic operations and increase reaction yields by reducing product loss due to purification. For a one-pot glycosylation to be successful, it is crucial that each reaction proceeds with high yield and stereoselectivity and the side products generated do not interfere with subsequent reactions. Over the past 20 years, many innovative one-pot glycosylation strategies have been successfully developed [1–6]. While the majority of one-pot syntheses are performed from the nonreducing end to the reducing end, it can also be carried out from the reducing end to the nonreducing

Glycochemical Synthesis: Strategies and Applications, First Edition.
Edited by Shang-Cheng Hung and Medel Manuel L. Zulueta.

end or through a combination of strategies (hybrid one-pot glycosylation) to expedite the production of the desired oligosaccharide.

6.2 ONE-POT GLYCOSYLATION FROM THE NONREDUCING END TO THE REDUCING END

To perform one-pot synthesis from the nonreducing end to the reducing end, the acceptor should bear, besides the free hydroxy group, an activatable aglycone moiety, which must be stable under the first glycosylation condition. Upon successful glycosylation, the newly formed glycoside should be able to undergo direct activation in the same reaction flask with the addition of a new promoter. Another critical consideration is the preferential activation of the glycosyl donor over the bifunctional acceptor. A variety of strategies have been developed to achieve this, which include designing building blocks with differential anomeric reactivities, chemoselective activation of different types of glycosyl donors, and preactivation of glycosyl donors.

6.2.1 Reactivity-Based One-Pot Glycosylation: Fine-Tuning of Anomeric Reactivities

Glycosyl building blocks can have different anomeric reactivities, which can be utilized in designing a reactivity-based one-pot synthesis. This type of synthesis starts by subjecting a mixture of donor **1** bearing the highest anomeric reactivity (armed donor) and the less reactive bifunctional acceptor **2** (disarmed) to a limiting amount of promoter (Scheme 6.1). Due to the anomeric reactivity disparity, **1** is preferentially activated and glycosylates the acceptor **2** producing one major glycoside product (**3**). This process can be repeated in the same reaction flask in the order of decreasing anomeric reactivities to produce the desired oligosaccharide **6**. A variety of factors can influence the anomeric reactivities, which include protecting groups, the anomeric leaving group, and solvents.

6.2.1.1 Reactivity Tuning through Protecting Groups Protecting groups play fundamental roles in carbohydrate chemistry. Besides controlling stereochemistry in the formation of glycosidic linkages, protecting groups significantly influence the

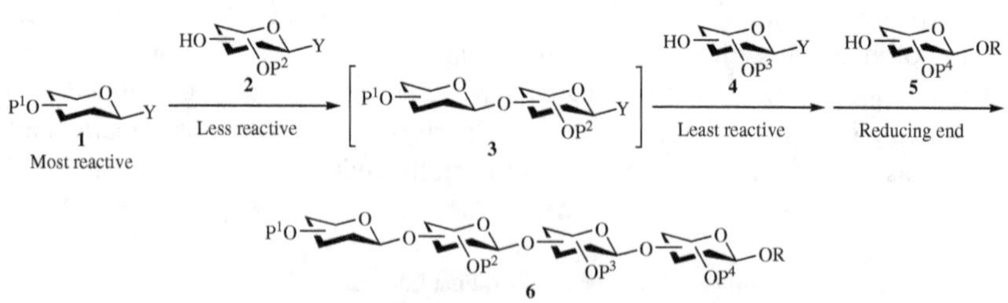

SCHEME 6.1 General scheme for reactivity-based one-pot glycosylation.

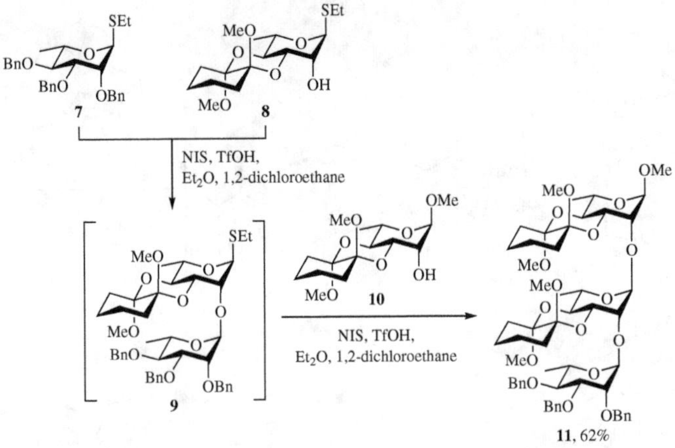

SCHEME 6.2 Ley's one-pot synthesis of trisaccharide **11**. Bn, benzyl.

anomeric reactivities of glycosyl building blocks through either electronic or conformational effects (see Section 4.3). Following the demonstration using *n*-pentenyl glycosides by Bert Fraser-Reid and coworkers [7, 8], the reactivity-based chemoselective glycosylation strategy arising from the reactivity tuning by protecting groups was found generally applicable to a variety of other glycosyl donor systems including glycals [9] and thioglycosides [10].

Steven Ley and coworkers reported the first reactivity-based one-pot glycosylation in their preparation of a trisaccharide unit from the polysaccharide antigen of group B streptococci [11]. They created the reactivity difference in rhamnose monosaccharides by using the protecting group cyclohexane-1,2-diacetal (CDA). A cyclic acetal such as CDA rigidifies the chair conformation of the glycan ring, rendering it more difficult to reach the twisted form adopted by oxocarbenium ions upon donor activation [12]. This phenomenon is referred to as conformational disarming. Thus, the conformationally flexible rhamnoside **7** can be activated faster than the CDA-bearing rhamnoside **8** (Scheme 6.2). Coupling of **7** and **8** in the presence of *N*-iodosuccinimide (NIS) and triflic acid (TfOH) afforded the disaccharide intermediate **9**, which was immediately treated with the acceptor **10** and the NIS/TfOH promoting system to give trisaccharide **11** in an excellent 62% yield [11].

Besides conformational disarming, one-pot methods also benefit from building blocks that are electronically disarmed. Biao Yu and coworkers utilized this effect in the one-pot assembly of the 19-membered macrolactone-containing tricolorin A (**16**) (Scheme 6.3) [13, 14]. The key design of their synthesis is the chemoselective activation and coupling of the perbenzylated donor **7** to the bis-acyl-containing disarmed bifunctional acceptor **12**. The subsequent glycosylation of macrolactone **14** in the same flask formed the protected tricolorin A **15** in 43% overall yield, which was deprotected to provide tricolorin A (**16**).

Analogously, Hirosato Kondo and coworkers carried out the one-pot assembly of Lewis X derivatives by taking advantage of the electronic disarming effect

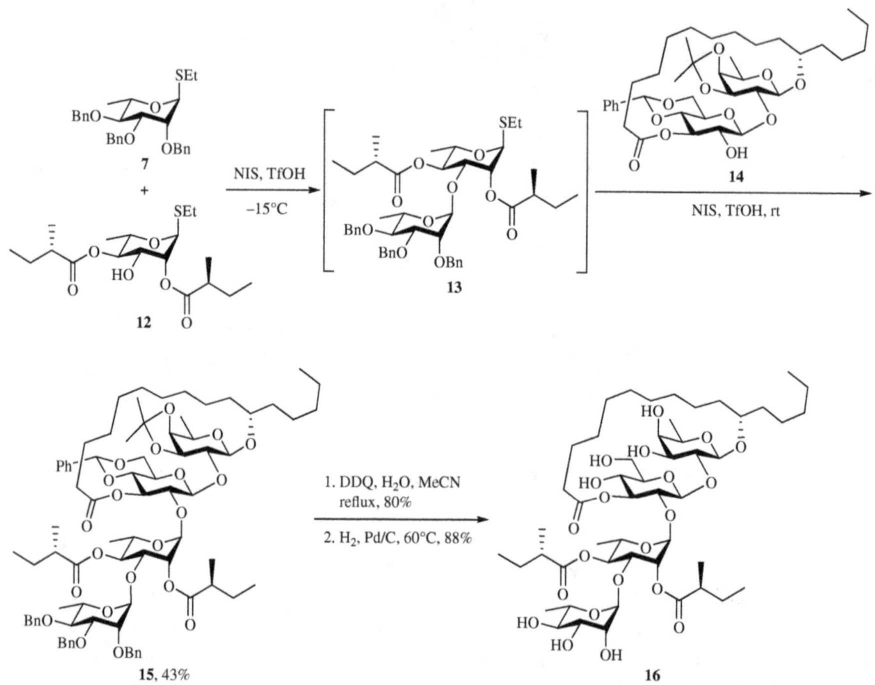

SCHEME 6.3 Yu's synthesis of tricolorin A. DDQ, 2,3-dichloro-5,6-dicyano-1-4,-benzoquinone.

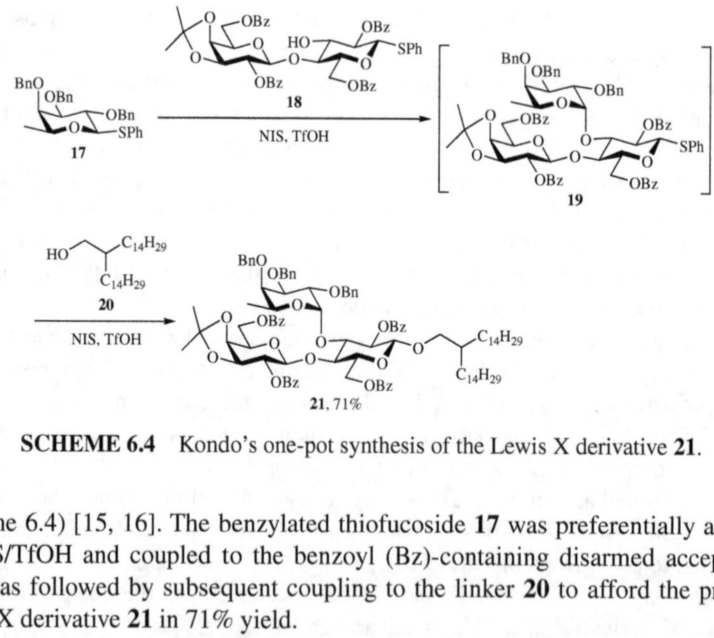

SCHEME 6.4 Kondo's one-pot synthesis of the Lewis X derivative **21**.

(Scheme 6.4) [15, 16]. The benzylated thiofucoside **17** was preferentially activated by NIS/TfOH and coupled to the benzoyl (Bz)-containing disarmed acceptor **18**. This was followed by subsequent coupling to the linker **20** to afford the protected Lewis X derivative **21** in 71% yield.

To synthesize longer oligosaccharide sequences using the armed–disarmed one-pot glycosylation approach, more than two consecutive glycosylation reactions are often required. As an aid in selecting suitable building blocks, it is desirable that the anomeric reactivity of the building blocks be quantified. Accordingly, Ley and coworkers explored the utility of the ^{1}H-NMR chemical shifts of anomeric protons as an indirect indicator of the anomeric reactivities [17]. This is based on the assumption that if a glycan contains multiple disarming electron-withdrawing protecting groups, its anomeric proton should be more downfield shifted. However, as chemical shifts can also be influenced by the magnetic anisotropy effect of the protecting groups, it is difficult to use chemical shift values to precisely predict the anomeric reactivities of a wide range of building blocks.

Alternatively, Chi-Huey Wong and coworkers developed an HPLC-based competition assay by subjecting two thioglycosides to substoichiometric amount of a promoter [18]. The relative anomeric reactivity values (RRVs) of building blocks were obtained from the integration of the respective glycoside products. Hundreds of monosaccharide/disaccharide (thioglycoside) building blocks were prepared and their respective RRVs were determined [6]. Monosaccharides were found to have different inherent reactivities with the following trend: fucose > galactose > glucose > mannose [18]. Both the structure and the location of protecting groups have significant impacts on anomeric reactivities. The RRVs were stored in a searchable database to help the selection of building blocks for the assembly of specific oligosaccharide structures. For example, the Lewis Y carbohydrate hapten **26** was prepared by this methodology (Scheme 6.5) [19]. Three advanced building blocks (**22**–**24**) with the respective RRVs of 7.2×10^4 and 1.2×10^4 and 0 were designed and synthesized. Sequential one-pot glycosylation afforded the fully protected Lewis Y hexasaccharide **25**

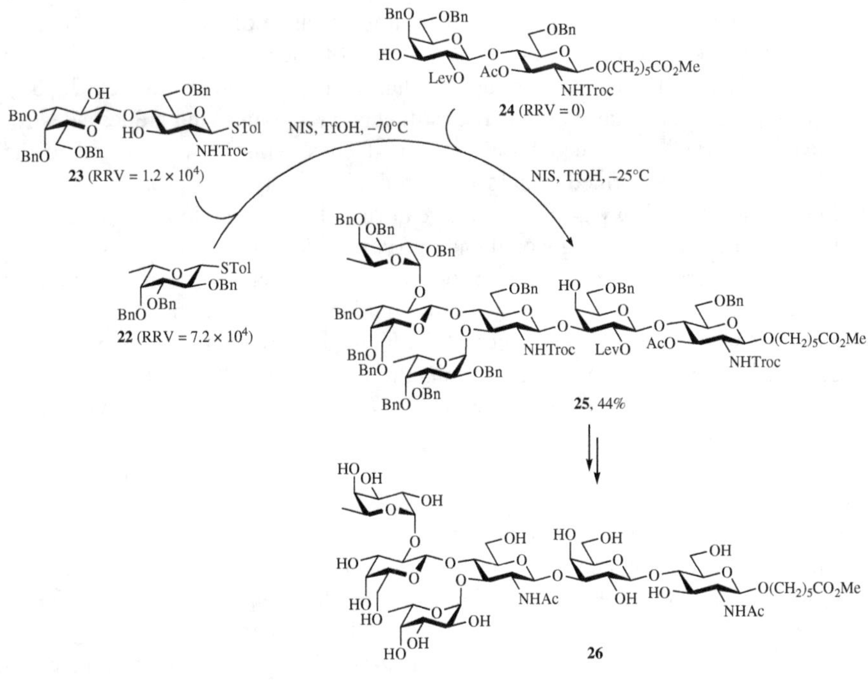

SCHEME 6.5 Wong's programmable one-pot synthesis of Lewis Y carbohydrate hapten. Ac, acetyl; Lev, levulinyl; Tol, 4-tolyl; Troc, trichloroethoxycarbonyl.

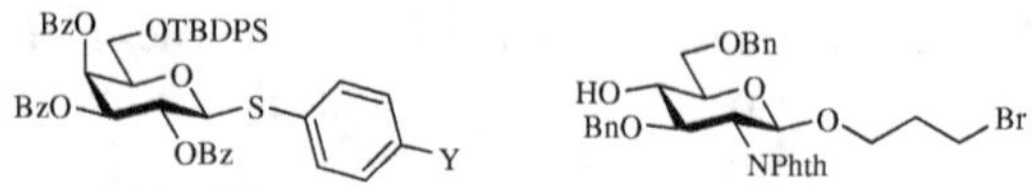

FIGURE 6.1 Building blocks employed by Huang in examining relative reactivities. Phth, phthaloyl; TBDPS, *tert*-butyldiphenylsilyl.

in 44% yield. Besides Lewis Y, the reactivity-based one-pot methodology has been applied to the assembly of an impressive array of complex oligosaccharides such as Globo H hexasaccharide [20], dimeric Lewis X [21], fucosyl GM1 [22], heparin-like oligosaccharide [23], tumor-associated antigen stage-specific embryonic antigen-4 (SSEA-4) [24], tumor-associated antigen N3 minor octasaccharide [25], KH-1 epitope [21], *N*-acetyllactosamine oligomers [26], vancomycin [27], oligomannoses [28], lactotetraose (Lc_4) [29], and oligosaccharide libraries [30].

The RRVs can provide guidance toward building block selections. However, these values were obtained through measuring the competitive glycosylation rate differences using methanol as the acceptor. Caution needs to be taken in extrapolating these values to other glycosylation conditions. Xuefei Huang and coworkers found that factors such as acceptor structure and solvents could significantly impact the relative reactivities [31]. For example, while donor **27** was 21.2 times more reactive than donor **28** with methanol as the acceptor, the difference dropped to only 5.95-fold when relative reactivities were measured with acceptor **29** (Fig. 6.1).

The success of the reactivity-based one-pot synthesis depends critically on the availability of glycosyl building blocks with a wide range of anomeric reactivities. Mikael Bols and coworkers performed a series of mechanistic investigations to understand the effect of hydroxy group orientation on anomeric reactivities [32, 33]. They discovered that axially oriented substituents on a glycan ring are less electronegative compared to the equatorially oriented ones. This knowledge led to the development of super-armed donors by introducing bulky silyl groups on neighboring equatorial hydroxy groups of a glycan ring. Due to the unfavorable gauche interactions, the pyranose ring would undergo conformational changes from the normally lowest energy ${}^4C_1$ conformation to a twisted boat conformation, thus placing the substituents into the less electronegative pseudoaxial orientations [34]. This conformational arming concept was successfully applied to the one-pot synthesis of trisaccharide **33** (Scheme 6.6) [35]. By mixing the super-armed donor **30**, armed donor **31**, and disarmed acceptor **32** with 2.1 equiv. of NIS and catalytic TfOH, they isolated trisaccharide **33** in 64% yield.

6.2.1.2 Reactivity Tuning through Anomeric Leaving Groups

Besides glycan ring protecting groups, the aglycone can significantly impact anomeric reactivity [36]. When the aglycone bears electron-donating moieties, it becomes more electron rich and thus can undergo faster electrophilic attack by the promoter. Richard Schmidt's trichloroacetimidate donor has been widely used for oligosaccharide synthesis [37]. As an alternative, Yu and coworkers developed glycosyl *N*-phenyltrifluoroacetimidate as

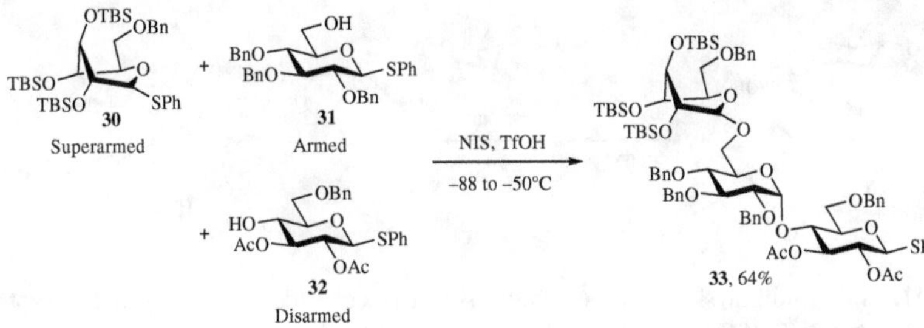

SCHEME 6.6 One-pot synthesis of trisaccharide **33** using the super-armed building block **30**. TBS, *tert*-butyldimethylsilyl.

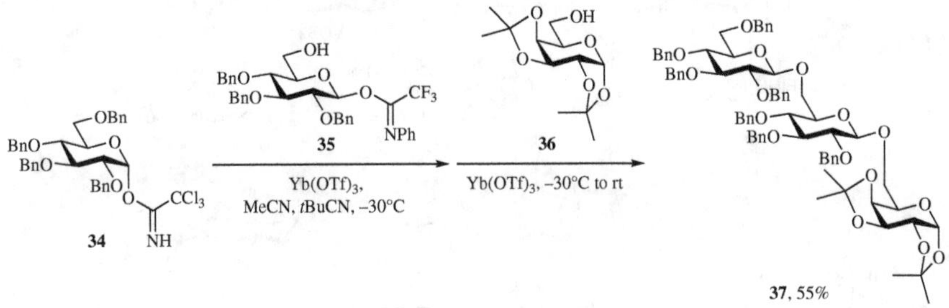

SCHEME 6.7 $Yb(OTf)_3$-mediated one-pot synthesis of trisaccharide **37**. Tf, triflyl.

glycosyl donors [38]. Alfonso Iadonisi et al. reported that glycosyl trichloroacetimidate can be selectively activated over the corresponding glycosyl trifluoroacetimidate presumably due to the electron-withdrawing property of the trifluoromethyl group [39, 40]. Hence, the trichloroacetimidate donor **34** was selectively activated by $Yb(OTf)_3$ under low temperature and glycosylated the trifluoroacetimidate **35** (Scheme 6.7), which was followed by coupling to acceptor **36** by warming to room temperature to produce trisaccharide **37** in 55% yield.

Gary Sulikowski et al. applied glycosyl phosphites for the one-pot assembly of oligosaccharides [41]. Their strategy relied on the differential activation rates of various dialkyl glycosyl phosphites. Because of the higher reactivity of the diethyl glycosyl phosphite **38** compared to the pinacol phosphite **39**, **38** was selectively activated to produce a disaccharide intermediate, which further underwent coupling with acceptor **40** to afford trisaccharide **41** in 50% overall yield (Scheme 6.8).

Instead of preparing each glycosyl building block bearing various aglycones separately, Huang and coworkers developed a divergent route to tune the anomeric reactivity by postsynthetic modification of the aglycone [31, 42]. The key thioglycosyl building block **42** bearing *p*-aminothiophenyl aglycone was prepared (Scheme 6.9a). To create donors with multiple levels of reactivities, the amine group was transformed into a variety of substituents through straightforward synthetic operations. In this manner, starting from an advanced intermediate, building blocks with multiple levels of anomeric reactivities (e.g., **27** and **43–47**) were obtained, thus reducing the amount of time required for building block preparation. Furthermore, as the building

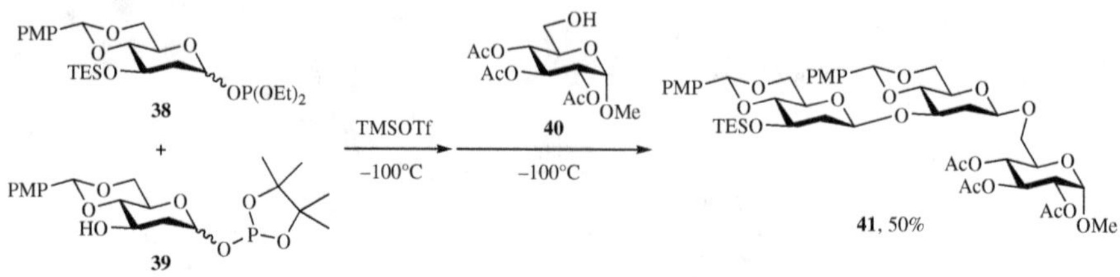

SCHEME 6.8 Sulikowski's one-pot synthesis of trisaccharide **41**. PMP, *p*-methoxyphenyl; TES, triethylsilyl; TMSOTf, trimethylsilyl triflate.

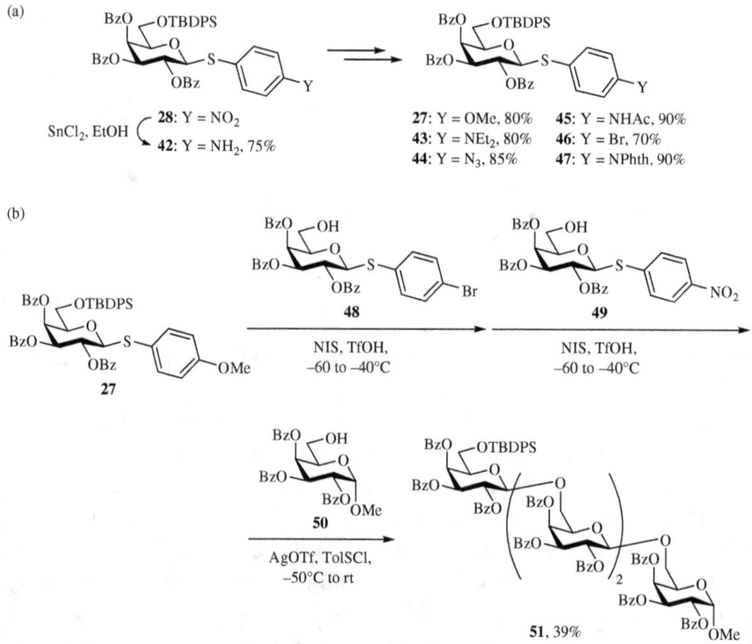

SCHEME 6.9 Huang's (a) postsynthetic aglycone modifications and (b) one-pot synthesis of tetrasaccharide **51**.

blocks only differ in aglycone substituents, the Hammett constant of the substituents was found useful for predicting the anomeric reactivities [31]. The synthetic utility of this method was demonstrated in several four-component one-pot syntheses. As an example, activation of the armed PMP-bearing donor **27** followed by sequential addition of **48**, **49**, and **50** upon completion of each coupling generated pentasaccharide **51** in 39% overall yield (Scheme 6.9b) [42]. The first two couplings were promoted by NIS and TfOH, while the last one was mediated by the more powerful promoter AgOTf/TolSCl due to the reduced reactivity of the growing saccharide chain.

Besides tuning the reactivity through aglycone substituents, different types of aglycone leaving groups can be used by taking advantage of their inherent reactivity

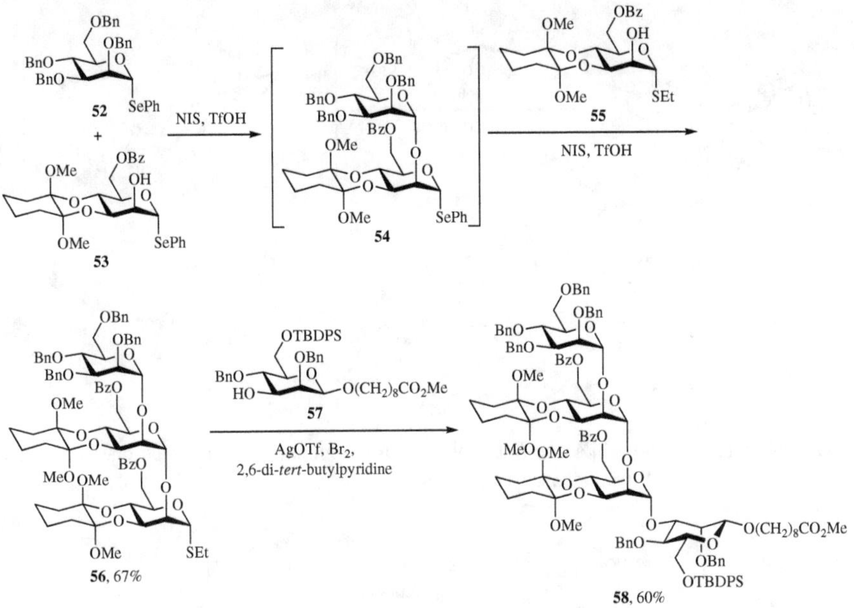

SCHEME 6.10 Synthesis of tetrasaccharide **58**.

differences. As an example, Ley and coworkers introduced selenoglycosides as glycosyl donors in one-pot synthesis because selenoglycosides are more reactive than their thioglycoside counterparts [17, 43, 44]. With the combination of the conformationally disarming CDA group, selenoglycosides, and thioglycosides, they were able to generate donors with four different levels of reactivities, which were applied to the three-component one-pot synthesis of a high mannose-type oligosaccharide—a part of the glycoprotein gp120 of the viral coat of HIV-1. As shown in Scheme 6.10, the coupling of the armed selenoglycoside **52** with the conformationally disarmed **53** gave disaccharide **54**, which was subjected to reaction with thioglycoside **55** producing trisaccharide **56** in one pot. NIS/TfOH failed to activate **56** to permit the final coupling with acceptor **57**, which might be due to the reduced anomeric reactivity of **56**. Using the more powerful promoter system AgOTf/Br_2 allowed the successful coupling of **56** with **57** to afford the tetramannoside **58**.

Instead of using selenoglycosides and thioglycosides, Yoshinori Kobayashi et al. explored the reactivity difference between glycosyl trichloroacetate and glycosyl *N*-trichloroacetylcarbamate building blocks [45]. They discovered that the *N*-trichloroacetylcarbamate donor **59** could be selectively activated over the bifunctional glycosyl trichloroacetate **60** by TMSOTf at 0°C to form disaccharide **61** (Scheme 6.11). Upon completion of the glycosylation, the third building block **62** was added to the mixture. Warming to room temperature led to the successful activation of **61** and the formation of trisaccharide **63**. However, the low levels of stereochemical control at the newly formed glycosidic linkages are a drawback of this synthesis.

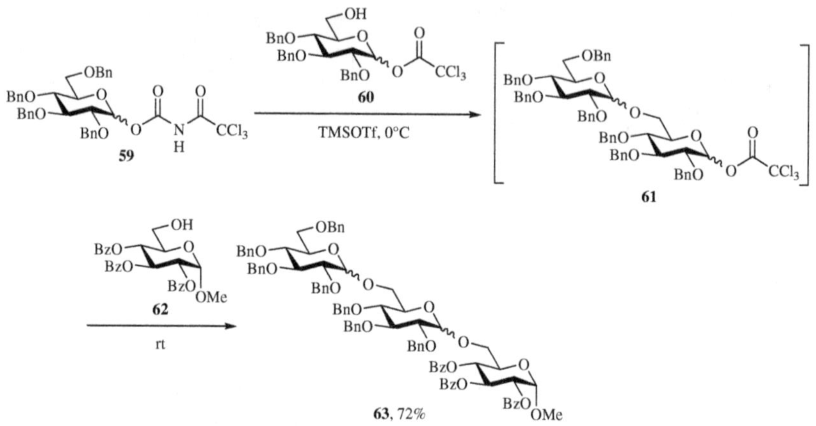

SCHEME 6.11 Kobayashi's one-pot synthesis of trisaccharide **63** through chemoselective activation of glycosyl *N*-trichloroacetylcarbamate.

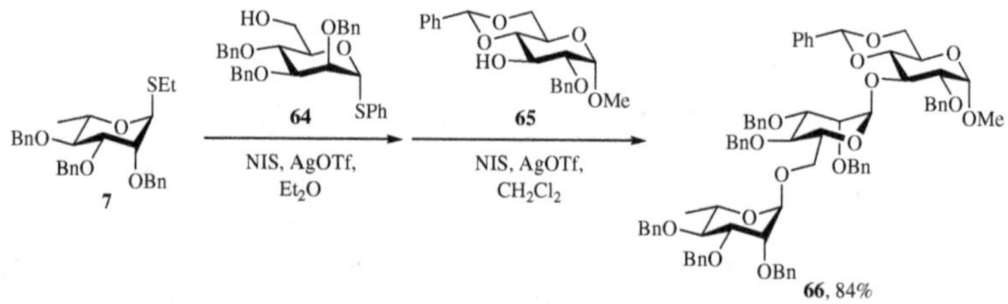

SCHEME 6.12 Oscarson's one-pot synthesis of trisaccharide **66**.

6.2.1.3 Solvent Tuning Solvents are well known to affect the outcome of glycosylation reactions. Many glycosylations are carried out in CH_2Cl_2 due to the higher reaction rate in this solvent, while coordinating solvents such as MeCN and Et_2O slow down the reactions. Based on this effect, Stefan Oscarson and coworkers designed a one-pot strategy by performing the first glycosylation step in Et_2O [46]. As the glycosylation reaction is slower in Et_2O, larger chemoselectivity could be potentially obtained for the preferential activation of the more reactive donor. Because the acceptor in the second step did not bear any activatable aglycone, CH_2Cl_2 was added to the reaction mixture to enhance the glycosylation rate. As shown in Scheme 6.12, donor **7** was selectively activated to glycosylate the acceptor **64** in Et_2O, followed by the addition of acceptor **65** in CH_2Cl_2 to produce trisaccharide **66** in an excellent 84% yield. When the first reaction was carried out in CH_2Cl_2, a complex product mixture was obtained, thus demonstrating the importance of the reaction solvent for the chemoselectivity.

Timor Baasov and coworkers reported a one-pot glycosylation approach, which combined the effects of glycone protecting groups, anomeric leaving group, and solvent tuning [47]. Two types of N-protecting groups (Troc and Phth) and two types of aglycone leaving groups (thioethyl and thiophenyl) were utilized. As NHTroc and

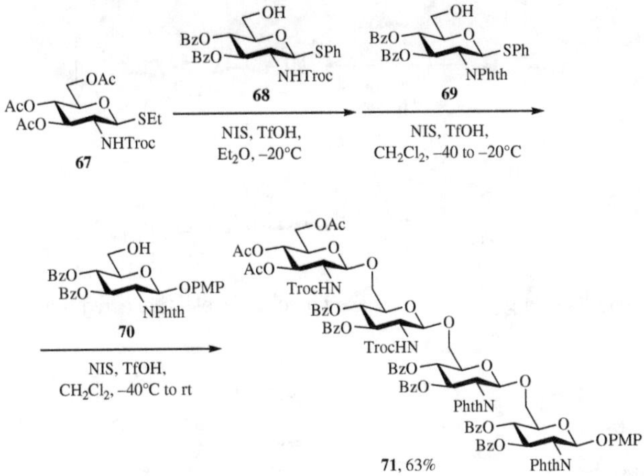

SCHEME 6.13 Baasov's one-pot synthesis of tetrasaccharide **71**.

thioethyl are more electron rich than NPhth and thiophenyl, respectively, the building blocks **67**, **68**, and **69** exhibit decreasing anomeric reactivities (Scheme 6.13). Selective activation of **67** over **68** was achieved in Et_2O, and upon completion of this coupling, **69** and **70** were added consecutively in CH_2Cl_2 to furnish the tetrasaccharide **71** in 63% yield.

6.2.2 One-Pot Glycosylation Based on Chemoselective Activation of Different Types of Glycosyl Donors

The chemoselectivity from the reactivity-based one-pot glycosylation comes from the higher anomeric reactivity of the glycosyl donor compared to the acceptor. Alternatively, chemoselective activation can be achieved by employing different types of building blocks bearing aglycones that require distinct promoters for activation. As shown in Scheme 6.14, selective activation of donor **72** using promoter 1 in the presence of **73** produces a disaccharide intermediate that can be selectively activated by promoter 2 to produce trisaccharide **76**. It is evident that selection of appropriate anomeric leaving groups is crucial for the success of this strategy.

6.2.2.1 Glycosyl Trichloroacetimidate and Thioglycoside Pair Trichloroacetimidate donors are typically activated with a Lewis acid such as TMSOTf, whereas thioglycosides are generally stable under acidic conditions. This can enable the selective activation of a trichloroacetimidate donor over thioglycosides by TMSOTf. Takashi Takahashi and coworkers demonstrated this principle in the synthesis of the elicitor-active hexaglucoside **80** via a two-step, one-pot glycosylation approach (Scheme 6.15) [48]. The trichloroacetimidate donor **77** was coupled to the thioglycoside acceptor **78** promoted by TMSOTf. The resulting tetrasaccharide intermediate was further activated by NIS and TfOH to glycosylate the reducing end disaccharide acceptor **79**, producing the hexasaccharide **80** in 50% yield.

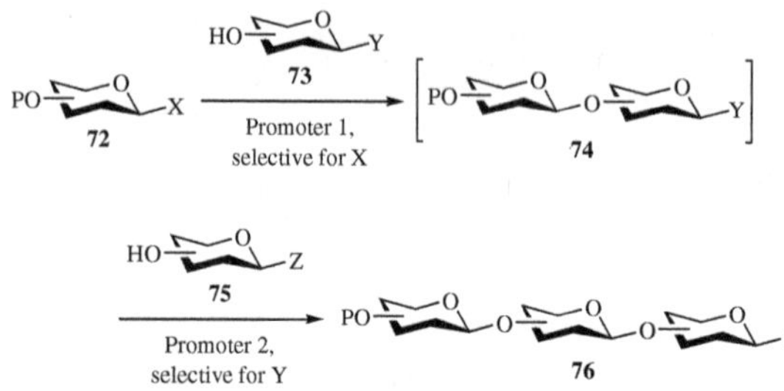

SCHEME 6.14 One-pot glycosylation based on chemoselective activation of different types of glycosyl donors.

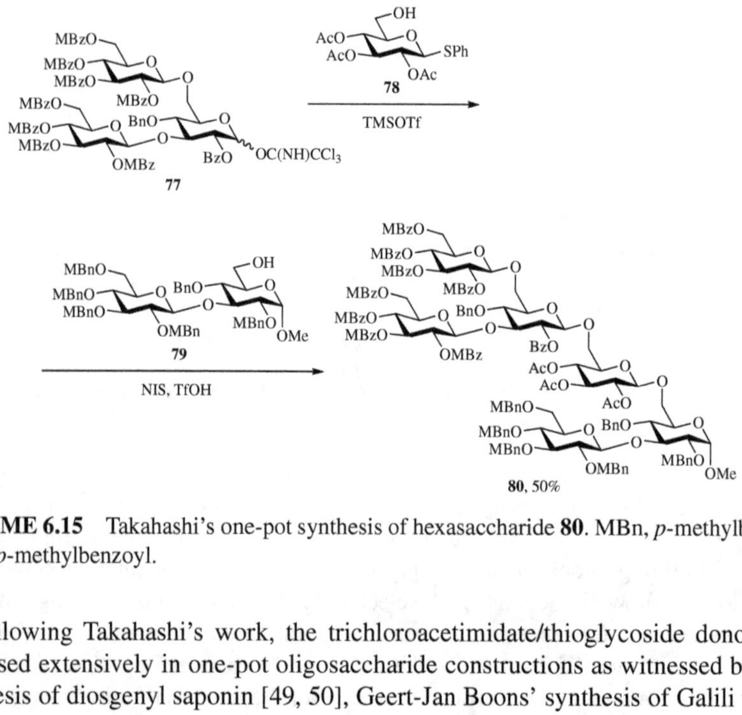

SCHEME 6.15 Takahashi's one-pot synthesis of hexasaccharide **80**. MBn, *p*-methylbenzyl; MBz, *p*-methylbenzoyl.

Following Takahashi's work, the trichloroacetimidate/thioglycoside donor pair was used extensively in one-pot oligosaccharide constructions as witnessed by Yu's synthesis of diosgenyl saponin [49, 50], Geert-Jan Boons' synthesis of Galili trisaccharide [51] and PSGL-1 hexasaccharide [52], Peter Seeberger's synthesis of a pentasaccharide against *Helicobacter pylori* [53], Todd Lowary's [54] and Jin-Song Yang's [55] synthesis of oligosaccharide fragments of the mycolyl-arabinogalactan complex, Gijsbert van der Marel's synthesis of hyaluronic acid oligomers [56], Yingxia Li's synthesis of flaccidoside II [57] and β-glucosyl chitobiose [58], and Feihong Bing's synthesis of the trisaccharide motif of natural saponin [59]. Among these, Boons' α-selective glycosylation approach using (1*S*)-phenyl-2-(phenylsulfanyl)ethyl moiety at O2 position as the participating neighboring group is worth mentioning (Scheme 6.16) [51]. Activation of the imidate donor **81** bearing (1*S*)-phenyl-2-(phenylsulfanyl)ethyl moiety led to the formation of the quasi-stable anomeric sulfonium ion **82** with a *trans*-decalin ring system (see Section 4.2.4).

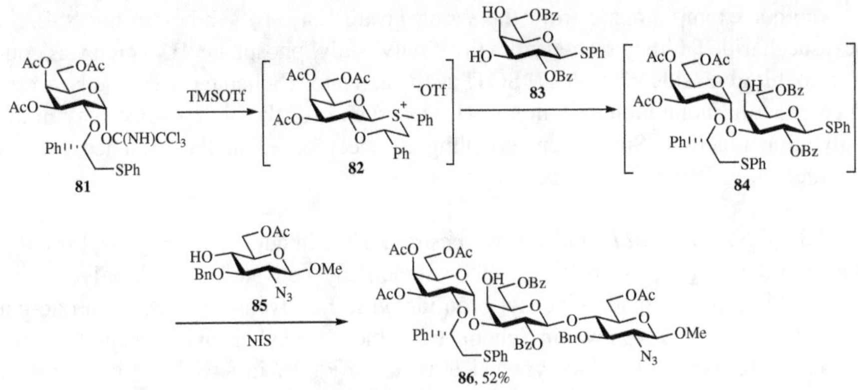

SCHEME 6.16 Stereoselective one-pot synthesis of trisaccharide **86**.

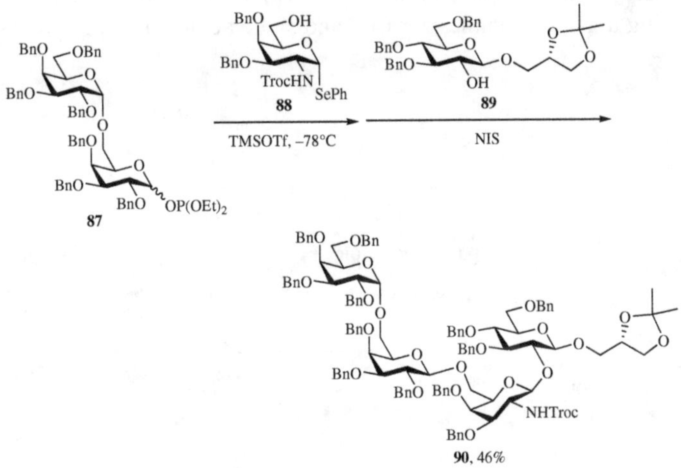

SCHEME 6.17 Wu's one-pot synthesis of tetrasaccharide **90**.

Displacement of the sulfonium ion by a hydroxy group of acceptor **83** in an S_N2-like fashion resulted in the formation of the α-glycoside **84**. This strategy was applied to the synthesis of trisaccharide **86** containing both α and β linkages in a one-pot fashion in 52% yield.

6.2.2.2 Glycosyl Phosphite/Phosphate and Seleno-/Thioglycoside Pair Similar to trichloroacetimidates, glycosyl phosphite/phosphate can be activated by a catalytic amount of TMSOTf, which in turn enables the glycosylation of selenoglycosyl/thioglycosyl acceptors. Shih-Hsiung Wu and coworkers reported the one-pot synthesis of a tetrasaccharide glycosyl glycerol by using this approach (Scheme 6.17) [60]. Activation of the glycosyl phosphite **87** over selenoglycoside **88** was achieved by TMSOTf. Further treatment with acceptor **89** and NIS in one pot produced trisaccharide **90** in 46% yield over the two steps.

Another example came from the Wong group [24]. To synthesize the SSEA-4 hexasaccharide **94**, they started from the dibutyl sialyl phosphate **91**, which was coupled to thioglycoside **92** with TMSOTf as the activator (Scheme 6.18). The N-acetylated oxazolidinone moiety in donor **91** was crucial for the high α selectivity in the sialylation reaction. Subsequent coupling to acceptor **93** in the presence of NIS afforded **94** in 78% yield for the two steps.

*6.2.2.3 **Glycosyl Halide and Thioglycoside Pair*** In addition to glycosyl trichloroacetimidates, glycosyl halides can be selectively activated over thioglycosides. Takahashi et al. combined glycosyl bromide with thioglycoside to perform one-pot glycosylation [61]. As shown in Scheme 6.19, the glycosyl bromide **95** and the thioglycoside **78** were utilized to synthesize trisaccharide **97** in 84% yield in a one-pot operation. Activation of the bromide **95** was achieved by AgOTf, and the resulting disaccharide donor was activated by the addition of NIS, forming trisaccharide **97**. This approach was later applied to the impressive assembly of a library of 72 trisaccharides using a manual synthesizer [62] and sialic acid-containing glycosyl amino acid [63].

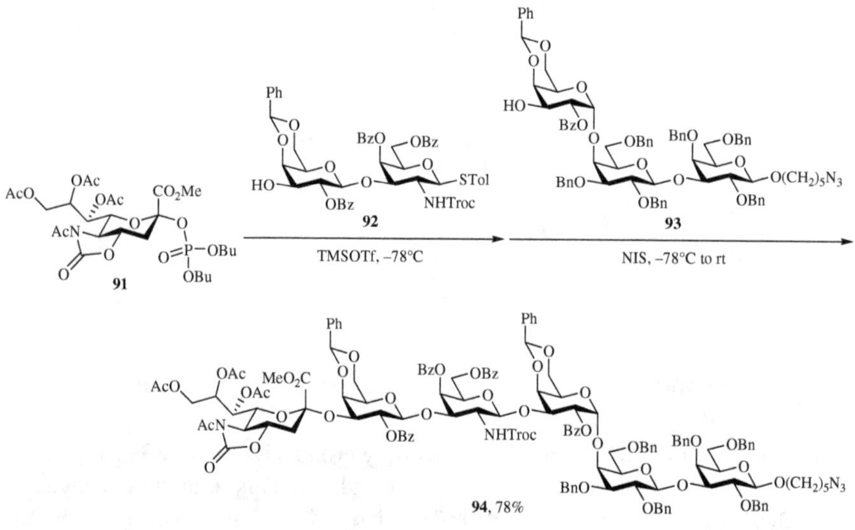

SCHEME 6.18 Wong's one-pot synthesis of the SSEA-4 hexasaccharide **94**.

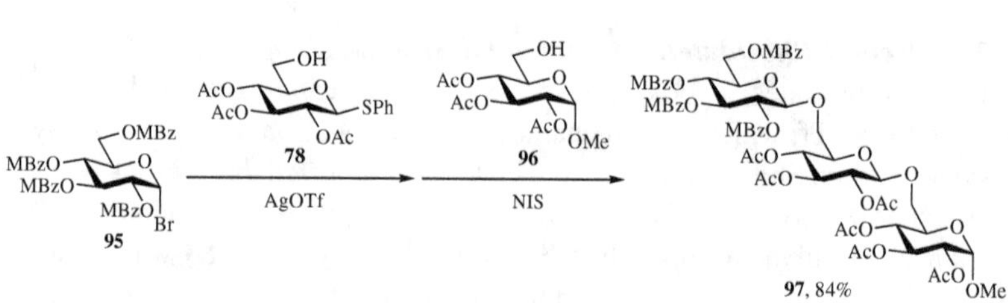

SCHEME 6.19 One-pot synthesis of trisaccharide **97**.

Methods for the stereoselective construction of the 1,2-*cis* linkage for 2-amino-2-deoxy sugars are in high demand because this linkage is very common in naturally existing oligosaccharides. With the 2,3-oxazolidinone group having been shown to significantly enhance α selectivity [64], Yukishige Ito and coworkers demonstrated that the *N*-benzyl-2,3-oxazolidinone group could be used to construct oligosaccharides with α linkages in one pot [65]. As shown in Scheme 6.20, the bromide donor **98** was activated by AgOTf to couple with **99**, producing a thioglycosyl disaccharide, which was immediately coupled to acceptor **100** to give trisaccharide **101** with two α linkages in 81% yield.

*6.2.2.4 **Glycosyl Thioimidates*** Thioglycosides such as ethyl thioglycoside, phenyl thioglycoside, and 4-tolyl thioglycoside are commonly used in glycosylation reactions [66]. Conversely, Demchenko and coworkers developed two novel types of glycosyl thioimidates: *S*-benzoxazolyl (SBox) [67–69] and *S*-thiazolyl (STaz) [70] thioglycosides, which have unique selectivities on activation compared to typical thioglycosides. They discovered that SBox could be activated over SEt by AgOTf, SEt could be activated over STaz by NIS/TfOH, and STaz could be activated by excess AgOTf. Based on this observation, a one-pot synthesis protocol was developed using building blocks containing different thio-leaving groups (Scheme 6.21). The coupling between **102** and **103** was promoted by AgOTf, and subsequent glycosylations were mediated by NIS/TfOH and AgOTf, respectively, leading to

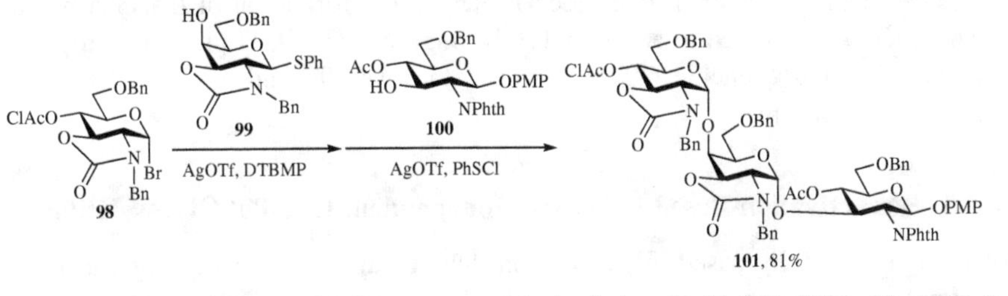

SCHEME 6.20 Ito's stereoselective one-pot synthesis of trisaccharide **101**. ClAc, chloroacetyl; DTBMP, di-*tert*-butylmethylpyridine.

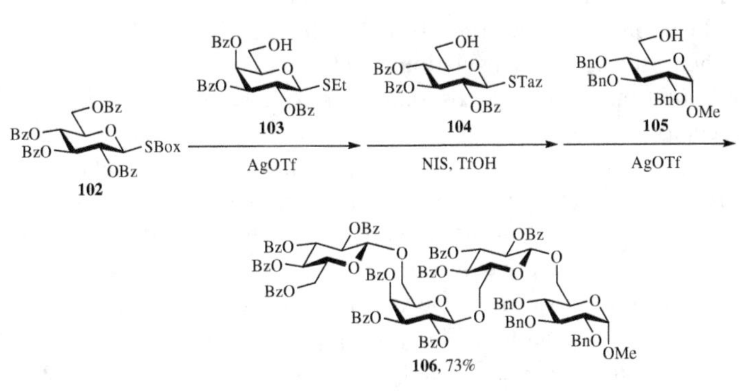

SCHEME 6.21 Demchenko's one-pot synthesis of tetrasaccharide **106**.

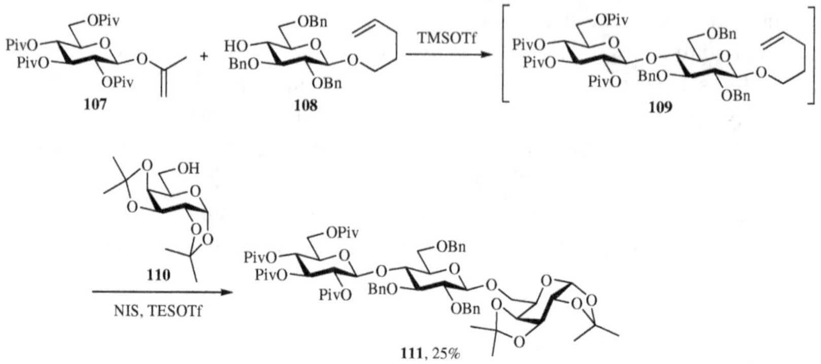

SCHEME 6.22 One-pot synthesis of trisaccharide **111**. Piv, pivaloyl.

tetrasaccharide **106** in 73% overall yield [71]. Encouraged by Demchenko's work, Takahashi synthesized α(2→9)-linked trisialic acid using the combination of SEt and SBox sialoside building blocks [72].

6.2.2.5 ***Isopropenyl and*** **n*-Pentenyl Glycoside Pair*** H. Keith Chenault and coworkers reported the selective activation of isopropenyl glycoside in the presence of an *n*-pentenyl glycoside [73]. Activation of donor **107** with TMSOTf in the presence of the *n*-pentenyl glycoside **108** led to the formation of the intermediate **109**, which was coupled to acceptor **110** through NIS/TESOTf promotion to afford trisaccharide **111** (Scheme 6.22). However, the overall yield is only 25%, which needs further optimization.

6.2.3 Preactivation-Based Reactivity-Independent One-Pot Glycosylation

Although reactivity-based glycosylation has greatly facilitated oligosaccharide assembly, extensive protecting group and aglycone manipulations have to be carried out to prepare the building blocks with desired anomeric reactivities. Furthermore, building blocks with suitably tuned anomeric reactivities need to be prepared depending on the location of the specific glycosidic linkage in the oligosaccharide chain [26]. The chemoselective glycosylation strategy based on different aglycones overcomes the limitation that the glycosyl donor must have higher anomeric reactivities than the acceptor. However, it requires the preparation of building blocks containing different types of aglycones. To further simplify the overall synthetic design, preactivation-based glycosylation was developed where one-pot glycosylation reactions can be performed independent of anomeric reactivities using building blocks containing the same aglycone. Preactivation refers to a glycosylation approach by which a glycosyl donor is activated in the absence of an acceptor to form a reactive intermediate followed by addition of the acceptor to couple with the activated species [74–77]. Unique chemoselectivity and stereoselectivity can be achieved with the preactivation approach.

van der Marel and coworkers reported a preactivation-based one-pot approach using glycosyl hemiacetal and thioglycosides [78]. They preactivated the hemiacetal donor **112** by using Ph_2SO/Tf_2O at −60°C followed by the addition of acceptor **113** (Scheme 6.23). The resulting mixture was warmed up to room temperature to afford a thioglycosyl disaccharide. Without separation, the reaction mixture was cooled back to −60°C followed by addition of Tf_2O, which generated a new thiophilic promoter. Addition of the final acceptor **114** to the reaction produced the protected hyaluronic acid trisaccharide **115** in 32% yield. This protocol has not been extended to more than two glycosylations in one pot because it is challenging to use the glycosyl hemiacetal as the bifunctional building block.

Xuefei Huang, Xin-Shan Ye, and coworkers established the concept of iterative one-pot synthesis of oligosaccharides based on the preactivation technique [75]. In this strategy, thioglycosides are used as the only type of donor and TolSOTf (prepared *in situ* from TolSCl/AgOTf) is used as the single promoter (Scheme 6.24). Preactivation of donor **116** generates the reactive intermediate **117**. Upon addition of acceptor **118**, nucleophilic attack of the reactive intermediate forms disaccharide **119**, which can be

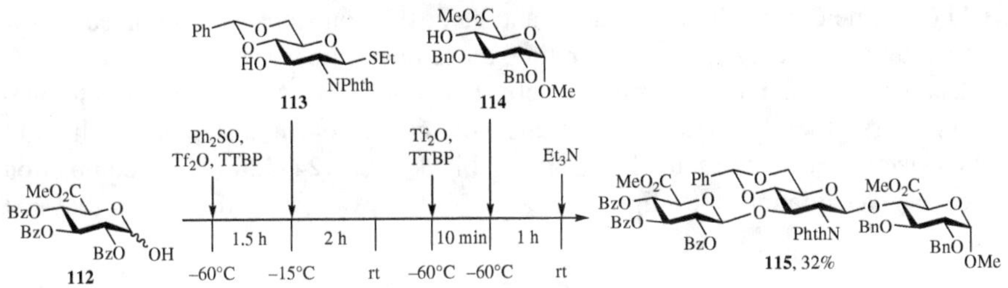

SCHEME 6.23 Preactivation-based one-pot synthesis of the hyaluronic acid trisaccharide **115**. TTBP, 2,4,6-tri-*tert*-butylpyrimidine.

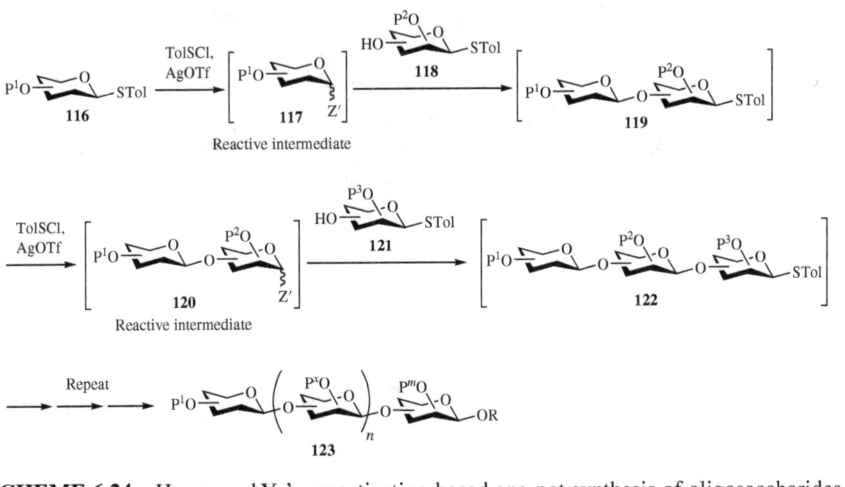

SCHEME 6.24 Huang and Ye's preactivation-based one-pot synthesis of oligosaccharides.

subjected to another round of preactivation and glycosylation. This process can be repeated in the same reaction flask until the desired oligosaccharide sequence is obtained. The success of this strategy is based on the following factors: (i) TolSOTf is stoichiometric in activating a wide range of glycosyl donors and is completely consumed by the donor to prevent the activation of the following building blocks and (ii) the intermediate generated after preactivation is stable in the absence of the acceptor and can rapidly react when the acceptor is added to the reaction mixture.

As the donor activation and acceptor glycosylation steps are separated, the preactivation-based glycosylation method can be performed without the donor bearing higher anomeric reactivities than the acceptor. The power of this strategy has been demonstrated by the successful assembly of many complex oligosaccharides and glycoconjugates such as chitotetraose [79], Globo H hexasaccharide [80], Man_5 oligosaccharide [81], iGb_3 oligosaccharide [82], Lewis X and dimeric Lewis X [83], hyaluronic acid oligosaccharides [84], influenza virus-binding sialyl trisaccharides [85], biantennary N-glycan dodecasaccharide [86], heparin/heparan sulfate oligosaccharide [87, 88], and syndecan-1 heparan sulfate glycopeptide [89]. As an example, the fully protected Globo H **127** was assembled from building blocks **22**, **124**, **125**, and **126** in a four-component one-pot manner in just a few hours with 47% overall yield (Scheme 6.25). All preactivation and glycosylation steps were initiated at low temperature (−78°C). After each coupling, the temperature was warmed up to −20°C to decompose the slight excess of the activated donor [80]. As a comparison to one-pot synthesis, the glycosylation reactions were carried out in a stepwise fashion to synthesize **127** using the same four building blocks (**22**, **124–126**). Upon completion of each glycosylation, the oligosaccharide product was purified prior to the next

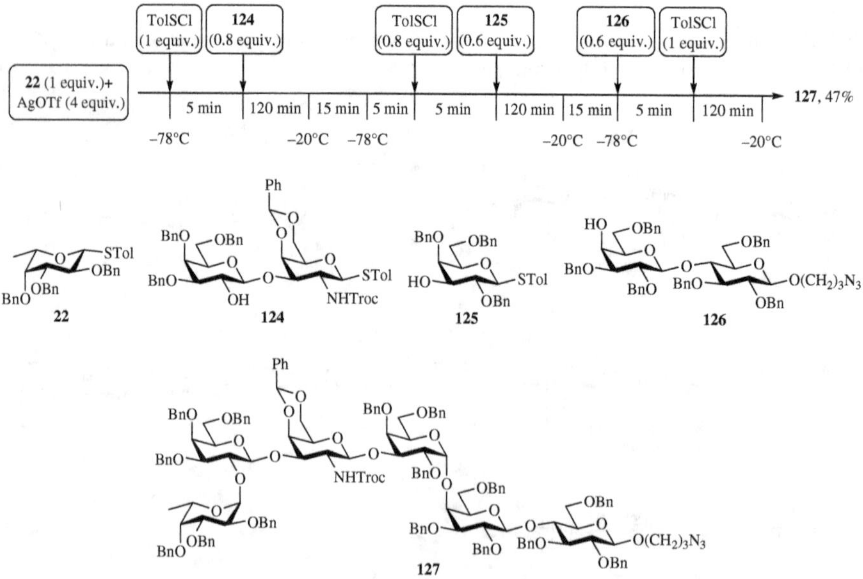

SCHEME 6.25 Huang's one-pot assembly of the Globo H construct **127**.

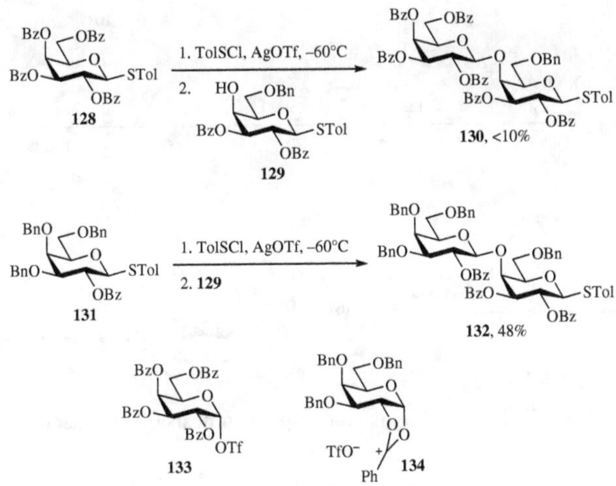

SCHEME 6.26 Coupling efficiency of donors bearing electron-withdrawing protecting groups compared with those containing electron-donating protecting groups under the preactivation strategy.

round of glycosylation. The overall yield for the stepwise synthesis was 36%. Therefore, one-pot synthesis not only shortened the amount of time spent but also improved the overall yield.

The reactivity-independent preactivation-based glycosylation method has a very general scope as evident from the diverse complex oligosaccharides prepared [75, 79–89]. However, in the case of sterically hindered unreactive acceptors (e.g., **129**), donors such as galactoside **128** failed to give productive coupling (Scheme 6.26) [90]. In contrast, the more electron-rich thiogalactosyl donor **131** successfully glycosylated the same acceptor. To better understand this phenomenon, the reactive intermediates formed after preactivation were examined by low-temperature NMR studies. Interestingly, the more electron-poor donor **128** formed the glycosyl triflate **133** as the major intermediate, while the electron-rich donor **131** generated the dioxolenium ion **134**. This differential outcome was presumably due to the large number of electron-withdrawing Bz moieties in **128**, which would disfavor the formation of the positively charged dioxolenium ion, thus producing the covalent glycosyl triflate intermediate. The dioxolenium ion **134** was found to be much more reactive than the glycosyl triflate **133** in glycosylation. Therefore, in challenging glycosylation reactions, donors bearing fewer electron-withdrawing protecting groups are preferred.

Common to all one-pot methodologies, silica gel column chromatography is utilized after the final glycosylation reaction to purify the desired oligosaccharide product from side products generated in the reaction. To expedite the purification of the final product from one-pot reactions, Huang and coworkers developed a fluorous separation protocol based on fluorous "catch and release" [91]. In this approach, the last acceptor in the one-pot synthesis bears a functionalized linker, which can react with a fluorous tag. After completion of the one-pot glycosylation, a fluorous tag is added into the reaction mixture to selectively "catch" the desired oligosaccharide,

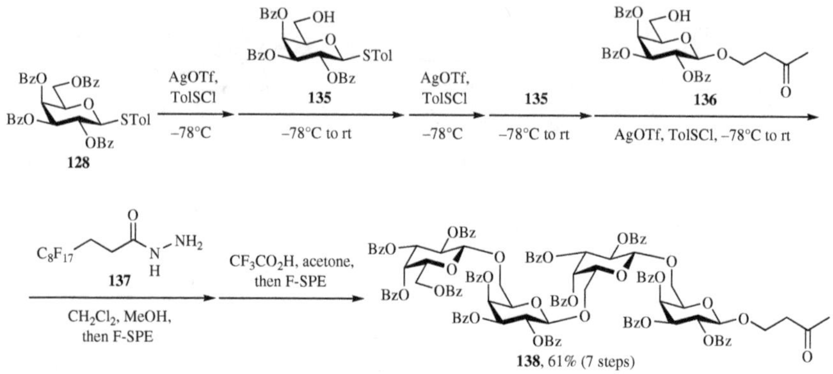

SCHEME 6.27 Fluorous-assisted one-pot synthesis of tetrasaccharide **138**.

which is easily separated from nonfluorous impurities by fluorous solid-phase extraction (F-SPE). Subsequent release of the tag and F-SPE leads to the desired pure oligosaccharide product without the need for any silica gel column purification. For example, tetrasaccharide **138** was synthesized from thioglycosides **128** and **135** and acceptor **136** bearing a ketone linker (Scheme 6.27). After the four-component one-pot glycosylation was complete, the fluorous-tagged hydrazine **137** was added into the reaction mixture, which selectively caught **138** by F-SPE. Release of the tag followed by another F-SPE afforded **138** in 61% yield with high purity. The whole glycoassembly and fluorous purification process took only a few hours.

TolSCl/AgOTf is a powerful promoter system for activating a wide range of thioglycosides. However, TolSCl is moisture sensitive, which requires careful handling for long-term storage. Several other promoters were utilized for preactivation of thioglycosides, which include benzenesulfinylpiperidine (BSP)/Tf_2O [92], benzenesulfinylmorpholine (BSM)/Tf_2O [82], diphenyl sulfoxide/Tf_2O [93], *p*-NO_2PhCl/AgOTf [94], and *O*,*O*-dimethylthiophosphonosulfenyl bromide (DMTPSB)/AgOTf [95]. However, many of these promoters are not suitable for one-pot synthesis because they generate side products that can quench the reactive intermediate or activate the product. In BSP/Tf_2O-promoted glycosylation, for example, it was necessary to add electrophiles such as triethyl phosphite to reduce the undesired product activation [96]. Therefore, further studies are necessary to develop more convenient and robust promoters for preactivation-based one-pot glycosylations.

High stereochemical control is crucial especially for one-pot glycosylation because the intermediate products generated are not purified. Typically, the formation of 1,2-*trans*-glycosidic linkages can be controlled through neighboring group participation of an acyl protecting group installed at the O2 position of the pyranoside ring. Generation of the 1,2-*cis* linkage in high stereoselectivity is a much more daunting task. For one-pot synthesis, one approach to bypass this difficulty is to preform the 1,2-*cis* linkage and use the glycoside containing the preexisting 1,2-*cis* linkage as building blocks [80]. To facilitate the direct formation of 1,2-*cis* linkage in one-pot glycosylations, a variety of conditions such as solvents and additives have been

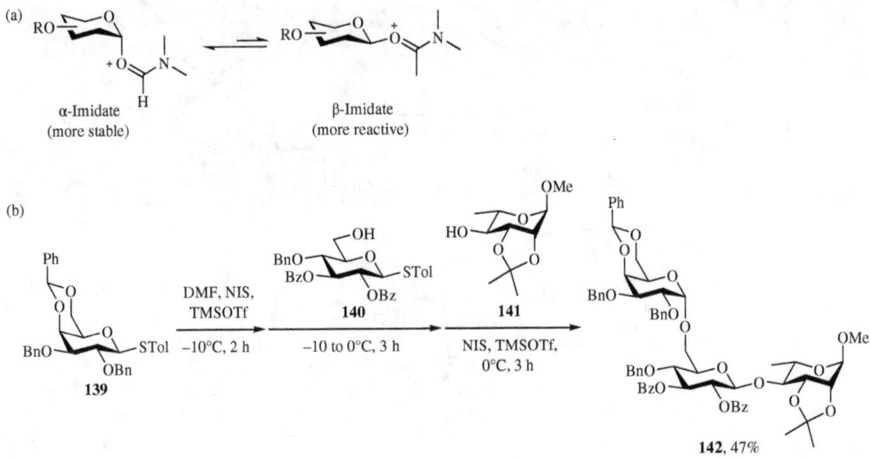

SCHEME 6.28 Mong's stereoselective one-pot synthesis using DMF as the novel additive to facilitate the 1,2-*cis*-glycoside formation. (a) The glycosyl imidates formed with DMF. (b) The one-pot synthesis of trisaccharide **142**.

investigated [97, 98]. Kwok-Kong Tony Mong and coworkers reported that *N,N*-dimethylformamide (DMF) is a novel additive to deliver α selectivity [99]. The rationale is that upon preactivation in the presence of DMF, α- and β-glycosyl imidate are formed through DMF addition to the oxocarbenium ion (Scheme 6.28a). These two imidates can exist in dynamic equilibrium with the α anomer being the major intermediate due to the anomeric effect, which was supported by low-temperature NMR studies. However, the β anomer is assumed to be more reactive, which, through an S_N2-like reaction pathway, would favor the formation of the α-glycoside product. The selective consumption of the β anomer intermediate would shift the imidate equilibrium toward the β anomer (Curtin–Hammett kinetics) [98, 99]. This knowledge was applied to the stereoselective one-pot synthesis of trisaccharide **142** containing both α and β linkages using monosaccharides **139**, **140**, and **141** (Scheme 6.28b). Preactivation of donor **139** by NIS/TMSOTf in the presence of 1.2 equiv. of DMF followed by the addition of acceptor **140** led to a disaccharide intermediate with excellent α selectivity. Subsequent addition of acceptor **141** together with NIS/TMSOTf afforded trisaccharide **142** in 47% overall yield [100].

6.3 REGIOSELECTIVE ONE-POT GLYCOSYLATION: CONSTRUCTION OF OLIGOSACCHARIDES FROM THE REDUCING END TO THE NONREDUCING END

Various innovative methodologies to carry out one-pot glycosylation from the nonreducing end to the reducing end have been discussed thus far. Alternatively, one-pot synthesis can be performed from the reducing end to the nonreducing end, which requires differentiation of the inherent reactivities of various hydroxy groups. For

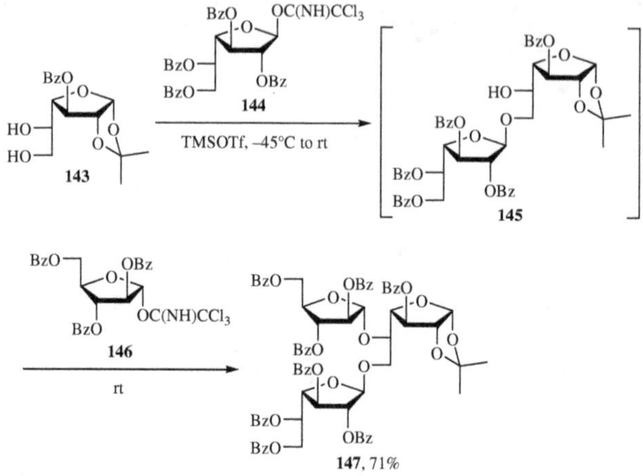

SCHEME 6.29 Ning's one-pot synthesis of trisaccharide **147**.

example, the primary 6-hydroxy group of galactofuranose is more reactive compared to hydroxy groups on other positions of the glycan ring. Jun Ning and coworkers took advantage of this reactivity difference to synthesize a trisaccharide fragment present in the *Mycobacterium tuberculosis* cell wall [101]. The diol acceptor **143** was first regioselectively glycosylated by the imidate donor **144** to give the disaccharide intermediate **145** (Scheme 6.29). The remaining free hydroxyl on **145** was coupled to the imidate donor **146** to afford trisaccharide **147** in 71% overall yield. Regioselective one-pot glycosylation based on the reactivity difference between primary and secondary hydroxyls was also demonstrated by other groups such as Takahashi's synthesis of 2,6-sialyl-T antigen [63, 102], Yang's synthesis of oligofuranose [103], and M. Carmen Galan's synthesis of branched model trisaccharides [104]. In Galan's work, the novel ionic liquid-based cosolvent/promoter [bmim][OTf], which promoted the activation of thioglycosides when combined with NIS, was developed.

Reactivity differences among secondary hydroxy groups were also utilized for the construction of branched oligosaccharides. Robert Field and coworkers reported that $HClO_4$ immobilized on silica gel could promote the one-pot synthesis of Lewis X trisaccharide [105]. Regioselective glycosylation of the diol acceptor **148** by donor **149** afforded the disaccharide intermediate **150** (Scheme 6.30). Addition of the second donor **151** led to trisaccharide **152** in 62% yield in one pot. The reactivity disparity between the 3-OH and 4-OH of **148** presumably came from the bulky TBDPS group at O6, which sterically hindered access to 4-OH. Interestingly, the reaction could be carried out on a column containing $HClO_4$-impregnated silica gel and the product could be directly purified using the same silica gel column.

Very recently, Junqiang Fang and coworkers reported the one-pot syntheses of timosaponin AIII and its analogues based on the reactivity differences of the hydroxy groups on different building blocks (Scheme 6.31) [106]. Because of the steric hindrance at the 2-OH of donor **153**, the alcohol **154** was more reactive and selectively coupled with the activated donor. Subsequent glycosylation by donor **155** afforded the conjugate **156** in 65% yield.

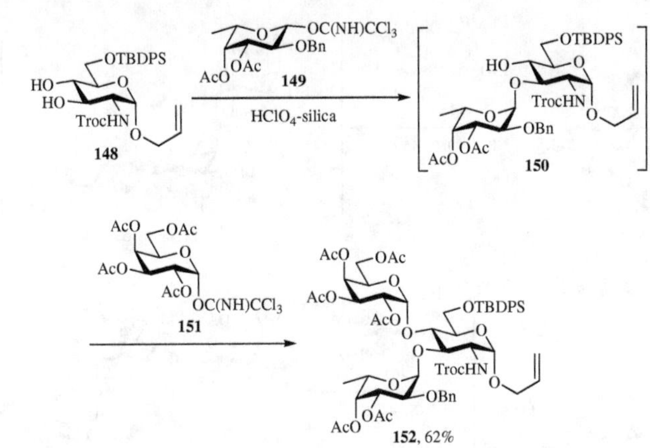

SCHEME 6.30 $HClO_4$–silica-promoted one-pot synthesis of trisaccharide **152**.

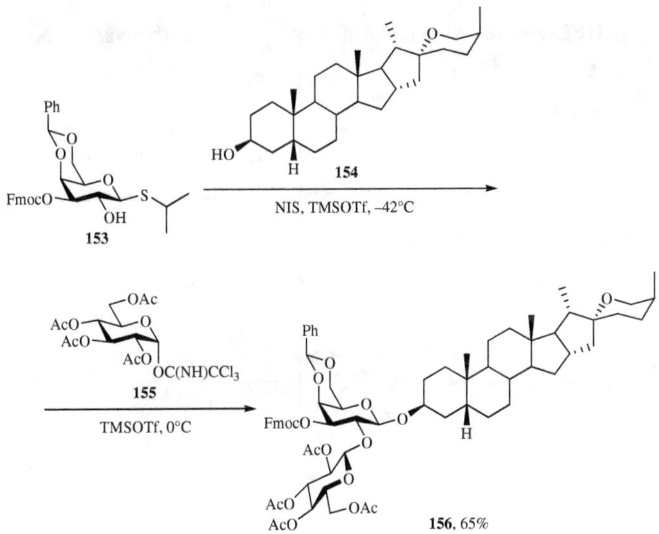

SCHEME 6.31 One-pot synthesis of a timosaponin AIII analogue.

The aforementioned examples show the application of differential reactivities of hydroxy groups for one-pot glycosylation. However, in many cases, the reactivity differences are not large enough to achieve high regioselectivities. Furthermore, depending on the structures of glycosyl donors, the regioselectivities can change or even reverse [107, 108]. To overcome this difficulty, protecting group strategies are designed to temporarily mask a reactive hydroxy group to ensure the location of the first coupling. The temporary protecting group is then removed *in situ* to initiate the next glycosylation. For example, Ulf Nilsson and coworkers demonstrated the one-pot synthesis of a globotetraose analogue by using *p*-methoxybenzyl (PMB) ether as the *in situ* removable protecting group [109]. As shown in Scheme 6.32, coupling of the thioglycoside donor **158** to acceptor **157** promoted by NIS/TfOH led to the

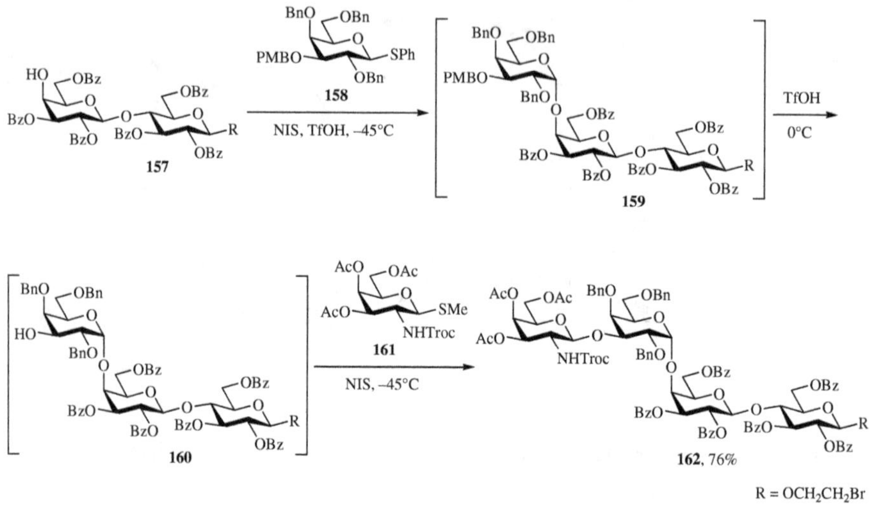

SCHEME 6.32 One-pot synthesis of a globotetraose analogue.

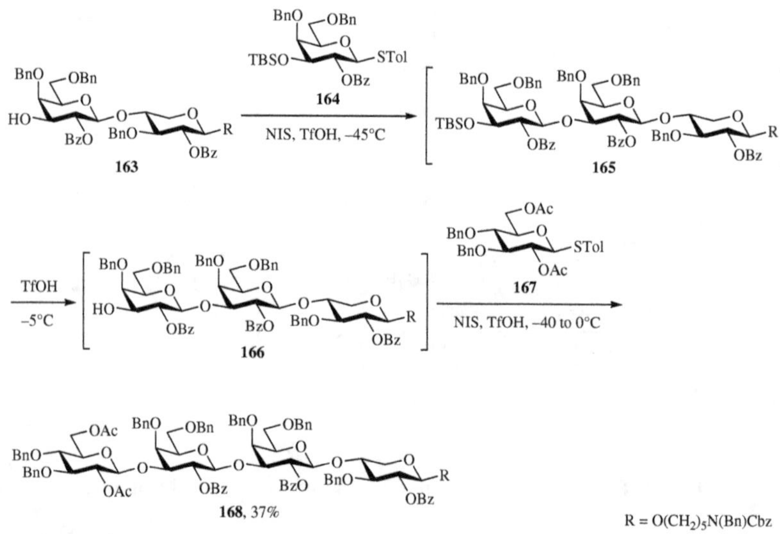

SCHEME 6.33 Hung's one-pot synthesis of tetrasaccharide **168**. Cbz, benzyloxycarbonyl.

formation of the trisaccharide intermediate **159**. The PMB moiety was stable at –45°C during the coupling but was cleaved by addition of TfOH and elevating the temperature to 0°C. Addition of thioglycoside **161** with NIS, again at –45°C, converted the trisaccharide acceptor **160** to tetrasaccharide **162** in 76% yield over the three steps.

Another example of *in situ* removal of temporary protecting groups is a novel one-pot synthesis of the tetrasaccharide linkage region of proteoglycans by Shang-Cheng Hung and coworkers [98]. In their synthesis, TBS group was used as the temporary masking group (Scheme 6.33). Glycosylation of acceptor **163** by the

TBS-bearing donor **164** afforded trisaccharide **165**, which was converted to acceptor **166** *in situ* by removal of the TBS protecting group with additional TfOH. Coupling of **166** with donor **167** produced tetrasaccharide **168** in 37% yield over three steps. In a similar manner, Carola Gallo-Rodriguez and coworkers utilized the TBS group as a temporary protecting group for the efficient synthesis of a branched mycobacterial arabinogalactan [110].

Besides the PMB and TBS moieties, other acid-labile protecting groups have been utilized in the one-pot assembly of oligosaccharides. These include benzylidene by Boons [111] and the trityl group by Li and coworkers in the synthesis of flaccidoside II, a bioactive component of Chinese traditional medicine [57].

6.4 HYBRID ONE-POT GLYCOSYLATION

To further enhance the overall synthetic efficiency, the approaches discussed in the previous sections can be combined to generate a hybrid-type one-pot glycosylation. For example, Daniel Kahne and coworkers reported the first one-pot glycosylation for the assembly of the protected ciclamycin trisaccharide **172** by using the glycosyl sulfoxides **169** and **170** and the thioglycosyl building block **171** (Scheme 6.34) [112]. TfOH was utilized as the promoter for the reaction as it could selectively activate glycosyl sulfoxides without affecting thioglycosides. Due to the *p*-methoxy moiety in the aglycone of donor **170**, it had higher anomeric reactivity than donor **169** and could be activated faster. Furthermore, the 4-OH of **170** was temporarily protected as a trimethylsilyl (TMS) ether, which was removed during the reaction to allow **170** to function both as a donor and an acceptor. Thus, by mixing the monosaccharide building blocks **169–171** together with TfOH, trisaccharide **172** was isolated as the major product in 25% yield. It was found that because phenylsulfenyl triflate was presumably formed as a side product of the glycosyl sulfoxide activation, it was necessary to add an electrophile such as methyl propiolate to prevent the undesired activation of the thioglycoside product by phenylsulfenyl triflate.

Ley and coworkers reported the one-pot synthesis of tetrasaccharide **176** by combining selective activation with reactivity-based armed–disarmed glycosylation [113]. As shown in Scheme 6.35, the first chemoselective coupling was achieved via selective activation of the fluoride donor **173** over the selenoglycoside acceptor **174**. The resulting disaccharide intermediate was more armed than the bifunctional

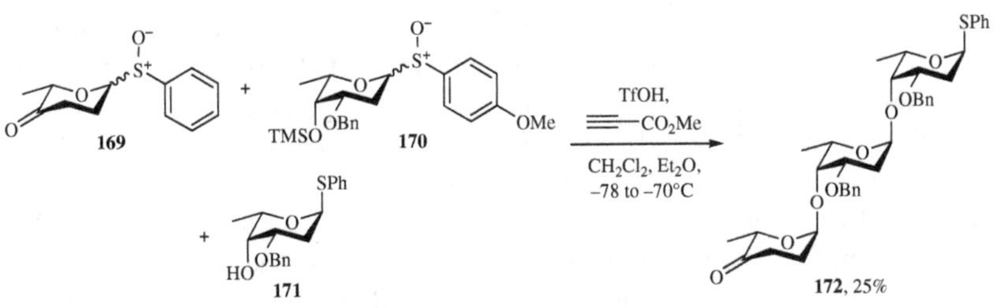

SCHEME 6.34 One-pot synthesis of protected ciclamycin trisaccharide **172**.

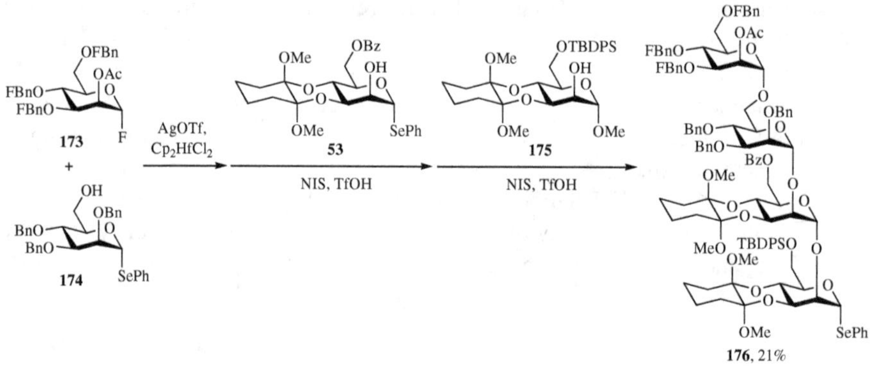

SCHEME 6.35 One-pot synthesis of tetrasaccharide **176**. FBn, 4-fluorobenzyl.

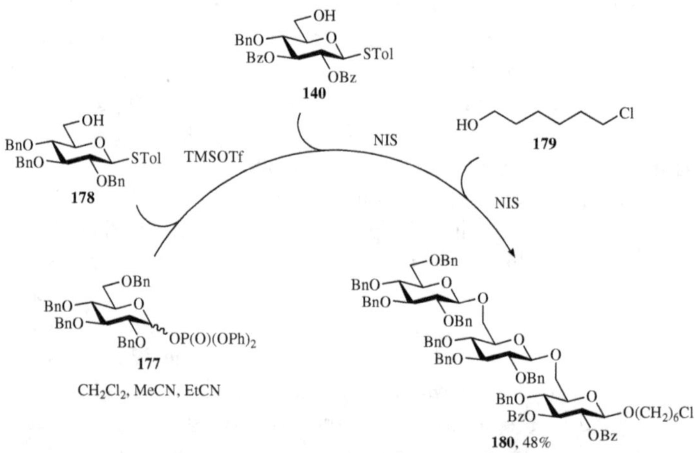

SCHEME 6.36 Mong's one-pot synthesis of trisaccharide **180** through combination of selective activation and armed–disarmed approach.

selenoglycoside **53** due to the conformational disarming effect of CDA in **53**. Therefore, the disaccharide intermediate was preferentially activated and subsequently glycosylated **53**. The last coupling to the reducing end acceptor **175** proceeded to afford tetrasaccharide **176** in 21% overall yield. This approach was later applied by the Ley group to the assemblies of penta-, hexa-, and heptasaccharides [114, 115].

Another example was shown by Mong and coworkers [116]. To synthesize trisaccharide **180**, they started from glucosyl diphenyl phosphate donor **177**, which chemoselectively coupled to thioglycoside **178** (Scheme 6.36). The next glycosylation to the thioglycoside acceptor **140** proceeded as expected due to the lower anomeric reactivity of **140**. This was followed by the subsequent coupling to linker **179**, producing trisaccharide **180** in 48% yield over the four steps. The 1,2-*trans* stereoselectivities of the first two glycosylations were achieved by the participation of the alkyl nitrile solvent.

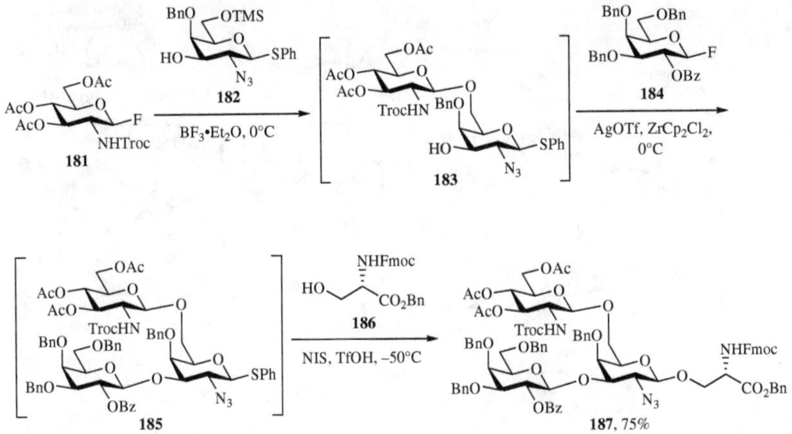

SCHEME 6.37 Takahashi's one-pot synthesis of the glycosyl amino acid **187** by combining selective activation and regioselective glycosylation.

Selective activation combined with regioselective glycosylation has been demonstrated by various groups. Takahashi and coworkers reported the one-pot synthesis of the core 2 class glycosyl amino acid **187** [117]. As shown in Scheme 6.37, the glycosyl fluoride **181** was selectively activated over the thioglycoside acceptor **182** through the fluorophilic promoter $BF_3{\cdot}Et_2O$. Interestingly, they discovered that it was crucial to protect the 6-OH as a TMS ether as the BF_3 complex with the TMS ether made it more nucleophilic to provide disaccharide **183** without the undesired overglycosylation. Reaction of the resulting disaccharide **183** with the glycosyl fluoride donor **184** afforded the trisaccharide **185**, which was immediately subjected to coupling with the serine acceptor **186** to give product **187** in an excellent 75% overall yield. Similar strategies have been applied by the same group for the assembly of branched trisaccharides [118], di-branched heptasaccharide with phytoalexin elicitor activity [119, 120], and a library of dimeric Lewis X derivatives [121]. Other groups such as those of Biao Yu [122], Bert Fraser-Reid [107], Asish Kumar Sen [123], and Jin-Song Yang [103] also demonstrated the usefulness of combining regioselective glycosylation with selective activation in the synthesis of branched oligosaccharides.

Regioselective glycosylation has been combined with the reactivity-based approach in a bidirectional one-pot synthesis of the hexasaccharide **193** by the Boons group (Scheme 6.38) [124]. The armed donor **188** was coupled to the disarmed acceptor **189** in the presence of NIS/TMSOTf to afford the nonreducing end trisaccharide intermediate **190**. At the same time, coupling of **189** to acceptor **191** in another flask gave the reducing end trisaccharide **192**. The regioselectivity achieved in the formation of **192** was presumably due to the multiple benzyl ether moieties on **191**, rendering its free hydroxy group more nucleophilic than that in **189**. Upon formation of **192**, the reaction mixture holding **190** was directly added to **192** together with NIS/TMSOTf to generate hexasaccharide **193** (53% overall yield). Other examples of one-pot synthesis combining regioselective glycosylation with the

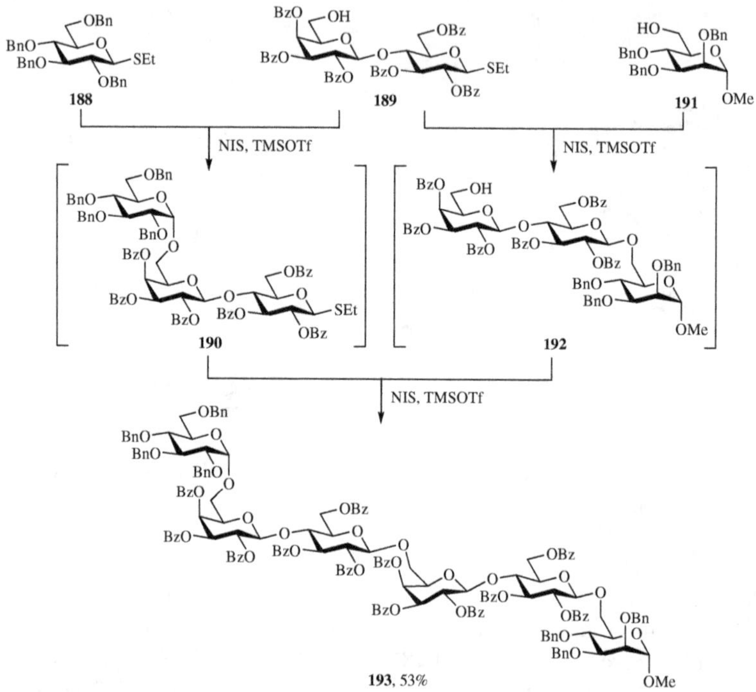

SCHEME 6.38 Bidirectional one-pot synthesis of hexasaccharide **193**.

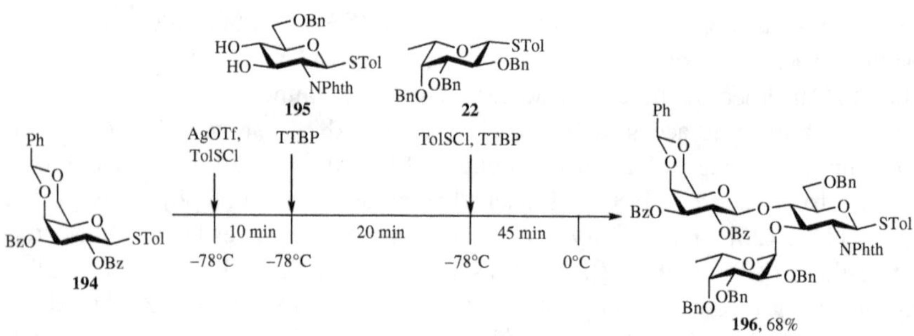

SCHEME 6.39 Huang's one-pot synthesis of the Lewis X trisaccharide **196**.

armed–disarmed approach include Fraser-Reid's double differential glycosylation [125] and Wong's sialyl Lewis X hexasaccharide synthesis [126].

Huang and coworkers combined regioselective glycosylation with their preactivation technique to assemble the Lewis X trisaccharide **196** (Scheme 6.39) [83]. Preactivation of the galactose donor **194** followed by addition of the glucosamine diol acceptor **195** resulted in the regioselective coupling at the 4-OH position. Subsequent addition of the fucosyl donor **22** produced trisaccharide **196** in 68% yield. The second chemoselective coupling was a result of the higher anomeric reactivity of **22** compared to the disaccharide product from the first glycosylation.

6.5 CONCLUSION

Tremendous progress has been made in the past two decades for the development of solution phase-based one-pot methodologies as witnessed from the complex structures of oligosaccharides that have been prepared. Much of the success is built upon careful experimental observations, in-depth thinking, and mechanistic understanding of the glycosylation reactions. However, to date, the majority of the synthesis involved two- to three-step reactions in one pot. The holy grail of the field is to develop a robust technology with the efficiency approaching that of automated peptide synthesis to prepare longer oligosaccharides for biological studies. Thus, there is still much room for improvement of one-pot methodologies, which will require further study and thorough understanding of the factors governing yield and stereoselectivity in glycosylation reactions.

ACKNOWLEDGMENTS

We would like to thank the general financial supports from the National Science Foundation (CHE 1111550) and the National Institute of General Medical Sciences, NIH (R01GM072667) toward our work.

REFERENCES

[1] Muthana, S.; Cao, H.; Chen, X. *Curr. Opin. Chem. Biol.* **2009**, *13*, 573–581.

[2] Wang, Y.; Ye, X.-S.; Zhang, L.-H. *Org. Biomol. Chem.* **2007**, *5*, 2189–2200.

[3] Yu, B.; Yang, Z.; Cao, H. *Curr. Org. Chem.* **2005**, *9*, 179–194.

[4] Codée, J. D. C.; Litjens, R. E. J. N.; van den Bos, L. J.; Overkleeft, H. S.; van der Marel, G. A. *Chem. Soc. Rev.* **2005**, *34*, 769–782.

[5] Bartolozzi, A.; Seeberger, P. H. *Curr. Opin. Struct. Biol.* **2001**, *11*, 587–592.

[6] Koeller, K. M.; Wong, C. H. *Chem. Rev.* **2000**, *100*, 4465–4493.

[7] Mootoo, D. R.; Konradsson, P.; Udodong, U.; Fraser-Reid, B. *J. Am. Chem. Soc.* **1988**, *110*, 5583–5584.

[8] Fraser-Reid, B.; Wu, Z.; Udodong, U. E.; Ottosson, H. *J. Org. Chem.* **1990**, *55*, 6068–6070.

[9] Friesen, R. W.; Danishefsky, S. J. *Tetrahedron* **1990**, *46*, 103–112.

[10] Veeneman, G. H.; van Boom, J. H. *Tetrahedron Lett.* **1990**, *31*, 275–278.

[11] Ley, S. V.; Priepke, H. W. M. *Angew. Chem. Int. Ed.* **1994**, *33*, 2292–2294.

[12] Fraser-Reid, B.; Wu, Z.; Andrews, C. W.; Skowronski, E. *J. Am. Chem. Soc.* **1991**, *113*, 1434–1435.

[13] Lu, S.-F.; O'Yang, Q.; Guo, Z.-W.; Yu, B.; Hui, Y.-Z. *Angew. Chem. Int. Ed.* **1997**, *36*, 2344–2346.

[14] Lu, S.-F.; O'Yang, Q.; Guo, Z.-W.; Yu, B.; Hui, Y.-Z. *J. Org. Chem.* **1997**, *62*, 8400–8405.

[15] Tsukida, T.; Yoshida, M.; Kurokawa, K.; Nakai, Y.; Achiha, T.; Kiyoi, T.; Kondo, H. *J. Org. Chem.* **1997**, *62*, 6876–6881.

[16] Yoshida, M.; Kiyoi, T.; Tsukida, T.; Kondo, H. *J. Carbohydr. Chem.* **1998**, *17*, 673–681.

[17] Douglas, N. L.; Ley, S. V.; Lucking, U.; Warriner, S. L. *J. Chem. Soc. Perkin Trans. 1* **1998**, 51–66.

[18] Zhang, Z.; Ollmann, I. R.; Ye, X.-S.; Wischnat, R.; Baasov, T.; Wong, C. H. *J. Am. Chem. Soc.* **1999**, *121*, 734–753.

[19] Mong, K.-K. T.; Wong, C.-H. *Angew. Chem. Int. Ed.* **2002**, *41*, 4087–4090.

[20] Burkhart, F.; Zhang, Z.; Wacowich-Sgarbi, S.; Wong, C. H. *Angew. Chem. Int. Ed.* **2001**, *40*, 1274–1277.

[21] Tsai, B.-L.; Han, J.-L.; Ren, C.-T.; Wu, C.-Y.; Wong, C.-H. *Tetrahedron Lett.* **2011**, *52*, 2132–2135.

[22] Mong, T. K.-K.; Lee, H. K.; Duron, S. G.; Wong, C.-H. *Proc. Natl. Acad. Sci. U. S. A.* **2003**, *100*, 797–802.

[23] Polat, T.; Wong, C. H. *J. Am. Chem. Soc.* **2007**, *129*, 12795–12800.

[24] Hsu, C.-H.; Chu, K.-C.; Lin, Y.-S.; Han, J.-L.; Peng, Y.-S.; Ren, C.-T.; Wu, C.-Y.; Wong, C.-H. *Chem. Eur. J.* **2010**, *16*, 1754–1760.

[25] Lee, J.-C.; Wu, C.-Y.; Apon, J. V.; Siuzdak, G.; Wong, C.-H. *Angew. Chem. Int. Ed.* **2006**, *45*, 2753–2757.

[26] Mong, T. K.-K.; Huang, C.-Y.; Wong, C.-H. *J. Org. Chem.* **2003**, *68*, 2135–2142.

[27] Ritter, T. K.; Mong, K.-K. T.; Liu, H.; Nakatani, T.; Wong, C.-H. *Angew. Chem. Int. Ed.* **2003**, *42*, 4657–4660.

[28] Lee, H.-K.; Scalan, C. N.; Huang, C. Y.; Chang, A. Y.; Calarese, D. A.; Dwek, R. A.; Rudd, P. M.; Burton, D. R.; Wilson, I. A.; Wong, C.-H. *Angew. Chem. Int. Ed.* **2004**, *43*, 1000–1003.

[29] Hsu, Y.; Lu, X.-A.; Zulueta, M. M. L.; Tsai, C.-M.; Lin, K.-I.; Hung, S.-C.; Wong, C.-H. *J. Am. Chem. Soc.* **2012**, *134*, 4549–4552.

[30] Ye, X.-S.; Wong, C.-H. *J. Org. Chem.* **2000**, *65*, 2410–2431.

[31] Li, X.; Huang, L.; Hu, X.; Huang, X. *Org. Biomol. Chem.* **2009**, *7*, 117–127.

[32] Jensen, H. H.; Nordstrom, L. U.; Bols, M. *J. Am. Chem. Soc.* **2004**, *126*, 9205–9213.

[33] McDonnell, C.; Lopez, O.; Murphy, P.; Fernandez Bolanos, J. G.; Hazell, R.; Bols, M. *J. Am. Chem. Soc.* **2004**, *126*, 12374–12385.

[34] Pedersen, C. M.; Nordstrøm, L. U.; Bols, M. *J. Am. Chem. Soc.* **2007**, *129*, 9222–9235.

[35] Pedersen, C. M.; Marinescu, L. G.; Bols, M. *Chem. Commun.* **2008**, 2465–2467.

[36] Lahmann, M.; Oscarson, S. *Can. J. Chem.* **2002**, *80*, 889–893.

[37] Schmidt, R. R.; Kinzy, W. *Adv. Carbohydr. Chem. Biochem.* **1994**, *50*, 21–123.

[38] Yu, B.; Tao, H. *Tetrahedron Lett.* **2001**, *42*, 2405–2407.

[39] Adinolfi, M.; Iadonisi, A.; Ravida, A. *Synlett* **2006**, 583–586.

[40] Valerio, S.; Pastore, A.; Adinolfi, M.; Iadonisi, A. *J. Org. Chem.* **2008**, *73*, 4496–4503.

[41] Pongdee, R.; Wu, B.; Sulikowski, G. A. *Org. Lett.* **2001**, *3*, 3523–3525.

[42] Huang, L.; Wang, Z.; Huang, X. *Chem. Commun.* **2004**, 1960–1961.

[43] Grice, P.; Ley, S. V.; Pietruszka, J.; Osborn, H. M. I.; Priepke, H. W. M.; Warriner, S. L. *Chem. Eur. J.* **1997**, *3*, 431–440.

[44] Grice, P.; Ley, S. V.; Pietruszka, J.; Priepke, H. W. M.; Walther, E. P. E. *Synlett* **1995**, 781–784.

[45] Shirahata, T.; Kojima, A.; Teruya, S.; Matsuo, J. I.; Yokoyama, M.; Unagiike, S.; Sunazuka, T.; Makino, K.; Kaji, E.; Omura, S.; Kobayashi, Y. *Tetrahedron* **2011**, *67*, 6482–6496.

[46] Lahmann, M.; Oscarson, S. *Org. Lett.* **2000**, *2*, 3881–3882.

[47] Fridman, M.; Solomon, D.; Yogev, S.; Baasov, T. *Org. Lett.* **2002**, *4*, 281–283.

[48] Yamada, H.; Harada, T.; Takahashi, T. *J. Am. Chem. Soc.* **1994**, *116*, 7919–7920.

[49] Yu, B.; Yu, H.; Hui, Y.; Han, X. *Tetrahedron Lett.* **1999**, *40*, 8591–8594.

[50] Yu, H.; Yu, B.; Wu, X.; Hui, Y.; Han, X. *J. Chem. Soc. Perkin Trans. 1* **2000**, 1445–1453.

[51] Kim, J.-H.; Yang, H.; Park, J.; Boons, G. J. *J. Am. Chem. Soc.* **2005**, *127*, 12090–12097.

[52] Vohra, Y.; Buskas, T.; Boons, G. J. *J. Org. Chem.* **2009**, *74*, 6064–6071.

[53] Wang, P.; Lee, H.; Fukuda, M.; Seeberger, P. H. *Chem. Commun.* **2007**, 1963–1965.

[54] Completo, G. C.; Lowary, T. L. *J. Org. Chem.* **2008**, *73*, 4513–4525.

[55] Liang, X.-Y.; Deng, L.-M.; Liu, X.; Yang, J.-S. *Tetrahedron* **2010**, *66*, 87–93.

[56] Dinkelaar, J.; Gold, H.; Overkleeft, H. S.; Codee, J. D. C.; van der Marel, G. A. *J. Org. Chem.* **2009**, *74*, 4208–4216.

[57] Liu, Q.; Zhang, L.; Li, X.; Guo, T.; Wang, P.; Li, Y. *J. Carbohydr. Chem.* **2009**, *28*, 506–519.

[58] Lu, Z.; Ding, N.; Zhang, W.; Wang, P.; Li, Y. *Tetrahedron Lett.* **2011**, *52*, 3320–3323.

[59] Huang, X.; Cheng, S.; Du, Y.; Bing, F. *Carbohydr. Res.* **2009**, *344*, 1153–1158.

[60] Ren, C.-T.; Tsai, Y.-H.; Yang, Y.-L.; Zou, W.; Wu, S.-H. *J. Org. Chem.* **2007**, *72*, 5427–5430.

[61] Yamada, H.; Harada, T.; Miyazaki, H.; Takahashi, T. *Tetrahedron Lett.* **1994**, *35*, 3979–3982.

[62] Takahashi, T.; Adachi, M.; Matsuda, A.; Doi, T. *Tetrahedron Lett.* **2000**, *41*, 2599–2603.

[63] Tanaka, H.; Adachi, M.; Takahashi, T. *Chem. Eur. J.* **2005**, *11*, 849–862.

[64] Benakli, K.; Zha, C.; Kerns, R. J. *J. Am. Chem. Soc.* **2001**, *123*, 9461–9462.

[65] Manabe, S.; Ishii, K.; Ito, Y. *J. Am. Chem. Soc.* **2006**, *128*, 10666–10667.

[66] Garegg, P. J. *Adv. Carbohydr. Chem. Biochem.* **1997**, *52*, 179–205.

[67] Demchenko, A. V.; Kamat, M. N.; De Meo, C. *Synlett* **2003**, 1287–1290.

[68] Demchenko, A. V.; Malysheva, N. N.; De Meo, C. *Org. Lett.* **2003**, *5*, 455–458.

[69] De Meo, C.; Kamat, M. N.; Demchenko, A. V. *Eur. J. Org. Chem.* **2005**, 706–711.

[70] Demchenko, A. V.; Pornsuriyasak, P.; De Meo, C.; Malysheva, N. N. *Angew. Chem. Int. Ed.* **2004**, *43*, 3069–3072.

[71] Pornsuriyasak, P.; Demchenko, A. V. *Tetrahedron: Asymmetry* **2005**, *16*, 433–439.

[72] Tanaka, H.; Tateno, Y.; Nishiura, Y.; Takahashi, T. *Org. Lett.* **2008**, *10*, 5597–5600.

[73] Chenault, H. K.; Castro, A. *Tetrahedron Lett.* **1994**, *35*, 9145–9148.

[74] Yamago, S.; Yamada, T.; Maruyama, T.; Yoshida, J. I. *Angew. Chem. Int. Ed.* **2004**, *43*, 2145–2148.

[75] Huang, X.; Huang, L.; Wang, H.; Ye, X.-S. *Angew. Chem. Int. Ed.* **2004**, *43*, 5221–5224.

[76] Crich, D.; Sun, S. *J. Am. Chem. Soc.* **1997**, *119*, 11217–11223.

[77] Kahne, D.; Walker, S.; Cheng, Y.; Van Engen, D. *J. Am. Chem. Soc.* **1989**, *111*, 6881–6882.

[78] Codée, J. D. C.; van den Bos, L. J.; Litjens, R. E. J. N.; Overkleeft, H. S.; Van Boom, J. H.; van der Marel, G. A. *Org. Lett.* **2003**, *5*, 1947–1950.

[79] Huang, L.; Wang, Z.; Li, X.; Ye, X.-S.; Huang, X. *Carbohydr. Res.* **2006**, *341*, 1669–1679.

[80] Wang, Z.; Zhou, L.; El-Boubbou, K.; Ye, X.-S.; Huang, X. *J. Org. Chem.* **2007**, *72*, 6409–6420.

[81] Teumelsan, N.; Huang, X. *J. Org. Chem.* **2007**, *72*, 8976–8979.

[82] Wang, C.; Wang, H.; Huang, X.; Zhang, L. H.; Ye, X.-S. *Synlett* **2006**, 2846–2850.

[83] Miermont, A.; Zeng, Y.; Jing, Y.; Ye, X.-S.; Huang, X. *J. Org. Chem.* **2007**, *72*, 8958–8961.

[84] Huang, L.; Huang, X. *Chem. Eur. J.* **2007**, *13*, 529–540.

[85] Wang, C.-C.; Lee, J.-C.; Luo, S.-Y.; Kulkarni, S. S.; Huang, Y.-W.; Lee, C.-C.; Chang, K.-L.; Hung, S.-C. *Nature* **2007**, *446*, 896–899.

[86] Sun, B.; Srinivasan, B.; Huang, X. *Chem. Eur. J.* **2008**, *14*, 7072–7081.

[87] Wang, Z.; Xu, Y.; Yang, B.; Tiruchinapally, G.; Sun, B.; Liu, R.; Dulaney, S.; Liu, J.; Huang, X. *Chem. Eur. J.* **2010**, *16*, 8365–8375.

[88] Lu, X.; Kamat, M. N.; Huang, L.; Huang, X. *J. Org. Chem.* **2009**, *74*, 7608–7617.

[89] Yang, B.; Yoshida, K.; Yin, Z.; Dai, H.; Kavunja, H.; El-Dakdouki, M. H.; Sungsuwan, S.; Dulaney, S. B.; Huang, X. *Angew. Chem. Int. Ed.* **2012**, *51*, 10185–10189.

[90] Zeng, Y.; Wang, Z.; Whitfield, D.; Huang, X. *J. Org. Chem.* **2008**, *73*, 7952–7962.

[91] Yang, B.; Jing, Y.; Huang, X. *Eur. J. Org. Chem.* **2010**, 1290–1298.

[92] Crich, D.; Smith, M. *J. Am. Chem. Soc.* **2001**, *123*, 9015–9020.

[93] Codée, J. D. C.; Litjens, R. E. J. N.; den Heeten, R.; Overkleeft, H. S.; van Boom, J. H.; van der Marel, G. A. *Org. Lett.* **2003**, *5*, 1519–1522.

[94] Crich, D.; Cai, F.; Yang, F. *Carbohydr. Res.* **2008**, *343*, 1858–1862.

[95] Peng, P.; Ye, X.-S. *Org. Biomol. Chem.* **2011**, *9*, 616–622.

[96] Codée, J. D. C.; van den Bos, L. J.; Litjens, R. E. J. N.; Overkleeft, H. S.; van Boeckel, C. A. A.; van Boom, J. H.; van der Marel, G. A. *Tetrahedron* **2004**, *60*, 1057–1064.

[97] Wasonga, G.; Zeng, Y.; Huang, X. *Sci. China: Chem.* **2011**, *54*, 66–73.

[98] Huang, T.-Y.; Zulueta, M. M. L.; Hung, S.-C. *Org. Lett.* **2011**, *13*, 1506–1509.

[99] Lin, Y.-H.; Ghosh, B.; Mong, K.-K. T. *Chem. Commun.* **2012**, *48*, 10910–10912.

[100] Liu, C.-Y. I.; Mulani, S.; Mong, K.-K. T. *Adv. Synth. Catal.* **2012**, *354*, 3299–3310.

[101] Wang, H.; Ning, J. *J. Org. Chem.* **2003**, *68*, 2521–2524.

[102] Adachi, M.; Tanaka, H.; Takahashi, T. *Synlett* **2004**, 609–614.

[103] Deng, L.-M.; Liu, X.; Liang, X.-Y.; Yang, J.-S. *J. Org. Chem.* **2012**, *77*, 3025–3037.

[104] Galan, M. C.; Tran, A. T.; Whitaker, S. *Chem. Commun.* **2010**, *46*, 2106–2108.

[105] Mukhopadhyay, B.; Maurer, S. V.; Rudolph, N.; Van Well, R. M.; Russell, D. A.; Field, R. A. *J. Org. Chem.* **2005**, *70*, 9059–9062.

[106] Fang, M.; Gu, L.; Gu, G.; Fang, J. *J. Carbohydr. Chem.* **2012**, *31*, 187–202.

[107] Jayaprakash, K. N.; Fraser-Reid, B. *Org. Lett.* **2004**, *6*, 4211–4214.

[108] Fraser-Reid, B.; Cristobal Lopez, J. C.; Radhakrishnan, K. V.; Mach, M.; Schlueter, U.; Gomez, A. M.; Uriel, C. *J. Am. Chem. Soc.* **2002**, *124*, 3198–3199.

[109] Bhattacharyya, S.; Magnusson, B. G.; Wellmar, U.; Nilsson, U. J. *J. Chem. Soc. Perkin Trans. 1* **2001**, 886–890.

[110] Gandolfi-Donadio, L.; Santos, M.; de Lederkremer, R. M.; Gallo-Rodriguez, C. *Org. Biomol. Chem.* **2011**, *9*, 2085–2097.

[111] Vohra, Y.; Vasan, M.; Venot, A.; Boons, G. J. *Org. Lett.* **2008**, *10*, 3247–3250.

[112] Raghavan, S.; Kahne, D. *J. Am. Chem. Soc.* **1993**, *115*, 1580–1581.

[113] Cheung, M. K.; Douglas, N. L.; Hinzen, B.; Ley, S. V.; Pannecoucke, X. *Synlett* **1997**, 257–260.

[114] Green, L.; Hinzen, B.; Ince, S. J.; Langer, P.; Ley, S. V.; Warriner, S. L. *Synlett* **1998**, 440–442.

[115] Baeschlin, D. K.; Green, L. G.; Hahn, M. G.; Hinzen, B.; Ince, S. J.; Ley, S. V. *Tetrahedron: Asymmetry* **2000**, *11*, 173–197.

[116] Chao, C.-S.; Yen, Y.-F.; Hung, W.-C.; Mong, K.-K. T. *Adv. Synth. Catal.* **2011**, *353*, 879–884.

[117] Tanaka, H.; Adachi, M.; Takahashi, T. *Tetrahedron Lett.* **2004**, *45*, 1433–1436.

[118] Yamada, H.; Kato, T.; Takahashi, T. *Tetrahedron Lett.* **1999**, *40*, 4581–4584.

[119] Tanaka, H.; Adachi, M.; Tsukamoto, H.; Ikeda, T.; Yamada, H.; Takahashi, T. *Org. Lett.* **2002**, *4*, 4213–4216.

[120] Yamada, H.; Takimoto, H.; Ikeda, T.; Tsukamoto, H.; Harada, T.; Takahashi, T. *Synlett* **2001**, 1751–1754.

[121] Tanaka, H.; Matoba, N.; Tsukamoto, H.; Takimoto, H.; Yamada, H.; Takahashi, T. *Synlett* **2005**, 824–828.

[122] Yu, B.; Xie, J.; Deng, S.; Hui, Y. *J. Am. Chem. Soc.* **1999**, *121*, 12196–12197.

[123] Sarkar, S.; Dutta, S.; Das, G.; Sen, A. K. *Tetrahedron* **2011**, *67*, 4118–4122.

[124] Boons, G. J.; Zhu, T. *Synlett* **1997**, 809–811.

[125] Fraser-Reid, B.; Lopez, J. C.; Radhakrishnan, K. V.; Nandakumar, M. V.; Gomez, A. M.; Uriel, C. *Chem. Commun.* **2002**, 2104–2105.

[126] Zhang, Z.; Niikura, K.; Huang, X.; Wong, C.-H. *Can. J. Chem.* **2002**, *80*, 1051–1054.

7

AUTOMATED OLIGOSACCHARIDE SYNTHESIS: TECHNIQUES AND APPLICATIONS

MATTAN HUREVICH[1], JEYAKUMAR KANDASAMY[2], AND PETER H. SEEBERGER[3]

[1] *Institute of Chemistry, The Hebrew University of Jerusalem, Jerusalem, Israel*

[2] *Department of Chemistry, Indian Institute of Technology (BHU)–Varanasi, Varanasi, India*

[3] *Department of Biomolecular Systems, Max Planck Institute of Colloids and Interfaces, Potsdam, Germany*

7.1 INTRODUCTION

Nucleic acids, proteins, and carbohydrates are important biopolymers responsible for most of the signal transduction processes in living organisms. Both genomics [1] and proteomics [2] have been extensively studied, utilizing synthetic constructs to elucidate nucleic acid and protein structures, functions, and interactions with other biomolecules. The well-established solid-phase syntheses of short peptides and oligonucleotides enable assembly to be carried out using automated synthesizers [3, 4]. By using the automated platform, nonchemists can synthesize well-defined oligonucleotides and peptides for biological investigations.

Carbohydrates mainly exist as glycoproteins and glycolipids (glycoconjugates) on cell surfaces. These glycoconjugates take part in many pathogenic and nonpathogenic cellular activities [5, 6] but are less studied than nucleic acids and proteins because the synthesis of oligosaccharides is more challenging than other biopolymers [5]. The increasing interest in the biological, chemical, and structural properties of carbohydrates demands a reliable and generalized method to obtain well-defined oligosaccharides [7, 8].

Glycochemical Synthesis: Strategies and Applications, First Edition.
Edited by Shang-Cheng Hung and Medel Manuel L. Zulueta.

Many strategies have been and are still being developed to access oligosaccharides by chemical synthesis. This chapter focuses on recent developments in the automated solid-phase synthesis of oligosaccharides and emphasizes the recent advances of novel microreactor techniques for carbohydrate synthesis.

7.2 CHALLENGES AND LIMITATIONS IN SOLUTION-PHASE OLIGOSACCHARIDE SYNTHESIS

Access to well-defined carbohydrates is crucial for biological, biochemical, and biophysical studies. Isolation of oligosaccharides from natural sources usually results in a heterogeneous mixture of glycans. Homogeneous glycan entities may be obtained by either chemical or enzymatic synthesis. Enzymatic preparation of carbohydrates is regio- and stereospecific [9] but is not flexible because not all enzymatic systems are general, scalable, and reproducible [10]. The chemical synthesis of oligosaccharides is challenging when compared to other linear biopolymers because carbohydrates are often branched. In addition, oligosaccharide synthesis requires the installation of a new stereocenter at each elongation step. Combinations of protecting groups and activation methods usually influence the selectivity and reactivity of each building block used for glycan synthesis.

Versatile synthetic procedures have been developed for the assembly of complex oligosaccharides [11–13]. Solution-phase oligosaccharide assembly relies on the coupling of a glycosylating agent (donor) with a nucleophile (acceptor) promoted by an activator (Scheme 7.1). Building blocks are synthesized with appropriate orthogonal permanent and temporary protecting groups along with an anomeric leaving group. Temporary protecting groups are removed from the newly synthesized molecule to uncover a new nucleophile for the next glycosylation reaction. Typically, after each glycosylation and deprotection step, the oligosaccharides are isolated from the reaction by chromatography before continuing to the next step. The assembly of

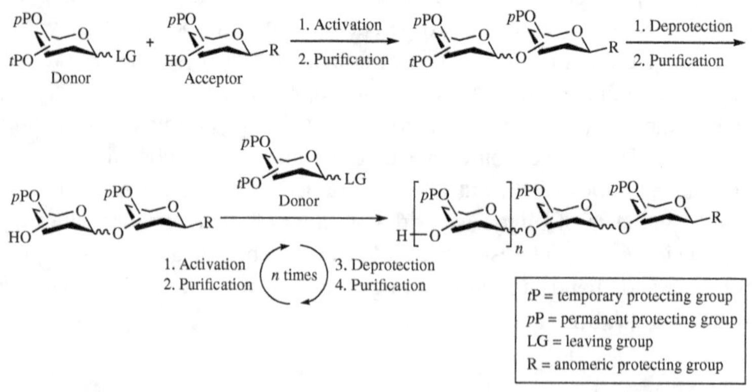

SCHEME 7.1 Solution-phase oligosaccharide synthesis strategy.

large oligosaccharides results in multiple purification steps that dramatically increase the overall synthesis time.

Stepwise oligosaccharide synthesis is a labor-intensive and time-consuming task and is usually carried out by synthetic experts [14–16]. Considerable efforts have been made to overcome classical stepwise oligosaccharide synthesis and have been reviewed recently [17]. A reactivity-based one-pot sequential glycosylation method that reduced the overall number of purification steps was pursued by Chi-Huey Wong et al. [18] (see Section 6.2.1). The anomeric reactivity (relative reactivity value (RRV)) of more than one hundred thioglycosides was analyzed using a computer program ("OptiMer") that enables the prediction of the optimal set of donors required to create a given oligosaccharide. The reactions are performed manually in solution phase, and the oligosaccharide chain is grown from the nonreducing to the reducing end based on thioglycoside reactivity. An automated solution-phase parallel synthesis of oligosaccharides was developed by Takashi Takahashi et al. based on the one-pot procedures [19]. However, restrictions on the overall length of the oligosaccharides prepared by one-pot methods currently hamper the more extensive use of these techniques.

7.3 SOLID-PHASE OLIGOSACCHARIDE SYNTHESIS

Solid-phase oligosaccharide synthesis is an appealing alternative to solution-phase strategies that dramatically reduce the purification process. Solid-phase synthesis of biopolymers was first established for oligopeptides in the 1960s [20, 21] and later extended to oligonucleotides [22]. Following these biopolymers, solid-phase oligosaccharide synthesis was first explored in the early 1970s [23]. The nature of the solid support and the linker are crucial factors that significantly influence the oligosaccharide assembly. The solid support should be chemically inert and should have high loading and good swelling properties in common organic solvents. An appropriate linker, on other hand, must be stable during oligosaccharide synthesis but readily cleavable from the resin after the assembly of the oligosaccharide. Building blocks have to be synthesized with orthogonal permanent and temporary protecting groups to facilitate the elongation at specific hydroxy groups. All the reactions performed on a solid support should be high yielding and stereoselective in order to avoid deletion sequences and the formation of complex mixtures.

Solid-phase synthesis starts with the coupling of the first monosaccharide building block to the linker-anchored solid support followed by removal of a temporary protecting group. The resulting free hydroxy group is glycosylated again and the iterative process continued for chain elongation. Excess reagents and building blocks are used at each glycosylation and deprotection step to avoid deletion sequences and are removed by simple washing after the reaction. Finally, cleavage of the solid support yields the desired oligosaccharides. In this way, the number of purification steps is minimized in comparison to conventional solution-phase methods.

7.3.1 Strategies for Solid-Phase Oligosaccharide Synthesis

Different strategies have been employed for the elongation of oligosaccharide chains on solid support (Fig. 7.1) [24, 25]. In the donor-bound strategy (Fig. 7.1a), a glycosylating agent is immobilized by attaching its nonreducing end to the resin via specific linkers and is elongated by reacting with a solution of the nucleophile [26–28]. In the acceptor-bound strategy (Fig. 7.1b), the glycosyl acceptor is immobilized by fixing the anomeric position to the solid support, and elongation is carried out by activating a donor [29]. The acceptor-bound strategy is superior to the donor-bound strategy because the resin-bound glycosylating agent is the limiting agent that tends to hydrolyze during glycosylation, leading to reduced yields [24, 25]. A combination of both strategies was developed to allow for elongation in both directions of a growing oligosaccharide [25]. In this approach, the glycosyl donor is attached to the solid support through an unmodified hydroxy group at the nonreducing region, and elongation proceeds in the reducing as well as the nonreducing ends (Fig. 7.1c). This bidirectional strategy is particularly useful for the preparation of branched oligosaccharides [30–32].

The first solid-phase synthesis of oligosaccharides was demonstrated in 1971 by using the acceptor-bound strategy (Scheme 7.2) [33–35]. The allyl alcohol-functionalized Merrifield resin **1** and the glucosyl bromide building block **2** were employed in the synthesis of trisaccharide **3** on solid support. Glycosylation and temporary protecting group removal were achieved by using 2,6-lutidine and NaOMe, respectively. Following oligosaccharide assembly, cleavage from the resin was accomplished by ozonolysis followed by reduction to furnish the protected 2-hydroxyethyl glycoside **4**.

This preliminary approach explored and highlighted the fundamental issues related to linkers, glycosylating agents, and temporary protecting groups essential to solid-phase oligosaccharide assembly. Long reaction times, high-temperature reactions, and incomplete stereocontrol were the major drawbacks associated with this synthesis. Since that time, numerous groups have utilized modified linker

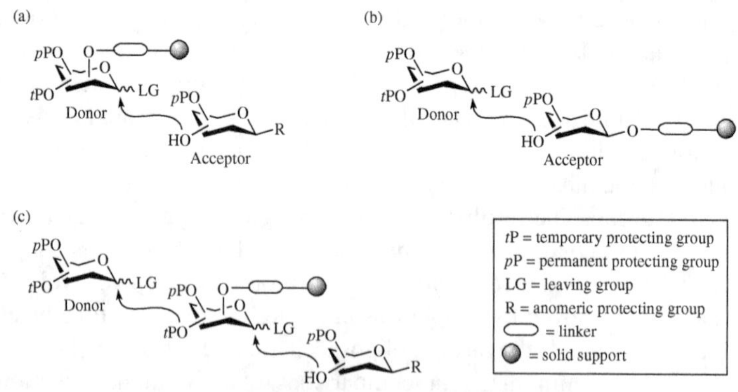

FIGURE 7.1 Synthetic strategies for oligosaccharide assembly on solid support. (a) Donor-bound strategy. (b) Acceptor-bound strategy. (c) Bidirectional strategy.

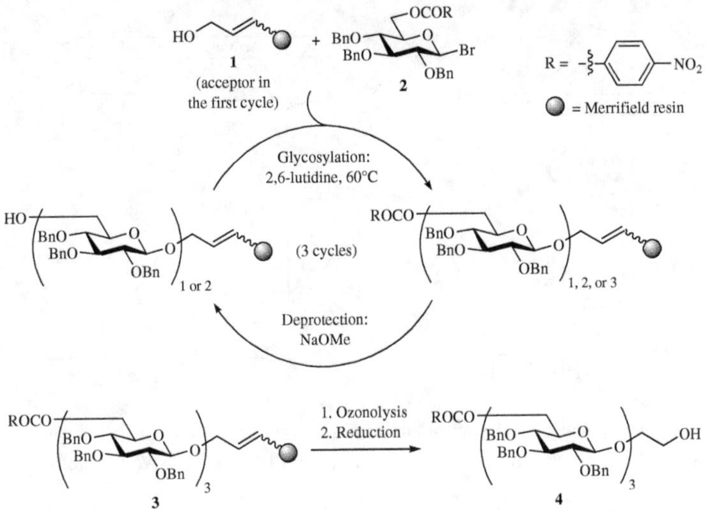

SCHEME 7.2 The first solid-supported oligosaccharide synthesis. Bn, benzyl.

systems, solid supports, and building blocks to accomplish the solid-phase synthesis of complex oligosaccharides. These reports have been reviewed [24, 25, 36, 37].

7.4 AUTOMATED OLIGOSACCHARIDE SYNTHESIS

Advances in solid-phase oligosaccharide synthesis led to the development of an automated platform that was first demonstrated by Peter Seeberger and coworkers in 2001 using a modified peptide synthesizer [38]. Automated washing, deprotection, and glycosylation steps dramatically reduced the time and labor usually attributed to solution- and solid-phase synthesis of oligosaccharides.

7.4.1 Technological Aspects of Automated Solid-Phase Oligosaccharide Synthesis

The complex structures of carbohydrates reflect also on the instruments designed to enable their synthesis. The variety of building blocks and conditions used for oligosaccharides synthesis demands a platform able to perform in a wide range of temperatures, efficient mixing of volatile and sometimes incompatible mixtures, and stability toward strong acids and bases. All parameters should be controlled and monitored by software. The three platforms that have been designed and used to date for automated solid-phase oligosaccharide synthesis are discussed in this section (Fig. 7.2). All three platforms utilize the acceptor-bound solid-phase oligosaccharide synthetic strategy.

First, a peptide synthesizer was modified to allow solid-phase oligosaccharide synthesis. This platform had the basic function of an automated synthesizer and allowed repeating cycles of glycosylation and deprotection to be mostly software controlled (Fig. 7.2a) [38]. This robust pressure-driven system relies on an established

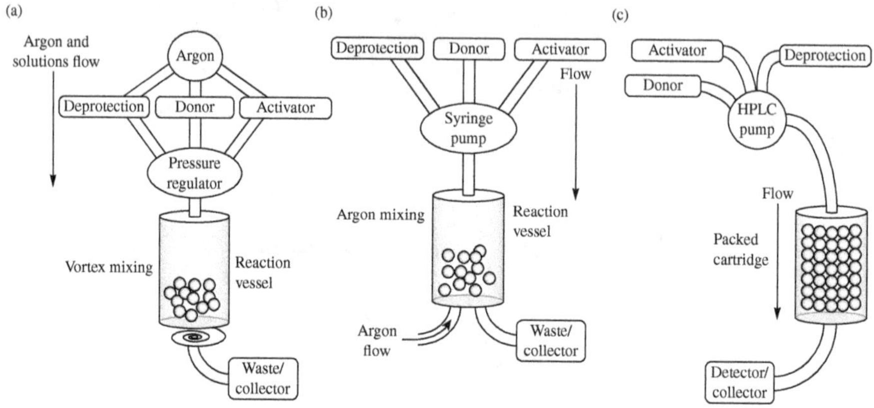

FIGURE 7.2 Three platforms for automated solid-phase oligosaccharide synthesis. (a) Pressure-driven system. (b) Syringe pump-driven system. (c) HPLC pump-driven system.

platform with years of success in peptide synthesis. In this synthesizer, reagents are delivered by regulating the argon pressure through a set of valves. To achieve efficient mixing, the resin is mixed by mechanical shaking ("vortexing") of the reaction vessel.

Second, the syringe pump-driven system represents a fully automated oligosaccharide synthesizer (Fig. 7.2b) [39]. A software program controls the reaction temperature as well as the addition of reagents via the syringe pump system that allows the accurate addition of reagents. The system is constantly under argon pressure, and resin is mixed by a flow of argon from the bottom of the reaction vessel, maintaining an inert atmosphere during the reaction.

Third, the HPLC pump-driven system represents a combination of solid-phase synthesis and continuous flow (Fig. 7.2c) [40]. The solid support is packed in an "HPLC column-like" cartridge, and reagents are delivered by an HPLC pump. In this case, mixing is replaced by the constant delivery of reagents by flow. The reagents continuously flow over the solid support and permeate in as they flush through the cartridge. Solid support washing is done in a similar way to clean the solid support before the next reaction. A computer-supported HPLC system equipped with an integral UV detector is used to monitor the reaction progress.

7.4.2 The First Decade of Automated Synthesis of Oligosaccharides

The automated solid-phase synthesis of oligosaccharides was first reported in 2001 [38]. A peptide synthesizer was modified to enable solid-phase oligosaccharide synthesis. The synthesizer was equipped with a cooling system for low-temperature glycosylations. As in the case with other automated platforms, a combination of new technologies and synthetic methods were developed, and distinctive breakthroughs were achieved. Many state-of-the-art synthetic and analytical methods were adopted and adjusted over the years. The first decade of automated solid-phase oligosaccharide synthesis relied on the development of new methods to access complex structures (Fig. 7.3).

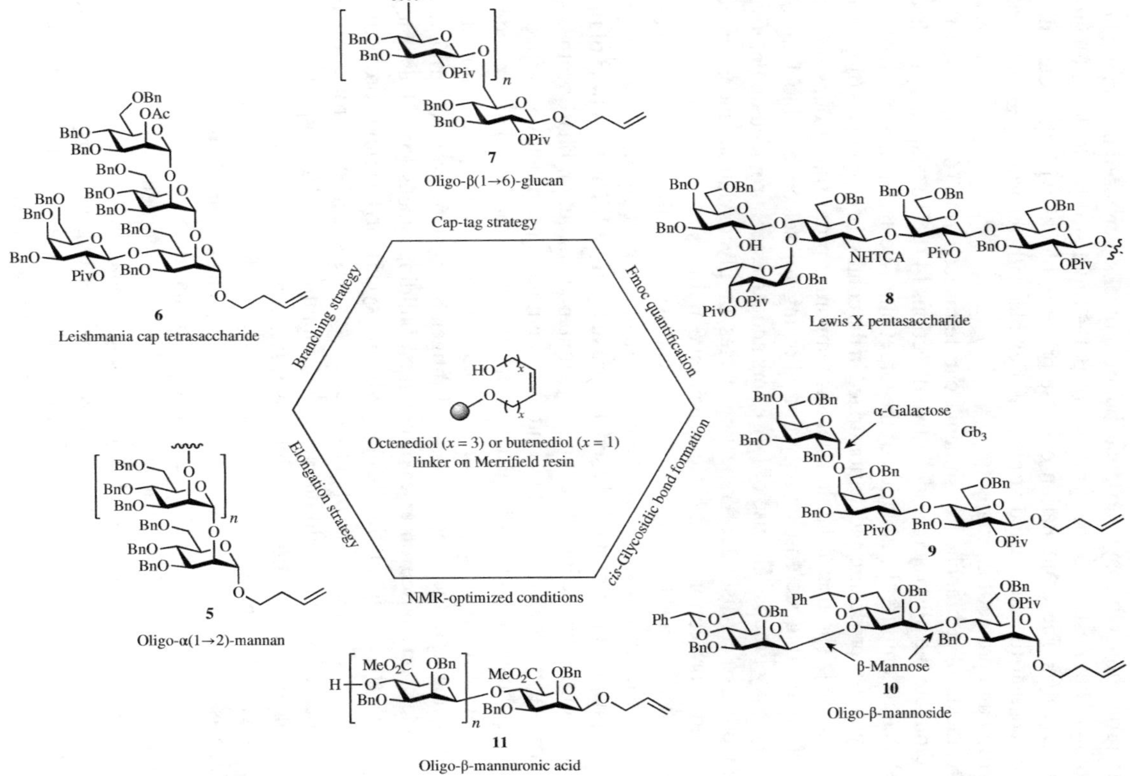

FIGURE 7.3 Oligosaccharides synthesized by automation using octenediol or butenediol linkers and the major milestones achieved. Piv, pivaloyl; TCA, trichloroacetyl.

The use of octenediol linker and Merrifield resin enabled the automated synthesis of various complex carbohydrates. A novel strategy to cleave an octenediol linker using olefin cross metathesis was developed to allow efficient removal of the oligosaccharides from the solid support [41–45]. Glycosyl trichloroacetimidates were used to synthesize a series of α(1→2)-mannosides (**5**) using an acetyl (Ac) ester for temporary protection. Glycosyl phosphate building blocks were used in combination with levulinyl (Lev) esters in the synthesis of β(1→3)/β(1→6) phytoalexin elicitor β-glucans [38, 46]. The combination of these building blocks with the octenediol linker showed that the automated strategy is flexible with regard to the anomeric leaving group and protecting groups.

The synthesis of a branched leishmania cap tetrasaccharide (**6**) illustrated the ability to elongate the oligosaccharide in two directions [47]. The "cap–tag" method was developed to simplify the isolation of the target oligosaccharide at the end of the automated synthesis procedure and demonstrated in the context of oligo-β(1→6)-glucan (**7**) syntheses [48]. The use of labile esters as temporary protecting groups proved valuable in controlling stereoselectivity and allowing for selective deblocking of particular hydroxy groups. The use of Lev esters and Ac esters provided an iterative automated process consisting of repeating cycles of glycosylation under acidic conditions and deprotection using basic or neutral conditions.

The automated approach to oligosaccharide synthesis was utilized for rapid access to carbohydrate-derived vaccine candidates as demonstrated by the synthesis of glycosylphosphatidylinositol (GPI)-based malaria antigen [49]. The total time of the GPI glycan assembly was reduced dramatically compared to solution-phase synthesis. The core hexasaccharide assembled by automated synthesis was readily modified in solution to the phosphorylated GPI malaria toxin [49]. The method was later extended to synthesize the core pentasaccharide of N-linked glycoproteins [50]. The automated solid-phase syntheses of the tumor-associated antigen **8** and blood group determinant oligosaccharides were based on a set of common building blocks [51]. This study highlighted the advantage of the 9-fluorenylmethoxycarbonyl (Fmoc) protecting group in automated solid-phase oligosaccharide synthesis. Fmoc deprotection is performed under mild basic conditions and can be monitored by UV/Vis spectroscopy to provide real-time information regarding glycosylation efficiency [52].

The *cis* glycosidic linkage formation is challenging in the context of complex carbohydrate assembly, because under the solid-phase synthesis paradigm, no real-time monitoring is possible. The automated syntheses of the tumor-associated carbohydrate antigens Gb_3 (**9**) and Globo-H demonstrated that the stereochemically challenging *cis* glycosylations can be performed in an automated manner (see Section 12.4.1) [53]. Optimizing the conditions for *cis* glycosylation using galactosyl phosphate building blocks enabled the formation of trisaccharides and hexasaccharides bearing α-galactosidic linkage. A strategy involving extended reaction times, lower reaction temperatures, and use of Et_2O as a participating solvent provided a significant increase in α selectivity. A similar approach was used for the automated synthesis of β-mannosidic linkages [54]. The anomeric phosphate triester was replaced by the stable carboxybenzyl (CB) leaving group that showed enhanced β selectivity for mannosylation in solution [55]. By careful placement of protecting groups, built in precedent by

David Crich [56], elongation of complex β-mannose-containing oligosaccharides (e.g., **10**) was demonstrated [54]. Apparently, new strategies for glycosidic bond formation can be readily transferred to automated oligosaccharide synthesis.

7.4.3 Recent Improvements in Automated Oligosaccharide Synthesis

For almost 10 years, automated solid-phase synthesis of oligosaccharides was performed on a modified peptide synthesizer. The first prototype of a specially made oligosaccharide synthesizer [39] features fully automated "tailor-made" syringe pump systems for accurate delivery of reagents.

With these instruments, Jeroen Codée and coworkers reported the automated synthesis of hyaluronan and oligo-β-mannuronic acid **11** (Fig. 7.3) using a butenediol linker [57, 58]. The synthesis of oligo-β(1→4)-mannuronic acid is notable because the stereoselective formation of the β(1→4)-linkage of mannuronic acid by chemical means is challenging. For the synthesis of oligo-β-mannuronic acid, low-temperature NMR spectroscopy studies were performed to optimize glycosylation conditions [59]. The stable *N*-phenyltrifluoroacetimidate leaving group [60], in combination with low temperatures, ensured complete selectivity of the glycosylation. These optimized conditions were implemented for the automated synthesis of the dodeca-β-mannuronic acid oligomer. The NMR technique for reaction optimization minimizes time, effort, and waste of precious building block usually involved in the optimization process.

The octenediol and butenediol linkers proved efficient for the synthesis of various oligosaccharides. The stability of the linker to harsh acidic conditions enabled an extensive use of the common glycosyl imidate and phosphate building blocks. However, olefin linkers have some drawbacks. Thioglycoside building blocks require electrophilic activators that may modify the double bond. In addition, several postassembly transformations of the linker are required to yield deprotected conjugation-ready oligosaccharides for biological studies.

7.4.4 Automated Synthesis of Conjugation-Ready Oligosaccharides

Linker selection has a major influence on glycan assembly because it dictates the type of activation strategy and protecting groups that can be used for oligosaccharide elongation. A ready-to-conjugate linker is very attractive because it facilitates conjugation to proteins, nanoparticles, and microarray systems to give a platform for biological evaluation of synthetic oligosaccharides.

A bifunctional linker system was developed by Seeberger et al. to render conjugation-ready oligosaccharides in a minimum of postautomation steps (Fig. 7.4) [39, 61]. The designed linker can be cleaved from the solid support by simple saponification along with the ester protecting groups and provides a conjugation-ready handle after final hydrogenolysis. Removal of the temporary protecting groups such as Fmoc, Lev, and silyl ethers, as well as the *N*-TCA groups, does not compromise linker integrity. The stability of the bifunctional linker to the presence of electrophiles has a major impact on building block selection. The thiol leaving group was preferred over glycosyl imidates and glycosyl phosphates because thioglycosides

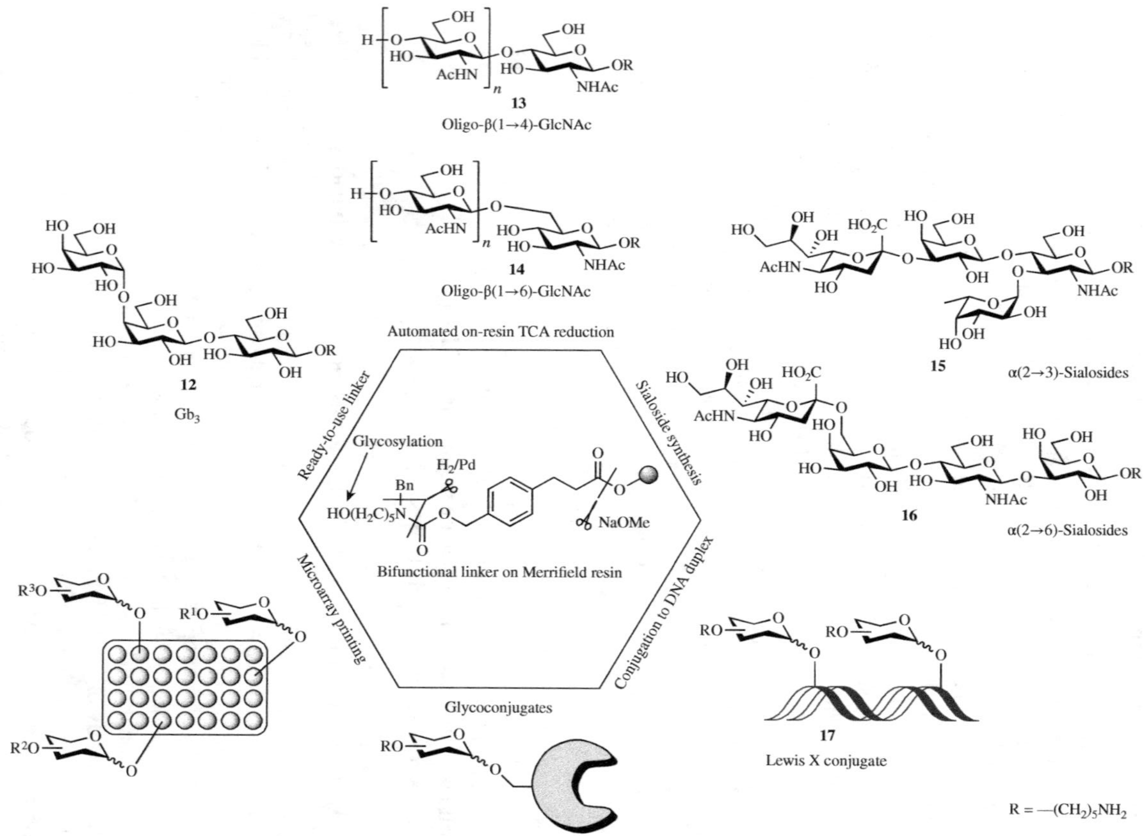

FIGURE 7.4 Automated synthesis utilizing the bifunctional linker gives access to ready-to-use oligosaccharide constructs that are useful for various biomedical applications.

are generally more stable than other glycosylating agents. The stable thioglycoside building blocks for use in automated synthesis were prepared by using both stepwise and one-pot procedures [62]. The new linker strategy in combination with the new automated platform enabled the synthesis of various oligosaccharides that can be used for biological applications without further modification.

The Gb_3 oligosaccharide **12** was synthesized as an example of a stereochemically challenging molecule (Fig. 7.4). The β-linked GalNAc- and GlcNAc-containing oligosaccharides were constructed with the help of the *N*-TCA group in C2 position of the corresponding thioglycosides. An automated radical reduction procedure was implemented to convert the *N*-TCA group to the native *N*-Ac on solid support before cleavage en route to the synthesis of the oligo-β(1→4)-GlcNAc **13** and the oligo-β(1→6)-GlcNAc **14** (Fig. 7.4) [39]. Fmoc and Lev ester protecting group manipulations proved efficient in the synthesis of linear as well as branched structures [39, 61].

Access to a series of α(2→3) (**15**) and α(2→6) (**16**) sialylated oligosaccharides (Fig. 7.4) required little manipulation and purification efforts [39]. The conjugation-ready oligosaccharides produced by automated synthesis were subsequently printed on microarrays or conjugated to proteins for immunological evaluation [39]. A combination of automated oligosaccharide and oligonucleotide synthesis was recently used to fix oligosaccharides on rigid scaffolds. Oligosaccharides such as the branched Lewis X trisaccharide can be synthesized and placed on a DNA duplex (Fig. 7.4, **17**) to allow for precise and defined presentation of the antigen [61]. The ability to merge automated procedures of different biopolymers to produce new hybrids is attractive for future applications.

7.4.5 HPLC-Assisted Automated Oligosaccharide Synthesis

An HPLC-assisted automated oligosaccharide synthesis system was recently reported by Alexei Demchenko and coworkers [40]. A modified HPLC system was used for the synthesis of various oligosaccharides on solid support. In this system, the anchored acceptor was packed in an Omnifit® chromatography column and was flushed with a mixed solution of building blocks and activator. *In situ* monitoring of Fmoc removal was achieved by using the HPLC-integrated UV detector. This solid-phase synthesis approach was used to create different oligosaccharides.

7.5 MICROFLUIDIC TECHNIQUES FOR OLIGOSACCHARIDE SYNTHESIS

Several technologies have been used in combination with or in addition to solid-phase oligosaccharide synthesis as an alternative to the classical carbohydrate synthetic approach [63–65]. Presented in the following text are summaries of important findings. More details concerning microfluidic systems and outcomes are described in Chapter 8.

Microfluidic-based devices are capable of conducting a wide range of organic transformations [66–69]. The small dimensions of microstructured reactors allow for the precise control of reaction variables such as reagent mixing, flow rates, reaction time, and heat and mass transfer [68, 70]. Microfluidic devices are also amenable to integrated real-time reaction monitoring techniques like UV/VIS, IR, NMR, mass spectrometry (MS), and LC/MS [69]. Microreactor reactions are easily scalable, representing a device capable of both analytical and semipreparative scales of production. Microfluidic systems are used in carbohydrate chemistry for optimization of glycosylation conditions as well as for the synthesis of oligosaccharides.

Microfluidic systems have initially been used for the optimization of glycosylation reactions because conditions can be carefully monitored even with small amounts of building blocks. In addition, microfluidic systems enable a wide range of reactions and conditions to be tested in a short time. A study to optimize α-mannosylation of the galactose acceptor **19** with the imidate **18** was investigated by a using a microfluidic system (Scheme 7.3a) [71]. Conditions were screened to optimize the formation of the desired disaccharide **20** over the orthoester by-product **21**. The glycosylations were carried out in a temperature range between −78 and 20°C, with different flow rates of glycosylating agents. The real-time progress of the reaction was monitored using HPLC. It was revealed that low temperatures (−78 to −70°C) and short reaction times (<1 min) favored the formation of the orthoester, while higher temperatures (−40°C) and longer reaction times (ca. 4 min) led to the formation of the desired α-linked product **20**.

A large number of variables influence the formation of the glycosidic linkage and make it difficult to predict the factors that will determine the outcome of a glycosylation. The influence of concentration, time, temperature, and solvent on the formation of the α(1→2)-dimannose **23** was investigated by comparative studies in a microreactor using the mannose donor **18** and acceptor **22** (Scheme 7.3b) [64]. A major influence of the solvent on the glycosylation outcome was revealed in this study. It was evident that, by changing CH_2Cl_2 to toluene, the desired disaccharide **23** was favored over the unwanted by-products **24** and **25**. A successful α-sialylation of the galactose nucleophile **27** with the sialic acid donor **26** to give the disaccharide **28** was optimized using microfluidic conditions (Scheme 7.3c) [65]. The use of a microfluidic system to control the effective concentration of the activated sialic acid building block **26** proved beneficial in increasing sialylation yields while maintaining the desired stereoselectivity. The preceding examples show that microfluidic systems enable the variation of several factors that influence the reaction outcome. Optimization in batch would be less efficient and demand more time and building blocks. The optimized conditions can be implemented easily for microfluidic scale up or batch reactions.

In addition to the glycosylation studies, oligosaccharides were prepared by combining a continuous-flow reactor and the fluorous-tag separation technique [63]. Reaction conditions for glycosylation and deprotection were first optimized and then scaled up to produce various oligosaccharides in excellent yield. Perfluorinated

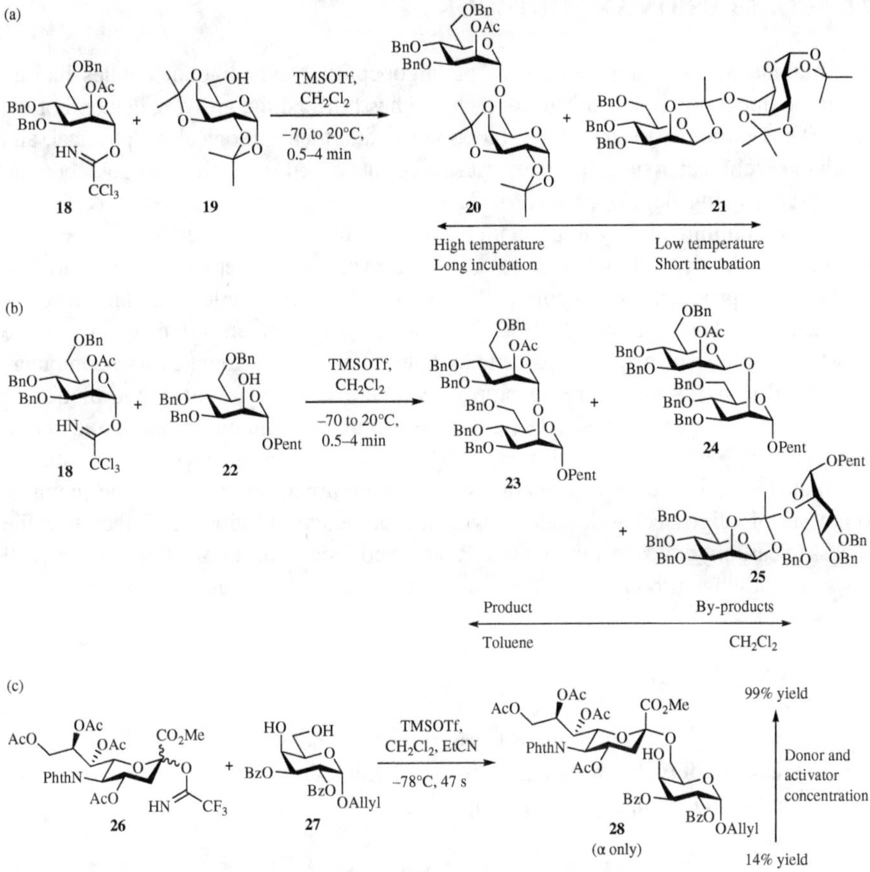

SCHEME 7.3 Optimization of glycosylation reactions in a microreactor. (a) Mannosylation optimization by varying temperature and incubation time. (b) Mannosylation optimization by solvent effect. (c) Sialylation optimization by donor and activator concentration.

linker was used to facilitate the isolation of the desired product by using fluorous solid-phase extraction (F-SPE) at the final step.

Koichi Fukase et al. reported an example of oligosaccharide synthesis by combining microreactors and an isolation strategy called synthesis based on affinity separation (SAS) [72, 73]. The glycosylation was carried out under continuous flow, and the mixture was directly sent to the stainless steel affinity column, and the flow channel was switched to wash the column with CH_2Cl_2. The labeled disaccharide was trapped in the column later to be eluted with a mixture of CH_2Cl_2 and methanol to obtain the isolated oligosaccharide product.

The use of microfluidic systems in oligosaccharide synthesis is in its infancy and far from being automated. More extensive use and development of such systems will open up new ways to access high-throughput and combinatorial oligosaccharide libraries.

7.6 CONCLUSION AND OUTLOOK

The automated solid-phase synthesis of oligopeptides and oligonucleotides had an immense impact on biochemical research and has fulfilled the growing demand of the scientific community to study these biopolymers and their biological, biophysical, and mechanical characteristics. In a similar fashion, automated solid-phase oligosaccharide synthesis is being developed to provide ready-to-use complex carbohydrates to the research community. Many important milestones have been achieved in the process of forming a reliable automated oligosaccharide synthesis system. The technological challenge in producing a suitable platform should not be overlooked. The progress from a modified pressure-driven system to a syringe pump-driven system gave exclusive access to previously inaccessible carbohydrates. The fully automated and computer-controlled systems that were developed in recent years offer attractive platforms for oligosaccharide synthesis. The combination of flow chemistry and solid-phase synthesis has not yet been fully explored and might present a viable solution for automated solid-phase oligosaccharide synthesis. Future developments in the chemical synthesis of oligosaccharides along with technological solutions will facilitate the access to carbohydrates by automation. Robust and accessible automation systems will assist in studying carbohydrates, their biological role, and their therapeutic potential.

ACKNOWLEDGMENTS

We thank the Max Planck Society for generous financial support. M.H. thanks the Max Planck Society for a Minerva postdoctoral fellowship. We thank Dr. C. L. Pereira and Dr. D. Kennedy for the help in editing this chapter.

REFERENCES

[1] Insight. *Nature* **2004**, *429*, 439–481.

[2] Insight. *Nature* **2003**, *422*, 191–237.

[3] Caruthers, M. H. *Science* **1985**, *230*, 281–285.

[4] Caruthers, M. H. *Acc. Chem. Res.* **1991**, *24*, 278–284.

[5] Varki, A., Cummings, R. D., Esko, J. D., Freeze, H. H., Stanley, P., Bertozzi, C. R., Hart, G. W., Etzler, M. E., Eds. *Essentials of Glycobiology*, 2nd ed.; Cold Spring Harbor Laboratory: Cold Spring Harbor, NY, **2009**.

[6] Varki, A. *Glycobiology* **1993**, *3*, 97–130.

[7] Ekelof, K.; Garegg, P. J.; Olsson, L.; Oscarson, S. *Pure Appl. Chem.* **1997**, *69*, 1847–1852.

[8] Seeberger, P. H.; Werz, D. B. *Nature* **2007**, *446*, 1046–1051.

[9] Koeller, K. M.; Wong, C. H. *Chem. Rev.* **2000**, *100*, 4465–4494.

[10] Seeberger, P. H.; Finney, N.; Rabuka, D.; Bertozzi, C. R. In *Essentials of Glycobiology*, 2nd ed.; Varki, A., Cummings, R. D., Esko, J. D., Freeze, H. H., Stanley, P., Bertozzi, C. R., Hart, G. W., Etzler, M. E., Eds.; Cold Spring Harbor Laboratory: Cold Spring Harbor, NY, **2009**, pp 691–704.

[11] Hudlicky, T.; Entwistle, D. A.; Pitzer, K. K.; Thorpe, A. J. *Chem. Rev.* **1996**, *96*, 1195–1220.

[12] Hanessian, S., Ed. *Preparative Carbohydrate Chemistry*; Marcel Dekker: New York, **1997**.

[13] Khan, S. H.; O'Neill, R. A., Eds. *Modern Methods in Carbohydrate Synthesis*; Harwood Academic: Amsterdam, **1996**.

[14] Mayer, T. G.; Kratzer, B.; Schmidt, R. R. *Angew. Chem. Int. Ed.* **1994**, *33*, 2177–2181.

[15] Seifert, J.; Lergenmuller, M.; Ito, Y. *Angew. Chem. Int. Ed.* **2000**, *39*, 531–534.

[16] Wang, Z. G.; Zhang, X. F.; Visser, M.; Live, D.; Zatorski, A.; Iserloh, U.; Lloyd, K. O.; Danishefsky, S. J. *Angew. Chem. Int. Ed.* **2001**, *40*, 1728–1732.

[17] Hsu, C. H.; Hung, S. C.; Wu, C. Y.; Wong, C. H. *Angew. Chem. Int. Ed.* **2011**, *50*, 11872–11923.

[18] Ye, X. S.; Wong, C. H. *J. Org. Chem.* **2000**, *65*, 2410–2431.

[19] Takahashi, T.; Adachi, M.; Matsuda, A.; Doi, T. *Tetrahedron Lett.* **2000**, *41*, 2599–2603.

[20] Merrifield, R. B. *J. Am. Chem. Soc.* **1963**, *85*, 2149–2154.

[21] Merrifield, R. B. *Angew. Chem. Int. Ed.* **1985**, *24*, 799–810.

[22] Matteucci, M. D.; Caruthers, M. H. *J. Am. Chem. Soc.* **1981**, *103*, 3185–3191.

[23] Frechet, J. M. In *Polymer-Supported Reactions in Organic Synthesis*; Hodge, P., Sherrington, D. C., Eds.; John Wiley & Sons Ltd: Chichester, UK, **1980**, pp 407–434.

[24] Weishaupt, M.; Eller, S.; Seeberger, P. H. In *Glycomics*; Fukuda, M., Ed.; Methods in Enzymology 478; Academic: San Diego, CA, **2010**, pp 463–484.

[25] Seeberger, P. H.; Haase, W. C. *Chem. Rev.* **2000**, *100*, 4349–4393.

[26] Roberge, J. Y.; Beebe, X.; Danishefsky, S. J. *Science* **1995**, *269*, 202–204.

[27] Danishefsky, S. J.; Mcclure, K. F.; Randolph, J. T.; Ruggeri, R. B. *Science* **1993**, *260*, 1307–1309.

[28] Danishefsky, S. J.; Bilodeau, M. T. *Angew. Chem. Int. Ed.* **1996**, *35*, 1380–1419.

[29] Wang, Z. G.; Douglas, S. P.; Krepinsky, J. J. *Tetrahedron Lett.* **1996**, *37*, 6985–6988.

[30] Zhu, T.; Boons, G. J. *Angew. Chem. Int. Ed.* **1998**, *37*, 1898–1900.

[31] Ito, Y.; Kanie, O.; Ogawa, T. *Angew. Chem. Int. Ed.* **1996**, *35*, 2510–2512.

[32] Kanie, O.; Ito, Y.; Ogawa, T. *J. Am. Chem. Soc.* **1994**, *116*, 12073–12074.

[33] Frechet, J. M.; Schuerch, C. *J. Am. Chem. Soc.* **1971**, *93*, 492–496.

[34] Frechet, J. M.; Schuerch, C. *Carbohydr. Res.* **1972**, *22*, 399–412.

[35] Frechet, J. M.; Schuerch, C. *J. Am. Chem. Soc.* **1972**, *94*, 604–609.

[36] Tanaka, K.; Fukase, K. In *Solid-Phase Organic Synthesis: Concepts, Strategies, and Applications*; Toy, P. H., Lam, Y., Eds.; John Wiley & Sons, Inc.: Hoboken, NJ, **2012**, pp 489–530.

[37] Tanaka, K.; Fukase, K. In *Glycoscience: Chemistry and Chemical Biology*, 2nd ed.; Fraser-Reid, B. O., Tatsuta, K., Thiem, J., Coté, G. L., Flitsch, S., Ito, Y., Kondo, H., Nishimura, S.-I., Yu, B., Eds.; Springer: Berlin, **2008**, pp 1241–1277.

[38] Plante, O. J.; Palmacci, E. R.; Seeberger, P. H. *Science* **2001**, *291*, 1523–1527.

[39] Kröck, L.; Esposito, D.; Castagner, B.; Wang, C. C.; Bindschadler, P.; Seeberger, P. H. *Chem. Sci.* **2012**, *3*, 1617–1622.

[40] Ganesh, N. V.; Fujikawa, K.; Tan, Y. H.; Stine, K. J.; Demchenko, A. V. *Org. Lett.* **2012**, *14*, 3036–3039.

[41] van Maarseveen, J. H.; den Hartog, J. A. J.; Engelen, V.; Finner, E.; Visser, G.; Kruse, C. G. *Tetrahedron Lett.* **1996**, *37*, 8249–8252.

[42] Miller, S. J.; Blackwell, H. E.; Grubbs, R. H. *J. Am. Chem. Soc.* **1996**, *118*, 9606–9614.

[43] Piscopio, A. D.; Miller, J. F.; Koch, K. *Tetrahedron Lett.* **1997**, *38*, 7143–7146.

[44] Nicolaou, K. C.; Winssinger, N.; Pastor, J.; Ninkovic, S.; Sarabia, F.; He, Y.; Vourloumis, D.; Yang, Z.; Li, T.; Giannakakou, P.; Hamel, E. *Nature* **1997**, *387*, 268–272.

[45] Andrade, R. B.; Plante, O. J.; Melean, L. G.; Seeberger, P. H. *Org. Lett.* **1999**, *1*, 1811–1814.

[46] Plante, O. J.; Andrade, R. B.; Seeberger, P. H. *Org. Lett.* **1999**, *1*, 211–214.

[47] Hewitt, M. C.; Seeberger, P. H. *Org. Lett.* **2001**, *3*, 3699–3702.

[48] Palmacci, E. R.; Hewitt, M. C.; Seeberger, P. H. *Angew. Chem. Int. Ed.* **2001**, *40*, 4433–4437.

[49] Hewitt, M. C.; Snyder, D. A.; Seeberger, P. H. *J. Am. Chem. Soc.* **2002**, *124*, 13434–13436.

[50] Ratner, D. M.; Swanson, E. R.; Seeberger, P. H. *Org. Lett.* **2003**, *5*, 4717–4720.

[51] Love, K. R.; Seeberger, P. H. *Angew. Chem. Int. Ed.* **2004**, *43*, 602–605.

[52] Carpino, L. A.; Han, G. Y. *J. Org. Chem.* **1972**, *37*, 3404–3409.

[53] Werz, D. B.; Castagner, B.; Seeberger, P. H. *J. Am. Chem. Soc.* **2007**, *129*, 2770–2771.

[54] Codée, J. D. C.; Kröck, L.; Castagner, B.; Seeberger, P. H. *Chem. Eur. J.* **2008**, *14*, 3987–3994.

[55] Kim, K. S.; Kim, J. H.; Lee, Y. J.; Lee, Y. J.; Park, J. *J. Am. Chem. Soc.* **2001**, *123*, 8477–8481.

[56] Crich, D. *Acc. Chem. Res.* **2010**, *43*, 1144–1153.

[57] Walvoort, M. T. C.; Volbeda, A. G.; Reintjens, N. R. M.; van den Elst, H.; Plante, O. J.; Overkleeft, H. S.; van der Marel, G. A.; Codée, J. D. C. *Org. Lett.* **2012**, *14*, 3776–3779.

[58] Walvoort, M. T. C.; van den Elst, H.; Plante, O. J.; Kröck, L.; Seeberger, P. H.; Overkleeft, H. S.; van der Marel, G. A.; Codée, J. D. C. *Angew. Chem. Int. Ed.* **2012**, *51*, 4393–4396.

[59] Walvoort, M. T. C.; Lodder, G.; Mazurek, J.; Overkleeft, H. S.; Codée, J. D. C.; van der Marel, G. A. *J. Am. Chem. Soc.* **2009**, *131*, 12080–12081.

[60] Yu, B.; Tao, H. C. *Tetrahedron Lett.* **2001**, *42*, 2405–2407.

[61] Schlegel, M. K.; Hutter, J.; Eriksson, M.; Lepenies, B.; Seeberger, P. H. *ChemBioChem* **2011**, *12*, 2791–2800.

[62] Wang, C. C.; Lee, J. C.; Luo, S. Y.; Kulkarni, S. S.; Huang, Y. W.; Lee, C. C.; Chang, K. L.; Hung, S. C. *Nature* **2007**, *446*, 896–899.

[63] Carrel, F. R.; Geyer, K.; Codée, J. D. C.; Seeberger, P. H. *Org. Lett.* **2007**, *9*, 2285–2288.

[64] Geyer, K.; Seeberger, P. H. *Helv. Chim. Acta* **2007**, *90*, 395–403.

[65] Tanaka, S. I.; Goi, T.; Tanaka, K.; Fukase, K. *J. Carbohydr. Chem.* **2007**, *26*, 369–394.

[66] Jahnisch, K.; Hessel, V.; Lowe, H.; Baerns, M. *Angew. Chem. Int. Ed.* **2004**, *43*, 406–446.

[67] Geyer, K.; Gustafsson, T.; Seeberger, P. H. *Synlett* **2009**, 2382–2391.

[68] Geyer, K.; Codée, J. D. C.; Seeberger, P. H. *Chem. Eur. J.* **2006**, *12*, 8434–8442.

[69] Mason, B. P.; Price, K. E.; Steinbacher, J. L.; Bogdan, A. R.; McQuade, D. T. *Chem. Rev.* **2007**, *107*, 2300–2318.

[70] deMello, A. J. *Nature* **2006**, *442*, 394–402.

[71] Ratner, D. M.; Murphy, E. R.; Jhunjhunwala, M.; Snyder, D. A.; Jensen, K. F.; Seeberger, P. H. *Chem. Commun.* **2005**, 578–580.

[72] Fukase, Y.; Zhang, S. Q.; Iseki, K.; Oikawa, M.; Fukase, K.; Kusumoto, S. *Synlett* **2001**, 1693–1698.

[73] Fukase, K.; Takashina, M.; Hori, Y.; Tanaka, D.; Tanaka, K.; Kusumoto, S. *Synlett* **2005**, 2342–2346.

8

SUGAR SYNTHESIS BY MICROFLUIDIC TECHNIQUES

KOICHI FUKASE[1], KATSUNORI TANAKA[2], YUKARI FUJIMOTO[3], ATSUSHI SHIMOYAMA[1], AND YOSHIYUKI MANABE[1]

[1] *Department of Chemistry, Graduate School of Science, Osaka University, Osaka, Japan*
[2] *Biofunctional Synthetic Chemistry Laboratory, RIKEN, Saitama, Japan*
[3] *Graduate School of Science and Technology, Keio University, Kanagawa, Japan*

8.1 INTRODUCTION

Continuous microflow synthesis is an innovative technology for chemical processes from small-scale synthesis to large-scale production [1–13]. This technology enables fast and efficient mixing, rapid heating and cooling, strict temperature control, precise residence time control, and effective mass transfer. Moreover, once the reaction conditions are optimized for a small-scale operation, the same conditions are directly applicable to large-scale synthesis under the flow process. For example, the alcohol dehydration process using microreactor was successfully developed and applied to the large-scale production of saturated terpenoid pristane, which has been used as an adjuvant for monoclonal antibody production in mice [12–14]. The applications of the advantageous features of the microfluidic systems to oligosaccharide synthesis have also been investigated [15–23]. Successful examples that employed this technique include α-sialylation [16–18], β-mannosylation [19], N-glycosylation on asparagine [20], α-selective glycosylation of 3-deoxy-D-*manno*-2-octulosonic acid (KDO) [21, 22], β-selective glucosaminylation [15], and reductive opening of the benzylidene acetal groups in sugars [23]. The improved procedure on the latter

---

Glycochemical Synthesis: Strategies and Applications, First Edition.
Edited by Shang-Cheng Hung and Medel Manuel L. Zulueta.

example under the microfluidic conditions enabled the preparation of key synthetic intermediates for oligosaccharides in multigram scales, eventually leading to the synthesis of the asparagine-linked oligosaccharide (N-glycan) [17] and *Helicobacter pylori* lipopolysaccharide (LPS) partial structures [21]. The details of these studies are described in the next section.

8.2 MICROFLUIDIC GLYCOSYLATION

Peter Seeberger and coworkers reported the first glycosylation reaction under microfluidic conditions by applying microflow synthesis to stereoselective α-mannosylation using a 2-*O*-acetylated mannosyl donor (see Section 7.5). Chemoselectivity was controlled by precise temperature and residence time control with silicon–glass reactor [24]. They then reported the microfluidic oligosaccharide synthesis integrated with fluorous separation technique. Stereoselective glycosylation was carried out in a microreactor by using neighboring group participation and iterative glycosylation, deprotection, and fluorous separation to afford the desired tetrasaccharide [25].

Koichi Fukase and coworkers also integrated a microreactor and the affinity separation system by using the interaction of the podand ether tag with solid-supported ammonium ion [15]. The procedure used a combination of an IMM micromixer (IMM stands for Institut für Mikrotechnik Mainz, a company based in Germany) and a stainless tube reactor (l=2.0 m, diameter (Φ)=500 μm) equipped with a stainless steel column as an affinity separation unit (Fig. 8.1). Glycosylation was carried out by mixing the solution of donor **1** and acceptor **2** in CH_2Cl_2 and the solution of trimethylsilyl triflate (TMSOTf) in CH_2Cl_2 under continuous flow. The reaction mixture was directly introduced to the stainless steel affinity column (4.6×250 mm), and then the flow channel was switched to wash the column with CH_2Cl_2. The desired disaccharide **3** was trapped in the column, whereas the other by-products were washed out. Elution with CH_2Cl_2/MeOH (1/1) gave the purified product **3** in 91% yield.

8.2.1 Microfluidic α-Sialylation

Sialic acids exist at the termini of glycans of glycoproteins and glycolipids of higher animals and are usually linked to galactose or *N*-acetylgalactosamine through α(2→3)- and α(2→6)-sialoside linkages. Because sialic acid-containing glycans play diverse and important roles in various recognition events such as inflammation, immunity, infection, tumor metastasis, and others, much effort has been devoted to the efficient and stereoselective synthesis of α-sialosides in order to investigate their biological functions.

α-Sialylation is recognized as one of the most difficult glycosylation in the field of oligosaccharide synthesis (see Chapter 5). Stereocontrol of sialylation is difficult and considerable amount of glycal is often formed because the α-sialoside is more thermodynamically unstable than the β-sialoside, no neighboring group participation

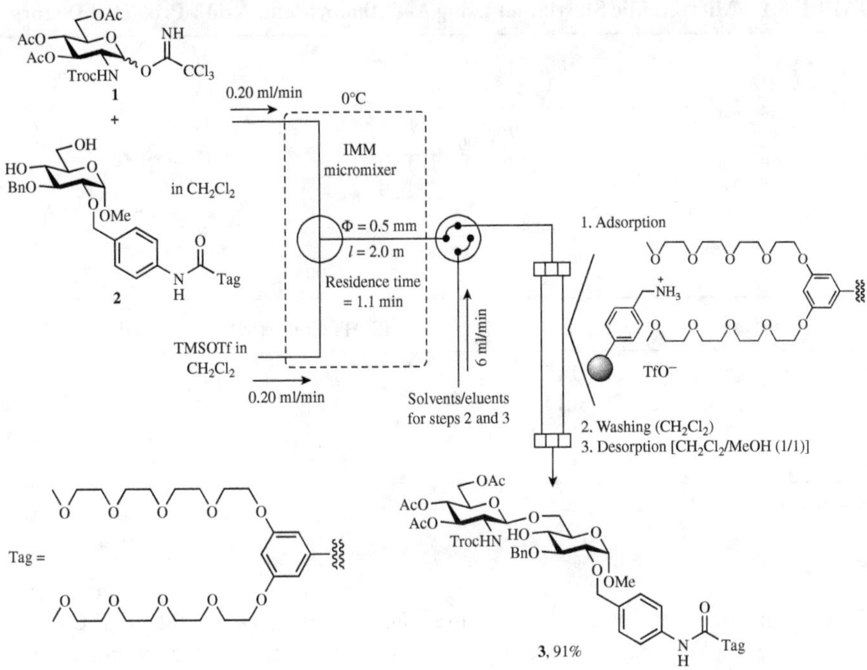

FIGURE 8.1 Continuous flow synthesis using microreactor and affinity separation system. Ac, acetyl; Bn, benzyl; TfO, triflate; Troc, trichloroethoxycarbonyl.

can be used because of the 3-deoxy structure, glycosylation should occur at a sterically hindered position, and the electron-withdrawing carboxylate promotes β-elimination of the leaving group. The breakthrough was the use of the kinetic solvent effect of nitrile for the α-selective glycosylation by Akira Hasegawa et al. [26]. Recent advances in glycosylation chemistry, such as the development of new leaving groups and protecting groups, enabled the efficient construction of α-sialoside linkages. Glycosyl *N*-phenyltrifluoroacetimidates proved to be efficient donors for various glycosylation including sialylation [16–22, 27–29]. In particular, sialylation using glycosyl donors having C5 substituents such as *N*-acetylacetamide [30], trifluoroacetamide [31, 32], NHTroc [32–34], and azide [35] groups improved the yields and α selectivities. Recently, the 4-*O*,5-*N*-cyclic carbamates have been utilized especially for the α(2→8)- and α(2→9)-sialylation case [36–41]. Sialylations using 4-*O*,5-*N*-cyclic carbamates afforded the desired α-sialosides with very high selectivity even without the nitrile solvent effect.

An efficient α-sialylation that took advantage of the highly reactive sialyl donors having C5 cyclic imides, especially phthalimide (NPhth), by virtue of the "fixed-dipole moment effects," was also reported [42]. For example, the use of the sialyl donor **4** and acceptor **6** at −78°C supplied the sialoside **7** as α only with 92% yield on 50 mg scale. The scale-up in batch process, however, significantly decreased the yield and stereoselectivity. This decrease in sialylation efficiency might be a result of

TABLE 8.1 Microfluidic Sialylation Using *N*-Phthaloyl- and Azido-Protected Donors[a]

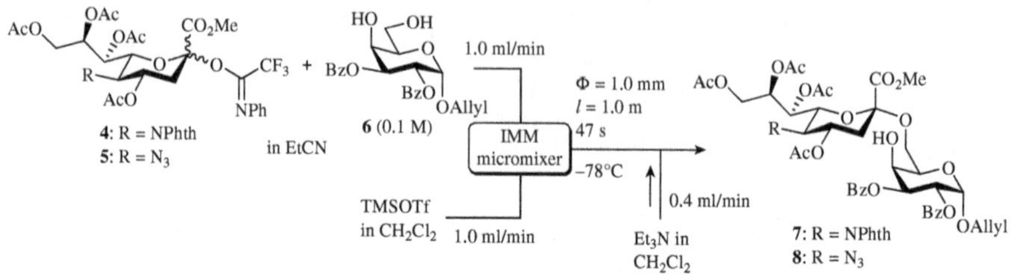

Entry	Donor	Donor Conc. (M)	TMSOTf Conc. (M)	Yield	α/β
1	**4**	0.15	0.08	14	α only
2	**4**	0.15	0.15	88	α only
3	**4**	0.2	0.15	>99	α only
4	**5**	0.2	0.15	>99	20/1

[a] Bz, benzoyl.

the local heating generated by the neutralization of the Lewis acid (TMSOTf) by Lewis bases and the reaction with the highly reactive **4**. In order to circumvent this problem, Fukase and coworkers used a continuous flow microreactor (Table 8.1) [16]. For the microfluidic sialylation, a solution of sialyl donor **4** and acceptor **6** in EtCN was mixed with a TMSOTf solution in CH_2Cl_2 at −78°C using an IMM micromixer with a channel width of 40 μm at a flow rate of 1.0 ml/min. After the reaction mixture was allowed to flow at −78°C for an additional 47 s through a reactor tube ($\Phi = 1.0$ mm, $l = 1.0$ m), the mixture was quenched by another flow of excess Et_3N dissolved in CH_2Cl_2 by using a T-shaped mixer at −78°C. The desired α-sialoside **7** was obtained in quantitative yield by using 0.2 M of donor **4** and 0.15 M of TMSOTf (entry 3). Thus, the microfluidic reaction successfully controlled the high reactivity of the sialyl donor **4** for α-sialylation. The efficient mixing of substrates with the high concentrations of the acid was responsible for the success of the microfluidic sialylation.

The 5-azido group was also employed because it should direct similar "fixed-dipole moment effects" and is easily converted to functional groups that naturally occur in neuraminic acids (i.e., *N*-acetyl or *N*-glycolyl groups) [17]. Under the typical batch process, the sialylation of the galactosyl acceptor **6** with the azide-containing donor **5** in the presence of TMSOTf in EtCN provided **8** in 90% yield with good α selectivity (α/β = 9/1 on 100 mg scale). The continuous microfluidic sialylation of **6** with **5** provided improved results as compound **8** was obtained quantitatively with near-perfect α selectivity (α/β = 20/1) (entry 4). The α and β stereoisomers were easily separated by chromatography on silica gel. The pure α isomer was readily converted to the desired imidate derivative **12** designed, along with building blocks **9–11**, for the solid-phase synthesis of the N-glycan **14** (Scheme 8.1). Compound **12** was successfully prepared on 5–10 g scale [17].

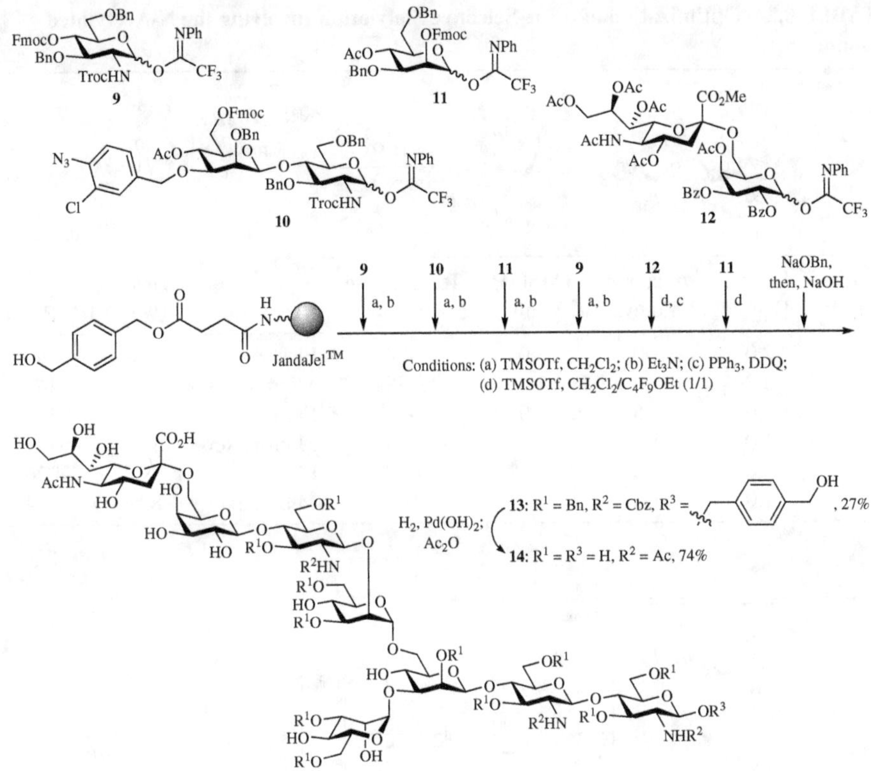

SCHEME 8.1 Solid-phase synthesis of an N-glycan. Cbz, benzyloxycarbonyl, DDQ, 2,3-dichloro-5,6-dicyano-1,4-benzoquinone; Fmoc, 9-fluorenylmethoxycarbonyl.

Sialylations using the 5-acetamido-containing donors usually afford modest yields and selectivity, although various glycosylation methods have been extensively studied [43–47]. The corresponding imidate donor shows high α selectivity in a few cases, such as the α(2→6)-sialylation of a galactose acceptor [27], although most reports described modest yields and α selectivities for the sialylation using the N-acetylated donors (about 50% and α/β = ~3/1, respectively) [42]. Recently, Leonid Kononov et al. reported that the "low-to-modest" efficiency in using the N-acetylated donors is caused by the hydrogen-bonding network. The supramolecular aggregation of the glycosyl donors, which is sensitively affected by the reaction concentrations, attenuates the nucleophilic attack and/or the α-face approach of the acceptor [48, 49].

With the realization that the efficiency and reproducibility in the conventional flask reaction is sensitive to the heat generated during syringe addition of the Lewis acid to the donor and acceptor, the reaction conditions of the sialylation using the N-acetylated donor was reinvestigated in pursuit of a practical route to α-sialylation. The initial optimization of the α-sialylation utilized the N-acetylated imidate **15** [27, 50, 51] and the galactose acceptor **6** in the presence of TMSOTf as an activator in

TABLE 8.2 Optimization of the α-Selective Sialylation Involving the N-Acetylated Donor 15

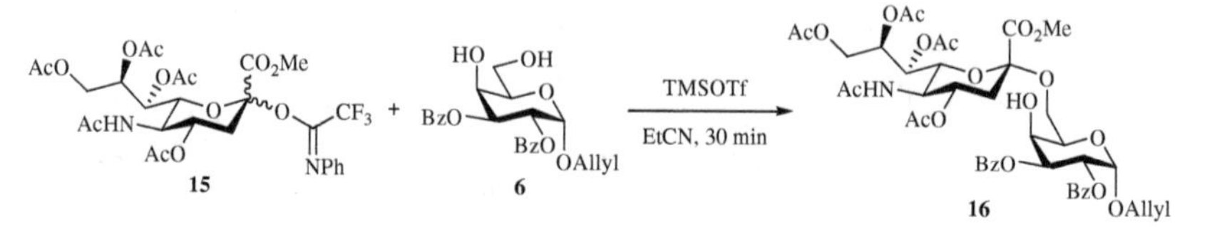

Entry	Donor Equiv.	Acceptor Equiv.	TMSOTf Equiv.	Temperature (°C)	Apparatus	Yield (%)	α/β Ratio
1	2.0	1.0	1.0	−78	Flask	77	96/4
2	2.0	1.0	1.0	−78	Micromixer	89	94/6
3	1.0	1.5	1.0	−78	Flask	86	93/7
4	1.0	1.5	1.0	−78	Micromixer	89	94/6
5	1.0	1.5	0.2	−20	Flask	89	74/26
6	1.0	1.5	1.0	−60	Micromixer	82	86/14

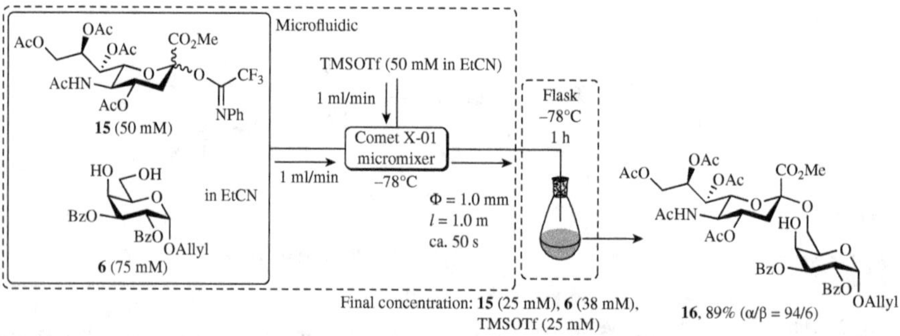

FIGURE 8.2 Microfluidic glycosylation to form the sialoside **16**.

EtCN at −78°C (Table 8.2). Assuming that temperature control is critical for sialylation in the presence of donor **15**, the reaction was initially performed in small scale (50–100 mg) by conventional syringe addition of the Lewis acid to a mixture containing the donor and the acceptor in a flask apparatus. Verification of the local heating effect to the glycosylation outcome was the one goal of this experiment. In the batch reaction, TMSOTf was carefully added in order to suppress the local heating, and the result was compared with the reaction performed under microfluidic conditions using a Comet X-01 apparatus, which produces efficient mixing and fast heat transfer (Fig. 8.2) [12–14, 16–20].

Efficient activation of the N-acetylated sialyl donor required 1.0 equiv. of the Lewis acid under −78°C, giving the disaccharide **16** in 77% yield with excellent α selectivity (α/β = 96/4, Table 8.2, entry 1). The conditions in entry 1 were reproducible under the microfluidic conditions (entry 2, 89% yield, α/β = 94/6). Because the excess donor **15** inhibited the purification of sialoside **16** from the glycal by-product due to similar polarities on silica gel, the ratio between the donor and the acceptor was reversed (entries 3–6). Similar conditions using 1.0 equiv. of TMSOTf successfully provided the α-sialoside **16** in 86% yield with good selectivity (α/β = 93/7, entry 3). The yield and selectivity were validated by performing the sialylation reaction under microfluidic conditions (89%, α/β = 94/6, entry 4). The concentrations of the substrates and Lewis acid did not affect the outcome of the sialylation efficiency under the conditions examined.

The microfluidic glycosylation with the N-acetylated donor **15** was carried out as follows: A solution of the sialyl donor **15** (50 mM) and the acceptor **6** (75 mM) in EtCN was mixed with TMSOTf in EtCN (50 mM) at −78°C using a Comet X-01 micromixer with a channel width of 500 μm at a flow rate of 1.0 ml/min (Fig. 8.2). The reaction mixture was allowed to flow with cooling at −78°C for an additional 50 s through the reactor tube (Φ = 1.0 mm, *l* = 1.0 m). The mixture was then introduced into a flask apparatus at −78°C to allow additional stirring to complete the reaction. For the gram-scale preparation of disaccharide **16**, the concentration of the substrates and Lewis acid were scaled down to half of those listed in Figure 8.2 to prevent solution blockage problems without decreasing efficiency.

It should be noted that the mixing temperature affected the α selectivity in a nontrivial way (Table 8.2, entries 5 and 6). The reaction proceeded smoothly by using 0.2 equiv. of TMSOTf under −20°C in a flask to give sialoside **16** in 89% yield, but the selectivity was not as good (α/β = 74/26) (entry 5). The inconsistent results may have resulted from temperature heterogeneities and heat generation during the inefficient mixing in the flask, which promoted the catalytic reaction at the expense of α selectivity. As noted earlier, most reported flask sialylation trials using the N-acetylated sialyl donors resulted in unsatisfactory results. Considering that a slight increase in the mixing temperature nontrivially affected the α selectivity, the low efficiency and reproducibility reported by various groups was not due to the inherent reactivity of the N-acetylated donors but was derived from the mixing inefficiencies associated with a flask apparatus.

After the α(2→6)-linked compound **16** was obtained efficiently by using the N-acetylated donor **15**, other important α-sialylation reactions forming the linkages of the natural glycans, including α-sialylation of the 3- and 6-hydroxyls of galactose and glucosamine acceptors, respectively, were examined (Scheme 8.2). Surprisingly, in contrast to the results in Table 8.2, entry 1, even catalytic amounts of the Lewis

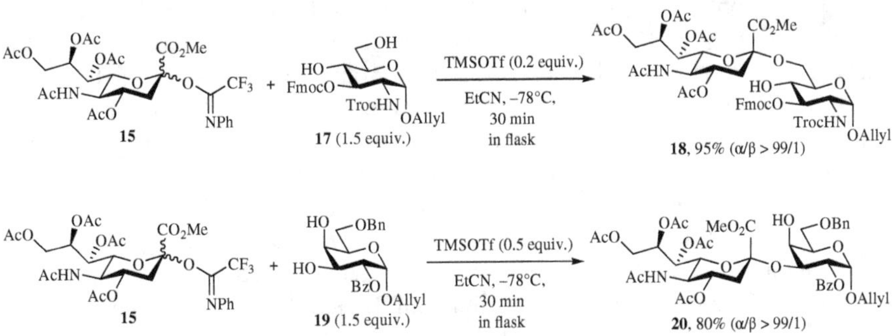

SCHEME 8.2 α-Sialylation with the 6-hydroxyl of glucosamine and the 3-hydroxyl of galactose acceptors.

acid promoted these glycosylation reactions. Hence, the amount of acid was optimized in each case to maximize the glycosylation efficiency. The reaction with the 6-hydroxyl of the *N*-Troc-protected glucosamine **17** in the presence of 0.2 equiv. of TMSOTf provided the corresponding sialoside **18** in 95% yield with perfect α selectively. Sialylation proceeded smoothly for the more sterically demanding 3-hydroxyl of the galactoside **19** in the presence of 0.5 equiv. of the acid to give **20** in 80% yield with high α selectivity.

8.2.2 Glycosylation with KDO

LPS is a glycoconjugate of Gram-negative bacteria composed of a polysaccharide and a glycolipid called lipid A (see Section 14.3). The acidic sugar KDO serves as a linker between the polysaccharide part and lipid A in LPS. LPS, also called endotoxin, has potent immunostimulatory and inflammatory activities upon recognition by the Toll-like receptor 4/myeloid differentiation factor-2 heterodimer, and the active entity is lipid A. On the other hand, LPS from parasitic bacteria such as *H. pylori* and *Porphyromonas gingivalis* show very low immunostimulatory activity compared to other enterobacterial preparations such as *Escherichia coli* LPS. The characteristic biological activities of *H. pylori* LPS are linked to the virulence and the innate immune response to infection. In addition, LPS from these parasitic bacteria has been implicated in chronic inflammation such as atherosclerosis.

In order to investigate the biological activity of *H. pylori* lipid A and LPS, synthetic studies are essential because LPS from natural sources are generally contaminated with other immunostimulative substances. The synthesis of *H. pylori* lipid A and KDO–lipid A have been investigated [21, 52–57]. For the synthesis of KDO–lipid A, glycosylation with KDO is a key reaction. Because KDO has a 2-keto-3-deoxy structure, the neighboring group effect is not available in glycosylations with KDO, and the presence of the 3-deoxy structure and a C1 carboxyl group easily lead to the β-hydrogen elimination to afford a glycal as the major by-product.

Glycosylation with KDO fluorides as glycosyl donors was developed by Fukase et al. for the synthesis of *E. coli* Re-type LPS, *H. pylori* KDO–lipid A, and their derivatives [54, 58–62]. High α selectivity was obtained by using 4,5-*O*-isopropylidene- or *tert*-butyldimethylsilyl (TBS)-protected KDO fluorides as donors. With these KDO donors, the undesirable β-side attack by the glycosyl acceptor is prevented by the presence of the bulky isopropylidene or TBS group. However, the cleavage or migration of acid-labile protecting groups sometimes occurs during the glycosylation, because strong Lewis acids are required for activation of the KDO fluorides. The synthetic strategy is therefore restricted, and longer reaction schemes are often needed for the synthesis of LPS partial structures composed of lipid A and KDO. Recently, KDO glycosylation with *N*-phenyltrifluoroacetimidate donor was developed and applied to the synthesis of *H. pylori* KDO–lipid A [21, 22]. The combination of TMSOTf as catalyst and cyclopentylmethyl ether (CPME) as solvent was found efficient for the glycosylation with the KDO *N*-phenyltrifluoroacetimidate **21** (Fig. 8.3). Glycosylation of the disaccharide acceptor **22** by using large

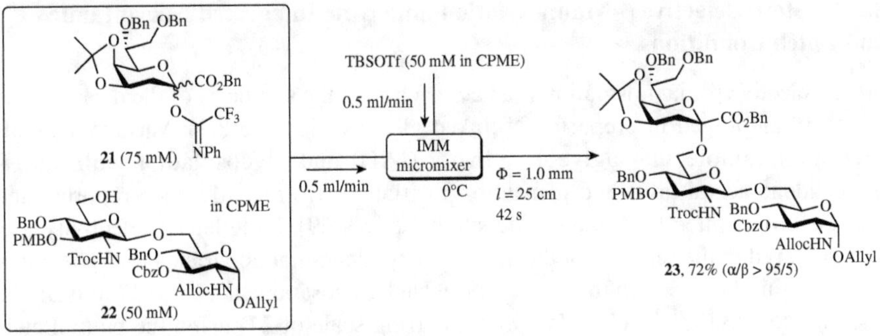

FIGURE 8.3 α-Selective glycosylation of KDO with a disaccharide acceptor using the microfluidic method. Alloc, allyloxycarbonyl; Cbz, benzyloxycarbonyl, PMB, *p*-methoxybenzyl.

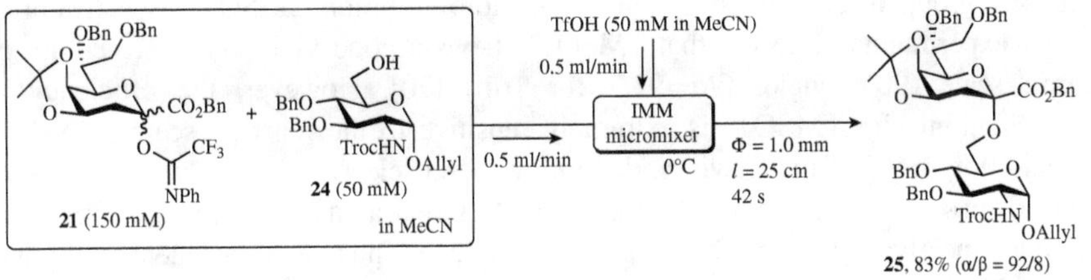

FIGURE 8.4 α-Selective glycosylation of monosaccharide acceptors with KDO using the microfluidic method.

excess of donor **21** (5.0 equiv.) in the presence of TBSOTf as activator in CPME gave trisaccharide **23** in 70% yield with high α selectivity (α/β>95/5). The low efficiency in the glycosylation might be due to the difficulty in temperature control. Donor **21** is highly reactive and should generate a significant amount of heat during the reaction. Microfluidic glycosylation was then applied expecting the suppression of glycal formation owing to the efficient removal of reaction heat as well as the promotion of intermolecular coupling reaction as a result of efficient mixing (Fig. 8.3). After optimization using TBSOTf as activator and CPME as solvent, the desired **23** was obtained in 72% yield with high stereoselectivity (α/β>95/5). The formation of the glycal was effectively suppressed under microfluidic conditions. Moreover, the amount of the KDO donor **21** was reduced to 1.5 equiv. from the 5.0 equiv. used in the batch reaction.

KDO glycosylations using monosaccharide acceptors were also examined [22]. Combination of TfOH as catalyst and CH_3CN as solvent gave better results than other catalysts (TMSOTf, TBSOTf) and solvents (CPME, CH_2Cl_2, toluene). The α selectivity was increased under microflow conditions, though the yields between batch (86%, α/β=85/15) and microflow conditions (83%, α/β=92/8) were comparable (Fig. 8.4).

8.2.3 Stereoselective β-Mannosylation under the Integrated Microfluidic and Batch Conditions

Stereoselective β-mannosylation is a key reaction in the synthesis of the disaccharide donor **10** employed in preparing N-linked glycans (Scheme 8.1). Various methods such as intramolecular aglycon delivery (IAD) and glycosylation with 4,6-*O*-benzylidene acetal-protected α-mannosyl triflates have recently been reported and successfully applied to β-mannoside synthesis [63–69]. Excellent β selectivity was also achieved in the reaction of the 4,6-*O*-benzylidene-mannopyranosyl-*N*-phenyltrifluoroacetimidate **26** with the *N*-Troc-protected glucosamine acceptor **27** to form the disaccharide **28** (93% yield, β/α=95/5 on 20mg scale of **27**) using the bulky Lewis acid $TMSB(C_6F_5)_4$ for 20min at −78°C (see Fig. 8.5 for the structures) [70].

Nevertheless, it is difficult to apply such β-mannosylation protocol to the gram-scale synthesis of compound **10** because the scaled-up glycosylation requires a large quantity of $TMSB(C_6F_5)_4$, which has limited commercial availability [70]. Therefore, Fukase's group refocused on applying the more common TMSOTf as activator as earlier experiments indicated that TMSOTf shows a good yield and β selectivity on 20mg scale (90% yield, β/α=93/7 for 16h) [70]. However, the efficiency of glycosylation with TMSOTf is extremely sensitive to the reaction scale as well as the addition speed of the Lewis acid, and low stereoselectivity was obtained on more than 50mg scale. Furthermore, slow addition of a Lewis acid in the large-scale reactions inhibited the glycosylation process at an earlier stage. Thus, the subsequent addition of the TMSOTf catalyst could not activate the glycosylation between the remaining starting materials. These results cannot be clearly explained based on presently available data. On the other hand, when the acid was added to the initial solution of the mannosyl donor **26** and acceptor **27** in one portion, mannosylation proceeded smoothly. However, the β selectivity decreased to 4.9/1 presumably due to the exothermic nature of the reaction (i.e., heat is generated while rapidly mixing), which leads to an overall decrease in the isolated yield of the β-disaccharide **28** (61% on 900mg scale).

Examination of the same mannosylation was performed by using the integrated microfluidic and batch systems as described earlier for microfluidic α-sialylation by the N-acetylated imidate **15** [19]. The Comet X-01 micromixer utilized for this case has a channel width of 500μm (Fig. 8.5) because the micromixing with IMM mixer

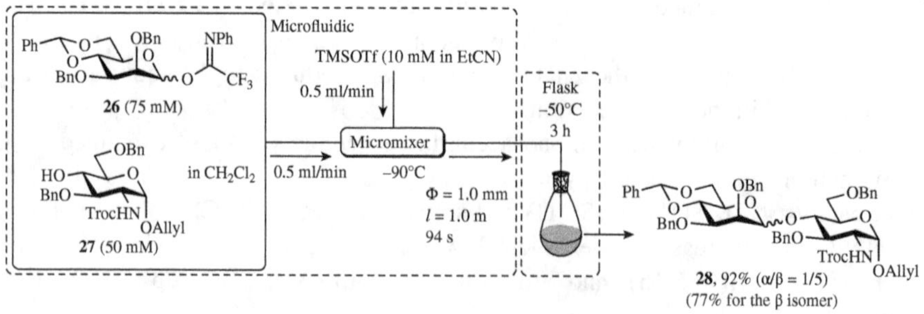

FIGURE 8.5 Stereoselective β-mannosylation.

(40 μm) caused significant solution blockage problems due to the low solubility of both **26** and **27** in CH_2Cl_2 at −78 to −90°C. The reaction solution after the micromixing was subsequently transferred to the flask and then was stirred for a few hours to complete the reaction. The optimal conditions in the integrated microfluidic/batch apparatus, that is, micromixing at −90°C and batch reaction at −50°C for 3 h, provided the α/β-mannoside **28** in 92% yield and with moderate β selectivity (β/α = 5.0/1). The coupling reaction toward compound **10** was performed in a similar manner as **28**.

8.2.4 Chemical N-Glycosylation of Asparagine under the Integrated Microfluidic and Batch Conditions

Chemical N-glycosylation of asparagine residues is a challenging topic in N-glycopeptide synthesis due to the inherently low nucleophilicity of the amide nitrogen toward glycosylation. Daniel Kahne and coworkers first reported N-glycosylation through amide functions by using an N-silylated acetamide to enhance the nucleophilicity of the amide nitrogen [71]. Takashi Takahashi and coworkers reported a more direct and efficient N-glycosylation by glycosyl imidates via protected asparagine derivatives [72]. They used *N*-phenyltrifluoroacetimidate as leaving group, TMSOTf as activator, and nitromethane as solvent. Glucosyl, glucosaminyl, mannosyl, and galactosyl imidates were successfully employed in N-glycosylations in 68–99% yields. Although Takahashi's protocol is very attractive, nitromethane is explosive and not suitable for large-scale synthesis. A reexamination of the reaction conditions by using more manageable solvents such as the nonflammable CH_2Cl_2 is warranted.

The N-glycosylation between the glucosaminyl *N*-phenyltrifluoroacetimidate **9** and the asparagine derivative **30** was initially examined under batch conditions [20]. Using 1.5 equiv. of the glycosyl donor **9** with 0.2 equiv. of TMSOTf in CH_2Cl_2, the desired N-glycoside **31** was afforded in 61% yield after 12 h at room temperature. This outcome is comparable to the reported results [72]. To adopt such conditions for the microfluidic N-glycosylation, the integrated microfluidic and batch system is employed (Table 8.3) because the N-glycosylation required long reaction times. Unlike the β-mannosylation case, the reaction was carried out at room temperature, and therefore, solution blockage was not a severe problem due to the relatively good solubility of both the donor and acceptor in CH_2Cl_2 at room temperature. An IMM micromixer with 40 μm width was used in order to obtain intense mixing. The flow rate and the concentration of TMSOTf were fixed at 1.0 ml/min and 43 mM, respectively. Batch stirring at room temperature for 12 h was required to complete the N-glycosylation. The condensation of the monosaccharide imidate **9** (86 mM) and acceptor **30** (172 mM) yielded the β-*N*-glycoside **31** in 60% (entry 1). A slight excess of donor **9** (130 mM) relative to **30** (86 mM) gave similar results (55%, entry 2). However, applying higher amounts of donor **9** (259 mM) completely consumed the asparagine acceptor, and the desired N-glycoside **31** was obtained in 85% yield (entry 3).

An efficient microfluidic N-glycosylation was also realized by using the disaccharide imidate **29** as donor. Nevertheless, the micromixing between 22 mM of **29** and

TABLE 8.3 N-Glycosylation of Asparagine under the Integrated Microfluidic and Batch Conditions

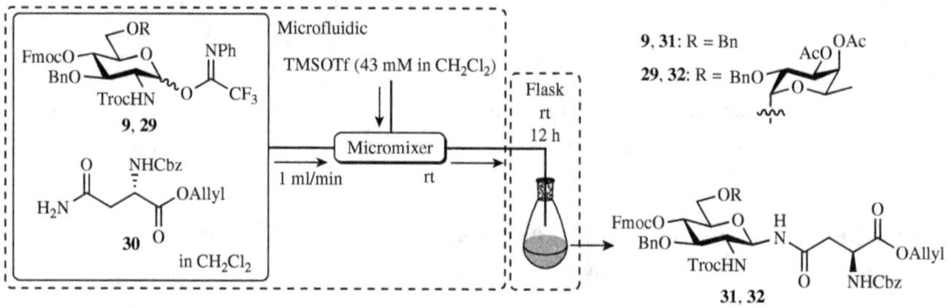

Entry	Donor (Conc. in mM)	Acceptor **30** Conc. (mM)	Product	Yield (%)
1	**9** (86)	172	**31**	60
2	**9** (130)	86	**31**	55
3	**9** (259)	86	**31**	85
4	**29** (22)	43	**32**	27
5	**29** (43)	86	**32**	84

43 mM of **30** provided the N-glycoside **32** only in 27% yield (entry 4). Fortunately, doubling the concentrations for both the reactants dramatically enhanced the yield (84%, entry 5). Thus, an efficient N-glycosylation involving glycosyl *N*-phenyltrifluoroacetimidates and asparagine amide under integrated microfluidic/batch conditions was established. The success of the protocol must be due to efficient mixing and fast heat transfer temperature control, which inhibit decomposition of the reactants and formation of complicated glycosyl intermediates during the batch mixing between the donor, acceptor, and activator.

8.3 CONCLUSION

As described, the microfluidic reactions can offer a practical route for stereoselective glycosylation. Microfluidic synthesis enables high mixing efficiency as well as precise temperature control and can achieve the ideal reaction kinetics. Microfluidic synthesis is therefore particularly useful for rapid and kinetically controlled reactions. On the other hand, the yields and selectivities sometimes vary depending on the addition and mixing methods and the researcher's skills. These effects are called mixing sensitivity. One major reason of such uncontrollability of the reaction is due to localized insufficient mixing that sometimes causes the local heating to enhance undesired reactions in the batch system. Microfluidic synthesis including the integrated microfluidic and batch synthesis is useful for such mixing-sensitive reactions.

REFERENCES

[1] Mason, B. P.; Price, K. E.; Steinbacher, J. L.; Bogdan, A. R.; McQuade, D. T. *Chem. Rev.* **2007**, *107*, 2300–2318.

[2] Fletcher, P. D. I.; Haswell, S. J.; Pombo-Villar, E.; Warrington, B. H.; Watts, P.; Wong, S. Y. F.; Zhang, X. *Tetrahedron* **2002**, *58*, 4735–4757.

[3] Geyer, K.; Codee, J. D. C.; Seeberger, P. H. *Chem. Eur. J.* **2006**, *12*, 8434–8442.

[4] Yoshida, J.; Nagaki, A.; Yamada, T. *Chem. Eur. J.* **2008**, *14*, 7450–7459.

[5] Ahmed-Omer, B.; Brandt, J. C.; Wirth, T. *Org. Biomol. Chem.* **2007**, *5*, 733–740.

[6] Pennemann, H.; Watts, P.; Haswell, S. J.; Hessel, V.; Loewe, H. *Org. Process Res. Dev.* **2004**, *8*, 422–439.

[7] Yoshida, J.; Nagaki, A.; Iwasaki, T.; Suga, S. *Chem. Eng. Technol.* **2005**, *28*, 259–266.

[8] Geyer, K.; Gustafsson, T.; Seeberger, P. H. *Synlett* **2009**, 2382–2391.

[9] Yoshida, J.; Saito, K.; Nokami, T.; Nagaki, A. *Synlett* **2011**, 1189–1194.

[10] Wegner, J.; Ceylan, S.; Kirschning, A. *Adv. Synth. Catal.* **2012**, *354*, 17–57.

[11] Suga, S.; Yamada, D.; Yoshida, J. *Chem. Lett.* **2010**, *39*, 404–406.

[12] Tanaka, K.; Fukase, K. *Org. Process Res. Dev.* **2009**, *13*, 983–990.

[13] Tanaka, K.; Fukase, K. *Beilstein J. Org. Chem.* **2009**, *5*, 40.

[14] Tanaka, K.; Motomatsu, S.; Koyama, K.; Tanaka, S.; Fukase, K. *Org. Lett.* **2007**, *9*, 299–302.

[15] Fukase, K.; Takashina, M.; Hori, Y.; Tanaka, D.; Tanaka, K.; Kusumoto, S. *Synlett* **2005**, 2342–2346.

[16] Tanaka, S.; Goi, T.; Tanaka, K.; Fukase, K. *J. Carbohydr. Chem.* **2007**, *26*, 369–394.

[17] Tanaka, K.; Fujii, Y.; Tokimoto, H.; Mori, Y.; Tanaka, S.; Bao, G.-M.; Siwu, E. R. O.; Nakayabu, A.; Fukase, K. *Chem. Asian J.* **2009**, *4*, 574–580.

[18] Uchinashi, Y.; Nagasaki, M.; Zhou, J.; Tanaka, K.; Fukase, K. *Org. Biomol. Chem.* **2011**, *9*, 7243–7248.

[19] Tanaka, K.; Mori, Y.; Fukase, K. *J. Carbohydr. Chem.* **2009**, *28*, 1–11.

[20] Tanaka, K.; Miyagawa, T.; Fukase, K. *Synlett* **2009**, 1571–1574.

[21] Shimoyama, A.; Saeki, A.; Tanimura, N.; Tsutsui, H.; Miyake, K.; Suda, Y.; Fujimoto, Y.; Fukase, K. *Chem. Eur. J.* **2011**, *17*, 14464–14474.

[22] Shimoyama, A.; Fujimoto, Y.; Fukase, K. *Synlett* **2011**, 2359–2362.

[23] Tanaka, K.; Fukase, K. *Synlett* **2007**, 164–166.

[24] Ratner, D. M.; Murphy, E. R.; Jhunjhunwala, M.; Snyder, D. A.; Jensen, K. F.; Seeberger, P. H. *Chem. Commun.* **2005**, 578–580.

[25] Carrel, F. R.; Geyer, K.; Codée, J. D. Seeberger, P. H. *Org. Lett.* **2007**, *9*, 2285–2288.

[26] Kanie, O.; Kiso, M.; Hasegawa, A. *J. Carbohydr. Chem.* **1988**, *7*, 501–506.

[27] Cai, S.; Yu, B. *Org. Lett.* **2003**, *5*, 3827–3830.

[28] Liu, Y.; Ruan, X.; Li, X.; Li, Y. *J. Org. Chem.* **2008**, *73*, 4287–4290.

[29] Bao, G.; Tanaka, K.; Ikenaka, K.; Fukase, K. *Bioorg. Med. Chem.* **2010**, *18*, 3760–3766.

[30] Demchenko, A. V.; Boons, G.-J. *Tetrahedron Lett.* **1998**, *39*, 3065–3068.

[31] De Meo, C.; Demchenko, A. V.; Boons, G.-J. *J. Org. Chem.* **2001**, *66*, 5490–5497.

[32] Ando, H.; Koike, Y.; Koizumi, S.; Ishida, H.; Kiso, M. *Angew. Chem. Int. Ed.* **2005**, *44*, 6759–6763.

[33] Ando, H.; Koike, Y.; Ishida, H.; Kiso, M. *Tetrahedron Lett.* **2003**, *44*, 6883–6886.

[34] Masaatsu, A.; Hiroshi, T.; Takashi, T. *Synlett* **2004**, 609–614.

[35] Yu, C.-S.; Niikura, K.; Lin, C.-C.; Wong, C.-H. *Angew. Chem. Int. Ed.* **2001**, *40*, 2900–2903.

[36] Tanaka, H.; Nishiura, Y.; Takahashi, T. *J. Am. Chem. Soc.* **2006**, *128*, 7124–7125.

[37] Tanaka, H.; Nishiura, Y.; Takahashi, T. *J. Am. Chem. Soc.* **2008**, *130*, 17244–17245.

[38] Farris, M. D.; De Meo, C. *Tetrahedron Lett.* **2007**, *48*, 1225–1227.

[39] Crich, D.; Li, W. *J. Org. Chem.* **2007**, *72*, 2387–2391.

[40] Hanashima, S.; Sato, K.; Ito, Y.; Yamaguchi, Y. *Eur. J. Org. Chem.* **2009**, 4215–4220.

[41] Hsu, C.-H.; Chu, K.-C.; Lin, Y.-S.; Han, J.-L.; Peng, Y.-S.; Ren, C.-T.; Wu, C.-Y.; Wong, C.-H. *Chem. Eur. J.* **2010**, *16*, 1754–1760.

[42] Tanaka, K.; Goi, T.; Fukase, K. *Synlett* **2005**, 2958–2962.

[43] Okamoto, K.; Goto, T. *Tetrahedron* **1990**, *46*, 5835–5857.

[44] DeNinno, M. P. *Synthesis* **1991**, 583–593.

[45] Schmidt, R. R.; Kinzy, W. *Adv. Carbohydr. Chem. Biochem.* **1994**, *50*, 21–123.

[46] Boons, G.-J..; Demchenko, A. V. *Chem. Rev.* **2000**, *100*, 4539–4566.

[47] Weishaupt, M.; Eller, S.; Seeberger, P. H. In *Glycomics*; Fukuda, M., Ed.; Methods in Enzymology 478; Academic: San Diego, CA, **2010**, pp 463–484.

[48] Kononov, L. O.; Malysheva, N. N.; Kononova, E. G.; Orlova, A. V. *Eur. J. Org. Chem.* **2008**, 3251–3255.

[49] Kononov, L. O.; Malysheva, N. N.; Orlova, A. V. *Eur. J. Org. Chem.* **2009**, 611–616.

[50] Yu, B.; Tao, H. *Tetrahedron Lett.* **2001**, *42*, 2405–2407.

[51] Adinolfi, M.; Barone, G.; Iadonisi, A.; Schiattarella, M. *Synlett* **2002**, 269–270.

[52] Sakai, Y.; Oikawa, M.; Yoshizaki, H.; Ogawa, T.; Suda, Y.; Fukase, K.; Kusumoto, S. *Tetrahedron Lett.* **2000**, *41*, 6843–6847.

[53] Ogawa, T.; Asai, Y.; Sakai, Y.; Oikawa, M.; Fukase, K.; Suda, Y.; Kusumoto, S.; Tamura, T. *FEMS Immunol. Med. Microbiol.* **2003**, *36*, 1–7.

[54] Fujimoto, Y.; Iwata, M.; Imakita, N.; Shimoyama, A.; Suda, Y.; Kusumoto, S.; Fukase, K. *Tetrahedron Lett.* **2007**, *48*, 6577–6581.

[55] Fujimoto, Y.; Tanaka, K.; Shimoyama, A.; Fukase, K. In *Glycomics*; Fukuda, M., Ed.; Methods in Enzymology 478; Academic: San Diego, CA, **2010**, pp 323–342.

[56] Fujimoto, Y.; Shimoyama, A.; Suda, Y.; Fukase, K. *Carbohydr. Res.* **2012**, *356*, 37–43.

[57] Fujimoto, Y.; Shimoyama, A.; Saeki, A.; Kitayama, N.; Kasamatsu, C.; Tsutsui, H.; Fukase, K. *Mol. BioSyst.* **2013**, *9*, 987–996.

[58] Imoto, M.; Kusunose, N.; Matsuura, Y.; Kusumoto, S.; Shiba, T. *Tetrahedron Lett.* **1987**, *28*, 6277–6280.

[59] Imoto, M.; Kusunose, N.; Kusumoto, S.; Shiba, T. *Tetrahedron Lett.* **1988**, *29*, 2227–2230.

[60] Kusumoto, S.; Kusunose, N.; Imoto, M.; Shimamoto, T.; Kamikawa, T.; Takada, H.; Kotani, S.; Rietschel, E. T.; Shiba, T. *Pure Appl. Chem.* **1989**, *61*, 461–464.

[61] Fukase, K.; Kurosawa, M.; Kusumoto, S. *J. Endotoxin Res.* **1994**, *1*, 149–163.

[62] Yoshizaki, H.; Fukuda, N.; Sato, K.; Oikawa, M.; Fukase, K.; Suda, Y.; Kusumoto, S. *Angew. Chem. Int. Ed.* **2001**, *40*, 1475–1480.

[63] Barresi, F.; Hindsgaul, O. *J. Am. Chem. Soc.* **1991**, *113*, 9376–9377.

[64] Stork, G.; Kim, G. *J. Am. Chem. Soc.* **1992**, *114*, 1087–1088.

[65] Ito, Y.; Ogawa, T. *Angew. Chem. Int. Ed. Engl.* **1994**, *33*, 1765–1767.

[66] Ishiwata, A.; Lee, Y. J.; Ito, Y. *Org. Biomol. Chem.* **2010**, *8*, 3596–3608.

[67] Crich, D.; Sun, S. *Tetrahedron* **1998**, *54*, 8321–8348.

[68] Crich, D.; Li, L. *J. Org. Chem.* **2007**, *72*, 1681–1690.

[69] Abdel-Rahman, A. A.; El Ashry, E. S. H.; Schmidt, R. R. *Carbohydr. Res.* **2002**, *337*, 195–206.

[70] Tanaka, S.; Takashina, M.; Tokimoto, H.; Fujimoto, Y.; Tanaka, K.; Fukase, K. *Synlett* **2005**, 2325–2328.

[71] Kahne, D.; Walker, S.; Cheng, Y.; Engen, D. V. *J. Am. Chem. Soc.* **1989**, *111*, 6881–6882.

[72] Tanaka, H.; Iwata, Y.; Takahashi, D.; Adachi, M.; Takahashi, T. *J. Am. Chem. Soc.* **2005**, *127*, 1630–1631.

9

CHEMOENZYMATIC SYNTHESIS OF CARBOHYDRATES

KASEMSIRI CHANDARAJOTI[1] AND JIAN LIU[2]

[1] *Department of Pharmaceutical Chemistry, Faculty of Pharmaceutical Sciences, Prince of Songkla University, Hat Yai, Thailand*
[2] *Division of Chemical Biology and Medicinal Chemistry, Eshelman School of Pharmacy, University of North Carolina at Chapel Hill, Chapel Hill, NC, USA*

9.1 INTRODUCTION

Utilizing enzymes to prepare carbohydrates has been known to chemists for a long time [1]. The inherently high stereoselectivity of enzymes can selectively convert a starting material to a product without the need for protection and deprotection steps, resulting in very high product yields. A number of studies on the biosynthesis of natural carbohydrates provide a platform for synthesizing complex carbohydrates that are considered difficult by chemical synthesis. A wide range of applications have been reported to synthesize individualized monosaccharides, oligosaccharides, and polysaccharides [2]. These approaches largely involve the use of glycosyltransferases (GTases) to build the specialized glycosidic bonds. Recently, the chemoenzymatic method was employed in the synthesis of complex carbohydrates, involving complicated sulfonations and epimerizations of saccharide residues [3] as well as the formation of the linkages between carbohydrates and polypeptides to engineer glycoproteins [4, 5]. These new chemoenzymatic syntheses evolved by using engineered enzymes and unnatural cofactors to complete the process. This chapter reviews the

Glycochemical Synthesis: Strategies and Applications, First Edition.
Edited by Shang-Cheng Hung and Medel Manuel L. Zulueta.

recent efforts for developing the chemoenzymatic approach to synthesize various oligosaccharides with emphasis on the strategy to overcome the difficulties in heparan sulfate (HS) and heparin syntheses. Other notable examples of chemoenzymatic synthesis can be found in Sections 3.4.3 and 19.5.11 of this book.

9.2 OLIGOSACCHARIDES AND POLYSACCHARIDES PRODUCED BY GTASES

GTases transfer saccharide donors to specific acceptors by utilizing nucleotide sugars (Leloir type), sugar phosphates, and polyprenol sugars (non-Leloir types). The acceptors of Leloir GTases can be nucleic acids, polysaccharides, proteins, lipids, and carbohydrates. To date, several mammalian GTases have been extensively studied including glucosyltransferases (GlcTs), glucuronosyltransferases (GlcATs), mannosyltransferases (ManTs), sialyltransferases (SiaTs), galactosyltransferases (GalTs), fucosyltransferases (FucTs), *N*-acetylglucosaminyltransferases (GlcNAcTs), and *N*-acetylgalactosaminyltransferases (GalNAcTs). The nucleotide sugar substrates of these enzymes are shown in Figure 9.1 [6]. The difficulties in the *Escherichia coli* expression of these mammalian GTases have drawn many interests in identifying bacterial GTase counterparts. Genetic engineering of these bacterial GTases is useful for large-scale synthesis to control the desired carbohydrate-containing products.

Three most useful GTases, namely, GlcATs, GlcNAcTs, and GalNAcTs, have been utilized in many glycan syntheses as the donor substrates of these synthases are key components in many glycans, especially glycosaminoglycans (GAGs). GAGs

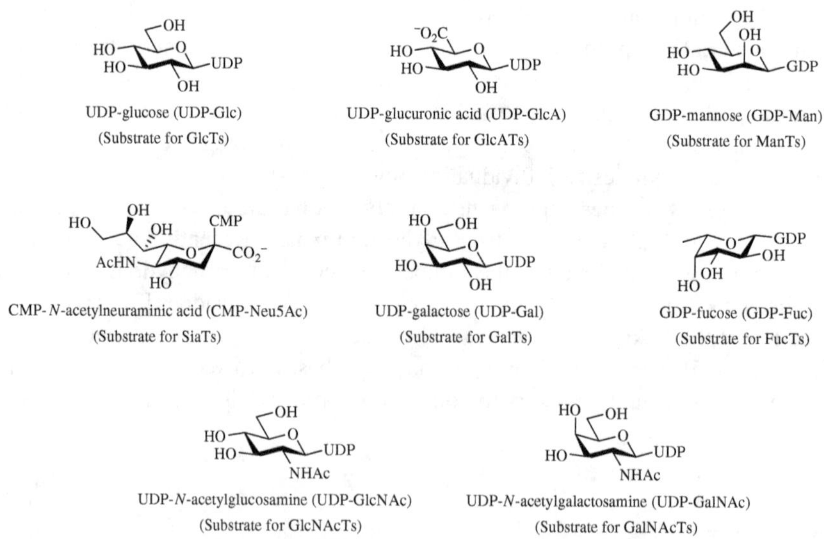

FIGURE 9.1 Nucleotide sugar substrates for GTases. CMP, cytidine monophosphate; GDP, guanidine diphosphate; UDP, uridine diphosphate.

are present in large quantities on the cell surface and in the extracellular matrix. These GAG polymers are formed by distinct disaccharide repeating units composed of a hexosamine and either a uronic acid for hyaluronan (HA), HS, heparin, chondroitin sulfate (CS), and dermatan sulfate or galactose for keratan sulfate (see Section 10.1 for the structures). Paul DeAngelis and colleagues have been investigating several GTases involved in the biosynthesis of GAGs [7]. HA is naturally synthesized without sulfation, while some of the GAG backbones such as HS and CS are modified by epimerase and sulfotransferases to gain unique biological functions. Typically, GTase is known for its one-enzyme–one-linkage property. However, the GTases that synthesize GAG backbones have two catalytic sites catering to both UDP-GlcA and UDP-GlcNAc/UDP-GalNAc [8]. Therefore, GTases can simultaneously elongate sugar residues to form GAG backbones. These bifunctional GTases are also found in bacteria producing capsular polysaccharides (e.g., *Pasteurella multocida*). The discovery of *P. multocida* hyaluronan synthase (PmHAS) [9], heparosan synthases (PmHS1 and PmHS2) [10], and chondroitin synthase (PmCS) [11] resulted in the synthesis of long-chain HA, unmodified HS, and unsulfated chondroitin, respectively. Typically, the synthesis of GlcA–GlcNAc/GalNAc repeating backbone can be carried out in two approaches: (i) sequential addition of UDP-sugars to the acceptors and (ii) simultaneous polymerization of the GAG backbones. In the first approach, a monofunctional synthase is employed to properly control the elongation. The latter approach needs a bifunctional synthase to efficiently synthesize long polymers. However, a stepwise reaction can be achieved by using a monofunctional mutant of which either the GlcA or HexNAc catalytic site is destroyed, resulting in better control for the synthesis of monodispersed oligosaccharides [10, 12].

Although the use of biosynthetic enzymes makes the synthesis of carbohydrates possible, the high cost of substrates and the difficulties in the synthesis of their derivatives hinder the ability to produce GAGs and related carbohydrates. The development of unnatural nucleotide analogues that can be recognized by bacterial GTases is beneficial for large-scale polysaccharide production. Robert Linhardt and coworkers developed a library of unnatural UDP-hexosamine analogues by chemoenzymatic synthesis [13]. The synthesis employed *N*-acetylglucosamine-1-phosphate uridyltransferase (GlmU) to catalyze the reaction between uridine triphosphate and GlcNAc/GalNAc-1-phosphate resulting in UDP-GlcNAc/GalNAc analogues. These analogues were tested for their incorporation into a polysaccharide by PmHAS, PmCS, and PmHSs. This study demonstrated the feasibility of using unnatural nucleotide sugars to synthesize polysaccharides. The most useful substrate produced by GlmU that is currently being used to synthesize HS and heparin is UDP-*N*-trifluoroacetylglucosamine (UDP-GlcNTFA) [14].

9.3 CHEMOENZYMATIC SYNTHESIS OF HS

HS is an essential glycan that carries a wide range of biological functions. Our knowledge of HS is far less developed than our knowledge of proteins and nucleic acids. One grand challenge in HS studies is to make HS oligosaccharides widely available to the scientific research community. HS oligosaccharides have been synthesized using purely chemical approaches (see Chapter 10). Although a number of successful examples have been reported [15–18], the method is unsuitable for a large number of oligosaccharides. The difficulties largely stem from low product yields for long oligosaccharides. In addition, the chemical method requires a large collection of precursors to synthesize the HS oligosaccharides with diverse structures. A chemoenzymatic approach has emerged as an alternative method. This method offers a completely new strategy to prepare HS oligosaccharides with extremely high efficiency [3].

9.3.1 Biosynthetic Pathway of HS and HS Biosynthetic Enzymes

HS is a highly sulfated polysaccharide with repeating sequences of glucuronic acid (GlcA) or iduronic acid (IdoA) linked to glucosamine. Both IdoA and glucosamine residues carry sulfate groups. HS participates in numerous biological processes, including embryonic development [19, 20], assisting viral and bacterial infections [21], and regulating blood coagulation [22]. The sulfation patterns and the location of IdoA residues play critical roles in determining the functions of HS. The biosynthesis of HS involves a series of specialized enzymes (Scheme 9.1) [20]. These enzymes include GTases, sulfotransferases, and an epimerase. Two HS polymerases, also known as EXTs, are responsible for building the backbone polysaccharide consisting of the GlcA–GlcNAc repeating unit [23]. Thus, EXTs belong to the GTase family. *N*-Deacetylase/*N*-sulfotransferase (NDST) is a dual-function enzyme capable of removing the acetyl group of GlcNAc residues and transferring a sulfonate group to the deacetylated glucosamine residue to form the N-sulfonated glucosamine (GlcNS) residue. 3′-Phosphoadenosine-5′-phosphosulfate (PAPS) acts as the sulfonate source. C5-epimerase (C5-epi) converts a GlcA residue to an IdoA residue. 2-*O*-Sulfotransferase (2-OST), 6-*O*-sulfotransferase (6-OST), and 3-*O*-sulfotransferase (3-OST) transfer a sulfonate group from PAPS to the 2-OH position of the IdoA residue, 6-OH position of the glucosamine residue, and 3-OH positions of the glucosamine residue, respectively.

The regioselectivities of HS biosynthetic enzymes are very high, offering the opportunity to conduct the enzymatic synthesis of HS. The critical reagents for the synthesis are the enzymes. To date, a majority of these biosynthetic enzymes have been expressed in *E. coli* with some exceptions, particularly EXTs and NDST [24]. For this reason, bacterial GTases and the *N*-sulfotransferase (NST) domain of NDST were used for the enzymatic synthesis of HS [10, 25].

9.3.2 Application of Biosynthetic Enzymes in HS and Heparin Oligosaccharide Synthesis

Chemoenzymatic synthesis of HS was first employed in the synthesis of structurally heterogeneous polysaccharides. In this method, a bacterial capsular polysaccharide isolated from the *E. coli* strain K5 is subjected to a series of enzymatic modifications, including N-sulfonation, epimerization, and 2-O-, 6-O-, and 3-O-sulfonations [26–28].

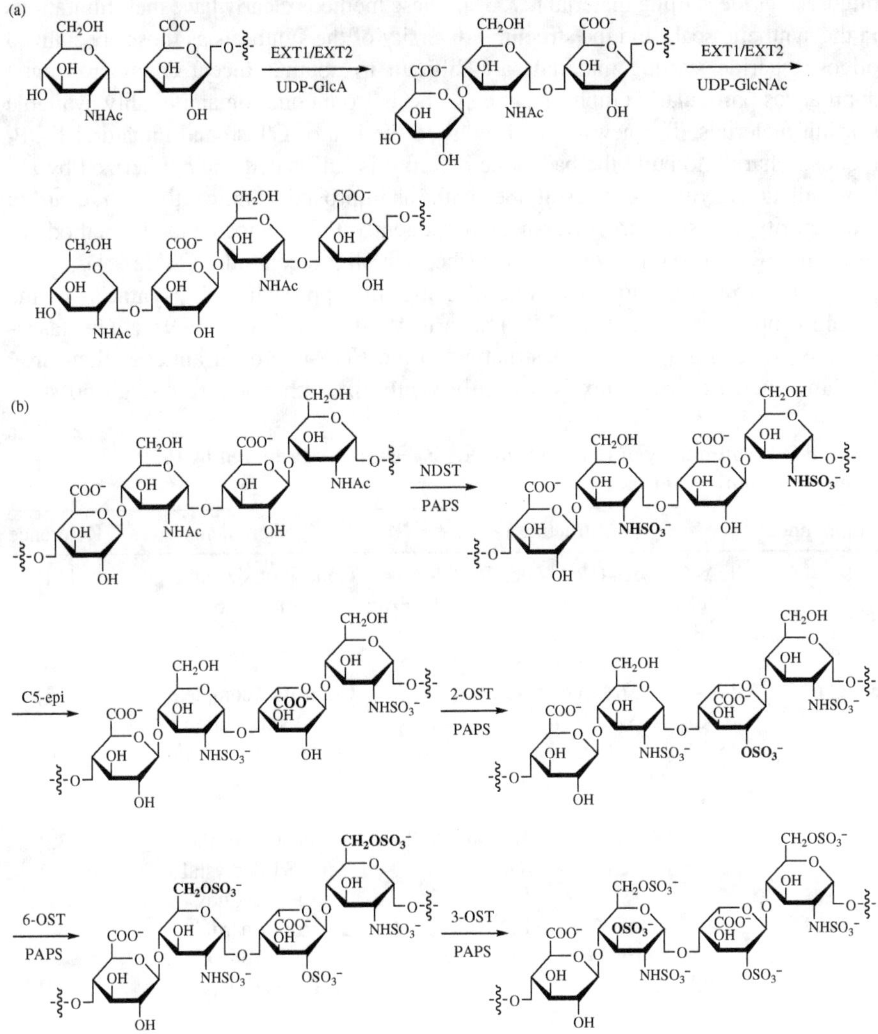

SCHEME 9.1 Biosynthesis of HS. (a) Elongation reactions to prepare the precursor polysaccharide backbone using both EXT1 and EXT2 enzymes. (b) Polysaccharide modifications.

As long as the appropriate sulfonation is performed, the products have the desired biological activities, including the anticoagulant activity often attributed to heparin, the structural analogue of HS. The method offers a new strategy to prepare synthetic heparin, with a goal to replace the heparins that are isolated from animal tissues in order to secure the safety of the heparin supply chain [29]. Recently, Linhardt and colleagues used this method to control the N-sulfonation steps in order to regulate the structures of synthetic heparins [30].

Another impetus is to use the chemoenzymatic approach to synthesize structurally defined oligosaccharides. Early attempts targeted a single product from a specialized oligosaccharide starting material [27, 31]. These methods clearly have their limitations in the synthetic scale and the structural diversity of the synthesis as those specialized oligosaccharide starting materials are difficult to obtain. Recent chemoenzymatic approaches particularly emphasize the synthesis from common and readily available starting materials. The new method required the use of GTase and unnatural UDP-monosaccharide to build the backbone before it is sulfonated and epimerized by HS biosynthetic enzymes. As a result, the synthesis improved significantly with regard to scale, purity, and structural diversity. Consequently, the chemoenzymatic method has been employed to synthesize several HS/heparin oligosaccharides (Table 9.1).

The most representative example utilizing this approach is the synthesis of the ultralow molecular weight (ULMW) heparin construct **27** (Scheme 9.2), a heptasaccharide that has a very similar structure to the FDA-approved anticoagulant drug fondaparinux. Fondaparinux is currently synthesized chemically through 50 steps

TABLE 9.1 Summary of HS/Heparin Oligosaccharides Prepared by the Chemoenzymatic Approach

Compounds	Structure	Comments	Reference
1–4	GlcA–GlcNR^1–GlcA–GlcNR^2–GlcA–GlcNAc–GlcA–AnMan (**1**, $R^1=R^2$=Ac; **2**, $R^1=SO_3^-$, R^2=Ac; **3**, R^1=Ac, $R^2=SO_3^-$; **4**, $R^1=R^2=SO_3^-$)	Control of size and N-sulfonation	[14]
5 and **6**	GlcA–GlcNS6S–(GlcA–GlcNS6S)$_n$–GlcA–AnMan (**5**, $n=3$; **6**, $n=4$)	Control of complete 6-O-sulfonation	[14]
7 and **8**	GlcNR6S–GlcA–GlcNS3S6S–IdoA2S–GlcNS6S–GlcA–AnMan (**7**, R=SO_3^-; **8**, R=Ac)	Synthesis of a complete heparin fragment	[3]
9	GlcNAc6S–GlcA–GlcNS6S–IdoA2S–GlcNS6S–GlcA–AnMan	Application of the 3-OST-1 crystal structure study	[32]
10 and **11**	GlcA–GlcNAc–GlcA–GlcNR6S–GlcA–AnMan (**10**, R=SO_3^-; **11**, R=Ac)	Control of partial 6-O-sulfonation	[33]
12–15	GlcNS–(GlcA–GlcNS)$_m$–(GlcA–GlcNAc)$_2$–(GlcA–GlcNS)$_2$–GlcA–AnMan (**12**, $m=2$; **13**, $m=3$; **14**, $m=4$; **15**, $m=5$)	Synthesis of large oligosaccharides up to henicosaccharide	[34]
16–21	GlcA–GlcNR^{1}6S–GlcA–GlcNR^{2}6S–GlcA–GlcNR^{3}6S–GlcA–GlcNR^{4}6S–GlcA–PNP (**16**, $R^1=R^2=R^3=R^4=SO_3^-$; **17**, R^1=Ac; $R^2=R^3=R^4=SO_3^-$; **18**, $R^1=R^2$=Ac, $R^3=R^4=SO_3^-$; **19**, $R^1=R^4$=Ac, $R^2=R^3=SO_3^-$; **20**, $R^1=R^2=R^3=SO_3^-$, R^4=Ac; **21**, $R^1=R^2$=Ac, $R^2=R^2=SO_3^-$)	Synthesis of a small library of N-sulfonated and 6-O-sulfonated nonasaccharides	[35]

2S, 2-*O*-SO_3^-; 3S, 3-*O*-SO_3^-; 6S, 6-*O*-SO_3^-; AnMan, 2,5-anhydromannitol; PNP, *p*-nitrophenyl; NS, *N*-SO_3^-.

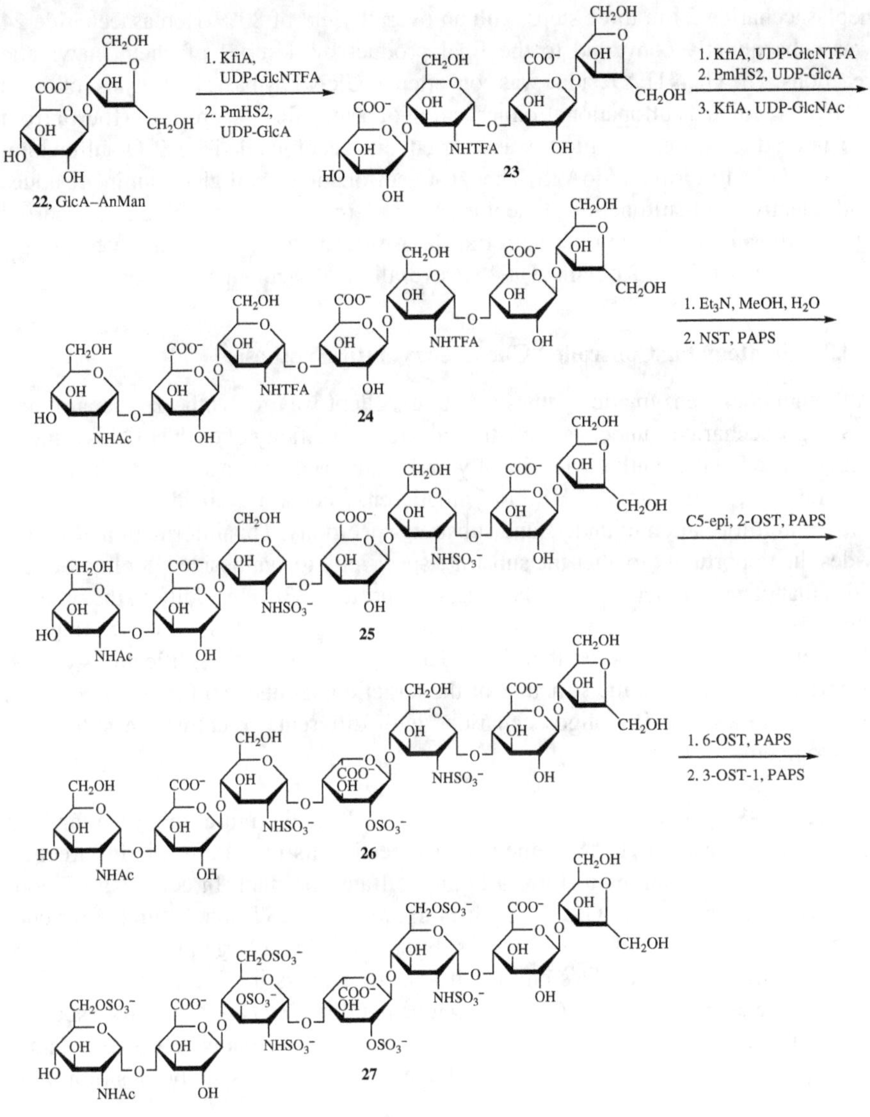

SCHEME 9.2 Chemoenzymatic synthesis of ULMW heparin **27**.

with an overall yield close to 0.1%. The chemoenzymatic synthesis performed by Jian Liu and coworkers has shortened the synthesis to about 10 steps. The entire process includes backbone elongation and saccharide modifications. Disaccharide **22** was elongated to tetrasaccharide **23** by using two bacterial GTases: the *N*-acetyl-glucosaminyltransferase from the *E. coli* strain K5 (KfiA) [25] and PmHS2 [10]. Tetrasaccharide **23** was designed with the unnatural monosaccharide *N*-trifluoro-acetylglucosamine (GlcNTFA) because the *N*-TFA group can be readily converted to an *N*-sulfonate group [14] in a later step. Tetrasaccharide **23** was elongated to heptasaccharide **24** in three steps with an overall yield of 80%. Heptasaccharide **24** was subsequently converted to the final product by a series of chemoenzymatic reactions. The GlcNTFA residue was converted to GlcNS by basic hydrolysis followed by NST-mediated sulfonation. Epimerization of a specific GlcA residue (fourth from the nonreducing end) to IdoA was carried out accompanied by 2-O-sulfonation using 2-OST to form an IdoA2S. Then, 6-O-sulfonation of all glucosamine residues and selective 3-O-sulfonation at the third residue from the nonreducing end occurred upon exposure to appropriate sulfotransferases and PAPS. After these ten steps, ULMW heparin **27** was obtained at 45% overall yield in about 50 mg scale.

9.3.3 Strategy for Controlled Chemoenzymatic Synthesis

Although chemoenzymatic synthesis is an excellent way to synthesize heparin and HS oligosaccharides, understanding the substrate specificity of each HS biosynthetic enzyme used in the synthesis is critically important to conduct a successful synthesis. Several crystal structures of HS biosynthetic enzymes are available, providing the structural information to understand substrate specificity. This information also provides the opportunity to alter the substrate specificity to synthesize HS oligosaccharides that cannot be achieved by the wild-type proteins [36]. In addition, the order of the enzymatic modifications can significantly reduce the side products to improve the yield. It should be noted that the order of the enzymatic modifications can be different depending on the structure of the target compound. So far, the procedures for the synthesis of HS oligosaccharides with different structures are still under development.

9.3.3.1 N-Sulfonation The formation of GlcNS is the initial modification step in the biosynthesis of HS (Scheme 9.1), suggesting its important role in directing subsequent modifications to form a highly sulfated product. In cells, NDST converts GlcNAc residues to GlcNS residues. However, NDST is not suitable for chemoenzymatic synthesis for two reasons. First, NDST is a large protein with more than 800 amino acid residues and is difficult to express in *E. coli* in high levels. Second, the conversion of the GlcNAc residue to GlcNS by NDST follows a specific pattern that depends on the isoform [37]. It is worth noting that NDST does not have the capacity to selectively generate the GlcNS residue at a specific position in an oligosaccharide.

Currently, the sulfamate groups in glucosamine residues are established by using KfiA coupled with an unnatural UDP-monosaccharide, that is, UDP-GlcNTFA (Scheme 9.2) [14]. KfiA recognizes UDP-GlcNTFA as a substrate and transfers GlcNTFA to the acceptor substrate holding a GlcA residue at the nonreducing end. Equally important, the GlcNTFA residue at the nonreducing end of the acceptor substrate is also an excellent substrate for PmHS2, this time for the transfer of a GlcA residue. Using these properties, one can synthesize an oligosaccharide with a GlcA–GlcNTFA disaccharide repeating unit. The conversion of the trifluoroacetamide to sulfamate is readily accomplished by chemical detrifluoroacetylation followed by N-sulfonation using NST. It should be noted that the NST domain of NDST is located at the C-terminal region. NST is only 260 amino acid residues long and can be expressed in *E. coli* at a very high level [38]. This method is highly effective in building oligosaccharides with designed distribution of GlcNAc and GlcNS. It has been used in the preparation of oligosaccharide targets at the size of up to 19 saccharide residues long [34].

9.3.3.2 C5-Epimerization The C5-epimerization step involves the conversion of a GlcA to an IdoA residue under the catalysis of C5-epi. In this reaction, C5-epi deprotonates the C5 position of the GlcA residue to form a putative carbanion intermediate (Scheme 9.3a). C5-epi is a unique enzyme that serves as a two-way catalyst, that is, C5-epi converts a GlcA residue to an IdoA residue and also converts an IdoA residue to a GlcA residue. N-Sulfonation of the residue (designated as residue −1) at the nonreducing side of GlcA (designated as residue 0) allows the

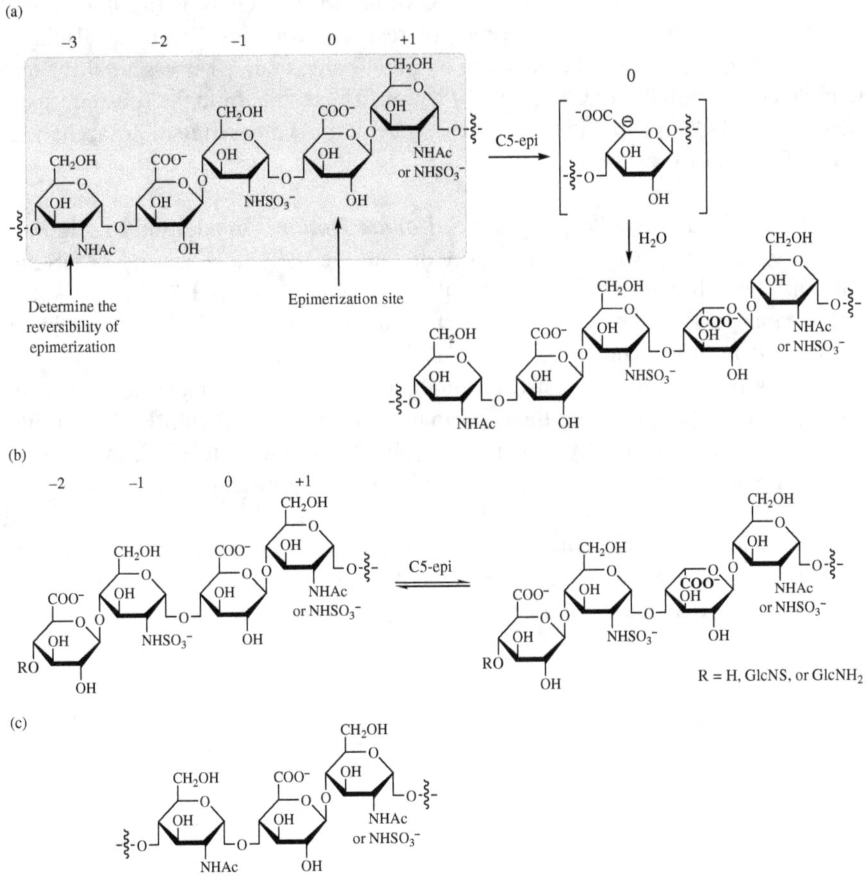

SCHEME 9.3 Substrate specificity of C_5-epi. (a) C5-epi recognition site resulting to irrevesible epimerization. (b) C5-epi recognition site resulting to revesible epimerization. (c) Site not acted upon by C5-epi.

identification of this GlcA residue as the epimerization site for the action of C5-epi (Scheme 9.3) [39]. The two-way catalytic property of C5-epi has complicated the process for the formation of IdoA using the chemoenzymatic method. Indeed, attempts to synthesize oligosaccharides containing multiple IdoA residues resulted in complex mixtures [32, 34, 40]. So far, introducing several IdoA residues in a structurally defined oligosaccharide remains a challenge.

A recent study on C5-epi reveals a very interesting aspect in its substrate specificity [41]. Liu's group discovered that C5-epi displays both reversible and irreversible catalytic modes by recognizing the structure of a pentasaccharide domain (Scheme 9.3) [41]. This pentasaccharide domain consists of the epimerization site and a residue that determines the mode of reaction at the epimerization site. The determinant residue is located three residues away at the nonreducing side (residue −3) of the epimerization site. The N-substitution status of residue −3 dictates the reversibility of the epimerization mediated by C5-epi. A GlcNAc residue directs C5-epi to display an irreversible reaction mode, while GlcNS, $GlcNH_2$ (an unmodified glucosamine), or the complete absence of residue −3 drives C5-epi to display a reversible reaction mode. The presence of GlcNS at residue −1 is essential for rendering the susceptibility to C5-epi modification. The results from the substrate specificity study of C5-epi could lead to a new strategy for synthesizing oligosaccharides with multiple IdoA residues.

9.3.3.3 Enzymatic Synthesis of the Sulfonate Donor In addition to oligosaccharide synthesis, sulfonations crucially dictate the biological activity of HS and heparin. The sulfonation steps by HS sulfotransferases require PAPS as the sulfonate donor. PAPS is expensive even for the milligram-scale synthesis in an academic laboratory. The availability of PAPS determines cost-effectiveness when performing large-scale synthesis. Although, a PAPS regeneration system has been employed to prepare heparin, the combination of this system with HS sulfotransferases yields unexpected by-products [26]. PAPS can be synthesized by an enzymatic approach in a one-pot format using adenosine triphosphate (ATP) sulfurylase (from *Kluyveromyces lactis*), APS kinase (from *Penicillium chrysogenum*), and pyrophosphatase (from *E. coli*). Bacterial expression of all three enzymes yielding crude extracts is efficient enough to convert ATP and sulfate to PAPS in gram scale as described in Scheme 9.4 [35].

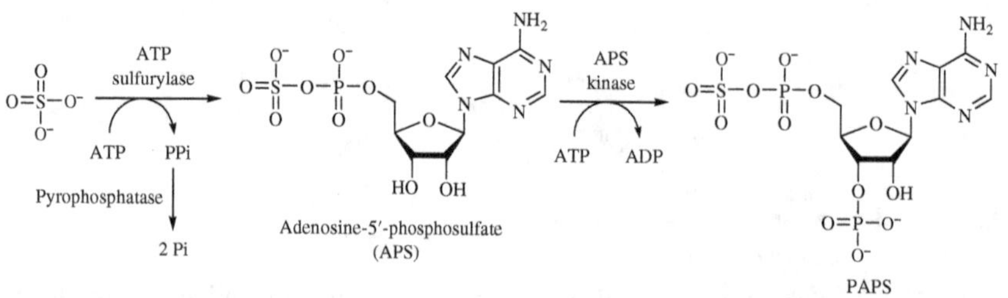

SCHEME 9.4 Enzymatic synthesis of PAPS starting from Na_2SO_4 and ATP. Pi, phosphate; PPi, pyrophosphate.

9.3.3.4 ***Product Purification*** The enzymatic conversion at each synthetic step is almost quantitative, and thus, no complicated purification procedures are necessary. The current method to separate HS oligosaccharides and enzymes is achieved by using a BioGel P-2 column, which is eluted with 0.1 M NH_4HCO_3. The sample is detected by mass spectrometry directly from the eluted fractions. Further purification also includes the use of a strong anion-exchange column (e.g., Q-sepharose). The anion-exchange column is able to further purify the sample if it contains incompletely sulfonated by-products. One technical issue about the use of Q-sepharose is the monitoring of the sample during purification. To overcome this problem, a UV tag should be introduced to the HS oligosaccharide products. To this end, GlcA-PNP may be used as starting material for the synthesis [42, 43]. The strong absorbance from the PNP tag provides easy detection during column purification, increasing the purity of the product to greater than 95%.

9.4 CONCLUSION

Chemoenzymatic approach served as an alternative method for the synthesis of complex carbohydrate-containing molecules. The use of biosynthetic enzymes not only provides high regioselectivity and stereoselectivity for the formation of glycosidic bonds but also solves the problems of protection and deprotection steps required by chemical synthesis. The discovery of bacterial counterparts and the advance in bioengineering technology led to breakthroughs in complex carbohydrate synthesis.

Taking advantage of high regioselectivity and quantitative conversion, the chemoenzymatic approach reduces the time frame for preparing HS oligosaccharides that is practical in an academic laboratory. The chemoenzymatic method is also highly flexible for synthesizing structurally diverse HS oligosaccharides. The method, involving eight enzymes and four cofactors, can prepare a wide range of oligosaccharides by rearranging the enzymatic modification steps. Such rearrangement is analogous to engineering the synthetic scheme for organic synthesis to prepare structurally complex natural products. It should be noted that the structures of the products are ultimately limited by the substrate specificity of the enzymes. Certain saccharide sequences, known as forbidden structures, are impossible to synthesize by the chemoenzymatic method because they are unnatural. To that end, a purely chemical approach is still the only choice for those "forbidden" oligosaccharides. It is conceivable that a combination of chemical and chemoenzymatic approaches may be the best solution to completely overcome the challenges in synthetic carbohydrate chemistry.

REFERENCES

[1] Gijsen, H. J. M.; Qiao, L.; Fitz, W.; Wong, C.-H. *Chem. Rev.* **1996**, *96*, 443–473.

[2] Muthana, S.; Cao, H.; Chen, X. *Curr. Opin. Chem. Biol.* **2009**, *13*, 573–581.

[3] Xu, Y.; Masuko, S.; Takieddin, M.; Xu, H.; Liu, R.; Jing, J.; Mousa, S. A.; Linhardt, R. J.; Liu, J. *Science* **2011**, *334*, 498–501.

[4] Zou, G.; Ochiai, H.; Huang, W.; Yang, Q.; Li, C.; Wang, L.-X. *J. Am. Chem. Soc.* **2011**, *133*, 18975–18991.

[5] Schwarz, F.; Huang, W.; Li, C.; Schulz, B. L.; Lizak, C.; Palumbo, A.; Numao, S.; Neri, D.; Aebi, M.; Wang, L.-X. *Nat. Chem. Biol.* **2010**, *6*, 264–266.

[6] Schmaltz, R. M.; Hanson, S. R.; Wong, C.-H. *Chem. Rev.* **2011**, *111*, 4259–4307.

[7] Tracy, B. S.; Avci, F. Y.; Linhardt, R. J.; DeAngelis, P. L. *J. Biol. Chem.* **2007**, *282*, 337–344.

[8] DeAngelis, P. L. *Glycobiology* **2002**, *12*, 9R–16R.

[9] DeAngelis, P. L.; White, C. L. *J. Bacteriol.* **2004**, *186*, 8529–8532.

[10] Sismey-Ragatz, A. E.; Green, E. D.; Otto, N. J.; Rejzek, M.; Field, R. A.; Deangelis, P. L. *J. Biol. Chem.* **2007**, *282*, 28321–28327.

[11] DeAngelis, P. L.; Padgett-McCue, A. J. *J. Biol. Chem.* **2000**, *275*, 24124–24129.

[12] Chavaroche, A. A. E.; Springer, J.; Kooy, F.; Boeriu, C.; Eggink, G. *Appl. Microbiol. Biotechnol.* **2010**, *85*, 1881–1891.

[13] Masuko, S.; Bera, S.; Green, D. E.; Weiwer, M.; Liu, J.; DeAngelis, P. L.; Linhardt, R. J. *J. Org. Chem.* **2012**, *77*, 1449–1456.

[14] Liu, R.; Xu, Y.; Chen, M.; Weïwer, M.; Zhou, X.; Bridges, A. S.; DeAngelis, P. L.; Zhang, Q.; Linhardt, R. J.; Liu, J. *J. Biol. Chem.* **2010**, *285*, 34240–34249.

[15] Arungundram, S.; Al-Mafraji, K.; Asong, J.; Leach III, F. E.; Amster, I. J.; Venot, A.; Turnbull, J. E.; Boons, G.-J. *J. Am. Chem. Soc.* **2009**, *131*, 17394–17405.

[16] de Paz, J. L.; Noti, C.; Seeberger, P. H. *J. Am. Chem. Soc.* 2006, *128*, 2766–2767.

[17] Hu, Y.-P.; Lin, S.-Y.; Huang, C.-Y.; Zulueta, M. M. L.; Liu, J.-Y.; Chang, W.; Hung, S.-C. *Nat. Chem.* **2011**, *3*, 557–563.

[18] Lu, L.-D.; Shie, C.-R.; Kulkarni, S. S.; Pan, G.-R.; Lu, X.-A.; Hung, S.-C. *Org. Lett.* **2006**, *8*, 5995–5998.

[19] Dejima, K.; Kanai, M. I.; Akiyama, T.; Levings, D. C.; Nakato, H. *J. Biol. Chem.* **2011**, *286*, 17103–17111.

[20] Esko, J. D.; Selleck, S. B. *Annu. Rev. Biochem.* **2002**, *71*, 435–471.

[21] Liu, J.; Thorp, S. C. *Med. Res. Rev.* **2002**, *22*, 1–25.

[22] Liu, J.; Pedersen, L. C. *Appl. Microbiol. Biotechnol.* **2007**, *74*, 263–272.

[23] Lind, T.; Tufaro, F.; McCormick, C.; Lindahl, U.; Lidholt, K. *J. Biol. Chem.* **1998**, *273*, 26265–26268.

[24] Peterson, S. P.; Frick, A.; Liu, J. *Nat. Prod. Rep.* **2009**, *26*, 610–627.

[25] Chen, M.; Bridges, A.; Liu, J. *Biochemistry* **2006**, *45*, 12358–12365.

[26] Chen, J.; Avci, F. Y.; Muñoz, E. M.; McDowell, L. M.; Chen, M.; Pedersen, L. C.; Zhang, L.; Linhardt, R. J.; Liu, J. *J. Biol. Chem.* **2005**, *280*, 42817–42825.

[27] Copeland, R. J.; Balasubramaniam, A.; Tiwari, V.; Zhang, F.; Bridges, A.; Linhardt, R. J.; Shukla, D.; Liu, J. *Biochemistry* **2008**, *47*, 5774–5783.

[28] Chen, J.; Jones, C. L.; Liu, J. *Chem. Biol.* **2007**, *14*, 986–993.

[29] Liu, H.; Zhang, Z.; Linhardt, R. J. *Nat. Prod. Rep.* **2009**, *26*, 313–321.

[30] Bhaskar, U.; Sterner, E.; Hickey, A. M.; Onishi, A.; Zhang, F.; Dordick, J. S.; Linhardt, R. J. *Appl. Microbiol. Biotechnol.* **2012**, *93*, 1–16.

[31] Balagurunathan, K.; Lech, M. Z.; Beeler, D. L.; Wu, Z. L.; Rosenberg, R. D. *Nat. Biotechnol.* **2003**, *21*, 1343–1346.

[32] Moon, A. F.; Xu, Y.; Woody, S.; Krahn, J. M.; Linhardt, R. J.; Liu, J.; Pedersen, L. C. *Proc. Natl. Acad. Sci. U. S. A.* **2012**, *109*, 5256–5270.

[33] Liu, R.; Liu, J. *Biochemistry* **2011**, *50*, 4382–4391.

[34] Xu, Y.; Pempe, E.; Liu, J. *J. Biol. Chem.* **2012**, *287*, 29054–29061.

[35] Zhou, X. X.; Chandarajoti, K.; Pham, T. Q.; Liu, R. P.; Liu, J. *Glycobiology* **2011**, *21*, 771–780.

[36] Liu, J.; Moon, A. F.; Sheng, J.; Pedersen, L. C. *Curr. Opin. Struct. Biol.* **2012**, *22*, 550–557.

[37] Sheng, Z.; Liu, R.; Xu, Y.; Liu, J. *J. Biol. Chem.* 2011, *286*, 19768–19776.

[38] Kakuta, Y.; Sueyoshi, T.; Negishi, M.; Pedersen, L. C. *J. Biol. Chem.* **1999**, *274*, 10673–10676.

[39] Li, J.-P.; Gong, F.; Darwish, K. E.; Jalkanen, M.; Lindahl, U. *J. Biol. Chem.* **2001**, *276*, 20069–20077.

[40] Pempe, E.; Xu, Y.; Gopalakrishnan, S.; Liu, J.; Harris, E. H. *J. Biol. Chem.* **2012**, *287*, 20774–20783.

[41] Sheng, J.; Xu, Y.; Dulaney, S. B.; Huang, X.; Liu, J. *J. Biol. Chem.* **2012**, *287*, 20996–21002.

[42] Peterson, S. B.; Liu, J. *J. Biol. Chem.* **2012**, *287*, 34836–34843.

[43] O'Leary, T.; Xu, Y.; Liu, J. *Glycobiology* **2012**, *23*, 132–141.

10

SYNTHESIS OF GLYCOSAMINOGLYCANS

MEDEL MANUEL L. ZULUETA, SHU-YI LIN, YU-PENG HU, AND SHANG-CHENG HUNG
Genomics Research Center, Academia Sinica, Taipei, Taiwan

10.1 INTRODUCTION

Glycosaminoglycans (GAGs) are a family of complex polyanionic polysaccharides best known for their hexosamine-containing disaccharide repeating units [1–3]. Based on structural features, GAGs can be classified into four major groups: hyaluronan or hyaluronic acid (HA), chondroitin sulfate (CS) and dermatan sulfate (DS), heparin and heparan sulfate (HS), and keratan sulfate (KS) (Table 10.1). HA is assembled as a copolymer of *N*-acetyl β-D-glucosamine (β-GlcNAc) and β-D-glucuronic acid (β-GlcA) by plasma membrane-bound glycosyltransferases before being transferred to the extracellular matrix. The degree of polymerization of this free sugar is the highest among GAGs, and it is the only type without sulfation. In contrast, the rest of the GAGs are assembled and processed by enzymes in the Golgi apparatus as components of sugar–protein macromolecules called proteoglycans [4, 5]. Proteoglycans are diverse and may include from one up to about 130 GAG chains alongside other non-GAG sugars.

The developing CS, DS, heparin, and HS chains are attached to particular serine residues of the core protein through a common tetrasaccharide bridge that starts with a D-xylose [6, 7]. Similar to HA, these GAGs are constructed with a β-GlcA residue in their repeating unit. The hexosamine unit in CS and DS is *N*-acetyl-β-D-galactosamine (β-GalNAc), whereas heparin and HS incorporate α-GlcNAc during chain

Glycochemical Synthesis: Strategies and Applications, First Edition.
Edited by Shang-Cheng Hung and Medel Manuel L. Zulueta.

TABLE 10.1 Structures of GAGs

GAG	Major Disaccharide Component	General Structure
HA	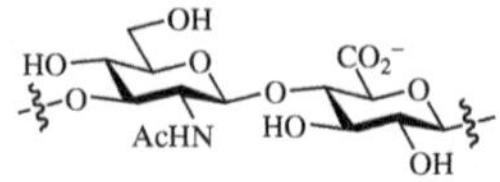	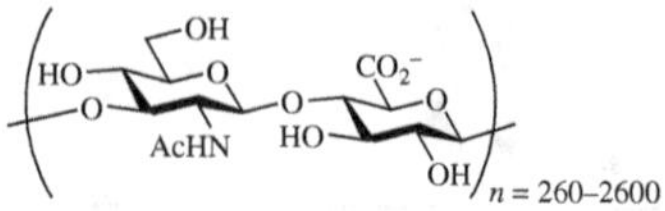
CS	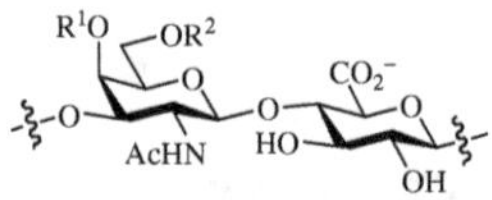 CS-A: $R^1 = SO_3^-$; $R^2 = H$ CS-C: $R^1 = H$; $R^2 = SO_3^-$	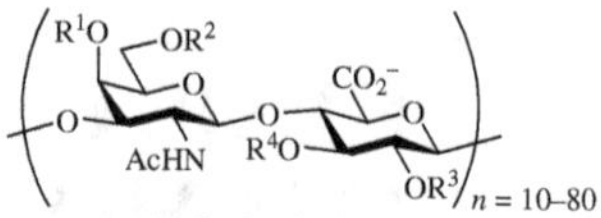 $R^1, R^2, R^3, R^4 = H$ or SO_3^-
DS	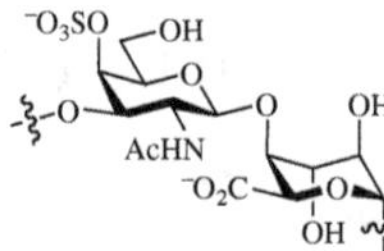	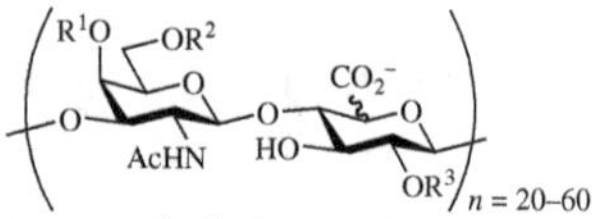 $R^1, R^2, R^3 = H$ or SO_3^-
Heparin	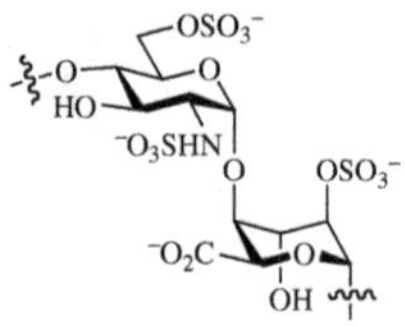	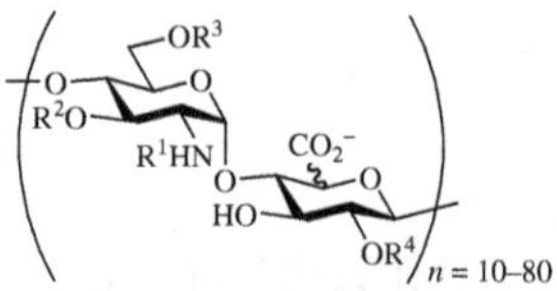 $R^1 = H$, Ac, or SO_3^-; $R^2, R^3, R^4 = H$ or SO_3^-
HS	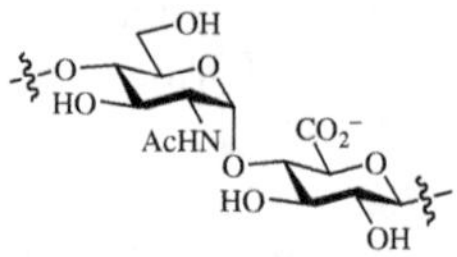	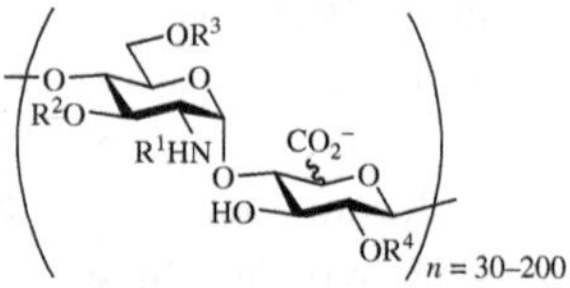 $R^1 = H$, Ac, or SO_3^-; $R^2, R^3, R^4 = H$ or SO_3^-
KS	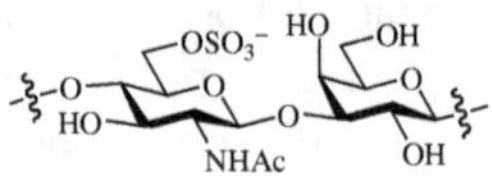	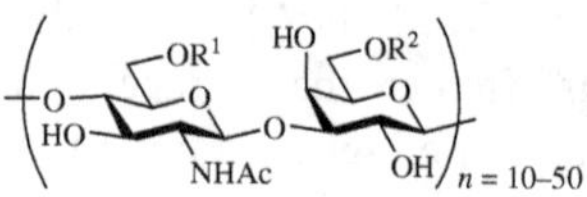 $R^1, R^2 = H$ or SO_3^-

elongation. Following assembly, these polysaccharides undergo incomplete and fairly tissue-specific modifications that add heterogeneity to their primary structures. DS, heparin, and HS chains are acted upon by C5-epimerases, which convert β-GlcA into α-L-iduronic acid (α-IdoA). In CS and DS, potential sulfonation sites include O2 of the uronic acid and O4 and O6 of β-GalNAc [8]. Rare 3-O-sulfonations are also observed at the β-GlcA residues of CS. The modification patterns in the disaccharide unit of CS chains were traditionally given alphabetical designations as A, C, D, E, and others, with B specifically assigned to DS [9]. Heparin and HS may be sulfonated at O2 of the uronic acid and at O3 and O6 of the hexosamine. Moreover, enzymes may cleave and further sulfonate the *N*-acetyl groups of heparin and HS, leading to the rare free amine and the prevalent sulfamate functionalities. The extent of epimerization, O-sulfonation, and N-sulfonation in heparin is far higher than in HS, which keeps about half of its residues unmodified [10]. Remarkably, alterations in HS tend to be localized in domains interspersed with stretches of unmodified regions. While other proteoglycans are ultimately transported intact to their site of action, serglycin—the heparin-containing proteoglycan—undergoes degradation in mastocytes and other hematopoietic cells to release unbound heparin fragments that are stored in the cell's secretory granules [11]. Heparin, in this case, is used to stabilize some basic proteins essential for inflammatory response.

The linkage region of KS varies and may resemble that of an N-glycan (KS I) or a mucin-type O-glycan (KS II), which starts from an α-GalNAc residue [12] (also see Section 11.2). KS attached to the protein core though an *O*-mannose link has also been identified. β-GlcNAc and β-Gal residues comprise the disaccharide repeating unit of KS, making it the only GAG type without a uronic acid. Sulfonations may occur at O6 positions of both component sugar units. Fucosylations at O3 of β-GlcNAc have been observed as well.

Most animal cells possess the ability to biosynthesize GAGs, which reflect to their ubiquitous presence in cell surfaces and the extracellular matrices of tissues [13]. Each tissue presents a unique array of GAGs that responds distinctively to hundreds of circulating free or bound proteins. Largely attracted to the high negative charge density of the sugars, these proteins have evolved to take advantage of the wealth of structural information in GAGs in affecting various aspects of development, physiology, and disease [2]. GAGs act as temporary storage sites of proteins while granting protection from proteases, increase local protein concentrations, mediate interaction with receptors and other proteins at the cell surface, and even function as receptors. In general, proteins bind to relatively short segments of the long GAG polymers. With the exemption of HA, the binding sites in GAGs often exhibit sequence heterogeneity. For example, the anticoagulant potency that heparin is long known for was traced to a specific and highly irregular 3-O-sulfonated pentasaccharide motif that tightly binds and activates antithrombin III, an inhibitor of the coagulation cascade [14]. This discovery led to the development of a synthetic pentasaccharide analogue, called fondaparinux, which is now an approved pharmaceutical. Conversely, the influence of length to the function of GAGs was highlighted by the observation that polymeric HA only captures CD44 (a nonproductive interaction), whereas short fragments of HA, upon binding to CD44, mediate the further activation of CD44 signaling [15].

Structure–activity relationship (SAR) evaluations are needed to understand the extent of binding and the specific nature of the binding sites within the GAG chain. This process requires pure GAG analogues of different lengths and structure variability to mimic the diversity of the natural polymer. By virtue of the intrinsic heterogeneity and polydispersity of the GAG chains, the enzymatic or chemical fragmentation of the natural polymers often leads to mixtures of compounds that are unsuitable for SAR work despite extensive purification. Considerable efforts in isolating the more exotic and sometimes biologically important sequences from the more abundant chain components are met with limited success. Conversely, several synthetic approaches have been reported over the years targeting the range of structures found in GAGs [16–25]. Given the complexity provided by modifications to the GAG chains, the chemical synthesis of oligomers encompassing all available structural possibilities is generally very tedious if not impossible. Biological studies using the nature-derived and chemoenzymatically modified materials can, nevertheless, give hints about the character of the active GAG chain, allowing for oligosaccharide targets to be narrowed down. With emerging chemoenzymatic methods in the horizon [26], chemical approaches have so far remained the most common and most reliable source of sequence-defined and homogeneous GAG-based compounds for SAR studies. Chemical methods also have the added flexibility of introducing tailor-made unnatural functional groups to the molecule. This chapter describes the chemical strategies that are currently utilized for the acquisition of GAG-based oligosaccharides and their mimics.

10.2 GENERAL STRATEGIES

The intricacy of the synthetic strategies aimed at GAGs depends on the uniformity of the repeating disaccharide, its functional group pattern, and the length of the target oligomers. GAG-based oligosaccharides with four residues or longer are typically retrosynthesized into individual disaccharide fragments, which in turn are generated from properly protected monosaccharide derivatives. With minimal building block and postassembly transformation requirements, compound targets having regular repeating units are easier to generate than irregular oligomers. Chain extensions may be carried out by iterative and modular additions of suitable disaccharide building blocks from the reducing end to the nonreducing end or vice versa. In the former, cleavage of a temporary protecting group at the nonreducing end permits the elongation process, whereas the latter employs selective and carefully executed leaving group activation. Generation of relatively long oligosaccharide donors and acceptors followed by convergent glycosylation may be performed to rapidly access lengthy oligomers. That aside, individual monosaccharide elongation [27] or a departure from even-numbered extensions is carried out as necessary.

Designing GAG building blocks demands balance between glycosylation efficiency and the need for the regioselective introduction of functionalities such as sulfate, carboxylate, acetamide, and sulfamate. A sophisticated orthogonal protecting group combination should be adopted to succeed in such process (see Fig. 10.1 [28–34]). Ester and amide or imide groups are positioned at C2 to take advantage of neighboring group participation in achieving 1,2-*trans* stereoselectivity. While the direct use of acetamide could minimize synthetic operations in these cases, N-acetylated sugar derivatives are often avoided because of their tendency to form the stable oxazoline intermediate when used as glycosyl donors and obstruct effective couplings when used as glycosyl acceptors [35–37]. Notably, the rigid benzylidene-type groups installed at O4 and O6 of the galactosamine donor (e.g., compound **4**) reduce the efficiency of β-glycosylation [38], but this protecting group is still employed in CS synthesis because of the convenience it provides in deprotections leading to

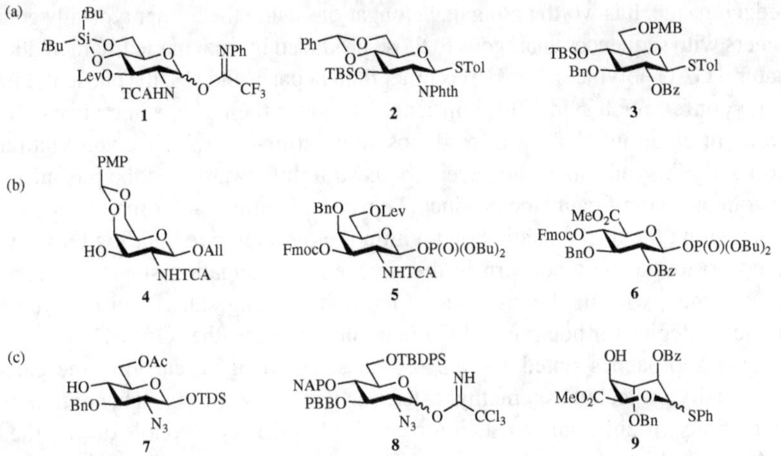

FIGURE 10.1 Some monosaccharide building blocks used in the assembly of (a) HA, (b) CS, and (c) heparin/HS showing the array of protecting groups. All, allyl; Bn, benzyl; Bz, benzoyl; Fmoc, 9-fluorenylmethoxycarbonyl; Lev, levulinyl; NAP, 2-naphthylmethyl; PBB, *p*-bromobenzyl; Phth, phthaloyl; PMB, *p*-methoxybenzyl; PMP, *p*-methoxyphenyl; TBDPS, *tert*-butyldiphenylsilyl; TBS, *tert*-butyldimethylsilyl; TCA, trichloroacetyl; TDS, dimethylthexylsilyl; Tol, 4-tolyl.

O-sulfonations. For the glucosamine building block in heparin and HS assembly, the nonparticipating azido group is utilized to permit 1,2-*cis* coupling mainly through the anomeric effect. The α stereoselectivity could be further enhanced by fine-tuning of protecting groups in order to take advantage of secondary factors such as remote participation and steric effect [33, 39]. An axially oriented nucleophilic center of the glycosyl acceptor was also found beneficial for this purpose [40]. Another set of temporary protecting groups is installed to guarantee regioselective O-sulfonation at the later stage of the synthesis. A number of stabilized sulfur trioxide reagents are currently utilized for chemical O- and N-sulfonation [41, 42]. Disaccharide building blocks with preinstalled sulfate as a trichloroethyl ester also exhibited good donor and acceptor properties and are promising alternative routes that could be explored further [43]. Permanent protecting groups are, then, assigned to mask the hydroxyls that would ultimately be free in the final product. The liberation of these groups should be tolerated by the base-labile uronic acid residues and acid-labile sulfonate groups. Additionally, noncarbohydrate linkers may be incorporated to the sugar to enable attachment to surfaces or proteins. Use of fluorous-based functionalities to aid various stages of purification has also been explored [44, 45]. At the end of the synthesis, the products are passed through a column of cation (usually Na^+)-exchange resin to afford a uniform countercation for the negative charges.

The sequence of residues in the disaccharide building blocks and the manner of access to the carboxyl group of the uronic acid demand further considerations. Depending on the intended target, the disaccharide building blocks may be constructed with the hexosamine unit positioned either at the reducing side or at the nonreducing side. It is worth noting that elongations using the former typically generate oligomers with sequences analogous to those produced by enzymatic fragmentations of the natural GAG polymers. For GAGs other than heparin and HS, the chosen sequence does not confer much synthetic significance because their glycosidic bonds all have 1,2-*trans* orientation. Control of α-glucosaminylation—a 1,2-*cis*-glycosylation—in heparin and HS synthesis is, however, not as straightforward, and the β isomer often comes out as a significant side product. This complicating factor may be minimized by establishing the α-glycosidic bond within the disaccharide building block, where it would no longer be a concern in the succeeding elongation steps. The resulting sugar sequence, with the hexosamine at the nonreducing side, is fairly common in synthetic strategies for heparin and HS in comparison to other GAGs.

Several approaches opted for the uronic acid surrogate carrying the carboxyl group, usually protected as a methyl ester, early in the synthesis. This option avoids the potentially troublesome oxidation of multiple primary alcohols during the late-stage functional group transformations. The electron-withdrawing carboxyl moiety reduces the reactivity of glycosyl donors [46], but various reports suggest that uronic acid donors are sufficiently viable for GAG synthesis [47, 48]. One more issue is the susceptibility of uronic acid derivatives to epimerization or elimination as a result of the elevated acidity of the C5 proton [43, 49]. On the other hand, successful oxidations by using 2,2,6,6-tetramethyl-1-piperidinyloxy (TEMPO) in the presence of a cooxidant, such as NaOCl or [bis(acetoxy)iodo]benzene (BAIB), made post-assembly carboxyl group generation feasible [50]. The effectiveness of this reagent combination for multiple oxidations has been demonstrated, for example, in the dodecasaccharide level [51]. Glycosyl donors with glucosyl or idosyl units at the reducing end can, therefore, be used for glycosylations without reactivity concerns. Here, another protecting group is needed to guarantee the selective liberation of the primary alcohol of the prospective uronic acid residue. As a result, the postassembly transformations are considerably longer than when a uronic acid derivative is used directly. Alternatively, the oxidation state adjustment may be performed at the disaccharide stage before chain elongation and after establishing the glycosidic bond between the uronic acid precursor and the hexosamine. Geert-Jan Boons and coworkers nicely demonstrated [32] this approach in the preparation of trichloroacetimidate **14** from the monosaccharide derivatives **10** and **7** (Scheme 10.1). TEMPO oxidation is selective for primary alcohols, allowing for the effective transformation from diol **12** to the methyl ester **13**. Nonetheless, this route is only advantageous if the uronic acid unit is positioned at the nonreducing side of the disaccharide donor.

10.3 SYNTHESIS OF DERIVATIVES OF L-IDOSE AND IDOA

Many proteins bind to motifs in DS, heparin, and HS chains that feature IdoA residues [2, 52]. Unfortunately, L-idose and IdoA are very expensive sugars, prompting chemists who wish to target IdoA-containing sequences to use cheaper alternatives for building block preparation [53]. Abundant D-glucose-based compounds are often

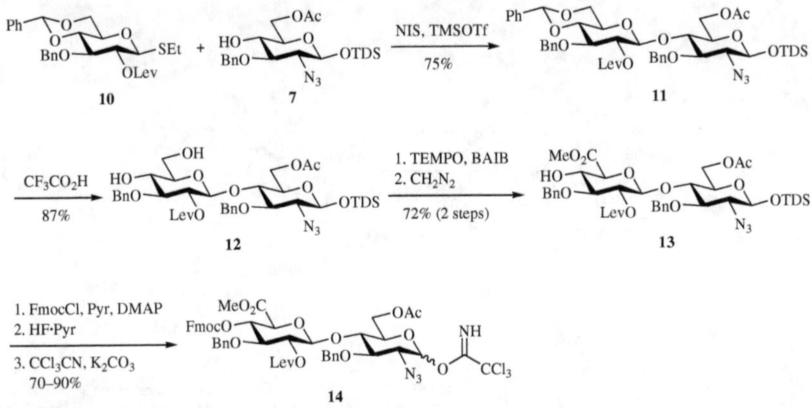

SCHEME 10.1 Preparation of a disaccharide building block for heparin and HS assembly that includes oxidation to form the uronic acid residue. DMAP, *N*,*N*-dimethyl-4-aminopyridine; NIS, *N*-iodosuccinimide; Pyr, pyridine; Tf, triflyl; TMS, trimethylsilyl.

used in these cases because they differ from the L-idose configuration only by the C5 stereochemistry. Taking advantage of this property, IdoA surrogates have been conveniently accessed through inversion of configuration at C5. The stereoselective C5 extension of D-xylose or D-*xylo*-dialdose, which could also be derived from D-glucose, is another approach used to arrive at IdoA derivatives [34, 54–57].

The S_N2 substitution of alkyl/aryl sulfonates is the typical method for C5 inversion [58–62], but stereoselective hydroboration of *exo*-glucals [63] and hydride-mediated inversion [64] were also reported. Scheme 10.2 illustrates some examples of S_N2 substitution involving D-glucurono-6,3-lactone (**15**) [60] and diacetone-D-glucose (**16**) [58] as starting materials. The triflation of the 5-alcohol **17**, obtained upon individual chemical manipulations of **15** and **16**, gave the intermediate sulfonate. The inversion was carried out by treatment with sodium levulinate, and hydrazine cleaved the resultant levulinate ester to fashion the L-*ido* alcohol **18**. The glycosyl acceptor **19** was subsequently generated in a two-step process. Starting from compound **16**, another route was also demonstrated toward the 1,6-anhydro-L-idose derivative **22** via the mesylate **20** [61]. Deprotection at O6 produces an alkoxide that subsequently attacks at the back side of C5 to form the epoxide **21** in the L-*ido* configuration. Acid reflux afforded the anhydroidose backbone, which was transformed into the 4-alcohol **22** after regioselective benzoylation. During assembly, the anhydro ring can act as a latent site that may be opened and functionalized with a leaving group if desired.

An example of stereoselective extension of a five-carbon compound toward an IdoA derivative is also shown in Scheme 10.2. The dialdose **23** was obtained from **16** through benzylation, selective 5,6-isopropylidene hydrolysis, and periodate oxidation [34, 56]. The $MgCl_2$-mediated attack by the cyanide ion on **23** produced the L-*ido*-configured cyanohydrin **24** as the major product. Further transformations led to the IdoA derivative **25**.

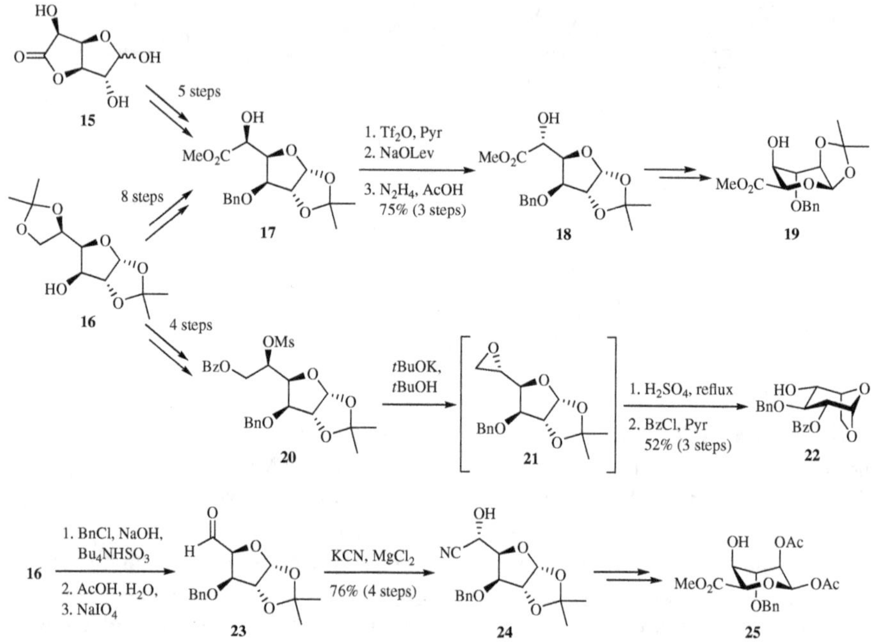

SCHEME 10.2 Preparations of IdoA and L-idose derivatives. Ms, mesyl.

10.4 SYNTHESIS VIA STEPWISE SOLUTION-PHASE ASSEMBLY AND COMPOUND DIVERSIFICATION

The stepwise assembly in solution phase is the most common method that is applied to the acquisition of GAG-based oligomers. This long approach, while considered tedious, is deemed highly reliable in producing clean target oligomers because the purification steps are carried out during each stage of the assembly. Thus, one-pot glycosylations are typically first tested for fidelity in the stepwise fashion before being implemented. The stepwise solution-phase assembly also offers the flexibility that, aside from the typical disaccharide elongation ($[2+n]$), other lengths of building blocks can be used to solve coupling problems or allow quicker access to long oligomers.

Frequent targets of chemical synthesis are heparin- and HS-based oligosaccharides of varying lengths that carry regular repeating disaccharide units, in particular those with the major component of heparin. John Gardiner and coworkers prepared such oligomers from heptasaccharide to dodecasaccharide using the $[2+n]$ route from the reducing end to the nonreducing end (Scheme 10.3) [65–67]. Bz and Bn groups protected the hydroxyls that would be sulfonated and free in the target compounds, respectively. Either PMB or TCA temporary protection was employed to permit extension at the nonreducing side of the chain. As the nonreducing end protection, the TCA group, however, gave way for sulfonation at O4, which is not found in nature. Such sulfonation may have minimal effects on a large oligosaccharide, but

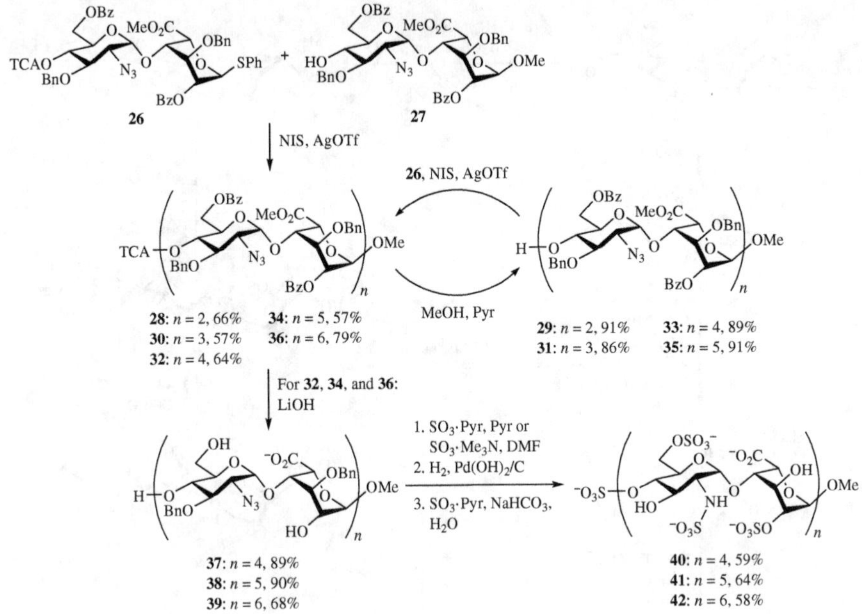

SCHEME 10.3 Preparation of oligosaccharides based on heparin and HS in a [2+*n*] manner. DMF, *N,N*-dimethylformamide.

if the natural activity is to be emulated, this functionalization is often avoided in shorter oligosaccharides by protecting group adjustment in the nonreducing end block. Iterative glycosylation–TCA deprotection cycles with the disaccharide donor **26** extended the disaccharide acceptor **27** to form several lengths of fully protected oligomers. Saponification of octamer **32**, decamer **34**, and dodecamer **36** freed the carboxylate and the hydroxy groups for subsequent sulfonation with $SO_3{\cdot}Me_3N$. After O-sulfonation, hydrogenolysis and selective N-sulfonation delivered the target compounds **40–42**.

The quest for long oligomers in solution phase can be augmented by convergent couplings of donors and acceptors that are both longer than disaccharides. Disaccharide building blocks, however, are still employed in making these donors and acceptors. Several octasaccharides and dodecasaccharides were suitably accessed in this manner [33, 45, 68–71]. David Bonnaffé's group, for instance, prepared a linker-attached heparin-based dodecasaccharide through iterative couplings of tetrasaccharide building blocks (i.e., the [4+4+4] assembly) (Scheme 10.4) [69]. To incorporate the amine-terminated linker, disaccharide **43** was first deallylated and converted into the trichloroacetimidate donor **44**. The condensation of **44** with the linker was followed by the acid-mediated PMB cleavage to allow extension at the nonreducing side. Thus, upon coupling again with donor **44** and PMB deprotection, the 4′-alcohol **45** was transformed into the tetrasaccharide acceptor **46**. Alternatively, acceptor **47** was obtained from **43** by DDQ treatment. Further glycosylation with donor **44** supplied the allylated tetrasaccharide, which underwent anomeric

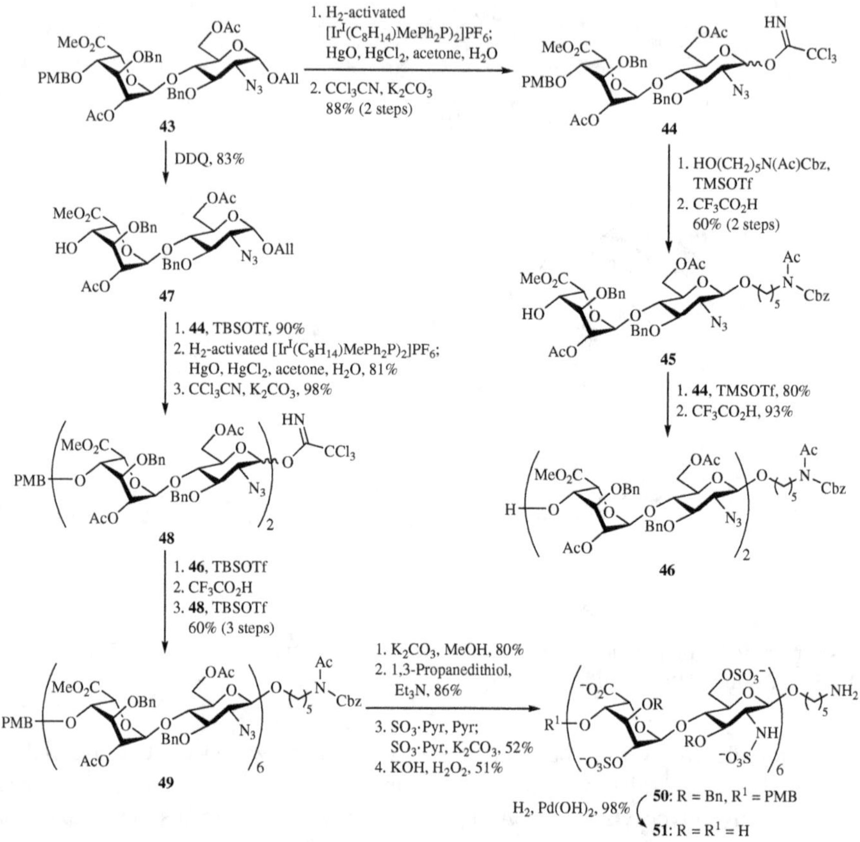

SCHEME 10.4 Preparation of a dodecasaccharide based on heparin and HS in a [4+4+4] manner. Cbz, benzyloxycarbonyl; DDQ, 2,3-dichloro-5,6-dicyano-1,4-benzoquinone.

deprotection and imidate formation. With the tetrasaccharide donor **48** and acceptor **46** in hand, two elongation cycles allowed the generation of the fully protected dodecasaccharide **49**. Exhaustive N- and O-sulfonations were carried out after exposure of the respective hydroxy and amino groups. Further hydrolysis and hydrogenolysis produced the desired compound **51**. This sugar construct was eventually conjugated to a CD4-mimetic peptide through the amine-terminated linker.

The preparation of various GAG oligosaccharides with sequence diversity and, in a broader sense, GAG sugar libraries demands an assortment of disaccharide building blocks that could be coupled in a modular fashion. Their structures can be built independently by coupling appropriately protected monosaccharides derivatives and, depending on the level of their functionalization, may be followed by simple conversions into donors and acceptors [32, 39, 51, 58, 59]. Moreover, the disaccharide precursors may be designed to permit elaborate divergent transformations into several building blocks [43, 70, 72–74]. This approach increases the utility of the precursors and significantly shortens the steps required for the synthesis of various

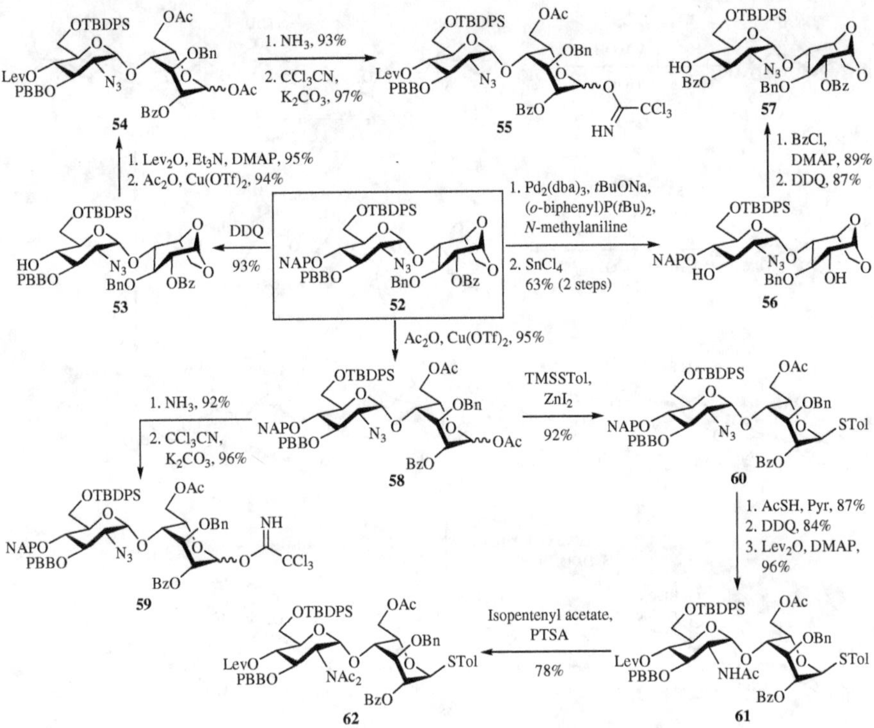

SCHEME 10.5 Hung's preparation of various disaccharide building blocks with compound **52** as the common precursor. PTSA, *p*-toluenesulfonic acid.

target compounds with irregular sequences. In this manner, Shang-Cheng Hung and coworkers derived a number of building blocks starting from disaccharide **52** (Scheme 10.5), which itself was acquired upon the condensation of compounds **8** and **22** [70]. Cleavage of the NAP group with DDQ formed acceptor **53**. Reprotection with Lev and anhydro-ring opening through acetolysis gave the diacetate **54**. The imidate donor **55** was acquired after further treatment with ammonia and then trichloroacetonitrile in the presence of a base. Palladium-catalyzed amination of the PBB group in **52** and subsequent exposure to $SnCl_4$ produced the diol **56**, which underwent dibenzoylation and denaphthylmethylation to furnish the acceptor **57**. Alternatively, acetolysis of the anhydro ring in **52** gave derivative **58**, which was transformed into the imidate donor **59** in two steps. The ZnI_2-mediated replacement of the anomeric acetate with thiotolyl group supplied the thioglycoside donor **60**. Access to the *N*,*N*-diacetyl derivative **62** was made possible by a four-step protecting group transformation via the acetamide **61**.

The disaccharide donors **55**, **59**, and **62** and the acceptors **53** and **57** were employed in the preparation of several oligosaccharides, including the octasaccharides **69–71** (Scheme 10.6) [33, 70]. Further elaborated in Scheme 10.6 is the synthesis of compound **69**—the most complicated of these structures—carrying the rare 3-O-sulfonation and several amino group functionalizations. The assembly of the

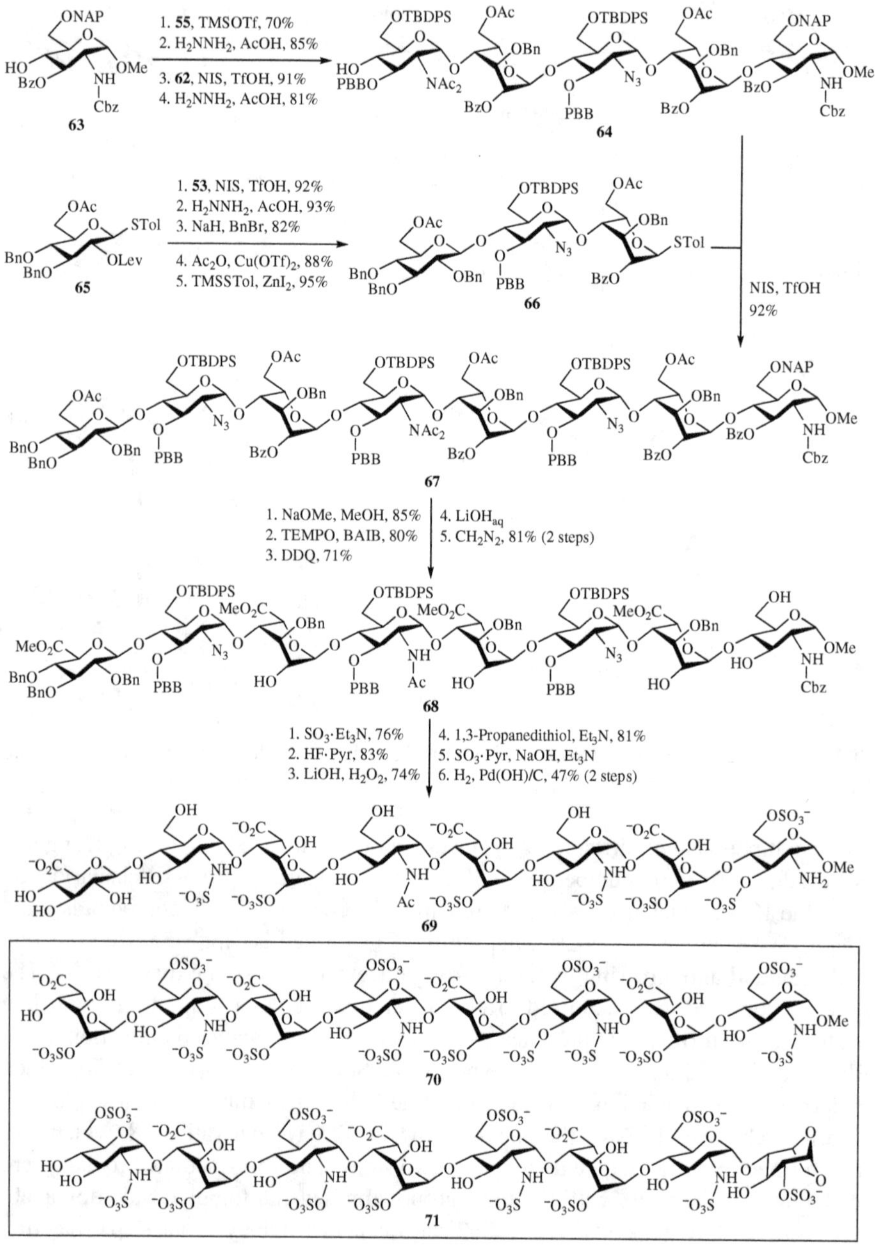

SCHEME 10.6 The heparin/HS-based octasaccharides prepared by Hung and coworkers.

octasaccharide occurred in a [3+5] manner. The reducing side pentasaccharide acceptor **64** was acquired by the iterative extension of the 4-alcohol **63** with donors **55** and **62**, whereas the nonreducing side trisaccharide donor was prepared from the monosaccharide **65** and the disaccharide acceptor **53**. The ensuing convergent glycosylation led to octasaccharide **67**, which holds a complex array of protecting groups. The acyl groups, including one of the Ac attached to the central amine, were first cleaved, followed by TEMPO oxidation of the primary hydroxyls to afford the uronate units. The succeeding deprotection enabled the exposure of all alcohols destined for O-sulfonation. Thus, the pentaol **68** was subjected to $SO_3{\cdot}Et_3N$ to access the compound, which underwent further removal of masking groups except the benzyl types (Bn, PBB, and Cbz). N-Sulfonation and final hydrogenolysis supplied the desired sugar **69**.

Instead of preparing CS oligomers from monosaccharide starting materials, Jean-Claude Jacquinet's group sourced their disaccharide building blocks directly from the CS polymers (Scheme 10.7) [72–74]. Controlled acid treatment of the natural polysaccharide results in the loss of sulfonate and *N*-acetyl groups and its fragmentation into chondrosine—the backbone disaccharide of CS. Methyl ester formation and TCA protection of the amino group in the bare disaccharide gave hemiacetal **72**, which was next benzylidenated, peracetylated, and transformed into the trichloroacetimidate derivative for condensation with (2-naphthyl)methanol. The so-formed NAP glycoside **73** was next converted in three steps to the diacetylated

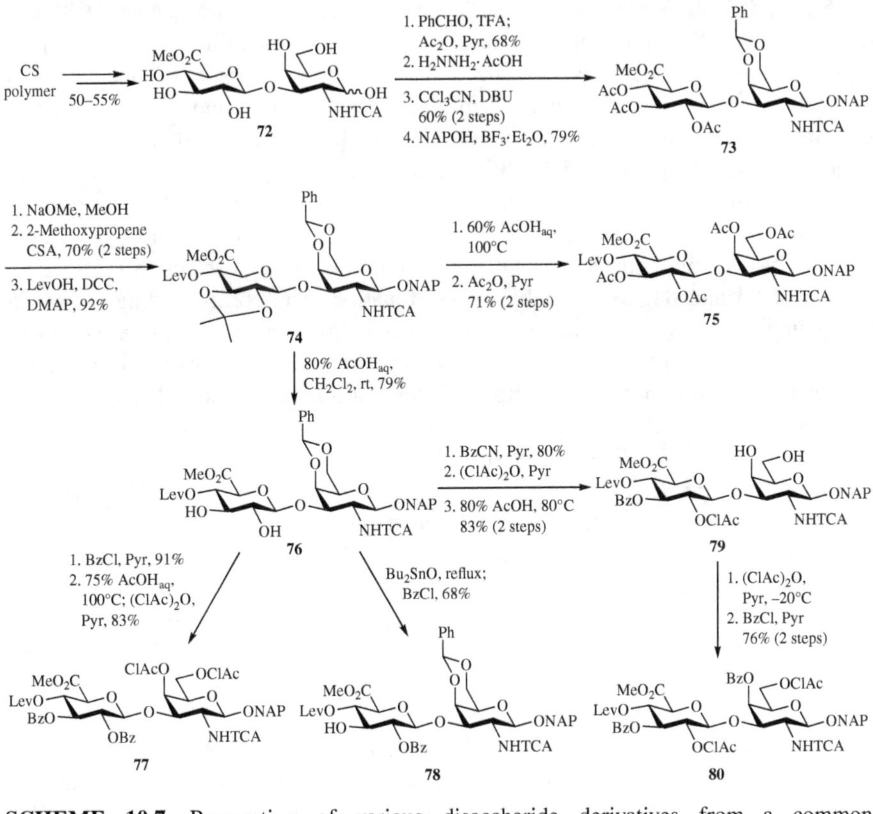

SCHEME 10.7 Preparation of various disaccharide derivatives from a common precursor by Jacquinet and coworkers. DBU, 1,8-diazabicyclo[5.4.0]undec-7-ene; DCC, *N,N'*-dicyclohexylcarbodiimide.

compound **74** designed for divergent transformation into various derivatives representing the sulfonation patterns found in CS. The tetra-acetylated disaccharide **75** was obtained after the full acetal cleavage by high-temperature acetic acid treatment followed by reacetylation of the freed hydroxyls. Meanwhile, the selective removal of the 2′,3′-*O*-isopropylidene groups in **74** at room temperature furnished diol **76**. Dibenzoylation and replacement of the benzylidene group with chloroacetyl (ClAc) groups gave compound **77**. Regioselective benzoylation of **76** was further exploited to increase the functional group diversity. Treatment with Bu_2SnO and then BzCl enabled 2′-O-benzoylation to form the 3′-alcohol **78**, while exposure to BzCN allowed the corresponding installation at O3′. With a Bz group at O3′, compound **79** was attained after chloroacetylation at O2′ and hydrolysis of the benzylidene acetal. The regioselective installation of ClAc and Bz groups at O6 and O4 positions, respectively, delivered compound **80**. The NAP and Lev groups in compounds **75**, **77**, **78**, and **80** are set to permit their straightforward transformations into glycosyl donors and acceptors.

Aside from the disaccharide building block preparations, postassembly divergent transformations are also profound means to increase compound diversity with minimal synthetic steps. Jacquinet and coworkers explored this approach in generating their CS oligomers. For instance, the fully protected tetrasaccharide **81** and hexasaccharide **82** originating from the disaccharide **77** were utilized to prepare three types of CS sulfation patterns (Scheme 10.8) [74]. The liberation of O4 and O6 under monochloroacetyl protection was performed using thiourea, and radical reduction gave the dechlorinated acetamide to form the polyols **83** and **84**. Exhaustive O-sulfonation supplied compounds **85** and **86**, which underwent saponification and hydrogenolysis to produce the 4,6-di-O-sulfonated CS oligomers **87** and **88**. The primary 6-OH of the GalNAc unit can be readily differentiated from the secondary and axial 4-OH. Thus, compounds **89** and **90** were achieved by regioselective 6-O-benzoylation of **83** and **84**, followed by 4-O-sulfonation. The corresponding target CS compounds **91** and **92** were attained after the usual three-step deprotection. Similarly, regioselective 6-O-sulfonation is possible to furnish the CS derivatives **93** and **94** and further the 6-O-sulfonated CS compounds **95** and **96**. Linda Hsieh-Wilson's group also achieved four CS-based tetrasaccharides from a single fully protected tetrasaccharide using a similar divergent strategy [30].

Likewise, oligosaccharides based on heparin and HS were achieved by divergent functional group transformations at the late stage of the synthesis to form part of compound libraries. Some examples of these are shown in Scheme 10.9. There are 48 theoretical disaccharide structures that could be found within the heparin and HS chains, and Hung's research group synthesized all of them starting from only two diversely protected disaccharide precursors [75]. The NAP, TBDPS, Bz, Ac, and N_3 groups were sufficiently transformed chemoselectively to allow the desired functionalization. Using four fully protected tetrasaccharides that differ only in the carboxyl orientation of the uronate unit, Boons and coworkers changed the manner of transformation in the amino functionality to differentially achieve acetamide and sulfamate groups [32]. After Staudinger reduction of the azido groups, treatment with Ac_2O in MeOH or SO_3·Pyr in the presence of NaOH and Et_3N supplied the

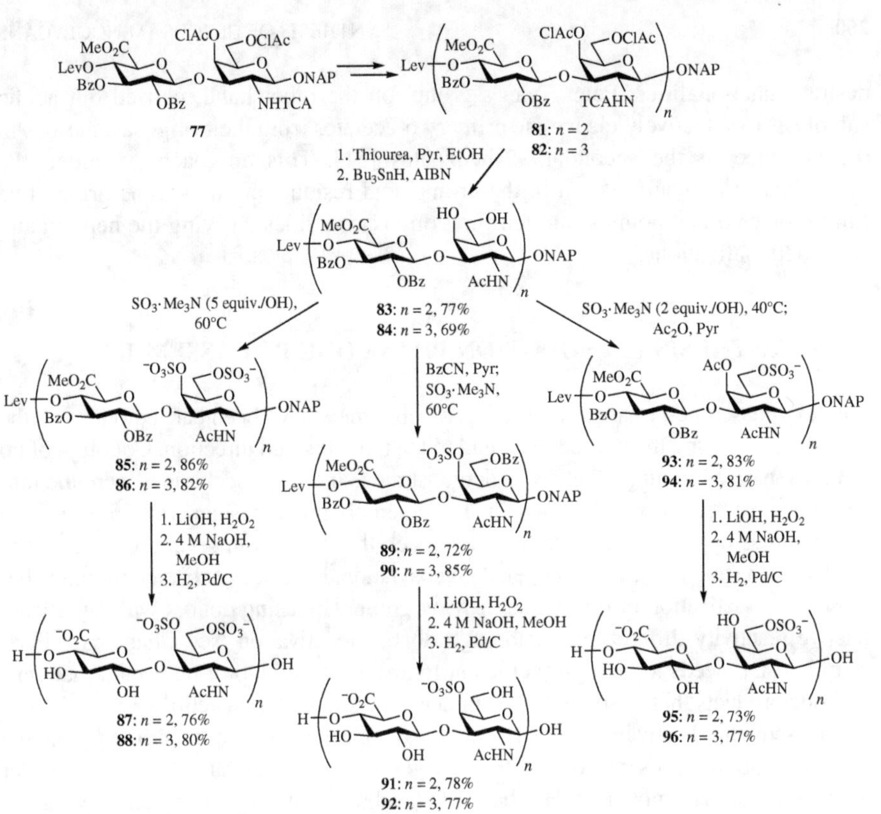

SCHEME 10.8 Divergent preparation of various CS oligomers by Jacquinet and coworkers. AIBN, azobisisobutyronitrile.

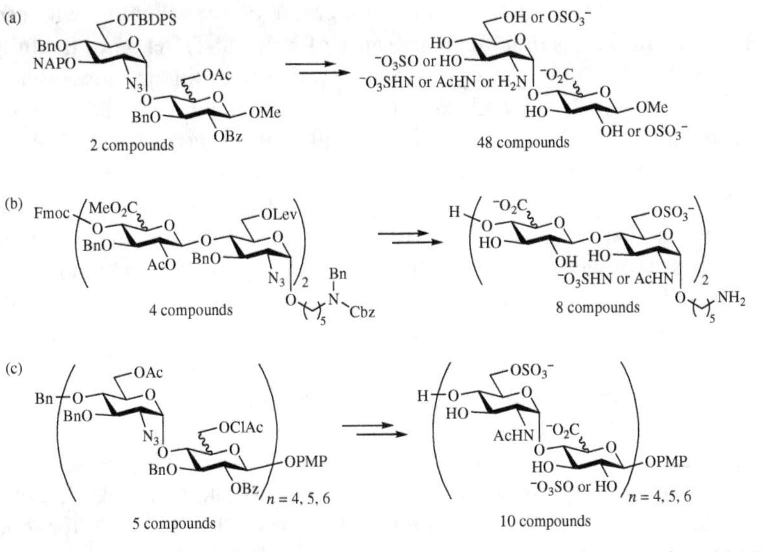

SCHEME 10.9 Various heparin and HS oligomers prepared by the laboratories of (a) Hung, (b) Boons, and (c) Tyler by postassembly divergent transformation.

desired functionalities. Peter Tyler's group, on the other hand, carried out acidic solvolysis to selectively cleave the primary 6-acetates from their oligosaccharides in the presence of the secondary 2-benzoates [51]. This approach permitted the discretionary 2-O-sulfonation in the uronic acid residue and thus transformed five fully protected compounds into ten final oligosaccharides carrying the heparin and HS functionalizations.

10.5 SYNTHESIS VIA SOLUTION-PHASE ONE-POT ASSEMBLY

Solution-phase one-pot assemblies of GAG oligomers have been carried out, with the extension following the nonreducing end to the reducing end direction. Coupling of up to four separate building blocks was demonstrated, and hexasaccharides were the longest that have been assembled thus far. The assembly process is generally finished in a few hours as opposed to the stepwise approach that lasts from several days to a few weeks. Such strategies call for targeted glycosylation that can be achieved through chemoselective activation of orthogonal leaving groups, by using donors with significant relative reactivity differences, or through donor preactivation (see Chapter 6). Each coupling step needs to be highly efficient to avoid the buildup of unreacted materials and side products that are bound to complicate the succeeding reactions.

The sequential activation of orthogonal leaving groups with compatible promoters presents opportunities for one-pot glycosylations. To illustrate, Gijsbert van der Marel's group assembled an HA-based trisaccharide in one pot by employing the trifluoroacetimidate donor **1** that could be activated in the presence of another donor—the thioglycoside **97** (Scheme 10.10) [29]. Thus, the addition of TMSOTf to the mixture of **1** and **97** only activated the imidate to provide the intermediate disaccharide donor (**99**) with a thiophenyl leaving group. Further coupling with acceptor **98** in the same flask, this time in the presence of NIS and TfOH, gave the fully protected trisaccharide **100** in good yield. With a properly positioned Lev group, additional iterative deprotections and extensions of the oligomer using the thioglycoside **99** provided a way to afford pentasaccharide **102** and heptasaccharide **104**. A four-step transformation supplied the desired oligomers **105–107**.

Tuning of the anomeric reactivity of glycosyl donors and one-pot coupling of building blocks that are guided by their relative reactivities are emerging means of gaining access to a wide variety of glycan structures [76]. The method requires the evaluation of the relative reactivity values (RRVs) of various thioglycoside donors and donor–acceptor hybrids and their proper selection to achieve targeted activation [77]. Chi-Huey Wong and coworkers prepared a tetrasaccharide and a pentasaccharide based on heparin and HS under this methodology [78]. The fully protected skeleton **111** corresponding to the pentasaccharide was generated from the initial activation of the more reactive donor **108** in the presence of the acceptor thioglycoside **109**, followed by a second coupling of the so-formed intermediate trisaccharide donor and the reducing end disaccharide acceptor **110** (Scheme 10.11). O-Sulfonation was performed after the cleavage of all ester groups in **111**. The target compound **112** was obtained after hydrogenolysis and selective N-sulfonation steps. This approach

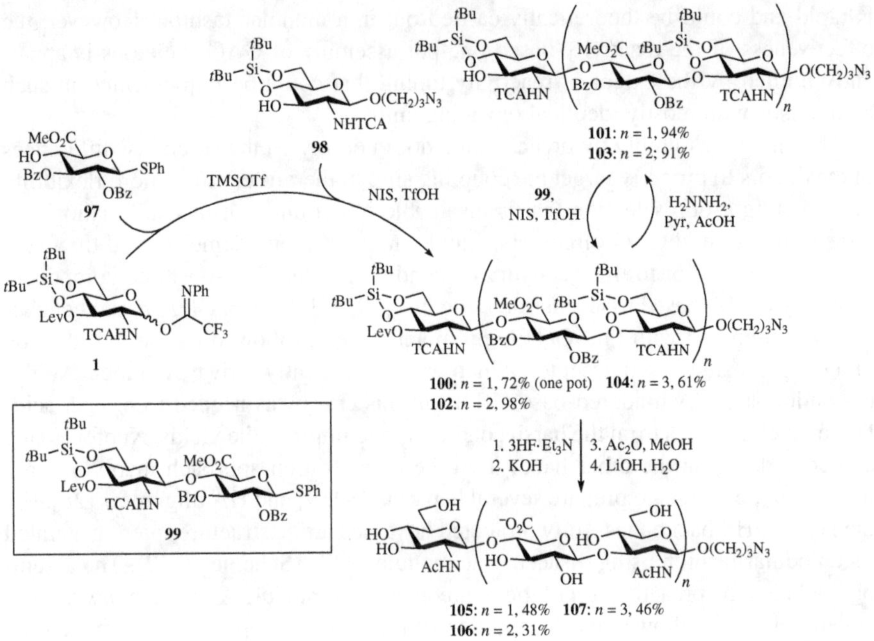

SCHEME 10.10 Synthesis of HA oligomers involving one-pot assembly to the trisaccharide level.

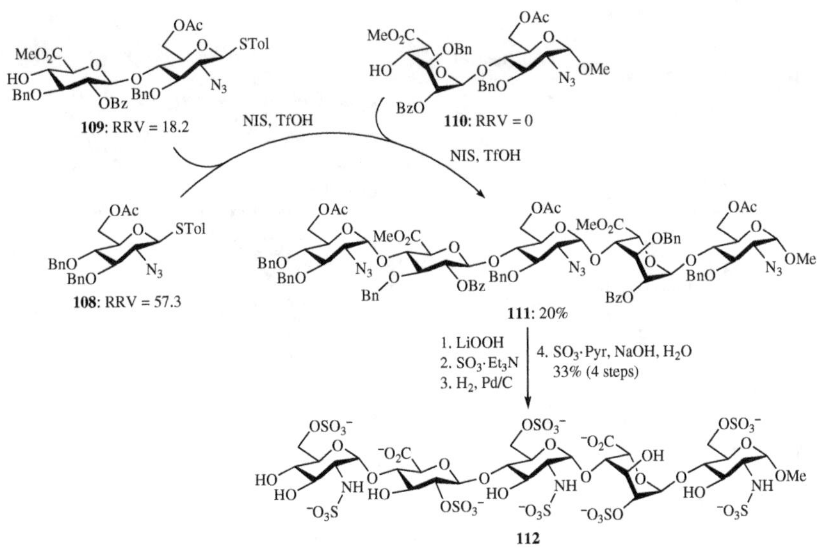

SCHEME 10.11 Anomeric reactivity-based one-pot assembly and preparation of a pentasaccharide based on heparin and HS.

is rapid and could be theoretically carried out in a modular fashion. However, the effectiveness of the reactivity-based one-pot assembly of GAG skeletons is apparently limited by the extent of reactivity tuning that could be implemented in such compounds with mostly identical repeating units.

The one-pot assembly by preactivation does not rely on the reactivity differences of the donors to properly target the coupling site. It, therefore, offers more flexibility in protecting group selection that is invaluable in defining various sulfonation patterns of the desired GAG oligomers. van der Marel's group demonstrated the feasibility of preactivation to affect dehydrative conditions in their one-pot three-component assembly of HA-based trisaccharide skeletons [79, 80]. Here, a GlcA-based hemiacetal was premixed with Ph_2SO/Tf_2O promoter system, followed by the introduction of a thioglycoside-based acceptor, which was subsequently activated in the next glycosylation stage. A hindered base (TTBP in this case) was needed to protect acid-labile functionalities from the harsh conditions and improve the yields. Xuefei Huang and coworkers, on the other hand, used the preactivation approach involving only thiotolyl-type donors to prepare several heparin/HS [39] and HA oligomers [28]. For the heparin/HS backbones, fully protected hexasaccharide structures were generated in a modular fashion using disaccharide modules **A–C** (Scheme 10.12). The assembly included the preactivation of the nonreducing end module **A** with the *in situ*-generated TolSOTf, followed by the addition of the donor–acceptor hybrid (module **B**) and TTBP. These steps ensured that module **B** would act only as glycosyl acceptor and not be activated as a donor. Once the coupling process is assumed completed,

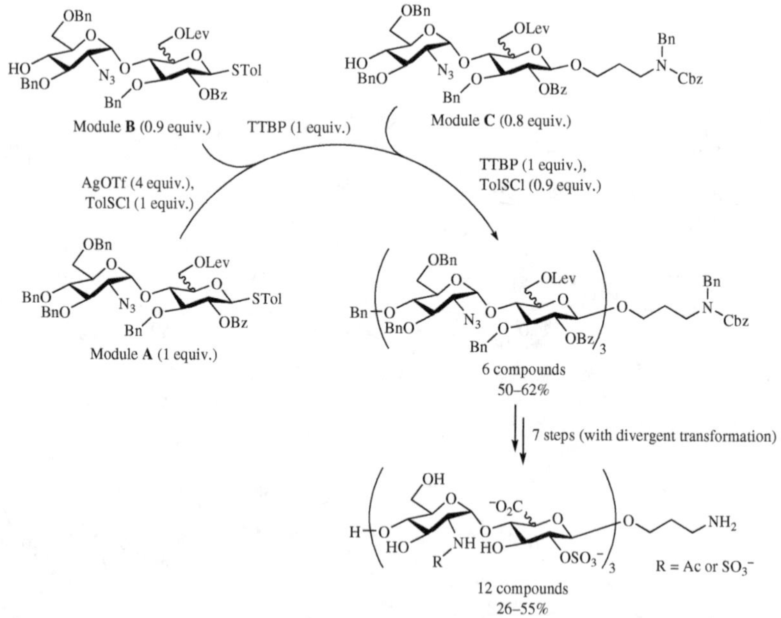

SCHEME 10.12 Preparation of hexasaccharides based on heparin and HS via preactivation. TTBP, 2,4,6-tri-*tert*-butylpyrimidine.

the reducing end module **C** was added together with the activator tandem to permit hexasaccharide formation. Application of divergent functional group transformation precisely aimed at preparing compounds that differ in the amino group functionalization on the six assembled derivatives delivered 12 heparin/HS-based oligomers. In the same manner, HA-based oligomers with four to six sugar units were assembled by Huang's group using three or four separate building blocks [28]. Preactivation was also utilized in the assembly of HA decasaccharides, but glycosylation issues were encountered, and the assembly in one pot was not pursued further [15].

10.6 POLYMER-SUPPORTED SYNTHESIS AND AUTOMATION

Dealing with the synthesis of GAGs oligomers through automated synthesis is currently an active area of experimentation. The linear and core repeating structures of GAG chains are enticing in these respects. Nevertheless, the relatively low reactivities of the building blocks used in GAG synthesis made the leap from solution phase to the polymer support required for automation quite a hurdle. Despite these issues, a few GAGs oligomers have been assembled using polymer support and automation by mainly employing acceptor-bound protocols [31, 81–86] (also see Chapter 7). The polymer support, by being readily separable from the reaction solution, permits repeated treatments with excess building blocks and reagents to force reactions to completion. Because sugar building blocks are expensive to produce, high yield and stereoselectivity must be ensured while keeping the number of glycosylation cycles as low as possible. The intricate postassembly functional group transformations inherent to GAG synthesis also remain a challenge that has only been partially addressed in polymer support.

The soluble polyethylene glycol (PEG) was employed during the early adoption of the polymer-supported synthesis of GAGs. It carries the advantages that the building blocks react in nearly similar efficiencies as regular solution phase, reaction monitoring can be performed using standard techniques, and the polymer with the bound sugars can be precipitated out of solution during workup. Nevertheless, precipitation could result in loss of materials, and the solubility of PEG is reduced at the low temperatures often required by chemical glycosylation. For the synthesis of heparin/HS-based oligosaccharides, Manuel Martín-Lomas and coworkers examined various attachment points through an ester-type linker with the soluble PEG ω-monomethyl ether (MPEG) and a PEG-grafted polystyrene (PS) resin [81, 82]. In the end, the MPEG attached to the carboxylate group of the reducing end IdoA unit through a succinoyl ester linker gave the best results in a $[2+n]$-type assembly (Scheme 10.13). The ethylene glycol transesterification with the methyl ester **113** that is necessary for MPEG attachment occurred under tin mediation. Acid hydrolysis of the polymer-bound disaccharide removed the benzylidene group, and further regioselective benzoylation produced the acceptor **114**. The extension of this acceptor with glycosyl donor **115** in two elongation cycles was followed by the capping of the unreacted acceptor with PS for subsequent removal by filtration. Similar processing as the disaccharide produced the subsequent acceptor for the next glycosylation steps.

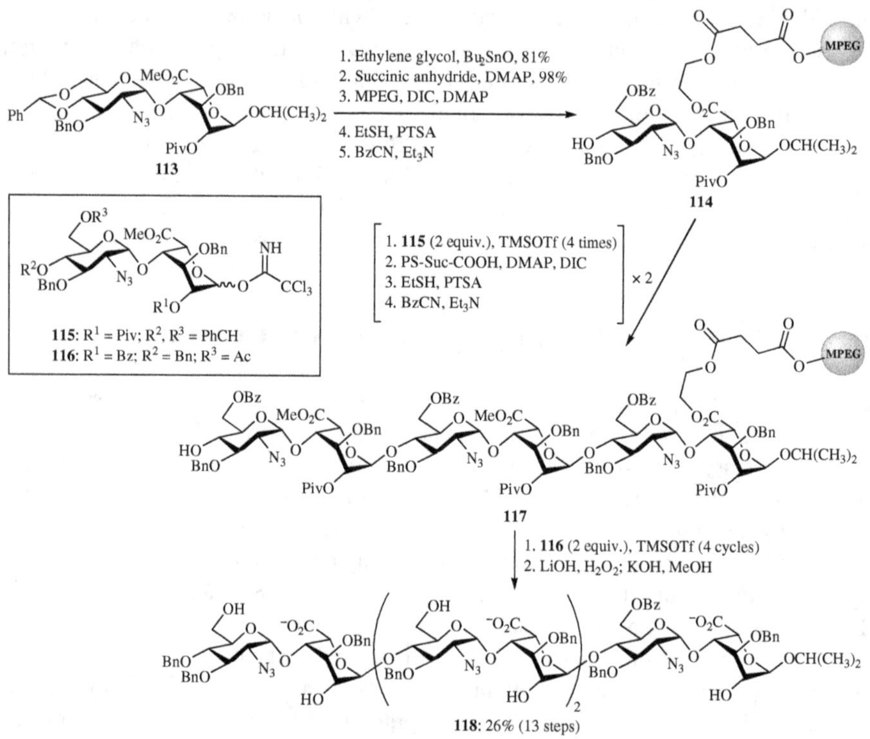

SCHEME 10.13 Polymer-supported assembly and release of an octasaccharide with a heparin/HS backbone. DIC, *N,N'*-diisopropylcarbodiimide; Piv, pivaloyl.

A different nonreducing end block (**116**) was employed for the final coupling to produce a hexasaccharide and an octasaccharide skeleton (**117**). Basic treatment enabled the release of the oligomers from the support and also liberated the hydroxyls for the O-sulfonation steps, which was performed according to the regular solution-phase procedure [87].

The power of automated solid-phase synthesis was demonstrated by Jeroen Codée et al. in the preparation of HA oligomers, the longest of which is a pentadecasaccharide (Scheme 10.14) [85]. Here, the Merrifield resin was first functionalized with a butenediol linker system anticipating a cross-metathesis cleavage from the solid support after chain assembly. The weakly reactive GlcA donor could not be coupled effectively to this resin-based acceptor (**119**) because acyl group migration to the reactive primary alcohol of the acceptor competed with donor–acceptor condensation. On the other hand, the glucosamine donor **1** gave favorable result in a coupling reaction repeated three times, paving its use as starting unit in the assembly. Deprotection of the Lev group with hydrazine exposes the hydroxy group needed for the extension. Several cycles of chain elongation with the disaccharide donor **120** produced various lengths of the resin-bound oligosaccharides. Grubbs' metathesis with trichloroacetamide as decoy substrate, followed by desilylation, produced the

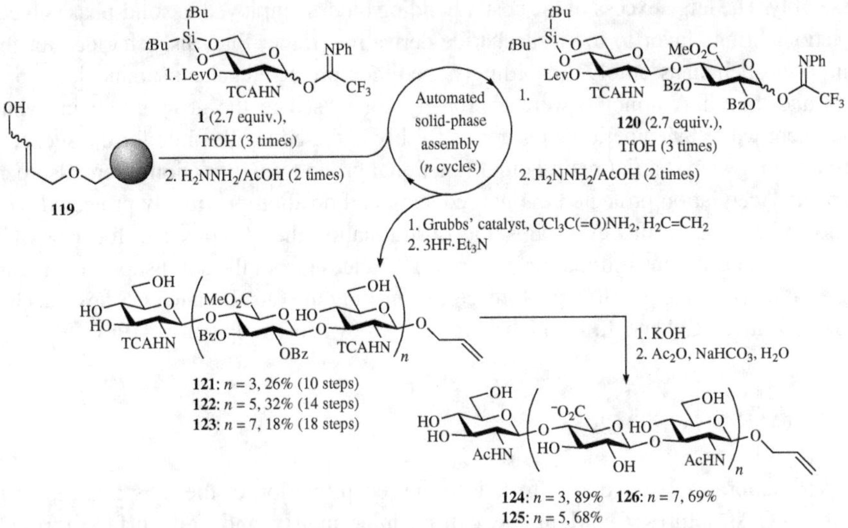

SCHEME 10.14 Automated solid-phase assembly and synthesis of HA oligomers.

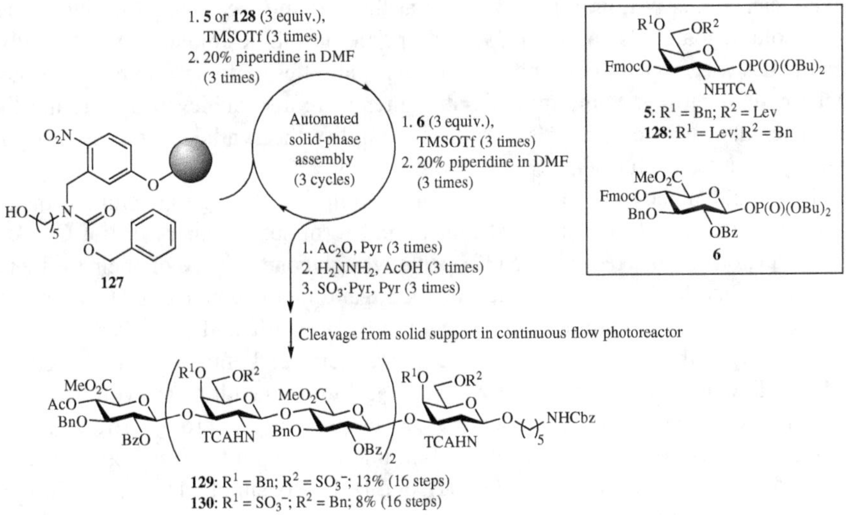

SCHEME 10.15 Automated solid-phase synthesis of CS-based hexasaccharides.

unbound polyols **121–123**. HPLC analysis detected only minor deletion sequences suggesting highly efficient coupling cycles. Subsequent basic hydrolysis of the ester and amide groups and selective N-acetylation led to the HA-based heptasaccharide **124**, undecasaccharide **125**, and pentadecasaccharide **126**.

For the automated synthesis of CS hexasaccharides, Peter Seeberger's group designed a photolabile nitrobenzyl ether-based linker to the Merrifield resin (Scheme 10.15) [31]. Their advanced strategy also differs by the use of monosaccharide donors for the assembly. The large excess of the costly building blocks employed in solid-phase glycosylations brings favor to monosaccharide derivatives rather than disaccharides for the purposes of chain assembly. Accordingly, the Fmoc-protected galactosamine donor **5** or **128** and the GlcA donor **6** were alternately condensed to the growing chain, while treatment with 20% piperidine ensured a freshly liberated nucleophile for the succeeding coupling step. Partial resin-bound functional group transformations were also performed. Acetylation protected the nonreducing end position previously protected with Fmoc. Cleavage of all Lev groups and O-sulfonation then followed to form two CS oligomers of different sulfonation patterns. The release from the solid support was carried out in a continuous-flow photoreactor furnishing the O-sulfonated CS hexasaccharide derivatives **129** and **130**, which were purified by using preparative HPLC.

10.7 GAG MIMETICS

Several laboratories devoted efforts into the recapitulation of the long and multivalent GAG structures found in the extracellular matrix and cell surface through synthetic chemistry. Current glycan assembly processes are still very limited in the length that could be achieved, and while attachments to metal surfaces through linkers can in a way mimic multivalency at the cell surface, generating individual freely soluble materials possessing such properties is still desirable. To this end, glycomimetics was tapped to expand the current capabilities to possess some semblance to the natural setting. Some mimetic structures were also achieved by altering the fundamental structure of GAGs in order to simplify the synthetic process, but this will be discussed no further.

GAG multivalency can be achieved, at its simplest, through attachment of the synthetic oligomers to bifunctional linkers and form sugar dimers. Elliot Chaikof et al. transformed the reducing end *n*-pentenyl aglycone moiety of their synthetic GAG disaccharides into an aldehyde, and the sugar dimer was generated by reductive amination to ethylenediamine [88, 89]. Similarly, Bonnaffé and coworkers relied on the reactivity of the allyl group installed at the reducing end of their synthetic heparin and HS oligomers to achieve dimerization [68]. Prior to the ultimate hydrogenolysis step, they conjugated the oligomers to α,ω-bis(thio)PEGs through irradiation at around 360 nm. The thioether functions that interfered with the last step were next oxidized to sulfones. Exposure to $Pd(OH)_2$ and H_2 finally supplied the target dimeric structure. With polyanionic sugars attached at opposite ends, the PEG spacer mimics the unmodified domain in HS. The strength of the interaction of the glycodimer with interferon-γ, a homodimeric protein with two HS binding sites, was found dependent on the length of the utilized spacer.

The generation of glycodendrimers presents reasonable means to produce multivalency. With amine-terminated heparin and HS oligomers in hand [90], Seeberger's group performed conjugation to a polyamidoamine (PAMAM) dendrimer holding 32 carboxylic acid end groups [91]. Amide formation was accomplished by activation with *N*-ethyl-*N*′-(dimethylaminopropyl)carbodiimide and *N*-hydroxysuccinimide. Only about 25% of the available sites were functionalized with the sugar, which was

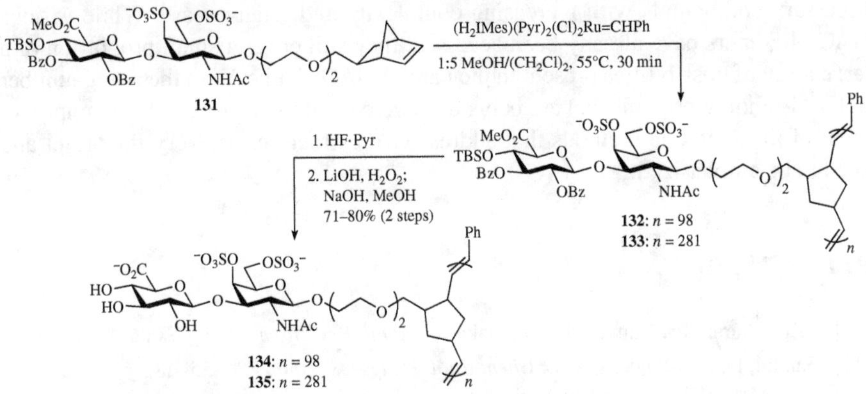

SCHEME 10.16 Synthesis of CS glycomimetic polymers through ROMP.

attributed to steric hindrance and electrostatic repulsion between the component sulfate and carboxylate groups in both PAMAM and the sugar. The resulting glycodendrimers, nevertheless, displayed better activity than their monovalent counterparts.

Ring-opening metathesis polymerization (ROMP) chemistry was employed by Hsieh-Wilson and coworkers to afford mimetic GAG glycopolymers [92–95]. Here, the reducing end linkers with terminal *cis*-cyclooctene and norbornene groups are manipulated as metathesis polymerization handles. Prior to the polymerization stage, the sugar component was properly sulfonated to ensure homogeneous functionalization in the growing polymer. Using this technique, disaccharides and tetrasaccharides based on CS, heparin, and HS were successfully polymerized. As illustrated in the example in Scheme 10.16, a fast-initiating Grubbs' catalyst facilitated the generation of CS polymers through the norbornyl function. The catalyst yields polymers having narrow polydispersity and with lengths that can be controlled by the catalyst concentration. For example, 1 mol% of the catalyst applied to compound **131** produced a polymer with about 98 CS disaccharide units (**132**), whereas about 281 monomeric units (**133**) were achieved by 0.5 mol% catalyst loading. Final deprotections such as desilylation, saponification, and hydrogenolysis were successfully implemented after the metathesis step. End functionalization with a biotin moiety for surface attachment was also made possible by the addition of a biotin-terminating agent to the ongoing reaction. A most interesting result of the bioactivity assay is that of the tetrasulfonated heparin-based disaccharide (the major repeating component plus 3-O-sulfonation) glycopolymer, which showed higher anticoagulant activity than the natural heparin and fondaparinux [95].

10.8 CONCLUSION

The utmost precision of chemical synthesis in defining the length and functionalization patterns ensures a prominent role in GAG acquisition for SAR studies. GAG syntheses remain formidable, but recent advances are instrumental in providing the necessary compounds with increasing complexity and relative ease. While various GAG oligomers have already been accessed, they still cover a small proportion of a large array of possibilities present in the natural GAG chains. With the rising number of GAG-binding proteins that are being discovered, continued research and improvement of the synthetic methods that address GAG structures are very important and need to be emphasized.

REFERENCES

[1] Sasisekharan, R.; Raman, R.; Prabhakar, V. *Annu. Rev. Biomed. Eng.* **2006**, *8*, 181–231.

[2] Gandhi, N. S.; Mancera, R. L. *Chem. Biol. Drug Res.* **2008**, *72*, 455–482.

[3] Kamhi, E.; Joo, E. J.; Dordick, J. S.; Linhardt, R. J. *Biol. Rev.* **2013**, *88*, 928–943.

[4] Kjellén, L.; Lindahl, U. *Annu. Rev. Biochem.* **1991**, *60*, 443–475.

[5] Couchman, J. R.; Pataki, C. A. *J. Histochem. Cytochem.* **2012**, *60*, 885–897.

[6] Silbert, J. E.; Sugumaran, G. *IUBMB Life* **2002**, *54*, 177–186.

[7] Sugahara, K.; Kitagawa, H. *IUBMB Life* **2002**, *54*, 163–175.

[8] Malavaki, C.; Mizumoto, S.; Karamanos, N.; Sugahara, K. *Connect. Tissue Res.* **2008**, *49*, 133–139.

[9] Trowbridge, J. M.; Gallo, R. L. *Glycobiology* **2002**, *12*, 117R–125R.

[10] Esko, J. D.; Selleck, S. B. *Annu. Rev. Biochem.* **2002**, *71*, 435–471.

[11] Rabenstein, D. L. *Nat. Prod. Rep.* **2002**, *19*, 312–331.

[12] Funderburgh, J. L. *Glycobiology* **2000**, *10*, 951–958.

[13] Yamada, S.; Sugahara, K.; Özbek, S. *Commun. Integr. Biol.* **2011**, *4*, 150–158.

[14] Petitou, M.; van Boeckel, C. A. A. *Angew. Chem. Int. Ed.* **2004**, *43*, 3118–3133.

[15] Lu, X.; Kamat, M. N.; Huang, L.; Huang, X. *J. Org. Chem.* **2009**, *74*, 7608–7617.

[16] Tamura, J. *Trends Glycosci. Glycotechnol.* **2001**, *13*, 65–88.

[17] Yeung, B. K. S.; Chong, P. Y. C.; Petillo, P. A. *J. Carbohydr. Chem.* **2002**, *21*, 799–865.

[18] Gama, C. I.; Hsieh-Wilson, L. C. *Curr. Opin. Chem. Biol.* **2005**, *9*, 609–619.

[19] Noti, C.; Seeberger, P. H. *Chem. Biol.* **2005**, *12*, 731–756.

[20] Bonnaffé, D. *C. R. Chim.* **2011**, *14*, 59–73.

[21] Bedini, E.; Parrilli, M. *Carbohydr. Res.* **2012**, *356*, 75–85.

[22] Dulaney, S. B.; Huang, X. *Adv. Carbohydr. Chem. Biochem.* **2012**, *67*, 95–136.

[23] Zulueta, M. M. L.; Lin, S.-Y.; Hung, S.-C. *Trends Glycosci. Glycotechnol.* **2013**, *25*, 141–158.

[24] Griffin, M. E.; Hsieh-Wilson, L. C. *Curr. Opin. Chem. Biol.* **2013**, *17*, 1014–1022.

[25] Zulueta, M. M. L.; Lin, S.-Y.; Hu, Y.-P.; Hung, S.-C. *Curr. Opin. Chem. Biol.* **2013**, *17*, 1023–1029.

[26] DeAngelis, P. L.; Liu, J.; Linhardt, R. J. *Glycobiology* **2013**, *23*, 764–777.

[27] Codée, J. D. C.; Stubba, B.; Schiattarella, M.; Overkleeft, H. S.; van Boeckel, C. A. A.; van Boom, J. H.; van der Marel, G. A. *J. Am. Chem. Soc.* **2005**, *127*, 3767–3773.

[28] Huang, L.; Huang, X. *Chem. Eur. J.* **2007**, *13*, 529–540.

[29] Dinkelaar, J.; Gold, H.; Overkleeft, H. S.; Codée, J. D. C.; van der Marel, G. A. *J. Org. Chem.* **2009**, *74*, 4208–4216.

[30] Gama, C. I.; Tully, S. E.; Sotogaku, N.; Clark, P. M.; Rawat, M.; Vaidehi, N.; Goddard, W. A.; Nishi, A.; Hsieh-Wilson, L. C. *Nat. Chem. Biol.* **2006**, *2*, 467–473.

[31] Eller, S.; Collot, M.; Yin, J.; Hahm, H. S.; Seeberger, P. H. *Angew. Chem. Int. Ed.* **2013**, *52*, 5858–5861.

[32] Arungundram, S.; Al-Mafraji, K.; Asong, J.; Leach, F. E., III; Amster, I. J.; Venot, A.; Turnbull, J. E.; Boons, G.-J. *J. Am. Chem. Soc.* **2009**, *131*, 17394–17405.

[33] Zulueta, M. M. L.; Lin, S.-Y.; Lin, Y.-T.; Huang, C.-J.; Wang, C.-C.; Ku, C.-C.; Shi, Z.; Chyan, C.-L.; Irene, D.; Lim, L.-H.; Tsai, T.-I.; Hu, Y.-P.; Arco, S. D.; Wong, C.-H.; Hung, S.-C. *J. Am. Chem. Soc.* **2012**, *134*, 8988–8995.

[34] Hansen, S. U.; Miller, G. J.; Baráth, M.; Broberg, K. R.; Avizienyte, E.; Helliwell, M.; Raftery, J.; Jayson, G. C.; Gardiner, J. M. *J. Org. Chem.* **2012**, *77*, 7823–7843.

[35] Crich, D.; Dudkin, V. *J. Am. Chem. Soc.* **2001**, *123*, 6819–6825.

[36] Enugala, R.; Carvalho, L. C. R.; Dias Pires, M. J.; Marques, M. M. B. *Chem. Asian J.* **2012**, *7*, 2482–2501.

[37] Bongat, A. F. G.; Demchenko, A. V. *Carbohydr. Res.* **2007**, *342*, 374–406.

[38] Bélot, F.; Jacquinet, J.-C. *Carbohydr. Res.* **2000**, *325*, 93–106.

[39] Wang, Z.; Xu, Y.; Yang, B.; Tiruchinapally, G.; Sun, B.; Liu, R.; Dulaney, S.; Liu, J.; Huang, X. *Chem. Eur. J.* **2010**, *16*, 8365–8375.

[40] Orgueira, H. A.; Bartolozzi, A.; Schell, P.; Seeberger, P. H. *Angew. Chem. Int. Ed.* **2002**, *41*, 2128–2131.

[41] Al-Horani, R. A.; Desai, U. R. *Tetrahedron* **2010**, *66*, 2907–2918.

[42] Gilbert, E. E. *Chem. Rev.* 1962, *62*, 549–589.

[43] Tiruchinapally, G.; Yin, Z.; El-Dakdouki, M.; Wang, Z.; Huang, X. *Chem. Eur. J.* **2011**, *17*, 10106–10112.

[44] Zong, C.; Venot, A.; Dhamale, O.; Boons, G.-J. *Org. Lett.* **2013**, *15*, 342–345.

[45] Tanaka, H.; Tateno, Y.; Takahashi, T. *Org. Biomol. Chem.* **2012**, *10*, 9570–9582.

[46] Guedes, N.; Czechura, P.; Echeverria, B.; Ruiz, A.; Michelena, O.; Martin-Lomas, M.; Reichardt, N.-C. *J. Org. Chem.* **2013**, *78*, 6911–6934.

[47] Tatai, J.; Osztrovszky, G.; Kajtár-Peredy, M.; Fügedi, P. *Carbohydr. Res.* **2008**, *343*, 596–606.

[48] Codée, J. D. C.; Christina, A. E.; Walvoort, M. T. C.; Overkleeft, H. S.; van der Marel, G. A. *Top. Curr. Chem.* **2011**, *301*, 253–289.

[49] Maza, S.; Macchione, G.; Ojeda, R.; Lopez-Prados, J.; Angulo, J.; de Paz, J. L.; Nieto, P. M. *Org. Biomol. Chem.* **2012**, *10*, 2146–2163.

[50] Bragd, P. L.; van Bekkum, H.; Besemer, A. C. *Top. Catal.* **2004**, *27*, 49–66.

[51] Schwörer, R.; Zubkova, O. V.; Turnbull, J. E.; Tyler, P. C. *Chem. Eur. J.* **2013**, *19*, 6817–6823.

[52] Volpi, N. *Carbohydr. Polym.* **2010**, *82*, 233–239.

[53] Zulueta, M. M. L.; Zhong, Y.-Q.; Hung, S.-C. *Chem. Commun.* **2013**, *49*, 3275–3287.

[54] Gavard, O.; Hersant, Y.; Alais, J.; Duverger, V.; Dilhas, A.; Bascou, A.; Bonnaffé, D. *Eur. J. Org. Chem.* **2003**, 3603–3620.

[55] Adibekian, A.; Bindschädler, P.; Timmer, M. S. M.; Noti, C.; Schützenmeister, N.; Seeberger, P. H. *Chem. Eur. J.* **2007**, *13*, 4510–4522.

[56] Hansen, S. U.; Baráth, M.; Salameh, B. A. B.; Pritchard, R. G.; Stimpson, W. T.; Gardiner, J. M.; Jayson, G. C. *Org. Lett.* **2009**, *11*, 4528–4531.

[57] Bindschädler, P.; Adibekian, A.; Grünstein, D.; Seeberger, P. H. *Carbohydr. Res.* **2010**, *345*, 948–955.

[58] Orgueira, H. A.; Bartolozzi, A.; Schell, P.; Litjens, R. E. J. N.; Palmacci, E. R.; Seeberger, P. H. *Chem. Eur. J.* **2003**, *9*, 140–169.

[59] Lu, L.-D.; Shie, C.-R.; Kulkarni, S. S.; Pan, G.-R.; Lu, X.-A.; Hung, S.-C. *Org. Lett.* **2006**, *8*, 5995–5998.

[60] Saito, A.; Wakao, M.; Deguchi, H.; Mawatari, A.; Sobel, M.; Suda, Y. *Tetrahedron* **2010**, *66*, 3951–3962.

[61] Hung, S.-C.; Lu, X.-A.; Lee, J.-C.; Chang, M. D.-T.; Fang, S.-L.; Fan, T.-C.; Zulueta, M. M. L.; Zhong, Y.-Q. *Org. Biomol. Chem.* **2012**, *10*, 760–772.

[62] Ojeda, R.; de Paz, J. L.; Martín-Lomas, M.; Lassaletta, J. M. *Synlett* **1999**, 1316–1318.

[63] Lee, J.-C.; Chang, S.-W.; Liao, C.-C.; Chi, F.-C.; Chen, C.-S.; Wen, Y.-S.; Wang, C.-C.; Kulkarni, S. S.; Puranik, R.; Liu, Y.-H.; Hung, S.-C. *Chem. Eur. J.* **2004**, *10*, 399–415.

[64] Yu, H. N.; Furukawa, J.-I.; Ikeda, T.; Wong, C.-H. *Org. Lett.* **2004**, *6*, 723–726.

[65] Cole, C. L.; Hansen, S. U.; Baráth, M.; Rushton, G.; Gardiner, J. M.; Avizienyte, E.; Jayson, G. C. *PLoS One* **2010**, *5*, e11644.

[66] Hansen, S. U.; Miller, G. J.; Jayson, G. C.; Gardiner, J. M. *Org. Lett.* **2013**, *15*, 88–91.

[67] Miller, G. J.; Hansen, S. U.; Avizienyte, E.; Rushton, G.; Cole, C.; Jayson, G. C.; Gardiner, J. M. *Chem. Sci.* **2013**, *4*, 3218–3222.

[68] Lubineau, A.; Lortat-Jacob, H.; Gavard, O.; Sarrazin, S.; Bonnaffé, D. *Chem. Eur. J.* **2004**, *10*, 4265–4282.

[69] Baleux, F.; Loureiro-Morais, L.; Hersant, Y.; Clayette, P.; Arenzana-Seisdedos, F.; Bonnaffé, D.; Lortat-Jacob, H. *Nat. Chem. Biol.* **2009**, *5*, 743–748.

[70] Hu, Y.-P.; Lin, S.-Y.; Huang, C.-Y.; Zulueta, M. M. L.; Liu, J.-Y.; Chang, W.; Hung, S.-C. *Nat. Chem.* **2011**, *3*, 557–563.

[71] Hansen, S. U.; Miller, G. J.; Cole, C.; Rushton, G.; Avizienyte, E.; Jayson, G. C.; Gardiner, J. M. *Nat. Commun.* **2013**, *4*, 2016.

[72] Lopin, C.; Jacquinet, J.-C. *Angew. Chem. Int. Ed.* **2006**, *45*, 2574–2578.

[73] Vibert, A.; Lopin-Bon, C.; Jacquinet, J.-C. *Chem. Eur. J.* **2009**, *15*, 9561–9578.

[74] Jacquinet, J.-C.; Lopin-Bon, C.; Vibert, A. *Chem. Eur. J.* **2009**, *15*, 9579–9595.

[75] Hu, Y.-P.; Zhong, Y.-Q.; Chen, Z.-G.; Chen, C.-Y.; Shi, Z.; Zulueta, M. M. L.; Ku, C.-C.; Lee, P.-Y.; Wang, C.-C.; Hung, S.-C. *J. Am. Chem. Soc.* **2012**, *134*, 20722–20727.

[76] Hsu, C.-H.; Hung, S.-C.; Wu, C.-Y.; Wong, C.-H. *Angew. Chem. Int. Ed.* **2011**, *50*, 11872–11923.

[77] Lee, J.-C.; Greenberg, W. A.; Wong, C.-H. *Nat. Protoc.* **2006**, *1*, 3143–3152.

[78] Polat, T.; Wong, C.-H. *J. Am. Chem. Soc.* **2007**, *129*, 12795–12800.

[79] Codée, J. D. C.; van den Bos, L. J.; Litjens, R. E. J. N.; Overkleeft, H. S.; van Boom, J. H.; van der Marel, G. A. *Org. Lett.* **2003**, *5*, 1947–1950.

[80] Dinkelaar, J.; Codée, J. D. C.; van den Bos, L. J.; Overkleeft, H. S.; van der Marel, G. A. *J. Org. Chem.* **2007**, *72*, 5737–5742.

[81] Ojeda, R.; de Paz, J.-L.; Martín-Lomas, M. *Chem. Commun.* **2003**, 2486–2487.

[82] Ojeda, R.; Terentí, O.; de Paz, J.-L.; Martín-Lomas, M. *Glycoconj. J.* **2004**, *21*, 179–195.

[83] Mar Kayser, M.; de Paz, J. L.; Nieto, P. M. *Eur. J. Org. Chem.* **2010**, 2138–2147.

[84] Czechura, P.; Guedes, N.; Kopitzki, S.; Vazquez, N.; Martin-Lomas, M.; Reichardt, N.-C. *Chem. Commun.* **2011**, *47*, 2390–2392.

[85] Walvoort, M. T. C.; Volbeda, A. G.; Reintjens, N. R. M.; van den Elst, H.; Plante, O. J.; Overkleeft, H. S.; van der Marel, G. A.; Codée, J. D. C. *Org. Lett.* **2012**, *14*, 3776–3779.

[86] Kandasamy, J.; Schuhmacher, F.; Hahm, H. S.; Klein, J. C.; Seeberger, P. H. *Chem. Commun.* **2014**, *50*, 1875–1877.

[87] de Paz, J.-L.; Angulo, J.; Lassaletta, J.-M.; Nieto, P. M.; Redondo-Horcajo, M.; Lozano, R. M.; Giménez-Gallego, G.; Martín-Lomas, M. *ChemBioChem* **2001**, *2*, 673–685.

[88] Iyer, S. S.; Rele, S. M.; Baskaran, S.; Chaikof, E. L. *Tetrahedron* **2003**, *59*, 631–638.

[89] Rele, S. M.; Iyer, S. S.; Baskaran, S.; Chaikof, E. L. *J. Org. Chem.* **2004**, *69*, 9159–9170.

[90] Noti, C.; de Paz, J. L.; Polito, L.; Seeberger, P. H. *Chem. Eur. J.* **2006**, *12*, 8664–8686.

[91] de Paz, J. L.; Noti, C.; Böhm, F.; Werner, S.; Seeberger, P. H. *Chem. Biol.* **2007**, *14*, 879–887.

[92] Rawat, M.; Gama, C. I.; Matson, J. B.; Hsieh-Wilson, L. C. *J. Am. Chem. Soc.* **2008**, *130*, 2959–2961.

[93] Lee, S.-G.; Brown, J. M.; Rogers, C. J.; Matson, J. B.; Krishnamurthy, C.; Rawat, M.; Hsieh-Wilson, L. C. *Chem. Sci.* **2010**, *1*, 322–325.

[94] Sheng, G. J.; Oh, Y. I.; Chang, S.-K.; Hsieh-Wilson, L. C. *J. Am. Chem. Soc.* **2013**, *135*, 10898–10901.

[95] Oh, Y. I.; Sheng, G. J.; Chang, S.-K.; Hsieh-Wilson, L. C. *Angew. Chem. Int. Ed.* **2013**, *52*, 11796–11799.

11

CHEMICAL GLYCOPROTEIN SYNTHESIS

YASUHIRO KAJIHARA[1], MASUMI MURAKAMI[1],
AND CARLO UNVERZAGT[2]

[1] *Department of Chemistry, Graduate School of Science, Osaka University, Osaka, Japan*
[2] *Bioorganische Chemie, Universität Bayreuth, Bayreuth, Germany*

11.1 INTRODUCTION

Posttranslational modifications of proteins are critical biosynthetic steps, which can enhance or alter protein functions [1]. These extensively studied protein modifications regulate important biological events. Phosphorylation is involved in signal transduction, ubiquitination mediates proteolysis, attachment of fatty acids leads to membrane anchoring, and glycosylation assists in protein folding, trafficking, and protein–protein interactions [1]. Among these modifications, glycosylation not only is the major pattern but also provides a complex system for protein modifications. N-Linked oligosaccharides consist of mannose-rich structures in the endoplasmic reticulum (ER), which are later converted into acidic oligosaccharides in the Golgi apparatus. The composition of these oligosaccharides changes depending on the biosynthetic status of the glycoprotein. Nearly all proteins on the mammalian cell surface as well as in body fluids are glycosylated, and these glycoproteins play central roles in many critical biological events such as cell–cell recognition, immunity, and development [2]. Recent studies have revealed the cross talk between the *N*-acetylglucosamine (GlcNAc) modification of serine (Ser) or threonine (Thr) and O-phosphorylation [3].

Glycochemical Synthesis: Strategies and Applications, First Edition.
Edited by Shang-Cheng Hung and Medel Manuel L. Zulueta.

In order to elucidate the many unknown functions of oligosaccharides in glycoproteins, numerous biological experiments have been carried out using glycoprotein substrates isolated from mammalian or eukaryotic sources. However, those isolated glycoproteins exhibit heterogeneous oligosaccharide structures, which frequently lead to ambiguous conclusions when trying to assign the contribution of an individual oligosaccharide structure to a biological event. This lack of substrates for research and therapy has stimulated chemists to prepare glycoprotein macromolecules exhibiting native three-dimensional protein structures combined with uniform oligosaccharides. This chapter summarizes the recent synthetic methodologies enabling the preparation of glycoproteins bearing homogeneous oligosaccharides.

11.2 OLIGOSACCHARIDE STRUCTURES

Proteins are synthesized in the cytoplasm, while glycoproteins are synthesized in the ER and the Golgi apparatus through complex biosynthetic processes. The oligosaccharides of glycoproteins are classified into two main types, namely, O-linked and N-linked derivatives [4, 5]. In the O-linked type, the alcohol of Ser or Thr is glycosylated with an *N*-acetyl-α-D-galactosamine (α-GalNAc) residue, which is subsequently elongated with additional sugar residues in the Golgi apparatus. O-Linked oligosaccharides are classified into eight subgroups depending on the sugar sequences (Fig. 11.1). In the N-linked glycans, a β-GlcNAc residue at the reducing end of the oligosaccharide is linked to the nitrogen of an asparagine (Asn) side chain. This series of oligosaccharides leads to larger and more branched glycans, which are divided into three main types, namely, complex, hybrid, and high mannose (Fig. 11.2). Complex-type N-glycans give rise to highly diverse structures, bearing di-, tri-, and tetra-branched extensions and terminal α(2→3)- or α(2→6)-linked sialic acid, poly-*N*-acetyllactosamine units, and α(2→8)-linked polysialic acid [6]. The combination of branching patterns and polymeric extensions results in the considerable heterogeneity

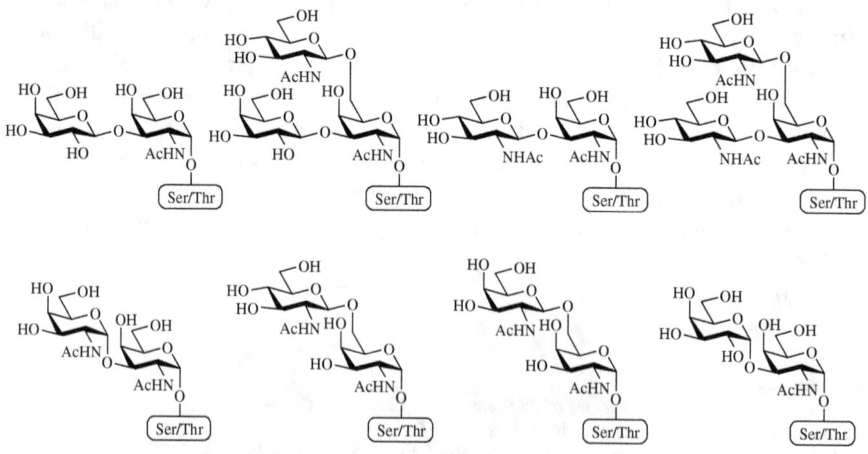

FIGURE 11.1 O-Linked oligosaccharide structures. Ac, acetyl.

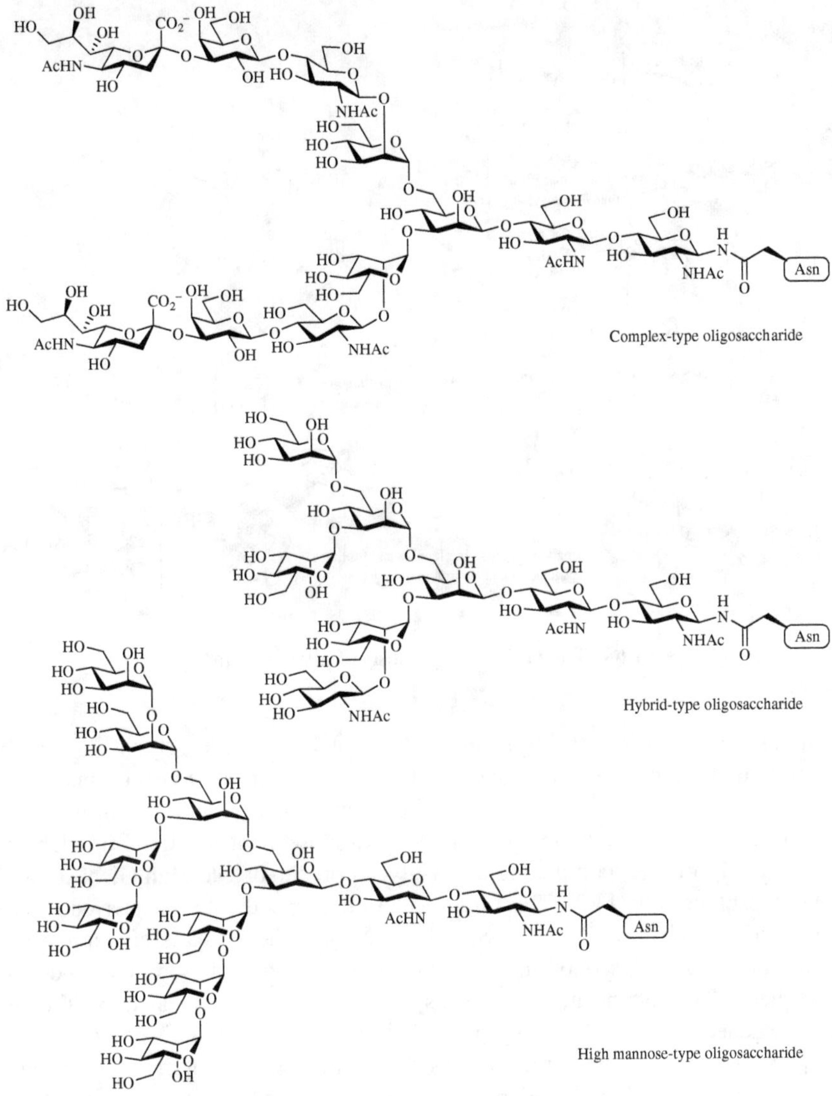

FIGURE 11.2 N-Linked oligosaccharide structures.

of the N-glycan structures. Less frequently, O-mannosides [7], O-GlcNAc [3], and C-mannosides [8] are found.

11.3 BIOSYNTHESIS OF GLYCOPROTEINS

Glycoproteins are synthesized by passing through two organelles, the ER and the Golgi apparatus [4]. The biosynthesis of N-linked oligosaccharides initiates with the cytosolic transfer of GlcNAc from its uridine diphosphate (UDP) form onto dolichol

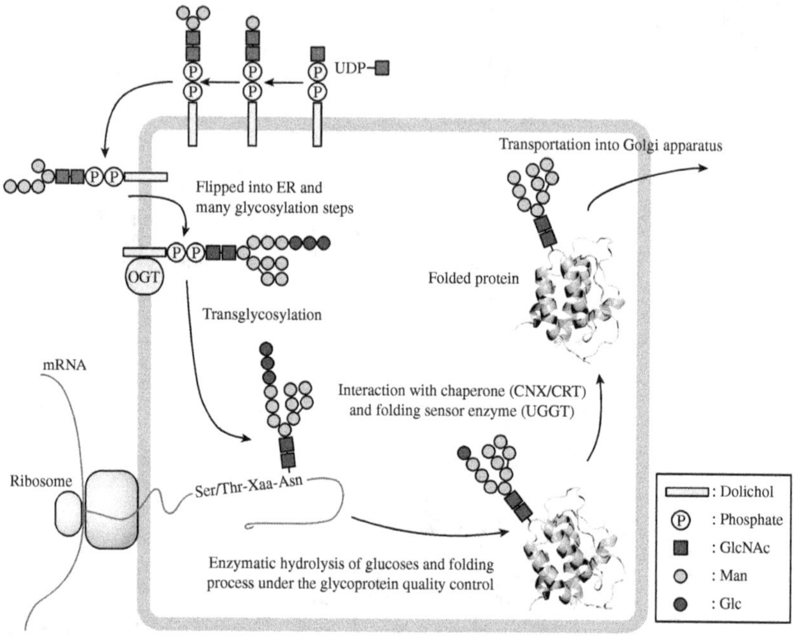

FIGURE 11.3 Biosynthesis of glycoproteins.

phosphate embedded in the lipid bilayer of the ER. Further glycosylations by *N*-acetylglucosaminyltransferase and additions of D-mannose (Man) by mannosyltransferases lead to a $Man_4GlcNAc_2$-dolichol diphosphate. After a flip of the oligosaccharide part to the luminal side of the ER, subsequent additions of Man and D-glucose (Glc) occur, completing the N-glycan precursor $Glc_3Man_9GlcNAc_2$-dolichol diphosphate (Fig. 11.3). This oligosaccharide is specifically transferred en bloc to the Asn side chain of Asn-Xaa-Ser/Thr sequons (where Xaa is any amino acid except for proline (Pro)) of the nascent peptide generated from the ribosome. Subsequently, the cotranslationally glycosylated peptides are elongated to ultimately yield the genetically encoded polypeptide chain. During the synthesis of the glycosylated polypeptide, its folding process starts. The folding process of the glycoprotein is further monitored and regulated by several enzymes and chaperones, providing a "glycoprotein quality control" system [9–20]. The 14-mer sugar attached to the protein is then shortened by glucosidases I and II to yield $Glc_1Man_9GlcNAc_2$, a key structure that can be recognized by the membrane-bound lectin chaperone calnexin (CNX) and its soluble orthologue calreticulin (CRT). The chaperones CNX and CRT assist in the folding of glycoproteins with the aid of the noncovalently associated oxidoreductase ERp57. Subsequently, glucosidase II removes the residual Glc residue to yield $Man_9GlcNAc_2$ (high mannose type; see Fig. 11.2). At this stage, any incompletely folded or misfolded glycoproteins can be reglucosylated by UDP-Glc/glycoprotein glucosyltransferase (UGGT) [14, 18]. UGGT works as the "folding sensor" in the ER quality control. Repetitive cycles of release and rebinding of glycoproteins to the chaperones CNX/CRT are regulated by the opposing actions of glucosidase II and UGGT that catalyze the removal and readdition of the terminal Glc residue, respectively. However, the discrimination of misfolded glycoproteins over the correctly folded glycoproteins by molecular recognition of the sensor proteins UGGT and CNX/CRT is still under investigation.

Properly folded glycoproteins are transported to the Golgi apparatus, while misfolded glycoproteins are translocated to the proteasome for digestion. In the Golgi apparatus, several glycosidases and glycosyltransferases convert the high mannose-type oligosaccharides into complex-type oligosaccharides or hybrid-type N-glycans. This process results in considerable heterogeneity in the structure of the final N-glycans. Thus, it is difficult to elucidate which oligosaccharide structures affect the trafficking, secretion, and bioactivities of glycoproteins in detail. In order to investigate how complex-type oligosaccharides influence protein bioactivity and how high mannose-type oligosaccharides interact with the glycoprotein quality control system, glycoproteins bearing homogeneous oligosaccharides are essential. Many individual oligosaccharide structures, especially the intermediates of high mannose-type oligosaccharide processed in the ER and intermediates of complex-type oligosaccharides in the Golgi apparatus, have a short lifetime in biosynthesis, and thus, conventional expression methodologies do not yield appropriate amounts of glycoproteins bearing many intermediate oligosaccharide structures. Due to these circumstances, access to glycoproteins by chemical synthesis was demonstrated in recent years, which gained further momentum due to the maturation of several key methodologies.

11.4 CHEMICAL PROTEIN SYNTHESIS

Chemical protein synthesis is now a practical approach based on the solid-phase peptide synthesis (SPPS) established by Robert Bruce Merrifield in 1963 [21]. Proteins exhibit complex three-dimensional folds including helices, sheets, and loop structures. Their activities depend on the amino acid sequence and the resulting complex protein conformations. Thus, proteins are fascinating to many scientists. In order to synthesize such large molecules, several routes have been investigated, but the total synthesis of the whole polypeptide chain of a protein with a single SPPS is still very difficult. As an alternative, peptide segment coupling strategies are favored as the major and more robust approach.

11.4.1 Native Chemical Ligation

For peptide segment couplings, several robust conditions are available. Saburo Aimoto and Hironobu Hojo demonstrated the use of partially side chain-protected peptide thioesters that can be activated with AgCl [22]. In 1994, the Stephen Kent group demonstrated native chemical ligation (NCL), which allows the chemoselective coupling of unprotected peptides leading to a native amide bond in solution (Scheme 11.1) [23]. NCL proceeds by reacting a C-terminal peptide thioester with a peptide containing a free cysteine (Cys) residue at the N-terminus via a thioester

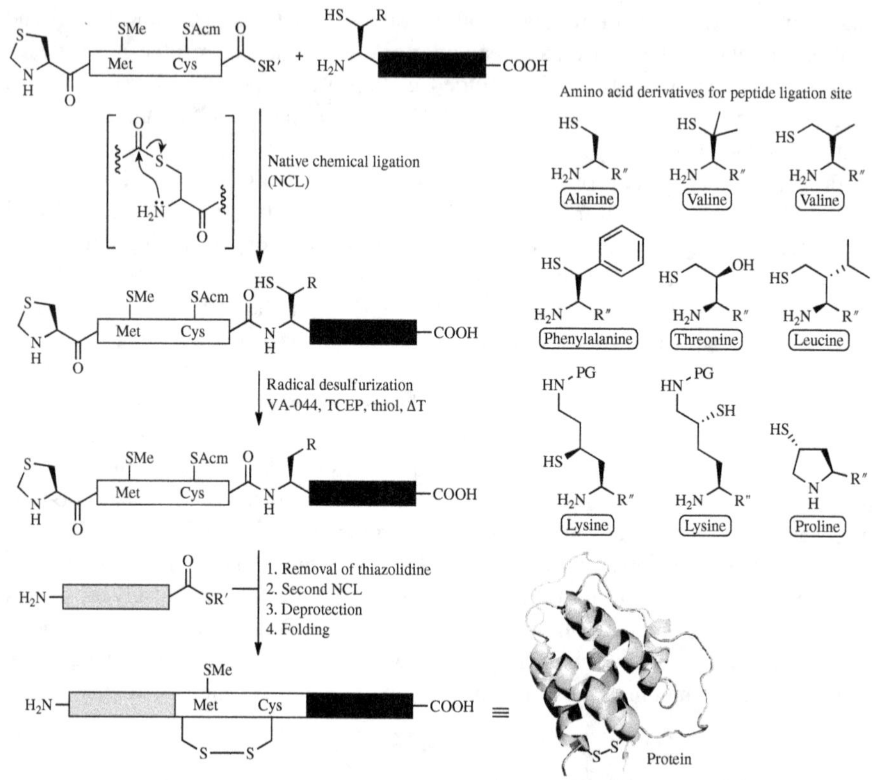

SCHEME 11.1 Native chemical ligation and a concept using amino acid having thiol at the β position. Acm, acetamidomethyl; PG, protecting group; VA-044, 2,2′- azobis[2-(2-imidazolin-2-yl)propane] dihydrochloride.

exchange reaction followed by intramolecular S-to-N acyl shift. The size of the used peptides is not limited and the reaction can yield even longer peptide chains through repetitive coupling. Apart from Cys and thioester peptides, selenocysteine and selenoesters are likewise applicable to NCL [24, 25]. By the use of thioesters, several proteins with molecular weights up to 20 kDa have been synthesized [26, 27]. However, conventional NCL can only be applied at the aa-Cys junction (aa, any amino acid). In order to overcome this general limitation, several additions have been developed.

11.4.2 NCL without the aa-Cys Junction

In order to perform NCL at any desired peptide-coupling junction, alternatives to the Cys side chain were investigated, such as auxiliary groups bearing a thiol [28–31], methionine ligation using homocysteine [32, 33], or sugar-assisted ligation by introducing a sulfhydryl group into a sugar residue of a glycopeptide [34]. Traceless Staudinger ligation couples phosphinothioesters with azides [35]. A very efficient concept is by using a dethiolation reaction after NCL, thereby converting Cys into alanine (Ala) residues (Scheme 11.1). The first dethiolation following this concept was the nickel-mediated reduction of the sulfhydryl group of Cys after conventional NCL, permitting the use of aa-Ala junctions for NCL [36, 37]. In addition to the aa-Ala junction, aa-phenylalanine [38], aa-lysine [39, 40], aa-valine [41], aa-leucine [42], aa-Pro [43], and aa-Thr [44] as the alternative ligation sites are now available based on the corresponding amino acids bearing sulfhydryl or selenol at the β position in most cases. Recently, metal-free reduction conditions (triscarboxyethylphosphine (TCEP)/radical starter/thiol) [45] were found, and these methods have already been widely used for the synthesis of long peptide chains. For selenocysteines, similar reduction conditions can also be applied [46]. With the postligation removal of thiols and selenols back to the parent amino acid residues, the NCL concept can now be applied to the chemical synthesis of nearly any type of protein sequence.

11.5 SYNTHESIS OF GLYCOPEPTIDES

After the first amino acyl sugar was synthesized, the synthesis of glycopeptide was performed in solution phase [47, 48]. The solution synthesis gradually shifted to solid-phase synthesis by using 9-fluorenylmethoxycarbonyl (Fmoc) protection for the α-amino group. In order to carry out glycopeptide synthesis, preparations of the corresponding O-linked-type threonyl- and seryl- and N-linked asparaginyl oligosaccharides are essential. The chemical synthesis of shorter O-linked-type oligosaccharide-Ser and Thr is well established. However, only recently, Jeffrey Gildersleeve and coworkers reported significant epimerization of *O*-glycosyl serines during SPPS [49, 50]. The preparation of the N-linked-type oligosaccharide-Asn derivatives requires more sophisticated synthetic techniques. Thus, the chemical synthesis of peptides bearing large complex-type and high mannose-type N-glycans is still a challenging task. Three useful synthetic strategies to prepare N-glycopeptides are shown in Scheme 11.2. N-Glycans obtained by chemical methods are converted into oligosaccharyl-β-amine derivatives [51] and then applied to convergent couplings leading to glycopeptides [52–54]. This convergent strategy has been improved by several groups and can be performed under solution- and solid-phase conditions (see Section 11.6). During the chemical synthesis of N-glycans, the core trisaccharide at the reducing end of the oligosaccharide requires the formation of the challenging β-mannosidic linkage [55]. Selective chemical elongation of this core scaffold can yield various branched oligosaccharides of the biantennary up to the pentaantennary type. Additions of core fucose, bisecting GlcNAc, and terminal sialic acids are also possible. As a convenient additional method, enzymatic incorporation of sialic acid or galactose with glycosyltransferases [56, 57] provides a facilitated and easily varied access to biorelevant structures.

In 1998, Morten Meldal and coworkers demonstrated the semisynthesis of Asn derivatives of complex-type bi- and triantennary asialo N-glycans [58]. After isolation of these oligosaccharides from the glycoproteins fetuin and ribonuclease (RNase) B by hydrazinolysis, the glycans were converted to Fmoc-Asn(oligosaccharide)-OH derivatives with an acetylated sugar part. In 2003, the Yasuhiro Kajihara group

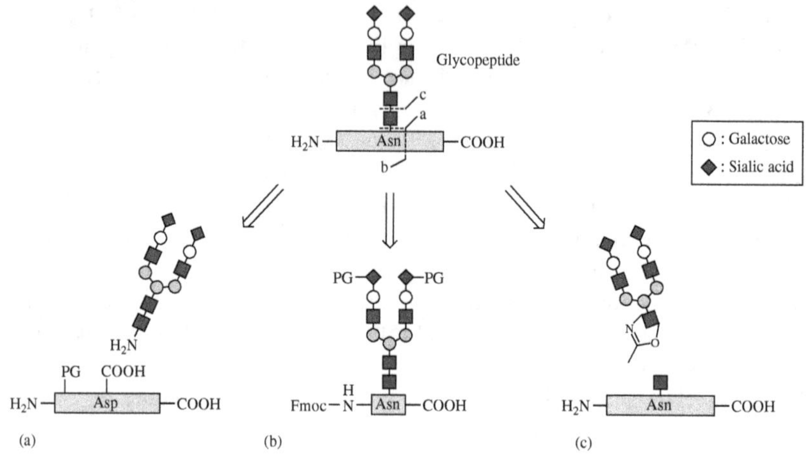

SCHEME 11.2 Synthetic strategies of glycopeptides. (a) Convergent coupling, (b) solid-phase glycopeptide synthesis, and (c) enzymatic coupling. Asp, aspartic acid.

demonstrated the semisynthesis of glycopeptides bearing complex-type disialyloligosaccharides with Fmoc-Asn(disialyloligosaccharide)-OH with unprotected sugar hydroxy groups [59]. A homogeneous asparaginyl complex-type sialyloligosaccharide, which can be isolated from egg yolk in gram scale, was used not only for the synthesis of several sialoglycopeptides by SPPS but also for the preparation of numerous valuable Fmoc-Asn building blocks by specific enzymatic degradation of Fmoc-Asn(disialylundecasaccharide)-OH [60]. Recently, a homogeneous high mannose-type oligosaccharide was also found to be available from egg yolk in practical amounts, which can be used in the synthesis of glycopeptides bearing high mannose-type oligosaccharides [61].

In addition to the chemical approaches, enzymatic transfer reactions can also be applied. An enzyme, which naturally hydrolyzes the chitobiose part of N-glycans, was used by the Kenji Yamamoto group to reversibly transfer the liberated oligosaccharide fragment onto peptidyl-GlcNAc acceptors [62]. Shin-ichiro Shoda et al. found that this enzymatic transfer reaction is accelerated when activated oligosaccharyl oxazolines are used. They subsequently developed a method for the preparation of the labile oligosaccharyl oxazolines [63]. Recently, the groups of Lai-Xi Wang, Benjamin Davis, and Antony Fairbanks demonstrated the enzymatic remodeling of several glycoproteins [64, 65].

11.6 SYNTHESIS OF GLYCOPEPTIDE-$^{\alpha}$THIOESTERS

In order to examine NCL, access to glycopeptide-$^{\alpha}$thioesters is crucial. In the last decade, many efficient thioesterification methods have been reported in the field of peptide chemistry. Most of these methodologies follow the Fmoc strategy using mild basic conditions (piperidine) to deprotect the Fmoc group of the growing chain. Because the thioester linkage between the peptide C-terminus and the solid support is labile to piperidine, C-terminal αthioesterification is best performed after completing the Fmoc-based peptide elongation. The following examples show the recent syntheses of glycopeptide-αthioesters.

11.6.1 Safety-Catch Linker

The Jonathan Ellman group demonstrated that safety-catch linkers yield peptide-αthioesters, and this method has been extensively used for peptides as well as for glycopeptide thioesters [66]. Carolyn Bertozzi and coworkers first examined the synthesis of glycopeptide-αthioesters for glycoprotein construction (Scheme 11.3a) [67]. The coupling of the first Fmoc-protected amino acid to the sulfonamide of the safety-catch linker and subsequent SPPS elongation yield peptides as well as glycopeptides on the safety-catch linker. After assembly of the desired glycopeptide on the safety-catch linker, alkylation of the sulfonamide bond followed by thiolysis smoothly generates the protected glycopeptide-αthioester. The final deprotection step with trifluoroacetic acid yields the glycopeptide-αthioester.

11.6.2 Thioesterification via Activation of C-Terminal Carboxylic Acids

A second straightforward thioesterification method employs simple condensation conditions of the C-terminal carboxylic acid and thiols by using benzotriazol-1-yl-oxytripyrrolidinophosphonium hexafluorophosphate (PyBOP) or other condensation

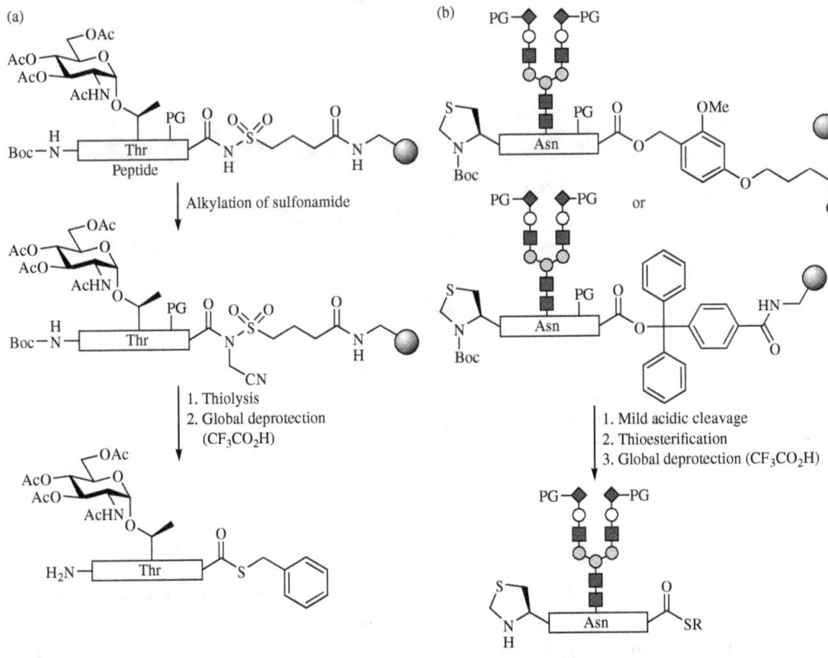

SCHEME 11.3 Synthesis of glycopeptide-αthioesters by (a) using safety-catch linker and (b) thioesterification of activated C-terminal carboxylic acids. Boc, *tert*-butoxycarbonyl.

reagents (Scheme 11.3b) [68, 69]. There are several suitable linkers for SPPS in order to prepare side chain-protected peptides with a C-terminal carboxylic acid. Typical examples employ acid-sensitive linkers such as trityl, chlorotrityl, and 4-(4-hydroxymethyl-3-methoxyphenoxy)butyric acid (HMPB). After the assembly of peptides under Fmoc conditions, mild acidic treatment yields the side chain-protected peptides. Subsequent activation of the C-terminal carboxylic acid with a condensation reagent and a thiol followed by deprotection with a trifluoroacetic acid cocktail affords the corresponding $^{\alpha}$thioester. However, the C-terminal activation easily results in epimerization of the C-terminal amino acid. In order to avoid this undesired epimerization, condensation with PyBOP at low temperature is employed [68]. Because the side chain-protected peptides are prone to interaction with the resin after mild acid treatment, repetitive washings should be performed in order to release the appropriate amount of protected peptide.

11.6.3 Convergent Methods for the Synthesis of Glycopeptide-$^{\alpha}$Thioesters

The convergent approach to N-glycopeptides was first demonstrated by Peter Lansbury et al. (Scheme 11.4) [52, 53]. Condensation of an oligosaccharyl amine with the Asp side chain of a suitably protected peptide was examined in solution phase but can also be carried out in solid phase. However, the drawback of this method is that the Asp unit meant to be coupled to an oligosaccharyl amine is prone to cyclic aspartimide formation with the amide nitrogen atom of the peptide backbone. Samuel Danishefsky and coworkers examined several conditions to overcome

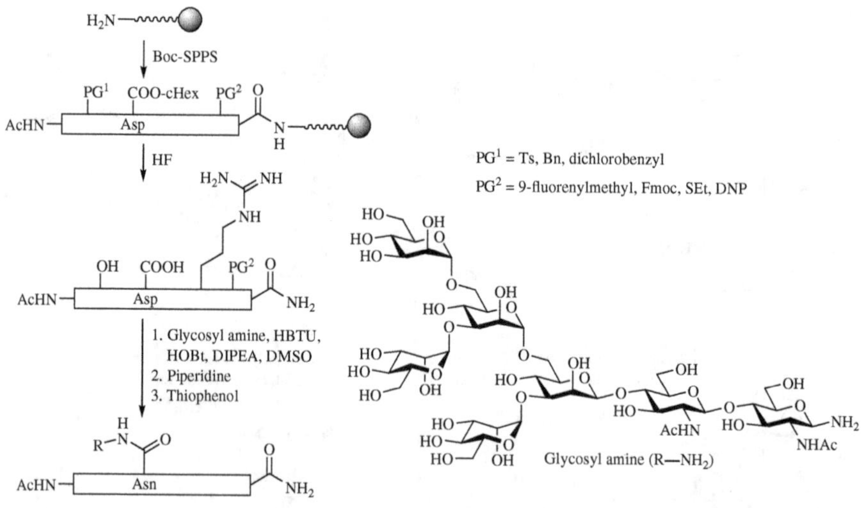

SCHEME 11.4 Convergent concept demonstrated by the Lansbury group. Bn, benzyl; cHex, cyclohexyl; DIPEA, diisopropylethylamine; DMSO, dimethyl sulfoxide; DNP, 2,4-dinitrophenyl; HBTU, 2-(1*H*-benzotriazol-1-yl)-1,1,3,3-tetramethyluronium hexafluorophosphate; HOBt, 1-hydroxybenzotriazole; Ts, tosyl.

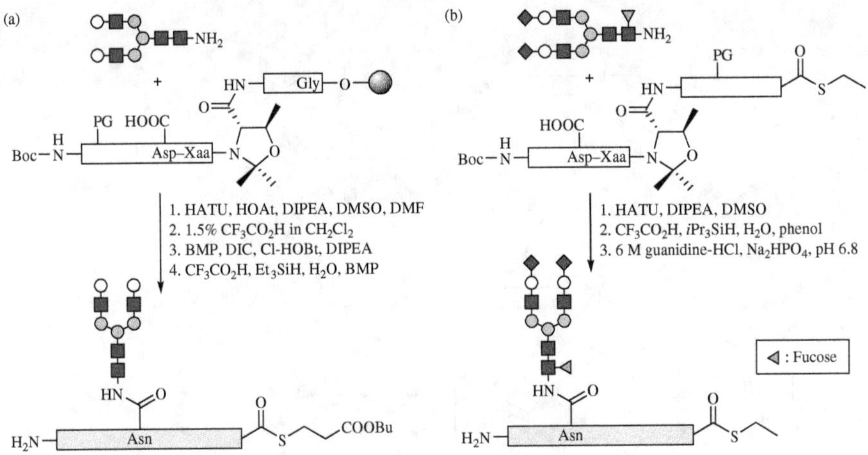

SCHEME 11.5 Efficient concept for convergent method using the pseudoproline scaffold reported by (a) Unverzagt et al. and (b) Danishefsky et al. BMP, butyl-3-mercaptopropionate; DIC, diisopropylcarbodiimide; DMF, *N*,*N*-dimethylformamide; Gly, glycine; HATU, *O*-(7-azabenzotriazol-1-yl)-*N*,*N*,*N*′,*N*′-tetramethyluronium hexafluorophosphate; HOAt, 1-hydroxy-7-azabenzotriazole.

these difficulties [70], and recently, the research groups of Carlo Unverzagt [71] (Scheme 11.5a) and Danishefsky [72] (Scheme 11.5b) independently demonstrated an efficient solution to reduce this side reaction by means of pseudoprolines at the Ser/Thr residues of the consensus Asp-Xaa-Ser/Thr sequence. This method enabled the synthesis of long glycopeptide thioesters of high complexity in solution and on the solid phase. The glycosylamines can be obtained by chemical approaches as well as through semisynthetic methods. Chemical synthesis of N-glycans allows the incorporation of an anomeric azide at an early stage in the synthetic route or as a final step [56]. In the semisynthetic approach, oligosaccharides with a free reducing end can be isolated from egg yolk [73]. The glycosyl amines can be obtained by reduction of the azide or via Kochetkov amination [51]. The oligosaccharyl amine can either be coupled directly to a peptide thioester or to a suitably protected peptide, which is later converted to an $^{\alpha}$thioester.

11.6.4 Thioesterification via O→S Transesterification for the Synthesis of Glycopeptide-$^{\alpha}$Thioesters

In addition to $^{\alpha}$thioesterifications employing the C-terminal condensation with alkylthiols, the Danishefsky group demonstrated an alternative type of $^{\alpha}$thioesterification (Scheme 11.6) [74]. A suitably side chain-protected peptide acid is condensed with an amino acid 2-dithioethyl phenyl ester. The 2-dithioethyl phenyl ester is prone to rearrangement to a thioester through O→S transesterification under reducing (NCL) conditions.

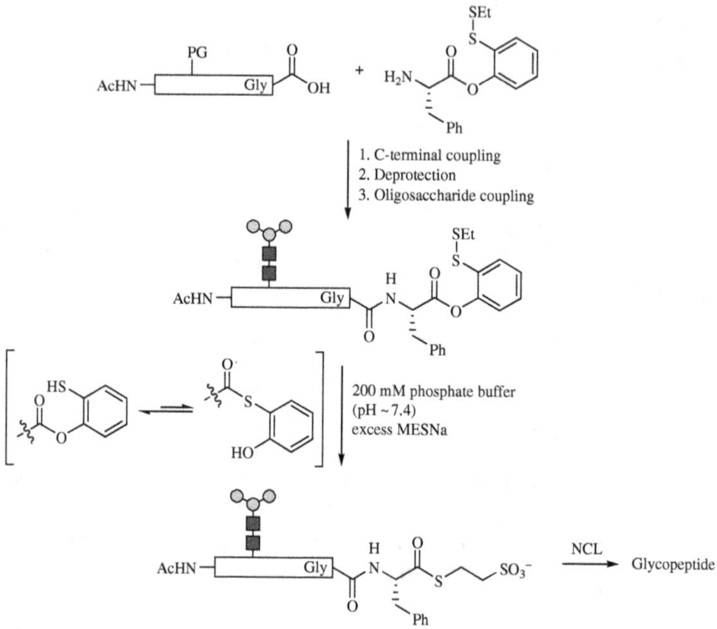

SCHEME 11.6 Glycopeptide-$^{\alpha}$thioesterification via O-to-S acyl shift. MESNa, sodium 2-mercaptoethanesulfonate.

11.6.5 Boc-SPPS for the Synthesis of Sialylglycopeptide-$^{\alpha}$Thioesters

Boc-based SPPS seems to be suitable for the preparation of practical amounts of peptide-$^{\alpha}$thioesters in a straightforward manner [75]. However, this approach requires strong acids such as HF in the final deprotection and the detachment of the peptide from the solid support. Therefore, Boc chemistry was thought to be unsuitable for the synthesis of the acid-labile sialylglycopeptide-$^{\alpha}$thioesters or the fucosylated glycopeptide-$^{\alpha}$thioesters that are important for the synthesis of glycoproteins. The lability of sialosides was thought to be attributed to the 3-deoxy function. However, Brian Capon, finding that acid hydrolysis of *o*-carboxyphenyl β-D-glucoside is 1000 times faster than that of phenyl β-D-glucoside [76], suggested that the carboxylic acid group of sialosides was also acting as an intramolecular acid catalyst facilitating the hydrolysis of the sialyl bond. This analogy was exploited by introducing a phenacyl (Pac) ester at the carboxylic acid function of sialosides [77]. The sialosides protected with a Pac ester were found to be very stable under the acidic conditions used for Boc-SPPS. This protecting group strategy enabled the synthesis of several sialylglycopeptide-$^{\alpha}$thioesters by means of Boc-SPPS *in situ* neutralization conditions, with the final deprotection step employing low acidic conditions (Scheme 11.7). In this approach, the glycopeptide thioesters were attached to the resin via a thiol linker (β-mercaptopropionamide). The sialylglycopeptide-$^{\alpha}$thioesters were not detached from the solid support during the final low acid deprotection, but subsequent thiolysis released the corresponding sialylglycopeptide thioesters. After construction of the sialylglycopeptide-$^{\alpha}$thioesters, repetitive short acid treatment

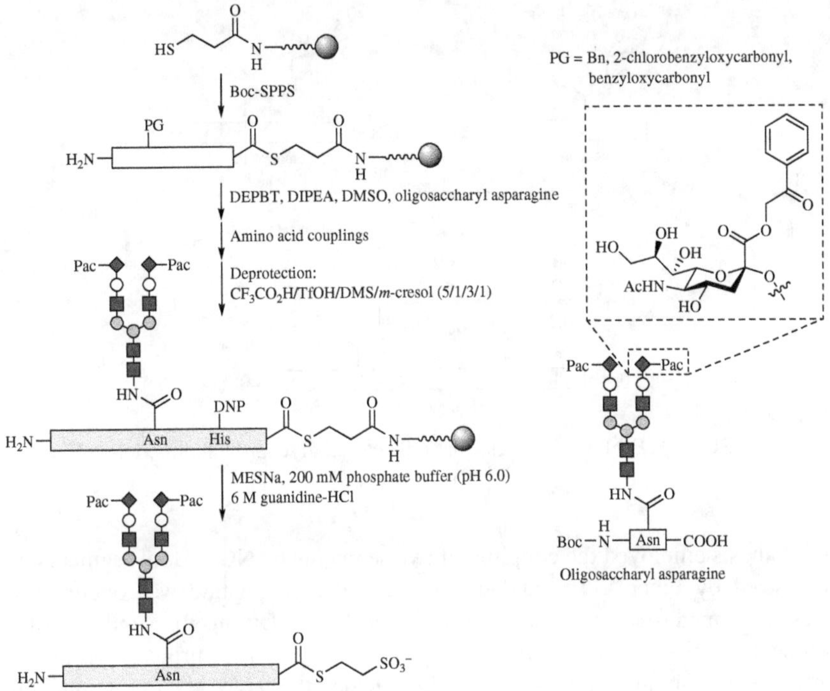

SCHEME 11.7 Glycopeptide-$^{\alpha}$thioesters by Boc-SPPS conditions. DEPBT, 3-(diethoxyphosphoryloxy)-1,2,3-benzotriazin-4(3*H*)-one.

using trifluoroacetic acid/triflic acid (TfOH)/dimethyl sulfide (DMS)/*m*-cresol (5/1/3/1) for 20–40 min and subsequent thiolysis performed in a sodium phosphate buffer (pH 7.0) containing 6 M guanidine–HCl and MESNa yielded the corresponding Pac-protected sialylglycopeptide-$^{\alpha}$thioesters [77]. The remaining Pac group can be removed during the subsequent NCL in the presence of high concentrations of alkylthiol or after NCL using piperidine.

11.7 CHEMICAL SYNTHESIS OF GLYCOPROTEINS

Several syntheses of notable glycoproteins have been reported in the literature, and some examples are briefly described in the sections that follow. The criteria for the following examples were the attainment of a correct fold, bioactivity, and large molecular weight of the target glycoproteins.

11.7.1 Antibacterial Glycoprotein Diptericin Bearing Two O-Linked GalNAc Residues

Using NCL, Bertozzi et al. first demonstrated the total chemical synthesis of the native antibacterial glycoprotein diptericin consisting of two O-linked GalNAc residues at positions Thr10 and Thr54 of an 82-amino acid chain (Scheme 11.8) [67].

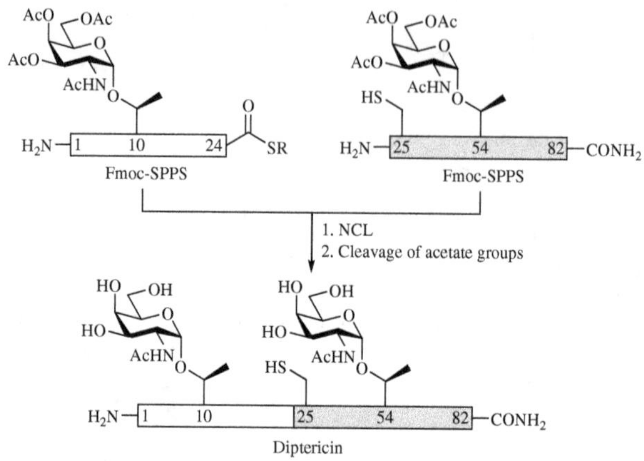

SCHEME 11.8 Synthesis of antibacterial glycoprotein diptericin.

The synthesis employed the coupling of two segments by NCL. Both segments were synthesized by Fmoc-SPPS, and the N-terminal glycopeptide was obtained as a thioester by means of the Ellman safety-catch linker. During the synthesis of this glycopeptide thioester, the hydroxy groups of the sugars were protected as acetates. The segment coupling with NCL smoothly yielded the target glycoprotein chain of diptericin in 55% isolated yield. Removal of the acetyl groups was achieved with 5% aqueous hydrazine containing excess dithiothreitol. The resultant purified glycoprotein exhibited antibacterial activity.

11.7.2 Lymphotactin

Lymphotactin is a glycoprotein consisting of 93 amino acids and eight O-linked GalNAc residues and was obtained by a two-segment coupling strategy (Scheme 11.9) [78]. The glycopeptide segment 48–93 was synthesized by Fmoc-SPPS by using temporary acetyl protection for the sugar part. Ligation with the peptide thioester 1–47 gave the glycoprotein 1–93, which was subjected to oxidative refolding, yielding the biologically active lymphotactin.

11.7.3 Bacterial Immunity Protein Im7

Im7 is a nonglycosylated protein and among the four homologous E colicin immunity proteins. The Barbara Imperiali group demonstrated the synthesis of the glycosylated Im7 variants bearing chitobiose (GlcNAc(β1→4)GlcNAc) linked to Asn residue and the nonglycosylated Im7 domain in order to evaluate the folding process and also the dependence of conformational properties on glycosylation (Scheme 11.10) [79, 80]. The Im7 domain consists of 87 amino acid residues and has four helix structures. The polypeptide chain was divided into two segments, namely, 2–28 and 29–87. The N-terminal peptide segment was synthesized as a glycopeptide thioester.

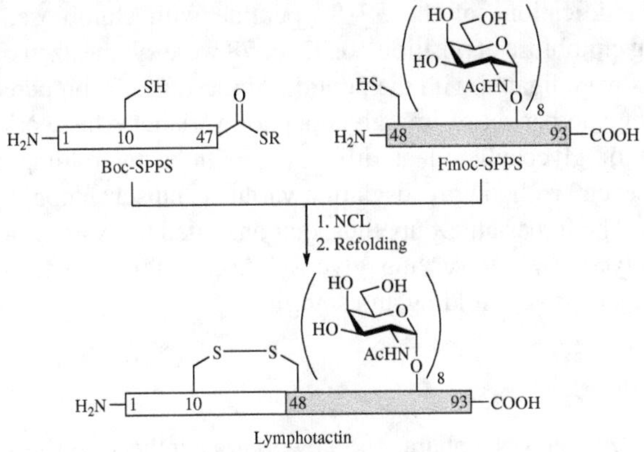

SCHEME 11.9 Synthesis of lymphotactin.

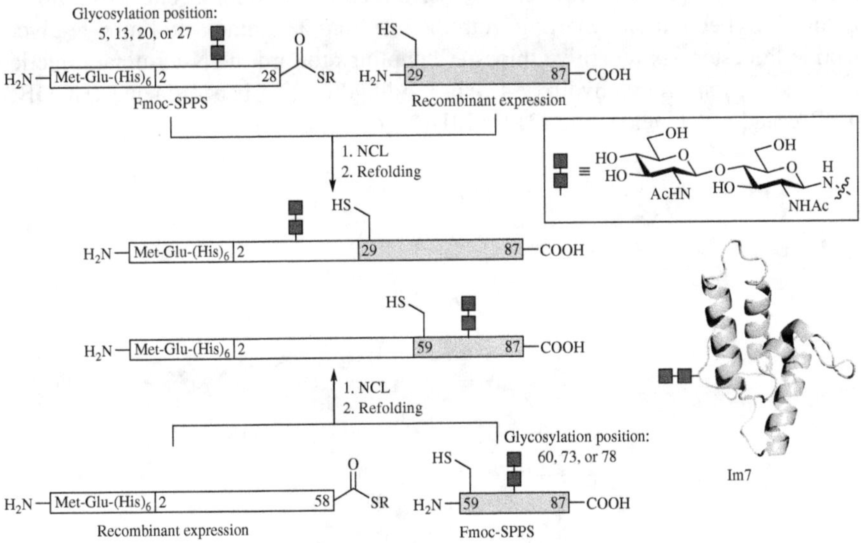

SCHEME 11.10 Synthesis of the Im7 domain of E colicin immunity protein. Glu, glutamic acid; His, histidine; Met, methionine.

Fmoc-SPPS on a trityl resin gave, after cleavage, the side chain-protected peptide acid, which was converted into a thioester by using PyBOP, DIPEA, and BnSH. During Fmoc-SPPS, one chitobiose-Asn was incorporated at position 5, 13, 20, or 27. The C-terminal peptide (29–87) bearing a Cys residue at position 29 was prepared from *Escherichia coli* by recombinant expression using the commercially available IMPACT system. The pH-dependent intein cleavage yields target peptides bearing a Cys at the N-terminus. NCL between the N-terminal glycopeptide-$^{\alpha}$thioester and Cys-peptide (29–87) successfully yielded the glycosylpolypeptide variants. For the selective modifications of the 59–87 peptide with chitobiose, glycopeptides (59–87) bearing chitobiose at position 60, 73, or 78 were synthesized by Fmoc-SPPS. These segments were ligated with the peptide-$^{\alpha}$thioester 2–58 prepared by the intein method, yielding another set of Im7 glycopeptide variants. This work further evaluated the effect of glycosylation at different positions by folding and unfolding kinetics using seven synthetic glycosylation variants and seven nonglycosylated reference proteins. The Imperiali group, thus, demonstrated by systematic synthesis and evaluation of glycoforms at varying glycosylation positions that N-glycosylation significantly affects protein folding and stability.

11.7.4 MUC-2

Hironobu Hojo, Yoshiaki Nakahara, and coworkers synthesized the 23 kDa MUC-2 glycoprotein (Scheme 11.11) [26]. A glycopeptide-$^{\alpha}$thioester bearing seven O-linked GalNAc residues was prepared by Fmoc-SPPS. Side chain deprotection and detachment of the glycopeptide thioester were performed with trifluoroacetic acid followed by low TfOH conditions. Starting from the resulting 23 amino acid Fmoc-N-glycopeptide thioester, six repetitive thioester couplings toward the N-terminal direction under activation with 3-hydroxy-4-oxo-3,4-dihydro-1,2,3-benzotriazine (HOOBt), DIPEA, and AgCl yielded the 23 kDa MUC-2 glycoprotein.

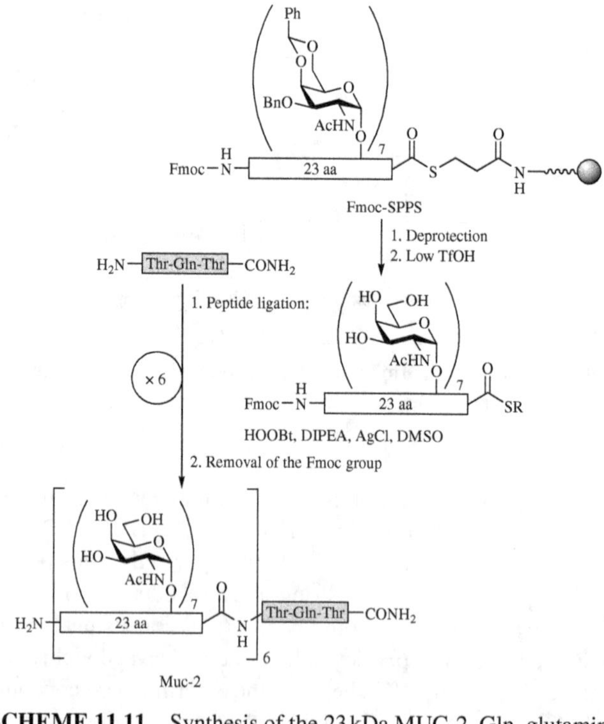

SCHEME 11.11 Synthesis of the 23 kDa MUC-2. Gln, glutamine.

11.7.5 GlyCAM-1

Derek Macmillan and Carolyn Bertozzi demonstrated the semisynthesis of three glycoforms of the O-linked glycoprotein GlyCAM-1 consisting of 132 amino acid residues (Scheme 11.12) [81]. The first glycoform contained O-linked GalNAc residues only in the N-terminal segment 1–40 of GlyCAM-1. Therefore, the corresponding N-terminal glycopeptide 1–40 was prepared by using Fmoc-SPPS, and the complementary Cys-polypeptide 41–132 was prepared as a peptide-$^{\alpha}$thioester by an intein expression system in *E. coli*. The N-terminal Cys residue of the expressed thioester was masked by a cleavable Ile-Glu-Gly-Arg tetrapeptide tag. This tetrapeptide extension prevented the reaction of the Cys residue with the intermediate peptide-$^{\alpha}$thioester formed during cleavage from the initial fusion protein. The Ile-Glu-Gly-Arg tetrapeptide was later cleaved by factor Xa. The resultant Cys site was used in NCL with the glycopeptide thioester 1–40 to afford GlyCAM-1 (15,550 Da) bearing seven GalNAc residues (Scheme 11.12a). The second glycoform was based on the same amino acid sequence, but with O-linked GalNAc residues only in the peptide segment 78–101, thus requiring the separate preparation of three segments (Scheme 11.12b). NCL was performed to link the Cys-glycopeptide 78–101 with the hydrophobic C-terminal segment 102–132, which was generated by using

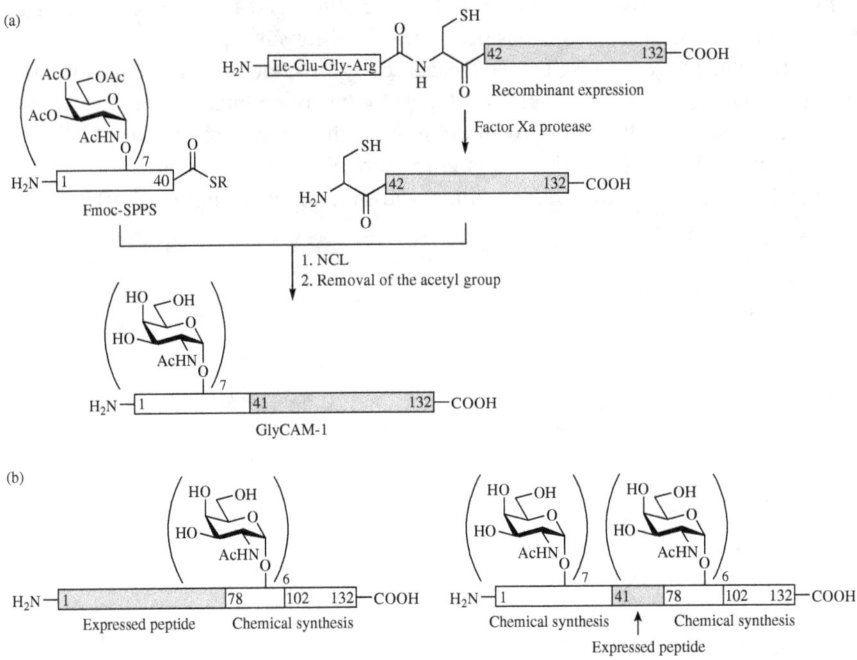

SCHEME 11.12 Semisynthesis of O-linked GlyCAM derivatives. (a) Preparation of a glycoform containing O-linked GalNAc residues only in the N-terminal segment. (b) Other assembled glycoforms with GalNAc residues at various positions. Arg, arginine; Ile, isoleucine.

Fmoc-SPPS. The acquired glycopolypeptide was then coupled to the bacteria-derived N-terminal region 1–77 in another round of NCL. The third variant contained O-linked GalNAc residues in the N- and C-terminal regions and required a four-segment coupling strategy that is a hybrid of the techniques used for the first two glycoforms. The central peptide thioester 41–77 was prepared recombinantly by an intein system.

11.7.6 Monocyte Chemotactic Protein-3

Monocyte chemotactic protein-3 (MCP-3), consisting of 76 amino acid residues, was synthesized by Kajihara et al. as the first native glycoprotein bearing a complex-type sialyloligosaccharide (Scheme 11.13) [82]. MCP-3 contains an N-terminal pyroglutamate (Glp) and an N-glycan at position 6. In addition, two disulfide bonds are present, forming the linkages Cys11–Cys36 and Cys12–Cys52. The chemical synthesis employed a three-segment coupling strategy. For the synthesis of the glycopeptide thioester 1–10, a complex-type biantennary sialyloligosaccharide was prepared from natural sources and converted into Fmoc-Asn(sialyloligosaccharide)-OH with unprotected hydroxy groups and sialic acids protected as Bn esters. Both the Boc and the Fmoc strategies successfully yielded the dibenzylated sialylglycopeptide-$^{\alpha}$thioester. Special Boc conditions were applied, where peptide elongation was performed by using Boc-amino acids with unprotected side chains. The final detachment of the peptide thioester from the resin was performed by thiolysis.

The first and the second NCL gave the desired glycosylpolypeptide and subsequent oxidative folding yielded MCP-3. The glycosylpolypeptide thus obtained was treated with 50 mM NaOH solution for 10 min to remove the Bn groups from the sialic acid residues to afford MCP-3, which was characterized by mass spectrometry, HPLC, circular dichroism (CD), disulfide mapping, and enzyme-linked immunosorbent assay.

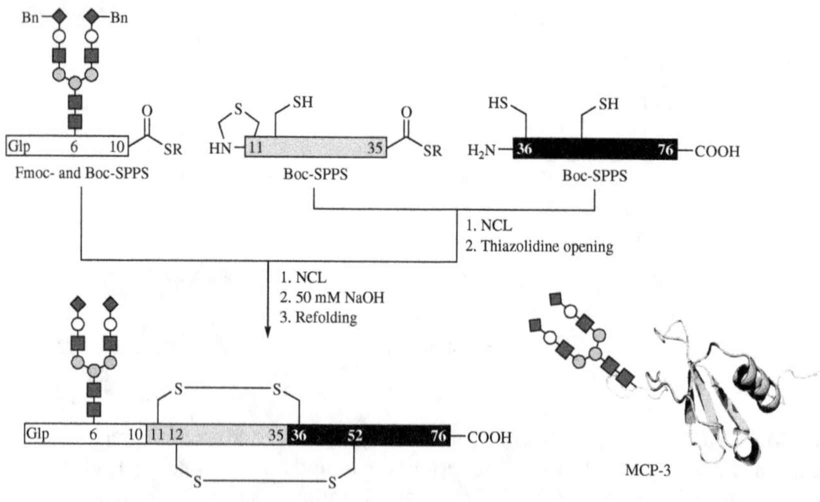

SCHEME 11.13 Chemical synthesis of the glycosylated MCP-3.

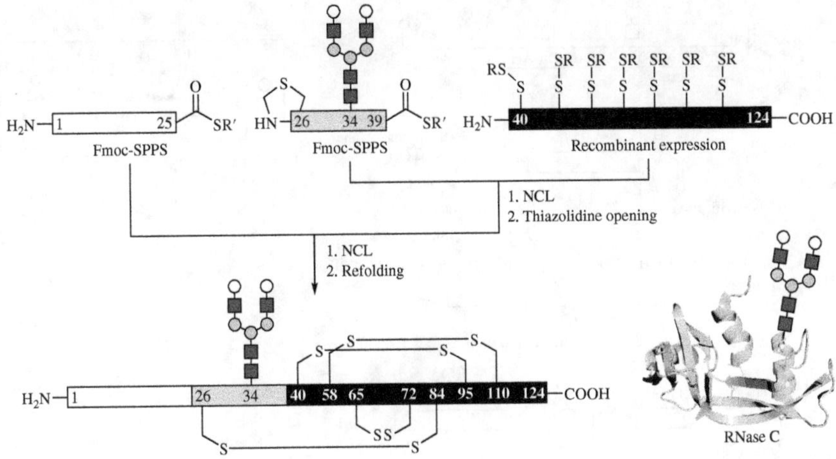

SCHEME 11.14 Semisynthesis of the glycosylated RNase.

11.7.7 Ribonuclease

Glycosylated RNase is the first synthetic glycoprotein enzyme obtained by NCL [83, 84]. The enzyme has 124 amino acid residues and contains eight Cys residues that form four disulfide bridges. The RNase C glycoform displaying a complex-type N-glycan was assembled by semisynthesis. A sequential ligation strategy was chosen by the Unverzagt group using recombinantly expressed fragments and synthetic thioesters bearing an N-glycan from natural sources (Scheme 11.14). The recombinant peptide 40–124 containing seven Cys residues could be obtained as a stable derivative containing mixed disulfides through the cleavage of an intein. Only after disulfide protection could the sensitive fragment be isolated and elongated by two consecutive ligations. The full-length glycoprotein was obtained by oxidative refolding of the crude mixture from the second ligation reaction. After purification by gel filtration, the synthetic RNase C was characterized by CD spectroscopy and enzymatic activity. In order to reduce the number of ligations, new synthetic methods for the glycosylated RNase 1–39 thioesters were developed.

11.7.8 Antifreeze Glycoproteins

Antifreeze glycoproteins are essential for fish in deep-sea Arctic and Antarctic waters. These glycoproteins prevent the growth of ice crystals, and this intriguing mechanism has been extensively investigated by chemical approaches. Shin-ichiro Nishimura and coworkers demonstrated a polymerization approach to obtain analogues of antifreeze glycoproteins and studied their inhibiting effect toward the growth of ice [85]. In 2012, the Richard Payne group synthesized glycosylated antifreeze glycopeptides of different lengths (Scheme 11.15) in order to evaluate the inhibiting ability on the growth of ice [86]. The synthetic disaccharide-modified amino acid (Gal(β1→3)GalNAc(α1→O))Thr was incorporated fourfold in a

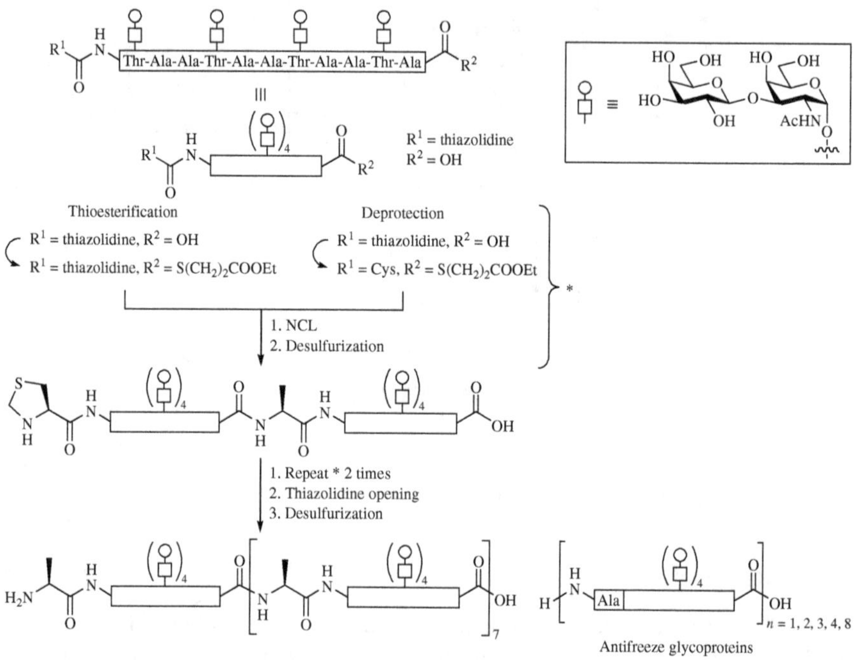

SCHEME 11.15 Synthesis of antifreeze glycoproteins.

12-amino acid consensus peptide by Fmoc-SPPS. C-terminal thioesterification was performed with PyBOP and an alkylthiol at −20°C. Repetitive NCLs and subsequent desulfurization successfully yielded glycopeptides ranging from a 12-amino acid peptide bearing four Gal(β1→3)GalNAc to a 96-amino acid peptide bearing 32 disaccharides. Out of these glycopeptides, the longest derivative exhibited potent thermal hysteresis.

11.7.9 Interleukin-8

Natural interleukin-8 (IL-8) is a nonglycosylated cytokine but was intentionally synthesized as a glycoprotein with a $Man_9GlcNAc_2$ modification, which is recognized by the glycoprotein quality control system in the ER. The Yukishige Ito and Yasuhiro Kajihara collaborative group synthesized the N-glycosylated IL-8 as a probe for misfolded proteins as well as correctly folded glycoproteins based on the well-known IL-8 structure (Scheme 11.16) [87]. The Asn(M9 high mannose-type oligosaccharide) building block was isolated from egg yolk [61] and incorporated by Fmoc chemistry close to the N-terminus of the 39-mer fragment. NCL between the 1–33 peptide thioester prepared by Boc chemistry and glycopeptide 34–76 bearing the $Man_9GlcNAc_2$ glycan followed by oxidative folding successfully yielded the correctly folded glycosylated IL-8, whereas three misfolded IL-8 glycoproteins were obtained under disulfide shuffle conditions. These model glycoproteins were characterized by disulfide mapping, CD spectroscopy, 1-anilinonaphthalene-8-sulfonic

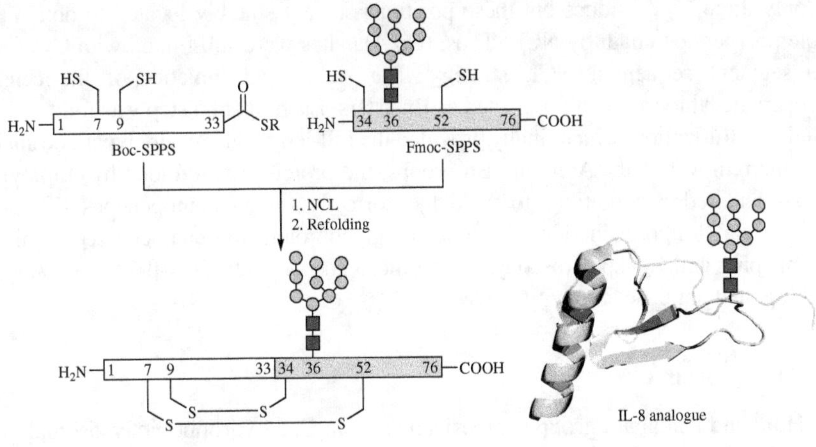

SCHEME 11.16 Synthesis of an IL-8 analogue having high mannose-type oligosaccharide.

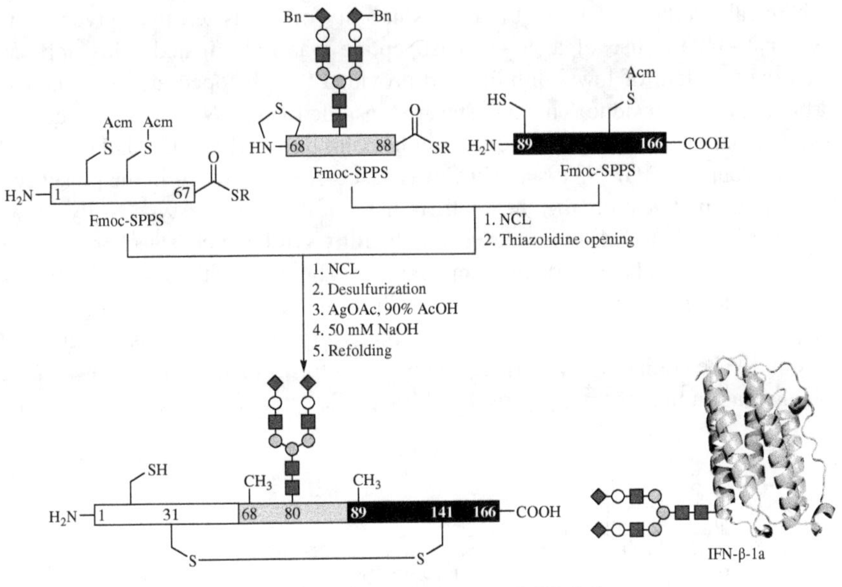

SCHEME 11.17 Synthesis of IFN-β-1a.

acid binding, and a functional assay by using the ER glucosyltransferase UGGT. This assay revealed that UGGT recognized the hydrophobic part of protein surfaces as a characteristic feature of misfolded proteins.

11.7.10 Interferon-β-1a

The glycoprotein interferon-β-1a (IFN-β-1a), an essential biotherapeutic for the treatment of multiple sclerosis, was synthesized by Kajihara and coworkers (Scheme 11.17) [88]. IFN-β-1a consists of 166 amino acids and has one N-glycosylation site. IFN-β-1a has only three Cys residues but these positions are not suitably located to obtain the whole polypeptide chain by NCL. Thus, two Ala sites were substituted with Cys for a three-segment sequential NCL strategy. The sialic acid moieties of the central glycopeptide-$^{\alpha}$thioester were protected as Bn esters. Each ligation step was followed by a mild desulfurization, which finally provided the full-length glycosylpolypeptide chain. After the removal of the Acm and Bn groups, the protein was refolded by employing dialysis and redox conditions to yield the correctly folded homogeneous IFN-β-1a bearing a sialyloligosaccharide. The synthetic glycoprotein was characterized by disulfide mapping and CD spectroscopy. Furthermore, the synthetic IFN-β-1a was found to exhibit potent anticancer activity *in vivo*.

11.7.11 Saposin C

The Hojo and Nakahara group synthesized saposin C, a hydrophobic lysosomal glycoprotein (Scheme 11.18) [89]. Saposin C consists of 80 amino acids and 6 Cys residues. The complex-type octasaccharide oxazoline used in this work was prepared by chemical synthesis. Due to the poor solubility of the N-terminal glycopeptide thioester 1–34, the use of an *O*-acyl isopeptide was investigated. This approach solved the problem of low solubility and provided the glycopeptide segment 1–34 bearing a GlcNAc residue. The thioester was installed via an *N*-ethyl-Cys precursor, which underwent N→S acyl shift under acidic conditions. After NCL and removal of the Acm groups, the GlcNAc-saposin C was successfully refolded in the presence of DMSO. After purification, the enzymatic transfer of the complex-type N-glycan was performed in buffer–DMSO mixture using the fully synthetic octasaccharide oxazoline and a glycosynthase derived from Endo-M. The enzymatic synthesis successfully yielded the glycosylated saposin C. When added to a lysosomal glucosidase assay, the biological activity of the differentially glycosylated forms of saposin C was found to be similar. However, the glycosylated forms of saposin C showed much less aggregation than the nonglycosylated lipophilic protein.

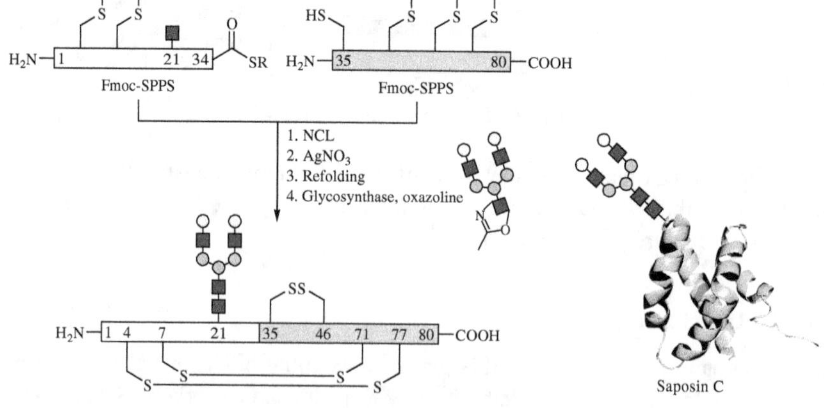

SCHEME 11.18 Synthesis of saposin C.

11.7.12 Erythropoietin

Erythropoietin (EPO) is the most important recombinant glycoprotein for the treatment of anemia. The Kajihara and Macmillan group demonstrated the synthesis of EPO analogues having two or three complex-type sialyloligosaccharides (Scheme 11.19) [90, 91]. EPO consists of 166 or 165 amino acid residues and has three N-linked complex-type oligosaccharides attached to Asn24, Asn38, and Asn83 and one O-linked oligosaccharide at Ser126. A synthetic strategy was designed to prepare a glycopeptide thioester bearing complex-type oligosaccharides attached through an analogous *N*-thioacetamide linkage. The synthetic glycopeptide 1–32 thioester having two complex-type sialyloligosaccharides was obtained by haloacetamide method. The ligation partner 33–166 was prepared by expression in *E. coli* followed by a BrCN cleavage. In the first synthetic EPO model, two analogous glycosylation sites at the natural position 24 and the unnatural position 30 were selected. The second model was prepared with three glycosylation sites at positions 24, 28, and 30. EPO contains two disulfide bonds (Cys7–Cys161 and Cys29–Cys33). The peptide-$^{\alpha}$thioester backbone was prepared by Fmoc-SPPS on an HMPB-poly-(ethylene glycol)–poly(dimethylacrylamide) copolymer resin (as shown Scheme 11.3b), and

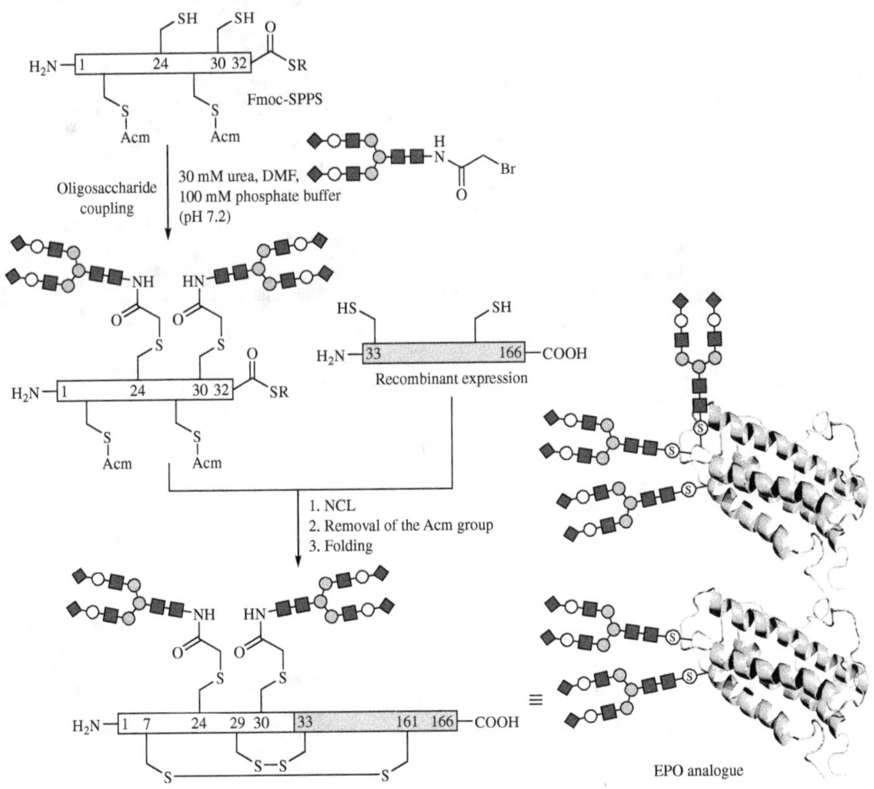

SCHEME 11.19 Semisynthesis of EPO analogues.

the thiol groups of Cys7 and Cys29 were protected by Acm groups. On the other hand, the polypeptide chain corresponding to residues 33–166 was expressed as His_6-tagged fusion protein where the His-tag was linked to the target peptide (33–166) through a Met residue. After expression and purification over a nickel–chelate column, the subsequent BrCN treatment cleaved the protein at the Met residue, yielding a homogeneous polypeptide chain with a Cys residue at the N-terminus. In this case, the native Met residues in the peptide sequence were substituted with leucine. NCL between the glycopeptide-$^{\alpha}$thioesters 1–32 and the 33–166 peptide prepared by expression in *E. coli* and subsequent refolding successfully yielded the biologically active EPO.

The function of the N-linked oligosaccharides of EPO has been elucidated, revealing that glycosylation at position 83 seems to exhibit a crucial role for the bioactivity of EPO. Chemical synthesis of one of the EPO glycoforms bearing a biantennary complex-type sialyloligosaccharide at position 83 was performed by the Kajihara group (Scheme 11.20) [77]. The native EPO sequence contains only four Cys residues, but the positions are placed too far apart in order obtain the full-length EPO polypeptide chain by well-accessible fragments. Therefore, the synthesis employed a metal-free desulfurization protocol that can convert Cys residues into Ala after NCL. Using this approach, a six-segment coupling strategy was planned

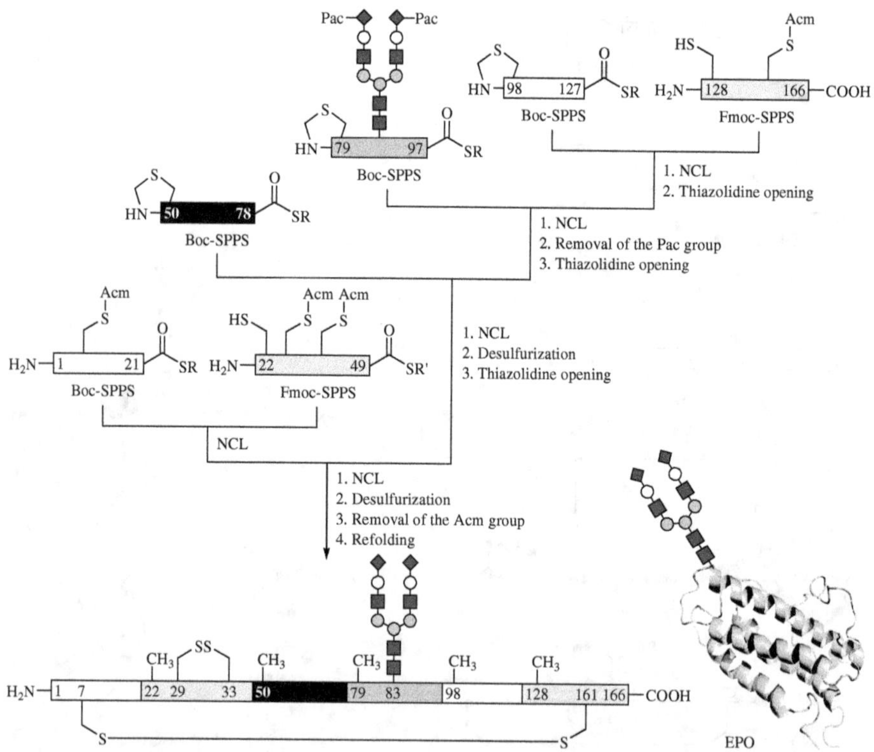

SCHEME 11.20 Chemical synthesis of EPO having a complex-type sialyloligosaccharide.

(mutation sites: Glu21Ala and Gln78Ala) anticipating a conversion of the Cys residues at the 22, 50, 79, 98, and 128 positions into Ala after NCL. Each segment was prepared by improved Boc conditions or Fmoc conditions (second and sixth segments). Sequential NCL reactions and desulfurizations successfully yielded the glycosylpolypeptide chain of EPO. The Pac groups required for the protection of the sialic acids were conveniently removed with piperidine after ligation of the glycopeptide segment. Final deprotection of the Acm groups with AgOAc and refolding under combined dialysis and redox conditions yielded the suitably folded EPO having a sialyloligosaccharide at position 83.

The Danishefsky group achieved the total chemical synthesis of an EPO glycoform with native amino acid sequence (Scheme 11.21) [92]. In their approach, the glycosylpolypeptide chain was divided into four segments (1–29, 30–78, 79–124, and 125–166), and chitobiosyl disaccharides were incorporated at the side chains of Asn24, Asn38, and Asn83. Furthermore, a chemically synthesized O-linked disialyltetrasaccharide was incorporated at the native Ser126 site. All four segments were designed for coupling by NCL and subsequent metal-free desulfurization intended to convert Cys to Ala. For the synthesis of the glycopeptide thioesters, Fmoc-based peptide elongation was employed. Selectively deprotected peptide-$^{\alpha}$thioesters were

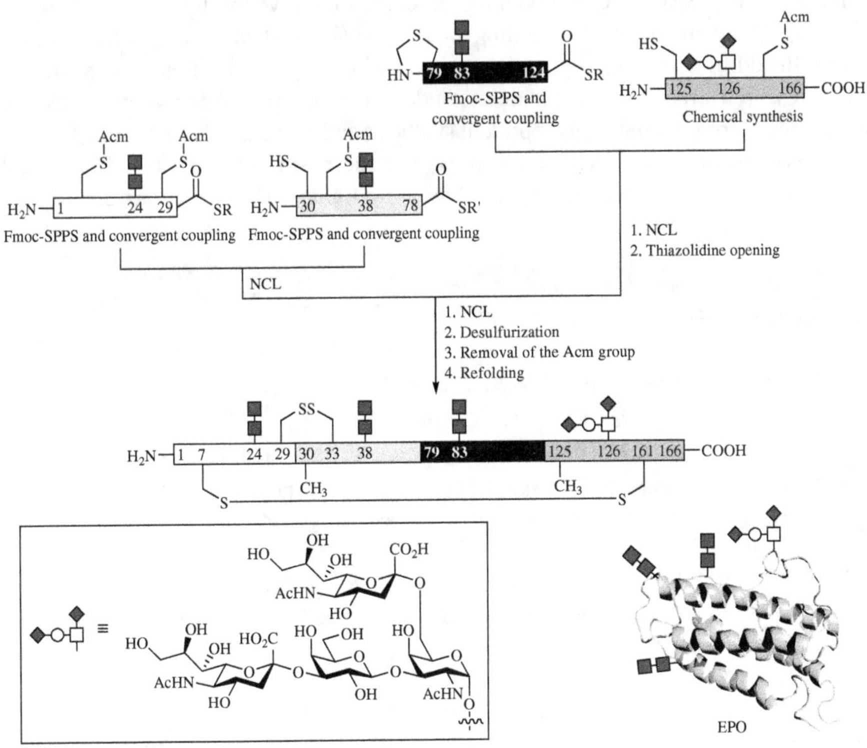

SCHEME 11.21 Synthesis of EPO bearing three N-linked chitobioses and an O-linked tetrasaccharide.

used as precursors for the N-glycosylated fragments. The chitobiose was introduced subsequently by the sophisticated pseudoproline method as described in Section 11.6.3 (Scheme 11.5b). All NCL-based couplings proceeded successfully, and the resultant glycosyl EPO full-length chain was subjected to refolding. Although typical folding conditions for EPO favor the guanidinium dialysis methods, in this case, a copper activation method was employed. This condition afforded the desired folded EPO glycoform bearing three N-linked chitobiosides and the O-linked disialyltetrasaccharide. When tested in a cell proliferation activity, moderate *in vitro* activity of the synthetic EPO was found. The same group later reported an improved synthetic strategy using five segments, providing the EPO glycoform with fucosylated biantennary N-glycans instead of the chitobiosides [93].

11.8 CONCLUSION

Chemical glycoprotein synthesis started with the synthesis of short glycopeptides. However, larger glycopeptides and glycoproteins became available only after the development of numerous advanced methodologies thriving on NCL and desulfurization concepts. Based on the expression of glycoproteins in yeast, *E. coli*, and other cell lines, the enzymatic remodeling of expressed glycoproteins has advanced [94–96]. These methods are suitable for large-scale production of pharmaceutical drugs. In addition to these applications, chemical glycoprotein synthesis can address basic research areas aiming at understanding the functions of glycoproteins. As mentioned before, chemical glycoprotein synthesis can conveniently vary glycosylation positions and oligosaccharide structures. The increasing momentum of chemical glycoprotein synthesis will allow shedding more light on the particular functions of oligosaccharides on proteins.

REFERENCES

[1] Walsh, G.; Jefferis, R. *Nat. Biotechnol.* **2006**, *24*, 1241–1252.
[2] Dwek, R. A. *Chem. Rev.* **1996**, *96*, 683–720.
[3] Rotty, J. D.; Hart, G. W.; Coulombe, P. A. *Nat. Cell Biol.* **2010**, *12*, 847–859.
[4] Varki, A.; Cummings, R. D.; Esko, J. D.; Freeze, H. H.; Stanley, P.; Bertozzi, C. R.; Hart, G. W.; Etzler, M. E., Eds. *Essentials of Glycobiology*, 2nd ed.; Cold Spring Harbor Laboratory: Cold Spring Harbor, NY, **2009**.
[5] Alley, W. R.; Mann, B. F.; Novotny, M. V. *Chem. Rev.* **2013**, *113*, 2668–2732.
[6] Egan, W.; Liu, T.-Y.; Dorow, D.; Cohen, J. S.; Robbins, J. D.; Gotschlich, E. C.; Robbins, J. B. *Biochemistry* **1977**, *16*, 3687–3692.
[7] Wells, L. *J. Biol. Chem.* **2013**, *288*, 6930–6935.
[8] Manabe, S.; Ito, Y. *J. Am. Chem. Soc.* **1999**, *121*, 9754–9755.
[9] Aebi, M.; Bernasconi, R.; Clerc, S.; Molinari, M. *Trends Biochem. Sci.* **2010**, *35*, 74–82.
[10] Baenziger, J. U. *FASEB J.* **1994**, *8*, 1019–1025.

[11] Cannon, K. S.; Helenius, A. *J. Biol. Chem.* **1999**, *274*, 7537–7544.

[12] Castro, O.; Chen, L. Y.; Parodi, A. J.; Abeijón, C. *Mol. Biol. Cell* **1999**, *10*, 1019–1030.

[13] Ceriotti, A.; Duranti, M.; Bollini, R. *J. Exp. Bot.* **1998**, *49*, 1091–1103.

[14] D'Alessio, C.; Caramelo, J. J.; Parodi, A. J. *Semin. Cell Dev. Biol.* **2010**, *21*, 491–499.

[15] Trombetta, E. S.; Parodi, A. J. *Annu. Rev. Cell Dev. Biol.* **2003**, *19*, 649–676.

[16] Ellgaard, L.; Molinari, M.; Helenius, A. *Science* **1999**, *286*, 1882–1888.

[17] Ellgaard, L.; Helenius, A. *Nat. Rev. Mol. Cell Biol.* **2003**, *4*, 181–191.

[18] Fanchiotti, S.; Fernández, F.; D'Alessio, C.; Parodi, A. J. *J. Cell Biol.* **1998**, *143*, 625–635.

[19] Fernández, F.; D'Alessio, C.; Fanchiotti, S.; Parodi, A. J. *EMBO J.* **1998**, *17*, 5877–5886.

[20] Ito, Y.; Takeda, Y. *Proc. Jpn. Acad. Ser. B Phys. Biol. Sci.* **2012**, *88*, 31–40.

[21] Merrifield, R. B. *J. Am. Chem. Soc.* **1963**, *85*, 2149–2154.

[22] Hojo, H.; Aimoto, S. *Bull. Chem. Soc. Jpn.* **1991**, *64*, 111–117.

[23] Dawson, P. E.; Muir, T. W.; Clark-Lewis, I.; Kent, S. B. *Science* **1994**, *266*, 776–779.

[24] Durek, T.; Alewood, P. F. *Angew. Chem. Int. Ed.* **2011**, *50*, 12042–12045.

[25] Quaderer, R.; Sewing, A.; Hilvert, D. *Helv. Chim. Acta* **2001**, *84*, 1197–1206.

[26] Hojo, H.; Matsumoto, Y.; Nakahara, Y.; Ito, E.; Suzuki, Y.; Suzuki, M.; Suzuki, A.; Nakahara, Y. *J. Am. Chem. Soc.* **2005**, *127*, 13720–13725.

[27] Kumar, K. S. A.; Bavikar, S. N.; Spasser, L.; Moyal, T.; Ohayon, S.; Brik, A. *Angew. Chem. Int. Ed.* **2011**, *50*, 6137–6141.

[28] Botti, P.; Carrasco, M. R.; Kent, S. B. H. *Tetrahedron Lett.* **2001**, *42*, 1831–1833.

[29] Canne, L. E.; Bark, S. J.; Kent, S. B. H. *J. Am. Chem. Soc.* **1996**, *118*, 5891–5896.

[30] Offer, J. *Pept. Sci.* **2010**, *94*, 530–541.

[31] Spetzler, J. C.; Hoeg-Jensen, T. *Bioorg. Med. Chem.* **2007**, *15*, 4700–4704.

[32] Pachamuthu, K.; Schmidt, R. R. *Synlett* **2003**, *5*, 659–662.

[33] Tam, J. P.; Xu, J.; Eom, K. D. *Pept. Sci.* **2001**, *60*, 194–205.

[34] Yang, Y.-Y.; Ficht, S.; Brik, A.; Wong, C.-H. *J. Am. Chem. Soc.* **2007**, *129*, 7690–7701.

[35] Nilsson, B. L.; Kiessling, L. L.; Raines, R. T. *Org. Lett.* **2001**, *3*, 9–12.

[36] Yan, L. Z.; Dawson, P. E. *J. Am. Chem. Soc.* **2001**, *123*, 526–533.

[37] Pentelute, B. L.; Kent, S. B. H. *Org. Lett.* **2007**, *9*, 687–690.

[38] Crich, D.; Banerjee, A. *J. Am. Chem. Soc.* **2007**, *129*, 10064–10065.

[39] Yang, R.; Pasunooti, K. K.; Li, F.; Liu, X.-W.; Liu, C.-F. *J. Am. Chem. Soc.* **2009**, *131*, 13592–13593.

[40] Yang, R.; Pasunooti, K. K.; Li, F.; Liu, X.-W.; Liu, C.-F. *Chem. Commun.* **2010**, *46*, 7199–7201.

[41] Haase, C.; Rohde, H.; Seitz, O. *Angew. Chem. Int. Ed.* **2008**, *47*, 6807–6810.

[42] Harpaz, Z.; Siman, P.; Kumar, K. S. A.; Brik, A. *ChemBioChem* **2010**, *11*, 1232–1235.

[43] Shang, S.; Tan, Z.; Dong, S.; Danishefsky, S. J. *J. Am. Chem. Soc.* **2011**, *133*, 10784–10786.

[44] Chen, J.; Wang, P.; Zhu, J.; Wan, Q.; Danishefsky, S. J. *Tetrahedron* **2010**, *66*, 2277–2283.

[45] Wan, Q.; Danishefsky, S. J. *Angew. Chem. Int. Ed.* **2007**, *46*, 9248–9252.

[46] Metanis, N.; Keinan, E.; Dawson, P. E. *Angew. Chem. Int. Ed.* **2010**, *49*, 7049–7053.

[47] Garg, H. G.; Jeanloz, R. W. *Adv. Carbohydr. Chem. Biochem.* **1985**, *43*, 135–201.

[48] Garg, H. G.; von dem Bruch, K.; Kunz, H. *Adv. Carbohydr. Chem. Biochem.* **1994**, *50*, 277–310.

[49] Zhang, Y.; Muthana, S. M.; Farnsworth, D.; Ludek, O.; Adams, K.; Barchi, J. J., Jr.; Gildersleeve, J. C. *J. Am. Chem. Soc.* **2012**, *134*, 6316–6325.

[50] Zhang, Y.; Muthana, S. M.; Barchi, J. J., Jr.; Gildersleeve, J. C. *Org. Lett.* **2012**, *14*, 3958–3961.

[51] Likhosherstov, L. M.; Novikova, O. S.; Derevitskaja, V. A.; Kochetkov, N. K. *Carbohydr. Res.* **1986**, *146*, C1–C5.

[52] Anisfeld, S. T.; Lansbury, P. T. *J. Org. Chem.* **1990**, *55*, 5560–5562.

[53] Cohen-Anisfeld, S. T.; Lansbury, P. T. *J. Am. Chem. Soc.* **1993**, *115*, 10531–10537.

[54] Chen, R.; Tolbert, T. J. *J. Am. Chem. Soc.* **2010**, *132*, 3211–3216.

[55] Crich, D.; Sun, S. *J. Am. Chem. Soc.* **1997**, *119*, 11217–11223.

[56] Prahl, I.; Unverzagt, C. *Angew. Chem. Int. Ed.* **2002**, *41*, 4259–4262.

[57] Weiss, H.; Unverzagt, C. *Angew. Chem. Int. Ed.* **2003**, *42*, 4261–4263.

[58] Meinjohanns, E.; Meldal, M.; Paulsen, H.; Dwek, R. A.; Bock, K. *J. Chem. Soc. Perkin Trans. 1* **1998**, 549–560.

[59] Yamamoto, N.; Ohmori, Y.; Sakakibara, T.; Sasaki, K.; Juneja, L. R.; Kajihara, Y. *Angew. Chem. Int. Ed.* **2003**, *42*, 2537–2540.

[60] Kajihara, Y.; Suzuki, Y.; Yamamoto, N.; Sasaki, K.; Sakakibara, T.; Juneja, L. R. *Chem. Eur. J.* **2004**, *10*, 971–985.

[61] Makimura, Y.; Kiuchi, T.; Izumi, M.; Dedola, S.; Ito, Y.; Kajihara, Y. *Carbohydr. Res.* **2012**, *364*, 41–48.

[62] Yamamoto, K.; Fujimori, K.; Haneda, K.; Mizuno, M.; Inazu, T.; Kumagai, H. *Carbohydr. Res.* **1998**, *305*, 415–422.

[63] Fujita, M.; Shoda, S.-i.; Haneda, K.; Inazu, T.; Takegawa, K.; Yamamoto, K. *Biochim. Biophys. Acta* **2001**, *1528*, 9–14.

[64] Goodfellow, J. J.; Baruah, K.; Yamamoto, K.; Bonomelli, C.; Krishna, B.; Harvey, D. J.; Crispin, M.; Scanlan, C. N.; Davis, B. G. *J. Am. Chem. Soc.* **2012**, *134*, 8030–8033.

[65] Wang, L.-X.; Lomino, J. V. *ACS Chem. Biol.* **2012**, *7*, 110–122.

[66] Backes, B. J.; Ellman, J. A. *J. Org. Chem.* **1999**, *64*, 2322–2330.

[67] Shin, Y.; Winans, K. A.; Backes, B. J.; Kent, S. B. H.; Ellman, J. A.; Bertozzi, C. R. *J. Am. Chem. Soc.* **1999**, *121*, 11684–11689.

[68] Kajihara, Y.; Yoshihara, A.; Hirano, K.; Yamamoto, N. *Carbohydr. Res.* **2006**, *341*, 1333–1340.

[69] Mezo, A. R.; Cheng, R. P.; Imperiali, B. *J. Am. Chem. Soc.* **2001**, *123*, 3885–3891.

[70] Wang, P.; Li, X.; Zhu, J.; Chen, J.; Yuan, Y.; Wu, X.; Danishefsky, S. J. *J. Am. Chem. Soc.* **2011**, *133*, 1597–1602.

[71] Ullmann, V.; Rädisch, M.; Boos, I.; Freund, J.; Pöhner, C.; Schwarzinger, S.; Unverzagt, C. *Angew. Chem. Int. Ed.* **2012**, *51*, 11566–11570.

[72] Wang, P.; Aussedat, B.; Vohra, Y.; Danishefsky, S. J. *Angew. Chem. Int. Ed.* **2012**, *51*, 11571–11575.

[73] Yamamoto, N.; Sakakibara, T.; Kajihara, Y. *Tetrahedron Lett.* **2004**, *45*, 3287–3290.

[74] Warren, J. D.; Miller, J. S.; Keding, S. J.; Danishefsky, S. J. *J. Am. Chem. Soc.* **2004**, *126*, 6576–6578.

[75] Schnölzer, M.; Alewood, P.; Jones, A.; Alewood, D.; Kent, S. B. H. *Int. J. Pept. Res. Ther.* **2007**, *13*, 31–44.

[76] Capon, B. *Tetrahedron Lett.* **1963**, *4*, 911–913.

[77] Murakami, M.; Okamoto, R.; Izumi, M.; Kajihara, Y. *Angew. Chem. Int. Ed.* **2012**, *51*, 3567–3572.

[78] Marcaurelle, L. A.; Mizoue, L. S.; Wilken, J.; Oldham, L.; Kent, S. B. H.; Handel, T. M.; Bertozzi, C. R. *Chem. Eur. J.* **2011**, *7*, 1129–1132.

[79] Chen, M. M.; Bartlett, A. I.; Nerenberg, P. S.; Friel, C. T.; Hackenberger, C. P. R.; Stultz, C. M.; Radford, S. E.; Imperiali, B. *Proc. Natl. Acad. Sci. U. S. A.* **2010**, *107*, 22528–22533.

[80] Hackenberger, C. P. R.; Friel, C. T.; Radford, S. E.; Imperiali, B. *J. Am. Chem. Soc.* **2005**, *127*, 12882–12889.

[81] Macmillan, D.; Bertozzi, C. R. *Angew. Chem. Int. Ed.* **2004**, *43*, 1355–1359.

[82] Yamamoto, N.; Tanabe, Y.; Okamoto, R.; Dawson, P. E.; Kajihara, Y. *J. Am. Chem. Soc.* **2008**, *130*, 501–510.

[83] Piontek, C.; Ring, P.; Harjes, O.; Heinlein, C.; Mezzato, S.; Lombana, N.; Pöhner, C.; Püttner, M.; Varón Silva, D.; Martin, A.; Schmid, F. X.; Unverzagt, C. *Angew. Chem. Int. Ed.* **2009**, *48*, 1936–1940.

[84] Piontek, C.; Varón Silva, D.; Heinlein, C.; Pöhner, C.; Mezzato, S.; Ring, P.; Martin, A.; Schmid, F. X.; Unverzagt, C. *Angew. Chem. Int. Ed.* **2009**, *48*, 1941–1945.

[85] Tachibana, Y.; Fletcher, G. L.; Fujitani, N.; Tsuda, S.; Monde, K.; Nishimura, S.-I. *Angew. Chem. Int. Ed.* **2004**, *43*, 856–862.

[86] Wilkinson, B. L.; Stone, R. S.; Capicciotti, C. J.; Thaysen-Andersen, M.; Matthews, J. M.; Packer, N. H.; Ben, R. N.; Payne, R. J. *Angew. Chem. Int. Ed.* **2012**, *51*, 3606–3610.

[87] Izumi, M.; Makimura, Y.; Dedola, S.; Seko, A.; Kanamori, A.; Sakono, M.; Ito, Y.; Kajihara, Y. *J. Am. Chem. Soc.* **2012**, *134*, 7238–7241.

[88] Sakamoto, I.; Tezuka, K.; Fukae, K.; Ishii, K.; Taduru, K.; Maeda, M.; Ouchi, M.; Yoshida, K.; Nambu, Y.; Igarashi, J.; Hayashi, N.; Tsuji, T.; Kajihara, Y. *J. Am. Chem. Soc.* **2012**, *134*, 5428–5431.

[89] Hojo, H.; Tanaka, H.; Hagiwara, M.; Asahina, Y.; Ueki, A.; Katayama, H.; Nakahara, Y.; Yoneshige, A.; Matsuda, J.; Ito, Y.; Nakahara, Y. *J. Org. Chem.* **2012**, *77*, 9437–9446.

[90] Hirano, K.; Macmillan, D.; Tezuka, K.; Tsuji, T.; Kajihara, Y. *Angew. Chem. Int. Ed.* **2009**, *48*, 9557–9560.

[91] Hirano, K.; Izumi, M.; Macmillan, D.; Tezuka, K.; Tsuji, T.; Kajihara, Y. *J. Carbohydr. Chem.* **2011**, *30*, 306–319.

[92] Wang, P.; Dong, S.; Brailsford, J. A.; Iyer, K.; Townsend, S. D.; Zhang, Q.; Hendrickson, R. C.; Shieh, J.; Moore, M. A. S.; Danishefsky, S. J. *Angew. Chem. Int. Ed.* **2012**, *51*, 11576–11584.

[93] Wang, P.; Dong, S.; Shieh, J.-H.; Peguero, E.; Hendrickson, R.; Moore, M. A. S.; Danishefsky, S. J. *Science* **2013**, *342*, 1357–1360.

[94] Schwarz, F.; Huang, W.; Li, C.; Schulz, B. L.; Lizak, C.; Palumbo, A.; Numao, S.; Neri, D.; Aebi, M.; Wang, L.-X. *Nat. Chem. Biol.* **2010**, *6*, 264–266.

[95] Valderrama-Rincon, J. D.; Fisher, A. C.; Merritt, J. H.; Fan, Y.-Y.; Reading, C. A.; Chhiba, K.; Heiss, C.; Azadi, P.; Aebi, M.; DeLisa, M. P. *Nat. Chem. Biol.* **2012**, *8*, 434–436.

[96] Hamilton, S. R.; Davidson, R. C.; Sethuraman, N.; Nett, J. H.; Jiang, Y.; Rios, S.; Bobrowicz, P.; Stadheim, T. A.; Li, H.; Choi, B.-K.; Hopkins, D.; Wischnewski, H.; Roser, J.; Mitchell, T.; Strawbridge, R. R.; Hoopes, J.; Wildt, S.; Gerngross, T. U. *Science* **2006**, *313*, 1441–1443.

12

SYNTHESIS OF GLYCOSPHINGOLIPIDS

SUVARN S. KULKARNI

Department of Chemistry, Indian Institute of Technology–Bombay, Mumbai, India

12.1 INTRODUCTION

Glycosphingolipids (GSLs) [1, 2] are a type of glycolipids ubiquitously distributed on the cell membranes of organisms from bacteria to man. They are major glycan components (~80%) of the vertebrate brain. In 1874, Johann Thudichum isolated cerebrosides from brain tissues [3]. Due to the difficulties in assigning the complex and unusual structure of the amino alcohol and their mysterious nature, the lipids were named "sphingolipids" after the mythological Egyptian sphinx [4]. Over the years, hundreds of unique GSL structures have been identified.

Structurally, these amphipathic molecules are composed of a hydrophilic sugar head group linked to a hydrophobic sphingolipid tail called ceramide (Fig. 12.1). Owing to its hydrophobic nature, the ceramide part remains embedded in the outer leaflet of the cell membrane and allows the complex carbohydrate component to protrude out to the glycocalyx. The sphingolipid section forms part in the dynamic clustering of microdomains in the cell membrane called lipid rafts and is implicated in various complex biochemical processes [5, 6]. Sphingolipids comprise of a sphingoid base, which is a long-chain amino alcohol, such as sphingosine, sphinganine, or phytosphingosine (Fig. 12.1a), usually acylated with a variety of fatty acids to form ceramides [7]. C18-sphingosine is most prevalent in GSLs, although large amounts of C20-sphingosine are also found in brain gangliosides along with C18-sphingosine [8]. The fatty acid attached to sphingosine varies in carbon chain lengths (C14 to C24),

Glycochemical Synthesis: Strategies and Applications, First Edition.
Edited by Shang-Cheng Hung and Medel Manuel L. Zulueta.

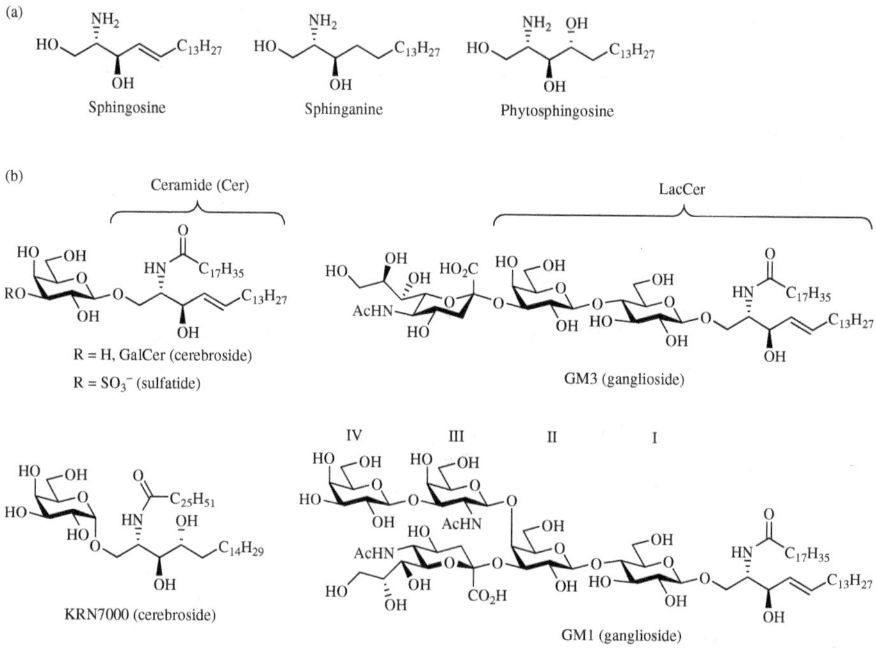

FIGURE 12.1 GSL structures. (a) Sphingoid bases. (b) Examples of GSLs and their classification based on the basic structural features in parentheses. Ac, acetyl.

degree of unsaturation, branching, and/or hydroxylation. This diversity and heterogeneity, together with their α- or β-glycosidic linkages, control the presentation of the glycan segment on the membrane and modulate their function. Moreover, GSLs show great structural diversity in their carbohydrate parts as well. Based on their basic glycan structure alone, GSLs are grouped into four classes, namely, cerebrosides (contain a single monosaccharide unit such as D-glucose (Glc) or D-galactose (Gal)), sulfoglycosphingolipids (sulfate-containing GSLs of which those containing only one sugar are called sulfatides), neutral GSLs (contain up to 50 neutral sugars), and gangliosides (contain one or more sialic acid residues). Some examples are shown in Figure 12.1b.

The most common sugars found in GSLs are Glc, Gal, L-fucose (Fuc), *N*-acetyl-D-galactosamine (GalNAc), *N*-acetyl-D-glucosamine (GlcNAc), and D-glucuronic acid (GlcA). In addition, gangliosides contain one or more sialic acids in the form of *N*-acetylneuraminic acid (Neu5Ac), *N*-glycolylneuraminic acid (Neu5Gc), 3-deoxy-D-*glycero*-D-*galacto*-non-2-ulosonic acid (KDN), and their derivatives (acetylated, sulfonated, or methylated). These acidic residues are connected to the neutral core sugar by an α-ketosidic linkage and are typically located at the nonreducing termini.

12.2 CLASSIFICATION AND NOMENCLATURE OF GSLs

Based on their carbohydrate structures, GSLs could be arranged in a series through sequential addition of single monosaccharide units. Typically, the first sugars linked to ceramide are β-linked Glc (GlcCer) and Gal (GalCer), which form the core unit of

TABLE 12.1 GSL Sugar Root Names, Abbreviations, and Structures

Root	Abbreviations	Root Structure (Sugar Residue Locant: IV–III–II–I)
Ganglio	Gg	Gal(β1→3)GalNAc(β1→4)Gal(β1→4)Glc(β1→
Isoganglio	iGg	Gal(β1→3)GalNAc(β1→3)Gal(β1→4)Glc(β1→
Lacto	Lc	Gal(β1→3)GlcNAc(β1→3)Gal(β1→4)Glc(β1→
Neolacto	nLc	Gal(β1→4)GlcNAc(β1→3)Gal(β1→4)Glc(β1→
Lactoganglio	LcGg	GalNAc(β1→4)[GlcNAc(β1→3)]Gal(β1→4)Glc(β1→
Globo	Gb	GalNAc(β1→3)Gal(α1→4)Gal(β1→4)Glc(β1→
Isoglobo	iGb	GalNAc(β1→3)Gal(α1→3)Gal(β1→4)Glc(β1→
Muco	Mc	Gal(β1→3)Gal(β1→4)Gal(β1→4)Glc(β1→
Gala	Ga	Gal(α1→4)Gal(α1→4)Gal(β1→4)Gal(β1→
Neogala	nGa	Gal(β1→6)Gal(β1→6)Gal(β1→
Mollu	Mu	GlcNAc(β1→2)Man(α1→3)Man(β1→4)Glc(β1→
Arthro	At	GalNAc(β1→4)GlcNAc(β1→3)Man(β1→4)Glc(β1→

GSLs. In higher animals, the GSLs start with GlcCer, and the Glc unit is extended with a β-linked Gal on the 4-hydroxyl of Glc to form lactosylceramide [Gal(β1→4)Glc(β1→O)Cer, LacCer]. Further elaboration of LacCer by GlcNAc, Gal, or GalNAc generates oligosaccharides of the lacto, globo, or ganglio series, respectively (Table 12.1). On the other hand, GSLs starting from GalCer are usually substituted by sialic acid or sulfonate, or they are appended by neutral sugars. In addition, gangliosides are found in all the carbohydrate series. The higher-order GSL sequences are usually capped with Fuc, GlcA, sialic acid, as well as the blood-group structures (e.g., sialyl Lewis X (sLe^x)) similar to those found in O- and N-glycans.

To simplify the nomenclature, the International Union of Pure and Applied Chemistry recommended abbreviations for the oligosaccharide chains of GSLs [9]. The neutral core (root) structures, their symbols, and commonly used abbreviations are shown in Table 12.1. The name of a particular GSL is composed of (root name) (root size) osylceramide. Thus, the sixth structure in Table 12.1, if linked to ceramide, is named globotetraosylceramide, which could be abbreviated as Gb_4Cer (the subscript 4 refers to the tetrasaccharide core).

The official nomenclature for GSLs is quite complex for common use. First, the sugar attached to ceramide is labeled as I, and the subsequent neutral core hexose units are labeled as II, III, IV, and so forth in a sequential manner. The prefix "iso" is used to denote the (1→3) versus (1→4) difference in the linkage position between the sugars III and II, while the term "neo" denotes the (1→4) versus (1→3) difference between residues IV and III. It should be noted that lacto (Lc) is different from lactose (Lac). Also, the root name "ganglio" should not be confused with gangliosides. All sialylated GSLs are referred to as "gangliosides" regardless of whether they are based on the ganglio-series neutral core structure. To form the abbreviations for the extended structures, the location of the core sugar carrying the substituents that extend or branch out from the neutral core structures are indicated by a Roman numeral, and a superscript denotes the position of the involved sugar hydroxyl. Thus, the common ganglioside Gal(β1→3)GalNAc(β1→4)[Neu5Ac(α2→3)]Gal(β1→4)Glc(β1→O)Cer (GM1, Fig. 12.1b) is shortened to II^3Neu5Ac-Gg_4Cer, which can be deciphered as a Neu5Ac unit attached to O3 of the second sugar residue (Gal in this case) in a ceramide-linked ganglio-series tetrasaccharide core.

To avoid this cumbersome method, gangliosides are more conveniently defined by a shorthand nomenclature system proposed by Lars Svennerholm [10]. For example, the previous structure (II^3Neu5Ac-Gg_4Cer) is simply referred to as "GM1." Here, "G" refers to gangliosides, the second letter refers to the number of sialic acid residues (mono (M), di (D), tri (T), etc.), and the number refers to the order of migration of the ganglioside on thin-layer chromatography (i.e., the order of migration of gangliosides is GM3 > GM2 > GM1). Svennerholm arranged the gangliosides into four series (asialo, a, b, and c) according to the number of sialic acid residues (0, 1, 2, and 3, respectively) "directly" linked to LacCer. Thus, GM1 belongs to "a" series as only one sialic acid is present on the Gal unit (O3′) of LacCer. A new series of gangliosides introduced much later—the so called "α-series"—comprise of gangliosides with α(2→6)-linked sialic acid residue attached to the GalNAc moiety of the gangliotetraose core structure [11].

12.3 BIOLOGICAL SIGNIFICANCE OF GSLs

GSLs play crucial roles in functions of the nervous system and skin, cell growth and differentiation, infections, cancer, and immune response [1, 2, 12]. Owing to their strategic position in membranes, they interact with toxins, bacteria, and viruses. They form membrane lipid rafts and present the attached carbohydrates as cell-surface receptors and, thus, serve as portals of entry for pathogens through carbohydrate–protein interactions [13]. For example, HIV entry is mediated by GalCer receptors of the host cells [14].

GSLs provide protection from the harsh extracellular environment, such as low pH and degrading enzymes. GSLs containing longer fatty acids and hydroxylated sphingolipids contribute to the water permeability barrier of the skin. In addition to their function as membrane components, GSLs are precursors of lipids such as sphingosines, ceramides, and their phosphorylated derivatives, which are involved in signal transduction [15–20]. GalCer and sulfatides are the major glycans in the brain and have demonstrated growth factor-like properties. They play vital roles in the structure and function of myelin. Gangliosides [21] are the major constituents of neuronal cell membranes and the endoplasmic reticulum. Acidic gangliosides influence the electrical field across the cellular membrane as well as the concentration of ions on the external surface of the cells. In addition, gangliosides may have a role in electrical insulation in myelin cells in the nervous system. Neurons of the central nervous system contain a variety of gangliosides in relatively high overall concentration. They are, thus, implicated in Alzheimer's disease, Parkinson's disease, and ischemic stroke.

Some gangliosides of the globo series (Globo-H, sLe^x hexasaccharyl ceramide, etc.) are useful cancer cell markers because they are elevated in the membranes of certain types of tumor cells such as melanomas and metastatic brain tumors. A number of GSLs play a role as tumor-associated carbohydrate antigens (TACAs) and in immunotherapy of individual cancer forms [22]. They may also serve as substrates and inhibitors of glycosidases and glycosyltransferases. Many GSLs are blood-group antigens.

Most of the GSLs discussed so far are β-linked (Table 12.1). However, a notable α-linked GSL is the α-galactosylceramide KRN7000 (Fig. 12.1b). This simplified analogue of agelasphins, originally isolated from the marine sponge *Agelas mauritianus*, shows remarkable immunological properties with great therapeutic potential. It is a powerful stimulator of the invariant natural killer T cells when presented as a ligand by CD1d molecules and produces various cytokines [23].

12.4 SYNTHESIS OF GSLs

The synthesis of GSLs has been categorically reviewed by Yashwant Vankar and Richard Schmidt in 2000 [2]. Recent reviews on cerebroside [24] and sulfatide syntheses [25] have also been published. The stereochemistry of the anomeric linkage of GSLs to the lipid component plays a crucial role in their mode and manifestation of biological activity. Thus, methods to control the formation of α- or β-glycosidic bonds have become pivotal [26, 27]. The synthesis of α- and β-GSLs with a special emphasis on glycosidic bond formation has been reviewed by Sergio Castillón and coworkers [28]. The unprecedented attention received by KRN7000 from the synthetic carbohydrate community has been covered in several comprehensive reviews [23, 29, 30]. This chapter focuses on the synthesis of globo- (e.g., Gb_3, Gb_5, Globo-H), isoglobo- (iGb_3), neolacto- (sLe^x hexasaccharide), and ganglio- (e.g., GM1–GM4) series GSLs as well as the complex GSLs, with a particular attention to stereoselective glycosylations.

In general, any GSL structure can be retrosynthetically divided into two main parts, that is, the oligosaccharide fragment and the ceramide unit. The ceramide unit can be further split into sphingosine and the fatty acid. Accordingly, most of the syntheses entail the procurement of a suitably protected oligosaccharide fragment followed by the stereoselective attachment of ceramide (or its precursor, which requires further manipulation) and global deprotection.

12.4.1 Synthesis of Globo- and Isoglobo-Series GSLs

GSLs of the globo series are associated with various types of cancers, including lung, breast, kidney, and ovarian cancers [22, 31]. Important members of this group include Globo-H ceramide, Gb_5Cer, and Gb_3Cer (Fig. 12.2). Synthetic Globo-H conjugates [32, 33] and a pentavalent construct containing Globo-H and other antigens [34, 35] showed very promising results as potential breast and prostate cancer vaccines. Gb_5Cer is abundant in testicular cancer. Because it is present during embryonic development, Gb_5 is often referred to as stage-specific embryonic antigen-3 (SSEA-3). Gb_3Cer is overexpressed in ovarian cancer, metastatic colon cancers [36], and Burkitt's B-cell lymphoma [37]; it also occurs in breast cancer and testicular carcinoma. It has been identified as P^k antigen in the P blood-group system [38], acts

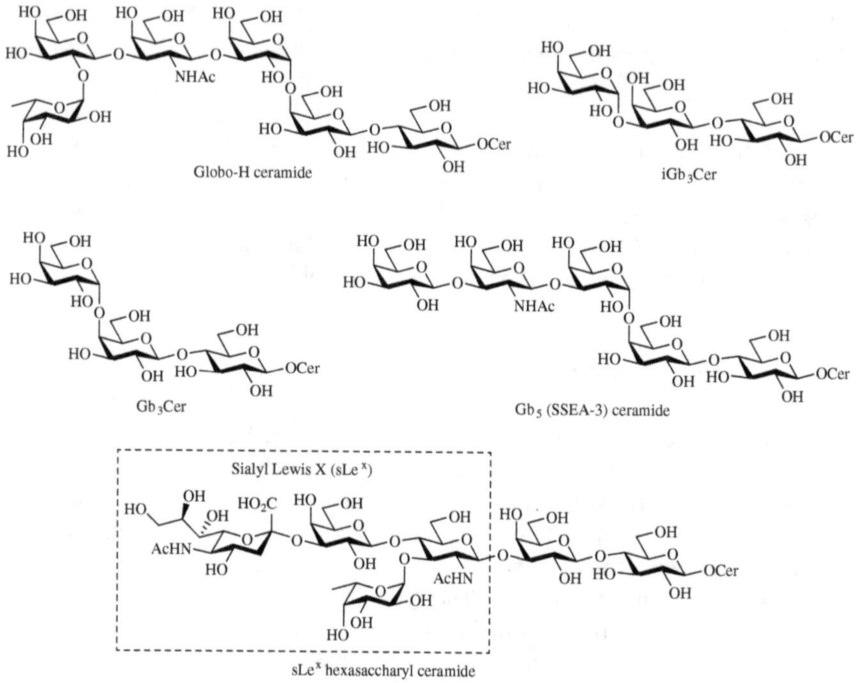

FIGURE 12.2 Important cancer-related GSLs.

as a receptor for Shiga-like toxins produced by *Escherichia coli* O-157:H-7 [39], is closely associated with Fabry's disease [40], and has been implicated in the entry of HIV-1 into cells [41]. The structurally related iGb$_3$Cer was found to be associated with malignant neoplasms. The Gal(α1→3)Gal epitope has been shown to serve as attachment sites of bacteria, toxins, and viruses. Recently, iGb$_3$ was disclosed as an endogenous ligand for human natural killer T cells [42]. Due to their importance as tumor markers and potential as anticancer vaccines, these antigens have been synthesized by several groups.

The globo series of carbohydrate antigens requires the selective formation of the challenging 1,2-*cis* galactosidic linkage on the axial 4-OH of Gal. α-Galactosidic linkages have been installed by using a variety of glycosylating donors and promoters. In fact, these complex glycoconjugates served as testing grounds for various glycosylation protocols. The α selectivity in the critical galactosylation is usually achieved by placing a nonparticipating ether-type (usually benzyl (Bn)) group at C2 and a participating ester-type group at C4 and/or C6 or the bulky di-*tert*-butylsilylidene (DTBS) acetal at O4 and O6 and/or by using participating ethereal solvents. The remaining 1,2-*trans* glycosidic linkages can be installed mainly by taking advantage of the neighboring group participation of the 2-*O*-acyl group. The formation of orthoester side product can be circumvented by using benzoyl (Bz) or pivaloyl (Piv) groups instead of Ac. Retrosynthetically, Gb$_3$ can be considered as a key intermediate for the synthesis of globo-series GSLs. Several syntheses of Gb$_3$Cer as well as its trisaccharide moiety are known, mostly starting from an appropriately protected 4′-OH lactose or 4-OH Gal acceptor.

*12.4.1.1 **Synthesis of Gb_3 and iGb_3*** The first total synthesis of Gb_3Cer was reported by David Shapiro and A. J. Acher [43] way back in 1978 using acetylated glycosyl halides and a ceramide acceptor via the classical Koenigs–Knorr method. This route suffered from the low selectivity of the α-galactosylation, Ac migration, and the low yield of the ceramide coupling. Several syntheses of Gb_3 appeared in the next 5 years [44–48]. Tomoya Ogawa and coworkers [49] reported a total synthesis of Gb_3Cer (Scheme 12.1) using the per-O-benzylated thiogalactoside 1 and the lactose-derived 4′-alcohol **2** using $CuBr_2$ and AgOTf as promoter to obtain trisaccharide **3** (94%, α/β = 3.4/1), which was converted to the glycosyl fluoride **4** and subsequently coupled to ceramide **5** to afford the fully protected Gb_3Cer (29%, β only). In a similar way, iGb_3 was synthesized using thioglycoside **1** and the corresponding lactosyl 3′,4′-diol derivative. The coupling gave a mixture of O3′ α-glycoside and 3′,4′-diglycosylated product (41% mono and 35% di). Further ceramide attachment was achieved in 33% yield via trichloroacetimidate coupling by using $BF_3{\cdot}Et_2O$ as promoter.

Ogawa's synthesis still called for improvement in the yield of the ceramide coupling step and the moderate α/β selectivity in galactosylation. The low glycosylation yields encountered with ceramide acceptors were attributed to the poor nucleophilicity of the ceramide due to intramolecular hydrogen bonding, its tendency toward orthoester formation, and its potential binding with the glycosylation catalyst. These problems were circumvented by using the azidosphingosine procedure developed by Richard Schmidt [50], which involves the use of azidosphingosine as the acceptor for

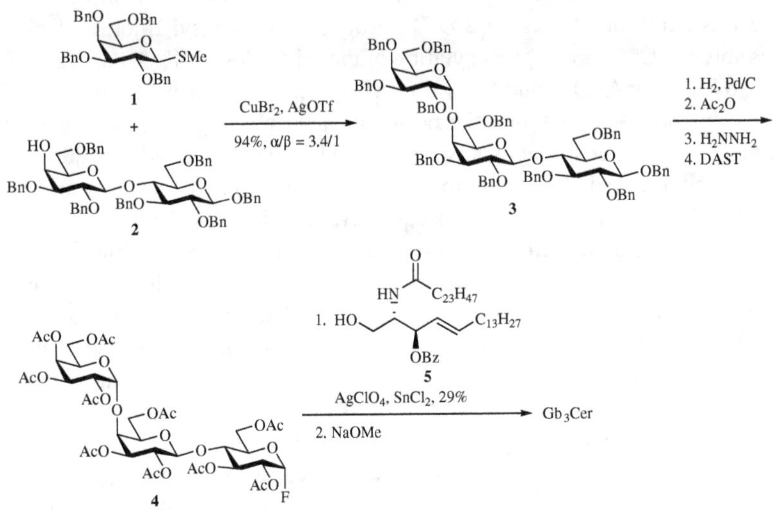

SCHEME 12.1 Ogawa's synthesis of Gb_3Cer. DAST, *N,N*-diethylaminosulfur trifluoride; Tf, triflyl.

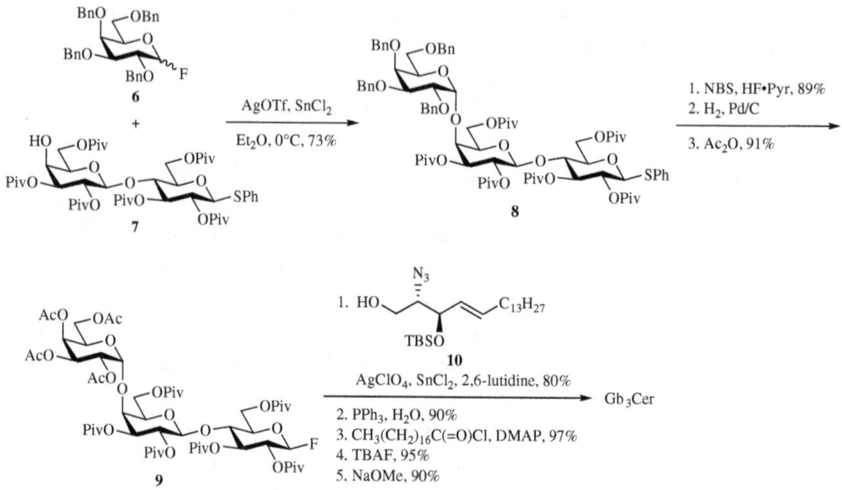

SCHEME 12.2 Nicolaou's synthesis of Gb_3Cer. DMAP, 4-dimethylaminopyridine; NBS, *N*-bromosuccinimide; TBAF, tetrabutylammonium fluoride.

glycosylation and the subsequent reduction of the azide to amine and its amidation. Further, the reaction course naturally allows one to synthesize the biologically important lysosphingolipid intermediate (glycosylsphingosine—one with free amino group). The azidosphingosine derivatives can be obtained either through chiral induction or chiral pool. Kyriacos Nicolaou synthesized Gb_3Cer by using this strategy (Scheme 12.2) [51, 52]. The azidosphingosine derivative **10** was first procured by chiral induction using an L-valine-derived oxazolidinone as chiral auxiliary. For the synthesis of the trisaccharide segment, the 1,2-*cis* stereoselectivity of the galactosyl linkage was controlled by using 2-*O*-Bn groups on galactosyl fluoride donors and Et_2O as solvent. Orthogonal glycosylation of the thiolactoside 4′-OH acceptor **7** with fluoride **6** by using AgOTf and $SnCl_2$ in Et_2O generated the α-linked trisaccharide **8** (73%). Transformation of **8** into the per-O-acylated glycosyl fluoride **9** and further coupling with the azidosphingosine acceptor **10** led to an intermediate (80%), which was subjected to Staudinger reduction, amidation, and global deprotection to afford Gb_3Cer in excellent overall yield. Subsequently, Schmidt [53] accomplished the total synthesis of Gb_3Cer via trichloroacetimidates by using the "inverted procedure" (donor is added to the mixture of promoter and acceptor) to achieve good α selectivity in galactosylation of the lactose-derived 4′-OH acceptor and the azidosphingosine protocol to install the β-linked ceramide moiety. iGb_3Cer was also synthesized through the same inverted procedure with the highly regioselective and stereoselective O3′ glycosylation of a lactose-derived 3′,4′-diol as the key step.

Shun-ichi Hashimoto and coworkers [54] carried out the total synthesis of Gb_3Cer by exploiting the armed–disarmed effect with glycosyl tetramethylphosphoroamidate donors (Scheme 12.3). The typical route (i.e., coupling of the galactosyl and lactosyl building blocks followed by ceramide attachment) suffered a drop in yield at the ceramide coupling stage partly due to the cleavage of the α-galactosyl linkage

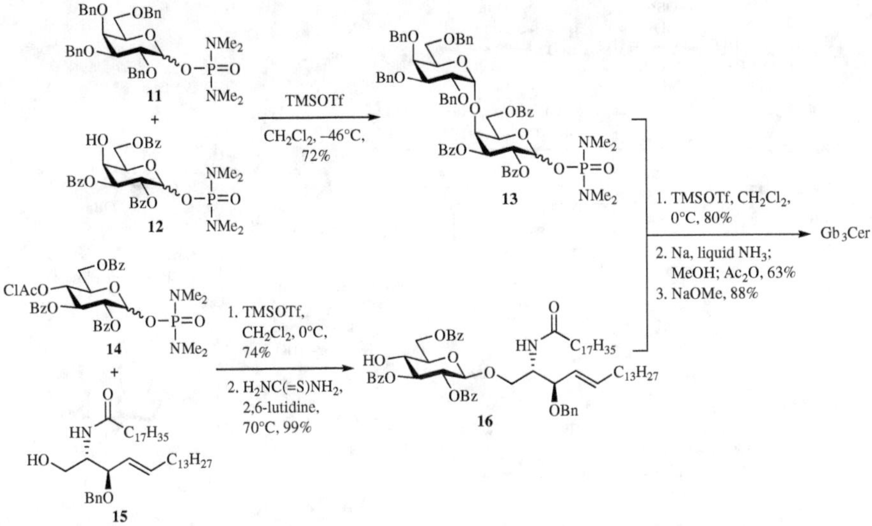

SCHEME 12.3 Hashimoto's synthesis of Gb_3Cer. TMS, trimethylsilyl.

under the glycosylation condition. Alternatively, a convergent route was tested. Chemoselective galactosylation of the disarmed galactosyl 4-alcohol **12** with the armed donor **11** fashioned the α-linked disaccharide **13** in 85% yield (α/β=97/3, 72% α after purification). In parallel, coupling of the glucosyl donor **14** with the ceramide acceptor **15** efficiently furnished the desired β-linked GlcCer derivative, which upon selective removal of the chloroacetyl (ClAc) group at O4 afforded the GlcCer acceptor **16**. Further coupling of **13** and **16** generated the desired Gb_3Cer derivative (80%), which was deprotected to deliver the target molecule. The convergent strategy has proven to be extremely efficient and provides an alternative solution to the problem of ceramide attachment. The high α selectivity observed in galactosylation was attributed to the steric and electronic disarming environment around the axial 4-OH of Gal. The retrosynthetic disconnection between Glc and Gal of LacCer, resulting in the use of GlcCer as acceptor, soon became known as the "ceramide cassette approach" and widely used in the synthesis of complex gangliosides.

A few syntheses of Gb_3, which are attached to different types of linkers (such as nitrophenyl, allyl, *n*-pentenyl, azidopropyl, 3-butene-1-ol) instead of ceramides, are also reported [55–58]. The critical α-galactosyl linkage was efficiently installed by the coupling of galactosyl chloride or α-methoxyphenylthioglycoside with linker-attached lactoside 4′-OH acceptor. These constructs have been used in creating multivalent assemblies as ligands for Shiga toxin. Fluorous protecting groups [59] have also been used effectively to expedite workup and purification in conjunction with the imidate donor and allyl lactoside as acceptor to synthesize a fluorescein-attached Gb_3 [60]. Enzymatic syntheses of Gb_3Cer and iGb_3Cer are also reported [61].

Xin-Shan Ye and coworkers [62] carried out the one-pot synthesis of Gb_3 and iGb_3 using thioglycoside donors with *N*-(phenylthio)-ε-caprolactam (**20**) and Tf_2O as

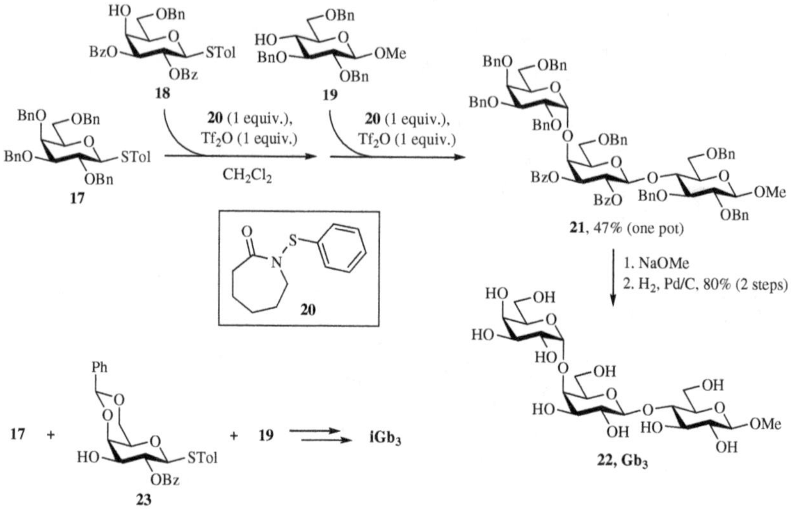

SCHEME 12.4 One-pot synthesis of Gb_3 and iGb_3. Tol, 4-tolyl.

promoter (Scheme 12.4) [63]. Consecutive coupling of **17**, **18**, and **19** in one pot afforded the fully protected Gb_3 **21** in 47% overall yield. Global deprotection gave Gb_3 as a methyl glycoside. For iGb_3, the 3-OH thiogalactoside **23** was used in place of **18**, and the related trisaccharide was obtained in 50% yield. All the couplings were done at room temperature, and promoter **20** is believed to be responsible for the α selectivity.

Tony Mong and coworkers [64] applied an operationally simple β-glycosylation method based on the well-known nitrile solvent effect to synthesize Gb_3 and iGb_3 derivatives (Scheme 12.5). This involves the use of thioglycoside donors without a C2 participating group and low substrate concentrations. Glycosylation of 6-chlorohexanol with the thiolactoside donor **24** in a nitrile-based solvent mixture at −70°C afforded the corresponding β-linked 4′-alcohol. Without any further purification, this acceptor was utilized in the coupling with the thiogalactoside donor **25** to obtain the α-linked trisaccharide **26** in 75% overall yield. The high α selectivity, in this case, is attributed to the steric effect of the DTBS acetal, which effectively blocks the β face during glycosylation and produces the α-glycoside regardless of the C2 protecting group [65–67]. This functional group was also successfully employed for the total synthesis of iGb_3Cer [68] by Makoto Kiso and coworkers. Removal of DTBS and Bn groups in **26** gave the Gb_3 derivative. A similar assembly sequence with the corresponding 3′-OH thiolactoside **27** afforded the protected iGb_3 derivative in 70% yield.

A report by Dapeng Zhou, Albert Bendelac, and coworkers [42] declaring that iGb_3Cer is the endogenous ligand for natural killer T cells stirred interest in this field. They synthesized iGb_3Cer through regioselective glycosylation of the lactosyl 3′,4′-diol **29** with bromide **28** to obtain trisaccharide **30** (61%), which was transformed into imidate **31** and coupled to the fully functionalized ceramide **32** in 45% yield

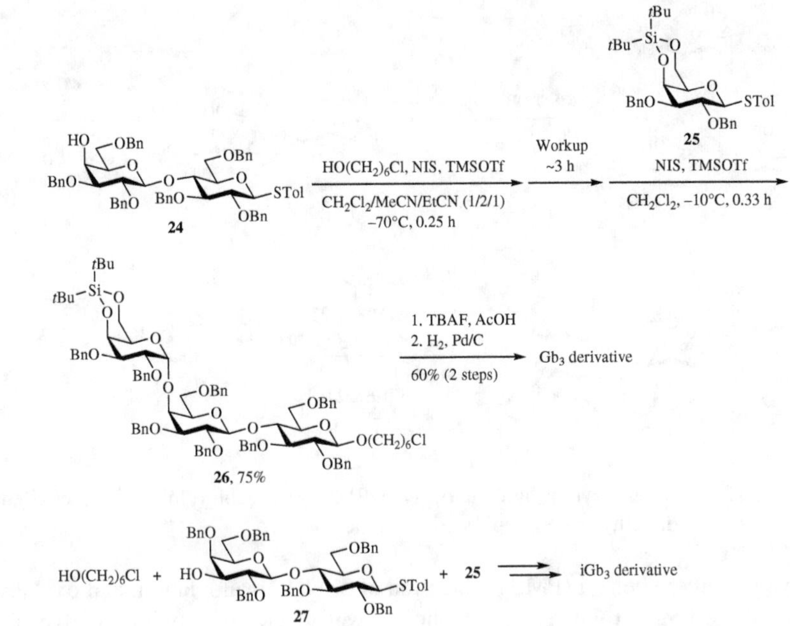

SCHEME 12.5 Mong's synthesis of Gb_3 and iGb_3. NIS, *N*-iodosuccinimide.

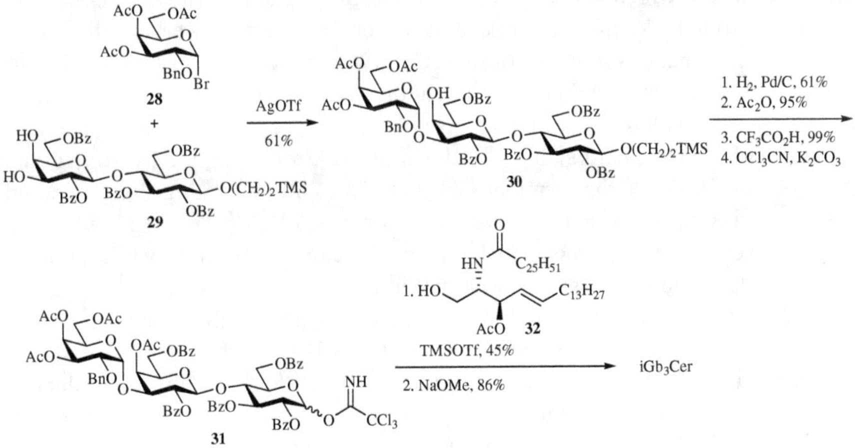

SCHEME 12.6 Synthesis of iGb_3Cer by Zhou et al.

(Scheme 12.6). The preparation of acceptor **29** involved isopropylidenation of the linker-attached lactoside followed by benzoylation and removal of isopropylidene. The route was short, although the glycosylation yield was moderate.

Peng George Wang and coworkers reported [69] the total synthesis of iGb_3Cer by using the azidosphingosine glycosylation procedure (Scheme 12.7). The lactosyl 3′-OH acceptor **34** was prepared from lactose via anomeric benzylation, tin-mediated

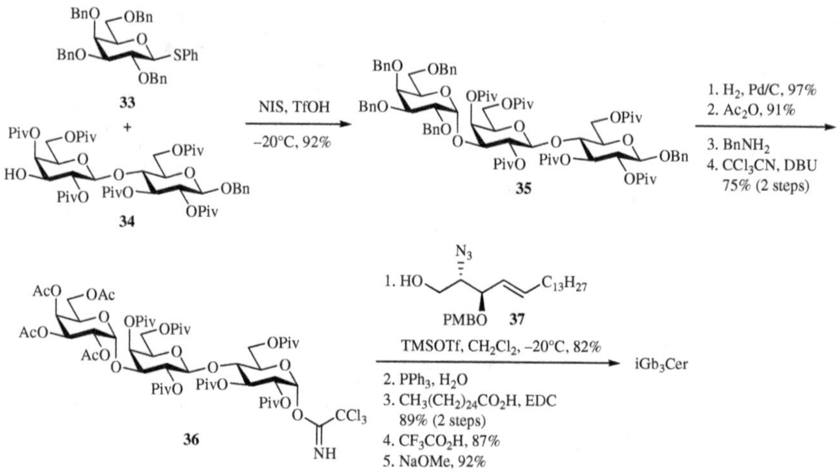

SCHEME 12.7 Wang's synthesis of iGb$_3$Cer. DBU, 1,8-diazabicyclo[5.4.0]undec-7-ene; EDC, 1-ethyl-3-(3-dimethylaminopropyl)carbodiimide.

selective *p*-methoxybenzyl (PMB) protection of 3′-OH, pivaloylation, and oxidative removal of the PMB group. The 3′-alcohol **34** was glycosylated with thioglycoside **33** under NIS/TfOH activation to obtain the desired α-linked trisaccharide **35** in high yield. Functional group manipulations were performed to fashion the trichloroacetimidate donor **36**, which was coupled to azidosphingosine **37**. Subsequent transformations led to iGb$_3$Cer in excellent overall yields. Wang and coworkers also carried out chemoenzymatic synthesis of iGb$_3$Cer [70]. Structure–activity relationship studies followed, which included modification of the ceramide moiety by reduction of the double bond or by using phytosphingosine [71], replacing the anomeric O-linkage of the terminal galabiose with sulfur [72], removing the hydroxy groups at various positions of the terminal Gal unit [73], and generating the α-ceramide (α-Gb$_3$Cer, α-iGb$_3$Cer) instead of the common β form [74]. Most of the modifications led to more effective analogues with better cytokine release profiles, while modifications at the C4″ and C6″ positions were not influential.

Gb$_3$ is also a key intermediate for the synthesis of structurally related antigens of the globo series. Stepwise addition of GlcNAc β(1→3), Gal β(1→3), and Fuc α(1→2) donors on a Gb$_3$ core 3″-OH acceptor allows access to Gb$_4$, Gb$_5$, and Globo-H, respectively. Ceramide attachment is usually done at a later stage. The first total synthesis of Gb$_5$Cer was reported by Ogawa and coworkers [75] using glycosyl chloride (for Gb$_3$) and imidate (for Gb$_5$) as donors. Ulf Nilsson and Göran Magnusson [76] used thioglycosides to make Gb$_5$ and attached it to bovine serum albumin and Sepharose for biological studies. Samuel Danishefsky and coworkers [77] employed thioglycoside coupling promoted by MeOTf to achieve the pentasaccharide sequence followed by glycal coupling for ceramide attachment to complete the total synthesis of Gb$_5$Cer. Peter Seeberger and coworkers [78] synthesized a series of cancer antigens in a linear fashion using glycosyl phosphate building blocks. The key α-galactosylation was achieved by temperature control (−78°C) and using Et$_2$O as solvent. Nikolay

Nifantiev and coworkers [79] synthesized the aminoethyl-linked Gb_3, Gb_4, and Gb_5 using stepwise addition of thioglycoside donors on a lactose 4′-OH acceptor. Alexei Demchenko's laboratory [80] carried out the synthesis of a linker-attached Gb_3, which was further extended to Gb_5 by a [2+3] strategy using an *n*-pentenyl disaccharide donor. They showed that thiazolinylthio (STaz) donors give excellent selectivity and efficiency in α-galactosylation with 4′-lactosyl acceptor to synthesize the Gb_3 construct. The enzymatic syntheses of Gb_3–Gb_5 glycans are also reported [81].

12.4.1.2 Synthesis of Globo-H Being an important TACA, Globo-H received much attention from chemists and biologists. The syntheses of Globo-H were achieved by Samuel Danishefsky [82–84], Richard Schmidt [85], Geert-Jan Boons [86], Chi-Huey Wong [87], Peter Seeberger [88], and Xuefei Huang [89] groups. The first total synthesis of Globo-H ceramide and its structure confirmation came from Danishefsky's laboratory by employing glycal assembly and sulfonamidation [82, 83]. As shown in Scheme 12.8, the hexasaccharide **40** was assembled from trisaccharides **38** and **39** by using MeOTf as promoter (70–85%) and further coupled to the azidosphingosine derivative **41** in 53% yield. Reduction of azide to amine followed by its amidation and removal of the protecting groups afforded Globo-H ceramide. For the installation of α linkages in **38** and **39**, the corresponding fucosyl and galactosyl fluoride donors were coupled to the respective acceptors by using $AgClO_4$ and $SnCl_2$ combination in approximately 50% yields. The obvious lacunae of this elegant synthesis, however, are the low yields encountered in the glycosylation and sulfonamidation steps.

Recently, Danishefsky's group came up [84] with a practical synthesis of Globo-H. The glycosylation efficiency was improved by replacing fluoride with thioglycosides and applying Huang's protocol [89] for preactivation-based glycosylation. As shown in Scheme 12.9, the thiofucosyl donor **43** was preactivated using TolSCl and AgOTf

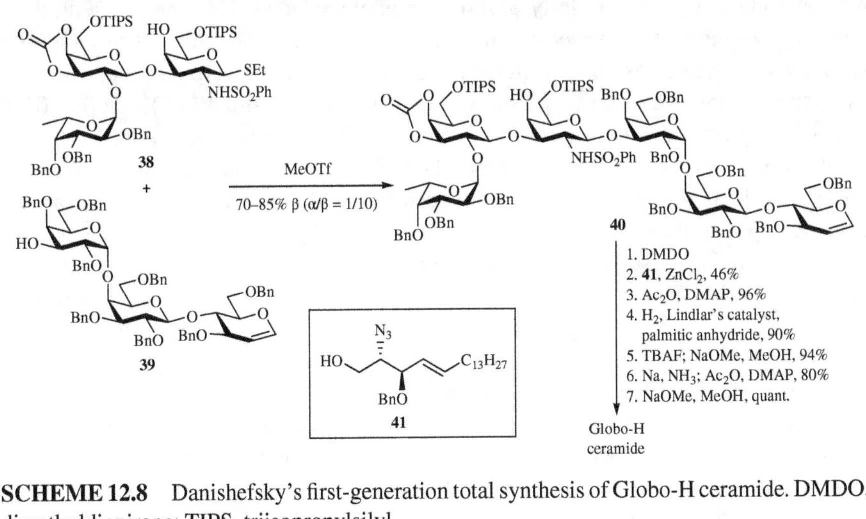

SCHEME 12.8 Danishefsky's first-generation total synthesis of Globo-H ceramide. DMDO, dimethyldioxirane; TIPS, triisopropylsilyl.

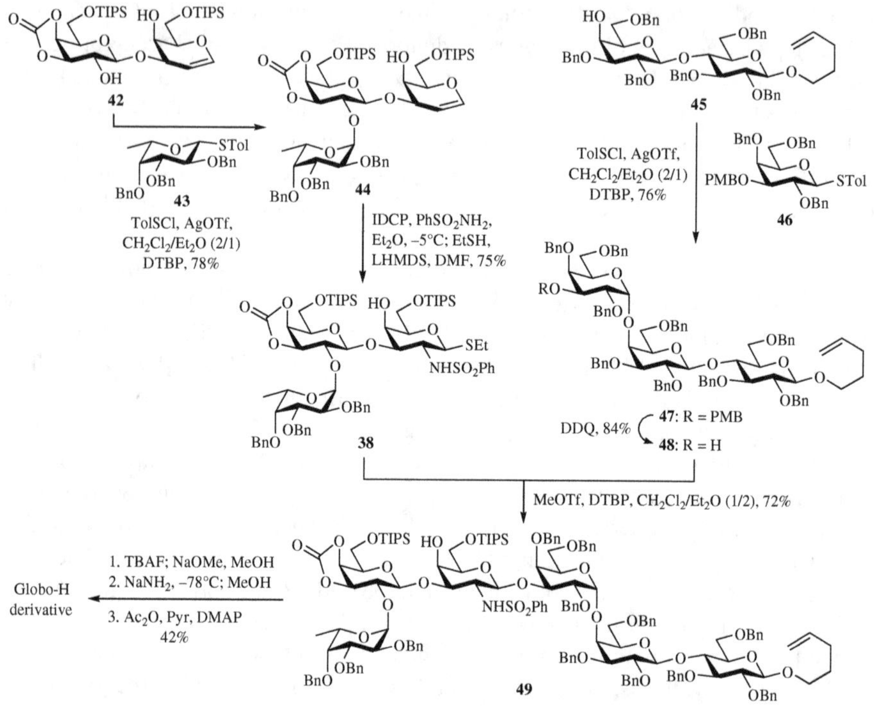

SCHEME 12.9 Danishefsky's second-generation synthesis of Globo-H. DMF, *N,N*-dimethylformamide; DTBP, 2,6-di-*tert*-butylpyridine; IDCP, iodonium di-*sym*-collidine perchlorate; LHMDS, lithium bis(trimethylsilyl)amide.

combination and selectively coupled to the 2′,4-diol **42** to obtain the desired α-linked trisaccharide **44** (78%). Similarly, glycosylation of the lactosyl acceptor **45** with thioglycoside **46** cleanly generated the α-linked trisaccharide **47** (76%). The yield and selectivity of the corresponding fluoride donor used earlier were also improved by using the strongly fluorophilic promoter $Cp_2Zr(OTf)_2$ to afford **47** (80%, α/β = 10/1). The PMB group in **47** was selectively removed by a brief treatment with 2,3-dichloro-5,6-dicyano-1,4-benzoquinone (DDQ). The glycal **44** was converted to the thioglycoside **38** through an improved iodosulfonamidation-rearrangement protocol. Finally, a [3+3] glycosylation of acceptor **48** with **38** generated the hexasaccharide **49** (72%), which was subjected to global deprotection–per-O-acetylation sequence to obtain Globo-H as an *n*-pentenyl glycoside. The olefinic handle could be oxidized to aldehyde for further coupling with proteins or with amino acids for vaccine studies.

José Lasaletta and Richard Schmidt [85] applied a stepwise relay glycosylation route to build Globo-H from the nonreducing end to the reducing end using the trichloroacetimidate method (Scheme 12.10). This synthesis applied the inverted procedure in Et_2O to achieve excellent α selectivity in two crucial glycosylations and used acetonitrile as solvent to afford β selectivity in the absence of a C2 participating group. Accordingly, the 3-alcohol **51** was glycosylated with galactosyl imidate **50**

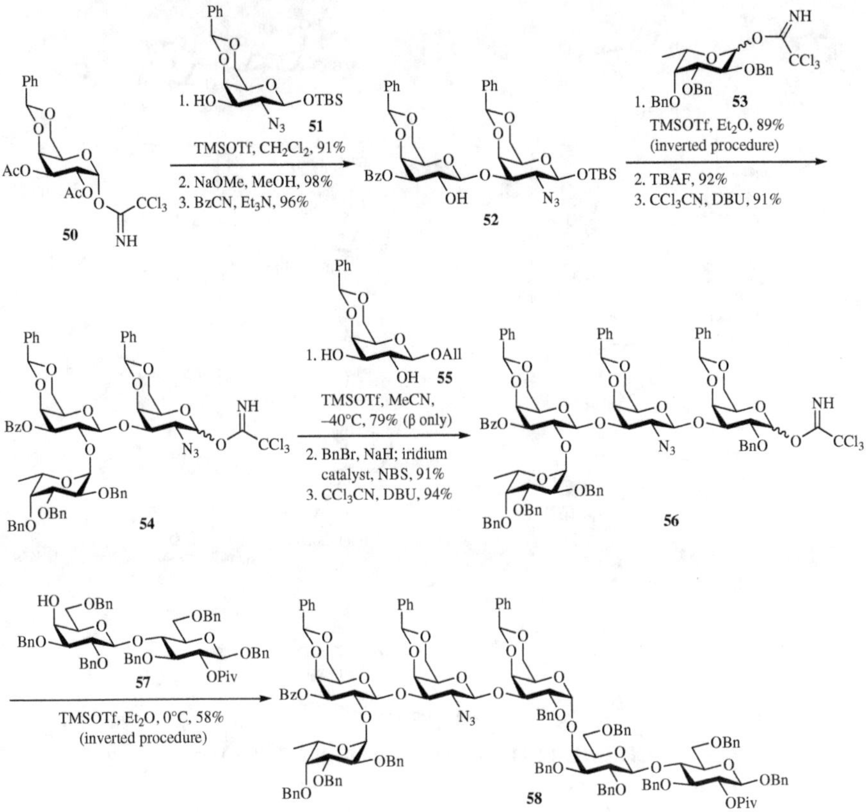

SCHEME 12.10 Schmidt's synthesis of Globo-H. All, allyl.

bearing a C2 acyl group using TMSOTf as promoter to obtain the requisite β-linked disaccharide (91%), which was subjected to deacetylation and regioselective 3-O-benzoylation to fashion the 2′-alcohol **52**. Stereoselective glycosylation of **52** with the fucosyl imidate **53** through the inverted procedure afforded the desired α-linked trisaccharide (89%). The anomeric *tert*-butyldimethylsilyl (TBS) group was removed, and the resulting hemiacetal was converted to imidate **54**. Regio- and stereoselective coupling with the galactosyl 2,3-diol **55** furnished the β-linked product (79%), which was again transformed into the corresponding hemiacetal and to imidate **56**. Finally, a [4+2] α-selective coupling of **56** with the lactosyl 4′-alcohol **57** using the inverted procedure offered the protected Globo-H **58**. This construct may be further attached to ceramide (or azidosphingosine) or fully deprotected to afford the bare Globo-H.

Tong Zhu and Geert-Jan Boons [86] reported a convergent and bidirectional approach by using chemoselective glycosylations for the assembly of Globo-H. Their route comprises five consecutive glycosylations without intermittent protecting group manipulations. This is achieved by exploiting the reactivity differences of the anomeric leaving groups and the subtle control of nucleophilicities of the sugar hydroxy groups and silyl ethers. The strategy allowed extensive use of Bn protecting

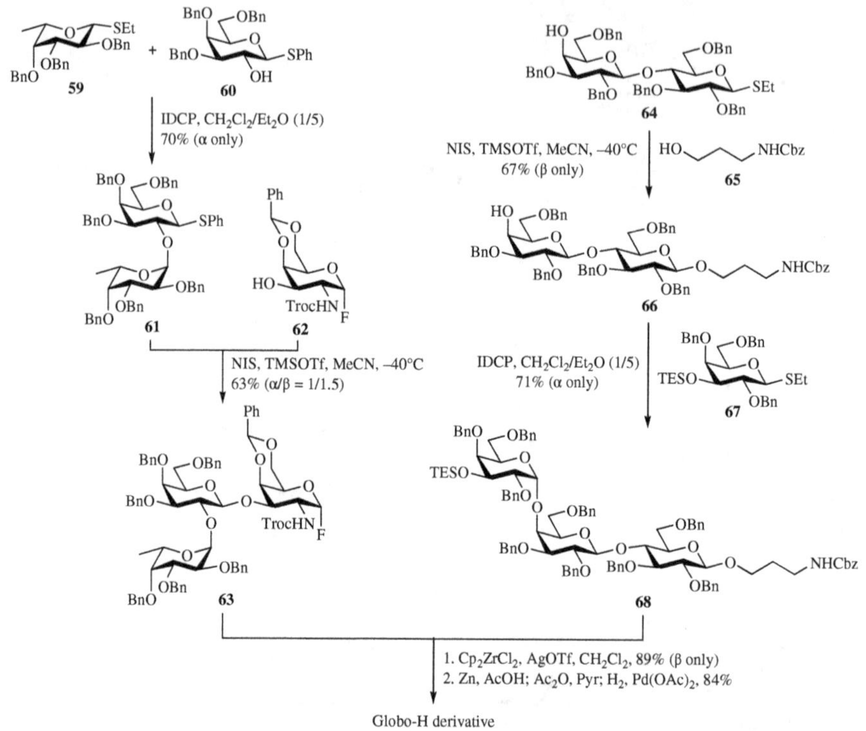

SCHEME 12.11 Boons' synthesis of Globo-H.

groups (Scheme 12.11). Thus, selective activation of the ethyl thiofucoside **59** over the phenyl thiogalactoside 2-alcohol **60** using the mild promoter IDCP afforded the α-linked disaccharide **61** (70%). Activation of the thiophenyl group in **61** by using the stronger promoter NIS/TMSOTf enabled the orthogonal glycosylation of the fluoride 3-OH acceptor **62** to generate trisaccharide **63** and its corresponding α isomer (63%, α/β = 1/1.5). In parallel, the primary alcohol **65** was glycosylated with the thiolactoside **64** using nitrile effect to fashion the β-linked lactosyl acceptor **66** (67%). Further stereoselective glycosylation with ethyl thioglycoside **67** by using IDCP in CH_2Cl_2/Et_2O solvent combination formed trisaccharide **68** (71%) with the triethylsilyl (TES) group intact. Finally, the 3″-*O*-TES-protected **68** was glycosylated with the trisaccharide donor **63** under strong activation by $Cp_2Zr(OTf)_2$/AgOTf to deliver the Globo-H construct in excellent yields (89%). Removal of the 2,2,2-trichloroethoxycarbonyl (Troc) group, *N*-acetylation, and hydrogenolysis gave the aminopropyl-linked Globo-H. The only shortcoming of this overall efficient approach was the difficulty in controlling β selectivity during the [2 + 1] glycosylation to synthesize compound **63**.

Wong and coworkers [87] synthesized Globo-H by applying their anomeric reactivity-based, programmable one-pot glycosylation of tolyl thioglycosides and using the computer program OptiMer to assist in synthesis planning. The reactivities of the

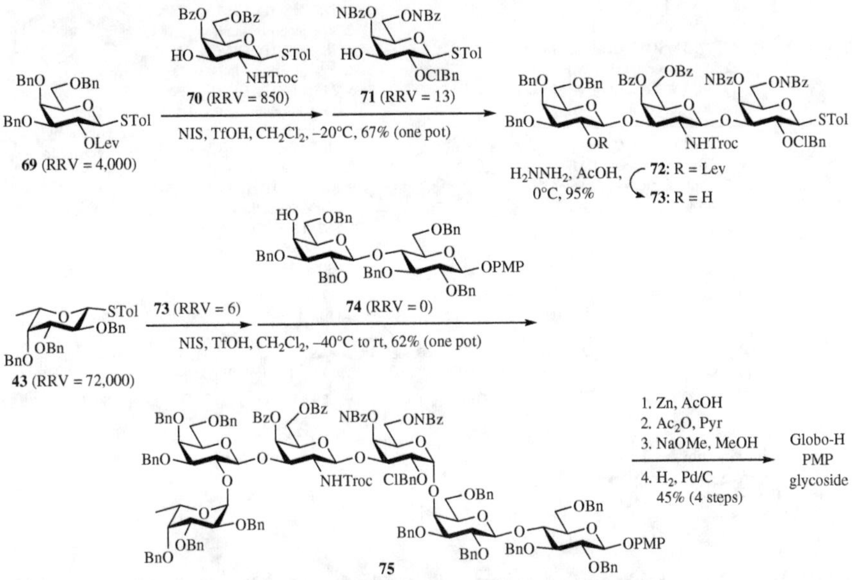

SCHEME 12.12 Wong's synthesis of Globo-H. ClBn, *o*-chlorobenzyl; NBz, *p*-nitrobenzoyl; PMP, *p*-methoxyphenyl; RRV, relative reactivity value.

building blocks were mainly tuned by the electron donating or withdrawing capacity of protecting groups to create sufficient reactivity difference between the coupling partners (Scheme 12.12). The sugar assembly involves two one-pot reactions separated by a selective deprotection step. Thus, the one-pot coupling of thioglycosides **69**, **70**, and **71** in the presence of NIS/TfOH generated the two β linkages to afford the trisaccharide **72** (67%). The levulinyl (Lev) protecting group was selectively removed to obtain acceptor **73** (95%). The second one-pot reaction between **43**, **73**, and **74** fixed the two α linkages and furnished the hexasaccharide **75** (62%). Global deprotection afforded Globo-H as a PMP glycoside (45%).

Seeberger and coworkers first streamlined a linear strategy for the solution-phase synthesis of Globo-H, SSEA-3, and Gb_3 by using glycosyl phosphates in anticipation of automating the synthesis on solid support [78]. Eventually, they carried out the automated solid-phase assembly of Globo-H and Gb_3 using building blocks **77–83** (Scheme 12.13) via iterative cycles of coupling and 9-fluorenylmethoxycarbonyl (Fmoc) deprotection [88]. Of the various donors employed, phosphate **79** and imidate **83** offered excellent α selectivity for the critical galactosylation. The glycans were detached from the resin using Grubbs' metathesis. Thus, Gb_3 was assembled on the linker-attached resin **76** in 12h by using building blocks **77**, **78**, and **83** in 46% overall yield. Similarly, Globo-H was assembled from building blocks **77–82** in 25h and 30% overall yield.

Yet another efficient assembly of Globo-H glycan was reported by Huang and coworkers [89] via a multicomponent one-pot synthesis. Their approach relies on preactivation-based iterative glycosylation of thiotolyl donors using TolSCl/AgOTf

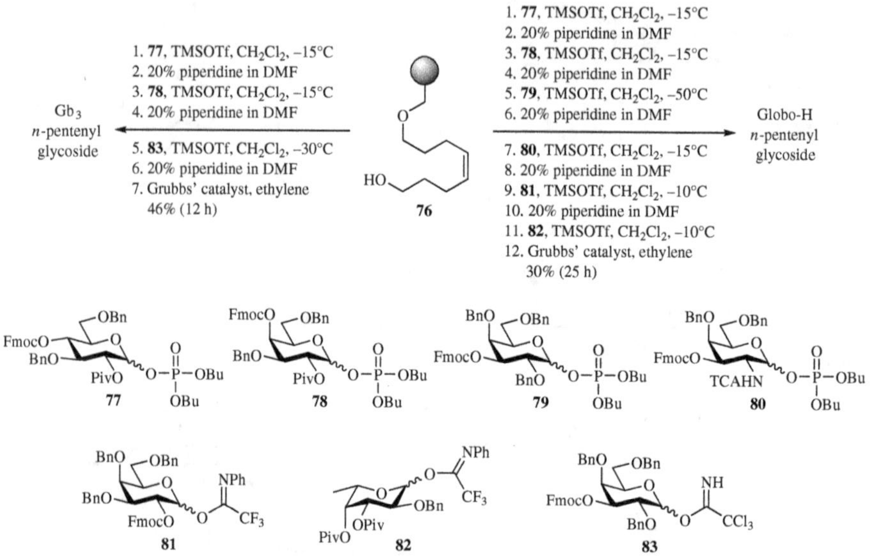

SCHEME 12.13 Seeberger's automated synthesis of Gb_3 and Globo-H. TCA, trichloroacetyl.

as promoter and, thus, obviates the necessity to control the anomeric reactivity of donors (see Section 6.2.3). Huang and coworkers also reported a chemoenzymatic synthesis of Globo-H by using their one-pot approach for the assembly of the reducing end pentasaccharide and a fucosyltransferase for the terminal α-fucosylation [90]. The nonreducing end tetrasaccharide sequence was also synthesized by Alfonso Iadonisi and coworkers [91] by using trifluoroacetimidate donors and $Yb(OTf)_3$ and acid-washed molecular sieves as activator.

12.4.2 Synthesis of Gangliosides

12.4.2.1 sLe^x In gangliosides, the sialic acid residue, which is a key player in many biological events including cancer [92, 93], is usually α-linked to either O3 or O6 of Gal, O8 of another sialic acid, or O6 of GalNAc. The tumor-associated antigen sLe^x (Fig. 12.2) has been implicated in inflammation and cancer metastasis [94, 95]. Structurally, the sLe^x hexasaccharide is based on the neolacto-series tetrasaccharide to which Fuc and sialic acid residues are appended to complete the sLe^x antigen. The chemical and chemoenzymatic synthesis of sLe^x structures has been recently reviewed by Horst Kunz [96]. The synthesis of sLe^x is made difficult by the steric hindrance encountered due to the close proximity of Fuc and Gal at O3 and O4 of GlcNAc, the low reactivity of the 4-OH group of GlcNAc toward galactosylations, the acid lability of the α-fucoside linkage, and the poor regio- and stereoselectivity as well as yields in sialylations.

The first total synthesis of sLe^x hexasaccharyl ceramide was reported by Akira Hasegawa's group in 1991 using thioglycoside donors, emphasizing on the utility

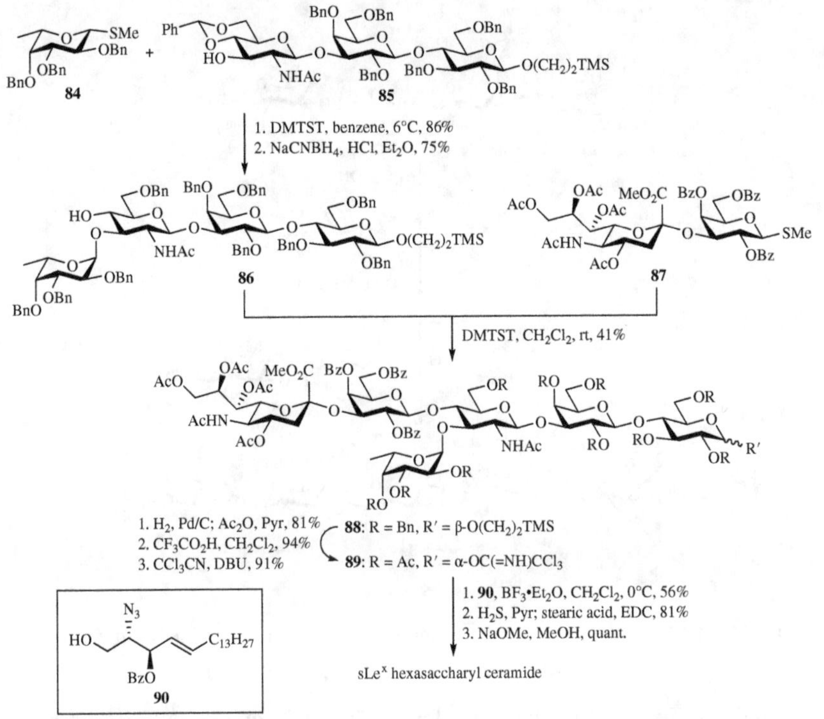

SCHEME 12.14 Hasegawa's total synthesis of sLe^x hexasaccharide ceramide.

of dimethyl(methylthio)sulfonium triflate (DMTST) as promoter (Scheme 12.14) [97]. The 3″-alcohol **85** was readily glycosylated with thiofucoside **84** using DMTST to obtain the α-linked tetrasaccharide (86%), which was subjected to regioselective 4-*O*-benzylidene ring opening to afford the 4″-alcohol **86** (75%). A [4+2] coupling of thioglycoside **87** with **86** under DMTST activation gave the hexasaccharide **88** (41%). Transformation into imidate **89** and further coupling with acceptor **90** using Schmidt's azidosphingosine protocol furnished the sLe^x hexasaccharyl ceramide construct. Similarly, Schmidt and coworkers employed the imidate methodology to synthesize fluorescence-labeled sLe^x GSLs for investigating microdomain formation in membranes [98–100]. These derivatives were prepared by replacing the natural amide-linked fatty acids with fluorescent analogues during the final steps.

Wong and coworkers tested their reactivity-based one-pot synthesis protocol to assemble the sLe^x hexasaccharide [101]. As shown in Scheme 12.15, sequential regioselective glycosylation of the phthalimide-protected 3,4-diol **92** with the thiotolyl disaccharide donor **91** at O4 and the fucosyl donor **43** at O3, followed by coupling with the lactosyl 4′-alcohol **93** afforded the sLe^x hexasaccharide **94** in 42% overall yield and in a one-pot manner. The most elegant part in this synthesis was the application of the promoter combination that circumvents the adventitious formation of the

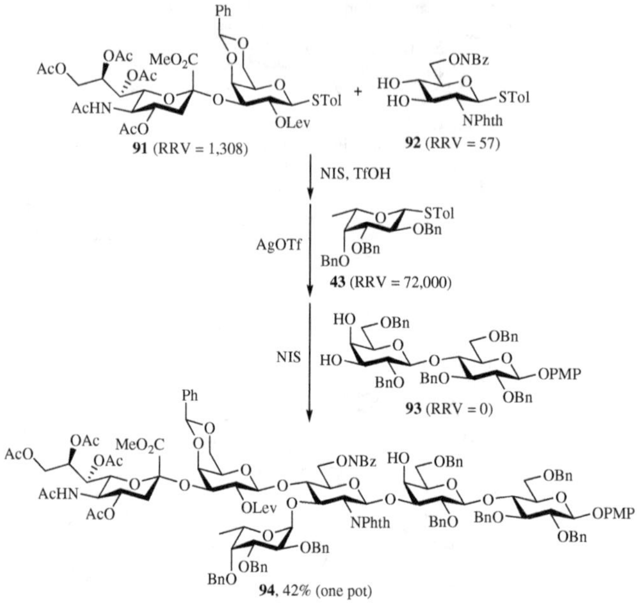

SCHEME 12.15 Wong's one-pot assembly of sLe^x hexasaccharide. Phth, phthaloyl.

succinimide derivative of the fucoside donor. First, 1.5 equiv. of NIS in combination with TfOH generates 1.5 equiv. of TolS-I in the first coupling step. Addition of AgOTf after adding thiofucoside **43** to this mixture *in situ* generates adequate amounts of TolSOTf, which further activates the fucoside for the second coupling.

Seeberger and coworkers also reported a synthesis of sLe^x hexasaccharide [102], which involved coupling the nonreducing end disaccharide segment at O4 of the glucosamine residue of the reducing end trisaccharide segment, followed by attachment of Fuc at O3. Most importantly, they reported a novel synthesis of the nonreducing end disaccharide donor (Scheme 12.16). In general, sialylation reaction has been extensively reviewed [103–105] and the synthesis of the terminal Neu5Ac ($\alpha 2 \rightarrow 3$) Gal has been the subject of intense investigation [106]. The 3-O-glycosylation of Gal is difficult mainly due to the low nucleophilicity of 3-OH and the steric hindrance of the O2 protecting group. Galactal is devoid of O2 function and its 3-OH, being allylic, is more nucleophilic. Thus, sialylation of the galactal acceptor **96** by the phosphite donor **95** by using TMSOTf as promoter and propionitrile as solvent cleanly generated the α-linked disaccharide **97** in excellent yield. The galactal portion of **97** was equipped with an *N*-phenyl trifluoacetimidate group to obtain **98**, which was used successfully as an efficient donor in the further assembly of sLe^x hexasaccharide.

A synthesis of two deoxy Lewis X pentaosyl GSLs is also reported [107]. Kyriacos Nicolaou [108], Samuel Danishefsky [109–111], and Khushi Matta [112] have reported the synthesis of the main sLe^x antigen. A fully automated enzymatic synthesis of sLe^x derivatives in 4 days and 16% overall yield from a simple GlcNac-linked aminooxy-functionalized G7 dendrimer is recently reported [113].

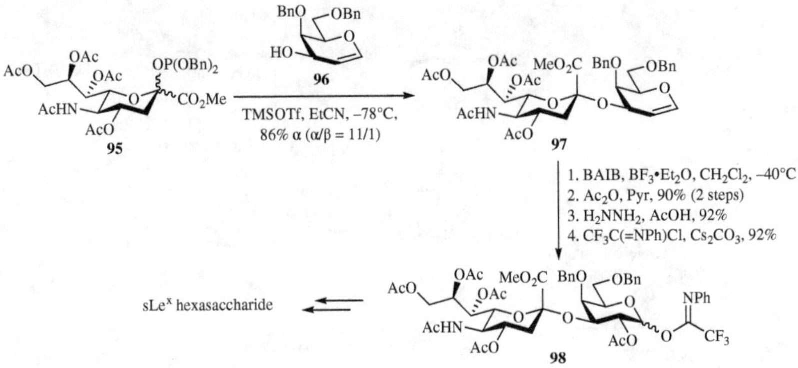

SCHEME 12.16 Seeberger's synthesis of the key sialo disaccharide. BAIB, [bis(acetoxy) iodo]benzene.

12.4.2.2 Conventional Gangliosides The chemical synthesis of gangliosides is an area of intense research. GM3 (Fig. 12.1) attracted great attention in recent years as an important TACA. It is also a key intermediate for the synthesis of higher gangliosides such as GM1, GM2, GD3, and so on. Various syntheses of GM3 have been reported by using chemical [114–124] and chemoenzymatic [125–128] means. The a-series (GM2 [129–131], GM1 [132–134], GD1a [132], and GT1a [135]) [124], b-series (GD2, GD3, GT1b, and GQ1b) [136–141], and α-series [142] GSLs have been synthesized by using various approaches. In addition, syntheses of a number of analogues of gangliosides have been reported in the last decade [143–156]. Kiso's group is very active in this area; an account of their efforts and recent developments over the past decade has been reviewed [106].

Progress in the synthesis of gangliosides has always been centered on solving the issues associated with the sialic acid moiety. Recently, Kiso and coworkers streamlined the synthesis of the a-series gangliosides by addressing the α-selective sialylation of the 3′-OH of lactose and its chromatographic separation to obtain the GM3 glycan, the β-selective glycosylation of the sterically hindered 4′-OH of the GM3 glycan, and the protecting group manipulation to convert the GM2 glycan into the corresponding acceptor [124]. To overcome these problems, they employed the *N*-Troc-protected sialyl lactose acceptor (GM3 acceptor) **101** and the *N*-Troc-protected galactosamine donor **102** capable of coupling with the unreactive 4′-OH group of **101** to form the GM2 unit (Scheme 12.17). Compound **101** was prepared by regioselective glycosylation of the lactosyl 3′,4′-diol **100** with the thiosialoside **99** promoted by NIS/TfOH at −50°C and using propionitrile as solvent. An earlier synthesis using the *N*-acetylated counterpart of **99** generated a difficulty to separate mixture (69%, α:β∼6:1), whereas **101** was easily separated on column chromatography and gave better yields (83%, α:β∼5:1). Acceptor **101** smoothly underwent stereoselective glycosylation with thiogalactoside **102** (76%, β only), followed by selective cleavage of *N*-Troc, acetylation, and desilylation to fashion the GM2 acceptor unit **103**. Further, **103** underwent facile glycosylations with thioglycoside

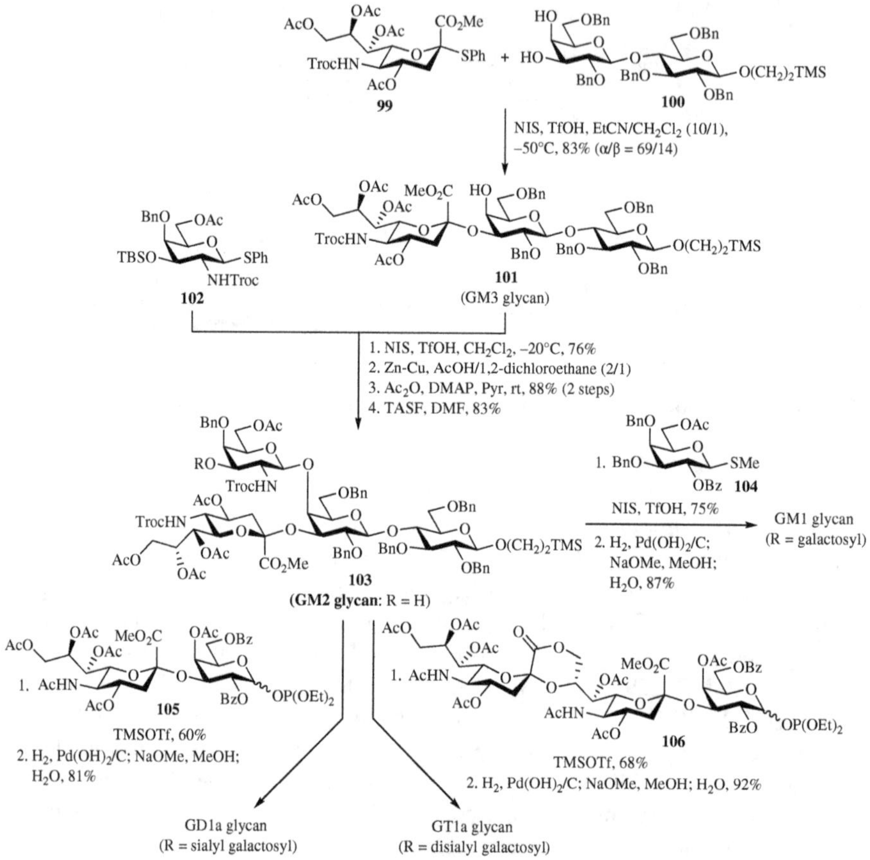

SCHEME 12.17 Synthesis of a-series ganglioside glycans. TASF, tris(dimethylamino)sulfonium difluorotrimethylsilicate.

104 and phosphites **105** and **106**, followed by global deprotections to furnish GM1, GD1a, and GT1a glycans, respectively.

Efforts have been devoted to expedite the glycan assembly. Particularly, Wong and coworkers [157, 158] extended their reactivity-based one-pot synthesis to synthesize the small-cell lung cancer marker fucosyl GM1 (GM1 bearing an α-fucosyl residue at O2 of the terminal Gal) using tolyl thioglycosides in 4 days and in 21% overall yield after global deprotection. Their strategy also involved a sialyl lactoside unit as a terminating building block, which was prepared via sialylation of the corresponding lactosyl 3′,4′,-diol by using sialyl dibenzyl phosphite donors.

Obviating the inefficient coupling of full-length glycans with ceramide or azidosphingosine, Hashimoto's ceramide cassette approach was used by Kiso and coworkers for the total synthesis of ganglioside GQ1b. As shown in Scheme 12.18, the heptasaccharide **107**, assembled using a common disialogalactose building block, was coupled to the 4-OH GlcCer acceptor **108** to obtain the GQ1b derivative **109** in an excellent 91% yield. PMB groups enhanced the reactivity of the acceptor **108** for coupling,

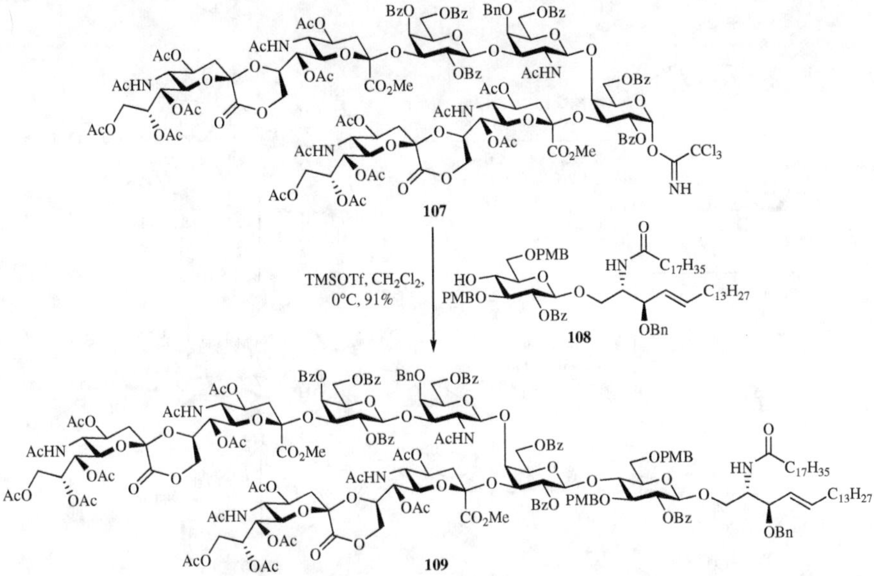

SCHEME 12.18 Kiso's total synthesis of GQ1b.

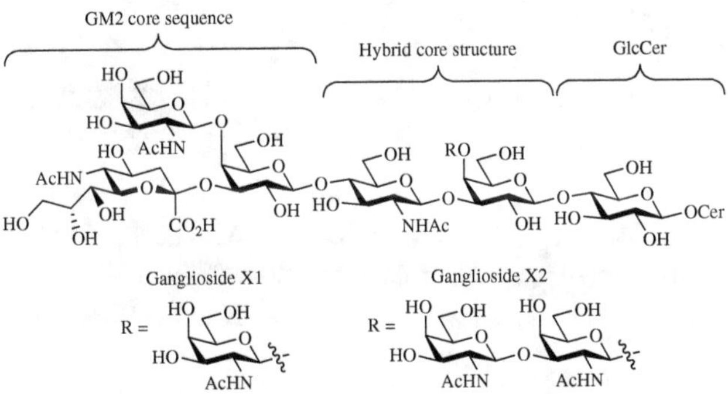

FIGURE 12.3 LcGg hybrid GSLs synthesized by Kiso and coworkers.

albeit the yield of coupling between the glucosyl donor and ceramide was only 48%. This problem was thought to arise from the self-aggregation of ceramide. It could be bypassed by tethering the ceramide to O6 of a similarly protected glucosyl donor through a succinoyl ester followed by glycosylation to afford a cyclic GlcCer in high yield (92%) [159]. Such cyclic ceramide derivatives showed similar high reactivity and efficiency in glycosylation compared to the acyclic **108** and were successfully used in the first total synthesis of GalNAc-GD1a [160]. Kiso and coworkers extended their studies to synthesize hybrid LcGg- and neolactoganglio-series gangliosides containing GM2 epitopes [161]. Recently, synthesis of such structurally related ganglioside X1 [162] and X2 [163] from LcGg series (Fig. 12.3) were accomplished by using the [3+4]+Cer and [3+4]+GlcCer (ceramide cassette) approaches, respectively.

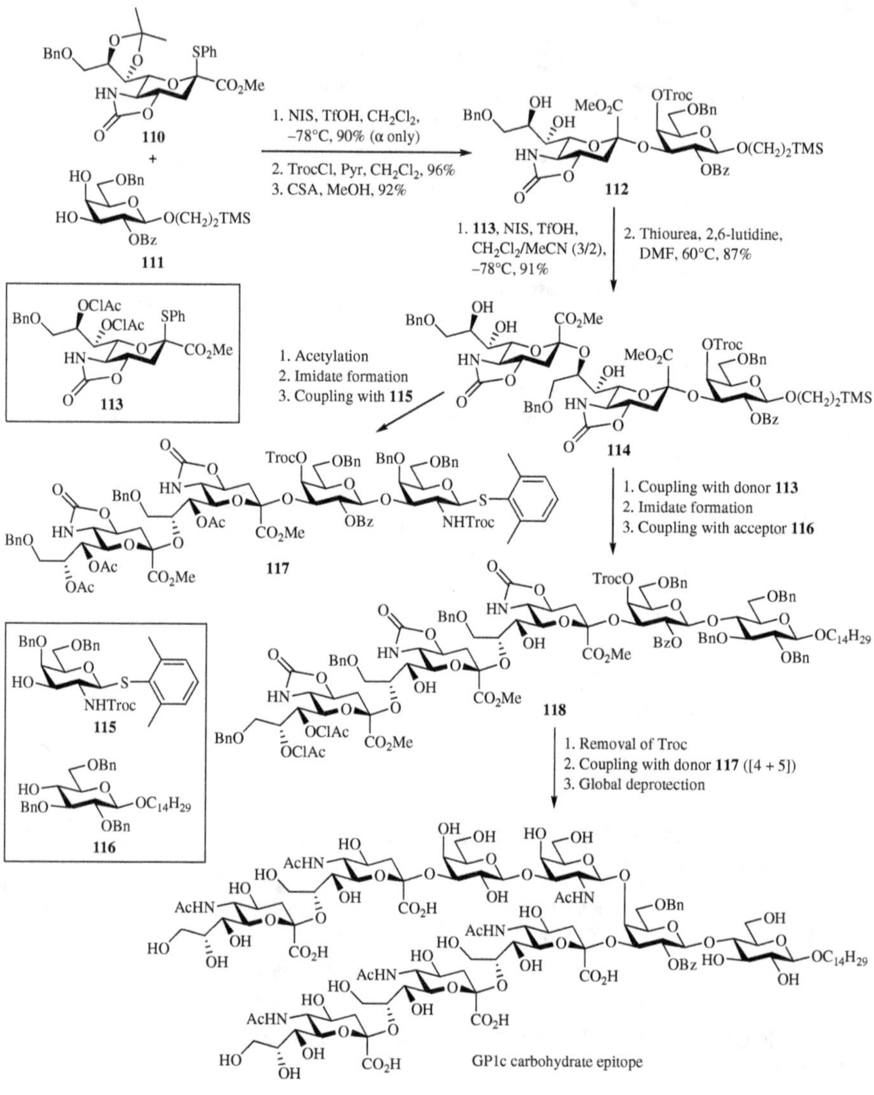

SCHEME 12.19 Takahashi's synthesis of GP1c. CSA, camphorsulfonic acid.

GP1c—one of the most complex c-series gangliosides—was recently synthesized by Takashi Takahashi and coworkers [164]. The GP1c nonasaccharide, bearing a C14 chain instead of ceramide, was synthesized by a [4+5] glycosylation strategy (Scheme 12.19). Highly regio- and stereoselective sialylation of the 4-*O*,5-*N*-oxazolidinone- and 7,8-*O*-isopropylidene-protected thiosialoside **110** with galactosyl 3,4-diol **111** using NIS/TfOH at −78°C exclusively provided the O3-linked α-disaccharide (90%). Subsequent O4 protection and isopropylidene removal provided the diol **112**. Glycosylation of **112** with the bis-ClAc-protected sialyl donor **113** generated the desired α-linked trisaccharide again as a single isomer, which upon selective removal of the ClAc groups furnished triol **114**. The couplings were highly specific; swiping the donors **110** and **113** between the first and second glycosylations caused a drop in selectivity in both the reactions. Again, **114** could be selectively α-glycosylated with **113** this time at the O8″ position to obtain the trisialylgalactoside, which was converted to the imidate and further coupled to the glucoside acceptor **116** to form the pentasaccharide **118** (TMSOTf, 94%). In parallel, triol **114** was acetylated, converted to the imidate, and chemoselectively activated in the presence of the thioglycoside acceptor **115** to obtain the corresponding tetrasaccharide **117** (TMSOTf, 95%). Here, the 2,6-dimethylphenylthio group was effective in chemoselective glycosylation without aglycone transfer. The Troc group in pentasaccharide **118** was removed (94%), and, finally, a [4+5]-glycosylation by donor **117** fashioned the GP1c construct (NIS/TfOH, 64%).

12.4.2.3 Gangliosides Containing Unusual Sialic Acid Residues Recently, echinodermatous gangliosides have attracted considerable attention due to their unusual structures and potent neurogenic activity, which greatly exceeds that of the mammalian ganglioside GM1 [165]. These gangliosides contain structural modifications, which are not seen in mammalian gangliosides. Their structures include a range of modified sialic acid residues connected by variety of linker groups, which are, at times, located in the inner portions of the molecule. To date, several series of echinodermatous gangliosides have been identified such as LLG, HLG, AG, HPG, SJG, and LMG. Kiso and coworkers developed an *N*-Troc-protected sialyl donor as key intermediate to access the modified sialic acid structures. They found that 1,5-lactam bridging in sialic acid imparts high reactivity to the C4 and C8 hydroxy groups and allows access to diasialic acid sequences such as Neu5Gc(α2→4)Neu5Ac and 8-*O*-SO_3H-Neu5Ac(α2→8)Neu5Ac in high yields [166]. They extended this work to accomplish the first total synthesis of the sea cucumber ganglioside HLG-2 [167].

The tetrasaccharide moiety of the starfish ganglioside LLG-3, which contains the 8-*O*-Me-Neu5Ac(α2→11)Neu5Gc structure, was first synthesized by Ken-ichi Sato and coworkers [168]. Kiso's group achieved the first total synthesis [169] of LLG-3 by combining their Neu5Troc chemistry and the ceramide cassette approach (Scheme 12.20). The stereoselective incorporation of benzylglycolate at the anomeric position of the thiosialoside donor **119** gave **120**. Selective 8-O-methylation from **120** is a challenge. For this, **120** was first treated with Zn under acidic conditions in DMF to afford the 8-hydroxy-*N*-acetyl sialic acid derivative through the concomitant O8-to-N5 acetyl group migration. Because methylation was not successful under a variety of conditions, a detour was followed. Chloracetylation at O8, N-acetylation of the 5-acetamide group (NAc→NAc_2), dechloroacetylation using 1-selenocarbamoylpiperidine and 2,6-lutidine as acid scavenger, methylation using Meerwein's reagent, and debenzylation afforded **121**. In parallel, coupling of **119** with diol **122** selectively offered the corresponding α(2→3) sialogalactoside. The free 4-OH of the disaccharide was acetylated and the *N*-Troc group was selectively removed using Zn and AcOH to obtain the amine **123**. Here, the use of acetonitrile in place of DMF suppressed the O8-to-N5 migration of the acetyl group during Troc removal.

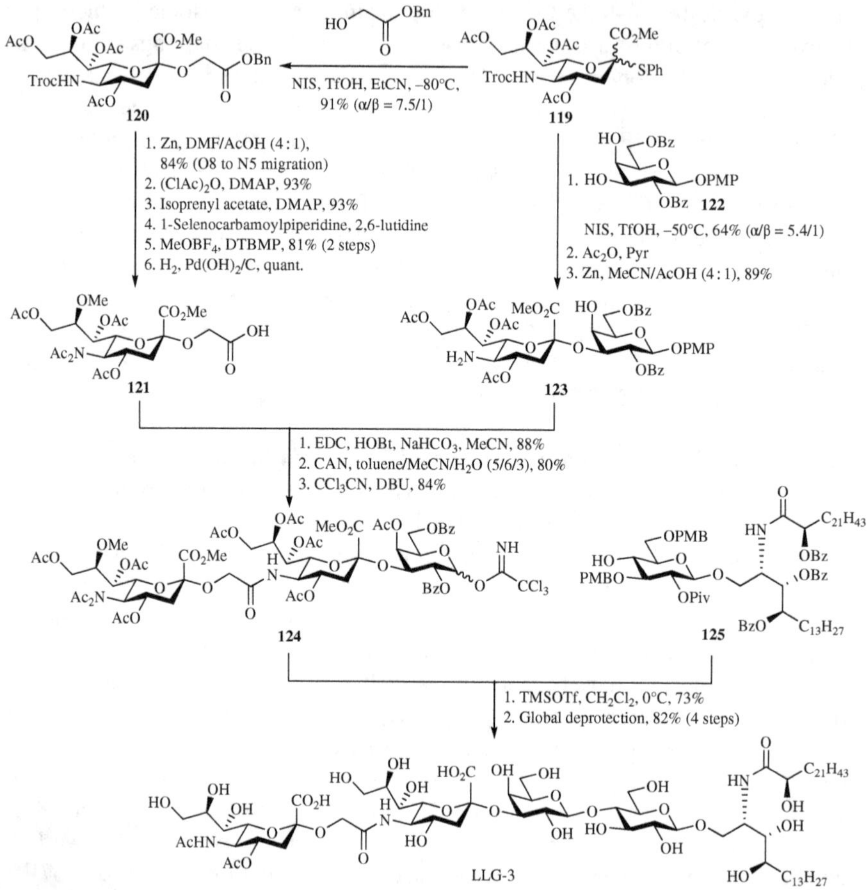

SCHEME 12.20 Total synthesis of starfish ganglioside LLG-3. CAN, ceric ammonium nitrate, DTBMP, 2,6-di-tert-butyl-4-methylpyridine, HOBt, 1-hydroxybenzotriazole.

Compounds **121** and **123** were coupled to get the amide. The PMP trisaccharide was converted to the imidate **124**, which was glycosylated with the GlcCer **125** containing a phytoceramide moiety via the ceramide cassette approach. Global deprotection delivered LLG-3. An interesting observation recorded during the formation of **125** was that the coupling of glucosyl donor with the ceramide acceptor was efficient only when a TBS group was present at O4 of Glc (81%), whereas a ClAc group at that position caused a drop in yield (23%). Very recently, Jamie Rich and Stephen Withers developed a chemoenzymatic synthesis of LLG-3 using an engineered and evolved synthase [170]. On similar lines, the 8-*O*-Me-Neu5Troc thioglycoside donor was successfully used to carry out the first synthesis of the pentasaccharide moiety of the ganglioside GAA-7 [171].

Apart from the unusual modifications, the atypical positioning of the sialic acid residue is among the noted structural variations. Yow-Fu Tsai and coworkers, for instance, reported the synthesis of the ganglioside Hp-s1 analogue in which the sialic acid is linked at O6 of GlcCer [172]. More complex structures result when the sialic acid moiety is located as the inner core residues. For example, Sato and coworkers reported the synthesis of the starfish ganglioside AG2 glycan containing sialic acid as residue III of the linear pentasaccharide structure [173]. Analogously, Kiso and coworkers reported a synthesis of the ganglioside HPG-1 glycan with embedded disialic acid by employing the 1,5-lactamized sialoside acceptor [174]. Their synthesis of the related saccharide moiety of HPG-7 containing an unusual trimer of sialic acid as inner sugar residues is shown in Scheme 12.21 [175]. The lactamized triol **126** was regioselectively glycosylated at O4 with thiosialoside **127** to obtain trisaccharide **128** (50%). The 1,5-lactam form of sialic acid raises the suitability of the 4-OH group toward sialylation. At this stage a Boc group was introduced to the lactam bridge, which was concomitantly opened by a methoxide ion to produce trimer **129**. Amide bond formation of the amine form of **129** and the acid unit **130** followed by multifunctional group transformations afforded the complex glycan moiety of HPG-7.

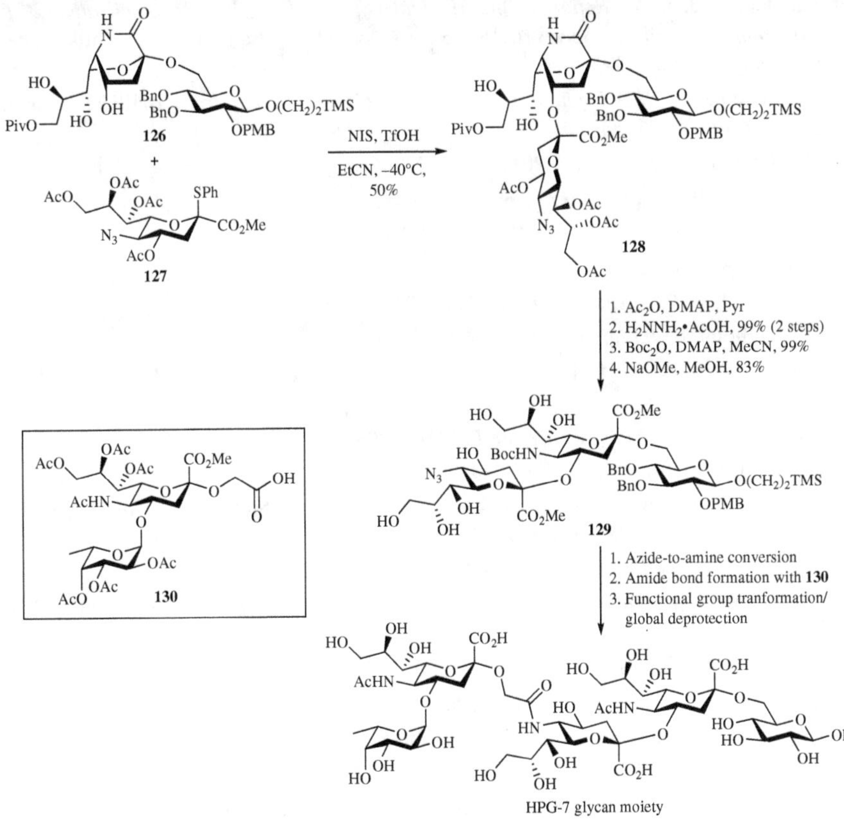

SCHEME 12.21 Synthesis of the glycan moiety of HPG-7. Boc, *tert*-butoxycarbonyl.

12.5 CONCLUSION

Several structurally complex and biologically important GSLs have been synthesized over the past two decades by employing various selective glycosylation protocols. The glycan assembly is further expedited and simplified by the advent of one-pot and automated glycosylations together with the one-pot protection of carbohydrates [176, 177] reported over the last decade. These developments are expected to speed up the biological evaluations of these GSLs and assist in the discovery of carbohydrate-based drugs and vaccines in years to come.

REFERENCES

[1] Schnaar, R. L.; Suzuki, A.; Stanley, P. In *Essentials of Glycobiology*; 2nd ed.; Varki, A.; Cummings, R. D.; Esko, J. D.; Freeze, H. H.; Stanley, P.; Bertozzi, C., Hart, G.,; Etzler, M. E., Eds.; Cold Spring Harbor Laboratory: Cold Spring Harbor, NY, **2009**, pp 129–141.

[2] Vankar, Y. D.; Schmidt, R. R. *Chem. Soc. Rev.* **2000**, *29*, 201–216.

[3] Thudichum, J. L. W. *Report of the Medical Officer of the Privy Council and Local Government Board New Series III*; Printed by George E. Eyre and William Spottiswoode: London, **1874**, pp 113–247.

[4] Thudichum, J. L. W. *A Treatise on the Chemical Constitution of Brain*; Bailliere, Tindall, and Cox: London, **1884**.

[5] Merrill, A. H., Jr. *Chem. Rev.* **2011**, *111*, 6387–6422.

[6] Simons, K.; Ikonen, E. *Nature* **1997**, *387*, 569–572.

[7] Kolter, T. *Chem. Phys. Lipids* **2011**, *164*, 590–606.

[8] Sonnino, S.; Chigorno, V. *Biochim. Biophys. Acta* **2000**, *1469*, 63–77.

[9] Chester, M. A. *Pure Appl. Chem.* **1997**, *69*, 2475–2487.

[10] Svennerholm, L. *J. Lipid Res.* **1964**, *5*, 145–155.

[11] Irie, F.; Hidari, K. I.-P. J.; Tai, T.; Li, Y.-T.; Seyama, Y.; Hirabayashi, Y. *FEBS Lett.* **1994**, *351*, 291–294.

[12] Zhang, X.; Kiechle, F. L. *Ann. Clin. Lab. Sci.* **2004**, *34*, 3–13.

[13] Fantini, J. *Curr. Med. Chem.* **2007**, *14*, 2911–2917.

[14] Harouse, J. M.; Bhat, S.; Spotalnik, S.; Laughin, M.; Stefano, K.; Silberberg, F.; Gonzalez-Scarano, F. *Science* **1991**, *253*, 320–323.

[15] Jenkins, G. M. *Cell. Mol. Life Sci.* **2003**, *60*, 701–710.

[16] Maggio, B.; Fanani, M. L.; Rosetti, C. M.; Wilke, N. *Biochim. Biophys. Acta* **2006**, *1758*, 1922–1944.

[17] Hakomori, S.-I. *Biochim. Biophys. Acta* **2008**, *1780*, 325–346.

[18] Bieberich, E. *Neurochem. Res.* **2011**, *36*, 1601–1611.

[19] Hyenne, V.; Labouesse, M. *Nat. Cell Biol.* **2011**, *13*, 1185–1187.

[20] Patwardhan, G. A.; Liu, Y.-Y. *Prog. Lipid Res.* **2011**, *50*, 104–114.

[21] Yu, R. K.; Tsai, Y.-T.; Ariga, T.; Yanagisawa, M. *J. Oleo Sci.* **2011**, *60*, 537–544.

[22] Yin, Z.; Huang, X., *J. Carbohydr. Chem.* **2012**, *31*, 143–186.

[23] Banchet-Cadeddu, A.; Hénon, E.; Dauchez, M.; Renault, J.-H.; Monneaux, F.; Haudrechy, A. *Org. Biomol. Chem.* **2011**, *9*, 3080–3104.

[24] Tan, R. X.; Chen, J. H. *Nat. Prod. Rep.* **2003**, *20*, 509–534.

[25] Compostella, F.; Panza, L.; Ronchetti, F. C. R. *Chimie* **2012**, *15*, 37–45.

[26] Pellissier, H. *Tetrahedron* **2005**, *61*, 2947–2993.

[27] Demchenko, A. V., Ed. *Handbook of Chemical Glycosylation*; Wiley: Weinheim, **2008**.

[28] Morales-Serna, J. A.; Boutureira, O.; Díaz, Y.; Matheu, M. I.; Castillón, S. *Carbohydr. Res.* **2007**, *342*, 1595–1612.

[29] Takuya, T.; Kenji, M. *Trends Glycosci. Glycotechnol.* **2010**, *22*, 280–295.

[30] Tashiro, T. *Biosci. Biotechnol. Biochem.* **2012**, *76*, 1055–1067.

[31] Hakomori, S.-I. *Glycoconj. J.* **2000**, *17*, 627–647.

[32] Slovin, S. F.; Ragupathi, G.; Adluri, S.; Ungers, G.; Terry, K.; Kim, S.; Spassova, M.; Bornmann, W. G.; Fazzari, M.; Dantis, L.; Olkiewicz, K.; Lloyd, K. O.; Livingston, P. O.; Danishefsky, S. J.; Scher, H. I. *Proc. Natl. Acad. Sci. U. S. A.* **1999**, *96*, 5710–5715.

[33] Huang, C.-Y.; Thayer, D. A.; Chang, A. Y.; Best, M. D.; Hoffmann, J.; Head, S.; Wong, C.-H. *Proc. Natl. Acad. Sci. U. S. A.* **2006**, *103*, 15–20.

[34] Ragupathi, G.; Koide, F.; Livingston, P. O.; Cho, Y. S.; Endo, A.; Wan, Q.; Spassova, M. K.; Keding, S. J.; Allen, J.; Ouerfelli, O.; Wilson, R. M.; Danishefsky, S. J. *J. Am. Chem. Soc.* **2006**, *128*, 2715–2725.

[35] Zhu, J.; Wan, Q.; Ragupathi, G.; George, C. M.; Livingston, P. O.; Danishefsky, S. J. *J. Am. Chem. Soc.* **2009**, *131*, 4151–4158.

[36] Kovbasnjuk, O.; Mourtazina, R.; Baibakov, B.; Wang, T.; Elowsky, C.; Choti, M. A.; Kane, A.; Donowitz, M. *Proc. Natl. Acad. Sci. U. S. A.* **2005**, *102*, 19087–19092.

[37] Farkas-Himsley, H.; Hill, R.; Rosen, B.; Arab, S.; Lingwood, C. A. *Proc. Natl. Acad. Sci. U. S. A.* **1995**, *92*, 6996–7000.

[38] Naiki, M.; Marcus, D. M. *Biochem. Biophys. Res. Commun.* **1974**, *60*, 1105–1111.

[39] Lindberg, A. A.; Brown, J. E.; Strömberg, N.; Westling-Ryd, M.; Schultz, J. E.; Karlsson, K.-A. *J. Biol. Chem.* **1987**, *262*, 1779–1785.

[40] Kint, J. A. *Science* **1970**, *167*, 1268–1269.

[41] Puri, A.; Hug, P.; Jernigan, K.; Barchi, J.; Kim, H.-Y.; Hamilton, J.; Wiels, J.; Murray, G. J.; Brady, R. O.; Blumenthal, R. *Proc. Natl. Acad. Sci. U. S. A.* **1998**, *95*, 14435–14440.

[42] Zhou, D.; Mattner, J.; Cantu, C., III; Schrantz, N.; Yin, N.; Gao, Y.; Sagiv, Y.; Hudspeth, K.; Wu, Y.-P.; Yamashita, T.; Teneberg, S.; Wang, D.; Proia, R. L.; Levery, S. B.; Savage, P. B.; Teyton, L.; Bendelac, A. *Science* **2004**, *306*, 1786–1789.

[43] Shapiro, D.; Acher, A. J. *Chem. Phys. Lipids* **1978**, *22*, 197–206.

[44] Jacquinet, J.-C.; Sinay, P. *Carbohydr. Res.* **1985**, *143*, 143–149.

[45] Garegg, P. J.; Hultberg, H. *Carbohydr. Res.* **1982**, *110*, 261–266.

[46] Cox, D. D.; Metzner, E. K.; Reist, E. J. *Carbohydr. Res.* **1978**, *63*, 139–147.

[47] Dahmén, J.; Frejd, T.; Magnusson, G.; Noori, G.; Carlström, A.-S. *Carbohydr. Res.* **1984**, *127*, 15–25.

[48] Paulsen, H.; Bünsch, A. *Carbohydr. Res.* **1982**, *100*, 143–167.

[49] Koike, K.; Sugimoto, M.; Sato, S.; Ito, Y.; Nakahara, Y.; Ogawa, T. *Carbohydr. Res.* **1987**, *163*, 189–208.

[50] Schmidt, R. R.; Zimmermann, P. *Angew. Chem. Int. Ed. Engl.* **1986**, *25*, 725–726.

[51] Nicolaou, K. C.; Caulfield, T.; Kataoka, H.; Kumazawa, T. *J. Am. Chem. Soc.* **1988**, *110*, 7910–7912.

[52] Nicolaou, K. C.; Caulfield, T. J.; Katoaka, H. *Carbohydr. Res.* **1990**, *202*, 177–191.

[53] Qiu, D.; Schmidt, R. R. *Liebigs Ann. Chem.* **1992**, 217–224.

[54] Hashimoto, S.-I.; Sakamoto, H.; Honda, T.; Abe, H.; Nakamura, S.-I.; Ikegami, S. *Tetrahedron Lett.* **1997**, *38*, 8969–8972.

[55] Dohi, H.; Nishida, Y.; Takeda, T.; Kobayashi, K. *Carbohydr. Res.* **2002**, *337*, 983–989.

[56] Miyagawa, A.; Kurosawa, H.; Watanabe, T.; Koyama, T.; Terunuma, D.; Matsuoka, K. *Carbohydr. Polym.* **2004**, *57*, 441–450.

[57] Matsuoka, K.; Terabatake, M.; Esumi, Y.; Terunuma, D.; Kuzuhara, H. *Tetrahedron Lett.* **1999**, *40*, 7839–7842.

[58] Kitov, P. I.; Bundle, D. R. *J. Chem. Soc. Perkin Trans. 1*, **2001**, 838–853.

[59] Miura, T.; Inazu, T. *Tetrahedron Lett.* **2003**, *44*, 1819–1821.

[60] Miura, T.; Tsujino, S.; Satoh, A.; Goto, K.; Mizuno, M.; Noguchi, M.; Kajimoto, T.; Node, M.; Murakami, Y.; Imai, N.; Inazu, T. *Tetrahedron* **2005**, *61*, 6518–6526.

[61] Adlercreutz, D.; Weadge, J. T.; Petersen, B. O.; Duus, J. Ø.; Dovichi, N. J.; Palcic, M. M. *Carbohydr. Res.* **2010**, *345*, 1384–1388.

[62] Wang, C.; Li, Q.; Wang, H.; Zhang, L.-H.; Ye, X.-S. *Tetrahedron* **2006**, *62*, 11657–11662.

[63] Durón, S. G.; Polat, T.; Wong, C.-H. *Org. Lett.* **2004**, *6*, 839–841.

[64] Chao, C.-S.; Li, C.-W.; Chen, M.-C.; Chang, S.-S.; Mong, K.-K. T. *Chem. Eur. J.* **2009**, *15*, 10972–10982.

[65] Imamura, A.; Ando, H.; Korogi, S.; Tanabe, G.; Muraoka, O.; Ishida, H.; Kiso, M. *Tetrahedron Lett.* **2003**, *44*, 6725–6728.

[66] Imamura, A.; Ando, H.; Ishida, H.; Kiso, M. *Org. Lett.* **2005**, *7*, 4415–4418.

[67] Imamura, A.; Kimura, A.; Ando, H.; Ishida, H.; Kiso, M. *Chem. Eur. J.* **2006**, *12*, 8862–8870.

[68] Kimura, A.; Imamura, A.; Ando, H.; Ishida, H.; Kiso, M. *Synlett* **2006**, 2379–2382.

[69] Xia, C.; Yao, Q.; Schümann, J.; Rossy, E.; Chen, W.; Zhu, L.; Zhang, W.; De Libero, G.; Wang, P. G. *Bioorg. Med. Chem. Lett.* **2006**, *16*, 2195–2199.

[70] Yao, Q.; Song, J.; Xia, C.; Zhang, W.; Wang, P. G. *Org. Lett.* **2006**, *8*, 911–914.

[71] Xia, C.; Schümann, J.; Emmanuel, R.; Zhang, Y.; Chen, W.; Zhang, W.; De Libero, G.; Wang, P. G. *J. Med. Chem.* **2007**, *50*, 3489–3496.

[72] Xia, C.; Zhou, D.; Liu, C.; Lou, Y.; Yao, Q.; Zhang, W.; Wang, P. G. *Org. Lett.* **2006**, *8*, 5493–5496.

[73] Chen, W.; Xia, C.; Wang, J.; Thapa, P.; Li, Y.; Nadas, J.; Zhang, W.; Zhou, D.; Wang, P. G. *J. Org. Chem.* **2007**, *72*, 9914–9923.

[74] Yin, N.; Long, X.; Goff, R. D.; Zhou, D.; Cantu, C., III; Mattner, J.; Mezard, P. S.; Teyton, L.; Bendelac, A.; Savage, P. B. *ACS Chem. Biol.* **2009**, *4*, 191–197.

[75] Nunomura, S.; Ogawa, T. *Tetrahedron Lett.* **1988**, *29*, 5681–5684.

[76] Nilsson, U.; Magnusson, G. *Carbohydr. Res.* **1995**, *272*, 9–16.

[77] Park, T. K.; Kim, I. J.; Danishefsky, S. J. *Tetrahedron Lett.* **1995**, *36*, 9089–9092.

[78] Bosse, F.; Marcaurelle, L. A.; Seeberger, P. H. *J. Org. Chem.* **2002**, *67*, 6659–6670.

[79] Cheshev, P. E.; Khatuntseva, E. A.; Gerbst, A. G.; Tsvetkov, Y. E.; Shashkov, A. S.; Nifantiev, N. E. *Russ. J. Bioorg. Chem.* **2003**, *29*, 372–381.

[80] Pornsuriyasak, P.; Demchenko, A. V. *Carbohydr. Res.* **2006**, *341*, 1458–1466.

[81] Randriantsoa, M.; Drouillard, S.; Breton, C.; Samain, E. *FEBS Lett.* **2007**, *581*, 2652–2656.

[82] Bilodeau, M. T.; Park, T. K.; Hu, S.; Randolph, J. T.; Danishefsky, S. J.; Livingston, P. O.; Zhang, S. J. *Am. Chem. Soc.* **1995**, *117*, 7840–7841.

[83] Park, T. K.; Kim, I. J.; Hu, S.; Bilodeau, M. T.; Randolph, J. T.; Kwon, O.; Danishefsky, S. J. *J. Am. Chem. Soc.* **1996**, *118*, 11488–11500.

[84] Jeon, I.; Iyer, K.; Danishefsky, S. J. *J. Org. Chem.* **2009**, *74*, 8452–8455.

[85] Lassaletta, J. M.; Schmidt, R. R. *Liebigs Ann.* **1996**, 1417–1423.

[86] Zhu, T.; Boons, G.-J. *Angew. Chem. Int. Ed.* **1999**, *38*, 3495–3497.

[87] Burkhart, F.; Zhang, Z.; Wacowich-Sgarbi, S.; Wong, C.-H. *Angew. Chem. Int. Ed.* **2001**, *40*, 1274–1277.

[88] Werz, D. B.; Castagner, B.; Seeberger, P. H. *J. Am. Chem. Soc.* **2007**, *129*, 2770–2771.

[89] Wang, Z.; Zhou, L.; El-Boubbou, K.; Ye, X.-S.; Huang, X. *J. Org. Chem.* **2007**, *72*, 6409–6420.

[90] Wang, Z.; Gilbert, M.; Eguchi, H.; Yu, H.; Cheng, J.; Muthana, S.; Zhou, L.; Wang, P. G.; Chen, X.; Huang, X. *Adv. Synth. Catal.* **2008**, *350*, 1717–1728.

[91] Adinolfi, M.; Iadonisi, A.; Ravidà, A.; Schiattarella, M. *J. Org. Chem.* **2005**, *70*, 5316–5319.

[92] Bobowski, M.; Cazet, A.; Steenackers, A.; Delannoy, P. In *Carbohydrate Chemistry: Chemical and Biological Approaches*; Rauter, A. P., Lindhorst, T. K., Eds.; Royal Society of Chemistry: Cambridge, UK, **2009**; Vol. *37*, pp 1–20.

[93] Ikami, T.; Ishida, H.; Kiso, M. In *Sphingolipid Metabolism and Cell Signaling Part A*; Merril, A. H., Jr.; Hannun, Y. A., Eds.; Methods in Enzymology 311; Academic: San Diego, CA, **1999**, pp 547–568.

[94] Tyrrell, D.; James, P.; Rao, N.; Foxall, C.; Abbas, S.; Dasgupta, F.; Nashed, M.; Hasegawa, A.; Kiso, M.; Asa, D.; Kidd, J.; Brandley, B. K. *Proc. Natl. Acad. Sci. U. S. A.* **1991**, *88*, 10372–10376.

[95] Kannagi, R.; Izawa, M.; Koike, T.; Miyazaki, K.; Kimura, N. *Cancer Sci.* **2004**, *95*, 377–384.

[96] Pudelko, M.; Bull, J.; Kunz, H. *ChemBioChem* **2010**, *11*, 904–930.

[97] Kameyama, A.; Ishida, H.; Kiso, M.; Hasegawa, A. *J. Carbohydr. Chem.* **1991**, *10*, 549–560.

[98] Gege, C.; Vogel, J.; Bendas, G.; Rothe, U.; Schmidt, R. R. *Chem. Eur. J.* **2000**, *6*, 111–122.

[99] Gege, C.; Oscarson, S.; Schmidt, R. R. *Tetrahedron Lett.* **2001**, *42*, 377–380.

[100] Gege, C.; Geyer, A.; Schmidt, R. R. *Chem. Eur. J.* **2002**, *8*, 2454–2463.

[101] Zhang, Z.; Niikura, K.; Huang, X.-F.; Wong, C.-H. *Can. J. Chem.* **2002**, *80*, 1051–1054.

[102] Hanashima, S.; Castagner, B.; Esposito, D.; Nokami, T.; Seeberger, P. H. *Org. Lett.* **2007**, *9*, 1777–1779.

[103] Boons, G.-J.; Demchenko, A. V. *Chem. Rev.* **2000**, *100*, 4539–4565.

[104] Halcomb, R. L.; Chappell, M. D. *J. Carbohydr. Chem.* **2002**, *21*, 723–768.

[105] Ando, H.; Imamura, A. *Trends Glycosci. Glycotechnol.* **2004**, *16*, 293–303.

[106] Ando, H.; Ishida, H.; Kiso, M. In *Glycomics*; Fukuda, M., Ed.; Methods in Enzymology 478; Academic Press: San Diego, CA, **2010**, pp 521–540.

[107] Luo, Y.; Gourmala, C.; Dong, D.; Barbault, F.; Fan, B.; Hu, Y.; Zhang, Y. *Glycoconj. J.* **2008**, *25*, 335–344.

[108] Nicolaou, K. C.; Hummel, C. W.; Bockovich, N. J.; Wong, C.-H. *J. Chem. Soc. Chem. Commun.* **1991**, 870–872.

[109] Danishefsky, S. J.; Gervay, J.; Peterson, J. M.; McDonald, F. E.; Koseki, K.; Oriyama, T.; Griffith, D. A.; Wong, C.-H.; Dumas, D. P. *J. Am. Chem. Soc.* **1992**, *114*, 8329–8331.

[110] Danishefsky, S. J.; Koseki, K.; Griffith, D. A.; Gervay, J.; Peterson, J. M.; McDonald, F. E.; Oriyama, T. *J. Am. Chem. Soc.* **1992**, *114*, 8331–8333.

[111] Danishefsky, S. J.; Gervay, J.; Peterson, J. M.; McDonald, F. E.; Koseki, K.; Griffith, D. A.; Oriyama, T.; Marsden, S. P. *J. Am. Chem. Soc.* **1995**, *117*, 1940–1953.

[112] Jain, R. K.; Vig, R.; Locke, R. D.; Mohammad, A.; Matta, K. L. *Chem. Commun.* **1996**, 65–67.

[113] Matsushita, T.; Nagashima, I.; Fumoto, M.; Ohta, T.; Yamada, K.; Shimizu, H.; Hinou, H.; Naruchi, K.; Ito, T.; Kondo, H.; Nishimura, S.-I. *J. Am. Chem. Soc.* **2010**, *132*, 16651–16656.

[114] Sugimoto, M.; Ogawa, T. *Glycoconj. J.* **1985**, *2*, 5–9.

[115] Murase, T.; Ishida, H.; Kiso, M.; Hasegawa, A. *Carbohydr. Res.* **1989**, *188*, 71–80.

[116] Numata, M.; Sugimoto, M.; Ito, Y.; Ogawa, T. *Carbohydr. Res.* **1990**, *203*, 205–217.

[117] Martin, T. J.; Brescello, R.; Toepfer, A.; Schmidt, R. R. *Glycoconj. J.* **1993**, *10*, 16–25.

[118] Tomoo, T.; Kondo, T.; Abe, H.; Tsukamoto, S.; Isobe, M.; Goto, T. *Carbohydr. Res.* **1996**, *284*, 207–222.

[119] Tietze, L. F.; Gretzke, D. *Eur. J. Org. Chem.* **1998**, 1895–1899.

[120] Sakamoto, H.; Nakamura, S.; Tsuda, T.; Hashimoto, S. *Tetrahedron Lett.* **2000**, *41*, 7691–7695.

[121] Xing, G.-W.; Chen, L.; Liang, F.-F. *Eur. J. Org. Chem.* **2009**, 5963–5970.

[122] Duclos, R. I., Jr. *Carbohydr. Res.* **2000**, *328*, 489–507.

[123] Liu, Y.; Ruan, X.; Li, X.; Li, Y. *J. Org. Chem.* **2008**, *73*, 4287–4290.

[124] Komori, T.; Imamura, A.; Ando, H.; Ishida, H.; Kiso, M. *Carbohydr. Res.* **2009**, *344*, 1453–1463.

[125] Ito, Y.; Paulson, J. C. *J. Am. Chem. Soc.* **1993**, *115*, 1603–1605.

[126] Liu, K. K.-C.; Danishefsky, S. J. *J. Am. Chem. Soc.* **1993**, *115*, 4933–4934.

[127] Yamada, K.; Fujita, E.; Nishimura, S.-I. *Carbohydr. Res.* **1998**, *305*, 443–461.

[128] Zehavi, U.; Tuchinsky, A. *Glycoconj. J.* **1998**, *15*, 657–662.

[129] Sugimoto, M.; Numata, M.; Koike, K.; Nakahara, Y.; Ogawa, T. *Carbohydr. Res.* **1986**, *156*, C1–C5.

[130] Hasegawa, A.; Nagahama, T.; Ohki, H.; Kiso, M. *J. Carbohydr. Chem.* **1992**, *11*, 699–714.

[131] Castro-Palomino, J. C.; Ritter, G.; Fortunato, S. R.; Reinhardt, S.; Old, L. J.; Schmidt, R. R. *Angew. Chem. Int. Ed.* **1997**, *36*, 1998–2001.

[132] Hasegawa, A.; Nagahama, T.; Kiso, M. *Carbohydr. Res.* **1992**, *235*, C13–C17.

[133] Stauch, T.; Greilich, U.; Schmidt, R. R. *Liebigs Ann.* **1995**, 2101–2111.

[134] Greilich, U.; Brescello, R.; Jung, K.-H.; Schmidt, R. R. *Liebigs Ann.* **1996**, 663–672.

[135] Ishida, H.-K.; Ando, H.; Ito, H.; Ishida, H.; Kiso, M.; Hasegawa, A. *J. Carbohydr. Chem.* **1997**, *16*, 413–428.

[136] Hasegawa, A.; Ishida, H.; Isogai, Y.; Ishida, H.; Kiso, M. *J. Carbohydr. Chem.* **1993**, *12*, 1217–1222.

[137] Ishida, H.; Ohta, Y.; Tsukada, Y.; Isogai, Y.; Ishida, H.; Kiso, M.; Hasegawa, A. *Carbohydr. Res.* **1994**, *252*, 283–290.

[138] Ishida, H.-K.; Ishida, H.; Kiso, M.; Hasegawa, A. *Carbohydr. Res.* **1994**, *260*, C1–C6.

[139] Ishida, H.-K.; Ishida, H.; Kiso, M.; Hasegawa, A. *Tetrahedron: Asymmetry* **1994**, *5*, 2493–2512.

[140] Castro-Palomino, J. C.; Simon, B.; Speer, O.; Leist, M.; Schmidt, R. R. *Chem. Eur. J.* **2001**, *7*, 2178–2184.

[141] Imamura, A.; Ando, H.; Ishida, H.; Kiso, M. *J. Org. Chem.* **2009**, *74*, 3009–3023.

[142] Ito, H.; Ishida, H.; Kiso, M. *J. Carbohydr. Chem.* **2001**, *20*, 207–225.

[143] Janssen, S.; Schmidt, R. R. *J. Carbohydr. Chem.* **2005**, *24*, 611–647.

[144] Andersen, S. M.; Ling, C.-C.; Zhang, P.; Townson, K.; Willison, H. J.; Bundle, D. R. *Org. Biomol. Chem.* **2004**, *2*, 1199–1212.

[145] Rich, J. R.; Wakarchuk, W. W.; Bundle, D. R. *Chem. Eur. J.* **2006**, *12*, 845–858.

[146] Otsubo, N.; Ishida, H.; Kiso, M. *Carbohydr. Res.* **2001**, *330*, 1–5.

[147] Kasuya, M. C.; Ito, A.; Hatanaka, K. *J. Fluorine Chem.* **2007**, *128*, 562–565.

[148] Hirai, G.; Watanabe, T.; Yamaguchi, K.; Miyagi, T.; Sodeoka, M. *J. Am. Chem. Soc.* **2007**, *129*, 15420–15421.

[149] Sodeoka, M.; Hirai, G.; Watanabe, T.; Miyagi, T. *Pure Appl. Chem.* **2009**, *81*, 205–215.

[150] Kurosu, M.; Kitagawa, I. *J. Carbohydr. Chem.* **2006**, *25*, 427–439.

[151] Zheng, M.; Ye, X.-S. *Tetrahedron* **2012**, *68*, 1475–1482.

[152] Komori, T.; Ando, T.; Imamura, A.; Li, Y.-T.; Ishida, H.; Kiso, M. *Glycoconj. J.* **2008**, *25*, 647–661.

[153] Pukin, A. V.; Weijers, C. A. G. M.; van Lagen, B.; Wechselberger, R.; Sun, B.; Gilbert, M.; Karwaski, M.-F.; Florack, D. E. A.; Jacobs, B. C.; Tio-Gillen, A. P.; van Belkum, A.; Endtz, H. P.; Visser, G. M.; Zuilhof, H. *Carbohydr. Res.* **2008**, *343*, 636–650.

[154] Fujikawa, K.; Imamura, A.; Ishida, H.; Kiso, M. *Carbohydr. Res.* **2008**, *343*, 2729–2734.

[155] Yu, H.; Cheng, J.; Ding, L.; Khedri, Z.; Chen, Y.; Chin, S.; Lau, K.; Tiwari, V. K.; Chen, X. *J. Am. Chem. Soc.* **2009**, *131*, 18467–18477.

[156] Pukin, A. V.; Florack, D. E. A.; Brochu, D.; van Lagen, B.; Visser, G. M.; Wennekes, T.; Gilbert, M.; Zuilhof, H. *Org. Biomol. Chem.* **2011**, *9*, 5809–5815.

[157] Mong, T. K.-K.; Lee, H.-K.; Durón, S. G.; Wong, C.-H. *Proc. Natl. Acad. Sci. U. S. A.* **2003**, *100*, 797–802.

[158] Lee, J.-C.; Greenberg, W. A.; Wong, C.-H. *Nat. Protoc.* **2006**, *1*, 3143–3152.

[159] Fujikawa, K.; Nohara, T.; Imamura, A.; Ando, H.; Ishida, H.; Kiso, M. *Tetrahedron Lett.* **2010**, *51*, 1126–1130.

[160] Fujikawa, K.; Nakashima, S.; Konishi, M.; Fuse, T.; Komura, N.; Ando, T.; Ando, H.; Yuki, N.; Ishida, H.; Kiso, M. *Chem. Eur. J.* **2011**, *17*, 5641–5651.

[161] Sawada, N.; Ito, M.; Ishida, H.; Kiso, M. *Tetrahedron Lett.* **2001**, *42*, 1745–1747.

[162] Nakashima, S.; Ando, H.; Imamura, A.; Yuki, N.; Ishida, H.; Kiso, M. *Chem. Eur. J.* **2011**, *17*, 588–597.

[163] Nakashima, S.; Ando, H.; Saito, R.; Tamai, H.; Ishida, H.; Kiso, M. *Chem. Asian J.* **2012**, *7*, 1041–1051.

[164] Tanaka, H.; Nishiura, Y.; Takahashi, T. *J. Am. Chem. Soc.* **2008**, *130*, 17244–17245.

[165] Kaneko, M.; Yamada, K.; Miyamoto, T.; Inagaki, M.; Higuchi, R. *Chem. Pharm. Bull.* **2007**, *55*, 462–463.

[166] Ando, H.; Koike, Y.; Koizumi, S.; Ishida, H.; Kiso, M. *Angew. Chem. Int. Ed.* **2005**, *44*, 6759–6763.

[167] Iwayama, Y.; Ando, H.; Ishida, H.; Kiso, M. *Chem. Eur. J.* **2009**, *15*, 4637–4648.

[168] Hanashima, S.; Ishikawa, D.; Akai, S.; Sato, K.-I. *Carbohydr. Res.* **2009**, *344*, 747–752.

[169] Tamai, H.; Ando, H.; Tanaka, H.-N.; Hosoda-Yabe, R.; Yabe, T.; Ishida, H.; Kiso, M. *Angew. Chem. Int. Ed.* **2011**, *50*, 2330–2333.

[170] Rich, J. R.; Withers, S. G. *Angew. Chem. Int. Ed.* **2012**, *51*, 8640–8643.

[171] Tamai, H.; Ando, H.; Ishida, H.; Kiso, M. *Org. Lett.* **2012**, *14*, 6342–6345.

[172] Tsai, Y.-F.; Shih, C.-H.; Su, Y.-T.; Yao, C.-H.; Lian, J.-F.; Liao, C.-C.; Hsia, C.-W.; Shui, H.-A.; Rani, R. *Org. Biomol. Chem.* **2012**, *10*, 931–934.

[173] Hanashima, S.; Yamaguchi, Y.; Ito, Y.; Sato, K.-I. *Tetrahedron Lett.* **2009**, *50*, 6150–6153.

[174] Shimizu, H.; Iwayama, Y.; Imamura, A.; Ando, H.; Ishida, H.; Kiso, M. *Biosci. Biotechnol. Biochem.* **2011**, *75*, 2079–2082.

[175] Iwayama, Y.; Ando, H.; Tanaka, H.-N.; Ishida, H.; Kiso, M. *Chem. Commun.* **2011**, *47*, 9726–9728.

[176] Wang, C.-C.; Lee, J.-C.; Luo, S.-Y.; Kulkarni, S. S.; Huang, Y.-W.; Lee, C.-C.; Chang, K.-L.; Hung, S.-C. *Nature* **2007**, *446*, 896–899.

[177] Wang, C.-C.; Kulkarni, S. S.; Lee, J.-C.; Luo, S.-Y.; Hung, S.-C. *Nat. Protoc.* **2008**, *3*, 97–113.

13

SYNTHESIS OF GLYCOSYLPHOSPHATIDYLINOSITOL ANCHORS

CHARLES JOHNSON AND ZHONGWU GUO
Department of Chemistry, Wayne State University, Detroit, MI, USA

13.1 INTRODUCTION

Glycosylphosphatidylinositol (GPI) anchors are a class of complex glycolipids that are responsible for linking a wide range of extracellular proteins and glycoproteins to the cell membrane of eukaryotic cells [1–4]. GPIs have been shown to take part in many biological events [5, 6], such as cell-surface adhesion and recognition [7], pathogenic infections [8, 9], signal transduction [10, 11], and cell-surface enzymatic reactions [12]. GPI anchoring of proteins was discovered in the 1970s [1], which was followed by the first full structural characterization of a *Trypanosoma brucei* GPI by Michael Ferguson's group in 1988 [13]. Since then, the structures of numerous GPIs have been determined and replicated synthetically. To date, all reported GPIs have the same pseudopentasaccharide core structure composed of *myo*-inositol, D-glucosamine ($GlcNH_2$), and three consecutive D-mannose (Man) residues capped by a phosphoethanolamine bridge (Fig. 13.1). The core Man residues are identified by Roman numerals in this chapter (i.e., the residue directly attached to $GlcNH_2$ is referred to as Man-I, etc.). A nonreducing end phosphoethanolamine bridge links to the C-terminus of the GPI-anchored proteins via an amide bond, while the lipids of phosphatidylinositol can insert into the cell membrane. The conserved core glycan of GPI anchors may include various functionalizations, typically additional sugar and phosphoethanolamine units. Lipid

Glycochemical Synthesis: Strategies and Applications, First Edition.
Edited by Shang-Cheng Hung and Medel Manuel L. Zulueta.

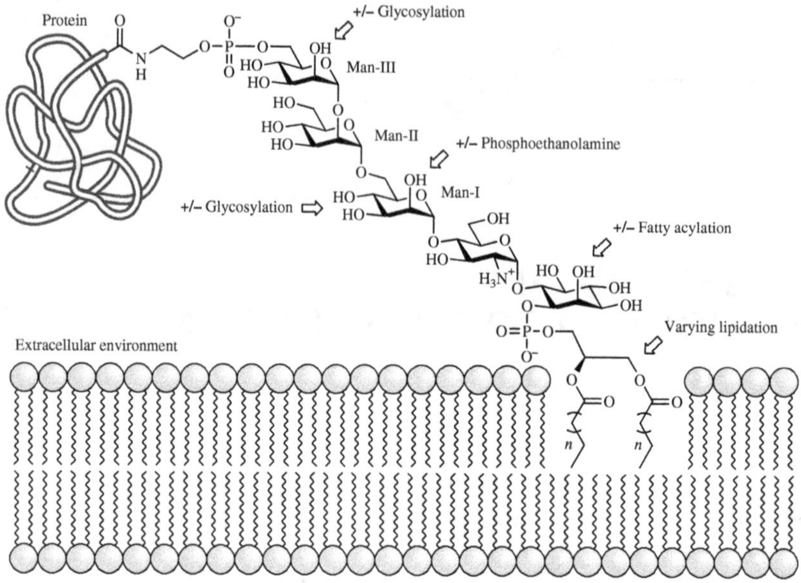

FIGURE 13.1 GPI core structure and modification sites.

composition may differ in terms of chain length and degrees of unsaturation. The O2 position of *myo*-inositol may also carry an acyl group. How the GPI core structure, as well as its many possible variations, affects its biological properties is largely unclear. To study such properties, access to homogeneous and structurally defined GPIs is necessary.

Since the first GPI total synthesis by Tomoya Ogawa's group in 1991 [14], many different synthetic strategies for GPIs have evolved. Nevertheless, GPI total synthesis remains a challenging feat as it requires careful planning and execution of carbohydrate, phosphate, inositol, and lipid chemistries. Synthesis of enantiomerically pure and properly differentiated *myo*-inositol derivatives is the first challenge in GPI synthesis. Many elegant syntheses of enantiomerically pure 1,6-*O*- and 1,2,6-*O*-differentiated inositol derivatives have been accomplished [15] and applied to GPI synthesis. Numerous partial syntheses of GPI structures (see reviews [16–19] and some articles [20–30]) greatly contributed to the strategies employed in GPI total synthesis. As this chapter is not intended to be an extensive survey of the whole field [18, 31], only selected total synthesis that is unique in certain aspects is presented to clearly show the evolution of the synthetic strategies in GPI anchor preparation.

13.2 SYNTHESIS OF THE *TRYP. BRUCEI* GPI ANCHOR

Commonly attached at O3 of Man-I of the *Tryp. brucei* variant surface glycoprotein GPI anchor is a digalactose branch, which may further carry additional D-galactose (Gal) units [13]. In 1991, the first total synthesis of the common region of the GPI anchor of *Tryp. brucei* was accomplished by the Ogawa group [14, 32]. The Gal

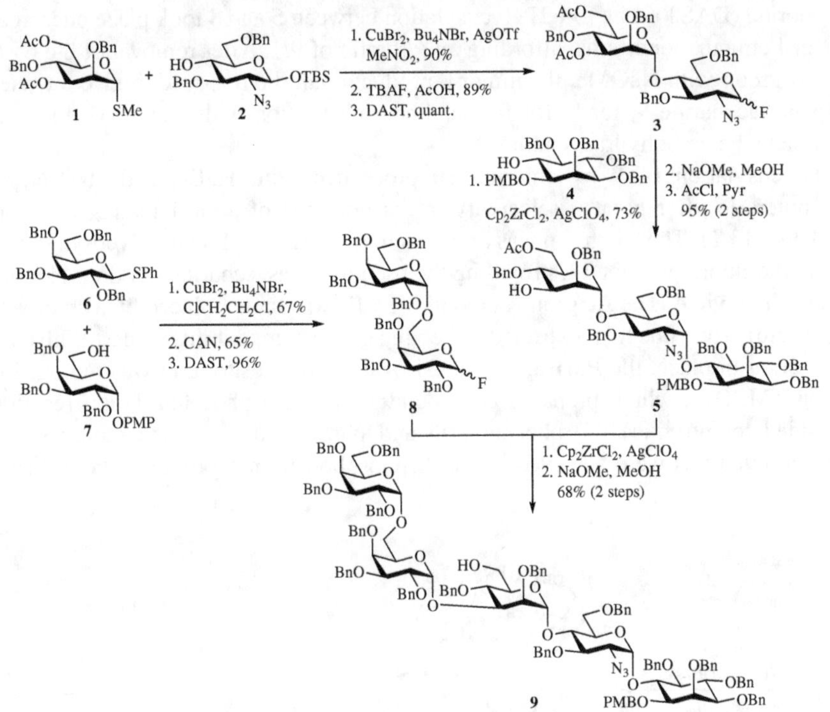

SCHEME 13.1 Synthesis of the pseudopentasaccharide intermediate **9**. CAN, ceric ammonium nitrate; PMB, *p*-methoxybenzyl; Pyr, pyridine; TBAF, tetrabutylammonium fluoride; Tf, triflyl.

modification required careful design and manipulation of protecting groups of the Man-I building block. Ogawa's synthetic strategy relied on benzyl (Bn) group for global hydroxyl protection and mainly linear elongation of the carbohydrate chain, except for the installation of the digalactose unit, by glycosylation using the Lemieux method [33], classical Koenigs–Knorr reaction [34], and Suzuki method [35, 36]. Phosphate groups were installed by the *H*-phosphonate method [37] at the final synthetic stage before global deprotection.

The sugar assembly began with the coupling of thiomannoside **1** and acceptor **2** by using modified Lemieux conditions to ensure α stereoselectivity, followed by removal of the anomeric *tert*-butyldimethylsilyl (TBS) group and transformation to the fluoride **3** (Scheme 13.1). Suzuki's method provided the glycosidic bond between donor **3** and asymmetric inositol-based acceptor **4**, which was acquired through chiral resolution, in a good yield. Such coupling is an inherently difficult task with respect to stereoselectivity due to the presence of the nonparticipating 2-azido group in **3**. Removal of the two acetyl (Ac) groups followed by selective acetylation at O6′ provided the pseudotrisaccharide acceptor **5**. A separate coupling of the thiogalactoside **6** and acceptor **7** under Lemieux conditions furnished the digalactosyl intermediate, which underwent oxidative removal of the anomeric *p*-methoxyphenyl (PMP) group followed by conversion to fluoride **8** upon treatment with *N*,*N*-diethylaminosulfur trifluoride (DAST). The [3+2] glycosylation between **5** and **8** took place once again under Lemieux conditions, affording an α/β ratio of 9/1. After removal of the 6″-*O*-acetyl group with NaOMe, the minor β product could be removed to give the pseudopentasaccharide **9**, ready for further linear elongation with mannosyl donors to complete the carbohydrate chain.

Glycosylation of **9** by chloride **10** proceeded with $HgBr_2$ and $Hg(CN)_2$ as promoters to offer great α selectivity in the presence of a participating Ac group (Scheme 13.2). The Ac group was removed to give way for Suzuki glycosylation by using the fluoride donor **11**, providing the pseudoheptasaccharide building block **12** in excellent yield. The Ac group at O6 of Man-III was then replaced by a chloroacetyl (ClAc) group due to unexpected protecting group compatibility issues. With ClAc protection in place, the PMB group at O1 of the *myo*-inositol unit was removed by using TMSOTf, which prepared compound **13** for two phosphorylation reactions intended to introduce the phospholipid and phosphoethanolamine moieties. The phosphorylations were achieved by the *H*-phosphonate method under the influence

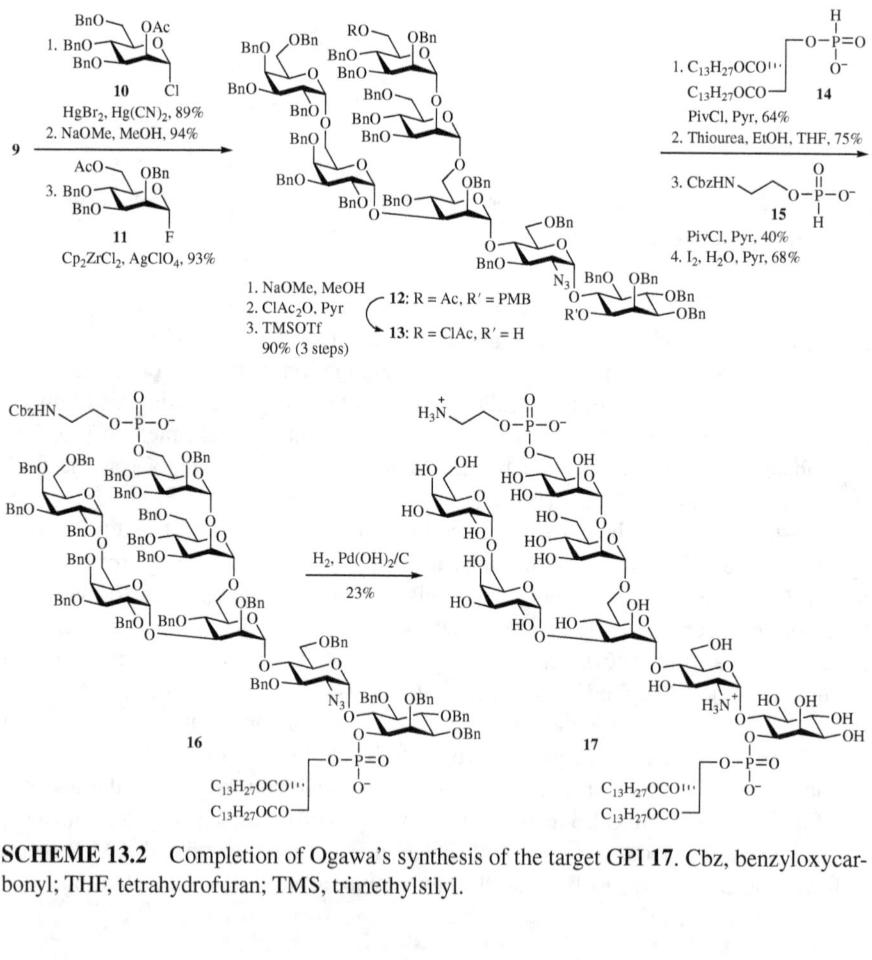

SCHEME 13.2 Completion of Ogawa's synthesis of the target GPI **17**. Cbz, benzyloxycarbonyl; THF, tetrahydrofuran; TMS, trimethylsilyl.

of pivaloyl chloride (PivCl) followed by one-pot oxidation of the trivalent phosphonate groups using I_2 and H_2O. In the process, removal of the ClAc protection was accomplished with thiourea. Finally, the phosphorylated intermediate **16** was subjected to Pd-catalyzed hydrogenolysis to remove all of the Bn groups, yielding the target GPI construct **17**.

Ogawa's landmark synthesis of the first natural GPI was based mainly on a linear glycosylation strategy, which was used less often in the future as more convergent approaches provided greater overall efficiency. However, the careful use of glycosyl halide donors to provide good α selectivity in the absence of participating groups was valuable to the overall efficiency of this synthesis.

In 1998, Steven Ley and coworkers reported the second synthesis of a *Tryp. brucei* GPI anchor [38, 39]. They used a convergent strategy, relying on a combination of Bn and butanediacetal (BDA) groups for global hydroxyl protection. Final-stage phosphorylation by using the mild phosphoramidite methodology [40] was adopted for this synthesis. The assembly of functionalized GPI pseudoheptasaccharide core was achieved in a [5+2] fashion, while the key pentasaccharide donor was quickly assembled using orthogonal selenoglycoside and thioglycoside donors.

The synthesis of the pentasaccharide donor **26** (Scheme 13.3) started with an efficient glycosylation between the "armed" selenogalactoside **18** and "disarmed" selenogalactoside **19**, a concept introduced by Bert Fraser-Reid [41]. Selenodigalactose **20** could then be used directly as donor under stronger reaction conditions for the next glycosylation with Man-I acceptor **21**. Removal of the Man-I TBS protection furnished acceptor **22**, which was ready for coupling with the dimannoside **25**. Disaccharide donor **25** was also prepared by the use of the armed–disarmed

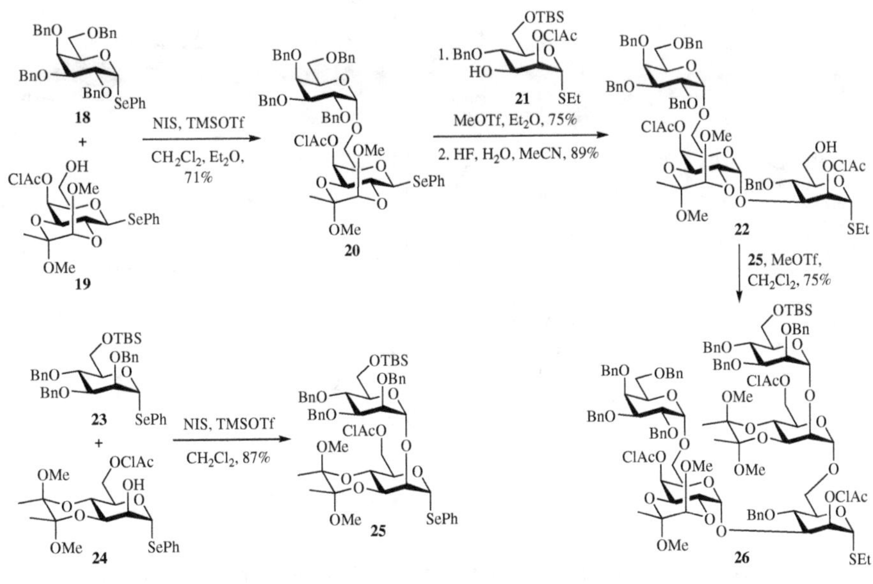

SCHEME 13.3 Synthesis of the pentasaccharide donor **26**.

glycosylation methodology, rendering it ready for the [3+2] seleno-selective coupling [42] to complete pentasaccharide **26**. The synthesis of **26** showcases an impressive display of anomeric control and rapid assembly of the desired product.

Following a short *myo*-inositol desymmetrization that employed a chiral bis(dihydropyran), the inositol building block **28** and glycosyl bromide **27** were joined by an α-selective glycosylation using Lemieux conditions (Scheme 13.4). After cleavage of the TBS group using TBAF, **29** was ready as glycosyl acceptor for the key [5+2] glycosylation with **26**, which occurred upon treatment with *N*-iodosuccinimide (NIS) and TMSOTf in a modest 51% yield. To permit the first phosphorylation reaction, the deprotection of the Man-III O6 position was accomplished with aqueous HF, giving the alcohol **30**. Reaction with phosphoramidite **31** and *in situ* oxidization of the trivalent phosphorous intermediate with *m*-chloroperoxybenzoic acid (mCPBA) provided the intermediate, which was further deallylated at the inositol unit and subjected to another round of phosphorylation to produce compound **33**. Pseudohexasaccharide **33** required three deprotection steps to yield the desired GPI anchor **17**.

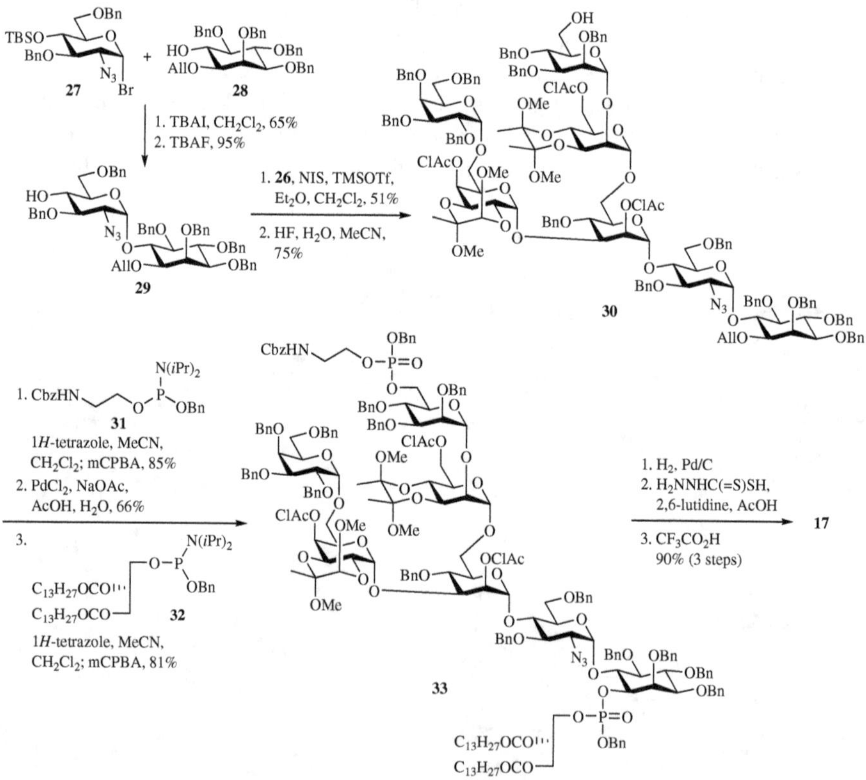

SCHEME 13.4 Completion of Ley's synthesis of the *Trypanosoma brucei* GPI anchor. All, allyl; TBAI, tetrabutylammonium iodide.

The Ley group's synthesis of a *Tryp. brucei* GPI showcased a great improvement, as far as efficiency is concerned, over the first report by Ogawa. These improvements came in the form of a more convergent strategy, a more rapid assembly of key polysaccharide building blocks, and also a much shorter synthesis of the inositol building block **28**.

13.3 SYNTHESIS OF THE YEAST GPI ANCHOR

The yeast GPI anchor notably carries a ceramide lipid section unlike the more common phospholipid moiety [43]. Moreover, an additional α-linked Man is found at O2 of Man-III. Richard Schmidt's group used a convergent approach for the synthesis of the yeast GPI anchor that they reported in 1994 [44, 45]. The key step of this synthesis is the glycosylation reaction between a tetramannose donor and a pseudodisaccharide acceptor. All glycosylation reactions relied on Schmidt's own trichloroacetimidate methodology [46], which produced good to excellent yields. Nearly all of the glycosylation reactions used donors that contained 2-*O*-acyl groups to take advantage of their neighboring participation property, which produced exclusive α selectivity. The Schmidt group also chose to carry out the phosphorylation reactions at a late stage.

The synthesis of the tetramannose donor (Scheme 13.5) began with the glycosylation of the Man-I acceptor **35** by the Man-II donor **34** in the presence of catalytic TMSOTf. Removal of the Man-II 2-*O*-acetyl group using NaOMe furnished the acceptor **36**, which was ready for glycosylation with Man-III **37** using the same condition. The last Man residue was introduced by the same procedures. At this point, the O1 and O2 positions of Man-I were transformed in order to obtain **38**, which was achieved through the use of Wilkinson's catalyst to remove both allyl groups, acetylation of the resulting two free hydroxyls, selective removal of the anomeric Ac group, and then imidate formation under mild basic conditions.

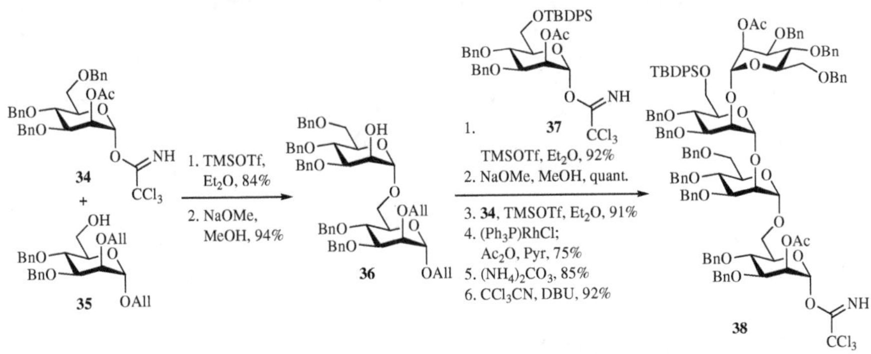

SCHEME 13.5 Synthesis of tetramannosyl donor **38**. DBU, 1,8-diazabicyclo[5.4.0]undec-7-ene; TBDPS, *tert*-butyldiphenylsilyl.

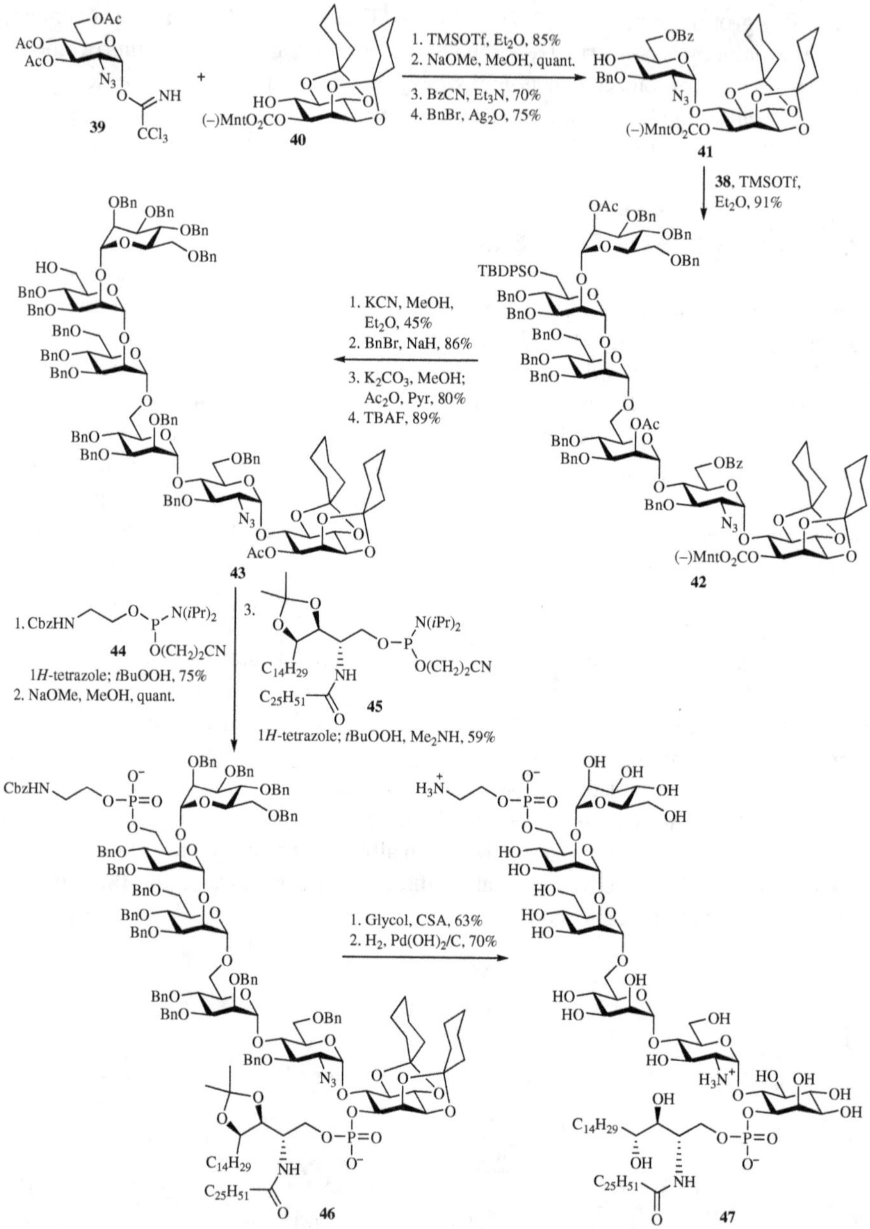

SCHEME 13.6 Completion of Schmidt's synthesis of the yeast GPI anchor. CSA, camphorsulfonic acid.

Glycosylation of the optically pure inositol derivative **40** with glucosamine donor **39** took place in 85% yield with complete α selectivity (Scheme 13.6). To prepare the proper protecting group arrangement for acceptor **41**, all Ac groups were removed and the O6′ position was regioselectively protected with a benzoyl (Bz) group. Then, selective benzylation at O3′ took place in the presence of Ag_2O. The key glycosylation between the pseudodisaccharide acceptor **41** and the tetramannosyl donor **38** proceeded smoothly with complete α selectivity, giving **42** in an excellent 91% yield. Pseudohexasaccharide **42** required several late-stage protecting group manipulations, beginning from the removal of Ac groups followed by Bn protection. Next, the inositol 1-*O*-(−)-menthyl (Mnt) carbonate group, which was employed as chiral auxiliary for enantiomeric resolution of **40**, was removed with K_2CO_3, and the resulting free hydroxy group was acetylated. After desilylation to give the alcohol **43**, installation of the phosphoethanolamine group on Man-III using **44** was carried out by using the phosphoramidite method [40]. Deacetylation at O1 of the inositol unit permitted the introduction of ceramide **45**, again using the phosphoramidite method. The phosphorylated intermediate **46** required two deprotection steps, starting from the removal of the ketal protecting groups under acidic conditions followed by debenzylation to give the target GPI **47**.

Schmidt's synthesis of GPI anchor **47** showcased the powerful glycosylation methodology of using trichloroacetimidate glycosyl donors along with proper protecting group selection. This allowed pseudohexasaccharide **42** to be assembled in a rapid and efficient manner with high yields and excellent stereocontrol for glycosidic bond formation. Unfortunately, several steps for late-stage protecting group manipulation were necessary, which slightly dampened the overall efficiency of this synthesis. Nonetheless, Schmidt's synthesis was a great step forward in the field of GPI total synthesis as various aspects of this synthesis were subsequently repeated by other groups.

13.4 SYNTHESIS OF THE RAT BRAIN THY-1 GPI ANCHOR

Fraser-Reid's synthesis of the rat brain Thy-1 GPI in 1995 marked the first total synthesis of a mammalian GPI anchor [47, 48]. The rat brain Thy-1 GPI is highly functionalized, containing two phosphoethanolamine groups (at O2 of Man-I and O6 of Man-III) as well as additional β-linked *N*-acetylgalactosamine (GalNAc) and α-linked Man units at O4 of Man-I and O2 of Man-III, respectively [49]. The assembly of the pseudoheptasaccharide backbone was achieved by a convergent approach using Fraser-Reid's own *n*-pentenyl glycoside chemistry [50, 51] as well as the classical Koenigs–Knorr glycosylation [34]. Phosphorylation reactions were carried out at a late stage using the phosphoramidite methodology [40].

The trimannosyl segment was achieved by two sequential Koenigs–Knorr glycosylation reactions with the help of 2-*O*-Bz groups in the donors to direct the desired stereochemistry (Scheme 13.7). Glycosylation between bromide **48** and 2-alcohol **49** followed by removal of the Bz group rendered the dimannosyl acceptor **50** ready for glycosylation with mannosyl donor **51**. Finally, the Bz and TBDPS groups were replaced with Bn and ClAc groups, respectively, to produce **52**. The glycosidic bond in GalNAc(β1→4) Man-I was realized through a β-selective Schmidt glycosylation between the GalNAc donor **53** and acceptor **54**. The resulting disaccharide **55** further underwent a series of transformations to obtain the suitable disaccharide donor **56**. Conversely, the synthesis of the pseudodisaccharide acceptor **59** was realized by

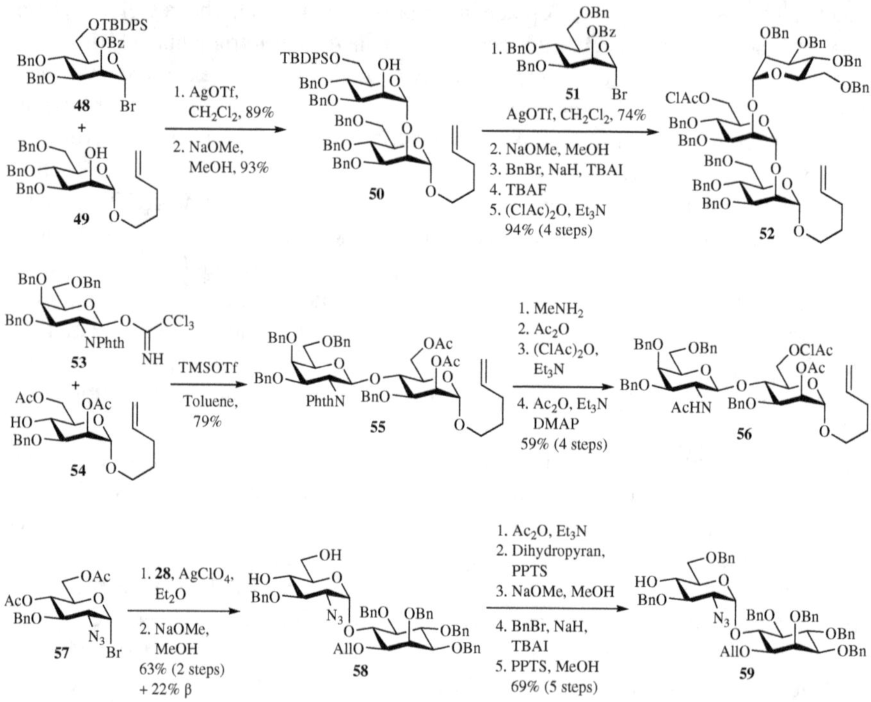

SCHEME 13.7 Fraser-Reid's synthesis of building blocks for the convergent assembly of the rat brain Thy-1 GPI anchor. DMAP, 4-dimethylaminopyridine; PPTS, pyridinium *p*-toluenesulfonate.

coupling the inositol acceptor **28** with glycosyl bromide **57** in the presence of $AgClO_4$, resulting in a modest α selectivity (α/β = 3/1). The mixture of anomers was subjected to NaOMe to remove the Ac groups, followed by separation of anomers to yield the pseudodisaccharide **58**. At this stage, the route to selective 6′-O-benzylation required five steps, including selective 6′-O-acetylation, protection of O4′ with a tetrahydropyran group, 6′-O-deacetylation, 6′-O-benzylation, and acidic methanolysis of the tetrahydropyran group.

The convergent assembly began with the [2+2] coupling of **56** and **59** promoted by NIS and TESOTf followed by removal of the Man-I 6-*O*-ClAc protection (Scheme 13.8). The resulting acceptor was then used in a [4+3] coupling with the trimannose donor **52**. Subsequent cleavage of the Man-III ClAc protection produced pseudoheptasaccharide **60**, which was ready for phosphorylation with phosphoramidite **31**. Thereafter, the Man-I O2 position was deacetylated and subjected to the same phosphorylation conditions. The final phospholipidation was carried out after deallylation at O1 of the inositol unit to afford the fully phosphorylated derivative **62**. Compound **62** underwent Pd-catalyzed hydrogenolysis, providing the target GPI **63** in 75% yield.

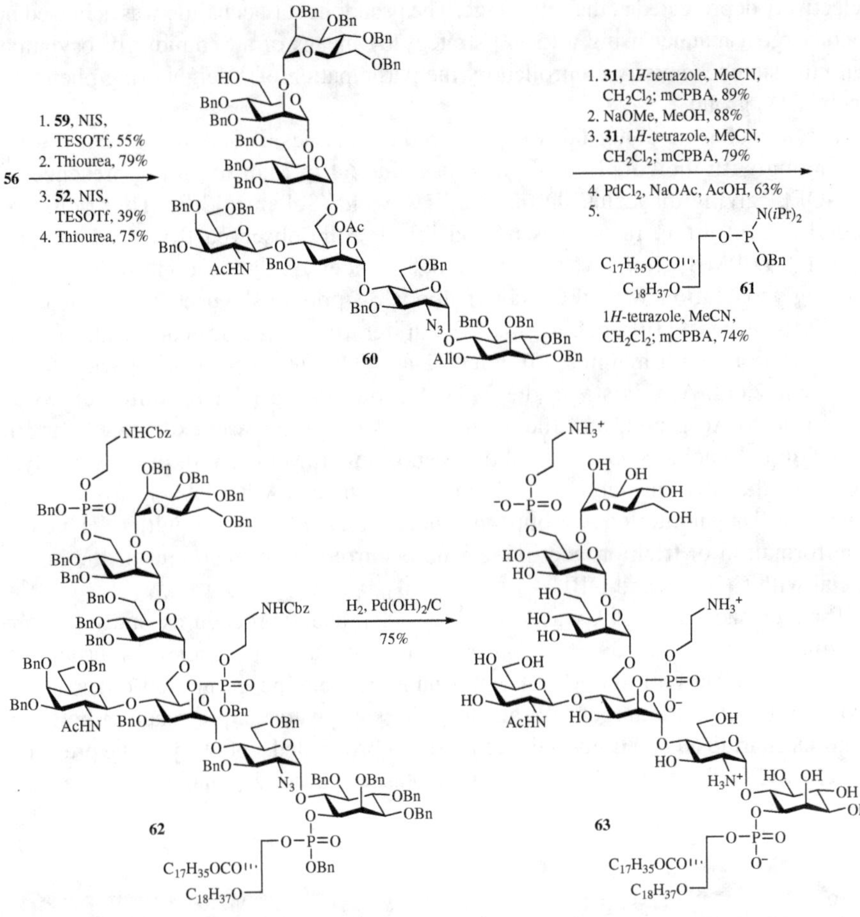

SCHEME 13.8 Completion of Fraser-Reid's synthesis of the rat brain Thy-1 GPI anchor. TES, triethylsilyl.

The first total synthesis of a mammalian GPI anchor by the Fraser-Reid group was highlighted by the use of a convergent strategy that allowed a highly functionalized rat brain Thy-1 GPI to be put together in an efficient manner. Fraser-Reid's own *n*-pentenyl glycoside glycosylation chemistry was in harmony with Schmidt's trichloroacetimidate chemistry and classical Koenigs–Knorr method for glycosylations, but this synthesis required a few late-stage protection–deprotection manipulations to reach the target GPI **63**.

The Schmidt group reported another synthesis of rat brain Thy-1 GPI in 1999 [52]. In 2003, they reported a rather similar synthesis [53] that involved differentially protected phosphoethanolamine groups useful for future regioselective protein/peptide coupling reactions. This differentiation was made possible by using Cbz and *tert*-butoxycarbonyl (Boc)-protected phosphoramidite reagents that could be selectively deprotected at the final stage. The pseudoheptasaccharide was achieved in a convergent manner using a [5+2] strategy by means of a Schmidt glycosylation with the stereochemistry controlled by the participation of a neighboring phenoxyacetyl (PA) group.

Pentasaccharide donor **68** was produced by a series of Schmidt glycosylation beginning with the coupling of galactosamine **64** to Man **65** in the presence of $BF_3 \cdot OEt_2$, giving disaccharide **66** in a 75% yield (Scheme 13.9). The trichloroacetyl (TCA) group in **64** was responsible for the observed β selectivity. The Man-I 6-*O*-PMB group was then removed under oxidative conditions before the α-glycosylation with the Man-II donor **34** previously used in the group's synthesis of yeast GPI anchor [44, 45]. An iterative series of deacetylations and glycosylations with mannosyl donors **37** and **34** offered pentasaccharide **67** in excellent yield. At this stage, the *N*-TCA group in the galactosamine unit was converted to Ac group, and the Man-IV 2-*O*-Ac group was exchanged for Bn protection. To achieve O1 and O2 differentiation at Man-I, Wilkinson's catalyst removed the allyl protections, which were replaced with PA groups at both positions. The anomeric PA group was, then, selectively cleaved with $(NH_4)_2CO_3$. The formation of trichloroacetimidate **68** occurred upon treatment of the hemiacetal with CCl_3CN and DBU.

Pseudodisaccharide **69** was achieved under usual procedures starting from building blocks **4** and **39**. The *N*-Boc protection in **69** is crucial for providing orthogonality between the Man-III phosphoethanolamine group and other amine groups present in the target GPI anchor. The key glycosylation reaction between pentasaccharide **68** and pseudodisaccharide **69** proceeded smoothly in the presence of TMSOTf (Scheme 13.10). Pseudoheptasaccharide **70** underwent the first

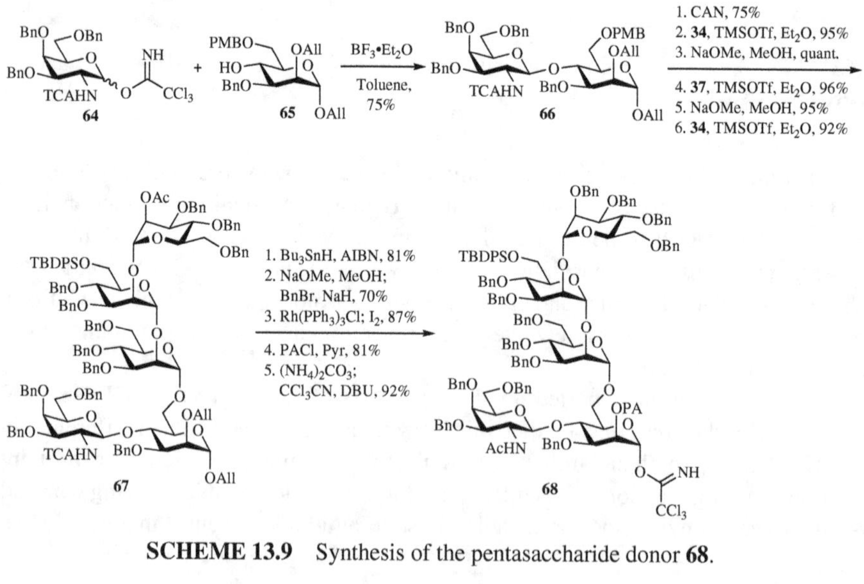

SCHEME 13.9 Synthesis of the pentasaccharide donor **68**.

phosphorylation event after the removal of the Man-III 6-*O*-TBDPS protection using TBAF. After phosphorylation with phosphoramidite **44**, the Man-I 2-*O*-PA group and phosphate cyanoethoxy protection were removed with $MeNH_2$, giving way for the second phosphorylation with phosphoramidite **71**, which contained amine protection orthogonal to that in **44**. Subsequent deacylation of the inositol O1 position using NaOMe enabled phospholipidation with phosphoramidite **72**. The fully phosphorylated intermediate **73** had the option to undergo two different deprotection sequences. To provide the rat brain Thy-1 GPI **63**, the Boc protections of the amino groups were removed first using acidic conditions followed by hydrogenolysis to remove all other protecting groups. The second deprotection protocol to provide the target GPI **74** only required hydrogenolysis to remove all of the

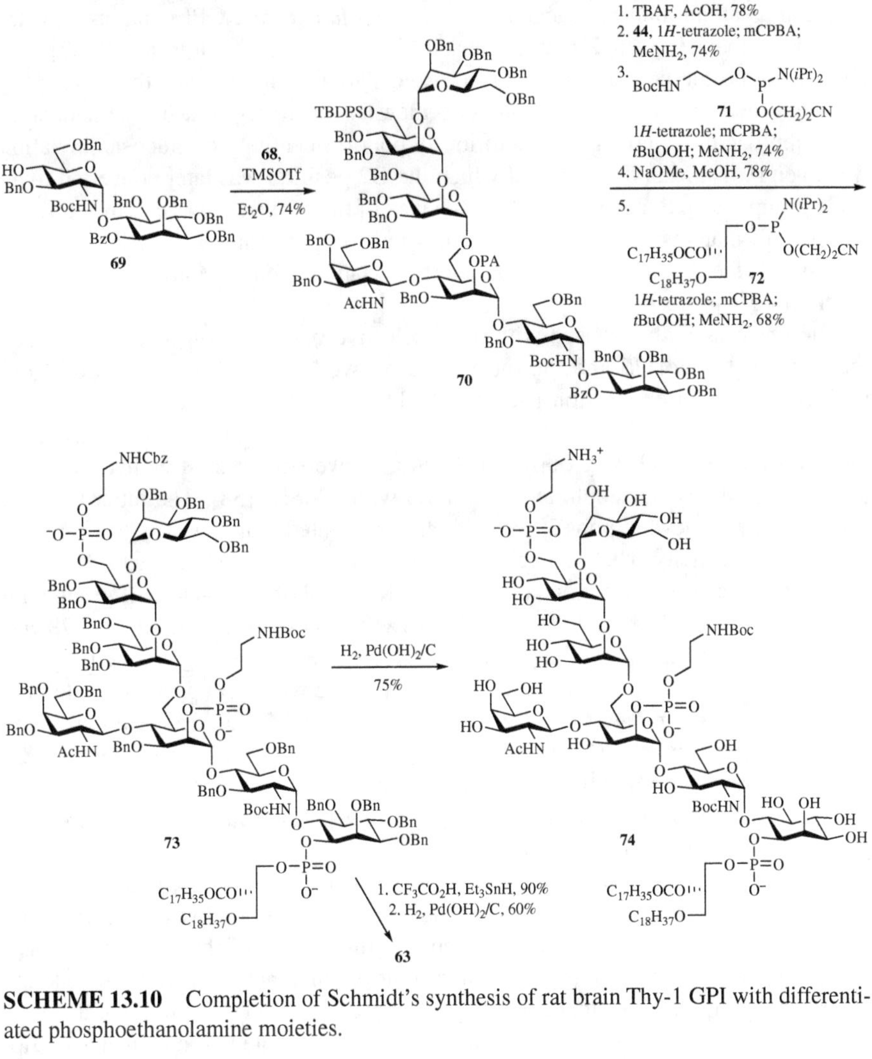

SCHEME 13.10 Completion of Schmidt's synthesis of rat brain Thy-1 GPI with differentiated phosphoethanolamine moieties.

Bn groups and Cbz protection, under which condition the amino Boc protection is stable. GPI **74** is the first fully phosphorylated synthetic GPI that was feasible for future coupling with peptides or proteins.

13.5 SYNTHESIS OF *PLASMODIUM FALCIPARUM* GPI ANCHOR

The Fraser-Reid group published the total synthesis of a *Plasmodium falciparum* GPI anchor prototype in 2004 [54] after a brief communication describing the synthesis of a *P. falciparum* GPI analogue with unnatural lipids [55]. *Plasmodium falciparum* uses free GPI anchors and GPI-anchored proteins to modulate the host immune system [56]. Therefore, a synthetic version of this GPI may have therapeutic potential as an antimalaria vaccine. *Plasmodium falciparum* GPI contains an ester lipid linked to the inositol 2-*O*-position and Man α(1→2)-linked to Man-III [57]. The GPI anchor prototype prepared by Fraser-Reid's group, which lacked the fourth Man residue, relied on using the *n*-pentenyl orthoester chemistry in a linear manner. A novel approach to overcome the notoriously poor α selectivity of glucosaminylation was to employ a Man derivative of which the O2 position was later converted to an azido group via an S_N2 reaction. The efficiency of the synthesis was enhanced by the fact that all mannosyl donors were synthesized from the same synthetic precursor. Like most other synthetic strategies, phosphorylation events occurred in late stage before final deprotection.

The linear assembly started with the α-selective glycosylation reaction between the mannosyl donor **75** and the inositol derivative **76**, which was acquired from D-glucose via Ferrier reaction (Scheme 13.11). The Man residue of the resulting pseudodisaccharide **77** was then converted to an azide-protected glucosamine by removal of the 2-*O*-Bz group with NaOMe, conversion of the free hydroxyl to a triflate, and finally nucleophilic substitution with $TMSN_3$ [58]. The latter reaction occurred in a somewhat poor 39% yield but represented a unique way to overcome the poor α selectivity. The relatively low yield of the glycosylation reaction was not a large loss because it gave good α selectivity. Removal of the 4′-*O*-PMB group with $BF_3·OEt_2$ provided the pseudodisaccharide **78** as a glycosyl acceptor. Alcohol **78** was coupled to mannoside **79** in the presence of NIS and $BF_3·OEt_2$. Again, the Bz group was removed, and the resulting alcohol was protected with a Bn group to match the global protection strategy. After removal of the 6″-*O*-TBS group, addition of the Man-II building block **81** proceeded to furnish the pseudotetrasaccharide **82**. Debenzoylation permitted the final glycosylation with mannosyl donor **83**. With all glycosidic bonds in place, a three-step protecting group adjustment delivered the pseudopentasaccharide **84**.

The first phosphorylation reaction took place using phosphoramidite **31** and mCPBA. The inositol 1,2-*O*-cyclohexylidineketal protection was then removed under acidic conditions to set the stage for the installation of the inositol 2-*O*-acyl group (Scheme 13.12). The acylation was achieved by reacting with **85** to generate an orthoester intermediate that could be rearranged using $Yb(OTf)_3$ to favor the 2-*O*-acylated product **86** (71% isolated yield) over the 1-*O*-acylated product in a 5/1 ratio.

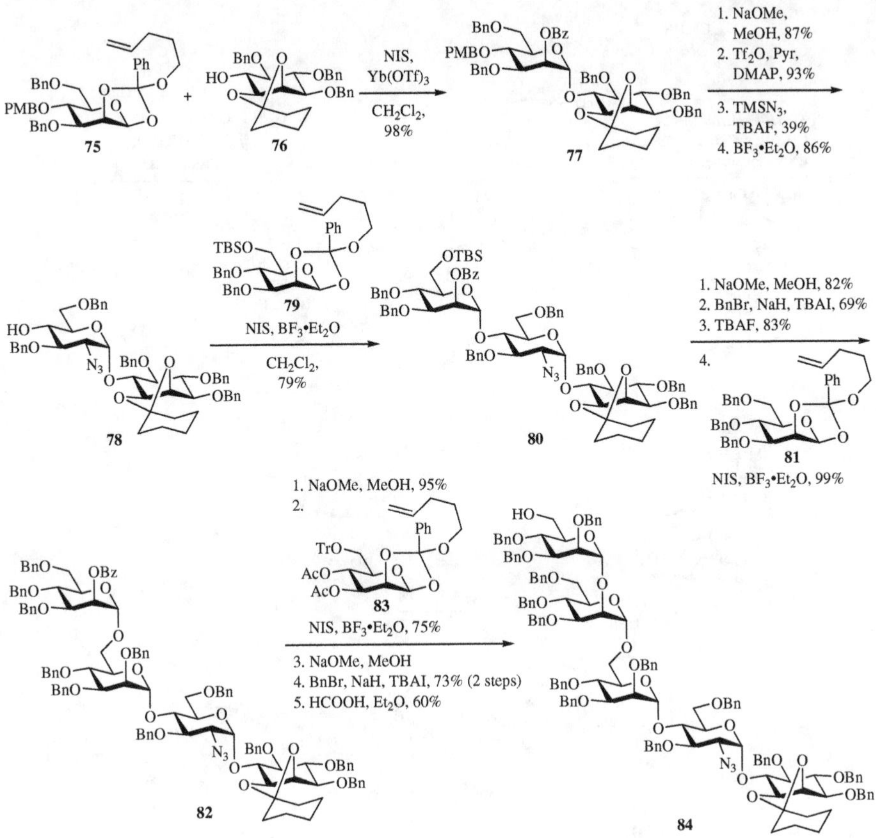

SCHEME 13.11 Synthesis of pseudopentasaccharide **84**. Tr, trityl.

Phospholipidation of the 1-alcohol **86** took place using phosphoramidite **87** followed by oxidation with mCPBA. Hydrogenation, catalyzed by Pd, reduced the azido group and removed the Bn protection to provide the target GPI **89**.

In addition to the interest in GPI-based antimalaria vaccines, Peter Seeberger's group also performed the total synthesis of some GPIs, including the *P. falciparum* GPI anchor [59, 60]. Unlike Fraser-Reid's version, the additional Man unit linked to Man-III was integrated in Seeberger's synthesis, which utilized Schmidt glycosylation to assemble the GPI core in a convergent manner. Seeberger also chose to have the inositol O1, O2, and O6 positions differentiated before its incorporation into the GPI core structure to allow a more efficient installation of the inositol 2-*O*-acyl group.

The synthesis of the tetramannose donor **93** (Scheme 13.13) resembled that of Schmidt's syntheses of yeast GPI and rat brain Thy-1 GPI [44, 53]. A series of glycosylations started with the formation of the Man(α1→6) bond between trichloroacetimidate **34** and 6-alcohol **90** in the presence of TMSOTf. Iterative removal of the O2 protecting group and glycosylations with mannosyl donors **91** and **34** completed the tetramannose **92**. To achieve the glycosyl donor **93**, the allyl group was removed with $PdCl_2$, followed by trichloroacetimidate formation.

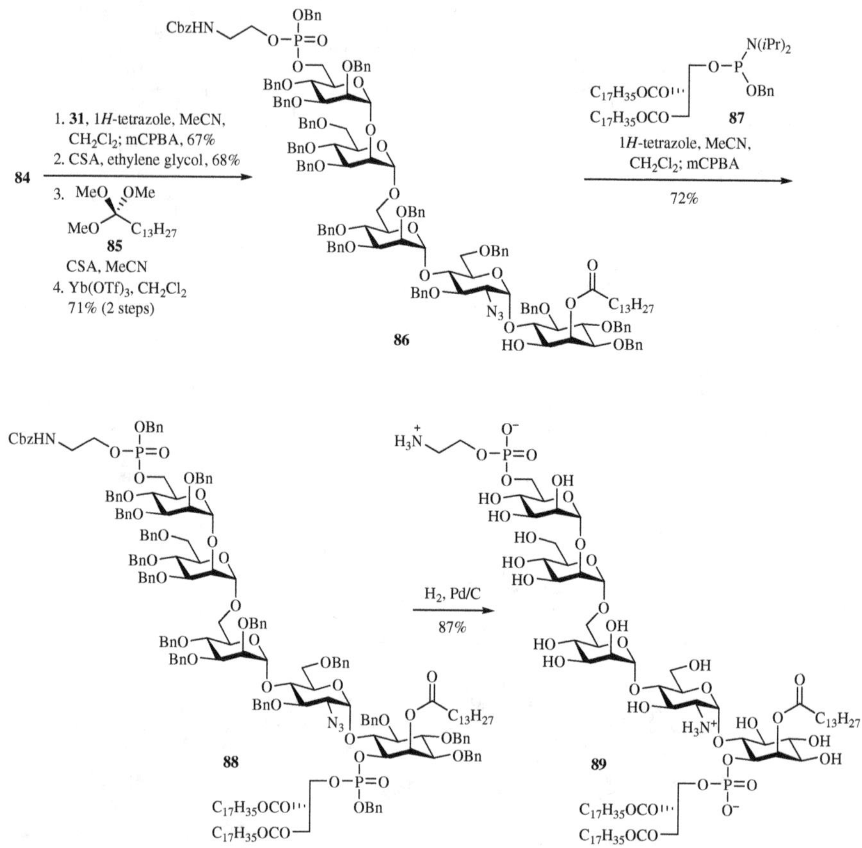

SCHEME 13.12 Completion of Fraser-Reid's synthesis of the *Plasmodium falciparum* GPI anchor.

En route to pseudodisaccharide **96** (Scheme 13.13), the 1,2,6-*O*-differentiated inositol derivative **94** was used to react with the trichloroacetimidate donor **39** in the presence of TMSOTf, which supplied **95** in an 89% yield ($\alpha/\beta=4/1$) after full deacetylation. The mixture of anomers carried forth to undergo 4′,6′-*O*-benzylidine protection, 3′-*O*-benzylation, and regioselective benzylidine ring opening, rendering the O4′ position free. At this point, the mixture of anomers could be separated, and the pseudodisaccharide acceptor **96** was ready for the key glycosylation step.

The glycosylation of **96** by **93** took place in an excellent yield (Scheme 13.14), which was followed by deacylation, benzylation of the two free hydroxy groups, and oxidative removal of the PMB protection in the inositol unit to afford the 2-alcohol **97**. Then, the free hydroxy group was coupled to palmitic acid with the aid of *N,N′*-dicyclohexylcarbodiimide (DCC). After 2-O-acylation, the inositol 1-*O*-allyl group was removed using $PdCl_2$ to produce the pseudohexasaccharide **98** ready for phospholipidation, which was achieved by the *H*-phosphonate method [37]. For this purpose, the Et_3NH^+ salt of **99** was reacted with **98** in the presence of PivCl, followed

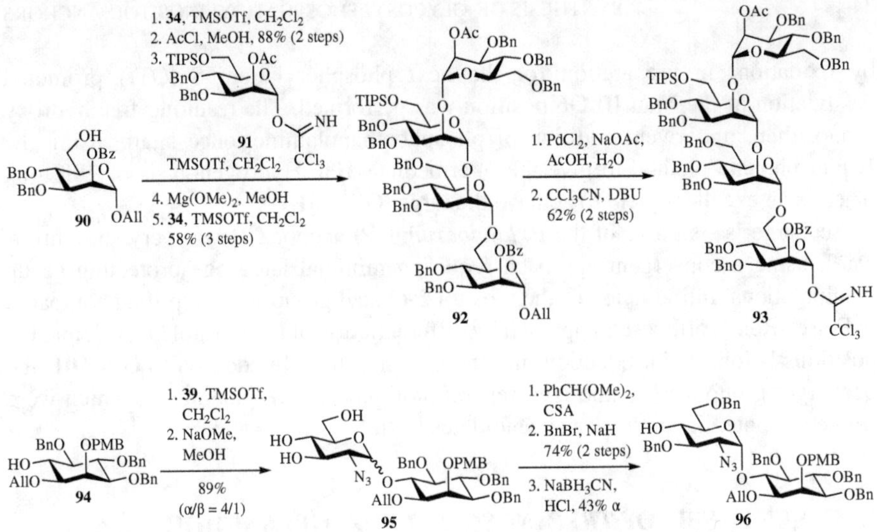

SCHEME 13.13 Synthesis of the key tetramannosyl donor **93** and pseudodisaccharide acceptor **96** for the assembly of the *Plasmodium falciparum* GPI anchor.

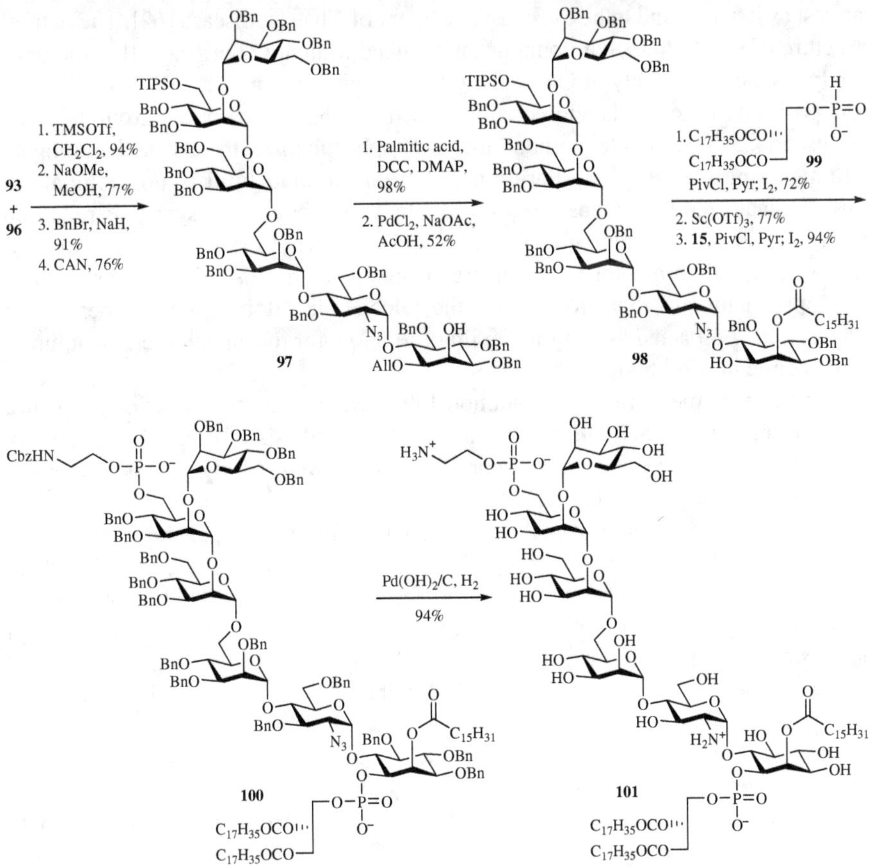

SCHEME 13.14 Completion of Seeberger's synthesis of the *Plasmodium falciparum* GPI anchor.

by oxidation. In preparation for the next phosphorylation, $Sc(OTf)_3$-promoted desilylation of the Man-III O6 position was performed. The resulting free hydroxy group, then, underwent addition of phosphoethanolamine, once again, using the *H*-phosphonate method to give the intermediate **100**. Hydrogenolysis of **100** took place in an excellent yield to achieve the target GPI **101**.

Seeberger's synthesis of the *P. falciparum* GPI anchor **101** was very straightforward, using a convergent approach requiring minimal late-stage protecting group manipulations. Introduction of the inositol 2-*O*-acyl group was simplified compared to Fraser-Reid's orthoester approach by differentiation of the inositol O1, O2, and O6 positions before its introduction into the GPI structure. In addition to GPI **101**, the Seeberger group also synthesized several nonlipidated GPI analogues, which were studied as potential antimalarial candidates [60].

13.6 SYNTHESIS OF *TRYPANOSOMA CRUZI* GPI ANCHOR

Ram Vishwakarma's group reported the first synthesis aiming for the *Trypanosoma cruzi* GPI anchor in 2005 [61]. The trypomastigote mucin GPI of *Tryp. cruzi* stirred interest as it was found to be the causative agent of Chagas' disease [62]. The natural structure of this GPI anchor contains an α-linked Man at O2 of Man-III, an aminoethylphosphonate moiety at O6 of the $GlcNH_2$ unit, and an unsaturated lipid at the *sn*-2 position of the alkylacylglycerolipid section. The Vishwakarma group targeted the pseudohexasaccharide without the aminophosphonate moiety and employed saturated lipids as the incorporation of an unsaturated lipid is not compatible with Pd-catalyzed hydrogenation used for Bn deprotection. The synthesis relied on a combination of Schmidt glycosylation and *n*-pentenyl orthoester chemistry for glycosidic bond formation in a convergent manner. The Vishwakarma group also developed a unique route to prepare the $GlcNH_2$–inositol segment, wherein the glucosamine unit acted as a chiral auxiliary to allow for the enantiomeric resolution of a racemic inositol building block.

Synthesis of the tetramannose donor **107** (Scheme 13.15) was conducted in a [2+2] fashion using Schmidt glycosylation. The synthesis of donor **104** began by reacting the trichloroacetimidate donor **102** with the Man-III glycosyl acceptor **103** in the presence of TMSOTf. The allyl group and the 4,6-*O*-benzylidine protection were then removed, and the three exposed hydroxy groups were acetylated. Subsequently, the anomeric Ac group was selectively removed using Me_2NH, which allowed for the formation of the trichloroacetimidate donor **104**. En route to the disaccharide glycosyl acceptor **106**, Man-I **105** was coupled with the *n*-pentenyl orthoester **81** in the presence of NIS and TESOTf. Removal of the Bz group rendered **106** ready for glycosylation with donor **104** in the presence of TMSOTf, resulting in 65% yield for the desired tetrasaccharide. To obtain trichloroacetimidate **107**, the anomeric allyl group was removed using $PdCl_2$, which was followed by reaction with CCl_3CN under basic conditions.

The synthesis of pseudodisaccharide **111** (Scheme 13.16) combined (+/−)-**4** with glucosamine donor **39** in the presence of TMSOTf in an α-selective glycosylation.

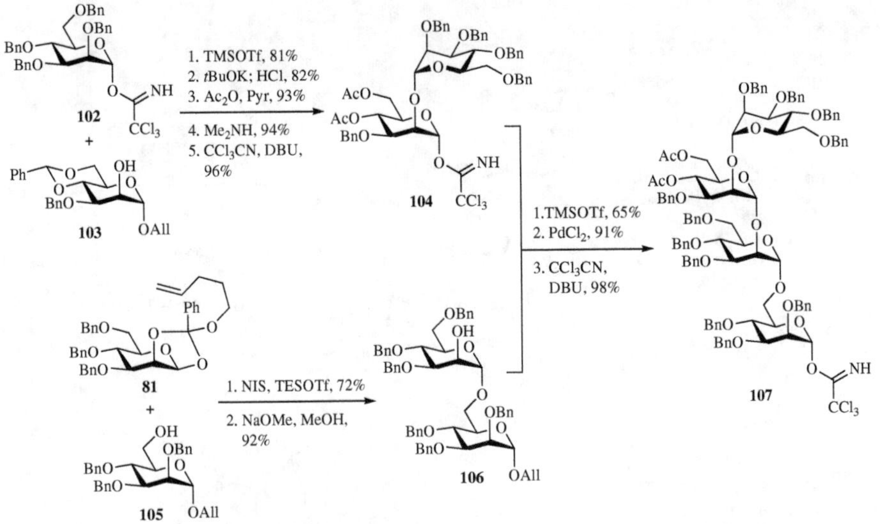

SCHEME 13.15 Synthesis of tetramannose donor **107**.

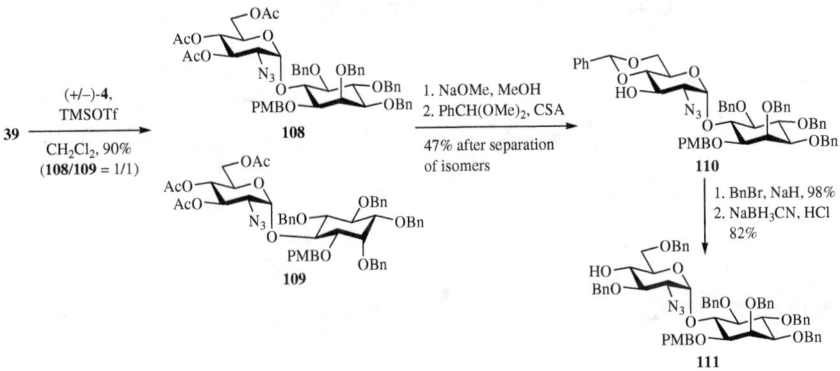

SCHEME 13.16 Synthesis of pseudodisaccharide acceptor **111**.

The resultant diastereomeric mixture was deacetylated before being subjected to 4′,6′-*O*-benzylidine formation, at which point diastereomeric separation occurred to provide **110**. Pseudodisaccharide **110**, then, underwent 3′-O-benzylation followed by regioselective benzylidine ring opening using $NaBH_3CN$ to afford the 4′-alcohol **111**.

The key step for this convergent synthesis was the coupling of tetramannose **107** and pseudodisaccharide **111**, which took place in the presence of TMSOTf (Scheme 13.17). Differentiation of the Man-III O4 and O6 positions required four steps, including deacetylation using NaOMe, selective protection at O6 with TBDPS group, 4-O-benzylation, and, finally, deprotection of O6 using TBAF. The resultant pseudohexasaccharide **112** was ready for phosphorylation. A Cbz-protected phosphoethanolamine group was added to the Man-III O6 position by the phosphoramidite method. The phosphoglycerolipid moiety was then introduced using **113** by the

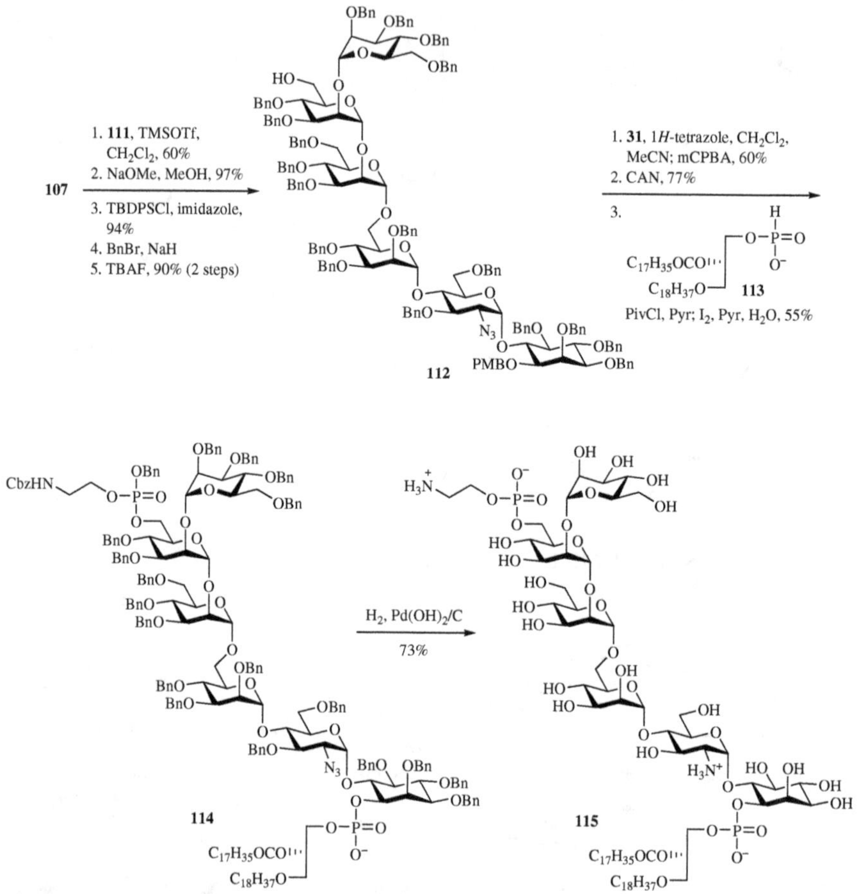

SCHEME 13.17 Completion of Vishwakarma's synthesis of the *Trypanosoma cruzi* GPI anchor.

H-phosphonate method after oxidative cleavage of the inositol 1-*O*-PMB group by CAN, yielding **114** that was subjected to global deprotection to obtain the desired GPI **115**.

Andrei Nikolaev's 2006 report on *Tryp. cruzi* GPI anchor preparation represented the first synthesis of a GPI anchor bearing unsaturated lipids [63]. In order to accomplish this feat, Bz group, which is labile under basic conditions, was chosen to replace the Bn group normally used for global protection of hydroxy groups. Although this is risky at first glance, as debenzoylation would seemingly interfere with acyl-linked lipids, the target GPIs were achieved. The GPI pseudohexasaccharide core was assembled in a [4+2] fashion by means of Schmidt glycosylation. Preceding the global deprotection, selective desilylations took place to give way for three late-stage phosphorylation reactions.

Construction of the pseudohexasaccharide core (Scheme 13.18) involved the sequential glycosylations with mannosyl trichloroacetimidates **117**, **118**, and **119**,

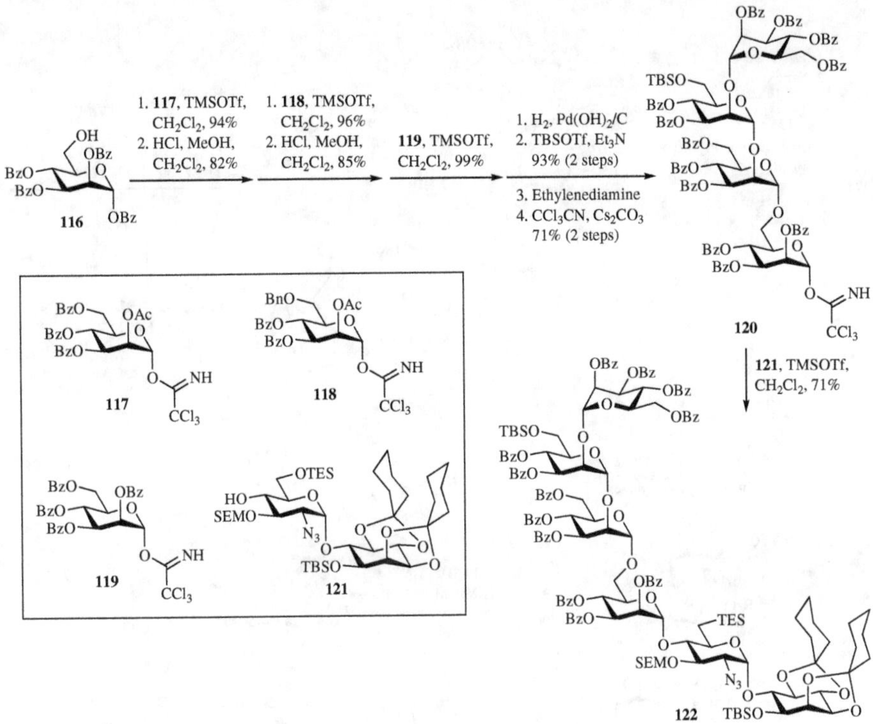

SCHEME 13.18 Synthesis of the pseudohexasaccharide core **122**.

with the 2-*O*-Ac groups removed selectively by acidic methanolysis in the intervening steps. The achieved tetramannose was subjected to Pd-catalyzed hydrogenolysis to remove the 6″-*O*-Bn group, which was replaced with a TBS group to provide orthogonality with future synthetic steps. At this point, the anomeric Bz protection was selectively removed by ethylenediamine, and the resulting hemiacetal was eventually transformed into the trichloroacetimidate to produce **120**, ready for the next glycosylation step with pseudodisaccharide **121**. Compound **121** was prepared in a number of steps following the coupling of glycosyl donor **39** and acceptor **40**. The [4+2] glycosylation involving **120** and **121**, thus, formed the desired pseudohexasaccharide structure **122**.

Relying on the 6′-*O*-TES group being more labile than the other silyl groups present in **122**, the TES group was selectively removed in a mild desilylation with TBAF (Scheme 13.19). The first phosphorylation occurred by using phosphonodichloridate **123** with 1*H*-tetrazole as promoter followed by displacement of the remaining phosphonochloride with methanol. To eliminate an extra final deprotection step, the azido group of glucosamine was reduced with PPh_3, and the resulting free amine was Boc protected. The primary Man-III TBS protection was selectively removed with $Et_3N{\cdot}3HF$ to expose the site for second phosphorylation with *H*-phosphonate **125** with PivCl as the promoter. The twice phosphorylated intermediate was then ready for the removal of the secondary inositol TBS protection to allow the introduction of the

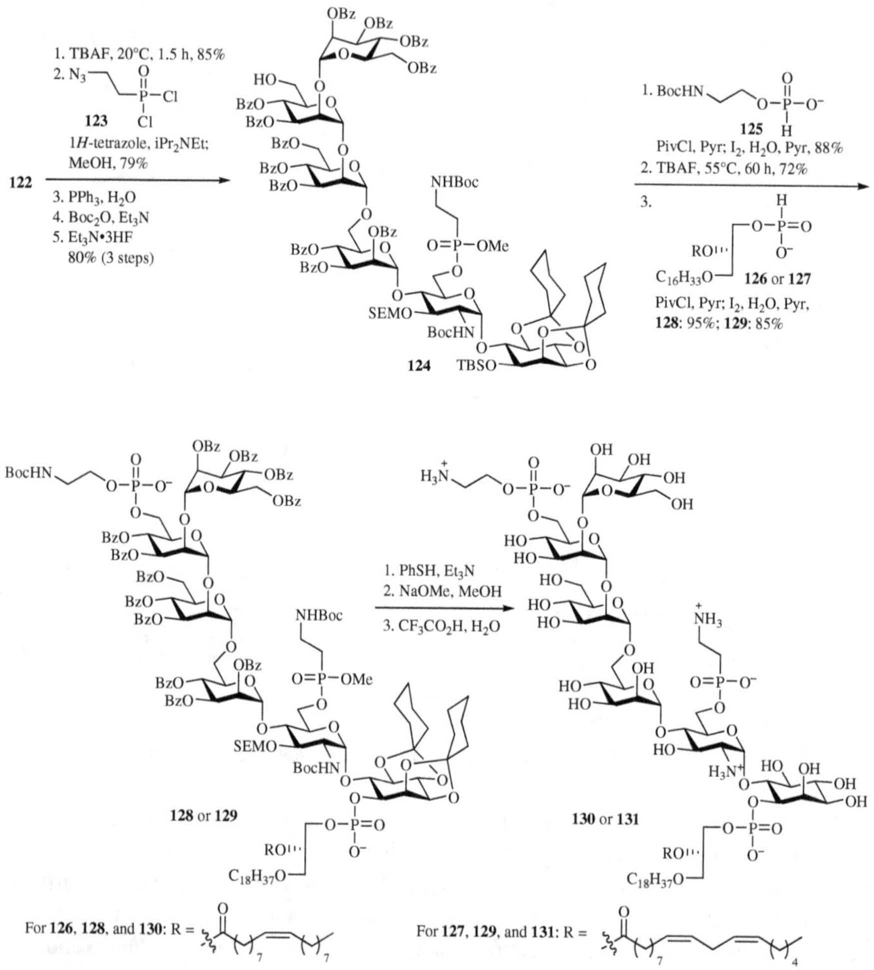

SCHEME 13.19 Completion of Nikolaev's synthesis of two *Trypanosoma cruzi* GPI anchors.

phosphoglycerolipid using **126** or **127**. The final deprotection started with demethylation of the methoxyphosphonate group at O6′. The critical debenzoylation step was achieved with 0.05 M NaOMe in methanol, and not surprisingly, only moderate yields (38–40%) were obtained. Finally, the acid-labile inositol cyclohexylidene protection and the 2-trimethylsilylethoxymethyl (SEM) group were removed with aqueous trifluoroacetic acid to provide the *Tryp. cruzi* GPIs **130** and **131** with varied lipid composition.

The Nikolaev group's first total synthesis of *Tryp. cruzi* GPI anchors carrying unsaturated lipids was made possible by using the Bz group for global protection for hydroxy groups to avoid Pd-catalyzed hydrogenation needed to deprotect the commonly used Bn group. The authors claim that during the final deacylation step, the amphiphilic nature of the GPI anchor would form aggregates based on lipid–lipid hydrophobic interactions, which would render the lipid acyl groups less prone to basic conditions. While this argument is reasonable, it may not hold true for the early stages of the final deacylation when the majority of the Bz groups would still be in place, giving less amphiphilic character to the molecule. This could be the reason for the moderate yields observed in the debenzoylation step. Nonetheless, the Nikolaev group reported a bold and novel strategy to a GPI that was unobtainable with the available methodology at the time. After this report, the Nikolaev group published an alternative strategy for the introduction of unsaturated lipids into a GPI construct by using acetal- and silyl-based global protection [18].

13.7 SYNTHESIS OF A HUMAN SPERM CD52 ANTIGEN GPI ANCHOR

In 2003, the Zhongwu Guo group reported the first total synthesis of a GPI anchor containing an acylated inositol, namely, the human sperm CD52 antigen [64]. Synthesis of a related structure was also reported in 2007, which used a similar synthetic methodology [65]. The human sperm CD52 GPI anchor is characterized by 2-O-palmitoylation at the *myo*-inositol unit, an *sn*-1-alkyl-2-lyso-glycerol structure instead of the regular diacylated anchor, and additional phosphoethanolamine at O2 of Man-I [66]. The pseudohexasaccharide core was obtained via a convergent [3+2] glycosylation involving a highly functionalized phospholipidated pseudodisaccharide, which was designed to increase convergency and to avoid a side reaction that was observed when a late-stage phospholipidation was attempted in the Guo group's 2003 report [64].

The synthesis of the requisite trimannosyl donor (Scheme 13.20) started with an orthogonal Koenigs–Knorr glycosylation between the thioglycoside acceptor **132** and mannosyl chloride **133**. Further deacetylation followed by glycosylation with mannosyl bromide **134** using Lemieux conditions assembled the trimannose intermediate, which underwent deacetylation at Man-III O6 position using NaOMe, and then, protection with a TBS group. Hydrolysis of the Man-I 1-*O*-thioacetal with NIS and TfOH under moisturized conditions followed by reaction with CCl_3CN in the presence of DBU provided the trichloroacetimidate donor **135**.

The preparation of the complementary pseudodisaccharide acceptor **139** was made possible due to an efficient synthesis of the 1,2,6-*O*-differentiated inositol derivative **136**, which was obtained from both enantiomers of protected *myo*-inositol [67]. The glycosidic bond in **137** (Scheme 13.20) was formed from **136** and glycosyl bromide **27** under Lemieux condition with only moderate yield but complete α selectivity. After the removal of the Ac protection by NaOMe, the inositol 2-*O*-palmitoleic acyl group was installed under the influence of DCC. The use of the palmitoleoyl group was in anticipation of the preparation of GPI conjugates through the ready transformation of the double bond into a carbonyl group. Oxidative cleavage of the inositol 1-*O*-PMB group allowed the installation of the phosphoglycerolipid using **138** by the phosphoramidite methodology. Pseudodisaccharide **139** was acquired after the removal of the 4′-*O*-TBS group upon treatment with $BF_3{\cdot}OEt_2$.

The glycosylation reaction between pseudodisaccharide **139** and trimannosyl donor **135** (Scheme 13.21) proceeded smoothly under Schmidt conditions. Simultaneous

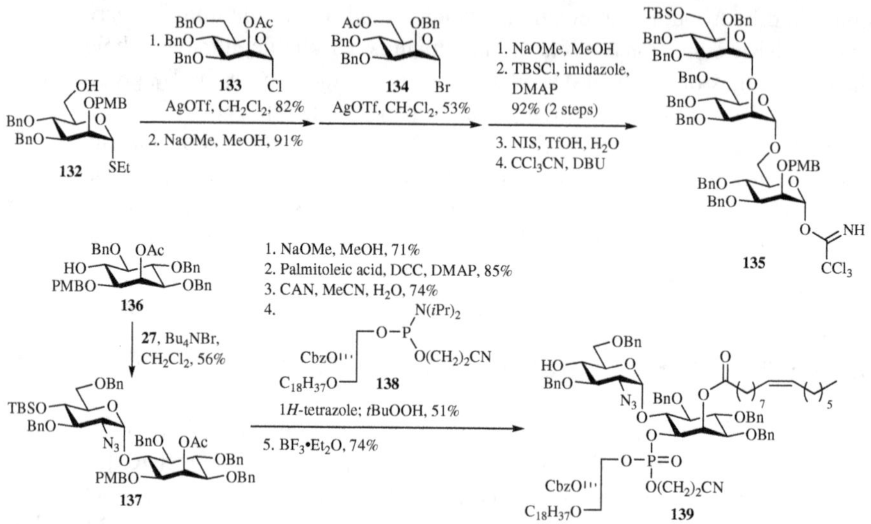

SCHEME 13.20 Synthesis of the tetramannose donor **135** and the pseudodisaccharide acceptor **139**.

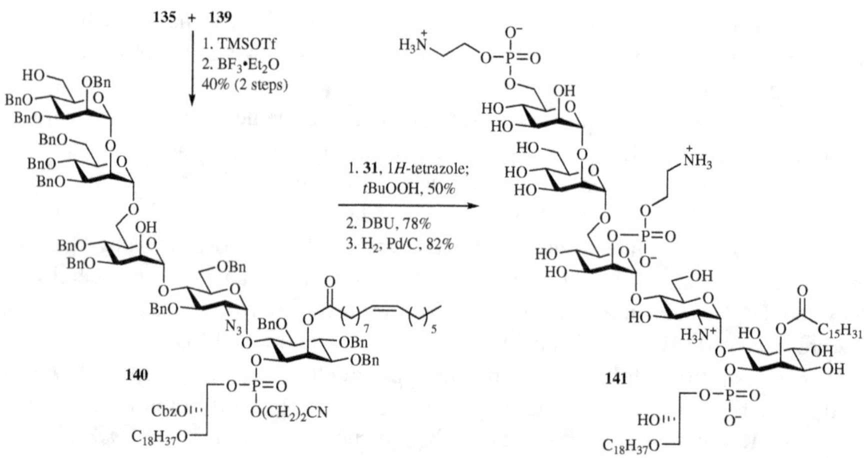

SCHEME 13.21 Completion of Guo's synthesis of the CD52 antigen GPI anchor **141**.

removal of the TBS and PMB groups with $BF_3{\cdot}OEt_2$ led to pseudopentasaccharide **140**, which was ready for two simultaneous phosphorylation reactions using phosphoramidite **31**. Following the phosphorylation step, removal of the phosphate cyanoethyl protection with DBU and Pd-catalyzed hydrogenation to remove the Bz protection and also reduce the carbon–carbon double bond eventually provided the target GPI **141**.

Guo's synthesis of several CD52 GPI anchors provided a straightforward convergent synthetic method for various GPI anchors, especially those with 2-*O*-acylated inositol. This synthesis is highlighted by the development of new efficient methods for the preparation of optically active 1,2,6-*O*-differented inositol derivatives and the installation of the phosphoglycerolipid moiety, which is one of the most difficult steps in GPI synthesis, at an early stage to avoid side reactions and difficult product purification at a later stage. Most significantly, this highly convergent synthetic strategy could enable the rapid assembly of GPI derivatives because glycosylation of a phosphoglycerolipidated pseudodisaccharide like **139** with various oligosaccharides can offer the desired GPI frameworks in one step. Indeed, it was used to prepare a number of GPI anchors and GPI derivatives [65, 68], some of which have been employed to explore biological problems related to GPI anchorage [68].

13.8 SYNTHESIS OF A HUMAN LYMPHOCYTE CD52 ANTIGEN GPI ANCHOR

In 2010, the Guo group reported a novel global protection strategy that employed the acid-labile PMB group for hydroxyl protection, allowing the incorporation of unsaturated lipids in the target GPI anchors [69]. This strategy was elaborated further in a 2011 report describing the synthesis of a human lymphocyte CD52 GPI anchor containing a rather labile polyunsaturated arachidonoyl group at the *sn*-2 position of the phospholipid moiety [70]. The human lymphocyte CD52 GPI anchor is also known to carry phosphoethanolamine at O2 of Man-I, an additional α-linked Man at O2 of Man-III, and 2-O-palmitoylation at the inositol unit [71]. The latter feature was not incorporated in Guo's target compound. Similar to Guo's report on the human sperm CD52 GPI anchor synthesis [64, 65], two simultaneous late-stage phosphorylation reactions took place after a convergent assembly of the phosphoglycerolipidated pseudohexasaccharide.

Tetramannose donor **146** (Scheme 13.22) was assembled in a straightforward fashion by using building blocks **142–144** and Schmidt methodology. The last glycosylation with **144** surprisingly resulted in a mixture of α/β anomers and were separated after the subsequent deallylation stage. Cha's deallylation protocol [72] was employed as other methods proved to be insufficient. This step was followed by silylation to produce **145**, which was converted to trichloroacetimidate **146** in two steps that include the hydrolysis of the thioglycoside, and then reaction with CCl_3CN in the presence of DBU.

Glycosylation between glucosaminyl donor **147** and inositol acceptor **148** followed by Ir-catalyzed deallylation provided the pseudodisaccharide core **149** (Scheme 13.22). It was coupled to the phosphoglycerolipid derivative **150** with the use of 1*H*-tetrazole as mild acidic promoter and then oxidized with *t*BuOOH at low temperature to avoid epoxidation of the unsaturated lipid. The phosphoglycerolipidated pseudodisaccharide then underwent TBS deprotection to afford the desired 4′-alcohol **151**.

The key [4+2] glycosylation followed by desilylation of the two phosphorylation sites resulted in a moderate 39% yield over two steps (Scheme 13.23). Phosphoramidite **153** was used to phosphorylate both of the resulting free hydroxy groups. A three-step protocol was applied for final deprotection, starting with azide reduction with Zn and acetic acid, then removal of cyanoethoxy phosphate protection with DBU, and finally

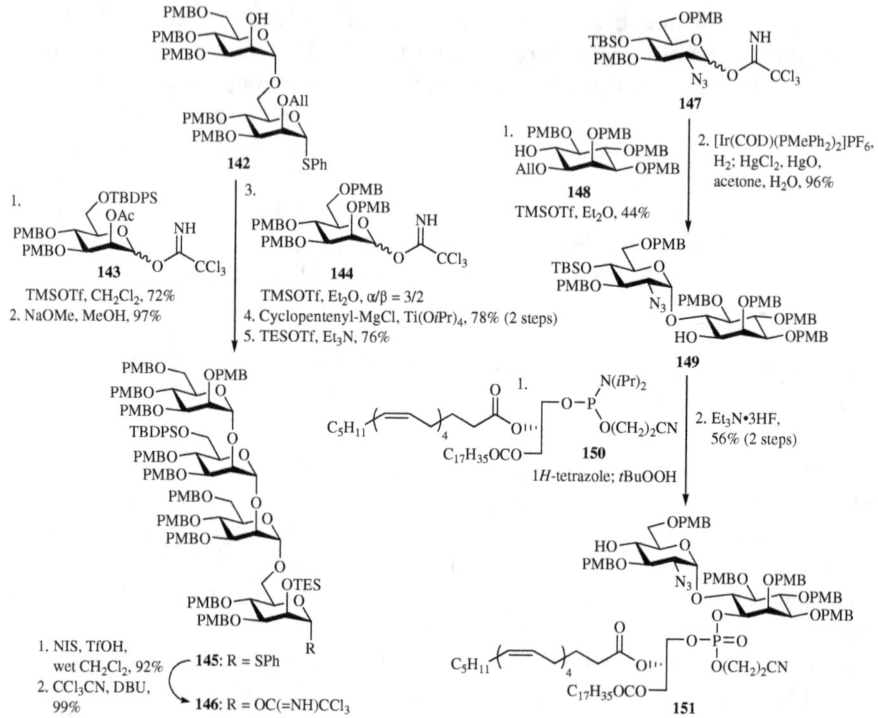

SCHEME 13.22 Synthesis of the tetramannose donor **146** and the pseudodisaccharide acceptor **151**. COD, 1,5-cyclooctadiene.

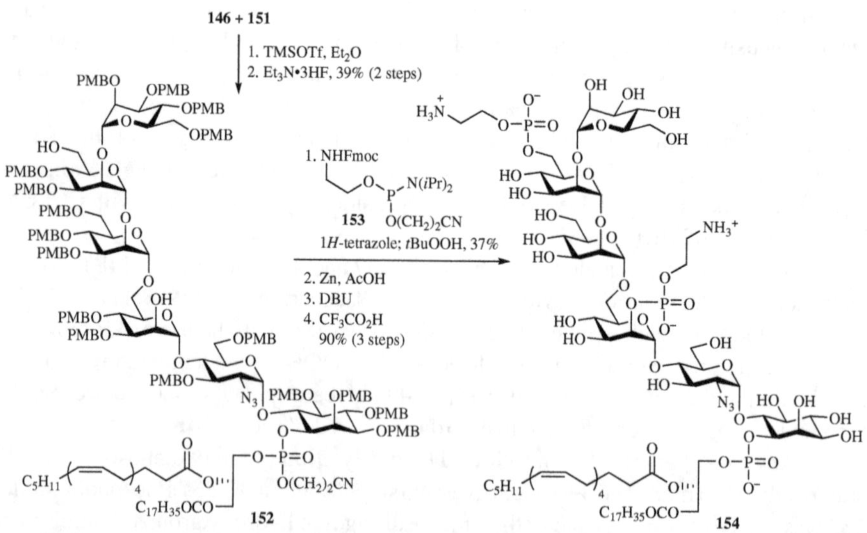

SCHEME 13.23 Completion of Guo's synthesis of the human lymphocyte CD52 antigen GPI anchor. Fmoc, 9-fluorenylmethoxycarbonyl.

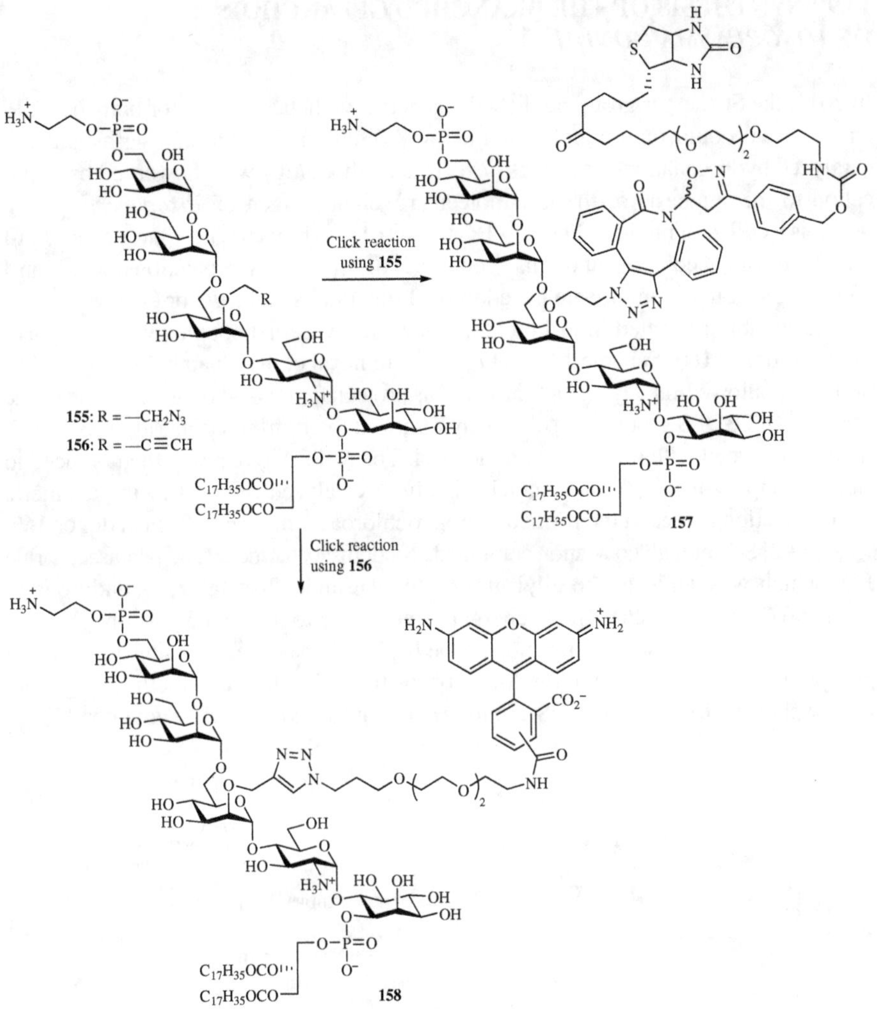

SCHEME 13.24 Functionalization of GPIs using click chemistry.

PMB removal with trifluoroacetic acid. All three steps took place quickly and efficiently to give an overall yield of 90% for the GPI anchor **154**.

Using PMB global protection allowed the synthesis of GPI anchors containing unsaturated lipids [69], including a GPI with the very labile polyunsaturated arachidonoyl lipid [70]. The PMB strategy proved advantageous as PMB could be removed under mild acidic conditions that did not affect the glycosidic linkages or acyl lipids and peptides linked to GPI. It can actually be used to prepare GPIs with other reduction-sensitive functional groups as demonstrated in Guo's synthesis of GPI derivatives **155** and **156** containing azide and alkyne functionalities, respectively [73]. These functionalities allowed the coupling of biologically useful appendages such as a fluorescent tag or a biotin molecule by click chemistry (Scheme 13.24) [74].

13.9 SYNTHESIS OF THE BRANCHED GPI ANCHOR OF *TOXOPLASMA GONDII*

In 2011, the Seeberger group published a general synthetic strategy for branched GPI anchors that could theoretically include any functionality that can withstand Pd-catalyzed hydrogenation conditions [75]. This methodology was showcased by application to the synthesis of the low-molecular-weight antigen of *Toxoplasma gondii*, which is a GPI containing a Glc(α1→4)GalNAc(β1→4) branch (Glc, glucose) at Man-I [76]. One critical component of this synthesis is the orthogonal protection of the Man-I building block to accommodate the additional sugar units at the O3 or O4 position.

The suitably protected Man-I building block **159** was first coupled with the galactosamine donor **160** (Scheme 13.25). Oxidative removal of the 2-napthylmethyl (NAP) protection allowed for glycosylation with Bn-protected glucosyl donor **162**. Next, the Man-I 6-*O*-TBDPS group was selectively removed to afford trisaccharide **163**, which was glycosylated with dimannosyl donor **164**. The *N*-TCA group was then reduced to an Ac group by using Zn and AcOH. Finally, Ir-catalyzed removal of the anomeric allyl protection preceded the formation of a trichloroacetimidate to furnish donor **165**.

A [5+2] Schmidt glycosylation combined pseudodisaccharide **29** and pentasaccharide **165**, which was subjected to allyl group removal under Ir-catalyzed conditions to form **166** (Scheme 13.26). The Seeberger group chose to install the phosphoglycerolipid moiety at the inositol O1 position by the *H*-phosphonate method using **99**. The next phosphorylation took place at the Man-III position with the same methodology and *H*-phosphonate **15** after the TIPS group was removed under Lewis acidic conditions.

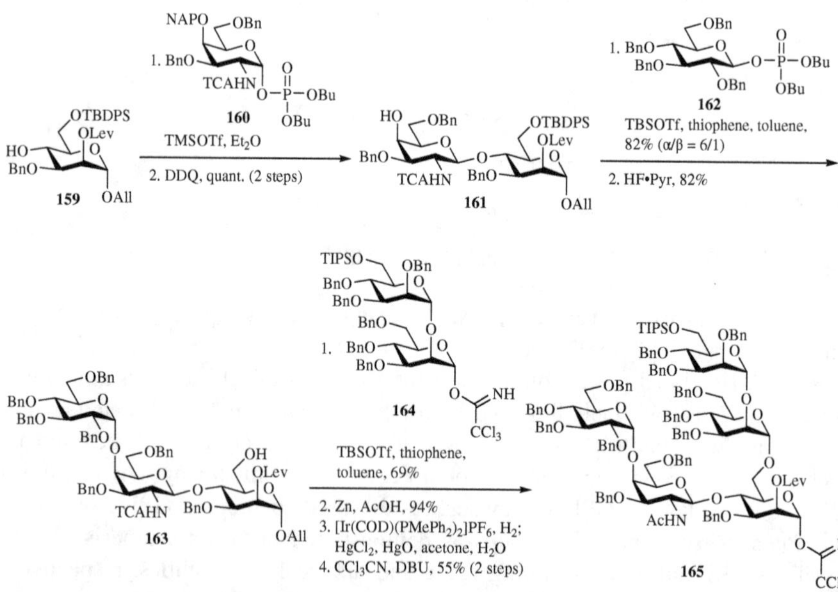

SCHEME 13.25 Synthesis of the pentasaccharide donor **165**. DDQ, 2,3-dichloro-5,6-dicyano-1,4-benzoquinone; Lev, levulinyl; TIPS, triisopropylsilyl.

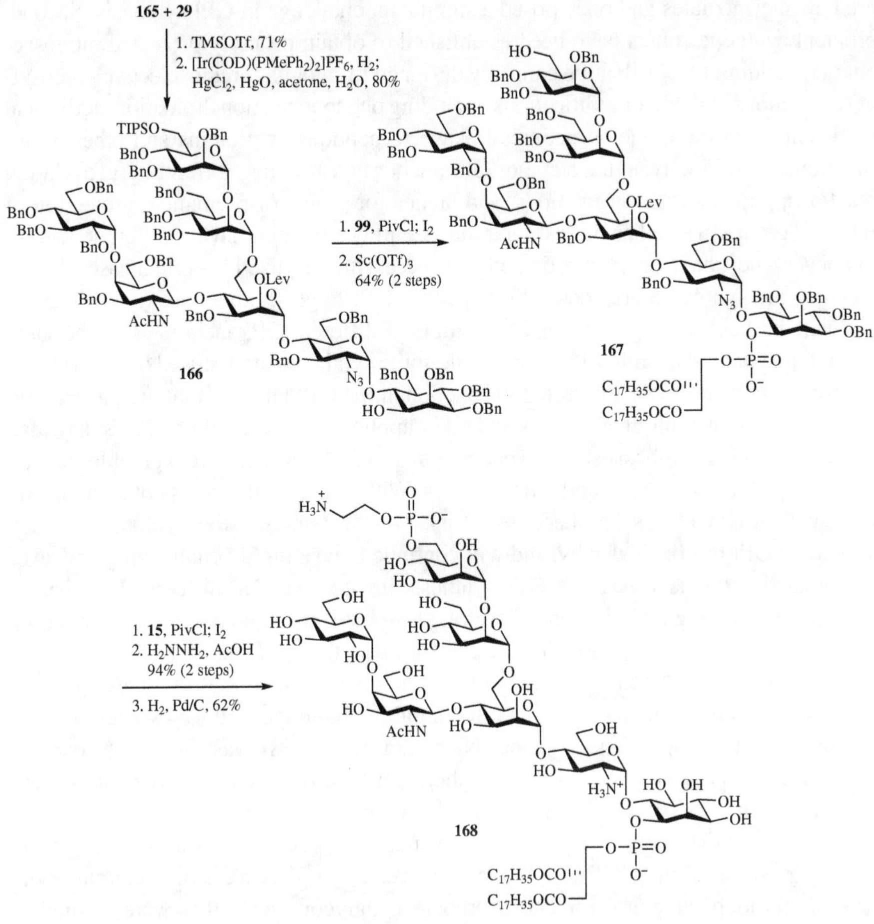

SCHEME 13.26 Completion of Seeberger's synthesis of a *Toxoplasma gondii* GPI anchor.

Phosphorylation would have been possible at the Man-I O2 position if desired as the Lev group could be selectively removed using hydrazine acetate before the Pd-catalyzed hydrogenolysis to remove all of the Bn groups to obtain the target GPI **168**.

The Seeberger group's synthetic strategy is focused on a general method for creating a wide range of highly functionalized GPIs. Another application of this methodology was reported in 2013, showcasing the synthesis of the *Tryp. brucei* variant surface glycoprotein GPI that contained additional sugars at the Man-I O3 position rather than the O4 position [77].

13.10 CONCLUSION

Since Ogawa's first report of a GPI total synthesis in 1991, the field has witnessed great multifaceted progress. First, a number of elegant synthetic approaches have been developed to access various position-differentiated inositol derivatives, which are key intermediates and once posed a significant challenge in GPI synthesis. Second, efficient synthetic routes have been established to obtain properly protected monosaccharide building blocks that are carefully designed to safeguard regioselective glycosylation reactions and other modifications, including phosphorylation, lipidation, additional sugar chain addition, and sometimes linkage to peptides or proteins and other useful functionalities. Finally, at the stage of GPI anchor construction, different glycosylation reactions, phosphorylation methods, and tactics for global or permanent protection of hydroxy groups, as well as oligosaccharide assembly strategies having different degrees of convergency, have been investigated and successfully utilized in certain cases. While some of the syntheses are focused on specific GPI targets, others are more diversity oriented, namely, aiming at the rapid assembly of different GPI anchors using the same or similar synthetic design. Moreover, while some syntheses are focused on natural GPI anchors, others are more interested in functionalized GPI anchors that are particularly useful for various biological studies and other applications. It should be fair to say that, at this stage, GPI synthesis starts to become mature, which will have a great impact on the study of GPI anchorage and related issues. With that said, this does not mean in any way that GPI synthesis has become trivial. On the contrary, the synthesis of each individual GPI has been, is now, and will continue to be a highly challenging endeavor for synthetic chemists, because GPI synthesis involves several different branches of chemistry, including carbohydrate, inositol, phosphate, and lipid chemistries, and each GPI synthesis must be individually designed to make the synthesis truly efficient.

Compared to the great progress in GPI anchor synthesis, GPI-anchored protein and glycoprotein synthesis remains a formidable challenge with only a few reports dealing with this topic. The Yoshiaki Nakahara group [78] was the first to explore GPI-linked peptide synthesis by native chemical ligation (NCL) between short peptides with simple GPI analogues. NCL was then utilized by the Carolyn Bertozzi group [29] and the Seeberger group [79] to successfully couple a fluorescent protein and the prion protein to GPI analogues, respectively. GPIs were also coupled with proteins through unnatural linkers to prepare neoglycoproteins that were studied as GPI-based vaccines [26]. Alternatively, the Guo group has reported the synthesis of GPI-linked peptides and glycopeptides via regiospecific chemical coupling reactions of partially protected GPIs and peptides/glycopeptides, followed by global deprotection [28, 80]. Recently, the Guo group also reported the use of sortase A for chemoenzymatic synthesis of GPI-linked peptides and glycopeptides [81–83], which may be employed to synthesize GPI-anchored proteins and glycoproteins. However, all the reported synthetic strategies are either difficult to apply to full-size proteins or result in products that contained unnatural amino acid or peptide residues. Therefore, a method for the synthesis of truly natural GPI-anchored proteins and glycoproteins is highly desirable.

ACKNOWLEDGMENT

The authors thank the National Science Foundation (CHE-0320878, 0715275, and 1053848) and National Institutes of Health (R01 GM090270) of the United States for their continuous support of research projects related to GPI anchors.

REFERENCES

[1] Ferguson, M. A. J.; Williams, A. F. *Annu. Rev. Biochem.* **1988**, *57*, 285–320.

[2] Englund, P. T. *Annu. Rev. Biochem.* **1993**, *62*, 121–138.

[3] Ikezawa, H. *Biol. Pharm. Bull.* **2002**, *25*, 409–417.

[4] Paulick, M. G.; Bertozzi, C. R. *Biochemistry* **2008**, *47*, 6991–7000.

[5] Menon, A. K.; Baumann, N. A.; Rancour, D. M. In *Carbohydrates in Chemistry and Biology*; Hart, G. W., Hart, G., Sinaÿ, P., Ernst, B., Sinaikin, P., Eds.; John Wiley & Sons: New York, **2000**; Vol. *4*, pp 757–769.

[6] Kasahara, K.; Sanai, Y. *Glycoconj. J.* **2000**, *17*, 153–162.

[7] He, H.-T.; Finne, J.; Goridis, C. *J. Cell Biol.* **1987**, *105*, 2489–2500.

[8] Cross, G. A. M. *Annu. Rev. Immunol.* **1990**, *8*, 83–110.

[9] Kinoshita, T.; Inoue, N.; Takeda, J. *Adv. Immunol.* **1995**, *60*, 57–103.

[10] Robinson, P. J.; Millrain, M.; Antoniou, J.; Simpson, E.; Mellor, A. L. *Nature* **1989**, *342*, 85–87.

[11] Eardley, D. D.; Koshland, M. E. *Science* **1991**, *251*, 78–81.

[12] Hwa, K. *Adv. Exp. Med. Biol.* **2001**, *491*, 207–214.

[13] Ferguson, M. A. J.; Homans, S. W.; Dwek, R. A.; Rademacher, T. W. *Science* **1988**, *239*, 753–759.

[14] Murakata, C.; Ogawa, T. *Tetrahedron Lett.* **1991**, *32*, 671–674.

[15] Sureshan, K. M.; Shashidhar, M. S.; Praveen, T.; Das, T. *Chem. Rev.* **2003**, *103*, 4477–4503.

[16] Gigg, R.; Gigg, J. In *Glycopeptides and Related Compounds: Synthesis, Analysis, and Applications*; Large, D. G., Warren, C. D., Eds.; Marcel Dekker: New York, **1997**, pp 327–392.

[17] Guo, Z.; Bishop, L. *Eur. J. Org. Chem.* **2004**, 3585–3596.

[18] Nikolaev, A. V.; Al-Maharik, N. *Nat. Prod. Rep.* **2011**, *28*, 970–1020.

[19] Tsai, Y.; Liu, X.; Seeberger, P. H. *Angew. Chem. Int. Ed.* **2012**, *51*, 11438–11456.

[20] Murakata, C.; Ogawa, T. *Tetrahedron Lett.* **1990**, *31*, 2439–2442.

[21] Reichardt, N.-C.; Martin-Lomas, M. *Angew. Chem. Int. Ed.* **2003**, *42*, 4674–4677.

[22] Lopez-Prados, J.; Martin-Lomas, M. *J. Carbohydr. Chem.* **2006**, *24*, 393–414.

[23] Pekari, K.; Tailler, D.; Weingart, R.; Schmidt, R. R. *J. Org. Chem.* **2001**, *66*, 7432–7442.

[24] Ruda, K.; Lindberg, J.; Garegg, P. J.; Oscarson, S.; Konradsson, P. *J. Am. Chem. Soc.* **2000**, *122*, 11067–11072.

[25] Hederos, M.; Konradsson, P. *J. Am. Chem. Soc.* **2006**, *128*, 3414–3419.

[26] Schofield, L.; Hewitt, M. C.; Evans, K.; Siomos, M.-A.; Seeberger, P. H. *Nature* **2002**, *418*, 785–789.

[27] Kwon, Y.; Soucy, R. L.; Snyder, D. A.; Seeberger, P. H. *Chem. Eur. J.* **2005**, *11*, 2493–2504.

[28] Xue, J.; Shao, N.; Guo, Z. *J. Org. Chem.* **2003**, *68*, 4020–4029.

[29] Paulick, M. G.; Wise, A. R.; Forstner, M. B.; Groves, J. T.; Bertozzi, C. R. *J. Am. Chem. Soc.* **2007**, *129*, 11543–11550.

[30] Paulick, M. G.; Forstner, M. B.; Groves, J. T.; Bertozzi, C. R. *Proc. Natl. Acad. Sci. U. S. A.* **2007**, *104*, 20332–20337.

[31] Swarts, B. M.; Guo, Z. *Adv. Carbohydr. Chem. Biochem.* **2012**, *67*, 137–219.

[32] Murakata, C.; Ogawa, T. *Carbohydr. Res.* **1992**, *235*, 95–114.

[33] Lemieux, R. U.; Hendriks, K. B.; Stick, R. V.; James, K. *J. Am. Chem. Soc.* **1975**, *97*, 4056–4062.

[34] Koenigs, W.; Knorr, E. *Chem. Ber.* **1901**, *34*, 957–981.

[35] Suzuki, K.; Maeta, H.; Matsumoto, T.; Tsuchihashi, G. *Tetrahedron Lett.* **1988**, *29*, 3571–3574.

[36] Suzuki, K.; Maeta, H.; Matsumoto, T. *Tetrahedron Lett.* **1989**, *30*, 4853–4856.

[37] Lindh, I.; Stawinski, J. *J. Org. Chem.* **1989**, *54*, 1338–1342.

[38] Baeschlin, D. K.; Chaperon, A. R.; Charbonneau, V.; Green, L. G.; Ley, S. V.; Lücking, U.; Walther, E. *Angew. Chem. Int. Ed.* **1998**, *37*, 3423–3428.

[39] Baeschlin, D. K.; Green, L. G.; Hahn, M. G.; Hinzen, B.; Ince, S. J.; Ley, S. V. *Tetrahedron: Asymmetry* **2000**, *11*, 173–179.

[40] Bannwarth, W.; Trzeciak, A. *Helv. Chim. Acta* **1987**, *70*, 175–186.

[41] Mootoo, D. R.; Konradsson, P.; Udodong, U.; Fraser-Reid, B. *J. Am. Chem. Soc.* **1988**, *110*, 5583–5584.

[42] Mehta, S.; Pinto, M. In *Preparative Carbohydrate Chemistry*; Hanessian, S., Ed.; Marcel Dekker: New York, **1997**, pp 106–129.

[43] Fankhauser, C.; Homans, S. W.; Thomas-Oates, J. E.; McConville, M. J.; Desponds, C.; Conzelmann, A.; Ferguson, M. A. *J. Biol. Chem.* **1993**, *268*, 26365–26374.

[44] Mayer, T. G.; Kratzer, B.; Schmidt, R. R. *Angew. Chem. Int. Ed. Engl.* **1994**, *33*, 2177–2181.

[45] Mayer, T. G.; Schmidt, R. R. *Eur. J. Org. Chem.* **1999**, 1153–1165.

[46] Schmidt, R. R.; Michel, J. *Angew. Chem. Int. Ed. Engl.* **1980**, *19*, 731–733.

[47] Udodong, U. E.; Madsen, R.; Roberts, C.; Fraser-Reid, B. *J. Am. Chem. Soc.* **1993**, *115*, 7886–7887.

[48] Campbell, A. S.; Fraser-Reid, B. *J. Am. Chem. Soc.* **1995**, *117*, 10387–10388.

[49] Homans, S. W.; Ferguson, M. A. J.; Dwek, R. A.; Rademacher, T. W.; Anand, R.; Williams, A. F. *Nature* **1988**, *333*, 269–272.

[50] Fraser-Reid, B.; Konradsson, P.; Mootoo, D. R.; Udodong, U. *J. Chem. Soc. Chem. Commun.* **1988**, 823–825.

[51] Mootoo, D. R.; Date, V.; Fraser-Reid, B. *J. Am. Chem. Soc.* **1988**, *110*, 2662–2663.

[52] Tailler, D.; Ferrieres, V.; Pekari, K.; Schmidt, R. R. *Tetrahedron Lett.* **1999**, *40*, 679–682.

[53] Pekari, K.; Schmidt, R. R. *J. Org. Chem.* **2003**, *68*, 1295–1308.

[54] Lu, J.; Jayaprakash, K. N.; Schlueter, U.; Fraser-Reid, B. *J. Am. Chem. Soc.* **2004**, *126*, 7540–7547.

[55] Lu, J.; Jayaprakash, K. N.; Fraser-Reid, B. *Tetrahedron Lett.* **2004**, *45*, 879–882.

[56] Azzouz, N.; Kamena, F.; Seeberger, P. H. *OMICS* **2010**, *14*, 445–454.

[57] Naik, R. S.; Branch, O. H.; Woods, A. S.; Vijaykumar, M.; Perkins, D. J.; Nahlen, B. L.; Lal, A. A.; Cotter, R. J.; Costello, C. E.; Ockenhouse, C. F.; Davidson, E. A.; Gowda, D. C. *J. Exp. Med.* **2000**, *192*, 1563–1576.

[58] Soli, E. D.; Manoso, A. S.; Patterson, M. C.; DeShong, P.; Favor, D. A.; Hirschmann, R.; Smith, A. B., III *J. Org. Chem.* **1999**, *64*, 3171–3177.

[59] Liu, X.; Kwon, Y.; Seeberger, P. H. *J. Am. Chem. Soc.* **2005**, *127*, 5004–5005.

[60] Kamena, F.; Tamborrini, M.; Liu, X.; Kwon, Y. U.; Thompson, F.; Pluschke, G.; Seeberger, P. H. *Nat. Chem. Biol.* **2008**, *4*, 238–240.

[61] Ali, A.; Gowda, D. C.; Vishwakarma, R. A., *Chem. Commun.* **2005**, 519–521.

[62] Almeida, I. C.; Camargo, M. M.; Procopio, D. O.; Silva, L. S.; Mehlert, A.; Travassos, L. R.; Gazzinelli, R. T.; Ferguson, M. A. J. *EMBO J.* **2000**, *19*, 1476–1485.

[63] Yashunsky, D. V.; Borodkin, V. S.; Ferguson, M. A. J.; Nikolaev, A. V. *Angew. Chem. Int. Ed.* **2006**, *45*, 468–474.

[64] Xue, J.; Guo, Z. *J. Am. Chem. Soc.* **2003**, *125*, 16334–16339.

[65] Wu, X.; Guo, Z. *Org. Lett.* **2007**, *9*, 4311–4313.

[66] Schröter, S.; Derr, P.; Conradt, H. S.; Nimtz, M.; Hale, G.; Kirchhoff, C. *J. Biol. Chem.* **1999**, *274*, 29862–29873.

[67] Xue, J.; Guo, Z. *Bioorg. Med. Chem. Lett.* **2002**, *12*, 2015–2018.

[68] Wu, X.; Shen, Z.; Zeng, X.; Lang, S.; Palmer, M.; Guo, Z. *Carbohydr. Res.* **2008**, *343*, 1718–1729.

[69] Swarts, B. M.; Guo, Z. *J. Am. Chem. Soc.* **2010**, *132*, 6648–6650.

[70] Burgula, S.; Swarts, B. M.;Guo, Z. *Chem. Eur. J.* **2012**, *18*, 1194–1201.

[71] Treumann, A.; Lifely, M. R.; Schneider, P.; Ferguson, M. A. J. *J. Biol. Chem.* **1995**, *270*, 6088–6099.

[72] Lee, J.; Cha, J. K. *Tetrahedron Lett.* **1996**, *37*, 3663–3666.

[73] Swarts, B. M.; Guo, Z. *Chem. Sci.* **2011**, *2*, 2342–2352.

[74] Sletten, E. M.; Bertozzi, C. R. *Acc. Chem. Res.* **2011**, *44*, 666–676.

[75] Tsai, Y.-H.; Gçtze, S.; Azzouz, N.; Hahm, H. K.; Seeberger, P. H.; Silva, D. V. *Angew. Chem. Int. Ed.* **2011**, *50*, 9961–9964.

[76] Striepen, B.; Zinecker, C. F.; Damm, J. B. L.; Melgers, P. A. T.; Gerwig, G. J.; Koolen, M.; Vliegenthart, J. F. G.; Dubremetz, J.-F.; Schwarz, R. T. *J. Mol. Biol.* **1997**, *266*, 797–813.

[77] Tsai, Y. H.; Gotze, S.; Vilotijevic, I.; Grube, M.; Silva, D. V.; Seeberger, P. H. *Chem. Sci.* **2013**, *4*, 468–481.

[78] Tanaka, Y.; Nakahara, Y.; Hojo, H.; Nakahara, Y. *Tetrahedron* **2003**, *59*, 4059–4067.

[79] Becker, C. F. W.; Liu, X.; Olschewski, D.; Castelli, R.; Seidel, R.; Seeberger, P. H. *Angew. Chem. Int. Ed.* **2008**, *47*, 8215–8219.

[80] Shao, N.; Xue, J.; Guo, Z. *Angew. Chem. Int. Ed.* **2004**, *43*, 1569–1573.

[81] Guo, X.; Wang, Q.; Swarts, B. M.; Guo, Z. *J. Am. Chem. Soc.* **2009**, *131*, 9878–9879.

[82] Wu, Z.; Guo, X.; Wang, Q.; Swarts, B. M.; Guo, Z. *J. Am. Chem. Soc.* **2010**, *132*, 1567–1571.

[83] Wu, Z.; Guo, X.; Guo, Z. *Chem. Commun.* **2010**, *46*, 5773–5774.

14

SYNTHESIS OF BACTERIAL CELL ENVELOPE COMPONENTS

Akihiro Ishiwata and Yukishige Ito
Synthetic Cellular Chemistry Laboratory, RIKEN, Saitama, Japan

14.1 INTRODUCTION

Glycans found in bacteria display a big variety of structural components [1]. As shown in Figure 14.1, bacterial cell-wall envelopes are quite complex. Different compositions can be seen in Gram-negative, Gram-positive, and mycobacterial organisms. Cell-wall components are critical for bacterial survival and, together with cell-wall biosynthetic enzymes, are the targets of antibiotics [2]. In most bacteria, peptidoglycan [3]—the major structural component of the cell wall conserved among species—encloses the cytoplasmic membrane (inner membrane) in the periplasm. The peptidoglycan layer is about 10 nm in Gram-negative and up to 80 nm in Gram-positive species [4]. The Gram-negative bacterial envelope has an asymmetric lipid bilayer outer membrane composed of inner phospholipid and outer lipopolysaccharide (LPS) leaflets [5]. Gram-positive bacteria have a thick peptidoglycan layer not only for protection but also as scaffold for surface proteins and secondary cell-wall biopolymers such as lipoteichoic acid, which is anchored into the inner membrane and extends through the peptidoglycan architecture up to the extracellular spaces [6]. In mycobacteria, peptidoglycans are linked to arabinogalactans through a linker disaccharide. Arabinogalactans, in turn, are connected to mycolic acids, which are long-chain β-hydroxy fatty acids that form a thick lipid barrier similar to the outer membrane in Gram-negative bacteria [7]. Lipoarabinomannan (LAM), also a cell

Glycochemical Synthesis: Strategies and Applications, First Edition.
Edited by Shang-Cheng Hung and Medel Manuel L. Zulueta.

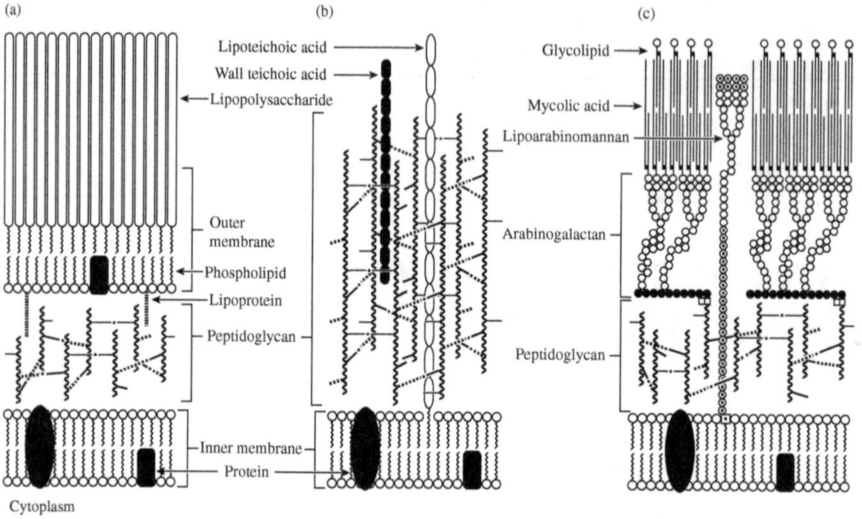

FIGURE 14.1 Cell envelope structures of (a) Gram-negative bacteria, (b) Gram-positive bacteria, and (c) mycobacteria.

envelope component in mycobacterial species, is anchored to the inner membrane through a phosphatidylinositol mannoside (PIM) moiety [8].

Approaches toward the preparation of bacterial cell envelope oligo- and polysaccharide components are described. The discussion includes the synthesis of peptidoglycans, LPS, arabinomannans, and other bacterial glycoconjugates. Recent synthetic studies targeting such complex molecules as well as the glycan of bacterial glycoproteins are covered as much as possible in this chapter, mainly focusing on the formation of glycosidic linkages in the molecules.

14.2 PEPTIDOGLYCAN AND RELATED GLYCOCONJUGATES

14.2.1 Lipid I, II, and IV Analogues

Peptidoglycan consists of repeating β(1→4)-linked *N*-acetylglucosamine (GlcNAc)-*N*-acetylmuramic acid (MurNAc) moieties with a pentapeptide (sequentially, L-alanine (L-Ala), γ-D-glutamic acid (γ-D-Glu), L-lysine (L-Lys) or *meso*-diaminopimelic acid (DAP), D-Ala, and D-Ala) attached to the D-lactyl moiety of each MurNAc. The pentapeptide stem participates in an interglycan cross-linking reaction, thus creating the cell-wall polymer. Lipid I is biologically synthesized from muramyl pentapeptide and undecaprenyl pyrophosphate. Glycosyl transfer from UDP-GlcNAc (UDP, uridine diphosphate) to lipid I leads to lipid II (Fig. 14.2). Further modifications by branched peptide chains such as pentaglycine [9], L-Ser-L-Ala (Ser, serine), and L-Ala-L-Ala dipeptides [10, 11] were also reported. The γ-D-Glu residue is converted to γ-D-glutamine (γ-D-Gln) in most Gram-negative species and mycobacteria [12]. A transporter protein presumably flip-flops lipid II monomers from the cytoplasmic

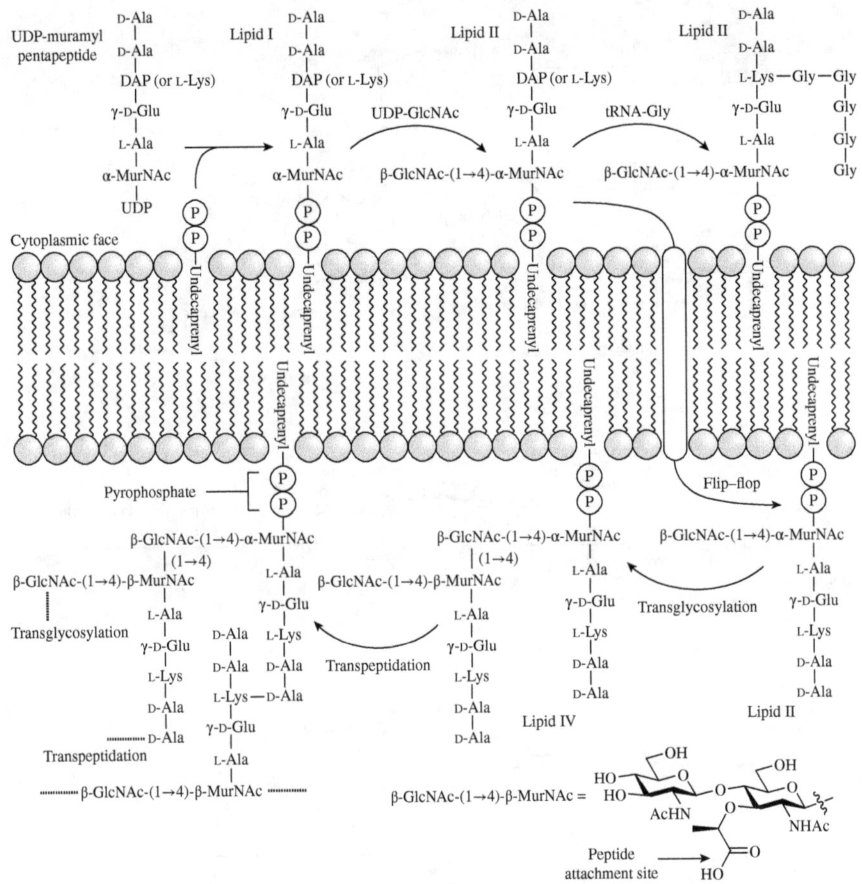

FIGURE 14.2 Peptidoglycan biosynthesis. Ac, acetyl; Gly, glycine.

face to the periplasmic surface. Peptidoglycan glycosyltransferases (PGTs) catalyze the polymerization of glycan chains in the bacterial cell wall by forming β(1→4) linkages between GlcNAc-MurNAc(peptide) units to produce the tetrasaccharide structure in lipid IV and the extended chains [13]. Transpeptidases subsequently form the durable cross-linked structure of the strands [14, 15].

Due to the structural complexity, low natural abundance, and difficult isolation of lipid I and lipid II, it is very important to chemically supply them for the study of peptidoglycan biosynthesis and the development of agents that may disrupt such process. Michael VanNieuwenhze et al. completed the synthesis of lipid I and lipid II having undecaprenyl pyrophosphate lipid carrier (**5**, **13**) [16–18] as well as the biosynthetic intermediate UDP-MurNAc-pentapeptide **7** (Scheme 14.1) [19]. Benjamin Schwartz's group also reported the synthesis of lipid II as well as the development of an appropriate assay system to observe both transglycosylation and transpeptidation catalyzed by penicillin-binding proteins (PBPs) [20]. Using VanNieuwenhze's scheme, Philip Collins and Gregory Weiss synthesized the derivatives of lipid I and

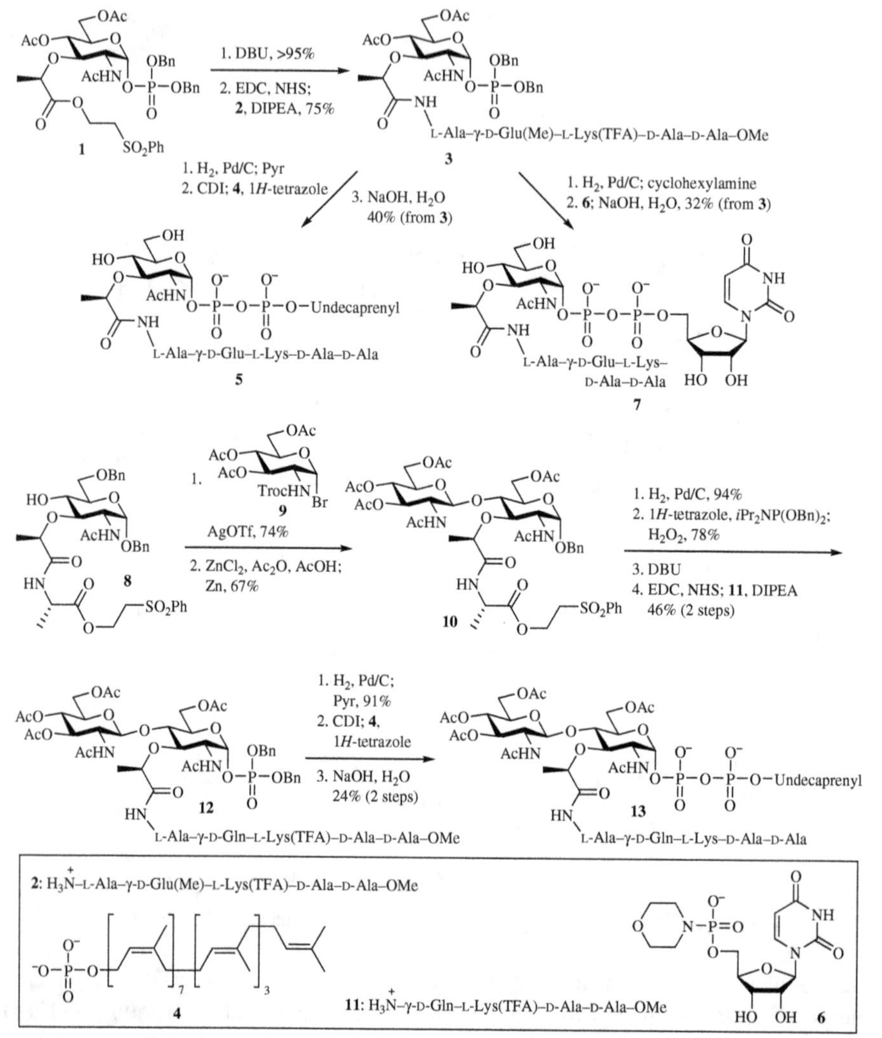

SCHEME 14.1 Synthesis of lipids I and II by VanNieuwenhze et al. Bn, benzyl; CDI, 1,1′-carbonyldiimidazole; DBU, 1,8-diazabicyclo[5.4.0]undec-7-ene; DIPEA, diisopropylethylamine; EDC, 1-ethyl-3-(3-dimethylaminopropyl)carbodiimide; NHS, *N*-hydroxysuccinimide; Pyr, pyridine; TFA, trifluoroacetyl; Troc, 2,2,2-trichloroethoxycarbonyl.

lipid II including a derivative of cross-linked lipid II for monitoring single-molecule lysozyme dynamics [21].

In order to find lead inhibitors against PGTs [15], Piet Herdewijn et al. [22] synthesized a series of substrate analogues of lipid I and lipid II with variations in lipid, pyrophosphate, and peptide moieties. Although many efforts were spent in designing and synthesizing lipid I- and lipid II-based inhibitors [23–26], the PGT inhibition assay performed with *Escherichia coli* PBP1B *in vitro* and antibacterial activity against *Bacillus subtilis* using lipid I and lipid II analogues indicated C16-phosphoglycerate-MurNAc(L-Ala-D-Glu)-GlcNAc as a promising lead compound for the design of new PGT inhibitors against multidrug-resistant bacteria.

PGTs, designated as class A PBPs, are potential antibacterial targets [27]. To afford bacterial PGT substrates, Chi-Huey Wong, Wei-Chieh Cheng, and coworkers synthesized lipid II and IV derivatives [28–30] with modifications to the peptide and lipid moieties and addition of fluorescent tags. According to the reactivity-based assembly strategy, coupling of two disaccharides (**17** and **18**) prepared from monosaccharides **14** and **15** afforded tetrasaccharide **19**, which was converted to the lipid IV **21** (Scheme 14.2). They also prepared the *N*-glycolyl lipid II **26** and analogues with fluorescent tags as mycobacterial PGT substrates (Scheme 14.3) [31].

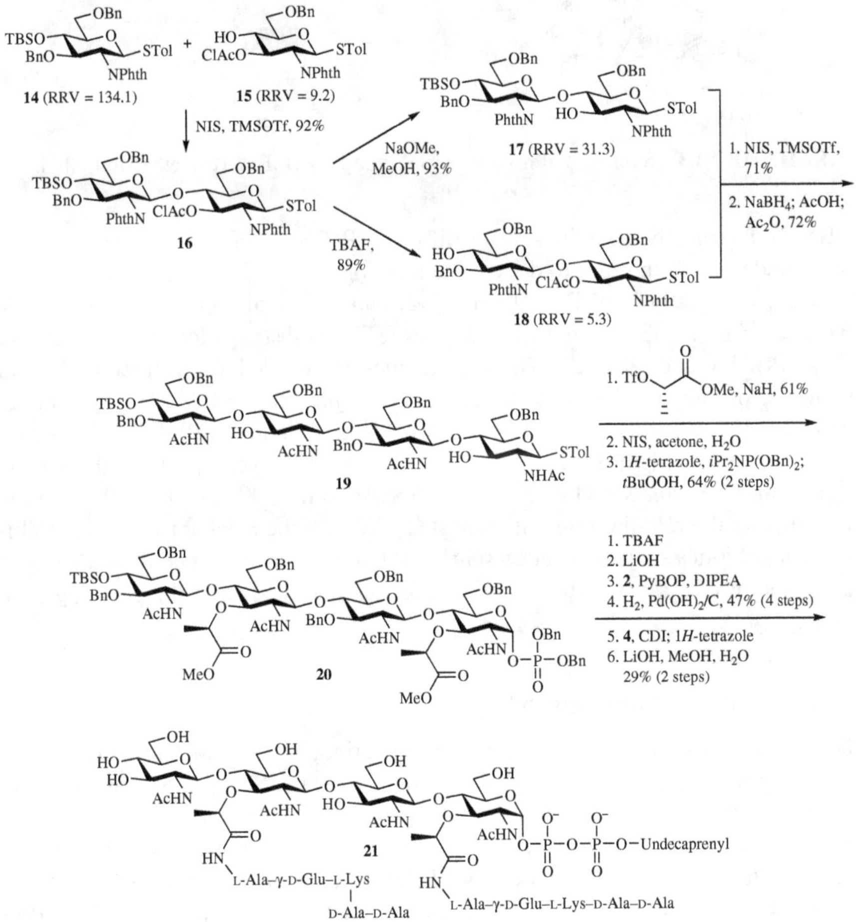

SCHEME 14.2 Synthesis of lipid IV by Wong, Cheng, and coworkers. ClAc, chloroacetyl; NIS, *N*-iodosuccinimide; Phth, phthaloyl; PyBOP, benzotriazol-1-yl-oxytripyrrolidinophosphonium hexafluorophosphate; RRV, relative reactivity value; TBAF, tetrabutylammonium fluoride; TBS, *tert*-butyldimethylsilyl; Tf, triflyl; TMS, trimethylsilyl; Tol, 4-tolyl.

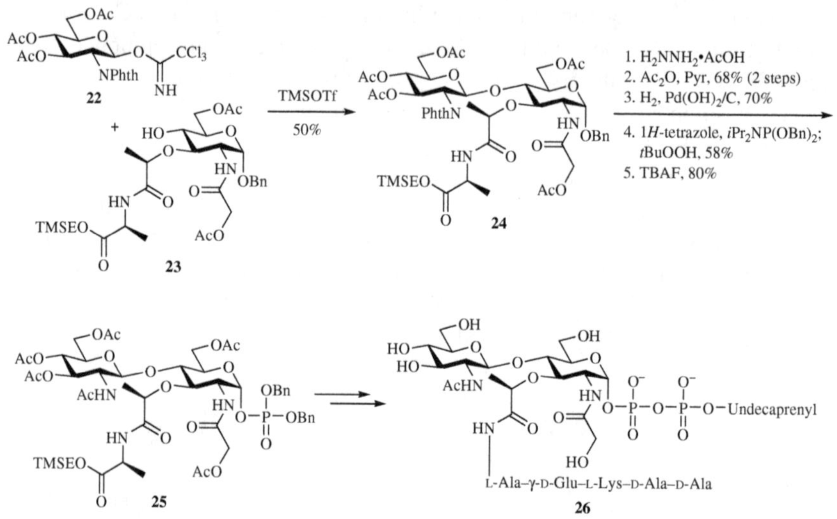

SCHEME 14.3 Synthesis of mycobacterial lipid II. TMSE, 2-(trimethylsilyl)ethyl.

Bacteria typically contain several different PGTs whose biological roles are poorly understood. In order to dissect the reaction, Daniel Kahne et al. synthesized chemoenzymatically lipid II analogues [32] and chemically heptaprenyl–lipid IV **36** (Scheme 14.4) [33, 34] and its analogues [35] as substrates for the major *E. coli* PGTs (PBP1A and PBP1B). These analogues were used for further biological researches on other PGTs, such as *Enterococcus faecalis* PBP2A, *Staphylococcus aureus* PBP2 [36], and *Aquifex aeolicus* PBP1A [37], and the cell division amidase AmiA required for hydrolysis of *E. coli* septal peptidoglycan [38], on the kinetic mechanism of *E. coli* MurG that couples GlcNAc to lipid I [39], on PGT activation by bypassing the rate-limiting initiation steps [40], on the mechanism of the inhibition of lantibiotics such as halodurascin α [41], and on the transpeptidase-mediated incorporation of a set of structurally diverse D-amino acids into the peptidoglycan fragment [42].

14.2.2 Peptidoglycan Fragments

The three-dimensional structure of bacterial peptidoglycan has been the subject of extensive investigation and speculation for over 40 years. Shahriar Mobashery et al. proposed the three-dimensional structure of peptidoglycans by using an aqueous solution structure obtained by NMR analysis of a synthetic GlcNAc-MurNAc(pentapeptide)-GlcNAc-MurNAc(pentapeptide) cell-wall fragment of the Gram-negative bacteria *E. coli* (**45**) (Scheme 14.5) [43–45] as well as that of the Gram-positive *Bacillus anthracis* [46]. Compound **45** was synthesized from disaccharide **39**, obtained from monosaccharides **37** and **38**, through two successive glycosylations with monosaccharide acceptor **40** and donor **42** followed by introduction of the lactate moiety to the resultant **43** and by coupling with pentapeptide **44**. The orientation of the

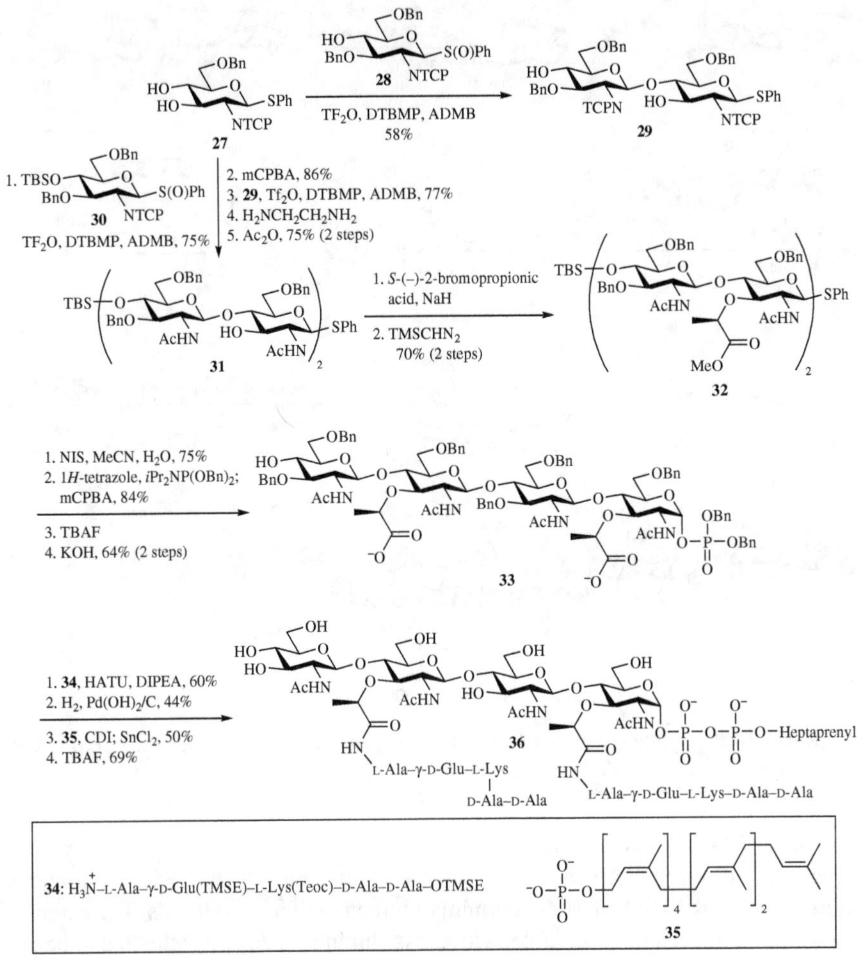

SCHEME 14.4 Synthesis of a lipid IV analogue by Khane and coworkers. ADMB, 4-allyl-1,2-dimethoxybenzene; DTBMP, 2,6-di-*tert*-butyl-4-methylpyridine; HATU, *O*-(7-azabenzotriazole-1-yl)-*N,N,N′,N′*-tetramethyluronium hexafluorophosphate; mCPBA, *m*-chloroperoxybenzoic acid; TCP, tetrachlorophthaloyl; Teoc, 2-trimethylsilylethoxycarbonyl.

peptidoglycan strand is originally presumed parallel to the bacterial outer membrane surface; however, Mobashery et al. proposed an orientation orthogonal to the membrane [47, 48].

Mobashery et al. identified the monomeric solution structure of the synthetic 2-kDa peptidoglycan fragment by NMR analysis and molecular modeling and then determined the oligomeric structure by computer prediction. Based on the average length of the *E. coli* strain W7 peptidoglycan, which consists of nine disaccharide repeats, they predicted the structure of longer peptidoglycans by using the defined and repeating glycosidic torsion angles of the synthetic segment. The result indicates a

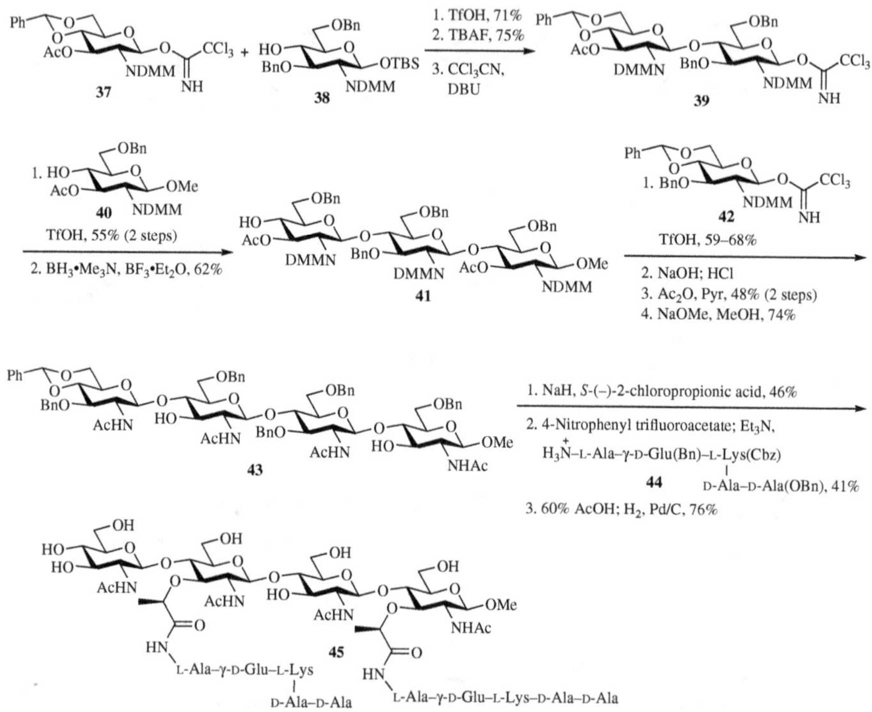

SCHEME 14.5 Synthesis of a peptidoglycan fragment by Mobashery and coworkers. Cbz, benzyloxycarbonyl; DMM, dimethylmaleoyl.

right-handed helix having three GlcNAc-MurNAc repeats per turn. The helix axis shows threefold symmetry with the peptide stems. Each peptidoglycan strand is assumed to be cross-linked to a maximum of three neighboring strands. The extent of cross-linking in bacteria is variable, with cross-linking for *E. coli* estimated to be less than 40%. *In silico* generation of the cell wall, assuming incomplete cross-links, resulted in a honeycomb pattern with larger pores due to missing cross-links.

Peptidoglycan recognition proteins (PGRPs) such as vancomycin are highly conserved pattern recognition molecules of the innate immune system [49]. Mobashery et al. reported the mechanism of binding of synthetic peptidoglycan fragments with human PGRPs by using X-ray, NMR, and molecular modeling methodologies [50]. On the basis of these analyses, it was proposed that PGRPs disrupt cell-wall maturation by sterically encumbering access of biosynthetic enzymes to the nascent peptidoglycan chains and by locking peptidoglycans into a conformation that prevents the formation of cross-links between peptide stems in the growing cell wall.

Geert-Jan Boons et al. also synthesized peptidoglycan fragments such as the disaccharide repeating unit [51] and muramyl peptide derivatives containing Lys and DAP residues [52–55] obtained through cross olefin metathesis [56]. The muramyl peptides, such as **48** prepared by solid-phase synthesis on Sieber amide resin (Scheme 14.6), were subjected to biological and structural analyses [50, 57, 58] of

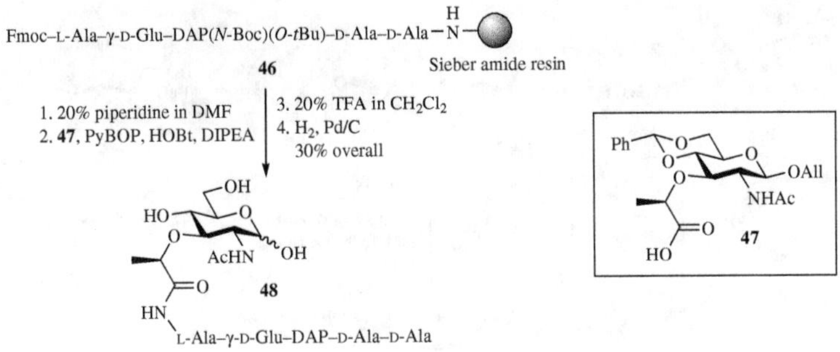

SCHEME 14.6 Synthesis of muramyl pentapeptide **48** with DAP residue by Boons et al. All, allyl; Boc, *tert*-butoxycarbonyl; HOBt, 1-hydroxybenzotriazole.

human PGRPs, which are bactericidal against pathogenic and nonpathogenic Gram-positive bacteria but not normal bacterial flora.

The mechanisms for the immunostimulatory and adjuvant properties of peptidoglycans and muramyl dipeptide, which is the minimum structural motif required for immunostimulation, are controversial [59, 60]. Koichi Fukase et al. synthesized peptidoglycan fragments including those that contain DAP and anhydroMurNAc residue [61–64], Lys [65, 66], and cross-links [67] to elucidate the immunostimulatory activity. To afford chains up to hexadecasaccharides, they employed a convergent approach through transformation of the fully protected oligosaccharide skeleton into glycosyl donor and acceptor followed by the respective coupling (Scheme 14.7). Using the peptidoglycan fragments, the recognition of peptidoglycans by the pattern recognition receptors Nod1 and Nod2 were suggested by Naohiro Inohara and Gabriel Nuñez [68, 69]. Peptidoglycan fragments induce Nod1-mediated immune responses through stimulation of chemokine production [70–72]. The native ligand of human Nod1 found in the most active fraction of *E. coli* K-12 culture supernatant is GlcNAc(β1→4)anhydroMurNAc(L-Ala-γ-D-Glu-DAP) [73].

For the observation of direct binding to Nod2, Catherine Grimes and coworkers developed a surface plasmon resonance (SPR) assay using 6-amino-muramyl dipeptide connected to carboxy-terminated alkanethiol self-assembled monolayers on chip by NHS/EDC activation of the carboxylic acid [74]. The 6-amino-muramyl dipeptide was synthesized through a modified Hasegawa protocol [75, 76]. The SPR experiments suggested that the γ-Gln stereochemistry is not a key recognition determinant and that Nod2 senses bacterial cell-wall fragments by binding directly to them.

The wild-type UDP-MurNAc-L-Ala-γ-D-Glu-L-Lys-D-Ala-D-Ala (UDP-MurNAc-pentapeptide) was first isolated from *S. aureus* in 1952 by James Park [77]. Nicolas Gisch et al. reported the synthesis of the biotinylated Lys-type and DAP-type muramyl tripeptides through their UDP derivatives [78]. Both compounds activated Nod2 in a cell-based assay, although the required minimal motif for the activation of Nod2 is muramyl dipeptide [79]. The chemical synthesis of the UDP-MurNAc-pentapeptide **7** (Scheme 14.1) [19] was followed by the chemoenzymatic synthesis by Chi-Huey Wong [80] and Michio Kurosu [81]. Wong et al. also synthesized the vancomycin-resistant UDP-MurNAc-L-Ala-γ-D-Glu-L-Lys-D-Ala-D-Lac

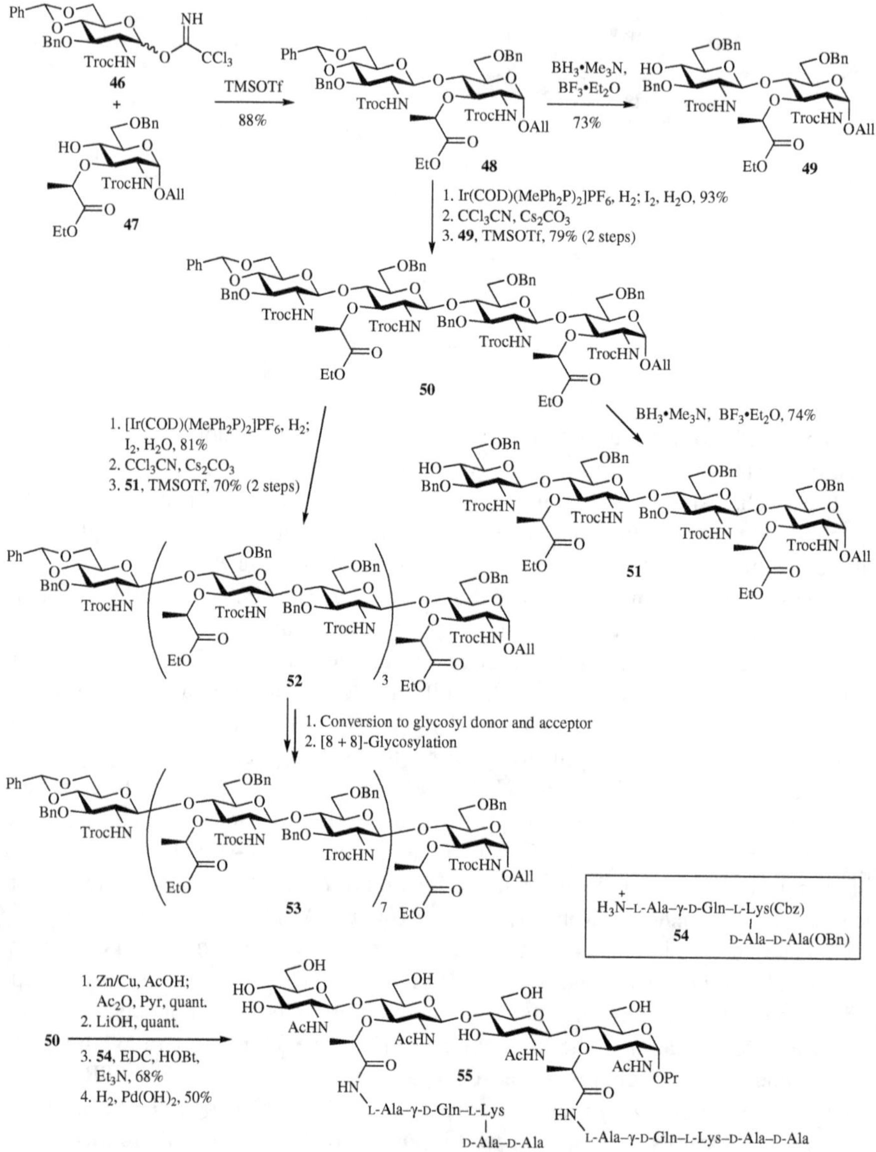

SCHEME 14.7 Synthesis of peptidoglycan fragments by Fukase and coworkers. COD, 1,5-cyclooctadiene.

(UDP-*N*-MurNAc-depsipentapeptide) derivatives including fluorescent-labeled derivatives. Shin-ichiro Nishimura et al. reported the synthesis of UDP-MurNAc-pentapeptide derivatives possessing a levulinyl (Lev) group on the side chain of the Lys residue [82, 83] for subsequent introduction of fluorescent label or oligosaccharide moieties by oxime formation after their incorporation through cell-wall biosynthesis. The 4-fluorinated UDP-MurNAc-pentapeptide was also synthesized as a stable biosynthesis inhibitor [84].

14.3 LPS AND RELATED GLYCOCONJUGATES

14.3.1 Lipid A

The LPS component of Gram-negative bacteria shows potent immunostimulatory activities upon recognition through Toll-like receptor (TLR)4/myeloid differentiation factor (MD)-2 heterodimer [85]. LPS has a glycolipid terminal anchor termed lipid A (endotoxin), which is the active principle of the endotoxic activity (Fig. 14.3). Lipid A builds up the outer monolayer of the outer membranes of most Gram-negative bacteria (Fig. 14.1) [86]. The LPS of *E. coli* consists of lipid A and 3-deoxy-D-*manno*-2-octulosonic acid (KDO) domains—the minimal structural requirement of growth—with additional core sugars and O-antigen sugars in wild-type strains, which help bacterial resistance against antibiotics, the complement system, and other environmental stresses.

Most lipid A has β(1→6)-linked glucosamine (GlcN) residues and fatty acids linked to the 2- and 2′-amino and 3- and 3′-hydroxy groups and one or two acidic groups at either or both the 1- and 4′-positions of the glycan moiety. In particular, *E. coli* lipid A possesses a hexaacylated structure for the common disaccharide motif, with two phosphate groups at the 1- and 4′-positions. Other parasitic bacteria may have longer and fewer fatty acids and lack one of the two phosphates found in *E. coli*. Binding of lipid A to MD-2 in the TLR4/MD-2 complex triggers the dimerization of the complex and leads to effective immunostimulation.

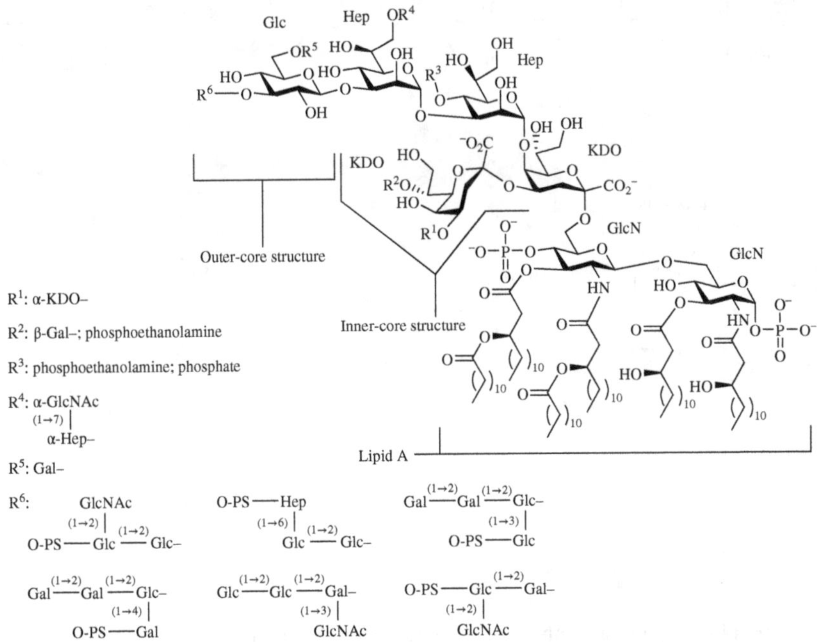

FIGURE 14.3 Structure of LPS with the core structure from *Escherichia coli* and the *Salmonella* species. Gal, galactose; Glc, glucose; Hep, L-*glycero*-D-*manno*-heptose; O-PS, O-antigen polysaccharide.

Tetsuo Shiba et al. were the first to synthesize *E. coli* lipid A [87, 88] as well as the biosynthetic precursor of lipid A with a phosphorylated tetraacyl disaccharide structure [89]. The synthetic compounds were applied for biological studies, which suggested that lipid A is the inflammation-inducing moiety of LPS [90]. Shoichi Kusumoto et al. also synthesized the simplest known bacterial LPS molecule, the Re-type LPS **64** (Scheme 14.8) [91] found in *E. coli* Re mutants. They used the fatty acid-modified glycosyl donor **56** and acceptor **57** for the selective introduction of three different fatty acids, followed by the introduction of the phosphonate group. For the *Helicobacter pylori* lipid A [92], Fukase et al. reported the synthesis of lipid A and its analogue using *p*-(trifluoromethyl)benzyl protecting group [93] and the

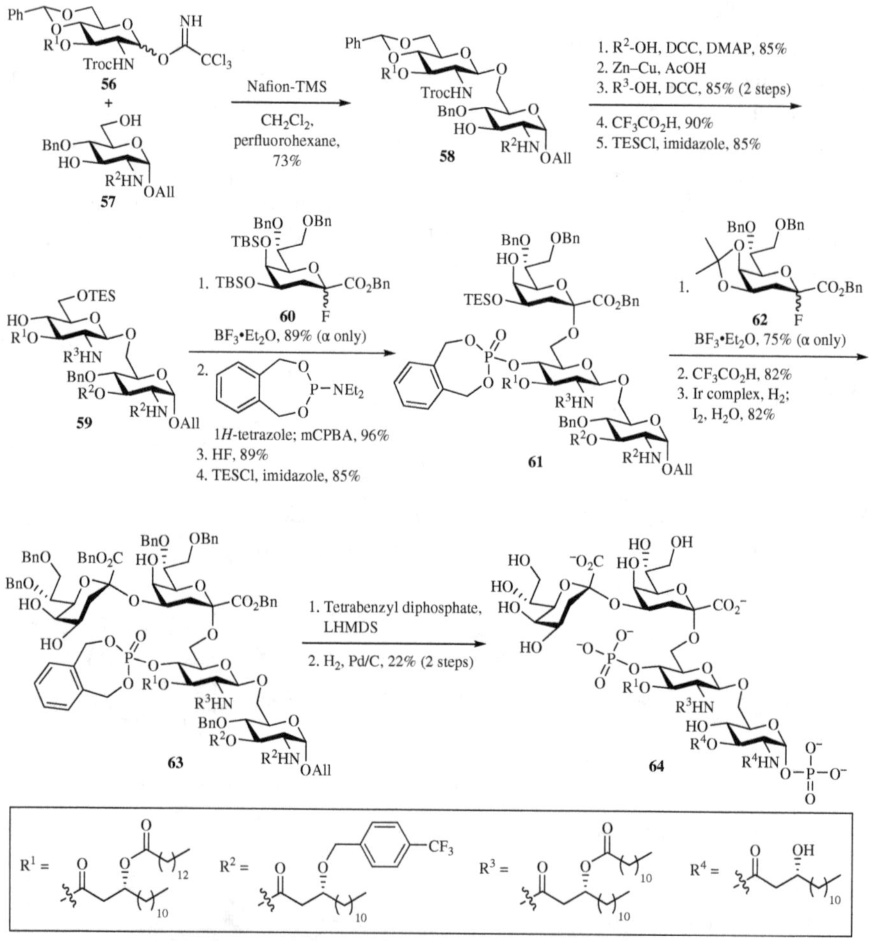

SCHEME 14.8 Synthesis of the Re-type lipid A from *Escherichia coli*. DCC, *N,N′*-dicyclohexylcarbodiimide; DMAP, 4-dimethylaminopyridine; LHMDS, lithium bis(trimethylsilyl) amide; TES, triethylsilyl.

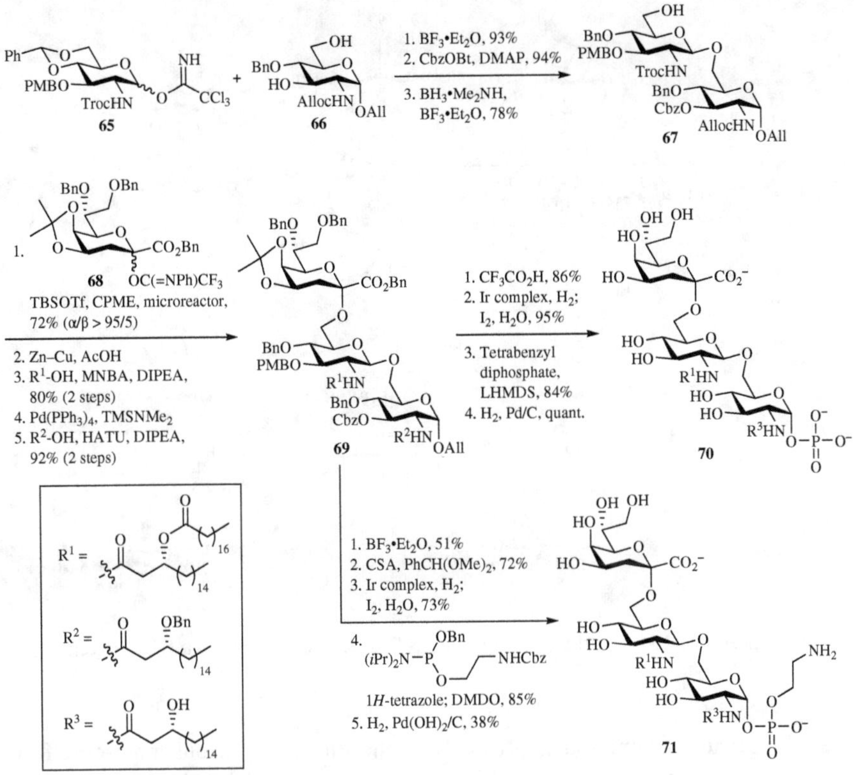

SCHEME 14.9 Synthesis of lipid A from *Helicobacter pylori*. Alloc, allyloxycarbonyl; CbzOBt, 1-(benzyloxycarbonyl)benzotriazole; CPME, cyclopentylmethyl ethel; CSA, camphorsulfonic acid; DMDO, dimethyldioxirane; MNBA, 2-methyl-6-nitrobenzoic anhydride; PMB, *p*-methoxybenzyl.

LPS partial structures **70** and **71** with KDO moiety (Scheme 14.9) [94, 95]. KDO fluoride (**60**, **62**) and *N*-phenyltrifluoroacetimidate (**68**) were used to selectively afford the α isomer. For donor **68**, the α-glycosylation was effectively achieved by using a microreactor (see Section 8.2.2 for details).

On the other hand, Boons et al. demonstrated the synthesis of *Rhizobium sin*-1 lipid A derivatives [96, 97], *E. coli* hexaacylated and *Salmonella typhimurium* heptaacylated lipid A [98], tetraacylated derivatives of lipid A from *Porphyromonas gingivalis* [99], and lipid A derivatives and KDO-linked lipid A of *Neisseria meningitidis* [100].

14.3.2 Oligo-KDO and Inner-Core Oligosaccharide

The outer membrane of *Chlamydia* contains a highly truncated and unique glycolipid composed of lipid A and KDO residues with a common KDO epitope [101, 102]. Because KDO is an important inner-core structural motif, many synthetic studies

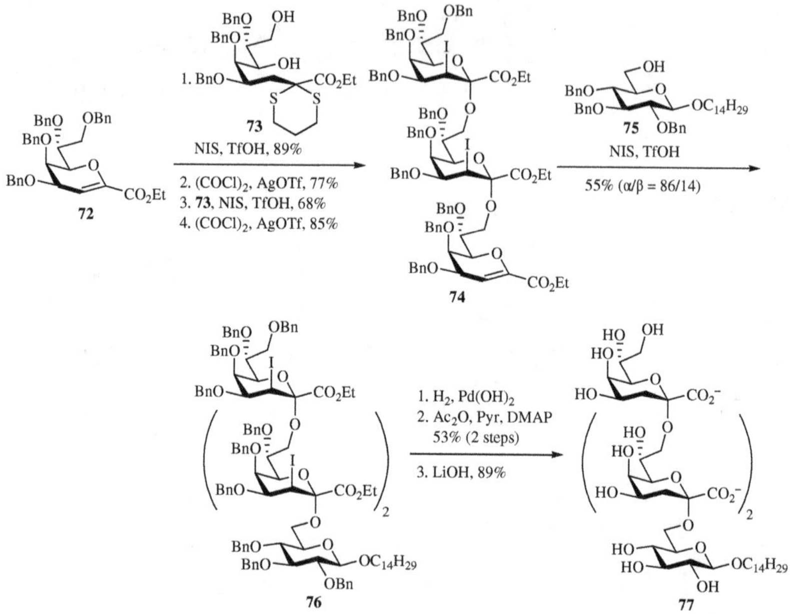

SCHEME 14.10 Synthesis of an oligo-KDO structure by using a KDO glycal.

have been carried out previously [103]. Synthetic studies of various inner-core KDO oligomeric structures including KDO dimers and trimers as well as KDO dimers with Hep were done by Hans Paulsen [104, 105], Paul Kosma [106–108], Shoichi Kusumoto [91], and Tsuyoshi Ichiyanagi [109] using KDO halide and imidate donors. The glycal derivative of KDO (**72**) has been used for the synthesis of oligo-KDO **77** by Takashi Takahashi et al. [110] through the 3-iodo-KDO derivative **76** followed by reduction to give the oligo-α(2→6)-KDO linkage (Scheme 14.10).

The inner-core glycans in LPS are very complex and species specific. These surface glycan components are antigenic and an important virulence determinant, which induces an adaptive immune response [111]. Vaccines based on these structural motifs have been targeted as alternatives to antibiotic therapy.

Francisella tularensis is a highly virulent pathogen with potential to induce several diseases such as tularemia in humans and animals [112]. Efforts to prevent infection by *F. tularensis* led to an attenuated live vaccine strain, which was expanded to a subunit vaccine composed of cell-surface components including LPS [113]. The LPS of *F. tularensis* contains lipid A, core oligosaccharide, and an O-antigen polysaccharide [114]. Boons et al. [115] reported the synthesis of an unusual inner-core oligosaccharide (**86**) having a single KDO moiety (Scheme 14.11). The respective constructions of β-mannoside and α linkage of galactosamine with donors **78** and **80** were carried out stereoselectively to give trisaccharide **81** after a further two-step conversion. Coupling of **81** and **82** followed by construction of the branch structure with imidates **83** and **85** afforded hexasaccharide **86** stereoselectively.

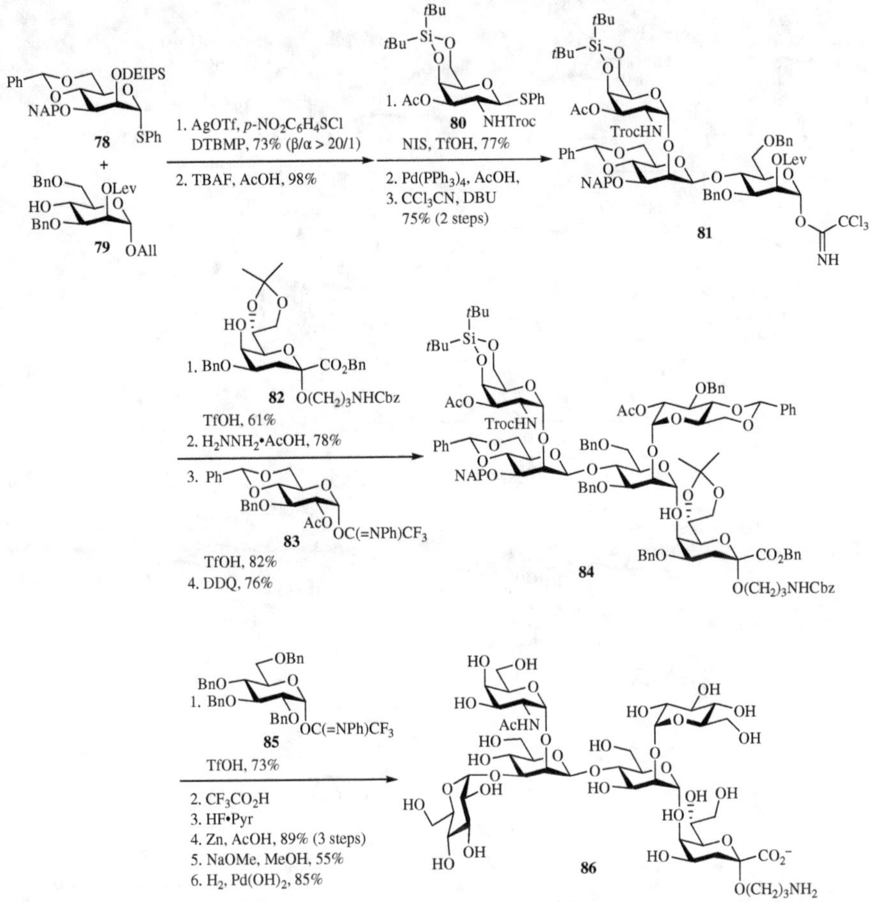

SCHEME 14.11 Synthesis of the LPS inner-core structure of *Francisella tularensis* by Boons et al. DDQ, 2,3-dichloro-5,6-dicyano-1,4-benzoquinone; DEIPS, diethylisopropylsilyl; NAP, 2-naphthylmethyl.

Bacterial meningitis is a severe infection of the meninges caused mainly by *N. meningitidis*, the only Gram-negative encapsulated bacterium responsible for large epidemics [116]. *Neisseria meningitidis* does not possess a polymeric O-antigen [117], and the core is exposed to be immunodominant. The core consists of a common Hep(α1→3)Hep(α1→5)KDO trisaccharide and a nonreducing α-GlcNAc terminal residue, which is specific to *N. meningitidis* and is shared by all clinically relevant meningococcal strains [118–120]. Although the inner-core oligosaccharide isolated from *N. meningitidis* elicited the production of antibodies that responded well to *N. meningitidis* infections [121], a well-defined, nontoxic LPS core structural motif obtained by chemical synthesis should be required to fully explore the effectiveness of LPS oligosaccharide as a vaccine candidate and to identify the most immunogenic epitopes. *Neisseria meningitidis* LPS-related oligosaccharides containing Hep and

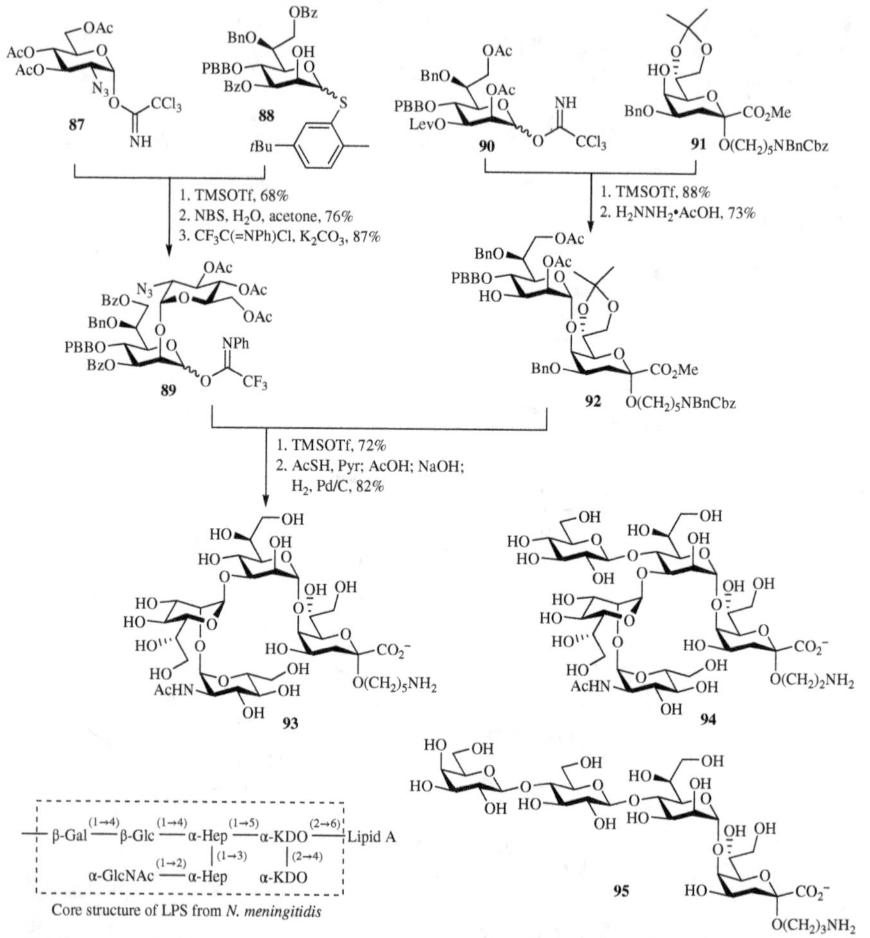

SCHEME 14.12 Synthesis of core structure of LPS from *Neisseria meningitidis*. Bz, benzoyl; NBS, *N*-bromosuccinimide; PBB, *p*-bromobenzyl.

KDO (**93–95**) have been synthesized by Peter Seeberger [122], Stefan Oscarson [123, 124], and Jacques van Boom [125, 126], respectively (Scheme 14.12). The backbone structure of tetramer **93** was afforded through the stereoselective [2+2] coupling of the α-GlcNAc-Hep donor **89** and Hep-KDO acceptor **92**.

14.3.3 Outer-Core Polysaccharides

The outer region of the core oligosaccharide affects the innate immune responses in the normal host. For example, patients with cystic fibrosis are hypersusceptible to chronic airway infections, particularly with *Pseudomonas aeruginosa* [127]. It was shown that the outer-core region of the LPS of *P. aeruginosa* influences a critical step

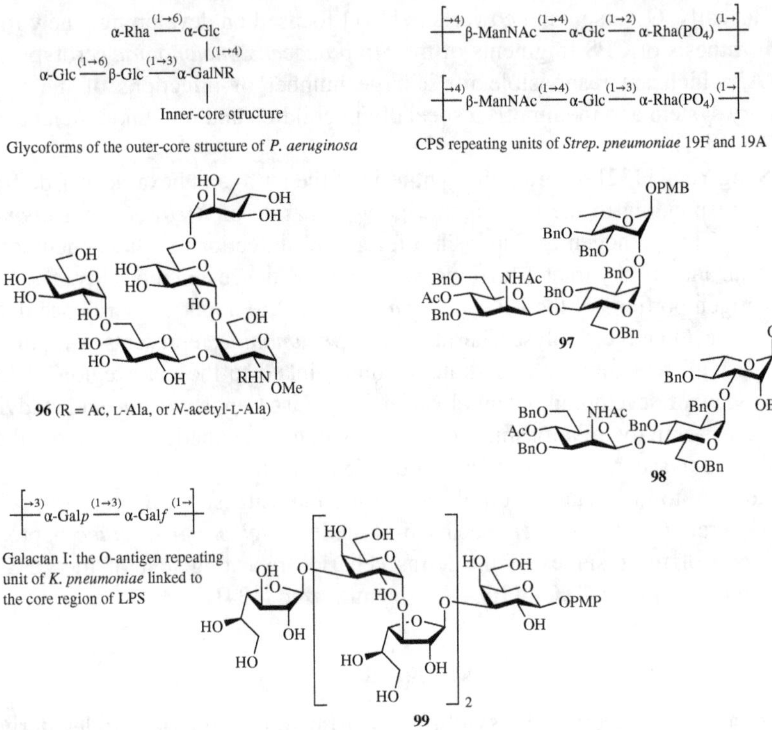

FIGURE 14.4 Outer-core polysaccharides and CPS as synthetic targets. Man, mannose; PMP, *p*-methoxyphenyl; Rha, rhamnose.

in the elimination of this bacterium from a normal host by binding to the cystic fibrosis transmembrane conductance regulator (CFTR), the protein that is missing or is dysfunctional in cystic fibrosis patients. To ascertain which naturally occurring glycoforms [128] of LPS outer core interact with the CFTR receptor, Nikolay Nifantiev et al. performed a systematic synthesis of the oligosaccharides representing these glycoforms (Fig. 14.4) [129].

14.3.4 Capsular Polysaccharide

Capsular polysaccharide (CPS) is present in encapsulated bacteria for protection against the host immune system and osmotic lysis and is responsible for virulence. The structure of CPS defines the serotype of bacterial cells. Approaches toward CPS-based vaccines have been reported, including the use of isolated fragments obtained after controlled lysis of the capsule, the CPS conjugated to a carrier protein, and synthetic vaccines bearing well-defined molecular structures [111]. The synthesis of the trisaccharide portion of the repeating unit of the *N. meningitidis* CPS was reported by Richard Schmidt et al. by using the glycosyl imidate chemistry

[130]. Recently, Oscarson and coworkers [131] focused on developing a new route for the synthesis of CPS fragments of the *Streptococcus pneumoniae* serotype 19F and 19A, which are responsible for a large number of infections of the upper respiratory system and meningitis, especially in children and immunodeficient subjects (Fig. 14.4).

Jin-Song Yang [132] reported the synthesis of the tetra- and hexasaccharide fragments corresponding to the O-antigenic polysaccharide of *Klebsiella pneumoniae* (Fig. 14.4). This pathogen is responsible for several infections such as pneumonia, bacteremia, and urinary tract infections, with high incidence and mortality. The CPS and O-antigen portions of the LPS of *K. pneumoniae* are important virulence determinants. The O-antigen polysaccharide of *K. pneumoniae* represents a family of molecules with repeating carbohydrate subunits linked to the core region of LPS [133]. In view of developing potential vaccines against *K. pneumoniae* and studying the structure–activity relationships of carbohydrates, synthetic studies on these antigenic *O*-polysaccharides are of considerable interest.

In addition, to understand the polysaccharide biosynthesis of a species-specific CPS in different *Campylobacter* species, 6-*O*-methyl-D-*glycero*-α-L-*gluco*-heptopyranose—one of the residues of the tetrasaccharide repeating unit of the CPS of *Campylobacter jejuni* NCTC 11168—was synthesized [134].

14.3.5 Secondary Cell-Wall Polysaccharide

Boons et al. [135] reported the synthesis of a range of oligosaccharides derived from the secondary cell-wall polysaccharides (SCWPs) [136] of *B. anthracis* (Scheme 14.13) [137], which causes anthrax and is a serious bioweapon [138]. SCWPs are ligands of cell-wall binding domains of endolysins expressed by bacteriophages to hydrolyze peptidoglycans. Endolysins offer an alternative to antibodies for detection of microbes such as *B. anthracis* [139]. The bindings to SCWPs are known to be essential for viability. It is very important to understand the molecular mechanisms that confer high affinity and selectivity to these carbohydrate-binding domains. The oligosaccharides contain an α-D-GlcNAc-(1→4)-β-D-ManNAc-(1→4)-β-D-GlcNAc backbone that is modified by various patterns of α-D-Gal and β-D-Gal branching points. Disaccharide **102**, synthesized through β-glucosylation followed by inversion as azide, was treated with imidate **103** to give the α-glycoside selectively. Introductions of β- and α-D-Gal branches to the resulting trisaccharide **104** by using donors **105** and **106** followed by functional group transformations gave the hexasaccharide **107**.

14.3.6 Zwitterionic Polysaccharide

Zwitterionic polysaccharides are candidate immunotherapeutic agents [140, 141] that can activate immune responses dependent on a major histocompatibility complex class II (MHCII)-mediated T cell in the absence of protein [142]. Polysaccharide

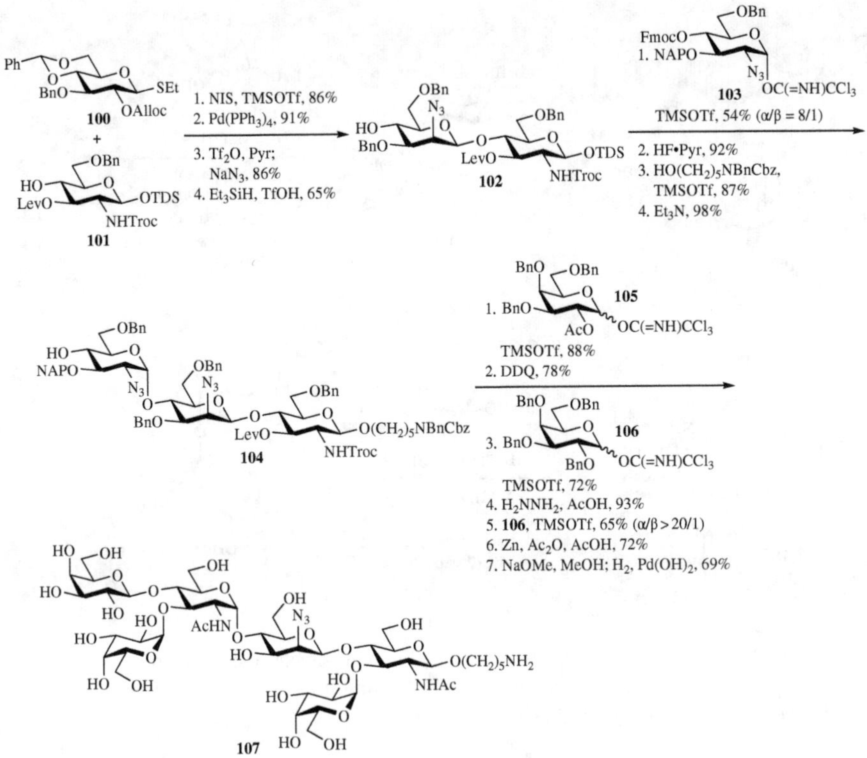

SCHEME 14.13 Synthesis of the SCWP fragment from *Bacillus anthracis* by Boons and coworkers. Fmoc, 9-fluorenylmethoxycarbonyl; TDS, dimethylthexylsilyl.

A1 (PS A1), found on the capsule of the commensal bacterium *Bacteroides fragilis* [143], is the best characterized natural zwitterionic polysaccharide. It strongly stimulates CD4+ T cells to produce cytokines and chemokines, initiates an inert immune response through TLR2 signaling [144], and shows anti-inflammatory activities [145–148]. The structure of native PS A1 is composed of pyruvated galactose, galactofuranose, and 2-acetamido-4-amino-2,4,6-trideoxy-D-galactose (AATDGal)—all novel residues often found in immunodominant epitopes. It has been suggested that PS A1 exhibits a helical structure [149].

Synthetic studies of the protected tetrasaccharide repeating unit of PS A1 have been reported by Gijsbert van der Marel et al. by using 1-thio, 1-seleno, and 1-hydroxy functionalized donor glycosides [150]. Seeberger et al. also studied the synthesis of PS A1 fragments and achieved the synthesis of PS A1 repeating tetrasaccharide unit **114** through a [3 + 1] glycosylation between the 3-alcohol **113** and the trisaccharide donor **112** obtained from the galactofuranose donor **111** and the α-AATDGal-GalN$_3$ disaccharide acceptor **110** (Scheme 14.14) [151, 152].

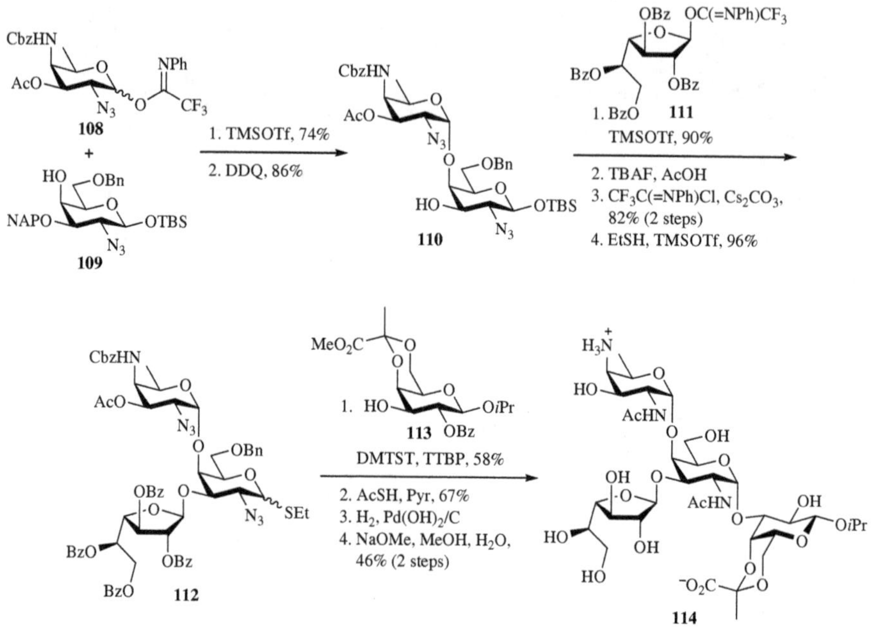

SCHEME 14.14 Synthesis of PS A1 repeating tetrasaccharide unit from *Bacteroides fragilis* by Seeberger and coworkers. DMTST, dimethyl(methylthio)sulfonium triflate; TTBP, 2,4,6-tri-*tert*-butylpyrimidine.

14.4 LIPOTEICHOIC ACID

Lipoteichoic acids are amphiphilic and negatively charged glycolipids that are abundant surface polymers in the envelope of most Gram-positive bacteria [153]. Majority of Gram-positive pathogens carry lipoteichoic acids with polyglycerophosphate backbones. Based on structure, lipoteichoic acids are classified into four types (types I–IV) (Fig. 14.5) [154, 155]. Synthetic studies on lipoteichoic acids having gentiobiosyl (Glc(β1→6)Glc) or kojibiosyl (Glc(α1→2)Glc) diacylglycerol moiety from *S. aureus*, *Streptococcus pyogenes*, and *Enterococcus hirae* have been carried out by van Boom [156, 157] and Kusumoto [158, 159]. Analysis of lipoteichoic acid from *S. aureus* suggested a crucial role attributed to preserved D-Ala residues at the polyglycerophosphate backbone [160–162]. *Staphylococcus aureus* lipoteichoic acid was synthesized by Schmidt et al. [163–165] to prove the significance of the D-Ala residues and to analyze structure–activity relationships [166]. They also synthesized polyglycerophosphate-type lipoteichoic acid with β-D-galactofuranosyl diacylglycerol moiety from *Streptococcus* species DSM 8747 [167]. The more complex *Strep. pneumoniae* lipoteichoic acid (type IV) (**123**) was completed also by Schmidt et al. (Scheme 14.15) [168, 169]. Trisaccharide **115** was assembled by α-glycosylation of

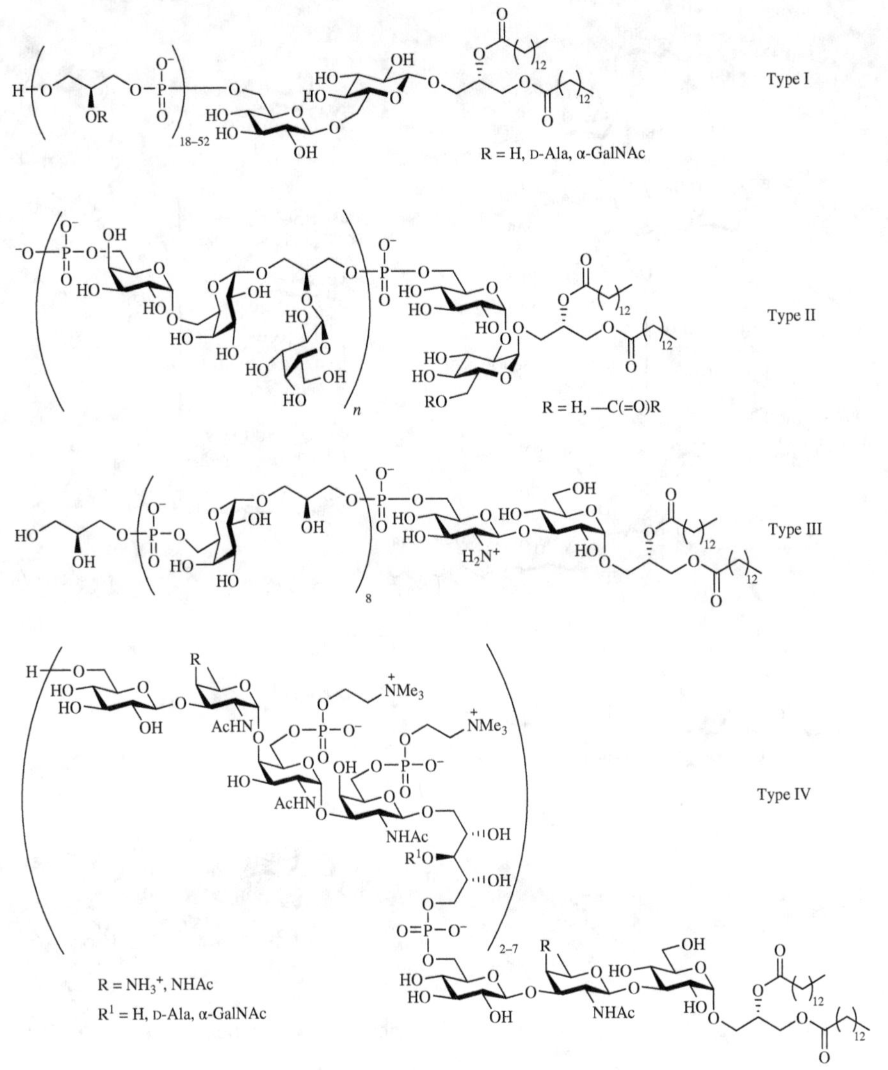

FIGURE 14.5 Structure of type I–IV lipoteichoic acid. GalNAc, *N*-acetylgalactosamine.

a nonparticipating azido donor and β-glucosylation with the help of nitrile solvent effect. Coupling with the galactosamine–ribitol derivative **116** followed by the installation of the phosphocholine units gave the ribitol-linked tetrasaccharide **118**. Compound **122**, derived from the Glc-diacylglycerol derivative **120** and imidate donors **119** and **121**, was connected to **118** through a phosphodiester linkage to give the desired lipoteichoic acid **123**.

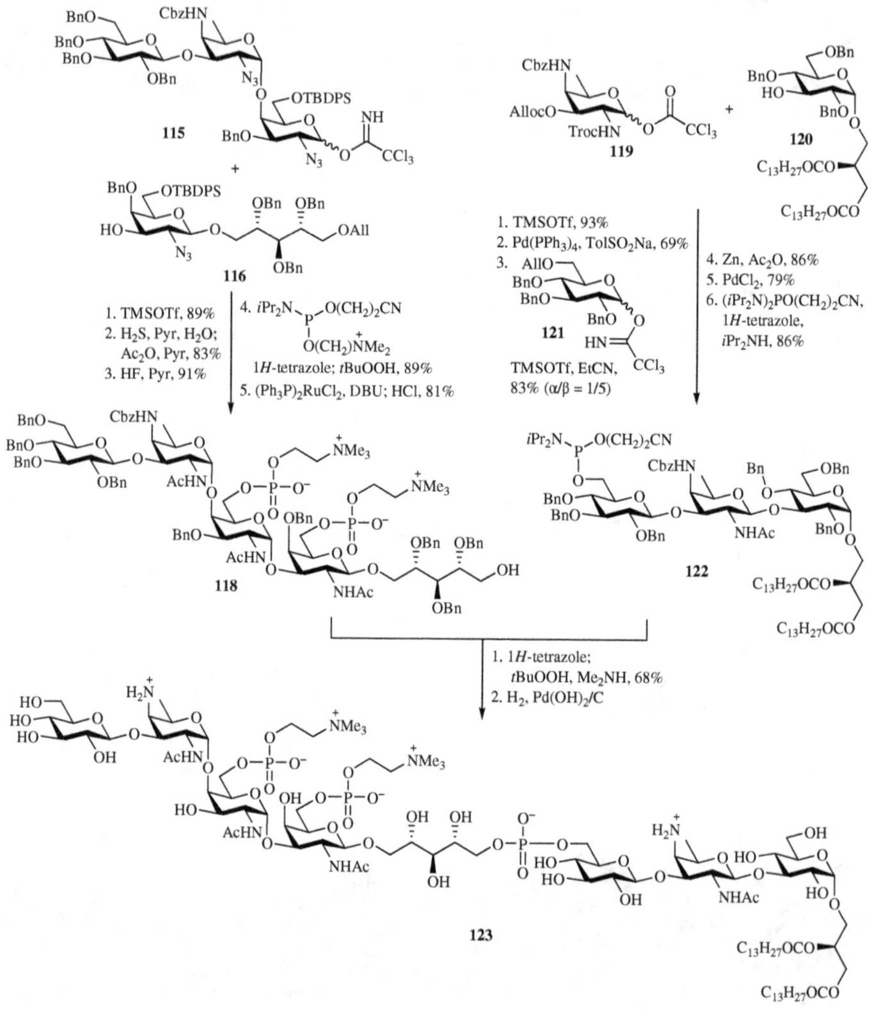

SCHEME 14.15 Synthesis of a type IV lipoteichoic acid by Schmidt et al. TBDPS, *tert*-butyldiphenylsilyl.

14.5 MYCOLYL ARABINOGALACTAN, LAM, AND RELATED GLYCOCONJUGATES

14.5.1 Arabinan, Galactan, and Related Glycoconjugates

Mycobacterium tuberculosis is the causative agent of tuberculosis, which is once again emerging in Southeast Asia and African and Western Pacific regions, as well as in HIV-infected patients. Although tuberculosis is a major public health concern, more serious is the discovery of multidrug-resistant and extensively drug-resistant tuberculosis, as well as extremely drug-resistant tuberculosis, which resists all first- and second-line drug treatments [170].

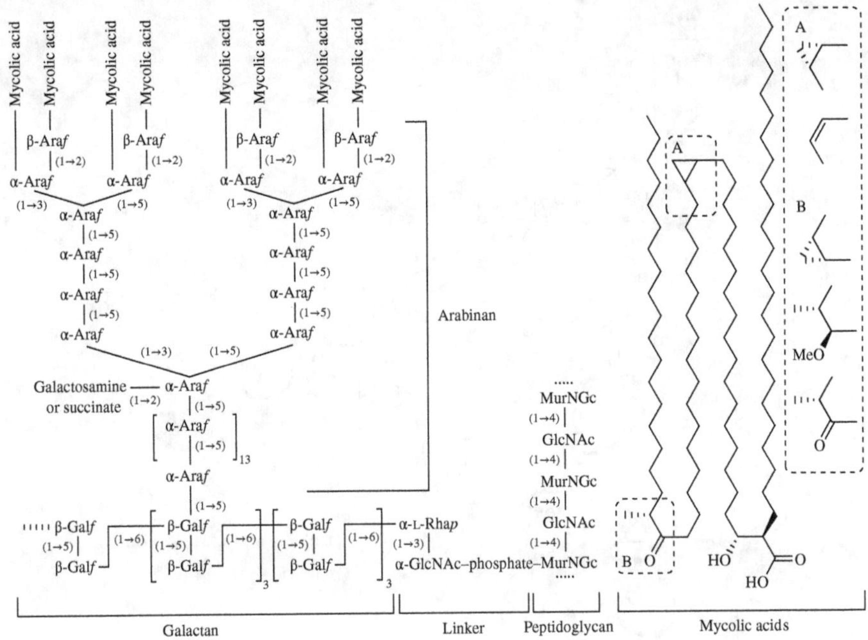

FIGURE 14.6 Structure of mycobacterial mycolyl arabinogalactan–peptidoglycan complex. Ara, arabinose; MurNGc, *N*-glycolylmuramic acid.

The mycobacterial cell-wall skeleton contains unique polyfuranosides, such as galactan (inner) and arabinan (outer), connected to peptidoglycan through a linker at the reducing end of the galactan moiety and to mycolic acid at the nonreducing end of arabinan. Mycobacterial galactan is an oligomer of β-D-galactofuranose, which is composed of Gal*f*(β1→5)Gal*f* repeating units linked by β(1→6) linkages (Fig. 14.6) [171].

Mycobacterial arabinan is a common constituent of both arabinogalactan and LAM. Arabinan, a challenging synthetic target, was originally proposed to be a docosasaccharide (22-mer) but was reclassified as a branched hentriacontasaccharide (31-mer) [172]. Two syntheses of β-Ara*f*-containing common arabinan docosasaccharide motif of mycobacterial cell wall (**124** and **125**) were achieved by the laboratories of Todd Lowary and Yukishige Ito (Fig. 14.7).

Stereocontrolled formation of β-Ara*f* is intrinsically problematic because of its 1,2-*cis*, nonaxial nature [173]. In order to solve this, several innovative methods have been developed. Of particular note were approaches based on S_N2-type displacement of α-triflates derived from 2,3-anhydro-modified and carboxybenzyl-substituted donors, which were developed by Todd Lowary [174, 175] and by Kwan Soo Kim [176], respectively. Alternatively, certain cyclic protections gave promising results by virtue of their abilities to bring conformational constrain to arabinofuranosyl donors. For instance, bicyclic donors such as those protected with 3,5-*O*-di-*tert*-butylsilylidene (DTBS) [177–179], 3,5-*O*-tetraisopropyldisiloxanylidene (TIPDS)

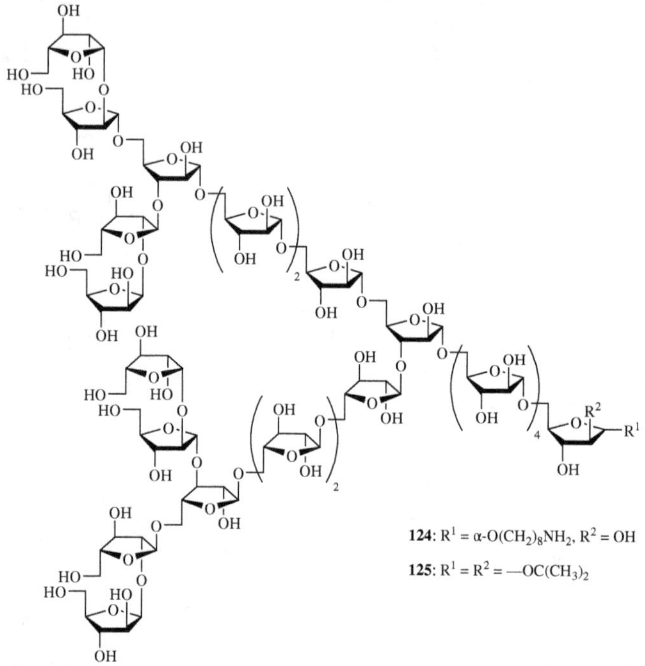

124: $R^1 = \alpha\text{-}O(CH_2)_8NH_2$, $R^2 = OH$

125: $R^1 = R^2 = \text{—}OC(CH_3)_2$

FIGURE 14.7 Structures of the docosasaccharide arabinan prepared by the laboratories of Lowary and Ito.

[180], and 2,3-*O*-xylylene [181] groups provided substantial β selectivity. To promote better selectivity in β-Ara*f* formation, approaches based on intramolecular aglycone delivery [182] have been investigated. These methods employed donors equipped with 2-*O*-PMB [183, 184], 2-*O*-NAP [185], or 5-*O*-NAP [186] groups. See Section 4.2 for discussions related to these approaches.

Lowary prepared monosaccharide arabinofuranose units [187], which were used as donors and acceptors for the synthesis of oligoarabinofuranosides [188–192] including hexasaccharides [193, 194]. The fragments were used to study the ligand specificity of the mycobacterial LAM-recognizing monoclonal antibody CS-35 [195, 196] and structural analysis of acylated arabinan fragments [197, 198]. For the synthesis of the docosasaccharide arabinan **124** (Fig. 14.7), three major fragments, specifically, the branched heptasaccharide **129**, the linear pentasaccharide **130**, and the branched pentasaccharide **134** with two β linkages selectively introduced by using the DTBS-protected **133**, were prepared first. Couplings of these fragments gave the desired docosasaccharide arabinan (Scheme 14.16) [199]. The highly branched arabinan structure could further involve galactosamine or succinate units [200–202]. Lowary et al. synthesized the branched structure having the galactosamine moiety to prove that this residue is attached via an α linkage in the natural glycan [203].

Ito's synthetic strategy toward the docosasaccharide arabinan **125** involves (i) the stereoselective β-arabinofuranosylation using both 3,5-*O*-TIPDS-protected and

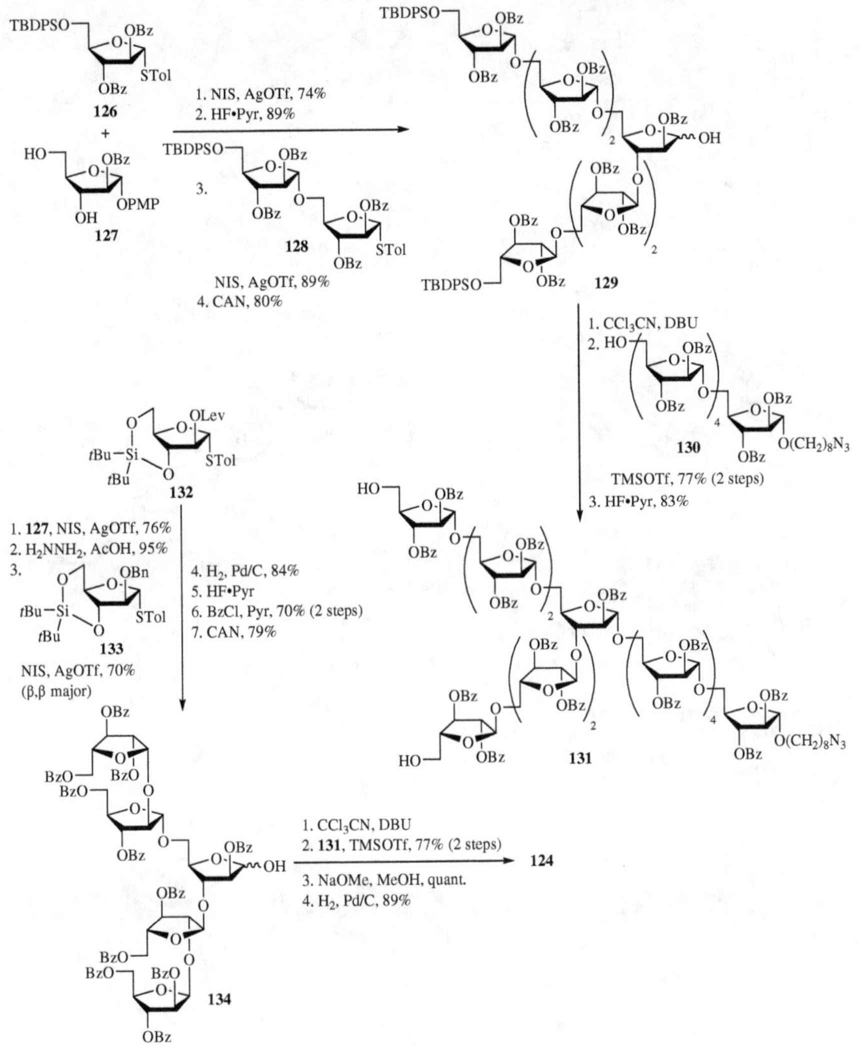

SCHEME 14.16 Synthesis of the docosasaccharide arabinan **124** by Lowary and coworkers. CAN, ceric ammonium nitrate.

NAP-protected arabinofuranosyl donors (**140, 141**) for straightforward intermolecular glycosylation and intramolecular aglycone delivery, respectively, and (ii) the convergent fragment coupling with branched fragments (branched octasaccharide **145** and branched heptasaccharide **142** with two β linkages) at the linear sequence using the thioglycoside donor obtained from the corresponding acetonide at the reducing terminal of each fragment through a three-step procedure (Scheme 14.17) [186]. Disaccharide **135** was converted to trisaccharide **137** and the branched pentasaccharides **139** and **144**. Because the acetonide installed at the reducing terminal

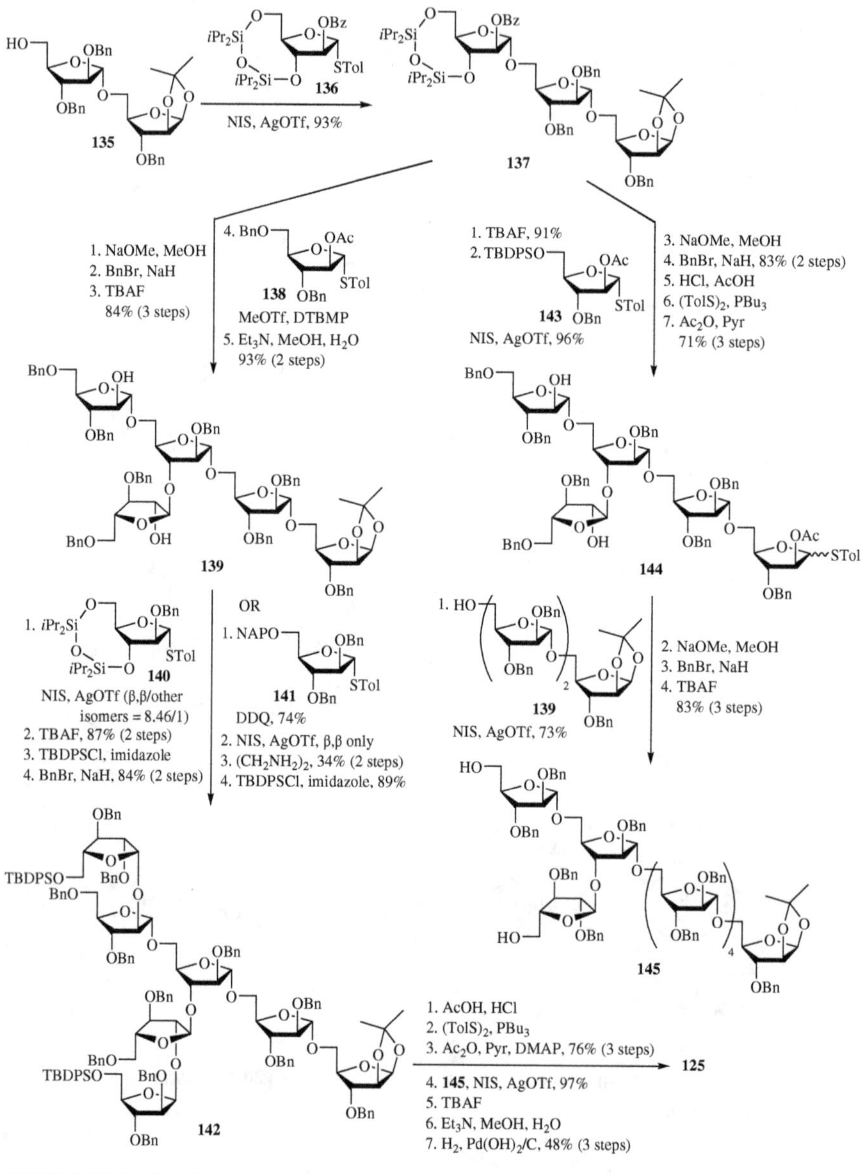

SCHEME 14.17 Synthesis of the docosasaccharide arabinan **125** by Ito and coworkers.

could be converted to thioglycoside and mainly Bn ether protections were used, this strategy should be readily applicable for the synthesis of more complex arabinan, arabinogalactan, and arabinomycolate derived from the mycobacterial cell-wall skeleton.

Additionally, the synthesis of mycobacterial cell-wall fragments were carried out by many groups. Oligoarabinofuranosides [183, 184, 204] and other fragments such as the capping motif of the LAM [205] and diacylated sulfoglycolipid [206] were synthesized by Jacques Prandi et al. For the synthesis of arabinan fragments [207, 208], Leonid Kononov et al. synthesized monosaccharide building blocks by using the 1,2,5-orthobenzoate as the key intermediate [209, 210]. One-pot strategy has been applied for the synthesis of oligoarabinofuranosides by using trichloroacetimidate and thioglycoside donors [211, 212]. Narayanaswamy Jayaraman et al. synthesized linear and branched oligoarabinofuranosides with lipid moieties [213, 214] as well as β-arabinofuranosylated [215] and mannosylated [216] derivatives for biological studies such as the interaction with surfactant protein A from *Mycobacterium smegmatis*.

Jun Ning et al. also reported a one-pot strategy for the synthesis of the arabinofuranose-branched galactan motif from *M. tuberculosis* [217] as well as a hexasaccharide galactan motif of *Bifidobacterium catenulatum* YIT 4016 [218]. The synthesis of the nonreducing end pentasaccharide of mycobacterial arabinan possessing 1,2-*cis* arabinofuranosides [219] has been reported by Mukund Gurjar et al. The β(1→5) digalactofuranose and branched arabinogalactan motifs [220, 221] were also synthesized. Lucia Gandolfi-Donadío et al. also reported the synthesis of linear galactan trisaccharide motifs [222] and di- and tetrasaccharides with an arabinofuranose branch [223–225]. Robert Reynolds et al. synthesized disaccharide analogues of the cell-wall fragment [226–233] as the substrate or inhibitor of glycosyltransferases in cell-wall biosynthesis and arabinofuranosyl phospholipid as the donor mimetic of arabinofuranosyltransferase [234].

Kim et al. studied the highly stereoselective synthesis of mycobacterial cell-wall fragments such as the arabinan octasaccharide [176] and cyclic oligogalactofuranosides with alternating β(1→5) and β(1→6) linkages found by enzymatic degradation of cell-wall arabinogalactan [235]. Analogues of the disaccharide linkage motif between arabinogalactan and peptidoglycan were synthesized by Philip Hultin [236] and K.P. Ravindranathan Kartha [237]. A tetrasaccharide motif containing this linkage region [β-Gal*f*-(1→5)-β-Gal*f*-(1→4)-α-Rha*p*-(1→3)-α-GlcNAc*p*-phosphate] was synthesized by Kim et al. [238]. Lowary et al. synthesized Gal*f*-containing glycans such as the fragment of mycobacterial galactan tetrasaccharide and β-Gal*f*-(1→6)-β-Gal*f*-(1→5)-β-Gal*f*-(1→4)-α-Rha*p*-(1→3)-α-GlcNAc*p* derivative [239] as well as the pentasaccharide repeating unit of the cell-wall polysaccharide of *Penicillium varians* [240].

14.5.2 Mycolates and Related Glycoconjugates

Trehalose 6,6′-dimycolate (TDM, cord factor) and 6-monomycolate (TMM) are important bacterial, especially mycobacterial, cell-surface components that show immunological responses [241]. Antigen 85, which interconverts TDM and TMM, is also an important alternative therapeutic target. Other surface-exposed glycolipids like glucose 6-monomycolate (GMM) [242, 243] produced by the mycolyltransferase antigen 85A [244] affect the regulation of TDM and T-cell activation.

The synthesis of mycolates including TDM, TMM [245–251], and GMM [252] as well as the mycolyl oligoarabinofuranoside **151** [253] (Scheme 14.18) was reported

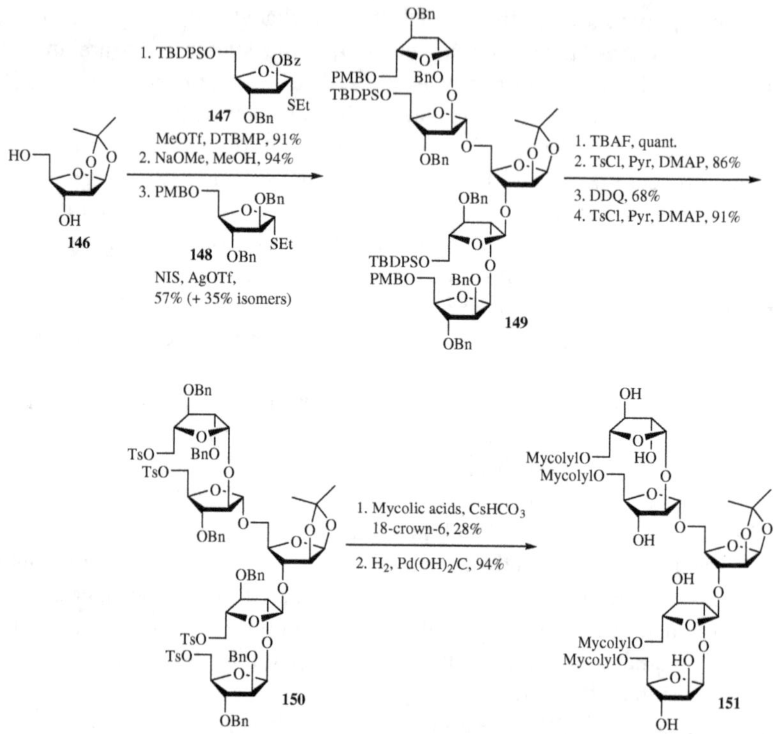

SCHEME 14.18 Synthesis of mycolyl arabinan **151**. Ts, tosyl.

mainly through alkylation using acid or through Keck esterification (DCC–DMAP) using protected acid. Because mycolic acids are β-hydroxy acids, ester formation is performed without protection of the hydroxy group of mycolic acid to avoid β-elimination. The pentasaccharide obtained from diol **146** and donors **147** and **148** was prepared as the tetratosylate **150** and converted to tetramycolate **151** by using the cesium salt of mycolic acid in the presence of 18-crown-6 (Scheme 14.18).

Azide-modified trehalose analogues were used by Carolyn Bertozzi et al. [254] for probing the mycobacterial trehalome through metabolic labeling in living organisms. Benjamin Davis et al. studied the kinetic analysis of antigen 85 by using synthetic trehalose derivatives and analogues through label-free high-throughput mass spectrometry monitoring assay [255, 256].

14.5.3 LAM and Related Glycoconjugates

The fine structure of mycobacterial LAM with a range of structural diversity is generally well understood (Fig. 14.8) [257, 258]. The core structure of LAM is composed of a phosphatidylinositol moiety connected to a mannan domain consisting of α(1→6)- and α(1→2)-linked mannopyranose residues. The mannan domain is linked to arabinan composed of α(1→5)-, α(1→3)-, and β(1→2)-linked arabinofuranose

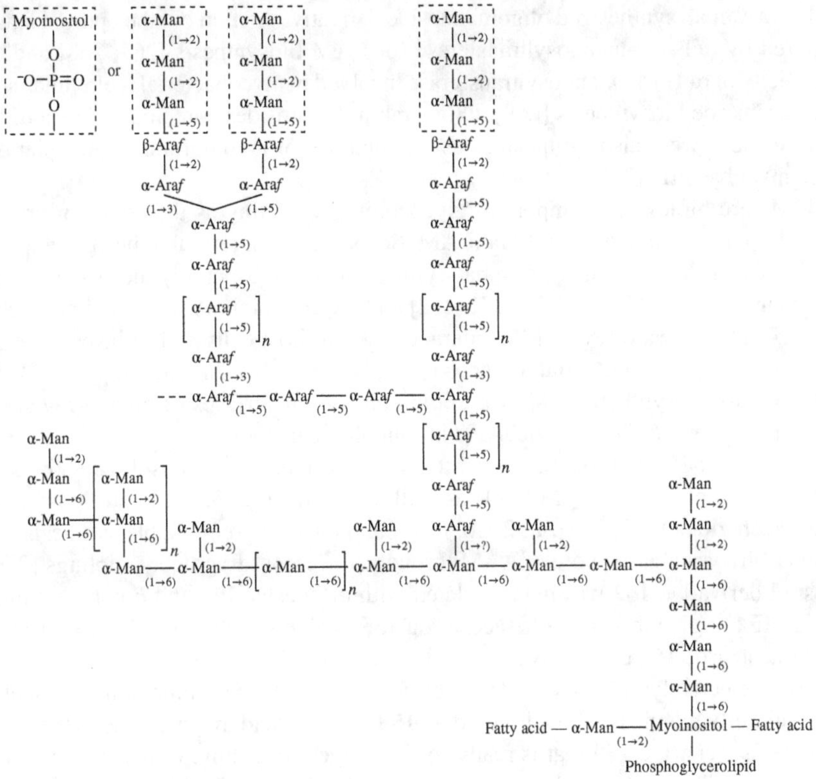

FIGURE 14.8 Structure of mycobacterial LAM.

residues. The arabinan segment is further functionalized at its nonreducing terminus by capping motifs such as oligo-α(1→2)-linked mannopyranosides to produce ManLAM in the cases of *M. tuberculosis*, *Mycobacterium bovis*, and *Mycobacterium avium* [259, 260] and inositol phosphate moieties to form a glycoconjugate called PILAM in the case of *M. smegmatis* [261]. Mycobacterial LAM with capping motifs has been implicated in a large and increasing number of important immunological events.

LAMs from laboratory strains (H37Rv and H37Ra) as well as clinical isolates (CSU20 and MT103) of *M. tuberculosis* [262, 263] contain an unusual 5-deoxy-5-methylthioxylofuranose (MTX) residue and its corresponding oxidized counterpart, 5-deoxy-5-methylsulfoxyxylofuranose (MSX) residue, which was also found in *Mycobacterium kansasii* attached not to the mannopyranose capping residues but rather to the mannan core [264, 265]. Lowary et al. synthesized the MTX/MSX residues [266] and established that the MTX/MSX residues in these glycoconjugates are of the D-configuration and that they are linked α(1→4) to a mannopyranose residue in the mannan portion of the glycan. They also performed conformational analysis of the MTX/MSX residue using NMR spectroscopy. Further assays suggested that this motif may play a role in the immune response arising from mycobacterial infection.

Lowary et al. synthesized oligomannoside derivatives to find out the optimal length required by α(1→6)-mannosyltransferase for LAM biosynthesis [267, 268] and the specificity of α(1→4)-mannosyltransferase involved in mycobacterial methylmannose polysaccharide biosynthesis [269]. Oligoarabinofuranosides containing C3-modified Ara*f* residues were also synthesized as the inhibitor of arabinofuranosyltransferases from mycobacteria [270].

PIMs are biologically important glycoconjugates acting as the common precursors of lipomannan, LAM, and ManLAM. Seeberger et al. reported the efficient synthesis of all PIMs including phosphatidylinositol and phosphatidylinositol mono- to hexamannoside (PIM_1 to PIM_6) [271, 272] and the core arabinomannan oligosaccharide [273] in order to reveal PIM interactions with host cells and to investigate the function of PIMs as potential antigens or adjuvants for vaccine development. Thiol linker-equipped synthetic PIMs were immobilized on surfaces of microarray slides and carrier proteins for biological and immunological studies.

Bert Fraser-Reid et al. used *n*-pentenyl orthoesters and glycosides for the synthesis of mycobacterial LAM [274] as well as arabinan [275, 276] and lipomannan oligosaccharides [277, 278]. In 2006, they completed the remarkable synthesis of a 28-mer oligosaccharide core of LAM through convergent fragment couplings [279]. Inositol derivative **152** was mannosylated with orthoester **153** and *n*-pentenyl mannoside **154** to give the pseudotrisaccharide **155**, which was elongated to the pseudoheptasaccharide **157** by iterative glycosylation with **156** (Scheme 14.19). The pentaol **157** was selectively mannosylated with orthoester **153** at the terminal primary alcohol and then fully mannosylated with imidate **158**. Further acid treatment led to the pseudododecasaccharide **159** that is ready for convergent coupling. Conversely, mannosylation of **157** with **158** followed by phosphorylation gave the lipomannan **160**. The branched hexasaccharide **164**, obtained from diol **161** and orthoesters **162** and **163**, was coupled to the branched arabinan **165** and then to compound **159** to give the 28-mer LAM core **166** (Scheme 14.20).

14.6 OLIGOSACCHARIDES OF BACTERIAL GLYCOPROTEIN AND RELATED GLYCOCONJUGATES

14.6.1 O-Linked Oligosaccharide from Bacillus Collagen-Like Protein of Anthracis

Glycoproteins are also found on the surface of the spores of bacteria such as *B. anthracis* [280, 281]. Bacillus collagen-like protein of anthracis (BclA) is a highly immunogenic glycoprotein that causes anthrax. The glycan part of BclA is an O-linked oligosaccharide containing the unusual nonreducing terminal sugar 2-*O*-methyl-4-(3-hydroxy-3-methylbutanamido)-4,6-dideoxy-D-glucopyranose, which is named anthrose (Ant) and is found β(1→3)-linked to an L-Rha residue. Ant does not occur in spores of other *Bacillus* species and is expected to be a species-specific biomarker for *B. anthracis*. The syntheses of the nonreducing terminal

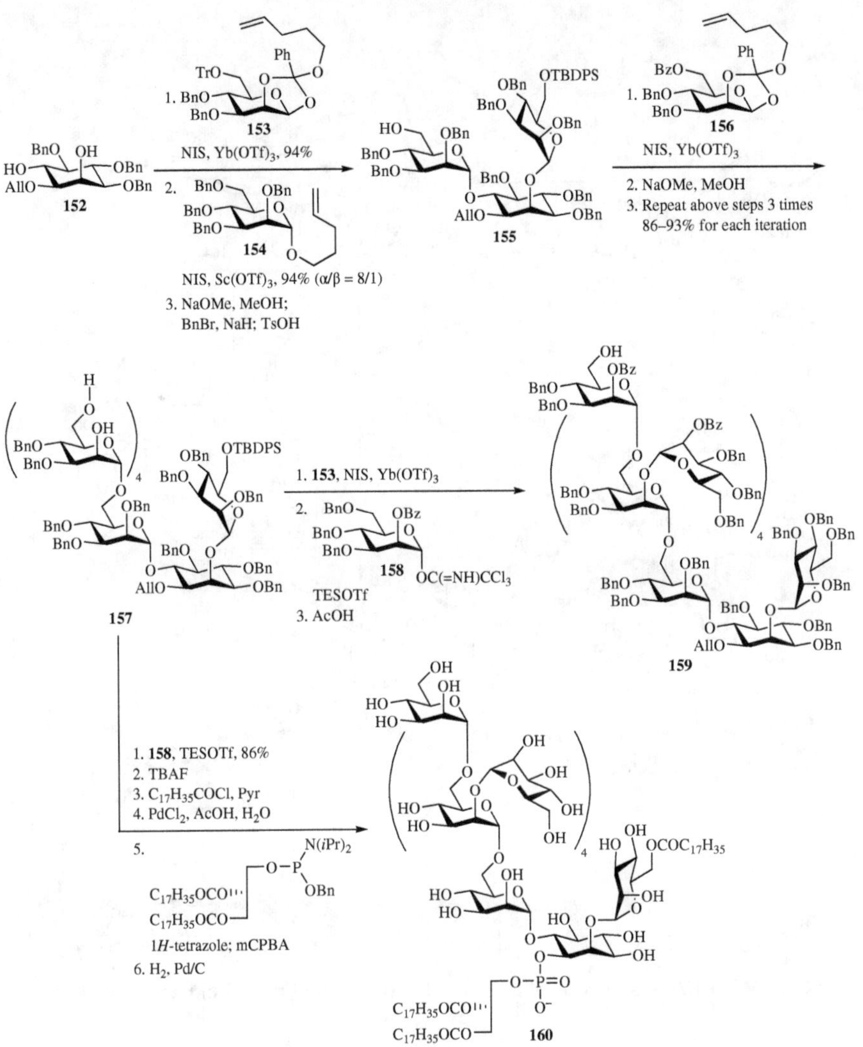

SCHEME 14.19 Lipomannan synthesis by Fraser-Reid and coworkers. Tr, trityl.

oligosaccharides of the BclA glycoprotein from *B. anthracis* exosporium were reported by Peter Seeberger [282–284], Pavol Kováč [285], Geert-Jan Boons [286], George O'Doherty [287], and David Crich [288]. To afford the terminal pentasaccharide segment, Seeberger et al. coupled the trisaccharide **170**, prepared from the GalNAc acceptor **167** and Rha donors **168** and **169**, and the disaccharide donor **173** obtained from Ant donor **171** and Rha acceptor **172** (Scheme 14.21). The target compound **175** was acquired after amidation with **174**.

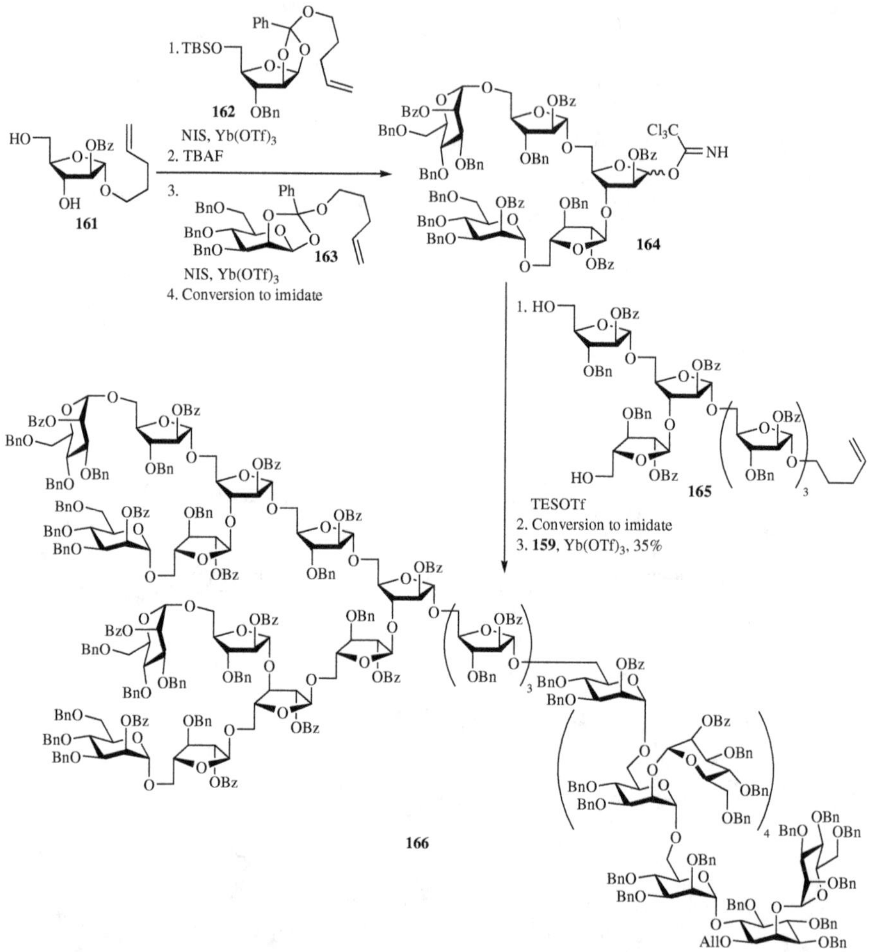

SCHEME 14.20 Synthesis of a 28-mer LAM core by Fraser-Reid and coworkers.

14.6.2 N-Linked Glycans from the Gram-Negative Bacterium *C. jejuni*

N-Glycans found in eukaryotic glycoproteins share a well-conserved core structure (i.e., $Man_3GlcNAc_2$) [289] (see Chapter 11). Contrary to previous belief, recent research has revealed that certain prokaryotes produce N-glycosylated proteins [290]. For instance, a major nonflagellin antigenic glycoprotein, designated as PEB3 or Cj0289c, identified in the pathogenic Gram-negative bacterium *C. jejuni* has multiple glycosylation sites, which carry N-linked glycans [291, 292]. The glycan chain is composed of an asparagine-linked rare sugar termed bacillosamine (2,4-diacetamido-2,4,6-trideoxy-D-glucose (Bac)), pentameric α(1→4)-linked GalNAc, and a β-linked Glc. The presence of the N-linked glycan on the surface of *C. jejuni* was shown to play a key role in enteric adhesion to host cells [293].

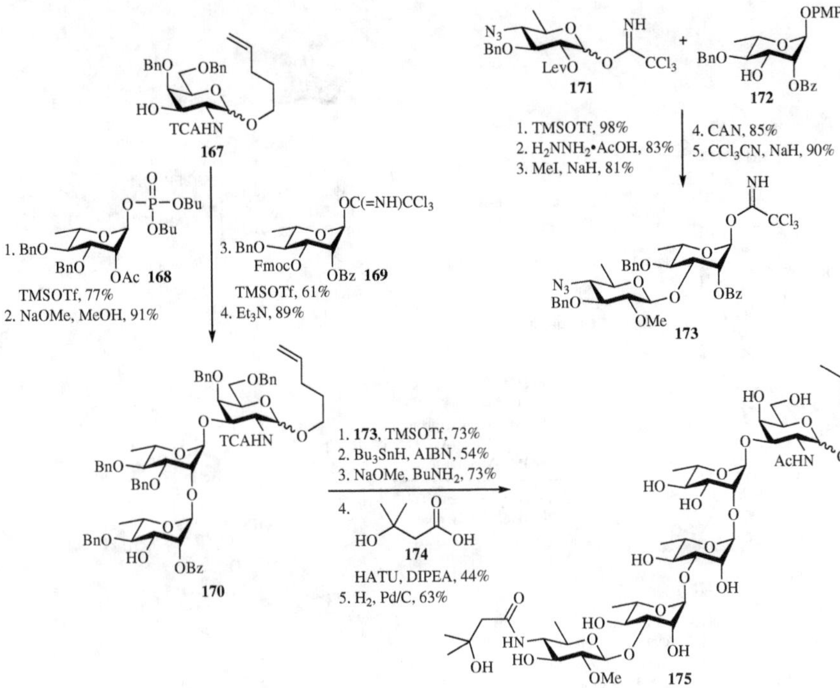

SCHEME 14.21 Synthesis of the O-linked oligosaccharide of BclA by Seeberger and coworkers. AIBN, azobisisobutyronitrile; TCA, trichloroacetyl.

The N-linked glycan consists of multiple α(1→4)-linked GalNAc repeats, requiring a glycosyl donor specifically protected at O4 position. Pentafluoropropionyl (PFP) group [294] was deemed suitable as a temporary protecting group for this purpose because of the mild deprotection conditions, the low volatility of the deprotection by-product ethyl pentafluoropropionate (b.p. 75–76°C), and its strongly electron-withdrawing nature. Based on stereoelectronic considerations, the neutralization of the strong dipole moment caused by the PFP ester at C4 in **177** is expected to enhance the intrinsic α selectivity of the 2-azido glycosyl donor (Scheme 14.22). Thus, glycosylation of the designed donor **177** under $AgClO_4/Cp_2HfCl_2$ activator system provided the desired α-glycoside in high yield and selectivity. In the cases of the glycosylation of the Bac acceptor **176** and by the Glc-containing disaccharide donor **181**, however, significant reduction in selectivity was observed even under optimized conditions ($AgClO_4/Cp_2HfCl_2$ in benzene). Even so, the corresponding desired α-glycosides were isolated in reasonable yields [295].

Starting with the Bac component **176**, which was designed for asparagine attachment [296], double-chain elongation with the galactosamine donor **177** afforded the trisaccharide **178**. Further couplings with disaccharide **181** and twice with donor **177** provided the full-length heptasaccharide in high yield. Simultaneous reduction of multiple azide groups followed by immediate N-acetylation and hydrogenolysis

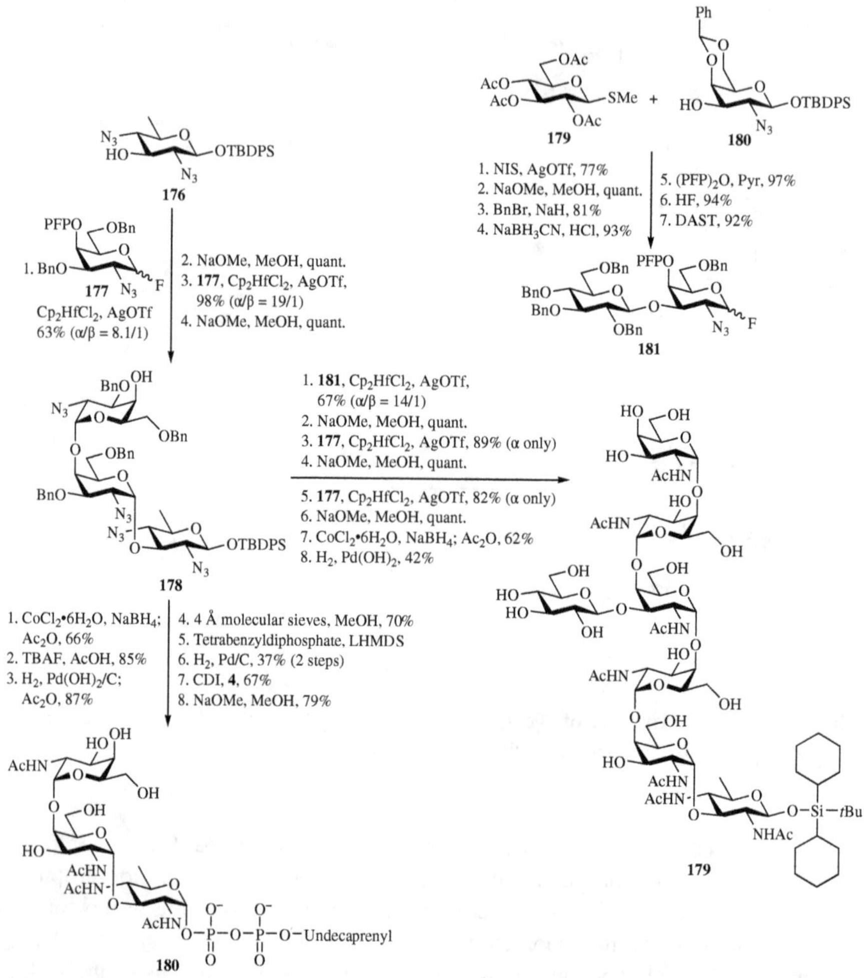

SCHEME 14.22 Synthesis of N-linked oligosaccharide and lipid-linked oligosaccharide from *Campylobacter jejuni* by Ito and coworkers. DAST, *N,N*-diethylaminosulfur trifluoride.

afforded the glycan construct **179** [297]. To supply a substrate for *C. jejuni* oligosaccharyltransferase, the undecaprenyl pyrophosphate-linked glycan **180** derived from trisaccharide **178** was also completed (Scheme 14.22) [298].

14.7 CONCLUSION

There is no doubt about the successful chemical syntheses of the glycan cell-wall components, and these approaches could be further applied to synthetic and biological studies on this area. After the discovery of antibiotics targeting the cell-wall biosynthesis, it became possible to control previously serious pathogenic bacteria, although some bacteria subsequently acquired antibiotic resistance. Because the cycle of control of pathogenic bacteria followed by development of resistance is likely to be repeated, continuous investigation of novel targets for new drugs is therefore required. As shown here, the bacterial cell wall contains quite a unique collection of glycan structural motifs that could be valuable in developing novel antibiotics and other therapeutic agents.

REFERENCES

[1] Ginsberg, C.; Brown, S.; Walker, S. In *Glycoscience: Chemistry and Biology*; Springer/Heidelberg: New York/Berlin, **2001**, pp 1535–1600.

[2] Bush, K.; Courvalin, P.; Dantas, G.; Davies, J.; Eisenstein, B.; Huovinen, P.; Jacoby, G. A.; Kishony, R.; Kreiswirth, B. N.; Kutter, E.; Lerner, S. A.; Levy, S.; Lewis, K.; Lomovskaya, O.; Miller, J. H.; Mobashery, S.; Piddock, L. J. V.; Projan, S.; Thomas, C. M.; Tomasz, A.; Tulkens, P. M.; Walsh, T. R.; Watson, J. D.; Witkowski, J.; Witte, W.; Wright, G.; Yeh, P.; Zgurskaya, H. I. *Nat. Rev. Microbiol.* **2011**, *9*, 894–896.

[3] Koch, A. L. *J. Mol. Microbiol. Biotechnol.* **2006**, *11*, 115–125.

[4] Cabeen, M. T.; Jacobs-Wagner, C. *Nat. Rev. Microbiol.* **2005**, *3*, 601–610.

[5] Kamio, Y.; Nikaido, H. *Biochemistry* **1976**, *15*, 2561–2570.

[6] Neuhaus, F. C.; Baddiley, J. *Microbiol. Mol. Biol. Rev.* **2003**, *67*, 686–723.

[7] Brennan, P. J. *Tuberculosis* **2003**, *83*, 91–97.

[8] Nigou, J.; Gilleron, M.; Puzo, G. *Biochimie* **2003**, *85*, 153-166.

[9] Rohrer, S.; Berger-Bächi, B. *Antimicrob. Agents Chemother.* **2003**, *47*, 837–846.

[10] Fiser, A.; Filipe, S. R.; Tomasz, A. *Trends Microbiol.* **2003**, *11*, 547–553.

[11] Filipe, S. R.; Severina, E.; Tomasz, A. *Microb. Drug Resist.* **2001**, *7*, 303-316.

[12] Vollmer, W.; Blanot, D.; de Pedro, M. A. *FEMS Microbiol. Rev.* **2008**, *32*, 149–167.

[13] van Heijenoort, J. *Glycobiology* **2001**, *11*, 25R–36R.

[14] Vollmer, W.; Bertsche, U. *Biochim. Biophys. Acta* **2008**, *1778*, 1714–1734.

[15] Sauvage, E.; Kerff, F.; Terrak, M.; Ayala, J. A.; Charlier, P. *FEMS Microbiol. Rev.* **2008**, *32*, 234–258.

[16] VanNieuwenhze, M. S.; Mauldin, S. C.; Zia-Ebrahimi, M.; Aikins, J. A.; Blaszczak, L. C. *J. Am. Chem. Soc.* **2001**, *123*, 6983–6988.

[17] Saha, S. L.; Van Nieuwenhze, M. S.; Hornback, W. J.; Aikins, J. A.; Blaszczak, L. C. *Org. Lett.* **2001**, *3*, 3575–3577.

[18] VanNieuwenhze, M. S.; Mauldin, S. C.; Zia-Ebrahimi, M.; Winger, B. E.; Hornback, W. J.; Saha, S. L.; Aikins, J. A.; Blaszczak, L. C. *J. Am. Chem. Soc.* **2002**, *124*, 3656–3660.

[19] Hitchcock, S. A.; Eid, C. N.; Aikins, J. A.; Zia-Ebrahimi, M.; Blaszczak, L. C. *J. Am. Chem. Soc.* **1998**, *120*, 1916–1917.

[20] Schwartz, B.; Markwalder, J. A.; Wang, Y. *J. Am. Chem. Soc.* **2001**, *123*, 11638–11643.

[21] Choi, Y.; Moody, I. S.; Sims, P. C.; Hunt, S. R.; Corso, B. L.; Seitz, D. E.; Blaszczak, L. C.; Blaszcazk, L. C.; Collins, P. G.; Weiss, G. A. *J. Am. Chem. Soc.* **2012**, *134*, 2032–2035.

[22] Dumbre, S.; Derouaux, A.; Lescrinier, E.; Piette, A.; Joris, B.; Terrak, M.; Herdewijn, P. *J. Am. Chem. Soc.* **2012**, *134*, 9343–9351.

[23] Hecker, S. J.; Minich, M. L.; Lackey, K. *J. Org. Chem.* **1990**, *55*, 4904-4911.

[24] Qiao, L.; Vederas, J. C. *J. Org. Chem.* **1993**, *58*, 3480–3482.

[25] Cudic, P.; Behenna, D. C.; Yu, M. K.; Kruger, R. G.; Szewczuk, L. M.; McCafferty, D. G. *Bioorg. Med. Chem. Lett.* **2001**, *11*, 3107–3110.

[26] Garneau, S.; Qiao, L.; Chen, L.; Walker, S.; Vederas, J. C. *Bioorg. Med. Chem.* **2004**, *12*, 6473–6494.

[27] Kahne, D.; Leimkuhler, C.; Lu, W.; Walsh, C. *Chem. Rev.* **2005**, *105*, 425–448.

[28] Shih, H.-W.; Chang, Y.-F.; Li, W.-J.; Meng, F.-C.; Huang, C.-Y.; Ma, C.; Cheng, T.-J. R.; Wong, C.-H.; Cheng, W.-C. *Angew. Chem. Int. Ed.* **2012**, *51*, 10123–10126.

[29] Shih, H.-W.; Chen, K.-T.; Cheng, T.-J. R.; Wong, C.-H.; Cheng, W.-C. *Org. Lett.* **2011**, *13*, 4600–4603.

[30] Liu, C.-Y.; Guo, C.-W.; Chang, Y.-F.; Wang, J.-T.; Shih, H.-W.; Hsu, Y.-F.; Chen, C.-W.; Chen, S.-K.; Wang, Y.-C.; Cheng, T.-J. R.; Ma, C.; Wong, C.-H.; Fang, J.-M.; Cheng, W.-C. *Org. Lett.* **2010**, *12*, 1608–1611.

[31] Meng, F.-C.; Chen, K.-T.; Huang, L.-Y.; Shih, H.-W.; Chang, H.-H.; Nien, F.-Y.; Liang, P.-H.; Cheng, T.-J. R.; Wong, C.-H.; Cheng, W.-C. *Org. Lett.* **2011**, *13*, 5306–5309.

[32] Ye, X. Y.; Lo, M. C.; Brunner, L.; Walker, D.; Kahne, D.; Walker, S. *J. Am. Chem. Soc.* **2001**, *123*, 3155–3156.

[33] Zhang, Y.; Fechter, E. J.; Wang, T.-S. A.; Barrett, D.; Walker, S.; Kahne, D. E. *J. Am. Chem. Soc.* **2007**, *129*, 3080–3081.

[34] Gampe, C. M.; Tsukamoto, H.; Wang, T.-S. A.; Walker, S.; Kahne, D. *Tetrahedron* **2011**, *67*, 9771–9778.

[35] Perlstein, D. L.; Zhang, Y.; Wang, T.-S.; Kahne, D. E.; Walker, S. *J. Am. Chem. Soc.* **2007**, *129*, 12674-12675.

[36] Wang, T.-S. A.; Manning, S. A.; Walker, S.; Kahne, D. *J. Am. Chem. Soc.* **2008**, *130*, 14068–14069.

[37] Perlstein, D. L.; Wang, T.-S. A.; Doud, E. H.; Kahne, D.; Walker, S. *J. Am. Chem. Soc.* **2010**, *132*, 48–49.

[38] Lupoli, T. J.; Taniguchi, T.; Wang, T.-S.; Perlstein, D. L.; Walker, S.; Kahne, D. E. *J. Am. Chem. Soc.* **2009**, *131*, 18230–18231.

[39] Chen, L.; Men, H.; Ha, S.; Ye, X.-Y.; Brunner, L.; Hu, Y.; Walker, S. *Biochemistry* **2002**, *41*, 6824–6833.

[40] Wang, T.-S. A.; Lupoli, T. J.; Sumida, Y.; Tsukamoto, H.; Wu, Y.; Rebets, Y.; Kahne, D. E.; Walker, S. *J. Am. Chem. Soc.* **2011**, *133*, 8528-8530.

[41] Oman, T. J.; Lupoli, T. J.; Wang, T.-S. A.; Kahne, D.; Walker, S.; van der Donk, W. A. *J. Am. Chem. Soc.* **2011**, *133*, 17544–17547.

[42] Lupoli, T. J.; Tsukamoto, H.; Doud, E. H.; Wang, T.-S. A.; Walker, S.; Kahne, D. *J. Am. Chem. Soc.* **2011**, *133*, 10748–10751.

[43] Hesek, D.; Lee, M.; Morio, K.-I.; Mobashery, S. *J. Org. Chem.* **2004**, *69*, 2137–2146.

[44] Hesek, D.; Suvorov, M.; Morio, K.-I.; Lee, M.; Brown, S.; Vakulenko, S. B.; Mobashery, S. *J. Org. Chem.* **2004**, *69*, 778–784.

[45] Lee, M.; Hesek, D.; Mobashery, S. In *Chemical Glycobiology*; Chen, X., Halcomb, R., Wang, P. G., Eds.; ACS Symposium Series 990; American Chemical Society: Washington, DC, **2008**, pp 54–78.

[46] Lee, M.; Hesek, D.; Shah, I. M.; Oliver, A. G.; Dworkin, J.; Mobashery, S. *ChemBioChem* **2010**, *11*, 2525–2529.

[47] Meroueh, S. O.; Bencze, K. Z.; Hesek, D.; Lee, M.; Fisher, J. F.; Stemmler, T. L.; Mobashery, S. *Proc. Natl. Acad. Sci. U. S. A.* **2006**, *103*, 4404–4409.

[48] Fisher, J. F.; Meroueh, S. O.; Mobashery, S. *Microbe* **2006**, *1*, 420–427.

[49] Fisher, J. F.; Mobashery, S. *J. Med. Chem.* **2010**, *53*, 4813–4829.

[50] Cho, S.; Wang, Q.; Swaminathan, C. P.; Hesek, D.; Lee, M.; Boons, G.-J.; Mobashery, S.; Mariuzza, R. A. *Proc. Natl. Acad. Sci. U. S. A.* **2007**, *104*, 8761–8766.

[51] Roy Chowdhury, A.; Siriwardena, A.; Boons, G.-J. *Tetrahedron Lett.* **2002**, *43*, 7805–7807.

[52] Siriwardena, A.; Jørgensen, M. R.; Wolfert, M. A.; Vandenplas, M. L.; Moore, J. N.; Boons, G.-J. *J. Am. Chem. Soc.* **2001**, *123*, 8145–8146.

[53] Guan, R.; Roychowdhury, A.; Ember, B.; Kumar, S.; Boons, G.-J.; Mariuzza, R. A. *Proc. Natl. Acad. Sci. U. S. A.* **2004**, *101*, 17168–17173.

[54] Roychowdhury, A.; Wolfert, M. A.; Boons, G.-J. *ChemBioChem* **2005**, *6*, 2088–2097.

[55] Kumar, S.; Roychowdhury, A.; Ember, B.; Wang, Q.; Guan, R.; Mariuzza, R. A.; Boons, G.-J. *J. Biol. Chem.* **2005**, *280*, 37005-37012.

[56] Roy Chowdhury, A.; Boons, G.-J. *Tetrahedron Lett.* **2005**, *46*, 1675–1678.

[57] Guan, R.; Brown, P. H.; Swaminathan, C. P.; Roychowdhury, A.; Boons, G.-J.; Mariuzza, R. A. *Protein Sci.* **2006**, *15*, 1199–1206.

[58] Kashyap, D. R.; Wang, M.; Liu, L.-H.; Boons, G.-J.; Gupta, D.; Dziarski, R. *Nat. Med.* **2011**, *17*, 676–683.

[59] Travassos, L. H.; Girardin, S. E.; Philpott, D. J.; Blanot, D.; Nahori, M.-A.; Werts, C.; Boneca, I. G. *EMBO Rep.* **2004**, *5*, 1000-1006.

[60] Dziarski, R.; Gupta, D. *Infect. Immun.* **2005**, *73*, 5212–5216.

[61] Kawasaki, A.; Karasudani, Y.; Otsuka, Y.; Hasegawa, M.; Inohara, N.; Fujimoto, Y.; Fukase, K. *Chem. Eur. J.* **2008**, *14*, 10318–10330.

[62] Fujimoto, Y.; Fukase, K. *J. Nat. Prod.* **2011**, *74*, 518–525.

[63] Fukase, K.; Fujimoto, Y.; Shimoyama, A.; Tanaka, K. *J. Synth. Org. Chem. Jpn.* **2012**, *70*, 113–130.

[64] Fujimoto, Y.; Pradipta, A. R.; Inohara, N.; Fukase, K. *Nat. Prod. Rep.* **2012**, *29*, 568-579.

[65] Inamura, S.; Fukase, K.; Kusumoto, S. *Tetrahedron Lett.* **2001**, *42*, 7613–7616.

[66] Inamura, S.; Fujimoto, Y.; Kawasaki, A.; Shiokawa, Z.; Woelk, E.; Heine, H.; Lindner, B.; Inohara, N.; Kusumoto, S.; Fukase, K. *Org. Biomol. Chem.* **2006**, *4*, 232–242.

[67] Fujimoto, Y.; Konishi, Y.; Kubo, O.; Hasegawa, M.; Inohara, N.; Fukase, K. *Tetrahedron Lett.* **2009**, *50*, 3631–3634.

[68] Chamaillard, M.; Hashimoto, M.; Horie, Y.; Masumoto, J.; Qiu, S.; Saab, L.; Ogura, Y.; Kawasaki, A.; Fukase, K.; Kusumoto, S.; Valvano, M. A.; Foster, S. J.; Mak, T. W.; Nuñez, G.; Inohara, N. *Nat. Immunol.* **2003**, *4*, 702–707.

[69] Inohara, N.; Ogura, Y.; Fontalba, A.; Gutierrez, O.; Pons, F.; Crespo, J.; Fukase, K.; Inamura, S.; Kusumoto, S.; Hashimoto, M.; Foster, S. J.; Moran, A. P.; Fernandez-Luna, J. L.; Nunez, G. *J. Biol. Chem.* **2003**, *278*, 5509–5512.

[70] Masumoto, J.; Yang, K.; Varambally, S.; Hasegawa, M.; Tomlins, S. A.; Qiu, S.; Fujimoto, Y.; Kawasaki, A.; Foster, S. J.; Horie, Y.; Mak, T. W.; Núñez, G.; Chinnaiyan, A. M.; Fukase, K.; Inohara, N. *J. Exp. Med.* **2006**, *203*, 203-213.

[71] Hasegawa, M.; Yang, K.; Hashimoto, M.; Park, J. H.; Kim, Y. G.; Fujimoto, Y.; Nunez, G.; Fukase, K.; Inohara, N. *J. Biol. Chem.* **2006**, *281*, 29054–29063.

[72] Hasegawa, M.; Kawasaki, A.; Yang, K.; Fujimoto, Y.; Masumoto, J.; Breukink, E.; Nunez, G.; Fukase, K.; Inohara, N. *J. Biol. Chem.* **2007**, *282*, 11757–11764.

[73] Pradipta, A. R.; Fujimoto, Y.; Hasegawa, M.; Inohara, N.; Fukase, K. *J. Biol. Chem.* **2010**, *285*, 23607–23613.

[74] Grimes, C. L.; Ariyananda, L. D. Z.; Melnyk, J. E.; O'Shea, E. K. *J. Am. Chem. Soc.* **2012**, *134*, 13535–13537.

[75] Grimes, C. L.; Podolsky, D. K.; O'Shea, E. K. *Bioorg. Med. Chem. Lett.* **2010**, *20*, 6061–6063.

[76] Okumura, H.; Tokushima, Y.; Saiki, I.; Azuma, I.; Kiso, M.; Hasegawa, A. *Carbohydr. Res.* **1983**, *122*, 87–98.

[77] Park, J. T. *J. Biol. Chem.* **1952**, *194*, 877–884.

[78] Gisch, N.; Buske, B.; Heine, H.; Lindner, B.; Zähringer, U. *Bioorg. Med. Chem. Lett.* **2011**, *21*, 3362–3366.

[79] Girardin, S. E.; Travassos, L. H.; Herve, M.; Blanot, D.; Boneca, I. G.; Philpott, D. J.; Sansonetti, P. J.; Mengin-Lecreulx, D. *J. Biol. Chem.* **2003**, *278*, 41702–41708.

[80] Liu, H.; Sadamoto, R.; Sears, P. S.; Wong, C.-H. *J. Am. Chem. Soc.* **2001**, *123*, 9916–9917.

[81] Kurosu, M.; Mahapatra, S.; Narayanasamy, P.; Crick, D. C. *Tetrahedron Lett.* **2007**, *48*, 799–803.

[82] Sadamoto, R.; Niikura, K.; Sears, P. S.; Liu, H.; Wong, C.-H.; Suksomcheep, A.; Tomita, F.; Monde, K.; Nishimura, S.-I. *J. Am. Chem. Soc.* **2002**, *124*, 9018–9019.

[83] Sadamoto, R.; Niikura, K.; Ueda, T.; Monde, K.; Fukuhara, N.; Nishimura, S.-I. *J. Am. Chem. Soc.* **2004**, *126*, 3755–3761.

[84] Ueda, T.; Feng, F.; Sadamoto, R.; Niikura, K.; Monde, K.; Nishimura, S.-I. *Org. Lett.* **2004**, *6*, 1753–1756.

[85] Park, B. S.; Song, D. H.; Kim, H. M.; Choi, B.-S.; Lee, H.; Lee, J.-O. *Nature* **2009**, *458*, 1191–1195.

[86] Raetz, C. R. H.; Whitfield, C. *Annu. Rev. Biochem.* **2002**, *71*, 635–700.

[87] Imoto, M.; Yoshimura, H.; Sakaguchi, N.; Kusumoto, S.; Shiba, T. *Tetrahedron Lett.* **1985**, *26*, 1545–1548.

[88] Imoto, M.; Yoshimura, H.; Shimamoto, T.; Sakaguchi, N.; Kusumoto, S.; Shiba, T. *Bull. Chem. Soc. Jpn.* **1987**, *60*, 2205-2214.

[89] Imoto, M.; Yoshimura, H.; Yamamoto, M.; Shimamoto, T.; Kusumoto, S.; Shiba, T. *Bull. Chem. Soc. Jpn.* **1987**, *60*, 2197–2204.

[90] Galanos, C.; Luderitz, O.; Rietschel, E. T.; Westphal, O.; Brade, H.; Brade, L.; Freudenberg, M.; Schade, U.; Imoto, M.; Yoshimura, H.; Kusumoto, S.; Shiba, T. *Eur. J. Biochem.* **1985**, *148*, 1–5.

[91] Yoshizaki, H.; Fukuda, N.; Sato, K.; Oikawa, M.; Fukase, K.; Suda, Y.; Kusumoto, S. *Angew. Chem. Int. Ed.* **2001**, *40*, 1475–1480.

[92] Fujimoto, Y.; Shimoyama, A.; Suda, Y.; Fukase, K. *Carbohydr. Res.* **2012**, *356*, 37–43.

[93] Sakai, Y.; Oikawa, M.; Yoshizaki, H.; Ogawa, T.; Suda, Y.; Fukase, K.; Kusumoto, S. *Tetrahedron Lett.* **2000**, *41*, 6843–6847.

[94] Fujimoto, Y.; Iwata, M.; Imakita, N.; Shimoyama, A.; Suda, Y.; Kusumoto, S.; Fukase, K. *Tetrahedron Lett.* **2007**, *48*, 6577–6581.

[95] Shimoyama, A.; Saeki, A.; Tanimura, N.; Tsutsui, H.; Miyake, K.; Suda, Y.; Fujimoto, Y.; Fukase, K. *Chem. Eur. J.* **2011**, *17*, 14464–14474.

[96] Demchenko, A. V.; Wolfert, M. A.; Santhanam, B.; Moore, J. N.; Boons, G.-J. *J. Am. Chem. Soc.* **2003**, *125*, 6103–6112.

[97] Santhanam, B.; Boons, G.-J. *Org. Lett.* **2004**, *6*, 3333–3336.

[98] Zhang, Y.; Gaekwad, J.; Wolfert, M. A.; Boons, G.-J. *J. Am. Chem. Soc.* **2007**, *129*, 5200–5216.

[99] Zhang, Y.; Gaekwad, J.; Wolfert, M. A.; Boons, G.-J. *Org. Biomol. Chem.* **2008**, *6*, 3371–3381.

[100] Zhang, Y.; Gaekwad, J.; Wolfert, M. A.; Boons, G.-J. *Chem. Eur. J.* **2008**, *14*, 558–569.

[101] Brade, H.; Brade, L.; Nano, F. E. *Proc. Natl. Acad. Sci. U. S. A.* **1987**, *84*, 2508–2512.

[102] Rund, S.; Lindner, B.; Brade, H.; Holst, O. *J. Biol. Chem.* **1999**, *274*, 16819–16824.

[103] Kosma, P.; Zamyatina, A. In *Bacterial Lipopolysaccharides*; Knirel, Y. A., Valvano, M. A., Eds.; Springer: Vienna, **2011**, pp 131–161.

[104] Paulsen, H.; Brenken, M. *Liebigs Ann. Chem.* **1991**, 1113–1126.

[105] Paulsen, H.; Wulff, A.; Brenken, M. *Liebigs Ann. Chem.* **1991**, 1127–1145.

[106] Kosma, P.; Schulz, G.; Unger, F. M.; Brade, H. *Carbohydr. Res.* **1989**, *190*, 191–201.

[107] Sixta, G.; Wimmer, K.; Hofinger, A.; Brade, H.; Kosma, P. *Carbohydr. Res.* **2009**, *344*, 1660-1669.

[108] Kosma, P.; Hofinger, A.; Müller-Loennies, S.; Brade, H. *Carbohydr. Res.* **2010**, *345*, 704–708.

[109] Ichiyanagi, T.; Fukunaga, M.; Tagashira, R.; Hayashi, S.; Nanjo, M.; Yamasaki, R. *Tetrahedron* **2011**, *67*, 5964–5971.

[110] Tanaka, H.; Takahashi, D.; Takahashi, T. *Angew. Chem. Int. Ed.* **2006**, *45*, 770-773.

[111] Lucas, A. H.; Apicella, M. A.; Taylor, C. E. *Clin. Infect. Dis.* **2005**, *41*, 705–712.

[112] Elkins, K. L.; Cowley, S. C.; Collazo, C. M. *Expert Rev. Vaccines* **2003**, *2*, 735–738.

[113] Pechous, R. D.; McCarthy, T. R.; Zahrt, T. C. *Microbiol. Mol. Biol. Rev.* **2009**, *73*, 684–711.

[114] Vinogradov, E.; Perry, M. B.; Conlan, J. W. *Eur. J. Biochem.* **2002**, *269*, 6112–6118.

[115] Boltje, T. J.; Zhong, W.; Park, J.; Wolfert, M. A.; Chen, W.; Boons, G.-J. *J. Am. Chem. Soc.* **2012**, *134*, 14255–14262.

[116] Pizza, M.; Rappuoli, R. *Curr. Opin. Microbiol.* **2015**, *23*, 68–72.

[117] Hansson, J.; Oscarson, S. *Curr. Org. Chem.* **2000**, *4*, 535–564.

[118] Rahman, M. M.; Kahler, C. M.; Stephens, D. S.; Carlson, R. W. *Glycobiology* **2001**, *11*, 703–709.

[119] Choudhury, B.; Kahler, C. M.; Datta, A.; Stephens, D. S.; Carlson, R. W. *Carbohydr. Res.* **2008**, *343*, 2971–2979.

[120] Mistretta, N.; Seguin, D.; Thiébaud, J.; Vialle, S.; Blanc, F.; Brossaud, M.; Talaga, P.; Norheim, G.; Moreau, M.; Rokbi, B. *J. Biol. Chem.* **2010**, *285*, 19874–19883.

[121] Gidney, M. A. J.; Plested, J. S.; Lacelle, S.; Coull, P. A.; Wright, J. C.; Makepeace, K.; Brisson, J. R.; Cox, A. D.; Moxon, E. R.; Richards, J. C. *Infect. Immun.* **2003**, *72*, 559–569.

[122] Yang, Y.; Martin, C. E.; Seeberger, P. H. *Chem. Sci.* **2012**, *3*, 896–899.

[123] Segerstedt, E.; Mannerstedt, K.; Johansson, M.; Oscarson, S. *J. Carbohydr. Chem.* **2004**, *23*, 443–452.

[124] Olsson, J. D. M.; Oscarson, S. *Tetrahedron: Asymmetry* **2009**, *20*, 875–882.

[125] Boons, G. J. P. H.; Overhand, M.; van der Marel, G. A.; van Boom, J. H. *Angew. Chem. Int. Ed. Engl.* **1989**, *28*, 1504–1506.

[126] Boons, G. J. P. H.; van Delft, F. L.; van der Klein, P. A. M.; van der Marel, G. A.; van Boom, J. H. *Tetrahedron* **1992**, *48*, 885–904.

[127] Campodónico, V. L.; Gadjeva, M.; Paradis-Bleau, C.; Uluer, A.; Pier, G. B. *Trends Mol. Med.* **2008**, *14*, 120–133.

[128] Bystrova, O. V.; Lindner, B.; Moll, H.; Kocharova, N. A.; Knirel, Y. A.; Zähringer, U.; Pier, G. B. *Carbohydr. Res.* **2003**, *338*, 1895–1905.

[129] Komarova, B. S.; Tsvetkov, Y. E.; Pier, G. B.; Nifantiev, N. E. *J. Org. Chem.* **2008**, *73*, 8411–8421.

[130] Kinzy, W.; Schmidt, R. R. *Liebigs Ann. Chem.* **1985**, 1537–1545.

[131] Bonaccorsi, F.; Catelani, G.; Oscarson, S. *Carbohydr. Res.* **2009**, *344*, 1442–1448.

[132] Zhu, S.-Y.; Yang, J.-S. *Tetrahedron* **2012**, *68*, 3795–3802.

[133] Shankar-Sinha, S.; Valencia, G. A.; Janes, B. K.; Rosenberg, J. K.; Whitfield, C.; Bender, R. A.; Standiford, T. J.; Younger, J. G. *Infect. Immun.* **2004**, *72*, 1423–1430.

[134] Peng, W.; Jayasuriya, A. B.; Imamura, A.; Lowary, T. L. *Org. Lett.* **2011**, *13*, 5290–5293.

[135] Mo, K.-F.; Li, X.; Li, H.; Low, L. Y.; Quinn, C. P.; Boons, G.-J. *J. Am. Chem. Soc.* **2012**, *134*, 15556–15562.

[136] Schäffer, C.; Messner, P. *Microbiology* **2005**, *151*, 643–651.

[137] Leoff, C.; Saile, E.; Rauvolfova, J.; Quinn, C. P.; Hoffmaster, A. R.; Zhong, W.; Mehta, A. S.; Boons, G.-J.; Carlson, R. W.; Kannenberg, E. L. *Glycobiology* **2009**, *19*, 665–673.

[138] Mock, M.; Fouet, A. *Annu. Rev. Microbiol.* **2001**, *55*, 647–671.

[139] Hermoso, J. A.; García, J. L.; García, P. *Curr. Opin. Microbiol.* **2007**, *10*, 461–472.

[140] Gallorini, S.; Berti, F.; Mancuso, G.; Cozzi, R.; Tortoli, M.; Volpini, G.; Telford, J. L.; Beninati, C.; Maione, D.; Wack, A. *Proc. Natl. Acad. Sci. U. S. A.* **2009**, *106*, 17481–17486.

[141] De Silva, R. A.; Wang, Q.; Chidley, T.; Appulage, D. K.; Andreana, P. R. *J. Am. Chem. Soc.* **2009**, *131*, 9622–9623.

[142] Avci, F. Y.; Kasper, D. L. *Annu. Rev. Immunol.* **2010**, *28*, 107–130.

[143] Baumann, H.; Tzianabos, A. O.; Brisson, J. R.; Kasper, D. L.; Jennings, H. J. *Biochemistry* **1992**, *31*, 4081–4089.

[144] Wang, Q.; McLoughlin, R. M.; Cobb, B. A.; Charrel-Dennis, M.; Zaleski, K. J.; Golenbock, D.; Tzianabos, A. O.; Kasper, D. L. *J. Exp. Med.* **2006**, *203*, 2853–2863.

[145] Ruiz-Perez, B.; Chung, D. R.; Sharpe, A. H.; Yagita, H.; Kalka-Moll, W. M.; Sayegh, M. H.; Kasper, D. L.; Tzianabos, A. O. *Proc. Natl. Acad. Sci. U. S. A.* **2005**, *102*, 16753–16758.

[146] Mazmanian, S. K.; Round, J. L.; Kasper, D. L. *Nature* **2008**, *453*, 620–625.

[147] Ochoa-Repáraz, J.; Mielcarz, D. W.; Wang, Y.; Begum-Haque, S.; Dasgupta, S.; Kasper, D. L.; Kasper, L. H. *Mucosal Immunol.* **2010**, *3*, 487–495.

[148] Mazmanian, S. K.; Liu, C. H.; Tzianabos, A. O.; Kasper, D. L. *Cell* **2005**, *122*, 107–118.

[149] Wang, Y.; Kalka-Moll, W. M.; Roehrl, M. H.; Kasper, D. L. *Proc. Natl. Acad. Sci. U. S. A.* **2000**, *97*, 13478–13483.

[150] van den Bos, L. J.; Boltje, T. J.; Provoost, T.; Mazurek, J.; Overkleeft, H. S.; van der Marel, G. A. *Tetrahedron Lett.* **2007**, *48*, 2697–2700.

[151] Pragani, R.; Seeberger, P. H. *J. Am. Chem. Soc.* **2011**, *133*, 102–107.

[152] Pragani, R.; Stallforth, P.; Seeberger, P. H. *Org. Lett.* **2010**, *12*, 1624–1627.

[153] Fischer, W. *Med. Microbiol. Immunol.* **1994**, *183*, 61–76.

[154] Greenberg, J. W.; Fischer, W.; Joiner, K. A. *Infect. Immun.* **1996**, *64*, 3318–3325.

[155] Schmidt, R. R.; Pedersen, C. M.; Qiao, Y.; Zähringer, U. *Org. Biomol. Chem.* **2011**, *9*, 2040–2052.

[156] Oltvoort, J. J.; van Boeckel, C. A. A.; de Koning, J. H.; van Boom, J. H. *Recl. Trav. Chim. Pays-Bas* **1982**, *101*, 87–91.

[157] Oltvoort, J. J.; Kloosterman, M.; van Boeckel, C. A. A.; van Boom, J. H. *Carbohydr. Res.* **1984**, *130*, 147–163.

[158] Fukase, K.; Matsumoto, T.; Ito, N.; Yoshimura, T.; Kotani, S.; Kusumoto, S. *Bull. Chem. Soc. Jpn.* **1992**, *65*, 2643–2654.

[159] Fukase, K.; Yoshimura, T.; Kotani, S.; Kusumoto, S. *Bull. Chem. Soc. Jpn.* **1994**, *67*, 473–482.

[160] Morath, S.; Geyer, A.; Hartung, T. *J. Exp. Med.* **2001**, *193*, 393–398.

[161] Morath, S.; Stadelmaier, A.; Geyer, A.; Schmidt, R. R.; Hartung, T. *J. Exp. Med.* **2002**, *195*, 1635–1640.

[162] Deininger, S.; Stadelmaier, A.; von Aulock, S.; Morath, S.; Schmidt, R. R.; Hartung, T. *J. Immunol.* **2003**, *170*, 4134–4138.

[163] Stadelmaier, A.; Morath, S.; Hartung, T.; Schmidt, R. R. *Angew. Chem. Int. Ed.* **2003**, *42*, 916-920.

[164] Figueroa-Perez, I.; Stadelmaier, A.; Deininger, S.; Aulock, S. V.; Hartung, T.; Schmidt, R. R. *Carbohydr. Res.* **2006**, *341*, 2901–2911.

[165] Stadelmaier, A.; Figueroaperez, I.; Deininger, S.; Vonaulock, S.; Hartung, T.; Schmidt, R. *Bioorg. Med. Chem.* **2006**, *14*, 6239–6254.

[166] Deininger, S.; Figueroa-Perez, I.; Sigel, S.; Stadelmaier, A.; Schmidt, R. R.; Hartung, T.; von Aulock, S. *Clin. Vaccine Immunol.* **2007**, *14*, 1629–1633.

[167] Qiao, Y.; Lindner, B.; Zähringer, U.; Truog, P.; Schmidt, R. R. *Bioorg. Med. Chem.* **2010**, *18*, 3696–3702.

[168] Pedersen, C. M.; Figueroa-Perez, I.; Boruwa, J.; Lindner, B.; Ulmer, A. J.; Zähringer, U.; Schmidt, R. R. *Chem. Eur. J.* **2010**, *16*, 12627–12641.

[169] Pedersen, C. M.; Figueroa-Perez, I.; Lindner, B.; Ulmer, A. J.; Zähringer, U.; Schmidt, R. R. *Angew. Chem. Int. Ed.* **2010**, *49*, 2585–2590.

[170] World Health Organization. Global Tuberculosis Control Report; World Health Organization: Geneva, **2010**.

[171] Besra, G. S.; Khoo, K.-H.; McNeil, M. R.; Dell, A.; Morris, H. R.; Brennan, P. J. *Biochemistry* **1995**, *34*, 4257–4266.

[172] Shi, L.; Berg, S.; Lee, A.; Spencer, J. S.; Zhang, J.; Vissa, V.; McNeil, M. R.; Khoo, K.-H.; Chatterjee, D. *J. Biol. Chem.* **2006**, *281*, 19512–19526.

[173] Imamura, A.; Lowary, T. *Trends Glycosci. Glycotechnol.* **2011**, *23*, 134–152.

[174] Gadikota, R. R.; Callam, C. S.; Wagner, T.; Del Fraino, B.; Lowary, T. L. *J. Am. Chem. Soc.* **2003**, *125*, 4155–4165.

[175] Hou, D.; Taha, H. A.; Lowary, T. L. *J. Am. Chem. Soc.* **2009**, *131*, 12937–12948.

[176] Lee, Y. J.; Lee, K.; Jung, E. H.; Jeon, H. B.; Kim, K. S. *Org. Lett.* **2005**, *7*, 3263–3266.

[177] Zhu, X.; Kawatkar, S.; Rao, Y.; Boons, G.-J. *J. Am. Chem. Soc.* **2006**, *128*, 11948–11957.

[178] Rao, Y.; Buskas, T.; Albert, A.; O'Neill, M. A.; Hahn, M. G.; Boons, G.-J. *ChemBioChem* **2008**, *9*, 381–388.

[179] Crich, D.; Pedersen, C. M.; Bowers, A. A.; Wink, D. J. *J. Org. Chem.* **2007**, *72*, 1553–1565.

[180] Ishiwata, A.; Akao, H.; Ito, Y. *Org. Lett.* **2006**, *8*, 5525–5528.

[181] Imamura, A.; Lowary, T. L. *Org. Lett.* **2010**, *12*, 3686–3689.

[182] Ishiwata, A.; Lee, Y. J.; Ito, Y. *Org. Biomol. Chem.* **2010**, *8*, 3596–3608.

[183] Sanchez, S.; Bamhaoud, T.; Prandi, J. *Tetrahedron Lett.* **2000**, *41*, 7447–7452.

[184] Marotte, K.; Sanchez, S.; Bamhaoud, T.; Prandi, J. *Eur. J. Org. Chem.* **2003**, 3587–3598.

[185] Ishiwata, A.; Munemura, Y.; Ito, Y. *Eur. J. Org. Chem.* **2008**, 4250–4263.

[186] Ishiwata, A.; Ito, Y. *J. Am. Chem. Soc.* **2011**, *133*, 2275–2291.

[187] D'Souza, F. W.; Cheshev, P. E.; Ayers, J. D.; Lowary, T. L. *J. Org. Chem.* **1998**, *63*, 9037–9044.

[188] Ayers, J. D.; Lowary, T. L.; Morehouse, C. B.; Besra, G. S. *Bioorg. Med. Chem. Lett.* **1998**, *8*, 437–442.

[189] D'Souza, F. W.; Ayers, J. D.; McCarren, P. R.; Lowary, T. L. *J. Am. Chem. Soc.* **2000**, *122*, 1251–1260.

[190] Han, J.; Gadikota, R. R.; McCarren, P. R.; Lowary, T. L. *Carbohydr. Res.* **2003**, *338*, 581–588.

[191] Cociorva, O. M.; Lowary, T. L. *Tetrahedron* **2004**, *60*, 1481–1489.

[192] Cociorva, O. M.; Lowary, T. L. *Carbohydr. Res.* **2004**, *339*, 853–865.

[193] D'Souza, F. W.; Lowary, T. L. *Org. Lett.* **2000**, *2*, 1493–1495.

[194] Yin, H.; D'Souza, F. W.; Lowary, T. L. *J. Org. Chem.* **2002**, *67*, 892–903.

[195] Rademacher, C.; Shoemaker, G. K.; Kim, H.-S.; Zheng, R. B.; Taha, H.; Liu, C.; Nacario, R. C.; Schriemer, D. C.; Klassen, J. S.; Peters, T.; Lowary, T. L. *J. Am. Chem. Soc.* **2007**, *129*, 10489–10502.

[196] Murase, T.; Zheng, R. B.; Joe, M.; Bai, Y.; Marcus, S. L.; Lowary, T. L.; Ng, K. K. S. *J. Mol. Biol.* **2009**, *392*, 381–392.

[197] Liu, C.; Richards, M. R.; Lowary, T. L. *J. Org. Chem.* **2010**, *75*, 4992–5007.

[198] Liu, C.; Richards, M. R.; Lowary, T. L. *Org. Biomol. Chem.* **2011**, *9*, 165–176.

[199] Joe, M.; Bai, Y.; Nacario, R. C.; Lowary, T. L. *J. Am. Chem. Soc.* **2007**, *129*, 9885–9901.

[200] Draper, P.; Khoo, K.-H.; Chatterjee, D.; Dell, A.; Morris, H. R. *Biochem. J.* **1997**, *327*, 519–525.

[201] Lee, A.; Wu, S.-W.; Scherman, M. S.; Torrelles, J. B.; Chatterjee, D.; McNeil, M. R.; Khoo, K.-H. *Biochemistry* **2006**, *45*, 15817–15828.

[202] Bhamidi, S.; Scherman, M. S.; Rithner, C. D.; Prenni, J. E.; Chatterjee, D.; Khoo, K. H.; McNeil, M. R. *J. Biol. Chem.* **2008**, *283*, 12992–13000.

[203] Peng, W.; Zou, L.; Bhamidi, S.; McNeil, M. R.; Lowary, T. L. *J. Org. Chem.* **2012**, *77*, 9826–9832.

[204] Sanchez, S.; Bamhaoud, T.; Prandi, J. *Eur. J. Org. Chem.* **2002**, 3864–3873.

[205] Désiré, J.; Prandi, J. *Carbohydr. Res.* **1999**, *317*, 110–118.

[206] Guiard, J.; Collmann, A.; Gilleron, M.; Mori, L.; De Libero, G.; Prandi, J.; Puzo, G. *Angew. Chem. Int. Ed.* **2008**, *47*, 9734–9738.

[207] Abronina, P. I.; Sedinkin, S. L.; Podvalnyy, N. M.; Fedina, K. G.; Zinin, A. I.; Torgov, V. I.; Kononov, L. O. *Tetrahedron Lett.* **2011**, *52*, 1794–1796.

[208] Fedina, K. G.; Abronina, P. I.; Podvalnyy, N. M.; Kondakov, N. N.; Chizhov, A. O.; Torgov, V. I.; Kononov, L. O. *Carbohydr. Res.* **2012**, *357*, 62–67.

[209] Podvalnyy, N. M.; Sedinkin, S. L.; Abronina, P. I.; Zinin, A. I.; Fedina, K. G.; Torgov, V. I.; Kononov, L. O. *Carbohydr. Res.* **2011**, *346*, 7–15.

[210] Abronina, P.; Podvalnyy, N.; Sedinkin, S.; Fedina, K.; Zinin, A.; Chizhov, A.; Torgov, V.; Kononov, L. *Synthesis* **2012**, *44*, 1219–1225.

[211] Liang, X.-Y.; Deng, L.-M.; Liu, X.; Yang, J.-S. *Tetrahedron* **2010**, *66*, 87–93.

[212] Deng, L.-M.; Liu, X.; Liang, X.-Y.; Yang, J.-S. *J. Org. Chem.* **2012**, *77*, 3025–3037.

[213] Naresh, K.; Bharati, B. K.; Jayaraman, N.; Chatterji, D. *Org. Biomol. Chem.* **2008**, *6*, 2388–2393.

[214] Naresh, K.; Bharati, B. K.; Avaji, P. G.; Chatterji, D.; Jayaraman, N. *Glycobiology* **2011**, *21*, 1237–1254.

[215] Naresh, K.; Avaji, P. G.; Maiti, K.; Bharati, B. K.; Syal, K.; Chatterji, D.; Jayaraman, N. *Glycoconj. J.* **2012**, *29*, 107–118.

[216] Naresh, K.; Bharati, B. K.; Avaji, P. G.; Jayaraman, N.; Chatterji, D. *Org. Biomol. Chem.* **2010**, *8*, 592–599.

[217] Wang, H.; Ning, J. *J. Org. Chem.* **2003**, *68*, 2521–2524.

[218] Zhang, G.; Fu, M.; Ning, J. *Tetrahedron: Asymmetry* **2005**, *16*, 733–738.

[219] Mereyala, H. B.; Hotha, S.; Gurjar, M. K. *Chem. Commun.* **1998**, 685–686.

[220] Gurjar, M. K.; Reddy, L. K.; Hotha, S. *Org. Lett.* **2001**, *3*, 321–323.

[221] Gurjar, M. K.; Reddy, L. K.; Hotha, S. *J. Org. Chem.* **2001**, *66*, 4657–4660.

[222] Gandolfi-Donadío, L.; Gallo-Rodriguez, C.; de Lederkremer, R. M. *J. Org. Chem.* **2003**, *68*, 6928–6934.

[223] Gandolfi-Donadío, L.; Santos, M.; de Lederkremer, R. M.; Gallo-Rodriguez, C. *Org. Biomol. Chem.* **2011**, *9*, 2085–2097.

[224] Gandolfi-Donadío, L.; Gallo-Rodriguez, C.; de Lederkremer, R. M. *Can. J. Chem.* **2006**, *84*, 486–491.

[225] Gandolfi-Donadío, L.; Gallo-Rodriguez, C.; de Lederkremer, R. M. *Carbohydr. Res.* **2008**, *343*, 1870–1875.

[226] Pathak, A. K.; Besra, G. S.; Crick, D.; Maddry, J. A.; Morehouse, C. B.; Suling, W. J.; Reynolds, R. C. *Bioorg. Med. Chem.* **1999**, *7*, 2407–2413.

[227] Pathak, A. K.; Pathak, V.; Bansal, N.; Maddry, J. A.; Reynolds, R. C. *Tetrahedron Lett.* **2001**, *42*, 979–982.

[228] Pathak, A. K.; Pathak, V.; Maddry, J. A.; Suling, W. J.; Gurcha, S. S.; Besra, G. S.; Reynolds, R. C. *Bioorg. Med. Chem.* **2001**, *9*, 3145–3151.

[229] Pathak, A. K.; Pathak, V.; Seitz, L.; Maddry, J. A.; Gurcha, S. S.; Besra, G. S.; Suling, W. J.; Reynolds, R. C. *Bioorg. Med. Chem.* **2001**, *9*, 3129–3143.

[230] Pathak, A. K.; Pathak, V.; Suling, W. J.; Gurcha, S. S.; Morehouse, C. B.; Besra, G. S.; Maddry, J. A.; Reynolds, R. C. *Bioorg. Med. Chem.* **2002**, *10*, 923–928.

[231] Pathak, A. K.; Pathak, V.; Riordan, J. R.; Suling, W. J.; Gurcha, S. S.; Besra, G. S.; Reynolds, R. C. *Bioorg. Med. Chem. Lett.* **2007**, *17*, 4527–4530.

[232] Pathak, A. K.; Pathak, V.; Seitz, L.; Gurcha, S. S.; Besra, G. S.; Riordan, J. M.; Reynolds, R. C. *Bioorg. Med. Chem.* **2007**, *15*, 5629–5650.

[233] Pathak, A. K.; Pathak, V.; Suling, W. J.; Riordan, J. R.; Gurcha, S. S.; Besra, G. S.; Reynolds, R. C. *Bioorg. Med. Chem.* **2009**, *17*, 872–881.

[234] Maddry, J. A.; Bansal, N.; Bermudez, L. E.; Comber, R. N.; Orme, I. M.; Suling, W. J.; Wilson, L. N.; Reynolds, R. C. *Bioorg. Med. Chem. Lett.* **1998**, *8*, 237–242.

[235] Kim, K. S.; Lee, B.-Y.; Yoon, S. H.; Jeon, H. J.; Baek, J. Y.; Jeong, K.-S. *Org. Lett.* **2008**, *10*, 2373–2376.

[236] Wen, X.; Crick, D. C.; Brennan, P. J.; Hultin, P. G. *Bioorg. Med. Chem.* **2003**, *11*, 3579–3587.

[237] Mugunthan, G.; Sriram, D.; Yogeeswari, P.; Kartha, K. P. R. *Carbohydr. Res.* **2011**, *346*, 2401–2405.

[238] Lee, Y. J.; Fulse, D. B.; Kim, K. S. *Carbohydr. Res.* **2008**, *343*, 1574–1584.

[239] Completo, G. C.; Lowary, T. L. *J. Org. Chem.* **2008**, *73*, 4513–4525.

[240] Bai, Y.; Lowary, T. L. *J. Org. Chem.* **2006**, *71*, 9672–9680.

[241] Ryll, R.; Kumazawa, Y.; Yano, I. *Microbiol. Immunol.* **2001**, *45*, 801–811.

[242] Moody, D. B.; Guy, M. R.; Grant, E.; Cheng, T. Y.; Brenner, M. B.; Besra, G. S.; Porcelli, S. A. *J. Exp. Med.* **2000**, *192*, 965–976.

[243] Moody, D. B.; Reinhold, B. B.; Guy, M. R.; Beckman, E. M.; Frederique, D. E.; Furlong, S. T.; Ye, S.; Reinhold, V. N.; Sieling, P. A.; Modlin, R. L.; Besra, G. S.; Porcelli, S. A. *Science* **1997**, *278*, 283–286.

[244] Matsunaga, I.; Naka, T.; Talekar, R. S.; McConnell, M. J.; Katoh, K.; Nakao, H.; Otsuka, A.; Behar, S. M.; Yano, I.; Moody, D. B.; Sugita, M. *J. Biol. Chem.* **2008**, *283*, 28835–28841.

[245] Toubiana, R.; Das, B. C.; Defaye, J.; Mompon, B.; Toubiana, M.-J. *Carbohydr. Res.* **1975**, *44*, 308–312.

[246] Liav, A.; Flowers, H. M.; Goren, M. B. *Carbohydr. Res.* **1984**, *133*, 53–58.

[247] Liav, A.; Goren, M. B. *Carbohydr. Res.* **1984**, *129*, 121–129.

[248] Liav, A.; Goren, M. B.; Yang, Y.-M.; Fales, H. M. *Carbohydr. Res.* **1986**, *155*, 223–228.

[249] Liav, A.; Goren, M. B. *Carbohydr. Res.* **1986**, *155*, 229–235.

[250] Azuma, I.; Sakurai, T.; Ishida, H.; Kitajima, T.; Yamazaki, M. *Carbohydr. Res.* **1991**, *212*, 47–53.

[251] Nishizawa, M.; Yamamoto, H.; Imagawa, H.; Barbier-Chassefière, V.; Petit, E.; Azuma, I.; Papy-Garcia, D. *J. Org. Chem.* **2007**, *72*, 1627–1633.

[252] Prandi, J. *Carbohydr. Res.* **2012**, *347*, 151–154.

[253] Ishiwata, A.; Akao, H.; Ito, Y.; Sunagawa, M.; Kusunose, N.; Kashiwazaki, Y. *Bioorg. Med. Chem.* **2006**, *14*, 3049–3061.

[254] Swarts, B. M.; Holsclaw, C. M.; Jewett, J. C.; Alber, M.; Fox, D. M.; Siegrist, M. S.; Leary, J. A.; Kalscheuer, R.; Bertozzi, C. R. *J. Am. Chem. Soc.* **2012**, *134*, 16123–16126.

[255] Backus, K. M.; Boshoff, H. I.; Barry, C. S.; Boutureira, O.; Patel, M. K.; D'Hooge, F.; Lee, S. S.; Via, L. E.; Tahlan, K.; Barry, C. E., III; Davis, B. G. *Nat. Chem. Biol.* **2011**, *7*, 228–235.

[256] Barry, C. S.; Backus, K. M.; Barry, C. E., III; Davis, B. G. *J. Am. Chem. Soc.* **2011**, *133*, 13232–13235.

[257] Briken, V.; Porcelli, S. A.; Besra, G. S.; Kremer, L. *Mol. Microbiol.* **2004**, *53*, 391–403.

[258] Mishra, A. K.; Driessen, N. N.; Appelmelk, B. J.; Besra, G. S. *FEMS Microbiol. Rev.* **2011**, *35*, 1126–1157.

[259] Nigou, J.; Gilleron, M.; Cahuzac, B.; Bounery, J. D.; Herold, M.; Thurnher, M.; Puzo, G. *J. Biol. Chem.* **1997**, *272*, 23094–23103.

[260] Khoo, K. H.; Tang, J. B.; Chatterjee, D. *J. Biol. Chem.* **2001**, *276*, 3863–3871.

[261] Khoo, K.-H.; Dell, A.; Morris, H. R.; Brennan, P. J.; Chatterjee, D. *J. Biol. Chem.* **1995**, *270*, 12380–12389.

[262] Treumann, A.; Xidong, F.; McDonnell, L.; Derrick, P. J.; Ashcroft, A. E.; Chatterjee, D.; Homans, S. W. *J. Mol. Biol.* **2002**, *316*, 89–100.

[263] Ludwiczak, P.; Gilleron, M.; Bordat, Y.; Martin, C.; Gicquel, B.; Puzo, G. *Microbiology* **2002**, *148*, 3029–3037.

[264] Turnbull, W. B.; Shimizu, K. H.; Chatterjee, D.; Homans, S. W.; Treumann, A. *Angew. Chem. Int. Ed.* **2004**, *43*, 3918–3922.

[265] Guérardel, Y.; Maes, E.; Briken, V.; Chirat, F.; Leroy, Y.; Locht, C.; Strecker, G.; Kremer, L. *J. Biol. Chem.* **2003**, *278*, 36637–36651.

[266] Joe, M.; Sun, D.; Taha, H.; Completo, G. C.; Croudace, J. E.; Lammas, D. A.; Besra, G. S.; Lowary, T. L. *J. Am. Chem. Soc.* **2006**, *128*, 5059–5072.

[267] Subramaniam, V.; Gurcha, S. S.; Besra, G. S.; Lowary, T. L. *Bioorg. Med. Chem.* **2005**, *13*, 1083–1094.

[268] Tam, P.-H.; Besra, G. S.; Lowary, T. L. *ChemBioChem* **2008**, *9*, 267–278.

[269] Xia, L.; Zheng, R. B.; Lowary, T. L. *ChemBioChem* **2012**, *13*, 1139–1151.

[270] Cociorva, O. M.; Gurcha, S. S.; Besra, G. S.; Lowary, T. L. *Bioorg. Med. Chem.* **2005**, *13*, 1369–1379.

[271] Boonyarattanakalin, S.; Liu, X.; Michieletti, M.; Lepenies, B.; Seeberger, P. H. *J. Am. Chem. Soc.* **2008**, *130*, 16791–16799.

[272] Liu, X.; Stocker, B. L.; Seeberger, P. H. *J. Am. Chem. Soc.* **2006**, *128*, 3638–3648.

[273] Hölemann, A.; Stocker, B. L.; Seeberger, P. H. *J. Org. Chem.* **2006**, *71*, 8071–8088.

[274] Fraser-Reid, B.; Lu, L.; Jayaprakash, K. N.; Chaudhuri, S. R. In *Carbohydrate-Based Vaccines*; Roy, R., Ed.; ACS Symposium Series 989; American Chemical Society: Washington, DC, **2008**, pp 199–238.

[275] Lu, J.; Fraser-Reid, B. *Org. Lett.* **2004**, *6*, 3051–3054.

[276] Lu, J.; Fraser-Reid, B. *Chem. Commun.* **2005**, 862–864.

[277] Jayaprakash, K. N.; Lu, J.; Fraser-Reid, B. *Angew. Chem. Int. Ed.* **2005**, *44*, 5894–5898.

[278] Fraser-Reid, B.; Chaudhuri, S. R.; Jayaprakash, K. N.; Lu, J.; Ramamurty, C. V. S. *J. Org. Chem.* **2008**, *73*, 9732–9743.

[279] Fraser-Reid, B.; Lu, J.; Jayaprakash, K. N.; López, J. C. *Tetrahedron: Asymmetry* **2006**, *17*, 2449–2463.

[280] Steichen, C. T.; Kearney, J. F.; Turnbough, C. L., Jr. *J. Bacteriol.* **2005**, *187*, 5868–5876.

[281] Daubenspeck, J. M.; Zeng, H.; Chen, P.; Dong, S.; Steichen, C. T.; Krishna, N. R.; Pritchard, D. G.; Turnbough, C. L., Jr. *J. Biol. Chem.* **2004**, *279*, 30945–30953.

[282] Werz, D. B.; Seeberger, P. H. *Angew. Chem. Int. Ed.* **2005**, *44*, 6315–6318.

[283] Werz, D. B.; Adibekian, A.; Seeberger, P. H. *Eur. J. Org. Chem.* **2007**, 1976–1982.

[284] Oberli, M. A.; Tamborrini, M.; Tsai, Y.-H.; Werz, D. B.; Horlacher, T.; Adibekian, A.; Gauss, D.; Möller, H. M.; Pluschke, G.; Seeberger, P. H. *J. Am. Chem. Soc.* **2010**, *132*, 10239–10241.

[285] Adamo, R.; Saksena, R.; Kováč, P. *Carbohydr. Res.* **2005**, *340*, 2579–2582.

[286] Mehta, A. S.; Saile, E.; Zhong, W.; Buskas, T.; Carlson, R.; Kannenberg, E.; Reed, Y.; Quinn, C. P.; Boons, G.-J. *Chem. Eur. J.* **2006**, *12*, 9136–9149.

[287] Guo, H.; O'Doherty, G. A. *Angew. Chem. Int. Ed.* **2007**, *46*, 5206–5208.

[288] Crich, D.; Vinogradova, O. *J. Org. Chem.* **2007**, *72*, 6513–6520.

[289] Dwek, R. A. *Chem. Rev.* **1996**, *96*, 683–720.

[290] Benz, I.; Schmidt, M. A. *Mol. Microbiol.* **2002**, *45*, 267–276.

[291] Wacker, M.; Linton, D.; Hitchen, P. G.; Nita-Lazar, M.; Haslam, S. M.; North, S. J.; Panico, M.; Morris, H. R.; Dell, A.; Wren, B. W.; Aebi, M. *Science* **2002**, *298*, 1790–1793.

[292] Young, N. M.; Brisson, J.-R.; Kelly, J.; Watson, D. C.; Tessier, L.; Lanthier, P. H.; Jarrell, H. C.; Cadotte, N.; St. Michael, F.; Aberg, E.; Szymanski, C. M. *J. Biol. Chem.* **2002**, *277*, 42530–42539.

[293] Szymanski, C. M.; Burr, D. H.; Guerry, P. *Infect. Immun.* **2002**, *70*, 2242–2244.

[294] Takatani, M.; Matsuo, I.; Ito, Y. *Carbohydr. Res.* **2003**, *338*, 1073–1081.

[295] Ishiwata, A.; Ohta, S.; Ito, Y. *Carbohydr. Res.* **2006**, *341*, 1557–1573.

[296] Amin, M. N.; Ishiwata, A.; Ito, Y. *Carbohydr. Res.* **2006**, *341*, 1922–1929.

[297] Amin, M. N.; Ishiwata, A.; Ito, Y. *Tetrahedron* **2007**, *63*, 8181–8198.

[298] Lee, Y. J.; Ishiwata, A.; Ito, Y. *Tetrahedron* **2009**, *65*, 6310–6319.

15

DISCOVERIES AND APPLICATIONS OF GLYCAN ARRAYS

Chung-Yi Wu and Shih-Huang Chang
Genomics Research Center, Academia Sinica, Taipei, Taiwan

15.1 INTRODUCTION

Carbohydrates, also called glycans, often exist on cell surfaces as glycoproteins or glycolipid conjugates and play important structural and functional roles in numerous biological recognition processes [1–4]. In addition, natural products often require glycosylation for biological activities, and many organisms diversify functions and detoxify harmful exogenous xenobiotics through glycosylation [5]. However, the study of carbohydrate–protein interactions still encounters much difficulty mainly because of the inaccessibility of such structurally complex molecules and their low-affinity interactions with glycan-binding proteins. The monomeric dissociation constant in a typical carbohydrate–protein interaction is in the millimolar range; thus, carbohydrate-mediated biological responses often occur through multivalent interactions on the cell surface in order to achieve high affinity and specificity [6].

15.2 DISCOVERIES OF GLYCAN ARRAYS

15.2.1 General

The traditional methods of evaluating the interactions of glycans and glycan-binding proteins are based on isothermal titration calorimetry, surface plasmon resonance (SPR), and enzyme-linked lectin assays. Recently, glycan arrays have been developed as a

Glycochemical Synthesis: Strategies and Applications, First Edition.
Edited by Shang-Cheng Hung and Medel Manuel L. Zulueta.

new technology for high-throughput evaluation of interactions between carbohydrates and proteins, antibodies, cells, or viruses [7–16]. Glycan arrays allow researchers to probe hundreds of receptor–ligand interactions in one experiment, providing a low-cost and high-throughput methodology for screening glycan interactions. Another main advantage is the multivalent display of arrayed glycans, which serve to mimic cell-surface display; thus, the array-based technique is particularly well suited for investigations of real glycan-mediated interactions in nature. Multiple proteins or glycans can be quantitatively and qualitatively analyzed in a monovalent or multivalent manner [17–26]. Because the glycan array system has numerous practical merits over traditional methods, it is expected to greatly contribute to pioneering new fields of glycoscience.

Due to the structural complexity of carbohydrates, preparation of structurally well-defined carbohydrates with broad diversity is a key issue for glycan array development. Because it is difficult to obtain sufficient quantities of glycans from natural sources, chemical or chemoenzymatic methods or even pure enzymatic methods have been used to address this deficiency. The recent advances in carbohydrate syntheses that are described in other chapters of this book have greatly reduced the labor for glycan preparation.

During the last decade, there has been an explosion of methods for creating glycan arrays. The most common solid support employed is the glass microscope slide due to its easy manipulation and low cost. Alternatively, gold, microtiter plates, aluminum, three-dimensional gel bead, and nitrocellulose membrane have been utilized as well. The three-dimensional substrate was constructed by polyacrylamide gel elements instead of two-dimensional dots on the array surface. As a result of the three-dimensional placement of the ligand, two orders of magnitude increase of its density is possible, providing principal signal improvement during fluorescent detection and increasing method sensitivity [27].

According to the method of immobilization, glycan array fabrications could be classified into two major groups: covalent-immobilized and noncovalent-immobilized glycan arrays. Almost all glycans are modified with distinct functional groups that then react with corresponding functional groups on the surface base substrate. With the modifications, glycans can be located on required positions and presented with required orientations.

15.2.2 Noncovalent Glycan Arrays

Noncovalent immobilization techniques are based on hydrophobic physical absorption or charge-based interaction to adsorb free or modified glycans on nonderivatized or derivatized solid surfaces. In the initially reported glycan array platform, unmodified glycan and array surface were used to construct a glycan array (Fig. 15.1a) [10]. This platform demonstrated the first microchip format with proteoglycan and neoglyproteins densely adsorbed to the chip surface through hydrophobic interaction. These methodologies are easy to operate as they do not require any chemical modifications. However, one disadvantage is the inability to retain low-mass glycans under washing. To improve the retention of glycans in washing steps, Ten Feizi and

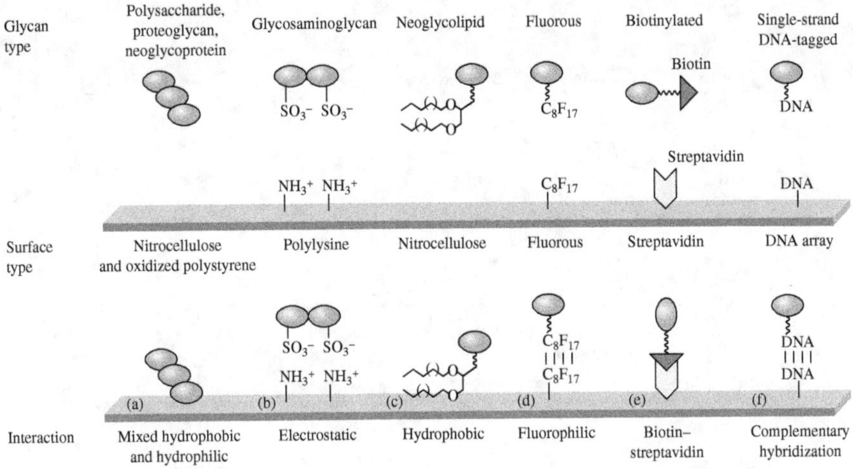

FIGURE 15.1 Noncovalent immobilization methods to construct a glycan array.

colleagues established an array system wherein the glycans are linked to lipids to generate neoglycolipids for arraying [11, 28]. Alternatively, Chi-Huey Wong's group reported the use of allyl- or azide-derivatized glycans for noncovalent immobilization on the free or alkyne-modified polystyrene surface of microtiter plates [29, 30]. Another noncovalent glycan microarray, developed by Nicola Pohl's group, utilized fluorophilic interaction, with glycans linked to fluorous tags and adsorbed avidly to slides coated with fluoroalkylsilane to form glycan array [31–34]. By taking advantage of the strong noncovalent fluorous–fluorous interaction, Wong's group developed a fluorous-based array that allowed convenient characterization of enzyme kinetics by laser desorption/ionization mass spectrometry without matrix [35]. Moreover, biotin–streptavidin interaction [36–39] or DNA complementary hybridization [40] has also been used for the construction of glycan arrays (Fig. 15.1b–f). All of these methods are stable after repeated washings. However, the hydrophobic moiety of the glycan derivative could be a hurdle for the experiment because glycan arrays are commonly used with aqueous solutions. Furthermore, presenting the glycans in the desired orientation is also hard to control under noncovalent immobilization.

15.2.3 Covalent Glycan Arrays

Covalently immobilized arrays are formatted by glycans with functionalized spacers that react with a complementary activated surface to form a covalent bond. Several different covalent interactions were reported to construct a specified glycan array. Amine chemistry and thiol chemistry are the two major methods to conjugate glycans to the reactive substrate in the array surface (Fig. 15.2). Thiol chemistry was first adapted by Injae Shin in 2002 to react with the maleimide functional group (Fig. 15.2a, b) [7, 41]. Disulfide bond formation was then reported for the fabrication

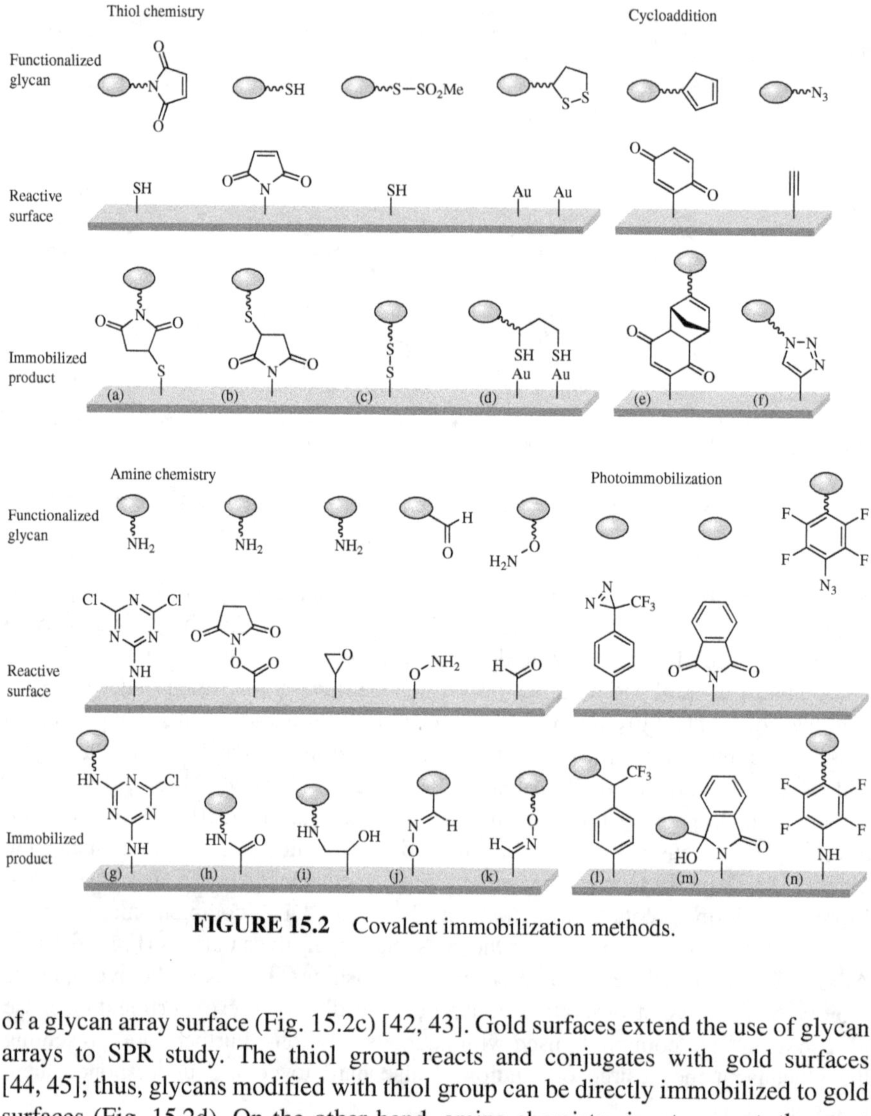

FIGURE 15.2 Covalent immobilization methods.

of a glycan array surface (Fig. 15.2c) [42, 43]. Gold surfaces extend the use of glycan arrays to SPR study. The thiol group reacts and conjugates with gold surfaces [44, 45]; thus, glycans modified with thiol group can be directly immobilized to gold surfaces (Fig. 15.2d). On the other hand, amine chemistry is, at present, the most widely used method, and several different kinds of array-based substrates are commercially available (Fig. 15.2g–i). The conjugation reaction between amine-linked glycans and *N*-hydroxysuccinimide (NHS) ester-modified glass slides served as the standard platform for the Consortium for Functional Glycomics (CFG) (Fig. 15.2h). In addition, several different functional groups, for example, cyanuric chloride, epoxide, and aldehyde, which could also react with amines, were utilized in the construction of glycan arrays [46–50]. Condensation between amines and other functional groups were also reported by several research groups (Fig. 15.2j, k) [51–54]. The minority methods to construct glycan array are cycloaddition reactions (Fig. 15.2e, f) and photoimmobilizations (Fig. 15.2l–n) [30, 55–57].

15.2.4 Quality Control of Glycan Arrays

Most glycan arrays used today are difficult to characterize and quantify and, thus, lack quality control. To tackle the problem of quality control, a special glycan array was developed by using 2,6-diaminopyridine as a fluorescent linker to sugars and printed on epoxy-activated slides [49, 50, 58]. Another approach is to combine the array method and mass spectrometry for label-free analysis. For example, Milan Mrksich's group developed a method using a combination of self-assembled monolayers and matrix-assisted laser desorption/ionization time-of-flight (MALDI-TOF) mass spectrometry to observe and optimize the synthesis of glycans on gold surface [59–61]. This strategy was further used by the Sabine Flitsch's group to monitor glycosyltransferases involved in protein O-glycosylation [62, 63]. To avoid nonspecific protein adsorption on the gold surface, a tri(ethylene glycol) spacer was used. Because the nature of the adducts and the degree of fragmentation of the parent molecular ions in MALDI-TOF are dependent on the laser fluence rate (laser energy) [64, 65], the addition of matrix is often required to reduce the fragmentation of the parent molecular ion. However, the signal of the matrix may interfere with the analyte spectrum, especially when the analyte is a small molecule. Wong's group recently developed a desorption/ionization on silicon mass spectrometry (DIOS-MS) method to characterize the oligosaccharides that are covalently bound to porous silicon with a built-in photocleavable linker [66]. In addition, a technique based on nanostructure-initiator mass spectrometry (NIMS) was also developed to characterize noncovalent glycan arrays by using fluorous-tagged substrates that are physically adhered to the fluorohydrocarbon-coated porous silicon surface [67]. However, incorporation of a photocleavable linker to individual glycans is quite tedious, and preparation of porous silicon plates requires corrosive acids, which may not generate uniform results. Another disadvantage is that covalent glycan arrays on conventional glass slides cannot be characterized directly by mass spectrometry. Furthermore, the research groups of Chi-Huey Wong and Chung-Yi Wu also recently developed a new type of aluminum oxide-coated glass slide, which can be functionalized by reacting with an alkyl monoethoxysilane and linked to a glycan containing a photocleavable linker [68]. The glycan array with a photocleavable linker on coated glass slide surfaces can be characterized by time-of-flight mass spectrometry without matrix. However, the array often gave relatively low signal-to-noise ratio most likely due to incomplete photocleavage. To tackle this problem, the same research groups used glycans with a polyfluorinated hydrocarbon ($-C_8F_{17}$) tail to spot robotically onto the aluminum oxide-coated glass slide surface containing a layer of polyfluorinated hydrocarbon terminated with phosphonic acid. The samples on noncovalent array subsequently produced quite high signal-to-noise ratio. A representative cellotetraose array was developed to study the activity and specificity of different cellulases and to differentiate exo- and endoglucanase activities.

15.2.5 Detection Methods of Glycan Arrays

Recent detection methods for glycan array include fluorescent assay, SPR, MALDI-TOF mass spectrometry, and nanoparticle assay. Fluorescence-based measurement is the prevalent principle for detecting binding to glycan microarrays. Rhodamine [9], indodicarbocyanine (Cy3 and Cy5) [13, 40, 69], or fluorescein isothiocyanate [7, 38, 70] are frequently coupled to proteins and antibodies to monitor binding interaction with glycan arrays. To enhance array sensitivity, secondary antibodies carrying fluorescent probes are used to identify the first antibodies that are bound to glycan array [47, 71]. Also, SPR has been developed as an advanced array platform to allow the measurement of association and dissociation constants on the array in a label-free manner in real time [9, 38, 42, 70–73]. With a single flow cell, multichannel SPR detects hundreds of spots independently [42, 73]. Moreover, SPR can recognize the relatively weak multivalent nature of carbohydrate interactions due to the elimination of the wash step in the procedure of array detection [8, 71]. Recently, coupling the glycan arrays with mass spectrometry [74–77] has become a key development in bioinformatics to decipher glycan sequences and control the quality of the glycan array. A nanoparticle-based detection for glycan array was developed by Wu et al. who synthesized iron oxide/gold core/shell nanoparticles conjugated with antibodies or proteins [78]. Core/shell nanoparticles consist of discrete domains of different materials and, thus, can exhibit the properties of different components in the same structure. Iron oxide/gold core/shell nanoparticles unite magnetic property for enrichment, surface modification, and signal enhancement in a single entity. This powerful combination enables researchers to quickly concentrate target biomolecules through an external magnetic field, easily conjugate biomolecules on gold surface, and amplify the signal by depositing silver on gold surface of core/shell nanoparticles. This nanoparticle-based assay can reach subattomole detection level and does not suffer the photochemical instability of fluorescent molecular dye label. Most importantly, no costly or complex instrumentation is needed. Recently, several alternative detection methods including detection of radioactivity [79, 80], electrochemiluminescence [81], cantilever-based detection [82], and oblique-incidence reflectivity difference microscopy [83] were developed.

15.3 APPLICATIONS OF GLYCAN ARRAY

15.3.1 Enzyme Activity/Inhibition Studies

Cell-surface glycans are synthesized by the sequential actions of glycoenzymes, which hydrolyze or transfer specific glycosidic linkages. A glycan array is ideal for glycoenzyme activity studies because it allows for the identification of enzyme specificity against a panel of immobilized potential glycans [69, 84–86]. Another important application of glycan array is the high-throughput screening of new inhibitors of glycan-processing enzymes that are involved in the biosynthesis of disease-related glycans. A chemoenzymatic synthesis of the sialyl Lewis X epitope was described by Wong's group [30]. Sialyl Lewis X, a potential inflammation drug target, was synthesized by fucosyltransferases, which mediates the transfer of fucose from guanosine diphosphate (GDP)-fucose to sialyl *N*-acetyllactosamine (LacNAc). To screen fucosyltransferase inhibitors, 85 synthetic compounds were incubated with $\alpha(1\rightarrow3)$-fucosyltransferase in the presence of GDP-fucose on a LacNAc-coated microtiter-type array. The fucosylation of the coupled LacNAc residues was detected by a fucose-specific lectin from *Tetragonolobus purpureas* conjugated to a peroxidase. Four potent inhibitors of the fucosyltransferase with nanomolar inhibition constants were identified [87]. The substrate specificities of galactosyl [69] and sialyltransferases [86] were examined in a high-throughput manner using carbohydrate microarrays. Shin et al. used *N*-acetylglucosamine (GlcNAc) immobilized on glass slides to build a sialyl Lewis X structure by three sequential enzymatic transformations [41]. Furthermore, the specificity of bovine β(1→4)-galactosyltransferase was identified by a glycan array with 20 different glycans. These results show that the enzyme binding affinity of the β-GlcNAc substrate is greater than that of the α-GlcNAc substrate [69]. Ola Blixt et al. used a biotinylated cytidine monophosphate (CMP)-*N*-acetylneuraminic acid (Neu5Ac) as an activated donor substrate for sialyltransferases. A substitution reaction was performed in several sialyltransferases that exchange a sialylated acceptor with a biotinylated sialic acid [86]. In fact, sialic acid-containing glycans are difficult to synthesize by chemical means. Therefore, the use of sialyltransferases to introduce the sialic acid moiety onto carbohydrates has attracted considerable attention. Chemoenzymatic synthesis of a diverse sialoside library in microtiter plates using a one-pot three-enzyme protocol was demonstrated by Xi Chen's group [88]. A high-throughput screening of various recombinant sialyltransferase acceptor specificities was recently demonstrated by Blixt et al. who produced a series of unnatural sialoside analogues via chemoenzymatic methods [89]. The fabricated glycan array was also used to evaluate the specificities of sialic acid-binding proteins.

Applications of glycan microarrays showcase the rapid screening of many interactions to deal with the complexity and heterogeneity of the glycome. Kinetic constants and IC_{50} values can be calculated using glycan microarrays [9, 69, 90, 91]. Thus, new specificities of glycosyltransferase and glycosyltransferase activities could be identified [79, 85]. Mrksich's group revealed the potential to combine on-chip synthesis and label-free assay to provide a nice approach in the study of glycoenzyme activity and glycan function [59]. Glycosyltransferase activity and reaction on the surface were monitored *in situ* by mass spectrometry. Mrksich also examined the importance of ligand density in enzymatic glycosylation [92]. With a similar concept, Flitsch et al. applied MALDI-TOF mass spectrometry to monitor the enzymatic synthesis of mucin-type glycopeptide arrays using a polypeptide *N*-acetylgalactosaminyltransferase [62]. Additionally, bovine β(1→4)-galactosyltransferase specificity was assessed in a glycan array [93]. Apart from mass spectrometry, lectins and antibodies are also widely used as detection tools for enzymatic glycosylation on glycan array. For example, Niels-Christian Reichardt's elegant construction of N-glycan microarrays used lectins to monitor the new glycans formed [94, 95]. Both the use of glycan array to explore enzyme activity and an enzymatic strategy to obtain arrays provide efficient ways to improve glycan array applications [96].

15.3.2 Glycan Array for Diseases Detection and Vaccine Development

Glycan array emerged as a powerful tool for glycobiology with significant potential for disease detection and vaccine development. The use of glycan arrays could accelerate screening procedures, making detection of cancer-related glycan epitopes simple and inexpensive. The initial achievements of the glycan array technology are discoveries of new glycan recognizing proteins, which have been used to evaluate immune responses to vaccine candidates [10, 12, 14, 24, 27, 47, 97–101].

15.3.2.1 AIDS The primary method to discover and characterize glycan antigens involves identifying disease-specific antibodies produced by the immune system. These antibodies are obtained, for example, by immunizing mice with the target cells, viruses, pathogens, or glycoproteins. Once an antibody has been obtained, the corresponding glycan antigen can be identified by glycan array. Because the structures of the relevant glycan antigens are not known, a series of potential glycans may need to be tested in order to find the antigen. The array format is ideally suited in identifying glycan antigens in a high-throughput manner. One example is identifying glycan antigens on HIV. Antibody 2G12 was found to have broadly neutralizing anti-HIV activity [102, 103] and was shown to protect against infection *in vivo* in monkey models [104]. However, the structure of 2G12 binding sites proved difficult to characterize. To solve this problem, Peter Seeberger's group provided a glycan array with high-mannose glycan and several synthetic mannose substructures. They found that 2G12 specifically recognizes Man(α1→2)Man (Man = mannose) linkages and binds the oligomannose-modified analogue; thus, they confirmed the structural requirements by 2G12 in order to bind to the HIV-1 envelope glycoprotein gp120 [70, 105]. Wong and coworkers further probed 2G12 with six synthetic oligomannosides bearing Man(α1→2)Man. The evaluation of 2G12 binding affinities was conducted on a microtiter plate. The results show that the antibody binds with higher affinity to smaller oligomannosides (Man_4 and Man_5) on the array compared with larger structures (Man_6, Man_7, Man_8, and Man_9). This suggests that the array surface enables better multivalent presentation of these smaller structures [106]. Recently isolated broadly neutralizing antibodies such as PG9, PG16, and PGT 141–PGT 145 neutralize 70–80% of circulating HIV-1 isolates, showing them to be both broad and potent [107] and suggesting that the targeted epitopes are highly conserved among the variants of the entire HIV-1 group M. The epitopes of these antibodies depend on the presence of N-linked glycosylation sites at certain positions in the V1/V2 or V3 variable loops [108–110]. By dissecting the binding specificities of different binding molecules, the molecular basis of protein interactions with glycan structures of gp120 was established, thereby identifying glycan antigens as HIV vaccine candidates.

15.3.2.2 Malaria To detect pathogen-specific antibodies, Seeberger and coworkers prepared a microarray comprising synthetic glycosylphosphatidylinositol (GPI) glycans [111]. The microarray was used to compare anti-GPI immunoglobulin G (IgG) levels in donors from malaria-endemic areas with nonexposed individuals. The results revealed that malaria exposure can have a deep effect on the anti-GPI antibody level and on the reactivity pattern. Furthermore, the epitope of the monoclonal antibody MG96 was identified by a glycan array displaying 46 oligosaccharides. The malaria vaccine approach is based on blocking the development of the parasite in mosquito by using the MG96 antibody. Successful identification of the MG96 epitope has increased the potential for the development of a transmission-blocking vaccine [112, 113]. Glycan array data shows several oligosaccharides as MG96 ligands and as components of the putative antigenic epitopes.

15.3.2.3 Anthrax Similarly, glycan arrays have also been exploited in studying the immune-dominant epitopes of *Bacillus anthracis*—the bacterium that causes anthrax. The major glycoprotein on *B. anthracis* is known as BclA. One component of the BclA glycoprotein is a tetrasaccharide made up of three rhamnose residues with anthrose—a newly characterized carbohydrate residue—at the nonreducing end [114]. Denong Wang's group constructed a glycan array with 35 unmodified carbohydrate structures, including anthrose-containing structures. The array was probed with rabbit polyclonal anti-anthrax antibodies. Strong signals were observed for anthrose-containing tri- and tetrasaccharides. These results demonstrated that the carbohydrates are immune-dominant epitopes and that a trisaccharide is the minimal structure required for binding, pointing to the feasibility of developing carbohydrate-based anthrax vaccines consisting of BclA carbohydrate structures [115].

15.3.2.4 Cancers Glycan arrays have been further applied in cancer antigen characterization and vaccine development. Recently, Wong's group reported the utility of a glycan array to measure antibody levels to the tumor-associated glycan antigen Globo-H and its related structures [47, 116]. Two monoclonal antibodies (MBr1 and VK-9) bound the terminal tetrasaccharide to the same degree as the full Globo-H hexasaccharide, and the fucose residue was found essential for monoclonal antibody recognition. These results implied that a smaller oligosaccharide analogue may have the same immunogenic properties as the Globo-H hexasaccharide. Moreover, when compared with normal individuals, breast cancer patients have higher levels of anti-Globo-H antibodies in their blood, which points toward glycan array application in diagnosis based on the Globo-H antigen. Recently, this method has been applied to different cancer studies, including hepatocellular carcinoma [117] and nonmucinous ovarian cancers [118]. However, due to the complex serum reactivity profile and nonspecific binding with autoantibodies [119], further sample processing before analysis is needed in order to increase the possibility of usage for cancer diagnosis [120]. In contrast, cancer-associated glycans displayed as glycopeptides on glycan array have shown more promise in detection of immunogenic glycopeptide epitopes [121–123]. These results indicated that the peptide parts also play an important role for the induction of anticancer-associated glycan antibodies.

To discover new antigens, glycan arrays have been used to identify signals of disease-specific antibody among the collection of antibodies found in individuals. Generally speaking, human serum contains a wide variety of glycan-binding antibodies, and the populations of these antibodies change as a result of disease, exposure to pathogens, and vaccination. A high-throughput glycan array comprising 37 glycans to profile serum antibodies of classical Hodgkin's lymphoma patients was reported by Charles Lawrie et al. [124]. Tn expression levels was found significantly elevated in classical Hodgkin's lymphoma patients by using HBTn1 antibody stain. Thus, a Tn-based vaccine may be useful for classical Hodgkin's lymphoma patients.

15.3.2.5 Severe Acute Respiratory Syndrome Glycan arrays were applied to monitor immune responses to an inactivated severe acute respiratory syndrome coronavirus (SARS-CoV) vaccine and discovered autoimmunogenic activity of this human viral pathogen. Using glycan microarrays comprising of 51 glycans, Wang's group characterized the carbohydrate-binding activity of SARS-CoV neutralizing antibodies and found that an undesired autoantibody reactivity is present in SARS-CoV neutralization antibodies [125]. Significant levels of IgG antibodies to a human serum glycoprotein, asialo-orosomucoid (ASOR), were detected, but there was no detectable antibody reactivity to agalacto-orosomucoid (AGOR). The results indicate that the vaccine may induce a glycan-dependent autoimmune response because AGOR differs from ASOR in that only the terminal galactose is absent. These observations raised concerns on the use of whole virus-based SARS vaccine on humans because the immunological property of the vaccine remains largely uncharacterized. The same idea has also been used for *Clostridium difficile* detection [126].

15.3.2.6 Influenza To analyze influenza viruses based on their binding preferences, glycan arrays have become a powerful tool to help identify virus strains by monitoring changes in the receptor-binding preference profile [127, 128]. Receptor specificity for the influenza virus is usually controlled by the glycoprotein hemagglutinin (HA) on the virus surface. These viral HAs bind to host cell receptors containing terminal sialic acids. A glycan array bearing 260 glycans including many sialylated structures was applied to screen the binding pattern of several human H3N2 and avian H5N1 HAs. The results show that avian H5N1 viruses prefer binding to Neu5Ac(α2→3)Gal residues (Gal = galactose), while human H3N2 viruses are usually specific for Neu5Ac(α2→6)Gal residues [129–131]. Because each protein has a characteristic binding pattern, glycan array should be feasible to analyze influenza viruses and to map the evolution of new avian influenza based on their binding preferences. Glycan array also has proven to be invaluable in the early identification of epidemic-causing viruses.

It has been widely accepted that adaptation of avian influenza viruses for replication and transmission in the human host requires mutations in HA. With a modest change of two amino acid mutations on HA, the 1918 influenza pandemic switched its binding preference from the human α(2→6) to the avian α(2→3) sialic acid receptor [132, 133]. The same mutations that made the 1918 influenza contagious do not apply to H5N1, as it has a different HA protein. However, when inserted into recombinant HAs or intact viruses of some recent avian H5N1 isolates, 1957 H2N2 and 1968 H3N2 pandemic viruses cause analogous specificity changes [130]. Thus, these avian strains should be considered high risk because of their significantly lower threshold for acquiring human receptor specificity. More recently, Terrence Tumpey's group examined the glycan-binding properties of the 2009 A(H1N1) HA by using the recombinantly expressed soluble HA and found it exhibited a dose-dependent binding to only longer α(2→6) glycan (sialylated di-LacNAc) structure [134]. The binding behavior is very different from seasonal influenza H1N1 virus HA. Glycan array combined with the zanamivir–biotin conjugate as a probe can also be used to differentiate different influenza virus strains with or without Tamiflu resistance [135].

15.3.2.7 Xenotransplants Glycan arrays have also been used to profile antibody responses in xenotransplants. Anticarbohydrate antibodies are implicated in hyperacute rejection resulting from transplantation and transfusion of organs and tissues displaying incompatible carbohydrate antigens. Strong antibody responses, including responses to the α-Gal determinant, occur following islet and other types of cellular xenotransplants. The α-Gal carbohydrate determinant of porcine and other mammalian organs is considered to be the major epitope responsible for rejection of xenotransplants to primates [136, 137]. However, it is believed that non-α-Gal antibodies are also involved in the rejection of pig-to-human organ xenotransplants. Using a glycan array of 200 different structures, Blixt et al. investigated antibody responses to porcine cells in transplant patients [138]. The anti-α-Gal antibodies are the predominant antibodies after porcine islet-like cell cluster transplantation, with some patients developing *N*-glycolylneuraminic acid-specific antibodies as well. Their clinical significance still needs to be established.

15.3.2.8 Siglec-8 The sialic acid-binding immunoglobulin-like lectins (siglecs) are receptors on cells of the hemopoietic, immune, and nervous systems that recognize sialylated glycans with differing preferences for sialic acid linkage and oligosaccharide backbone sequence (see Chapter 18 for more discussion). Siglec-8 is expressed only on human mast cells, basophils, and eosinophils, giving it a unique expression pattern on effector cells of allergic disease. Activation via Siglec-8 could potentially be used to inhibit eosinophil survival *in vivo*, providing a strategy for reducing or inhibiting these cells in allergic and other diseases [139]. Beside a preference for sialic acid-bearing compounds, little was known about the function of Siglec-8. By screening a microarray containing 172 carbohydrates including 40 sialylated carbohydrates, the binding preference of Siglec-8 was examined [38]. One high-affinity-binding sugar, 6′-sulfo-sialyl Lewis X, was revealed. Furthermore, five siglecs, namely, human Siglec-7, Siglec-8, Siglec-9, and murine Siglec-2 and Siglec F, were determined using glycan arrays bearing 190 different carbohydrates [140]. Each siglec exhibited a specific binding pattern. It was further revealed that sulfation is an important modulator of siglec–sugar interactions.

15.3.3 Consortium for Functional Glycomics

The CFG—one of the best glycan array resources—was funded by the National Institutes of General Medical Sciences (http://www.functionalglycomics.org). The success of the CFG approach is a result of the combined efforts and generous input from many investigators in the field. Scientists can utilize the arrays in their own research and access the CFG database for viewing CFG raw and summary data. The CFG array has over 600 highly diverse glycans attached to NHS-derivatized glass microscope slides via amino-functionalized linkers. Some examples include the screening of the monoclonal antibody KEN-5 [141], Siglec-8 ligand [142], glycosyltransferase and sulfotransferase gene expression [143], and galectin-1 that recognizes high mannose on HIV-1 [144]. The mechanism of carbohydrate-dependent cancer metastasis was studied by screening the carbohydrate-binding activity and binding specificity of Sfrs1 protein, which may be responsible for fucosylated carbohydrate-dependent lung metastasis of epithelial cancers [145]. Comparison of the receptor-binding capabilities of natural and mutated human and avian influenza viruses was also conducted by CFG array [146]. The CFG arrays have become important tools in antigen detection and vaccine development and provided a wealth of information on ligand specificities for glycan-binding proteins.

15.4 CONCLUSION

The glycan array technology has become more mature and drawn a lot of scientific attention. However, several points are worth noting. First, cross-reactivity may cause misinterpretations during the analysis of biological samples such as serum. Second, the complexity of glycans on carbohydrate epitope environment is far beyond the glycans on the array. And finally, arraying pure glycans per spot may not be able to represent the real epitope environment. Development of methods in designing a glycan array that closely mimic the natural system is expected to attract more attention in the near future. Although there are big advancements in complex oligosaccharide synthesis by using programmable one-pot and automated solid-phase methods, no automated machine is available in the market now. To design a dream machine for oligosaccharide synthesis is a long-term interest in this field and will benefit the integrity of glycan arrays.

Employing mass spectrometry for glycan array quality control and monitoring the enzyme activity and specificity on the array surface will greatly improve glycan array applications. In addition, development of new inexpensive and user-friendly personal machines for glycan array detection will improve the capability of glycan arrays in early identification of emerging viruses to avoid new epidemics. When glycan arrays become commonly used tools for the serodiagnostic detection of pathogen-specific antibodies or human cancers, the information obtained from these assays will be invaluable in anti-infectious diseases or anticancer vaccine development and drug discovery.

REFERENCES

[1] Apweiler, R.; Hermjakob, H.; Sharon, N. *Biochim. Biophys. Acta* **1999**, *1473*, 4–8.

[2] Raman, R.; Raguram, S.; Venkataraman, G.; Paulson, J. C.; Sasisekharan, R. *Nat. Methods* **2005**, *2*, 817–824.

[3] Fukuda, M.; Hindsgaul, O., Eds. *Molecular and Cellular Glycobiology*, 2nd ed.; Frontiers in Molecular Biology 30; Oxford University: Oxford, **2000**.

[4] Seeberger, P. H. *Nature* **2005**, *437*, 1239.

[5] Thorson, J. S.; Vogt, T. In *Carbohydrate-Based Drug Discovery*; Wong, C.-H., Ed.; Wiley: Weinheim, **2003**, pp 685–711.

[6] Weis, W. I.; Drickamer, K. *Annu. Rev. Biochem.* **1996**, *65*, 441–473.

[7] Park, S.; Shin, I. *Angew. Chem. Int. Ed.* **2002**, *41*, 3180–3182.

[8] Shin, I.; Park, S.; Lee, M.-R. *Chem. Eur. J.* **2005**, *11*, 2894–2901.

[9] Houseman, B. T.; Mrksich, M. *Chem. Biol.* **2002**, *9*, 443–454.

[10] Wang, D.; Liu, S.; Trummer, B. J.; Deng, C.; Wang, A. *Nat. Biotechnol.* **2002**, *20*, 275–281.

[11] Fukui, S.; Feizi, T.; Galustian, C.; Lawson, A. M.; Chai, W. *Nat. Biotechnol.* **2002**, *20*, 1011–1017.

[12] Love, K. R.; Seeberger, P. H. *Angew. Chem. Int. Ed.* **2002**, *41*, 3583–3586.

[13] Willats, W. G. T.; Rasmussen, S. E.; Kristensen, T.; Mikkelsen, J. D.; Knox, J. P. *Proteomics* **2002**, *2*, 1666–1671.

[14] Ortiz Mellet, C.; García Fernández, J. M. *ChemBioChem* **2002**, *3*, 819–822.

[15] Paulson, J. C.; Blixt, O.; Collins, B. E. *Nat. Chem. Biol.* **2006**, *2*, 238–248.

[16] Feizi, T.; Fazio, F.; Chai, W.; Wong, C.-H. *Curr. Opin. Struct. Biol.* **2003**, *13*, 637–645.

[17] Liang, P.-H.; Wu, C.-Y.; Greenberg, W. A.; Wong, C.-H. *Curr. Opin. Chem. Biol.* **2008**, *12*, 86–92.

[18] Wu, C.-Y.; Liang, P.-H.; Wong, C.-H. *Org. Biomol. Chem.* **2009**, *7*, 2247–2254.

[19] Liang, C.-H.; Wu, C.-Y. *Expert Rev. Proteomics* **2009**, *6*, 631–645.

[20] Shang, R.; Fu, Y.; Wang, Y.; Xu, Q.; Yu, H.-Z.; Liu, L. *Angew. Chem. Int. Ed.* **2009**, *48*, 9350–9354.

[21] Oyelaran, O.; Gildersleeve, J. C. *Curr. Opin. Chem. Biol.* **2009**, *13*, 406–413.

[22] Campbell, C.; Gildersleeve, J. In *Carbohydrate Recognition: Biological Problems, Methods, and Applications*; Wang, B., Boons, G.-J., Eds.; Wiley: Weinheim, **2011**, pp 205–227.

[23] Wu, C.-Y.; Wong, C.-H. *Chem. Commun.* **2011**, *47*, 6201–6207.

[24] Lepenies, B.; Seeberger, P. H. *Immunopharmacol. Immunotoxicol.* **2010**, *32*, 196–207.

[25] Katrlík, J.; Švitel, J.; Gemeiner, P.; Kožár, T.; Tkac, J. *Med. Res. Rev.* **2010**, *30*, 394–418.

[26] Park, S.; Gildersleeve, J. C.; Blixt, O.; Shin, I. *Chem. Soc. Rev.* **2013**, *42*, 4310–4326.

[27] Dyukova, V. I.; Dementieva, E. I.; Zubtsov, D. A.; Galanina, O. E.; Bovin, N. V.; Rubina, A. Y. *Anal. Biochem.* **2005**, *347*, 94–105.

[28] Feizi, T.; Chai, W. *Nat. Rev. Mol. Cell Biol.* **2004**, *5*, 582–588.

[29] Bryan, M. C.; Plettenburg, O.; Sears, P.; Rabuka, D.; Wacowich-Sgarbi, S.; Wong, C.-H. *Chem. Biol.* **2002**, *9*, 713–720.

[30] Fazio, F.; Bryan, M. C.; Blixt, O.; Paulson, J. C.; Wong, C.-H. *J. Am. Chem. Soc.* **2002**, *124*, 14397–14402.

[31] Ko, K.-S.; Jaipuri, F. A.; Pohl, N. L. *J. Am. Chem. Soc.* **2005**, *127*, 13162–13163.

[32] Mamidyala, S. K.; Ko, K.-S.; Jaipuri, F. A.; Park, G.; Pohl, N. L. *J. Fluorine Chem.* **2006**, *127*, 571–579.

[33] Chen, G.-S.; Pohl, N. L. *Org. Lett.* **2008**, *10*, 785–788.

[34] Zhu, X.; Schmidt, R. R. *Angew. Chem. Int. Ed.* **2009**, *48*, 1900–1934.

[35] Chang, S.-H.; Han, J.-L.; Tseng, S. Y.; Lee, H.-Y.; Lin, C.-W.; Lin, Y.-C.; Jeng, W.-Y.; Wang, A. H. J.; Wu, C.-Y.; Wong, C.-H. *J. Am. Chem. Soc.* **2010**, *132*, 13371–13380.

[36] Galanina, O. E.; Mecklenburg, M.; Nifantiev, N. E.; Pazynina, G. V.; Bovin, N. V. *Lab Chip* **2003**, *3*, 260–265.

[37] Guo, Y.; Feinberg, H.; Conroy, E.; Mitchell, D. A.; Alvarez, R.; Blixt, O.; Taylor, M. E.; Weis, W. I.; Drickamer, K. *Nat. Struct. Mol. Biol.* **2004**, *11*, 591–598.

[38] Bochner, B. S.; Alvarez, R. A.; Mehta, P.; Bovin, N. V.; Blixt, O.; White, J. R.; Schnaar, R. L. *J. Biol. Chem.* **2005**, *280*, 4307–4312.

[39] Godula, K.; Bertozzi, C. R. *J. Am. Chem. Soc.* **2010**, *132*, 9963–9965.

[40] Chevolot, Y.; Bouillon, C.; Vidal, S.; Morvan, F.; Meyer, A.; Cloarec, J.-P.; Jochum, A.; Praly, J.-P.; Vasseur, J.-J.; Souteyrand, E. *Angew. Chem. Int. Ed.* **2007**, *46*, 2398–2402.

[41] Park, S.; Lee, M.-R.; Pyo, S.-J.; Shin, I. *J. Am. Chem. Soc.* **2004**, *126*, 4812–4819.

[42] Smith, E. A.; Thomas, W. D.; Kiessling, L. L.; Corn, R. M. *J. Am. Chem. Soc.* **2003**, *125*, 6140–6148.

[43] Harris, L. G.; Schofield, W. C. E.; Doores, K. J.; Davis, B. G.; Badyal, J. P. S. *J. Am. Chem. Soc.* **2009**, *131*, 7755–7761.

[44] Bain, C. D.; Biebuyck, H. A.; Whitesides, G. M. *Langmuir* **1989**, *5*, 723–727.

[45] Love, J. C.; Estroff, L. A.; Kriebel, J. K.; Nuzzo, R. G.; Whitesides, G. M. *Chem. Rev.* **2005**, *105*, 1103–1170.

[46] Schwarz, M.; Spector, L.; Gargir, A.; Shtevi, A.; Gortler, M.; Altstock, R. T.; Dukler, A. A.; Dotan, N. *Glycobiology* **2003**, *13*, 749–754.

[47] Wang, C.-C.; Huang, Y.-L.; Ren, C.-T.; Lin, C.-W.; Hung, J.-T.; Yu, J.-C.; Yu, A. L.; Wu, C.-Y.; Wong, C.-H. *Proc. Natl. Acad. Sci. U. S. A.* **2008**, *105*, 11661–11666.

[48] Wang, C.-C.; Chen, J.-R.; Tseng, Y.-C.; Hsu, C.-H.; Hung, Y.-F.; Chen, S.-W.; Chen, C.-M.; Khoo, K.-H.; Cheng, T.-J.; Cheng, Y.-S. E.; Jan, J.-T.; Wu, C.-Y.; Ma, C.; Wong, C.-H. *Proc. Natl. Acad. Sci. U. S. A.* **2009**, *106*, 18137–18142.

[49] Song, X.; Xia, B.; Stowell, S. R.; Lasanajak, Y.; Smith, D. F.; Cummings, R. D. *Chem. Biol.* **2009**, *16*, 36–47.

[50] Xia, B.; Kawar, Z. S.; Ju, T.; Alvarez, R. A.; Sachdev, G. P.; Cummings, R. D. *Nat. Methods* **2005**, *2*, 845–850.

[51] Lee, M.-R.; Shin, I. *Org. Lett.* **2005**, *7*, 4269–4272.

[52] de Paz, J. L.; Spillmann, D.; Seeberger, P. H. *Chem. Commun.* **2006**, 3116–3118.

[53] Tully, S. E.; Rawat, M.; Hsieh-Wilson, L. C. *J. Am. Chem. Soc.* **2006**, *128*, 7740–7741.

[54] Gama, C. I.; Tully, S. E.; Sotogaku, N.; Clark, P. M.; Rawat, M.; Vaidehi, N.; Goddard, W. A., III; Nishi, A.; Hsieh-Wilson, L. C. *Nat. Chem. Biol.* **2006**, *2*, 467–473.

[55] Bryan, M. C.; Fazio, F.; Lee, H.-K.; Huang, C.-Y.; Chang, A.; Best, M. D.; Calarese, D. A.; Blixt, O.; Paulson, J. C.; Burton, D.; Wilson, I. A.; Wong, C.-H. *J. Am. Chem. Soc.* **2004**, *126*, 8640–8641.

[56] Michel, O.; Ravoo, B. J. *Langmuir* **2008**, *24*, 12116–12118.

[57] Wendeln, C.; Heile, A.; Arlinghaus, H. F.; Ravoo, B. J. *Langmuir* **2010**, *26*, 4933–4940.

[58] Song, X.; Xia, B.; Lasanajak, Y.; Smith, D. F.; Cummings, R. D. *Glycoconj. J.* **2008**, *25*, 15–25.

[59] Ban, L.; Mrksich, M. *Angew. Chem. Int. Ed.* **2008**, *47*, 3396–3399.

[60] Mrksich, M. *ACS Nano* **2008**, *2*, 7–18.

[61] Gurard-Levin, Z. A.; Mrksich, M. *Annu. Rev. Anal. Chem.* **2008**, *1*, 767–800.

[62] Laurent, N.; Voglmeir, J.; Wright, A.; Blackburn, J.; Pham, N. T.; Wong, S. C. C.; Gaskell, S. J.; Flitsch, S. L. *ChemBioChem* **2008**, *9*, 883–887.

[63] Laurent, N.; Haddoub, R.; Flitsch, S. L. *Trends Biotechnol.* **2008**, *26*, 328–337.

[64] Trevor, J. L.; Lykke, K. R.; Pellin, M. J.; Hanley, L. *Langmuir* **1998**, *14*, 1664–1673.

[65] Gong, W.; Elitzin, V. I.; Janardhanam, S.; Wilkins, C. L.; Fritsch, I. *J. Am. Chem. Soc.* **2001**, *123*, 769–770.

[66] Lee, J.-C.; Wu, C.-Y.; Apon, J. V.; Siuzdak, G.; Wong, C.-H. *Angew. Chem. Int. Ed.* **2006**, *45*, 2753–2757.

[67] Northen, T. R.; Lee, J.-C.; Hoang, L.; Raymond, J.; Hwang, D.-R.; Yannone, S. M.; Wong, C.-H.; Siuzdak, G. *Proc. Natl. Acad. Sci. U. S. A.* **2008**, *105*, 3678–3683.

[68] Tseng, S. Y.; Wang, C.-C.; Lin, C.-W.; Chen, C.-L.; Yu, W.-Y.; Chen, C.-H.; Wu, C.-Y.; Wong, C.-H. *Chem. Asian J.* **2008**, *3*, 1395–1405.

[69] Park, S.; Shin, I. *Org. Lett.* **2007**, *9*, 1675–1678.

[70] Ratner, D. M.; Adams, E. W.; Su, J.; O'Keefe, B. R.; Mrksich, M.; Seeberger, P. H. *ChemBioChem* **2004**, *5*, 379–383.

[71] Zhi, Z.-L.; Powell, A. K.; Turnbull, J. E. *Anal. Chem.* **2006**, *78*, 4786–4793.

[72] de Boer, A. R.; Hokke, C. H.; Deelder, A. M.; Wuhrer, M. *Glycoconj. J.* **2008**, *25*, 75–84.

[73] Karamanska, R.; Clarke, J.; Blixt, O.; MacRae, J. I.; Zhang, J. Q.; Crocker, P. R.; Laurent, N.; Wright, A.; Flitsch, S. L.; Russell, D. A.; Field, R. A. *Glycoconj. J.* **2007**, *25*, 69–74.

[74] Ban, L.; Pettit, N.; Li, L.; Stuparu, A. D.; Cai, L.; Chen, W.; Guan, W.; Han, W.; Wang, P. G.; Mrksich, M. *Nat. Chem. Biol.* **2012**, *8*, 769–773.

[75] Guan, W.; Ban, L.; Cai, L.; Li, L.; Chen, W.; Liu, X.; Mrksich, M.; Wang, P. G. *Bioorg. Med. Chem. Lett.* **2011**, *21*, 5025–5028.

[76] Sanchez-Ruiz, A.; Serna, S.; Ruiz, N.; Martin-Lomas, M.; Reichardt, N.-C. *Angew. Chem. Int. Ed.* **2011**, *50*, 1801–1804.

[77] Beloqui, A.; Sanchez-Ruiz, A.; Martin-Lomas, M.; Reichardt, N.-C. *Chem. Commun.* **2012**, *48*, 1701–1703.

[78] Liang, C.-H.; Wang, C.-C.; Lin, Y.-C.; Chen, C.-H.; Wong, C.-H.; Wu, C.-Y. *Anal. Chem.* **2009**, *81*, 7750–7756.

[79] Shipp, M.; Nadella, R.; Gao, H.; Farkas, V.; Sigrist, H.; Faik, A. *Glycoconj. J.* **2008**, *25*, 49–58.

[80] Pedersen, H. L.; Fangel, J. U.; McCleary, B.; Ruzanski, C.; Rydahl, M. G.; Ralet, M. C.; Farkas, V.; von Schantz, L.; Marcus, S. E.; Andersen, M. C. F.; Field, R.; Ohlin, M.; Knox, J. P.; Clausen, M. H.; Willats, W. G. T. *J. Biol. Chem.* **2012**, *287*, 39429–39438.

[81] Han, E.; Ding, L.; Jin, S.; Ju, H. *Biosens. Bioelectron.* **2011**, *26*, 2500–2505.

[82] Gruber, K.; Horlacher, T.; Castelli, R.; Mader, A.; Seeberger, P. H.; Hermann, B. A. *ACS Nano* **2011**, *5*, 3670–3678.

[83] Fei, Y. Y.; Schmidt, A.; Bylund, G.; Johansson, D. X.; Henriksson, S.; Lebrilla, C.; Solnick, J. V.; Borén, T.; Zhu, X. D. *Anal. Chem.* **2011**, *83*, 6336–6341.

[84] Park, S.; Lee, M.-R.; Shin, I. *Nat. Protocols* **2007**, *2*, 2747–2758.

[85] Seibel, J.; Hellmuth, H.; Hofer, B.; Kicinska, A.-M.; Schmalbruch, B. *ChemBioChem* **2006**, *7*, 310–320.

[86] Blixt, O.; Allin, K.; Bohorov, O.; Liu, X.; Andersson-Sand, H.; Hoffmann, J.; Razi, N. *Glycoconj. J.* **2008**, *25*, 59–68.

[87] Bryan, M. C.; Lee, L. V.; Wong, C.-H. *Bioorg. Med. Chem. Lett.* **2004**, *14*, 3185–3188.

[88] Chokhawala, H. A.; Huang, S.; Lau, K.; Yu, H.; Cheng, J.; Thon, V.; Hurtado-Ziola, N.; Guerrero, J. A.; Varki, A.; Chen, X. *ACS Chem. Biol.* **2008**, *3*, 567–576.

[89] Blixt, O.; Han, S.; Liao, L.; Zeng, Y.; Hoffmann, J.; Futakawa, S.; Paulson, J. C. *J. Am. Chem. Soc.* **2008**, *130*, 6680–6681.

[90] Liang, P.-H.; Wang, S.-K.; Wong, C.-H. *J. Am. Chem. Soc.* **2007**, *129*, 11177–11184.

[91] de Paz, J. L.; Noti, C.; Böhm, F.; Werner, S.; Seeberger, P. H. *Chem. Biol.* **2007**, *14*, 879–887.

[92] Houseman, B. T.; Mrksich, M. *Angew. Chem. Int. Ed.* **1999**, *38*, 782–785.

[93] Zhi, Z. L.; Laurent, N.; Powell, A. K.; Karamanska, R.; Fais, M.; Voglmeir, J.; Wright, A.; Blackburn, J. M.; Crocker, P. R.; Russell, D. A.; Flitsch, S.; Field, R. A.; Turnbull, J. E. *ChemBioChem* **2008**, *9*, 1568–1575.

[94] Serna, S.; Etxebarria, J.; Ruiz, N.; Martin-Lomas, M.; Reichardt, N.-C. *Chem. Eur. J.* **2010**, *16*, 13163–13175.

[95] Serna, S.; Yan, S.; Martin-Lomas, M.; Wilson, I. B. H.; Reichardt, N.-C. *J. Am. Chem. Soc.* **2011**, *133*, 16495–16502.

[96] Gray, C. J.; Weissenborn, M. J.; Eyers, C. E.; Flitsch, S. L. *Chem. Soc. Rev.* **2013**, *42*, 6378–6405.

[97] Shivatare, S. S.; Chang, S.-H.; Tsai, T.-I.; Ren, C.-T.; Chuang, H.-Y.; Hsu, L.; Lin, C.-W.; Li, S.-T.; Wu, C.-Y.; Wong, C.-H. *J. Am. Chem. Soc.* **2013**, *135*, 15382–15391.

[98] Chuang, H.-Y.; Ren, C.-T.; Chao, C.-A.; Wu, C.-Y.; Shivatare, S. S.; Cheng, T.-J. R.; Wu, C.-Y.; Wong, C.-H. *J. Am. Chem. Soc.* **2013**, *135*, 11140–11150.

[99] Wang, C.-H.; Li, S.-T.; Lin, T.-L.; Cheng, Y.-Y.; Sun, T.-H.; Wang, J.-T.; Cheng, T.-J. R.; Mong, K. K. T.; Wong, C.-H.; Wu, C.-Y. *Angew. Chem. Int. Ed.* **2013**, *52*, 9157–9161.

[100] Huang, Y.-L.; Hung, J.-T.; Cheung, S. K. C.; Lee, H.-Y.; Chu, K.-C.; Li, S.-T.; Lin, Y.-C.; Ren, C.-T.; Cheng, T.-J. R.; Hsu, T.-L.; Yu, A. L.; Wu, C.-Y.; Wong, C.-H. *Proc. Natl. Acad. Sci. U. S. A.* **2013**, *110*, 2517–2522.

[101] Liao, S.-F.; Liang, C.-H.; Ho, M.-Y.; Hsu, T.-L.; Tsai, T.-I.; Hsieh, Y. S.-Y.; Tsai, C.-M.; Li, S.-T.; Cheng, Y.-Y.; Tsao, S.-M.; Lin, T.-Y.; Lin, Z.-Y.; Yang, W.-B.; Ren, C.-T.; Lin, K.-I.; Khoo, K.-H.; Lin, C.-H.; Hsu, H.-Y.; Wu, C.-Y.; Wong, C.-H. *Proc. Natl. Acad. Sci. U. S. A.* **2013**, *110*, 13809–13814.

[102] Burton, D. R.; Pyati, J.; Koduri, R.; Sharp, S. J.; Thornton, G. B.; Parren, P. W. H. I.; Sawyer, L. S. W.; Hendry, R. M.; Dunlop, N.; Nara, P. L.; Lamacchia, M.; Garratty, E.; Stiehm, E. R.; Bryson, Y. J.; Cao, Y.; Moore, J. P.; Ho, D. D.; Barbas, C. F., III *Science* **1994**, *266*, 1024–1027.

[103] Trkola, A.; Purtscher, M.; Muster, T.; Ballaun, C.; Buchacher, A.; Sullivan, N.; Srinivasan, K.; Sodroski, J.; Moore, J. P.; Katinger, H. *J. Virol.* **1996**, *70*, 1100–1108.

[104] Mascola, J. R.; Stiegler, G.; VanCott, T. C.; Katinger, H.; Carpenter, C. B.; Hanson, C. E.; Beary, H.; Hayes, D.; Frankel, S. S.; Birx, D. L.; Lewis, M. G. *Nat. Med.* **2000**, *6*, 207–210.

[105] Adams, E. W.; Ratner, D. M.; Bokesch, H. R.; McMahon, J. B.; O'Keefe, B. R.; Seeberger, P. H. *Chem. Biol.* **2004**, *11*, 875–881.

[106] Calarese, D. A.; Lee, H. K.; Huang, C. Y.; Best, M. D.; Astronomo, R. D.; Stanfield, R. L.; Katinger, H.; Burton, D. R.; Wong, C. H.; Wilson, I. A. *Proc. Natl. Acad. Sci. U. S. A.* **2005**, *102*, 13372–13377.

[107] Walker, L. M.; Huber, M.; Doores, K. J.; Falkowska, E.; Pejchal, R.; Julien, J.-P.; Wang, S.-K.; Ramos, A.; Chan-Hui, P.-Y.; Moyle, M.; Mitcham, J. L.; Hammond, P. W.; Olsen, O. A.; Phung, P.; Fling, S.; Wong, C.-H.; Phogat, S.; Wrin, T.; Simek, M. D.; Protocol G Principal Investigators; Koff, W. C.; Wilson, I. A.; Burton, D. R.; Poignard, P. *Nature* **2011**, *477*, 466–470.

[108] McLellan, J. S. et al. *Nature* **2011**, *480*, 336–343.

[109] Doores, K. J.; Burton, D. R. *J. Virol.* **2010**, *84*, 10510–10521.

[110] Pancera, M.; Shahzad-ul-Hussan, S.; Doria-Rose, N. A.; McLellan, J. S.; Bailer, R. T.; Dai, K.; Loesgen, S.; Louder, M. K.; Staupe, R. P.; Yang, Y.; Zhang, B.; Parks, R.; Eudailey, J.; Lloyd, K. E.; Blinn, J.; Alam, S. M.; Haynes, B. F.; Amin, M. N.; Wang, L.-X.; Burton, D. R.; Koff, W. C.; Nabel, G. J.; Mascola, J. R.; Bewley, C. A.; Kwong, P. D. *Nat. Struct. Mol. Biol.* **2013**, *20*, 804–813.

[111] Kamena, F.; Tamborrini, M.; Liu, X.; Kwon, Y.-U.; Thompson, F.; Pluschke, G.; Seeberger, P. H. *Nat. Chem. Biol.* **2008**, *4*, 238–240.

[112] Dinglasan, R. R.; Fields, I.; Shahabuddin, M.; Azad, A. F.; Sacci, J. B., Jr. *Infect. Immun.* **2003**, *71*, 6995–7001.

[113] Dinglasan, R. R.; Valenzuela, J. G.; Azad, A. F. *Insect Biochem. Mol. Biol.* **2005**, *35*, 1–10.

[114] Daubenspeck, J. M.; Zeng, H.; Chen, P.; Dong, S.; Steichen, C. T.; Krishna, N. R.; Pritchard, D. G.; Turnbough, C. L. *J. Biol. Chem.* **2004**, *279*, 30945–30953.

[115] Wang, D.; Carroll, G. T.; Turro, N. J.; Koberstein, J. T.; Kováč, P.; Saksena, R.; Adamo, R.; Herzenberg, L. A.; Herzenberg, L. A.; Steinman, L. *Proteomics* **2007**, *7*, 180–184.

[116] Huang, C.-Y.; Thayer, D. A.; Chang, A. Y.; Best, M. D.; Hoffmann, J.; Head, S.; Wong, C.-H. *Proc. Natl. Acad. Sci. U. S. A.* **2006**, *103*, 15–20.

[117] Wu, C.-S.; Yen, C.-J.; Chou, R.-H.; Li, S.-T.; Huang, W.-C.; Ren, C.-T.; Wu, C.-Y.; Yu, Y.-L. *PLoS One* **2012**, *7*, e39466.

[118] Jacob, F.; Goldstein, D. R.; Bovin, N. V.; Pochechueva, T.; Spengler, M.; Caduff, R.; Fink, D.; Vuskovic, M. I.; Huflejt, M. E.; Heinzelmann-Schwarz, V. *Int. J. Cancer* **2012**, *130*, 138–146.

[119] Bovin, N.; Obukhova, P.; Shilova, N.; Rapoport, E.; Popova, I.; Navakouski, M.; Unverzagt, C.; Vuskovic, M.; Huflejt, M. *Biochim. Biophys. Acta* **2012**, *1820*, 1373–1382.

[120] Obukhova, P.; Piskarev, V.; Severov, V.; Pazynina, G.; Tuzikov, A.; Navakouski, M.; Shilova, N.; Bovin, N. *Glycoconj. J.* **2011**, *28*, 501–505.

[121] Wandall, H. H.; Blixt, O.; Tarp, M. A.; Pedersen, J. W.; Bennett, E. P.; Mandel, U.; Ragupathi, G.; Livingston, P. O.; Hollingsworth, M. A.; Taylor-Papadimitriou, J.; Burchell, J.; Clausen, H. *Cancer Res.* **2010**, *70*, 1306–1313.

[122] Blixt, O.; Cló, E.; Nudelman, A. S.; Sørensen, K. K.; Clausen, T.; Wandall, H. H.; Livingston, P. O.; Clausen, H.; Jensen, K. J. *J. Proteome Res.* **2010**, *9*, 5250–5261.

[123] Blixt, O.; Bueti, D.; Burford, B.; Allen, D.; Julien, S.; Hollingsworth, M.; Gammerman, A.; Fentiman, I.; Taylor-Papadimitriou, J.; Burchell, J. M. *Breast Cancer Res.* **2011**, *13*, R25.

[124] Lawrie, C. H.; Marafioti, T.; Hatton, C. S. R.; Dirnhofer, S.; Roncador, G.; Went, P.; Tzankov, A.; Pileri, S. A.; Pulford, K.; Banham, A. H. *Int. J. Cancer* **2006**, *118*, 3161–3166.

[125] Wang, D. *Physiol. Genomics* **2004**, *18*, 245–248.

[126] Oberli, M. A.; Hecht, M.-L.; Bindschädler, P.; Adibekian, A.; Adam, T.; Seeberger, P. H. *Chem. Biol.* **2011**, *18*, 580–588.

[127] Lao, W.-I.; Wang, Y.-F.; Kuo, Y.-D.; Lin, C.-H.; Chang, T.-C.; Su, I.-J.; Wang, J.-R.; Chang, C.-F. *Future Med. Chem.* **2011**, *3*, 283–296.

[128] Liu, Y.; Childs, R. A.; Palma, A. S.; Campanero-Rhodes, M. A.; Stoll, M. S.; Chai, W.; Feizi, T. In *Carbohydrate Microarrays: Methods and Protocols*; Chevolot, Y., Ed.; Methods in Molecular Biology 808; Springer: New York, **2012**, pp 117–136.

[129] Stevens, J.; Blixt, O.; Glaser, L.; Taubenberger, J. K.; Palese, P.; Paulson, J. C.; Wilson, I. A. *J. Mol. Biol.* **2006**, *355*, 1143–1155.

[130] Stevens, J.; Blixt, O.; Chen, L.-M.; Donis, R. O.; Paulson, J. C.; Wilson, I. A. *J. Mol. Biol.* **2008**, *381*, 1382–1394.

[131] Stevens, J.; Blixt, O.; Paulson, J. C.; Wilson, I. A. *Nat. Rev. Microbiol.* **2006**, *4*, 857–864.

[132] Tumpey, T. M.; Maines, T. R.; Van Hoeven, N.; Glaser, L.; Solorzano, A.; Pappas, C.; Cox, N. J.; Swayne, D. E.; Palese, P.; Katz, J. M.; Garcia-Sastre, A. *Science* **2007**, *315*, 655–659.

[133] Srinivasan, A.; Viswanathan, K.; Raman, R.; Chandrasekaran, A.; Raguram, S.; Tumpey, T. M.; Sasisekharan, V.; Sasisekharan, R. *Proc. Natl. Acad. Sci. U. S. A.* **2015**, *105*, 2800–2805.

[134] Maines, T. R.; Jayaraman, A.; Belser, J. A.; Wadford, D. A.; Pappas, C.; Zeng, H.; Gustin, K. M.; Pearce, M. B.; Viswanathan, K.; Shriver, Z. H.; Raman, R.; Cox, N. J.; Sasisekharan, R.; Katz, J. M.; Tumpey, T. M. *Science* **2009**, *325*, 484–487.

[135] Cheng, T.-J. R.; Wang, S.-Y.; Wen, W.-H.; Su, C.-Y.; Lin, M.; Huang, W.-I.; Liu, M.-T.; Wu, H.-S.; Wang, N.-S.; Cheng, C.-K.; Chen, C.-L.; Ren, C.-T.; Wu, C.-Y.; Fang, J.-M.; Cheng, Y.-S. E.; Wong, C.-H. *Angew. Chem. Int. Ed.* **2013**, *52*, 366–370.

[136] Galili, U. R. I.; Tibell, A.; Samuetsson, B.; Rydberg, L.; Groth, C. G. *Transplantation* **1995**, *59*, 1549–1556.

[137] Korbutt, G. S.; Aspeslet, L. J.; Rajotte, R. V.; Warnock, G. L.; Ao, Z.; Ezekowitz, J.; Malcolm, A. J.; Koshal, A.; Yatscoff, R. W. *Xenotransplantation* **1996**, *3*, 207–216.

[138] Blixt, O.; Kumagai-Braesch, M.; Tibell, A.; Groth, C. G.; Holgersson, J. *Am. J. Transplant.* **2009**, *9*, 83–90.

[139] Nutku, E.; Aizawa, H.; Hudson, S. A.; Bochner, B. S. *Blood* **2003**, *101*, 5014–5020.

[140] Campanero-Rhodes, M. A.; Childs, R. A.; Kiso, M.; Komba, S.; Le Narvor, C.; Warren, J.; Otto, D.; Crocker, P. R.; Feizi, T. *Biochem. Biophys. Res. Commun.* **2006**, *344*, 1141–1146.

[141] Pospisil, R.; Kabat, J.; Mage, R. G. *Mol. Immunol.* **2009**, *46*, 2456–2464.

[142] Hudson, S. A.; Bovin, N. V.; Schnaar, R. L.; Crocker, P. R.; Bochner, B. S. *J. Pharmacol. Exp. Ther.* **2009**, *330*, 608–612.

[143] Trottein, F.; Schaffer, L.; Ivanov, S.; Paget, C.; Vendeville, C.; Cazet, A.; Groux-Degroote, S.; Lee, S.; Krzewinski-Recchi, M.-A.; Faveeuw, C.; Head, S. R.; Gosset, P.; Delannoy, P. *Glycoconj. J.* **2009**, *26*, 1259–1274.

[144] Krishnamoorthy, L.; Bess, J. W.; Preston, A. B.; Nagashima, K.; Mahal, L. K. *Nat. Chem. Biol.* **2009**, *5*, 244–250.

[145] Hatakeyama, S.; Sugihara, K.; Nakayama, J.; Akama, T. O.; Wong, S.-M. A.; Kawashima, H.; Zhang, J.; Smith, D. F.; Ohyama, C.; Fukuda, M.; Fukuda, M. N. *Proc. Natl. Acad. Sci. U. S. A.* **2009**, *106*, 3095–3100.

[146] Yen, H.-L.; Aldridge, J. R.; Boon, A. C. M.; Ilyushina, N. A.; Salomon, R.; Hulse-Post, D. J.; Marjuki, H.; Franks, J.; Boltz, D. A.; Bush, D.; Lipatov, A. S.; Webby, R. J.; Rehg, J. E.; Webster, R. G. *Proc. Natl. Acad. Sci. U. S. A.* **2009**, *106*, 286–291.

16

SYNTHESIS AND APPLICATIONS OF GLYCONANOPARTICLES, GLYCODENDRIMERS, AND GLYCOCLUSTERS IN BIOLOGICAL SYSTEMS

Avijit Kumar Adak, Ching-Ching Yu, and Chun-Cheng Lin
Department of Chemistry, National Tsing Hua University, Hsinchu, Taiwan

16.1 INTRODUCTION

Understanding the central role that glycans play in a wide range of biological events is crucial for developing new drugs and more effective therapeutic agents to treat human diseases [1, 2]. Glycans (free carbohydrates or carbohydrate fragments of glycoproteins and glycolipids) and their conjugates are of considerable importance for high-throughput diagnostics including the detection of viral and bacterial infections, cellular recognition events in proliferation and differentiation, and various complex biological processes such as immune response [3, 4]. The initial stages of a large number of biological and pathological processes are mediated by multiple ligand–receptor binding [5, 6]. Multivalency is ubiquitous and especially essential in the specific interactions between carbohydrate ligands and carbohydrate-binding proteins (lectins) [7, 8]. Unlike univalent binding, which is generally weak and unspecific, the interactions of glycan epitopes displayed in a multivalent array is remarkably enhanced in terms of their binding capacity and selectivity by the multipoint contacts between the clusters of glycans and proteins. Therefore, covalently attaching glycans to various macromolecular backbones to generate

Glycochemical Synthesis: Strategies and Applications, First Edition.
Edited by Shang-Cheng Hung and Medel Manuel L. Zulueta.

clusters of glycans (glycoclusters) was developed to mimic biological carbohydrate–lectin interactions. Optimization of the biological potency of glycoclusters is greatly influenced by several factors such as the number of glycan copies, sizes, and their architecture [7]. Consequently, well-defined multivalent glycomimetics of various architectures such as small glycoclusters, neoglycoproteins, glycopolymers, or glycodendrimers became valuable tools in glycobiology, especially in the context of multivalency, which is important for the investigation of carbohydrate–protein interactions.

16.2 SIGNIFICANCE OF MULTIVALENT BINDING INTERACTIONS IN BIOLOGICAL SYSTEMS

Because they are easily accessible, glycans displayed on the surface of mammalian cells provide enormous opportunities to bind to many microbial pathogens, ranging from viruses to molecular toxins and from pathogenic bacteria to parasites. In multivalent binding, multiple interactions between ligands and various receptors are common (Fig. 16.1). One representative example is ricin—a versatile and durable A–B-type toxin—in which one of the protein chains (the B chain) is a lectin that interacts and binds terminal galactose (Gal) on the surface of eukaryotic cells with multivalent interactions to facilitate entry of the other peptide chain (the A chain) into the cell to cause cellular death via the catalytic

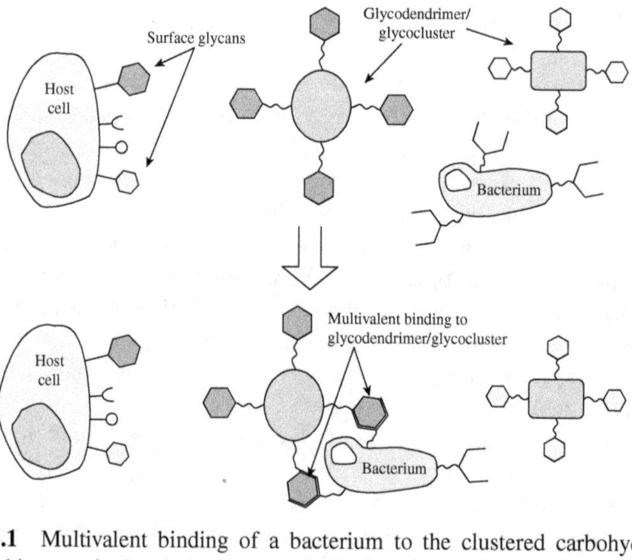

FIGURE 16.1 Multivalent binding of a bacterium to the clustered carbohydrate ligand carrying matching terminal carbohydrate residue(s) found in the host cell surface. The multivalent interaction prevents the attachment of the bacterium to the host cell surface. The functional role of the linker is the optimal presentation of the glycan into the binding site of the receptor, but it may also be involved in additional hydrophobic contacts.

deactivation of ribosomes. Viruses also specifically exploit the glycans of the host cell surface as binding epitopes in the first step of infection. For instance, in viral infection of influenza, the trimeric hemagglutinin protein selectively binds to multiple copies of sialic acid residues on the surface of the target cell [5]. Many pathogenic bacteria also utilize glycan recognition to adhere to the host cells prior to infection. The adhesion of *Escherichia coli* to gastrointestinal epithelial cells or the urinary tract is mediated by lectins present on its fimbriae. These lectins preferentially bind to the terminal carbohydrates of cell-surface glycoproteins and glycolipids. Also, among other biospecific interactions, antibodies divalently bind to specific antigens.

Although the affinity and specificity required for protein–carbohydrate interactions in such biological processes are essentially high, the monovalent interaction between a glycan and a protein typically has weak binding affinity (K_d values in the millimolar to micromolar range) and apparent relatively low selectivity. However, nature overcomes this issue by presenting multiple copies of the glycan on the glycoprotein scaffold or in the glycolipid, and the oligomer of the lectin or a lectin with multiple binding sites is displayed at the cell surface to achieve high-affinity binding. The binding affinities achieved by this multivalent effect are orders of magnitude higher than a monovalent binding event [5]. As a consequence of such multivalent binding sites, multiple distinct ligands often allow for numerous low binding events to occur simultaneously with the receptors, resulting in a high overall avidity; thus, such ligands might serve as (ant)agonists for multiple receptors. This enhancement in the binding affinity exhibited by a multivalent carbohydrate ligand is known as the "glycoside cluster effect" and was first recognized by Yuan C. Lee [7, 8].

In light of the aforementioned factors, a variety of diverse neoglycoconjugates with variable valency, topology, and modes of ligand presentation such as glyconanoparticles [9–11], glycoclusters, and glycodendrimers [12–16] have been introduced for amplified binding and improved carbohydrate–protein interactions. Due to their inherent high surface area-to-volume ratio, nanoparticles allow for greater contact surface area to improve multivalent interactions. These features made the self-assembly of carbohydrate monolayers on the surface of gold nanoparticles (AuNPs), termed glyco-AuNPs, as the most studied multivalent nanocarriers because of their distinct physical and chemical attributes [17]. Many excellent reviews have recently surveyed the chemistry and biology of these glyconanoparticles and provided in-depth discussions of the topic [10, 11]. The fundamental and classical aspects of nanomaterials, including iron oxide magnetic nanoparticles (MNPs) and AuNPs, have also been covered previously [17–21] and are not considered explicitly in this chapter. In this chapter, recent applications of glyconanoparticles, including MNPs and AuNPs, as probes for protein recognition and detection, cell targeting, and imaging are illustrated. Additionally, recent advances in the synthesis of glycodendrimers and glycoclusters, emphasizing on their impact on glycobiology, are briefly discussed. Finally, some specific examples of biological applications of these important classes of multivalent glycoconjugates are presented, focusing on achievements made in the last decade.

16.3 GLYCONANOPARTICLES, GLYCODENDRIMERS, AND GLYCOCLUSTERS: GENERAL OVERVIEW

Glyconanoparticles consist of an inorganic core of nanoscale dimensions (below 100 nm in size), and anchored in their surface is a flexible organic layer (linker or polymer) linked either covalently or noncovalently to glycans. The typical approach is to place the reducing end of terminal glycan residues onto the inorganic cores. Depending on the applications, different materials could be employed for the core, such as noble metal nanoparticles [17] or MNPs [21]. However, the conjugation chemistries used in the surface modification of nanoparticles are crucial in achieving ligand assembly. In addition, covalent coupling via amide bond formations and Cu(I)-catalyzed 1,2,3-triazole formations have also been employed for the surface modification of iron oxide MNPs [22, 23]. Traditional procedures for fabricating AuNPs from gold salts employ mild reducing agents and capping agents by modification of the Turkevich or Brust method [24, 25]. While the former used sodium citrate as both the reducing and capping agent in hot aqueous solution to produce colloids of 12–20 nm particle sizes, the latter utilized $NaBH_4$ in the presence of thiolated molecules as the capping ligands to yield AuNPs of 1–4 nm size. Thus, by changing the reaction conditions, the size, shape, and morphology of the AuNPs can be tuned. The synthesized AuNPs can be further modified by employing a ligand-exchange process using thiolated molecules, benefiting from the easy formation of the Au—S bond. Assembling the target molecules on the surface of AuNPs allows the resulting functionalized AuNPs to be dispersed in the appropriate solvent of interest.

By using nanomaterials as scaffold, multivalent interactions between glycan ligands and proteins have been demonstrated. The average size of MNPs is typically two orders of magnitude smaller than a bacterium, which leads to multiple nanoparticle attachments onto the cell surface. Moreover, MNPs have unique high magnetization via their superparamagnetic properties. These features make them easy to separate via magnets, avoiding contamination from samples. In addition to their high aqueous solubility, biocompatibility, and facile chemical modification, AuNPs have characteristic surface plasmon resonance (SPR) in the visible region (400–700 nm), which makes AuNPs useful as tools to study or monitor biological binding events. Soledad Penadés et al. were the first to use carbohydrate (lactose (Lac) and Lewis X)-encapsulated AuNPs in the study of carbohydrate–carbohydrate interactions [9].

Dendrimers are a unique class of multivalent macromolecules. In 1993, René Roy first coined the term "glycodendrimers" [26] to refer to glycosylated dendrimers, a versatile new class of compounds (Scheme 16.1). The attachment of terminal carbohydrate residues onto the dendrimer scaffolds generates a multivalent display of sugars that mimic cell-surface glycans. Glycodendrimers were aimed to fill the gap between glycopolymers and small glycoclusters in the study of multivalent carbohydrate–protein interactions. These glycodendrimers can be classified into three major types: carbohydrate-coated, carbohydrate-centered, and carbohydrate-based dendrimers (Fig. 16.2) [8, 12–16, 27, 28]. Glycoclusters are medium-sized glycoconjugates that are nonlinear but show fewer elements of symmetry than glycodendrimers (Scheme 16.1). Clustered carbohydrate arrays induce strong and

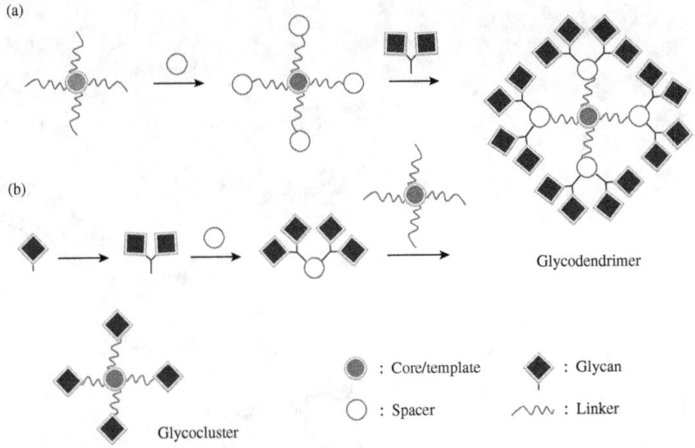

SCHEME 16.1 Structures of multivalent architectures and their general synthetic strategies. (a) Convergent approach. (b) Divergent approach.

specific carbohydrate recognition processes because of their multivalent carbohydrate arrays. Thisbe Lindhorst was the first to realize their potential in the synthesis of dendrimers and as cores for the construction of polyamidoamine (PAMAM)-type compounds [29].

The typical syntheses of glycodendrimers adopt two different strategies. In the convergent approach, the glyco-coated dendrons are synthesized first and then assembled on an oligofunctional core molecule (Scheme 16.1a). In contrast, in the divergent approach, the branching units are attached successively to the core/template, followed by linkages to the glycan moieties at the periphery (Scheme 16.1b). Given the instability associated with the linker and the requirements for the protection of glyco-coated dendrons, it is apparent why the convergent strategy is preferred. Numerous excellent review articles and book chapters covering synthetic advances in glycodendrimers and glycoclusters have been recently published [8, 12–16, 27, 28]. The representative examples [14, 30] of core-modified glycodendrimers for each class as mentioned earlier and the schematic illustration of the synthesis of multivalent architectures as shown in Figure 16.2 and Scheme 16.1, respectively, should provide a general idea for the wealth of possibilities.

Glyconanomaterials, which are nanoparticles carrying multiple glycan ligands, provide an excellent platform for studying carbohydrate-mediated biological processes. Because of their high surface-to-volume ratio, glyco-MNPs and glyco-AuNPs can display multiple copies of glycans to maximize the multivalent or glycoside cluster effects. Glycodendrimers and glycoclusters are also appealing candidates to obtain relevant affinities for studying carbohydrate–protein interactions. Furthermore, the size of the glycodendrimers can be varied in a controlled manner depending on how the dendrimers are generated. Most notably, neoglycoconjugates, which are composed of natural and synthetic glycan sequences representing the major glycans from cell-surface glycoproteins and glycolipids, were examined with several plant and human

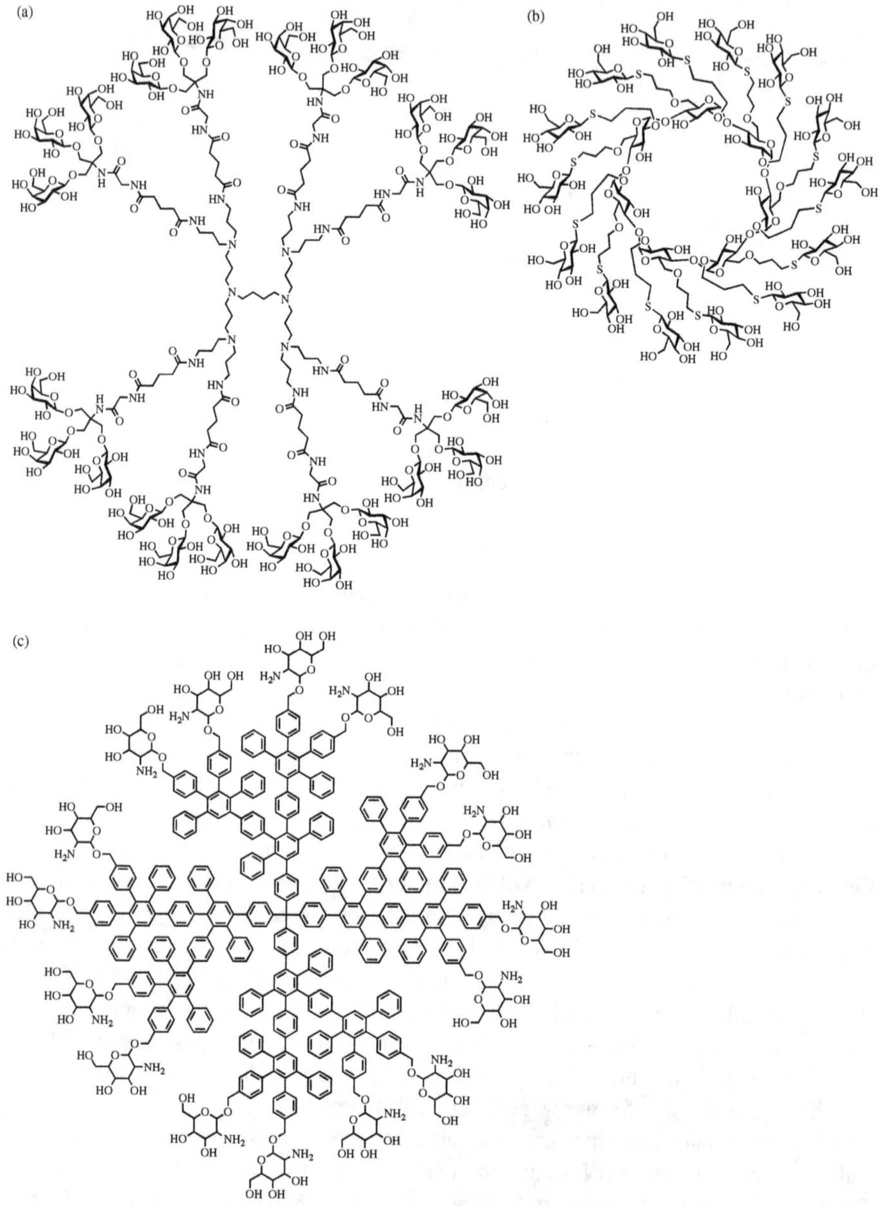

FIGURE 16.2 Molecular structures of different glycodendrimers. (a) A carbohydrate-coated dendrimer with 24 pendant monosaccharide residues. (b) A carbohydrate-centered dendrimer with a β-cyclodextrin (βCD) core and 14 copies of a glycosyl thiol residue. (c) A β-D-glucosamine-based glycodendrimer comprising 16 monosaccharide residues arranged around a tetravalent polyphenylene core.

lectins, glycan-specific antibodies, and bacterial and viral glycan-binding proteins. The collective data showed a wide variety of binding specificities, and these neoglycoconjugates have functioned as useful tools in studying carbohydrate-mediated recognition events.

16.4 PLANT LECTINS

Lectins are proteins known to interact with carbohydrates noncovalently but in a reversible and highly specific manner. Most plant lectins are simple and are formed as homo- or heterooligomers. The sizes of the subunits are moderate (molecular weight is usually below 40 kDa) and may contain more than one carbohydrate-binding site. Because many lectins have been well studied and are available in large quantities, they are invaluable tools for chemical biology and functioned as model systems in the study of carbohydrate–protein interactions [31, 32].

16.4.1 Concanavalin A

Pure concanavalin A (Con A) was first isolated from jack bean with binding specificity to α-mannosides and a related weak affinity to glucosides. At neutral pH, Con A is a homotetramer with binding sites separated by approximately 72 Å. Kimoon Kim et al. functionalized cucurbit[*n*]urils (CB[*n*], $n=5$–10) with various thioglycosides by the thiol–ene reaction [33]. Matrix-assisted laser desorption/ionization time-of-flight (MALDI-TOF) mass spectrometry analyses revealed that approximately 9–12 monosaccharides were attached to the CB[6] core, which has a diameter and thickness of 2.9 and 1.8 nm, respectively. The binding abilities of the mannosylated CB[6]-based glycocluster to Con A was investigated by turbidimetric assay and isothermal titration calorimetry (ITC). The latter method confirmed that the mannosylated glycocluster behaves predominantly as a trivalent ligand to the lectin with binding (association) constant (K_a) of $1.9\times10^5\,M^{-1}$, which is 25-fold higher than that with methyl α-D-mannopyranoside to Con A. Protein clustering is important in many cellular pathways and has been implicated in infection and proliferation of tumors and pathogens [5]. Mary Cloninger et al. discovered that an intrinsic fluorescence lifetime measurement could be used to characterize glycodendrimer-mediated protein aggregation in solution (Fig. 16.3) [34]. They employed mannose (Man)-coated PAMAM dendrimers with various degrees of functionalization to mediate unlabeled Con A aggregation. The changes in fluorescence were believed to be most likely associated with Con A–Con A interactions mediated by binding to the glycodendrimer framework.

Recently, Nicoleta Bogdan et al. demonstrated a quantitative analysis of the binding constant between a labeled lectin and glycodendrimer-coated AuNPs using the surface energy transfer (SET) process [35]. A water-soluble glycodendrimer (Man-functionalized PAMAM-based [G0] cystamine core) was coated onto 1.5 nm AuNPs (Au-Man) and used as a biosensor in SET optical signaling to monitor protein–carbohydrate interactions (Fig. 16.4a). The signal resulting from the SET

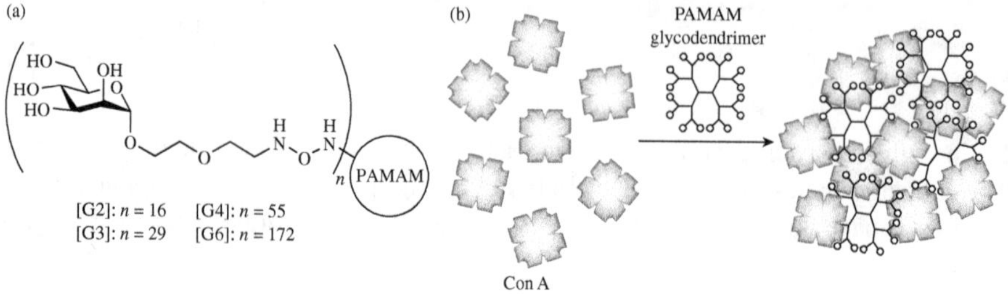

FIGURE 16.3 Glycodendrimer-mediated protein aggregation in solution. (a) Structures of the Man-coated PAMAM dendrimers with G values referring to various dendrimer generations. (b) Transformation of uncomplexed Con A to the cross-linked state in the presence of a glycodendrimer.

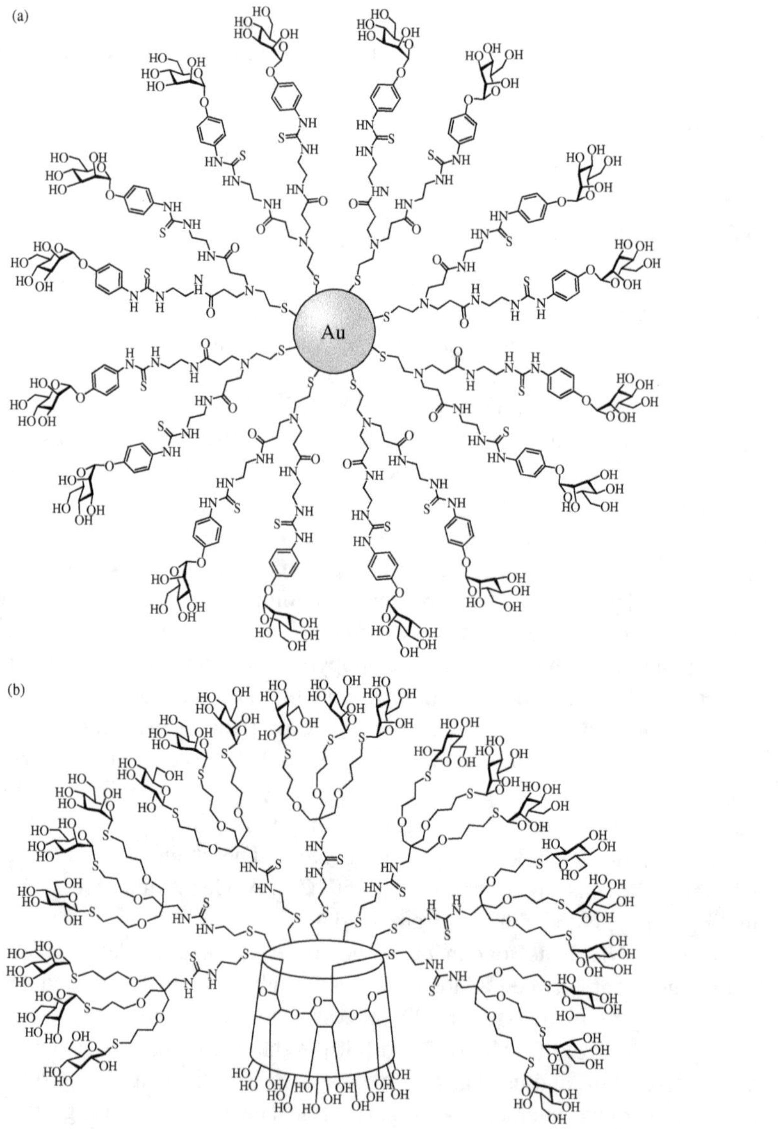

FIGURE 16.4 (a) Glycodendrimer-coated gold nanoparticles used for protein detection based on the SET process. (b) A representative heteroglycocluster on a βCD core used to investigate comparative binding on lectin–carbohydrate interactions.

between the fluorescein isothiocyanate (FITC)-labeled Con A and the nanometal surface of Au-Man was used to quantify the binding constant. In this assay, Con A bind to Au-Man with $K_a = 5.6 \times 10^6\,M^{-1}$, which is 100-fold higher than that with glycodendrimer [G2]. Alternatively, Peter Seeberger et al. used fluorescent Ru(II) complex-conjugated glycodendrimers to sense unlabeled lectins with the photoinduced electron transfer process [36]. The results demonstrated that a hexamannose dendrimer bearing a highly fluorescent derivative of tris(bipyridine)ruthenium(II) ($[Ru(bipy)_3]^{2+}$) can sense Con A at concentrations as low as 25 nM.

In addition, José Blanco et al. studied the heterocluster effect by comparing the interactions of low- and high-density homo- and heteroglycoclusters with Con A acting as model lectin [37]. A series of molecularly well-defined homo- and heteromultivalent glycoclusters containing 7–21 α-mannopyranosides and β-glucopyranosides at the periphery of βCD was synthesized (Fig. 16.4b). The binding affinities were assessed using enzyme-linked lectin assay (ELLA), ITC, and SPR. The results showed that high-density glycoclusters amplified the Con A binding strength compared to their low-density counterparts. Furthermore, highly dense functionalized homo- and heteroglycoclusters similarly enhanced the affinity for Con A binding, supporting the heterocluster effect (synergic effect) caused by the secondary interactions of the nonactive structural motifs.

Sébastien Gouin and others used click chemistry to conjugate several alkynated Man ligands on azido-terminated carbohydrate scaffolds to form glycoclusters that contain valencies ranging from 1 to 4 and differential spatial arrangements [38]. The binding affinities of the synthetic glycoclusters with Con A evaluated by ELLA revealed a remarkable cluster effect that enhances the relative binding potency, varying from 8.1 to 9.1 per Man unit. They also performed molecular dynamics simulation to rationalize the results obtained that was found to be consistent with the ELLA observations.

Although the binding affinities of glycoconjugates toward lectins were routinely measured in solution by agglutination inhibition assays, the affinities are predicted to be different by several orders of magnitude for surface-based multivalent ligand–receptor interactions relative to those involving disperse soluble species. Ricardo Riguera et al. highlighted comparative studies of carbohydrate–lectin binding in solution and surface-bound direct measurements by using SPR (Fig. 16.5) [39]. The results indicated that the affinities of Man to Con A increased with the generation of the dendrimer and [G3]-Man was the best binder with an affinity enhancement of 13.8-fold and 3100-fold on a per sugar basis relative to methyl α-D-mannoside for the competitive (solution-based) binding and SPR direct binding (surface-based) experiments, respectively.

As mentioned earlier, glyco-AuNPs exhibit strong glycoside cluster effect that enhanced the binding affinity of carbohydrate to lectin. However, the quantitative analysis of multivalent affinities at nanoparticle surfaces still represents a challenge. Mingdi Yan et al. measured the binding affinity of Man-AuNPs with Con A by fluorescent competition binding assay [40]. Unmodified glycans were photochemically conjugated onto AuNPs, followed by quantitative evaluation of the effect of ligand density, linker length, and nanoparticle size on the affinity enhancement

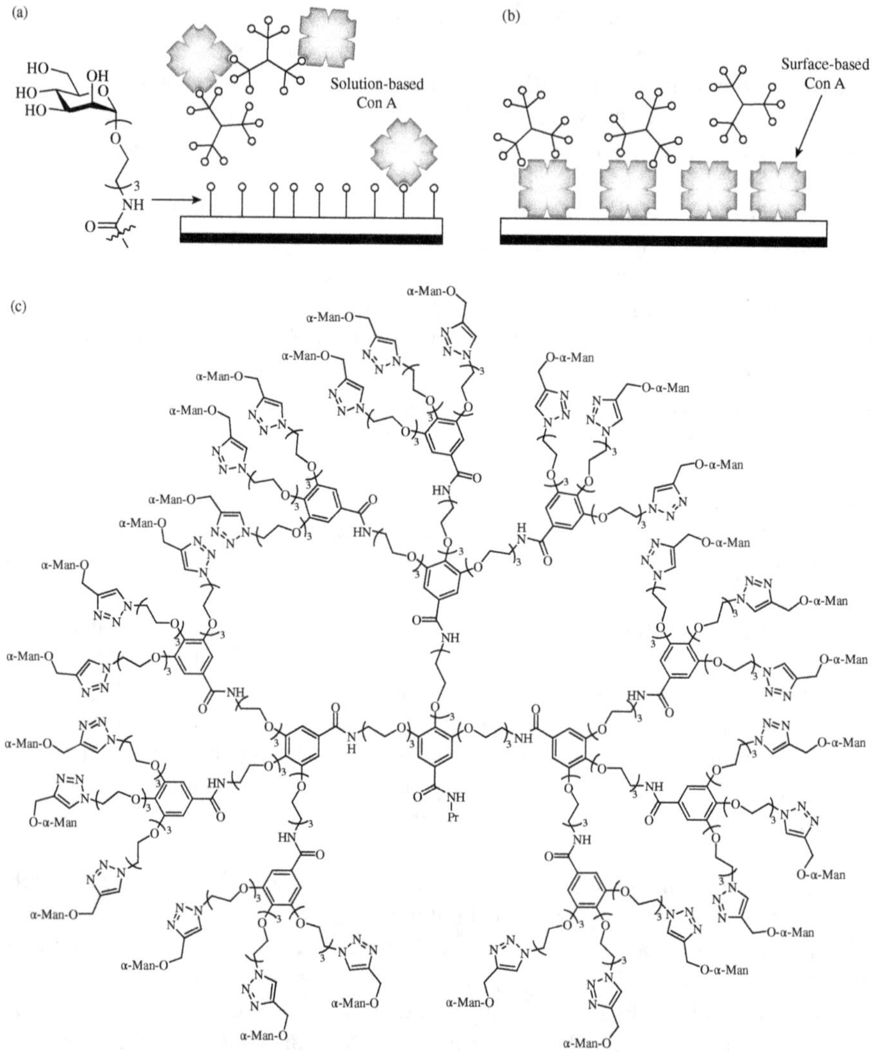

FIGURE 16.5 A schematic representation of (a) SPR competitive and (b) direct binding assays of Con A–Man recognition. (c) The structure of clicked [G3]-Man glycodendrimer used for studying the glycoside cluster effect.

toward Con A. The specific interactions between Con A and α-linked Man and glucose (Glc) derivatives are well defined and constitute an excellent model system for investigating the multivalent effect. In this context, Chun-Cheng Lin et al. quantitatively measured the binding affinities of glyco-AuNPs with Con A using the SPR technique [41]. Various types of carbohydrate-functionalized (Man, Glc, and Gal) AuNPs of different particle sizes and linker lengths were synthesized as multivalent Con A ligands. The SPR studies indicated that AuNP of ca. 20 nm diameter with a relatively short five-carbon spacer decorated with ca. 680 Man residues at the periphery was found to be the most potent and was about 128-fold stronger inhibitor than methyl α-D-mannopyranoside on the surface.

The high surface area-to-volume ratio of AuNPs allowed for improved SPR sensitivity and colorimetric changes [20]. Chao-Tsen Chen et al. developed a competitive colorimetric assay using Man-AuNPs to identify the binding interaction with Con A [42]. The assay sensitivity is excellent, and a Con A concentration of 5 nM could be detected within 2 min by the naked eye. Similarly, David Russell et al. utilized Man-functionalized silver nanoparticles and AuNPs for the colorimetric detection of Con A with a superior sensitivity of the latter at 0.04 μM [43]. Recently, Huan-Tsung Chang et al. prepared fluorescent nanometer-sized (~1.8 nm) Au nanodots stabilized with Man residues and utilized them as probes in lectin sensing [44]. The results showed that Man-Au nanodots can provide an impressive sensitivity for Con A at concentrations as low as 75 pM. Robert Field et al. introduced thioctic acid at the reducing end of the α-mannoside residue to prepare Man-AuNPs [45]. The use of thioctic acid as a linker conferred low nonspecific protein binding and increased the stability of the glyco-AuNPs in solution. Moreover, Huangxian Ju et al. demonstrated the aggregation of glyco-AuNPs as a label-free scanometric method for the *in situ* monitoring of cell-surface carbohydrates [46]. As shown in Figure 16.6, in the presence of Con A as a model lectin, the lectin-induced Man-AuNP aggregation could be inhibited by the specific binding of the lectin to the cell-surface carbohydrates. As a result, there was a net decrease in the free lectin concentration in the glyconanoparticle recognition system. They used the silver enhancement technique to amplify the aggregation signal. Thus, the method could be used to determine the cell-surface densities of different carbohydrates without the need for pretreatment and labeling.

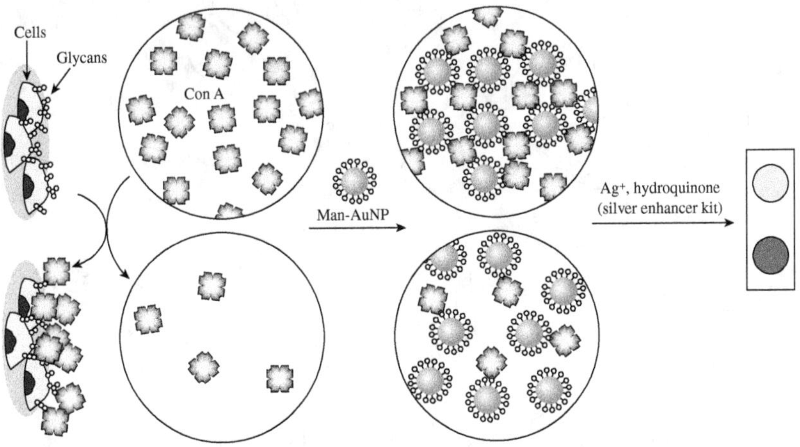

FIGURE 16.6 Scanometric strategy for the in situ detection of Man groups on living cells.

16.4.2 Peanut Agglutinin

Peanut agglutinin (PNA) is a Gal-specific lectin composed of four identical subunits [31]. This lectin has an asymmetric structure, and the distances between the two binding sites are 54Å and 74Å, respectively. Recently, a βCD-scaffold appended with α-Man and β-Lac antennae was synthesized as heterogeneous glycocluster in a controlled manner and used to probe carbohydrate–protein recognition with two different lectins in heterogeneous environments [47]. The highly dense sugar patches orient toward the same face of the βCD cone, and the resulting architecture is very compact with a diameter of ca. 6 nm and covered by 21 sugar valency (α-Man + β-Lac) (Fig. 16.7). The ELLA results indicated that the β-Lac units in the heterogeneous glycocluster strengthen the binding of α-Man to Con A, although a control experiment showed that homogeneous β-Lac clusters are not recognized by this lectin. Conversely, the presence of α-Man motifs resulted in a substantial enhancement in the binding of β-Lac toward PNA, a lectin known to have strong affinity to lactosides but with no detectable affinity for α-Man. The results of the turbidimetric assay are also in agreement with the previous observations. These data clearly suggest that the heterocluster effect is not only restricted to Con A but can also influence the binding mechanism of other lectins.

Similarly, a turbidity assay can be used to monitor the formation of cross-linked complexes in real time. Recently, Xiao-Liu Li et al. used Lac-functionalized perylene bisimide (PBI-Lac) to form a right-handed supramolecular stacking induced by the chiral Lac moieties. In the turbidity assay, the PNA lectin was applied to investigate the formation of three-dimensional chiral aggregates [48]. Upon addition of PBI-Lac

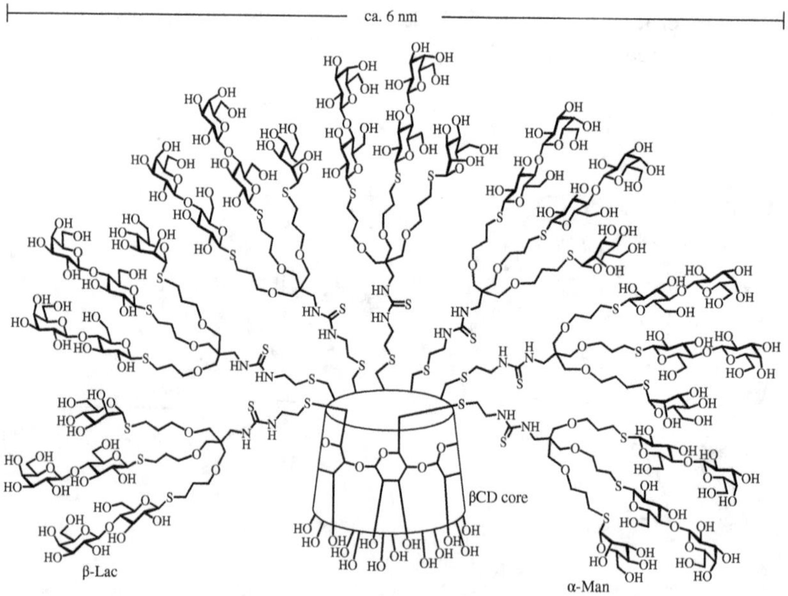

FIGURE 16.7 Structure of the βCD-scaffolded heteroglycocluster used to study the heterocluster effect.

to a solution of PNA in phosphate-buffered saline, an immediate increase in turbidity was observed, which is an indication of carbohydrate–protein interaction. Lokesh Joshi et al. measured carbohydrate–lectin interactions by using label-free electrochemical impedance spectroscopy [49]. The PNA lectin was covalently immobilized on the electrodes of a layered Cu/Ni/Au printed circuit board, and subsequently, TF-antigen-coated AuNPs were run through the PNA-immobilized electrodes. The detection sensitivity of the PNA-immobilized electrode was found to be 1 pg/ml for the TF-AuNPs.

16.4.3 Wheat Germ Agglutinin

Wheat germ agglutinin (WGA) is a dimeric protein that specifically recognizes *N*-acetylglucosamine (GlcNAc). This lectin has eight binding sites for GlcNAc, and the carbohydrate recognition sites are separated by distances varying from approximately 14 to 53 Å. Daiyo Terunuma et al. synthesized lactotriaose (GlcNAc(β1→3)Gal(β1→4)Glc)-containing carbosilane dendrimers with 3, 4, 6, and 12 valences and evaluated their WGA binding by fluorescence methods [50]. The hexavalent glycodendrimer binds most strongly with a maximum K_a of $2.8 \times 10^6 M^{-1}$, which is 2500-fold higher than the binding affinity of free lactotriaose.

In addition to functioning glycodendrimers as inhibitors for WGA, glycoclusters have also been used to investigate the glycoside cluster effect. Alessandro Dondoni et al. discovered that an octasilsesquioxane scaffold appended with GlcNAc units showed a striking glycoside cluster effect toward WGA [51]. They synthesized the glycoclusters by reacting glycosyl thiols with octasilsesquioxane derivatives, which display eight polyethylene glycol chains, using thiol–ene coupling. The specificities of the resulting glycoclusters to lectins were evaluated by ELLA. The results indicated a modest binding to Con A when Man and Glc clusters were used as ligands; however, a high affinity toward WGA was observed with an IC_{50} at 3.0 nM using an octasilsesquioxane scaffold featuring eight GlcNAc units. The relative inhibition potency per number of sugar units was measured as 10^6, a value that supports the occurrence of glycoside cluster effect.

16.4.4 *Ricinus communis* Agglutinin 120

Ricinus communis agglutinin 120 (RCA_{120}) is a plant lectin isolated from the seeds of castor bean. RCA_{120} is related to ricin, but unlike ricin, it does not deactivate protein synthesis and is therefore less cytotoxic. However, RCA_{120} is composed of two ricin-like subunits (A chain and B chain) and is therefore a good surrogate for developing an assay for ricin. Both RCA_{120} and ricin specifically recognize and bind to Lac and Gal. However, both lectins show a low monovalent binding affinity to Gal with a K_a value between 10^3 and $10^4 M^{-1}$ and $2.2 \times 10^3 M^{-1}$ for ricin and RCA_{120}, respectively [31]. Kazunori Kataoka et al. described carbohydrate–lectin binding measurements using colorimetric assay [52]. Lac-stabilized AuNPs were prepared, and then, the aggregation of particles was measured by adding RCA_{120} to induce a color change. Russell, Field, and coworkers applied the same colorimetric method but used

Gal-AuNPs for RCA_{120} detection [53]. The Gal-AuNPs could detect RCA_{120} concentrations as low as 9 nM. In addition, Lee et al. prepared water-soluble and RCA_{120}-recognizable carbohydrate-centered glycoclusters by using the Huisgen 1,3-dipolar cycloaddition reaction between an alkynated β-Gal core and azido-Lac [54]. The lectin binding affinities of the glycoclusters examined by using capillary affinity electrophoresis revealed that the glycocluster showed a 400-fold stronger inhibitory effect on RCA_{120} compared with monovalent Lac.

16.4.5 Other Plant Lectins

A hexadecavalent cyclopeptide-based glycocluster, RK-αMan_{16}, was shown to be a more potent inhibitor (IC_{50} = 0.7 μM) to Con A over the monovalent binder as determined by ELLA [55]. Recently, Taichi Usui et al. synthesized tetravalent sialoglycoclusters on an ethylene glycol bis(β-aminoethyl ether)-*N,N,N′,N′*-tetraacetate backbone as ligands for the sialic acid-binding lectin *Sambucus sieboldiana* agglutinin [56]. The tetravalent sialoglycocluster showed a 16-fold greater inhibitory effect than that of Neu5Ac(α2→6)*N*-acetyllactosamine (Neu5Ac = *N*-acetylneuraminic acid) as determined by the hemagglutinin inhibition assay (HIA). Mary Cloninger et al. used Man-functionalized PAMAM dendrimers to study the lectin–carbohydrate interaction using a mitogenic pea (*Pisum sativum*) lectin [57]. HIA measurements indicated binding of the dendrimer in a monovalent fashion to the pea lectin, with no measurable increase in binding affinity as the dendrimer size increases up to the sixth generation. The unfavorable shape complementarity between the dendrimer and the pea lectin was the suggested reason for the monovalency as the glycoside cluster effect is clearly observable in the case of the same dendrimers and Con A.

16.5 AB_5 TOXINS

The AB_5 toxins are important virulence factors for various major bacterial pathogens. Their catalytic A subunit is responsible for the disruption of essential host functions. The B subunits recognize and bind the glycan receptors displayed on the surface of eukaryotic cells and facilitate toxin transport into the host cell—the vital first step of their infection [58]. Inhibitors that block this interaction can serve as therapeutic agents for bacterial infections (Fig. 16.1). Despite sharing similar structural architectures, AB_5 toxins vary from one family to another. Within the same family, they can differ in their host cell-surface receptor specificity. The B pentamers of the cholera toxin produced by *Vibrio cholerae* is closely related to the heat-labile enterotoxins (LT) produced by the enterotoxigenic *E. coli* and specifically recognize the oligosaccharide moiety of GM1, a ganglioside present on the surface of eukaryotic cells [58]. However, there are other serogroups of the LT toxin, namely, LT-IIa and LT-IIb, the B subunits of which show affinity for gangliosides GD1b and GD1a, respectively [59]. Furthermore, the Shiga toxins produced by *Shigella dysenteriae* and the Shiga-like toxins (SLT-I and SLT-II) generated by *E. coli* bind the oligosaccharide component of the glycosphingolipid globotriaosyl (Gb_3) ceramide [60]. Receptor specificity is critical for the pathogenic process because it determines host susceptibility. Because the AB_5 toxins can bind to multiple glycans simultaneously, they are ideal targets for multivalent ligand interaction. By appending glycans on the surface of dendrimers, the resulting glycodendrimers can present antimicrobial activities [61].

16.5.1 Cholera Toxin

Upon entry of *V. cholerae* into the epithelial cells lining the intestine, the secreted cholera toxin causes diarrhea, massive dehydration, and, occasionally, death of the affected individual if left untreated. LT from *E. coli* is structurally similar to cholera toxin and causes traveler's diarrhea upon exposure to cells in the intestine. Roland Pieters et al. investigated the affinity of the GM1-coated dendrimer against cholera toxin using enzyme-linked immunosorbent assay (ELISA). The oligo-GM1-linked glycodendrimers showed an unprecedented binding enhancement that is 380,000-fold stronger than that observed by using the corresponding monovalent analogue [62]. Russell et al. prepared Lac-AuNPs and then combined with colorimetric changes to detect cholera toxin [63]. The B subunit of cholera toxin binds to the Lac segment and then induces a color change of the Lac-AuNPs due to the aggregation of the nanoparticles. The Lac-AuNP-based sensor was able to detect cholera toxin at the low concentration of 54 nM within 10 min of incubation.

16.5.2 Shiga-Like Toxins

David Bundle et al. reported the most striking examples of affinity enhancement with a designed multivalent ligand for the lectin subunits of verotoxin from *E. coli* O157 [64]. The STARFISH ligand, which represents two copies of Gb_3 at the ends of each of its five arms emanating from a glucose core, bridges the binding sites in all five B subunits of the toxin complex simultaneously (Fig. 16.8a). Although the STARFISH ligand was not described as a dendrimer, conceptually, it has all of the structural features of a dendrimer. The ligand also appears to interact at the secondary binding sites with a subnanomolar K_d that is more than a millionfold higher than the mono-P^k-trisaccharide ligand. Moreover, Lin et al. prepared Gb_3-coated AuNPs (Gb_3-AuNPs) with different particle sizes ranging from 4 to 20 nm and containing from 60 to 2000 sugars on the surface (Fig. 16.8b) [65]. These Gb_3-AuNPs were examined as inhibitors of the B subunit of SLT-I by using SPR. The results revealed that Gb_3-AuNPs are very potent inhibitors with the relative potencies of each sugar ranging from 1300 for the 4 nm particle with shorter linker to 228,000 for the 20 nm particle with longer linker. The strong affinity of Gb_3-AuNP with the B subunit of SLT-I was further employed to develop a chip-based detection of the toxin with detection sensitivity at 1 μg/ml.

Later, Suri Iyer et al. synthesized similar Gb_3-AuNPs but with different length of linkers and surface glycan densities [66]. They conducted luciferase assay to measure the toxin-mediated inhibition of protein synthesis and found that Gb_3-AuNPs were nontoxic to Vero monkey kidney cells and protected Vero cells from Shiga toxin in a dose-dependent manner.

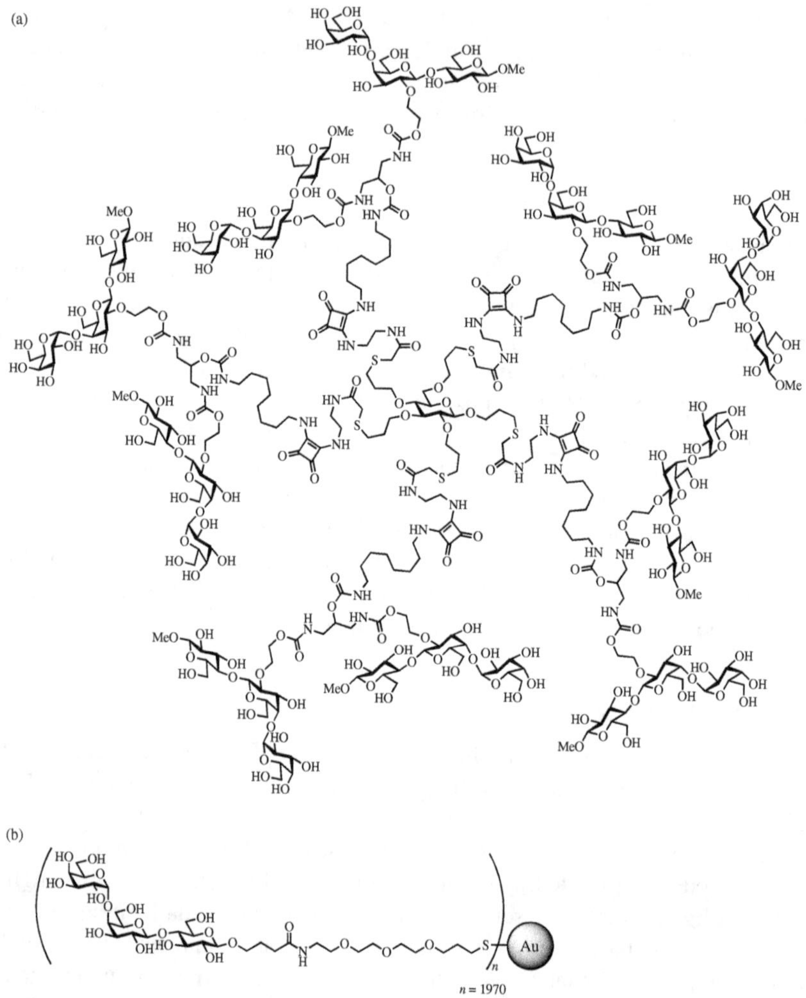

FIGURE 16.8 Gb_3 coated on (a) a carbohydrate core and (b) a gold nanoparticle used to study the binding interactions with Shiga-like toxin.

16.6 BACTERIAL ADHESION LECTINS

Microbial pathogens utilize different types of lectins for targeting the glycans on the surface of host cells. Many bacteria are covered with pili or fimbriae that contain a very special class of lectins known as adhesins because they play a role in attachment to epithelial cells. These lectins are monomeric and comprise only one binding site. Because the adhesins are repeated on the pilus, a larger number of adhesins on the bacterial surface create multivalent interaction with the host glycans.

(a)

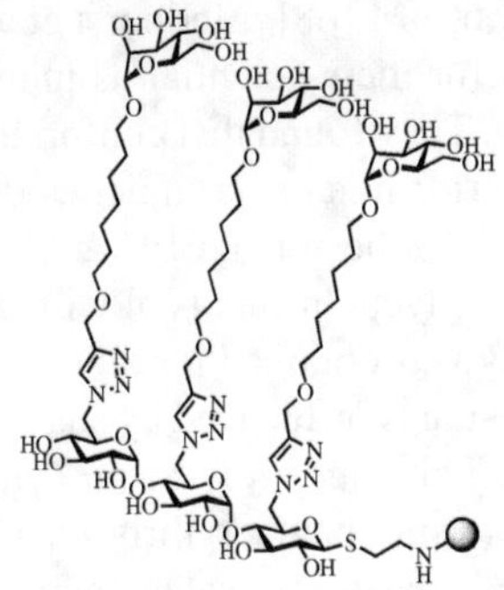

(b)

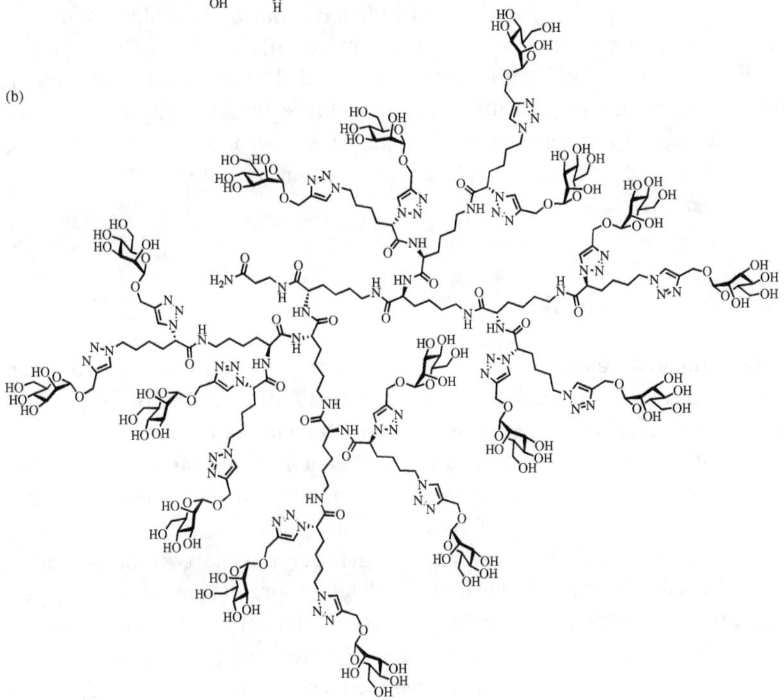

FIGURE 16.9 Selected mannosylated glycocluster based on (a) a carbohydrate and (b) an L-lysine core used as a potential antagonist against *Escherichia coli* FimH.

16.6.1 FimH Adhesin

The adhesion protein FimH mediates the attachment of uropathogenic *E. coli* strains with the host cell glycocalyx and specifically recognizes mannosylated structures. In 2002, Lin et al. first demonstrated that the glyco-AuNP can be used as a probe for staining the binding protein on the cell surface through carbohydrate–receptor interactions [67]. Man-AuNPs were used to visualize the FimH adhesins on the type I pili of *E. coli* via transmission electron microscopy.

Gouin et al. described a heptavalent α-Man-based glycocluster (Fig. 16.9a) as a strong inhibitor to block the hemagglutination of type 1 fimbrial adhesins of *E. coli*

(UTI89) to the bladder cells with a titer value of 60 nM [68]. Lindhorst et al. evaluated spacer-modified oligomannoside mimetics for their potential as inhibitors of type 1 fimbriae-mediated bacterial adhesion [69]. They found that conformationally flexible glycodendrons decorated with four exterior mannosyl moieties displayed on thiahexyl spacers are good inhibitors of bacterial adhesion in ELISA. The studies were later extended to cysteine (Cys)-based Man glycoclusters [70] with 2–4 Man units. Although only moderate inhibitory potency was obtained for all of the glycoclusters tested, they suggested that bacterial adhesion is influenced by sugar valency, sugar scaffolding, and nature of the glycosidic aglycone. Roy et al. synthesized potential inhibitors for type 1 fimbria of bacteria using pentaerythritol and triazole linkages [71]. The preliminary results indicated that a tetravalent mannosylated glycocluster was approximately a 100-fold more efficient in inhibiting agglutination of *E. coli* strain × 7122 with a titer of 3.96 mM. They also demonstrated that the multiantennary mannopyranoside bearing a triazole functionality close to the anomeric position was a potent antagonist against *E. coli* FimH with an HIA titer of 1 μM, which is a 32-fold better binding affinity when corrected on a per Man residue basis (Fig. 16.9b) [72].

16.6.2 PA-IL and PA-IIL

The opportunistic human pathogen *Pseudomonas aeruginosa*, a Gram-negative bacterium, produces two soluble lectins, namely, PA-IL and PA-IIL (also referred to as LecA and LecB) with binding specificities for Gal and L-fucose (Fuc), respectively. *Pseudomonas aeruginosa* is a causative agent of mortality-related infections and can be especially pernicious in immunocompromised and cystic fibrosis patients often exhibiting resistance to antibiotic that is partially due to the formation of biofilm. Synthetic multivalent carbohydrate ligands for these two lectins have been reported. Recently, carbohydrate-functionalized resorcin[4]arene-based glycoclusters prepared by Cu-catalyzed azide–alkyne click chemistry was described [73]. In ELLA binding studies, the tetravalent galactosylated glycocluster was identified as an effective ligand for PA-IL and demonstrated a strong affinity with IC_{50} of 0.7 μM. Furthermore, impressively high-affinity multivalent carbohydrate-coated gold glycoclusters were synthesized by Serge Pérez et al. (Fig. 16.10a), and their interactions with PA-IL were studied [74]. The results of HIA showed that the most effective gold glycocluster exhibited a K_d of 50 nM for each monosaccharide.

Carbohydrate–protein interaction is not only dependent on the valency of the sugar ligand but also on the topology of the glycoclusters. In studies using different core structures to append Gal residues, Sébastien Vidal et al. revealed that the calix[6]arene-based hexavalent glycocluster showed high affinity to PA-IL (Fig. 16.10b) [75]. In the ITC measurement of the glycocluster/PA-IL interaction, the hexavalent glycocluster displayed a K_d value of 136 nM. The group later discovered that the dodecavalent fullerene-based glycoclusters (Fig. 16.10c) displayed a significant glycoside cluster effect with up to a 12,000-fold increase in binding to PA-IL compared with the monovalent ligand [76].

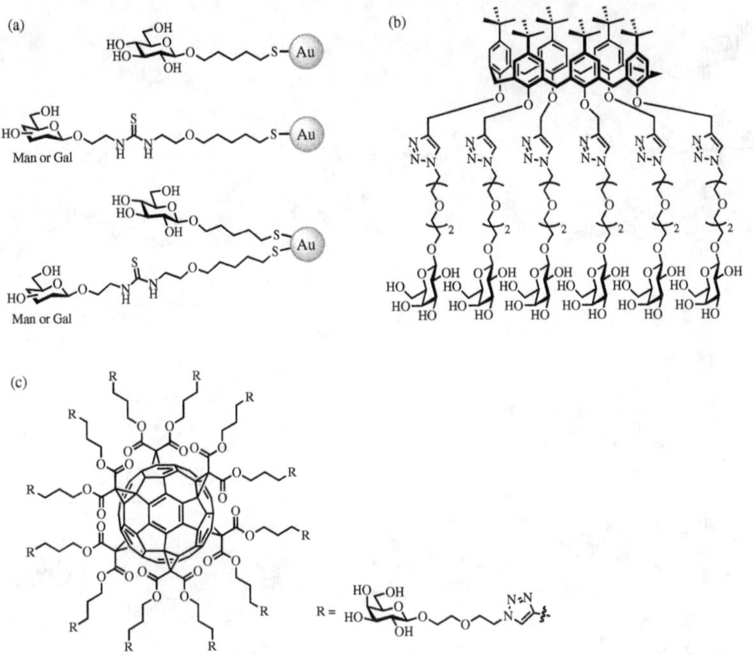

FIGURE 16.10 High-affinity glycocluster ligands for PA-IL. (a) Gold glycoclusters. (b) Calix[6]arene-based hexavalent glycoclusters. (c) Fullerene-based dodecavalent galactosylated glycoclusters synthesized using azide–alkyne click chemistry.

Roy et al. synthesized divalent and trivalent glycoclusters bearing the Fuc(α1→4) GlcNAc disaccharide unit—a high-affinity ligand for PA-IIL—conjugated with different lengths and core geometries by 1,2,3-triazole formation [77]. Among them, a dimeric Fuc(α1→4)GlcNAc glycocluster bearing a flexible hexa(ethylene glycol) linker displayed the highest inhibition affinity (K_d=90 nM) to PA-IIL as tested by microcalorimetry. The group also constructed a heterobifunctional glycodendrimer bearing four Gal (for PA-IL) and four Fuc (for PA-IIL) residues to prevent the adhesion behavior of *P. aeruginosa* (Fig. 16.11a) [78]. Turbidimetric assays demonstrated that the heteroglycodendrimer recognized and linked both binding site domains of PA-IL and PA-IIL simultaneously.

In addition, glycodendrimers were found to inhibit the formation of biofilms by *P. aeruginosa*. Jean-Louis Reymond et al. synthesized a neoglycopeptide dendrimer ending with a C-linked fucoside (Fig. 16.11b) [79]. The strongest binding to PA-IIL was observed with a tetravalent dendrimer (IC_{50}=0.6 μM) having positive guanidine residues in close proximity to the carbohydrate residue. The group later identified FD2, a highly potent peptide-based C-linked fucoside dendrimer ligand for PA-IIL, which improved the IC_{50} value to 0.14 μM (Fig. 16.11c) [80]. Because a *P. aeruginosa* mutant that lacks PA-IIL protein shows impaired biofilm formation [81], blocking this lectin with Fuc residues should therefore inhibit biofilm formation.

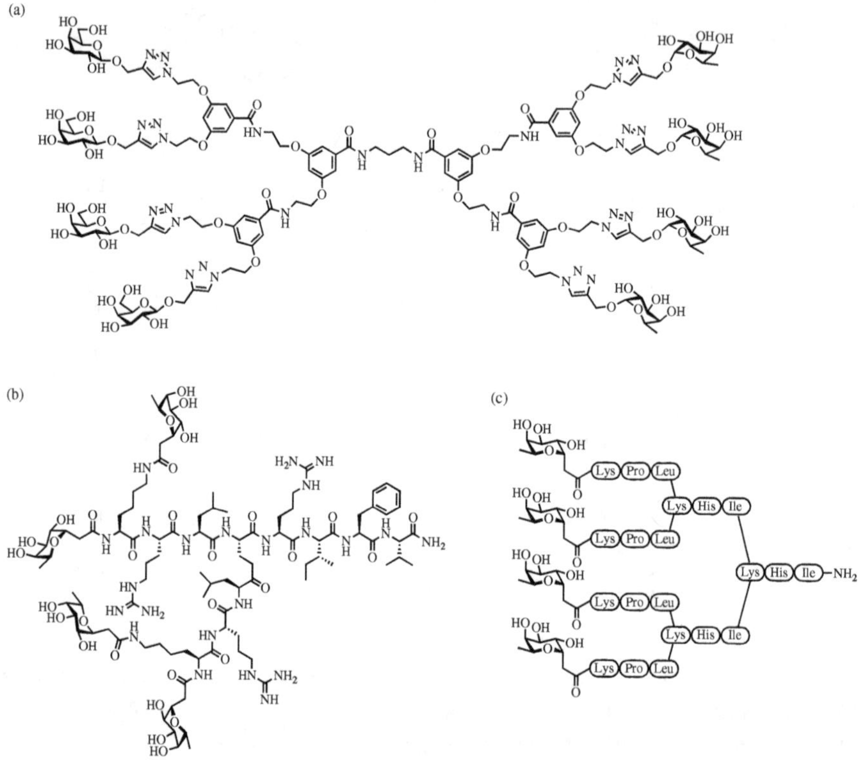

FIGURE 16.11 (a) The heterobifunctional glycodendrimer bearing both Gal and Fuc residues for simultaneous binding to PA-IL and PA-IIL, respectively. (b) A tetravalent C-fucoside neoglycopeptide and (c) FD2 are the best ligand candidates for PA-IIL as selected from a library of glycopeptidomimetics. His, histidine; Ile, isoleucine; Leu, leucine; Lys, lysine; Pro, proline.

Later, Reymond et al. found that FD2 completely inhibited the formation of *P. aeruginosa* biofilms at 50 μM concentration (IC_{50} = 10 μM) and caused a complete dispersion of established biofilms for both the wild-type strain and clinical isolates [82]. Moreover, the fucosylated glycopeptide dendrimer showed no cytotoxicity against human kidney embryonic cells (293T) [82], suggesting that these dendrimers may be used in the treatment of *P. aeruginosa* infections in humans. The same group later showed that the D-FD2 dendrimer, which contains the D- instead of L-amino acids, retained the inhibitory effect against the formation of *P. aeruginosa* biofilms but was resistant to proteolysis indicating greater stability [83].

Glyco-AuNPs have been shown to simultaneously enrich and isolate proteins from a very dilute solution with minimal sample handling [84]. AuNPs offer an additional advantage in that bound proteins can be identified directly by mass spectrometry without the elution of the captured protein because of the electrical conductivity of the nanoparticles. Lin et al. previously prepared Gal-AuNP and P^{k}-AuNP and demonstrated their use as PA-IL probes for specific protein capture and purification [84]. The capture of the target protein was also demonstrated at low femtomole levels and directly analyzed by MALDI-TOF mass spectrometry.

16.7 INFLUENZA VIRUS

Koji Matsuoka et al. developed sialic acid-conjugated carbosilane glycodendrimers as novel sialidase inhibitors [85]. Biological evaluation of these sialodendrimers against human influenza sialidases (H1N1 and H3N2) showed that ether- and amine-elongated sialodendrimers with an anomeric thioether linkage have inhibitory activity against sialidase. In particular, the most potent inhibition was provided by the ether- and amine-elongated carbosilane glycodendrimers displaying six Neu5Ac residues at the six terminal ends for the H3N3 and H1N1 sialidase, respectively. The same research group later prepared carbosilane dendrimers that were uniformly functionalized with Neu5Ac(α2→3)Gal(β1→4)Glc segments [86], and the HIA results showed 1000-fold better inhibition using the same dendrimer core. Furthermore, sialic acids were linked at either one or both rims of calix[4]arene through click chemistry to yield tetra- and octavalent sialoclusters with a hydrolytically stable carbon–sulfur bond at the anomeric position [87]. These glycoclusters showed moderate glycoside cluster effects when evaluated as inhibitors for the BK and influenza A viruses.

16.8 DETECTION OF BACTERIA

Carbohydrate-functionalized MNPs (glyco-MNPs) have also been used in rapid and accurate pathogen detection. Xuefei Huang et al. prepared MNPs conjugated with Man residues using the [2+3] Huisgen cycloaddition reaction. The resulting Man-MNP was used to detect *E. coli* strain ORN178 at a concentration of 10^4 cells/mL (Fig. 16.12) [88]. In addition, the rapid magnetic separation significantly reduced the operational time and contamination of samples, a common serious limitation in conventional separation processes.

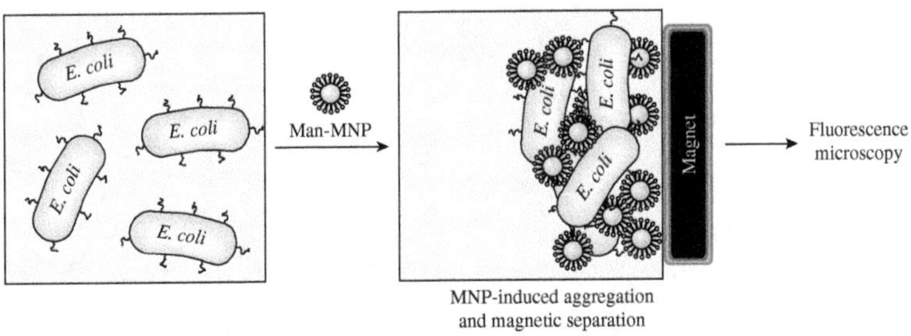

FIGURE 16.12 Schematic representation of the glyco-MNP-based pathogen detection.

Subsequently, Yan et al. incorporated a perfluorophenyl azide group on the surface of the MNP through a phosphate linkage [89]. The MNP was then used to covalently link Man residues via photoirradiation, and the resulting glyconanoparticle showed binding affinity to *E. coli* strain ORN178. Furthermore, Seeberger et al. synthesized glycodendronized poly-L-lysine polymers, which carry Man-bearing dendrons [90]. In combination with Con A-FITC, the glycodendronized polylysine decorated with nine Man residues was used to detect the presence of *E. coli* by confocal microscopy. The group later used Man-functionalized Ru(II)-glycodendrimers to detect *E. coli* [91]. The Ru(II)-containing complexes were observable by virtue of their inherent red fluorescence. In addition, Lin et al. developed an integrated MNP-based cell sorting system that can effectively separate bacterial cells from a complex mixture [92]. The microfluidic system consisted of Man-MNPs, which are capable of sorting *E. coli* ORN178 cells at concentrations as low as 10^3 cells/mL as rapidly as 1 min with an efficiency level of more than 70%.

16.9 GLYCO-MNPs AS NANOPROBES FOR LABELING CELLS AND MAGNETIC RESONANCE IMAGING AGENTS

MNPs have great potential as probes in clinical diagnostics and therapeutic techniques [93, 94]. Due to the high surface area-to-volume ratio, glyco-MNPs can contain multiple copies of glycans on their surface, resulting in significant enhancement in their binding affinities to the target lectins. Designed glyco-MNPs were also demonstrated to specifically target the asialoglycoprotein receptor (ASGP-R) on the surface of HepG2 cells (Fig. 16.13) [95]. Preassembled triantennary galactosyl residues were immobilized on the fluorescent MNP. The resulting T-Gal-l-Cy3@MNPs were used to target ASGP-R on the HepG2 cell surface, and through receptor-mediated endocytosis, the fluorescent glyco-MNPs were delivered into the cell. Moreover, these

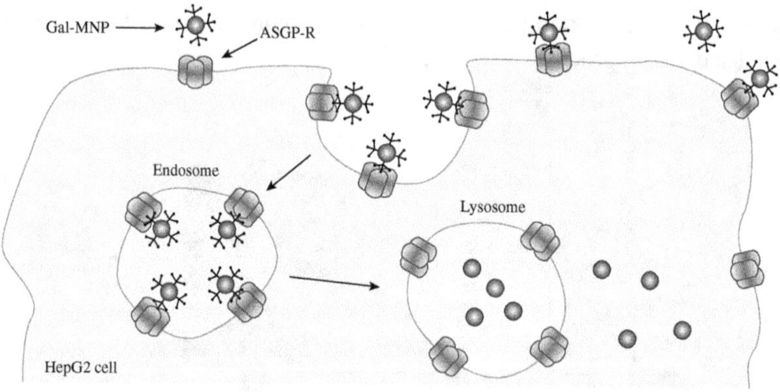

FIGURE 16.13 Well-defined multifunctional fluorescent Gal-MNP was used to specifically target HepG2 cells.

synthetic carbohydrate-functionalized MNPs (Gal-Cy3@MNPs) did not show any apparent cytotoxicity. Because of the ease of surface functionalization and the absence of cytotoxicity of the multifunctional MNPs, glyco-MNPs could serve as nanoprobes in biomedical applications.

Glyco-MNPs have also been recently developed as contrast agents for magnetic resonance imaging (MRI). Benjamin Davis et al. reported the *in vivo* imaging of an inflamed brain using sialyl Lewis X-functionalized MNPs (sLe^x-MNPs) [96]. In the synthesis of the glyco-MNPs, the presence of a masked *S*-cyanomethyl group at the linker of glycans was successfully reacted with an amino group on the MNP at the later stage of surface functionalization. The sLe^x-MNP successfully targeted the endothelial marker E-selectin in a model of acute inflammation inside a rat brain. Moreover, Huang et al. used glyco-MNPs to detect and differentiate normal and cancer cell lines [97]. Five different types of glyco-MNPs, namely, Man-, Gal-, Fuc-, Neu5Ac-, and GlcNAc-MNPs, were prepared, and their binding abilities to 10 different cell lines were examined. Analysis of the statistical data showed that all 10 cell lines could be distinguished based on their respective MRI signatures. In addition, the antiadhesive properties of glyco-MNPs were also used to establish potential antiadhesive cancer agents.

16.10 CYCLOPEPTIDE-BASED GLYCOCLUSTERS AS VACCINE ADJUVANTS

Cyclic peptides are good candidates for the production of template-assembled glycoclusters because they can be traced as adjuvants and carriers to direct other molecules to immunocompetent cells [98]. This design is appealing in its simplicity and involves the formation of two faces with regioselectively functionalizable amino acid side chains. Because of the presence of these accessible side chains, the templates are also referred to as regioselectively addressable functionalized templates (RAFT). Numerous variations of these themes can be conceived by simply changing the geometry or amino acid of the cyclic peptide. Of special interest are synthetic RAFT-based glycoclusters as vaccine constructs that may enhance cluster-recognizing antibody responses [99]. This section focuses specifically on commendable examples of vaccine adjuvants conducted with a cyclic peptide template.

In the design of carbohydrate-based vaccine constructs, the carbohydrate-specific antibody response may be influenced greatly by the orientation in which the antigen is displayed in its natural environments (i.e., on the surface of the transformed cell). Pascal Dumy et al. assembled the tumor-associated carbohydrate antigen (TACA) Tn by forming oxime bonds on a cyclic decapeptide template that contained two Pro-Gly (Gly = glycine) as β-turn inducers to form a stable conformation in solution (Fig. 16.14a) [100]. The cyclic template was amenable to providing the appropriate spatial orientation and number of the Tn antigens in the cluster. The antibodies elicited in mice immunized with Tn-conjugated RAFT recognized the native form of the Tn epitope expressed on the human breast cancer cell line MCF-7. Lai-Xi Wang et al. also selected a similar template to mimic the dimension of the glycoepitope on the

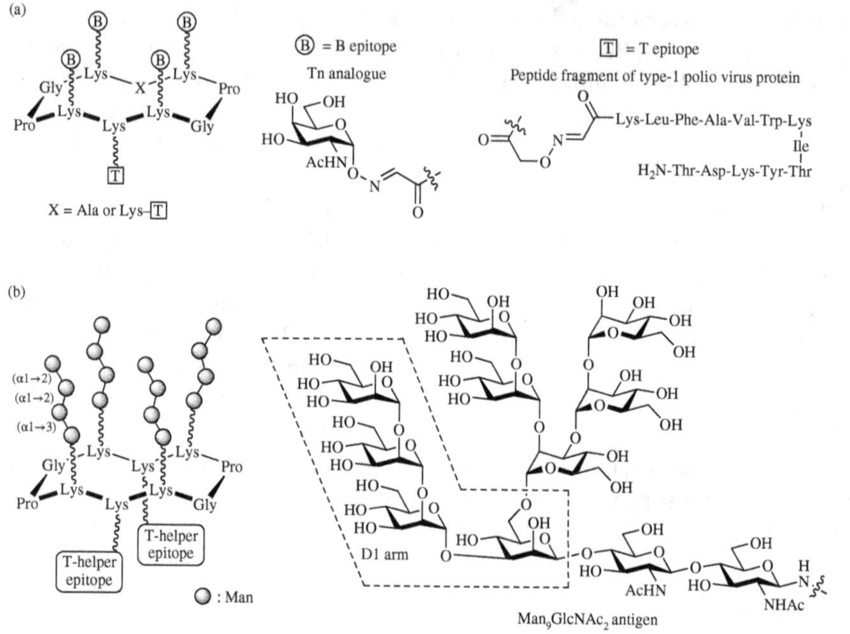

FIGURE 16.14 Template-assembled cyclopeptide vaccine candidates containing (a) the Tn antigen and (b) the D1 arm of the $Man_9GlcNAc_2$ antigen. Ala, alanine; Asp, aspartic acid; Phe, phenylalanine; Thr, threonine; Trp, tryptophan; Tyr, tyrosine; Val, valine.

HIV envelope glycoprotein gp120. The rigidity and multiple functionality of RAFT allows for tetravalent glycosylation with oligomannoside and the attachment of two T-helper peptide epitopes [101]. Cu(I)-catalyzed cycloaddition reaction was employed to incorporate the D1 arm of the $Man_9GlcNAc_2$ antigen presented on gp120 at one face of the RAFT scaffold (Fig. 16.14b), and the binding affinity of the resulting glyco-RAFT to human antibody 2G12 was evaluated. The results suggested that an appropriate spatial orientation of sugar chains in the cluster is crucial for high-affinity binding to 2G12. This approach is suitable for further designs of mimics for understanding the glyco-binding epitope of 2G12.

Samuel Danishefsky et al. capitalized in this interesting strategy by modifying the peptide scaffold with a pair of additional β-turn inducers—a D-Pro-L-Pro sequence at both ends of the macrocycle—and assembled $Man_9GlcNAc_2$ antigens to develop an anti-HIV vaccine (Fig. 16.15a) [102]. The optimal presentation of the glycan structures on the cyclopeptide template can be modulated by changing the numbers of Asp residues (4 or 6) (Fig. 16.15b) [103]. They also assembled multiple copies of two TACAs, namely, Tn and STn, on a single cyclic peptide RAFT scaffold. The vaccines generated by conjugating the antigen-RAFT complexes with carrier protein were shown to induce higher antibody titers against carbohydrate epitopes [92]. It is anticipated that the higher molecular ratio of antigen versus carrier protein could offer additional advantages in terms of immunogenicity.

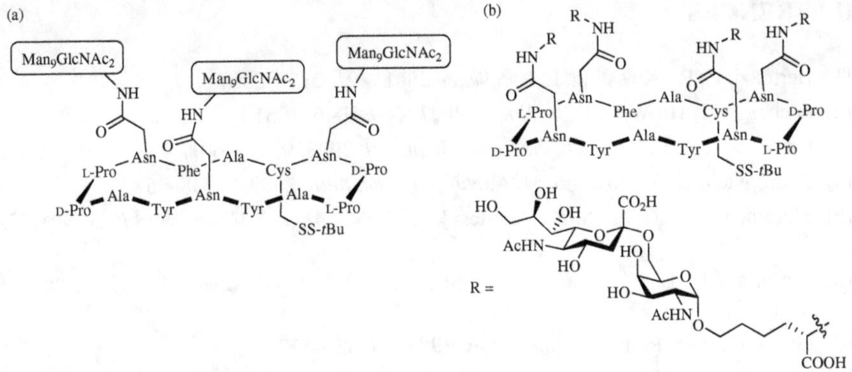

FIGURE 16.15 RAFT-antigen constructs by Danishefsky et al. (a) HIV vaccine construct with three $Man_9GlcNAc_2$ structures for gp120 recognition. (b) RAFT scaffold-based STn antigen construct.

16.11 CONCLUSION

Various types of multivalent neoglycoconjugates have been synthesized over the years. The primary focus was frequently to enhance the affinity or amplify the inhibitory potency of these systems. Glycodendrimers were found to have binding constants that were orders of magnitude higher than monosaccharides for lectins. Glyconanoparticles are excellent biomaterials because they offer a multivalent display of glycans similar to the glycocalyx structures covering cell surfaces. In addition, glyconanoparticles provide several advantages, including water solubility, ease of synthesis, stability, and, most importantly, absence of toxicity.

Furthermore, because of the fine specificity of microbial pathogens for the glycans present on only few species, tissues, or cell types, it has therefore been possible to design glycomimetics. Thus, the designed neoglycoconjugates are capable of targeting bacterial cells and bacterial toxins, cancer imaging, targeted delivery, and antiadhesion. Due to the globular shape and orientation of the ligands displayed on the surface of the designed glycoconjugates, they are well suited to engage in multivalent interactions to block the adhesion of bacterial cells. This chapter also attempted to generate a snapshot of some potential future approaches of glycans to treat diseases. By optimizing the multivalent effects in ligand–receptor recognition events, the designed and synthesized glyconanoparticles, glycodendrimers, and glycoclusters have provided many useful implications and shown their potential uses as therapeutic agents or delicate sensors for diseases.

ACKNOWLEDGMENTS

The authors thank the National Science Council in Taiwan, National Tsing Hua University, and Academia Sinica for financial support.

REFERENCES

[1] Bertozzi, C. R.; Kiessling, L. L. *Science* **2001**, *291*, 2357–2364.

[2] Seeberger, P. H.; Werz, D. B. *Nature* **2007**, *446*, 1046–1051.

[3] van Kooyk, Y.; Rabinovich, G. A. *Nat. Immunol.* **2008**, *9*, 593–601.

[4] Rabinovich, G. A.; Toscano, M. A. *Nat. Rev. Immunol.* **2009**, *9*, 338–352.

[5] Mammen, M.; Choi, S.-K.; Whitesides, G. M. *Angew. Chem. Int. Ed.* **1998**, *37*, 2754–2794.

[6] Kiessling, L. L.; Gestwicki, J. E.; Strong, L. E. *Angew. Chem. Int. Ed.* **2006**, *45*, 2348–2368.

[7] Lee, Y. C.; Lee, R. T. *Acc. Chem. Res.* **1995**, *28*, 321–327.

[8] Lundquist, J. J.; Toone, E. J. *Chem. Rev.* **2002**, *102*, 555–578.

[9] de la Fuente, J. M.; Penadés, S. *Biochim. Biophys. Acta* **2006**, *1760*, 636–651.

[10] Wei-Ting, C.; Ching-Ching, Y.; Chien-Fu, L.; Chian-Hui, L.; Po-Chiao, L.; Chun-Cheng, L. In *Petite and Sweet: Glyco-Nanotechnology as a Bridge to New Medicines*; Huang, X., Barchi, J. J., Jr., Eds.; ACS Symposium Series 1091; American Chemical Society: Washington, DC, **2011**, pp 15–36.

[11] Lin, P.-C.; Adak, A.; Lin, C.-C. In *The Molecular Immunology of Complex Carbohydrates-3*; Wu, A. M., Ed.; Advances in Experimental Medicine and Biology 705; Springer: New York, **2011**, pp 727–755.

[12] Cloninger, M. J. *Curr. Opin. Chem. Biol.* **2002**, *6*, 742–748.

[13] Roy, R.; Baek, M.-G. *Rev. Mol. Biotechnol.* **2002**, *90*, 291–309.

[14] Turnbull, W. B.; Stoddart, J. F. *Rev. Mol. Biotechnol.* **2002**, *90*, 231–255.

[15] Röckendorf, N.; Lindhorst, T. K. In *Dendrimers IV*; Vögtle, F., Schalley, C. A., Eds.; Topics in Current Chemistry 217; Springer: Berlin, **2001**, pp 201–238.

[16] Lindhorst, T. K. In *Host-Guest Chemistry*; Penadés, S., Ed.; Topics in Current Chemistry 218; Springer: Berlin, **2002**, pp 201–235.

[17] Saha, K.; Agasti, S. S.; Kim, C.; Li, X.; Rotello, V. M. *Chem. Rev.* **2012**, *112*, 2739–2779.

[18] Katz, E.; Willner, I. *Angew. Chem. Int. Ed.* **2004**, *43*, 6042–6108.

[19] Rosi, N. L.; Mirkin, C. A. *Chem. Rev.* **2005**, *105*, 1547–1562.

[20] Murphy, C. J.; Gole, A. M.; Stone, J. W.; Sisco, P. N.; Alkilany, A. M.; Goldsmith, E. C.; Baxter, S. C. *Acc. Chem. Res.* **2008**, *41*, 1721–1730.

[21] Algar, W. R.; Prasuhn, D. E.; Stewart, M. H.; Jennings, T. L.; Blanco-Canosa, J. B.; Dawson, P. E.; Medintz, I. L. *Bioconjug. Chem.* **2011**, *22*, 825–858.

[22] Lin, P.-C.; Ueng, S.-H.; Yu, S.-C.; Jan, M.-D.; Adak, A. K.; Yu, C.-C.; Lin, C.-C. *Org. Lett.* **2007**, *9*, 2131–2134.

[23] Lin, P.-C.; Chou, P.-H.; Chen, S.-H.; Liao, H.-K.; Wang, K.-Y.; Chen, Y.-J.; Lin, C.-C. *Small* **2006**, *2*, 485–489.

[24] Enustun, B. V.; Turkevich, J. *J. Am. Chem. Soc.* **1963**, *85*, 3317–3328.

[25] Brust, M.; Walker, M.; Bethell, D.; Schiffrin, D. J.; Whyman, R. *J. Chem. Soc. Chem. Commun.* **1994**, 801–802.

[26] Roy, R.; Zanini, D.; Meunier, S. J.; Romanowska, A. *J. Chem. Soc. Chem. Commun.* **1993**, 1869–1872.

[27] Niederhafner, P.; Šebestík, J.; Ježek, J. *J. Pept. Sci.* **2007**, *14*, 2–43.

[28] Lahmann, M. In *Glycoscience and Microbial Adhesion*; Lindhorst, T. K., Oscarson, S., Eds.; Topics in Current Chemistry 288; Springer: Berlin, **2009**, pp 17–65.

[29] Dubber, M.; Lindhorst, T. K. *Chem. Commun.* **1998**, 1265–1266.

[30] Sakamoto, J.; Müllen, K. *Org. Lett.* **2004**, *6*, 4277–4280.

[31] Lis, H.; Sharon, N. *Chem. Rev.* **1998**, *98*, 637–674.

[32] Ambrosi, M.; Cameron, N. R.; Davis, B. G. *Org. Biomol. Chem.* **2005**, *3*, 1593–1608.

[33] Kim, J.; Ahn, Y.; Park, K. M.; Kim, Y.; Ko, Y. H.; Oh, D. H.; Kim, K. *Angew. Chem. Int. Ed.* **2007**, *46*, 7393–7395.

[34] Schlick, K. H.; Lange, C. K.; Gillispie, G. D.; Cloninger, M. J. *J. Am. Chem. Soc.* **2009**, *131*, 16608–16609.

[35] Bogdan, N.; Roy, R.; Morin, M. *RSC Adv.* **2012**, *2*, 985–991.

[36] Kikkeri, R.; Grünstein, D.; Seeberger, P. H. *J. Am. Chem. Soc.* **2010**, *132*, 10230–10232.

[37] Gómez-García, M.; Benito, J. M.; Gutiérrez-Gallego, R.; Maestre, A.; Mellet, C. O.; Fernández, J. M. G.; Blanco, J. L. J. *Org. Biomol. Chem.* **2010**, *8*, 1849–1860.

[38] Gouin, S. G.; Vanquelef, E.; García Fernández, J. M.; Ortiz Mellet, C.; Dupradeau, F.-Y.; Kovensky, J. *J. Org. Chem.* **2007**, *72*, 9032–9045.

[39] Munoz, E. M.; Correa, J.; Fernandez-Megia, E.; Riguera, R. *J. Am. Chem. Soc.* **2009**, *131*, 17765–17767.

[40] Wang, X.; Ramström, O.; Yan, M. *Anal. Chem.* **2010**, *82*, 9082–9089.

[41] Lin, C.-C.; Yeh, Y.-C.; Yang, C.-Y.; Chen, G.-F.; Chen, Y.-C.; Wu, Y.-C.; Chen, C.-C. *Chem. Commun.* **2003**, 2920–2921.

[42] Tsai, C.-S.; Yu, T.-B.; Chen, C.-T. *Chem. Commun.* **2005**, 4273–4275.

[43] Schofield, C. L.; Haines, A. H.; Field, R. A.; Russell, D. A. *Langmuir* **2006**, *22*, 6707–6711.

[44] Huang, C.-C.; Chen, C.-T.; Shiang, Y.-C.; Lin, Z.-H.; Chang, H.-T. *Anal. Chem.* **2009**, *81*, 875–882.

[45] Karamanska, R.; Mukhopadhyay, B.; Russell, D. A.; Field, R. A. *Chem. Commun.* **2005**, 3334–3336.

[46] Ding, L.; Qian, R.; Xue, Y.; Cheng, W.; Ju, H. *Anal. Chem.* **2010**, *82*, 5804–5809.

[47] Gómez-García, M.; Benito, J. M.; Butera, A. P.; Mellet, C. O.; Fernández, J. M. G.; Blanco, J. L. J. *J. Org. Chem.* **2012**, *77*, 1273–1288.

[48] Wang, K.-R.; An, H.-W.; Wu, L.; Zhang, J.-C.; Li, X.-L. *Chem. Commun.* **2012**, *48*, 5644–5646.

[49] La Belle, J. T.; Gerlach, J. Q.; Svarovsky, S.; Joshi, L. *Anal. Chem.* **2007**, *79*, 6959–6964.

[50] Yamada, A.; Hatano, K.; Koyama, T.; Matsuoka, K.; Takahashi, N.; Hidari, K. I. P. J.; Suzuki, T.; Suzuki, Y.; Terunuma, D. *Bioorg. Med. Chem.* **2007**, *15*, 1606–1614.

[51] Lo Conte, M.; Staderini, S.; Chambery, A.; Berthet, N.; Dumy, P.; Renaudet, O.; Marra, A.; Dondoni, A. *Org. Biomol. Chem.* **2012**, *10*, 3269–3277.

[52] Otsuka, H.; Akiyama, Y.; Nagasaki, Y.; Kataoka, K. *J. Am. Chem. Soc.* **2001**, *123*, 8226–8230.

[53] Schofield, C. L.; Mukhopadhyay, B.; Hardy, S. M.; McDonnell, M. B.; Field, R. A.; Russell, D. A. *Analyst (Cambridge, U. K.)* **2008**, *133*, 626–634.

[54] Gao, Y.; Eguchi, A.; Kakehi, K.; Lee, Y. C. *Bioorg. Med. Chem.* **2005**, *13*, 6151–6157.

[55] Bossu, I.; Šulc, M.; Křenek, K.; Dufour, E.; Garcia, J.; Berthet, N.; Dumy, P.; Křen, V.; Renaudet, O. *Org. Biomol. Chem.* **2011**, *9*, 1948–1959.

[56] Ogata, M.; Yano, M.; Umemura, S.; Murata, T.; Park, E. Y.; Kobayashi, Y.; Asai, T.; Oku, N.; Nakamura, N.; Matsuo, I.; Usui, T. *Bioconjug. Chem.* **2012**, *23*, 97–105.

[57] Schlick, K. H.; Udelhoven, R. A.; Strohmeyer, G. C.; Cloninger, M. J. *Mol. Pharm.* **2005**, *2*, 295–301.

[58] Fan, E.; Merritt, E. A.; Verlinde, C. L. M. J.; Hol, W. G. J. *Curr. Opin. Struct. Biol.* **2000**, *10*, 680–686.

[59] Fukuta, S.; Magnani, J. L.; Twiddy, E. M.; Holmes, R. K.; Ginsburg, V. *Infect. Immun.* **1988**, *56*, 1748–1753.

[60] Ling, H.; Boodhoo, A.; Hazes, B.; Cummings, M. D.; Armstrong, G. D.; Brunton, J. L.; Read, R. J. *Biochemistry* **1998**, *37*, 1777–1788.

[61] Imberty, A.; Chabre, Y. M.; Roy, R. *Chem. Eur. J.* **2008**, *14*, 7490–7499.

[62] Pukin, A. V.; Branderhorst, H. M.; Sisu, C.; Weijers, C. A. G. M.; Gilbert, M.; Liskamp, R. M. J.; Visser, G. M.; Zuilhof, H.; Pieters, R. J. *ChemBioChem* **2007**, *8*, 1500–1503.

[63] Schofield, C. L.; Field, R. A.; Russell, D. A. *Anal. Chem.* **2007**, *79*, 1356–1361.

[64] Kitov, P. I.; Sadowska, J. M.; Mulvey, G.; Armstrong, G. D.; Ling, H.; Pannu, N. S.; Read, R. J.; Bundle, D. R. *Nature* **2000**, *403*, 669–672.

[65] Chien, Y.-Y.; Jan, M.-D.; Adak, A. K.; Tzeng, H.-C.; Lin, Y.-P.; Chen, Y.-J.; Wang, K.-T.; Chen, C.-T.; Chen, C.-C.; Lin, C.-C. *ChemBioChem* **2008**, *9*, 1100–1109.

[66] Kulkarni, A. A.; Fuller, C.; Korman, H.; Weiss, A. A.; Iyer, S. S. *Bioconjug. Chem.* **2010**, *21*, 1486–1493.

[67] Lin, C.-C.; Yeh, Y.-C.; Yang, C.-Y.; Chen, C.-L.; Chen, G.-F.; Chen, C.-C.; Wu, Y.-C. *J. Am. Chem. Soc.* **2002**, *124*, 3508–3509.

[68] Almant, M.; Moreau, V.; Kovensky, J.; Bouckaert, J.; Gouin, S. G. *Chem. Eur. J.* **2011**, *17*, 10029–10038.

[69] Heidecke, C. D.; Lindhorst, T. K. *Chem. Eur. J.* **2007**, *13*, 9056–9067.

[70] Schierholt, A.; Hartmann, M.; Schwekendiek, K.; Lindhorst, T. K. *Eur. J. Org. Chem.* **2010**, 3120–3128.

[71] Touaibia, M.; Shiao, T. C.; Papadopoulos, A.; Vaucher, J.; Wang, Q.; Benhamioud, K.; Roy, R. *Chem. Commun.* **2007**, 380–382.

[72] Papadopoulos, A.; Shiao, T. C.; Roy, R. *Mol. Pharm.* **2012**, *9*, 394–403.

[73] Soomro, Z. H.; Cecioni, S.; Blanchard, H.; Praly, J.-P.; Imberty, A.; Vidal, S.; Matthews, S. E. *Org. Biomol. Chem.* **2011**, *9*, 6587–6597.

[74] Reynolds, M.; Marradi, M.; Imberty, A.; Penadés, S.; Pérez, S. *Chem. Eur. J.* **2012**, *18*, 4264–4273.

[75] Cecioni, S.; Faure, S.; Darbost, U.; Bonnamour, I.; Parrot-Lopez, H.; Roy, O.; Taillefumier, C.; Wimmerová, M.; Praly, J.-P.; Imberty, A.; Vidal, S. *Chem. Eur. J.* **2011**, *17*, 2146–2159.

[76] Cecioni, S.; Oerthel, V.; Iehl, J.; Holler, M.; Goyard, D.; Praly, J.-P.; Imberty, A.; Nierengarten, J.-F.; Vidal, S. *Chem. Eur. J.* **2011**, *17*, 3252–3261.

[77] Marotte, K.; Préville, C.; Sabin, C.; Moumé-Pymbock, M.; Imberty, A.; Roy, R. *Org. Biomol. Chem.* **2007**, *5*, 2953–2961.

[78] Deguise, I.; Lagnoux, D.; Roy, R. *New J. Chem.* **2007**, *31*, 1321–1331.

[79] Johansson, E. M. V.; Kolomiets, E.; Rosenau, F.; Jaeger, K.-E.; Darbre, T.; Reymond, J.-L. *New J. Chem.* **2007**, *31*, 1291–1299.

[80] Kolomiets, E.; Johansson, E. M. V.; Renaudet, O.; Darbre, T.; Reymond, J.-L. *Org. Lett.* **2007**, *9*, 1465–1468.

[81] Tielker, D.; Hacker, S.; Loris, R.; Strathmann, M.; Wingender, J.; Wilhelm, S.; Rosenau, F.; Jaeger, K.-E. *Microbiology* **2005**, *151*, 1313–1323.

[82] Johansson, E. M. V.; Crusz, S. A.; Kolomiets, E.; Buts, L.; Kadam, R. U.; Cacciarini, M.; Bartels, K.-M.; Diggle, S. P.; Cámara, M.; Williams, P.; Loris, R.; Nativi, C.; Rosenau, F.; Jaeger, K.-E.; Darbre, T.; Reymond, J.-L. *Chem. Biol.* **2008**, *15*, 1249–1257.

[83] Johansson, E. M. V.; Kadam, R. U.; Rispoli, G.; Crusz, S. A.; Bartels, K.-M.; Diggle, S. P.; Cámara, M.; Williams, P.; Jaeger, K.-E.; Darbre, T.; Reymond, J.-L. *Med. Chem. Commun.* **2011**, *2*, 418–420.

[84] Chen, Y.-J.; Chen, S.-H.; Chien, Y.-Y.; Chang, Y.-W.; Liao, H.-K.; Chang, C.-Y.; Jan, M.-D.; Wang, K.-T.; Lin, C.-C. *ChemBioChem* **2005**, *6*, 1169–1173.

[85] Sakamoto, J.-I.; Koyama, T.; Miyamoto, D.; Yingsakmongkon, S.; Hidari, K. I. P. J.; Jampangern, W.; Suzuki, T.; Suzuki, Y.; Esumi, Y.; Nakamura, T.; Hatano, K.; Terunuma, D.; Matsuoka, K. *Bioorg. Med. Chem.* **2009**, *17*, 5451–5464.

[86] Oka, H.; Onaga, T.; Koyama, T.; Guo, C.-T.; Suzuki, Y.; Esumi, Y.; Hatano, K.; Terunuma, D.; Matsuoka, K. *Bioorg. Med. Chem.* **2009**, *17*, 5465–5475.

[87] Marra, A.; Moni, L.; Pazzi, D.; Corallini, A.; Bridi, D.; Dondoni, A. *Org. Biomol. Chem.* **2008**, *6*, 1396–1409.

[88] El-Boubbou, K.; Gruden, C.; Huang, X. *J. Am. Chem. Soc.* **2007**, *129*, 13392–13393.

[89] Liu, L.-H.; Dietsch, H.; Schurtenberger, P.; Yan, M. *Bioconjug. Chem.* **2009**, *20*, 1349–1355.

[90] Laurino, P.; Kikkeri, R.; Azzouz, N.; Seeberger, P. H. *Nano Lett.* **2011**, *11*, 73–78.

[91] Grünstein, D.; Maglinao, M.; Kikkeri, R.; Collot, M.; Barylyuk, K.; Lepenies, B.; Kamena, F.; Zenobi, R.; Seeberger, P. H. *J. Am. Chem. Soc.* **2011**, *133*, 13957–13966.

[92] Shih, P.-H.; Shiu, J.-Y.; Lin, P.-C.; Lin, C.-C.; Veres, T.; Chen, P. *J. Appl. Phys.* **2008**, *103*, 07A316.

[93] Jun, Y. W.; Lee, J. H.; Cheon, J. *Angew. Chem. Int. Ed.* **2008**, *47*, 5122–5135.

[94] Gao, J.; Gu, H.; Xu, B. *Acc. Chem. Res.* **2009**, *42*, 1097–1107.

[95] Lai, C.-H.; Lin, C.-Y.; Wu, H.-T.; Chan, H.-S.; Chuang, Y.-J.; Chen, C.-T.; Lin, C.-C. *Adv. Funct. Mater.* **2010**, *20*, 3948–3958.

[96] van Kasteren, S. I.; Campbell, S. J.; Serres, S.; Anthony, D. C.; Sibson, N. R.; Davis, B. G. *Proc. Natl. Acad. Sci. U. S. A.* **2008**, *106*, 18–23.

[97] El-Boubbou, K.; Zhu, D. C.; Vasileiou, C.; Borhan, B.; Prosperi, D.; Li, W.; Huang, X. *J. Am. Chem. Soc.* **2010**, *132*, 4490–4499.

[98] Boturyn, D.; Defrancq, E.; Dolphin, G. T.; Garcia, J.; Labbe, P.; Renaudet, O.; Dumy, P. *J. Pept. Sci.* **2008**, *14*, 224–240.

[99] Shiao, T. C.; Roy, R. *New J. Chem.* **2012**, *36*, 324–339.

[100] Grigalevicius, S.; Chierici, S.; Renaudet, O.; Lo-Man, R.; Dériaud, E.; Leclerc, C.; Dumy, P. *Bioconjug. Chem.* **2005**, *16*, 1149–1159.

[101] Wang, J.; Li, H.; Zou, G.; Wang, L.-X. *Org. Biomol. Chem.* **2007**, *5*, 1529–1540.

[102] Krauss, I. J.; Joyce, J. G.; Finnefrock, A. C.; Song, H. C.; Dudkin, V. Y.; Geng, X.; Warren, J. D.; Chastain, M.; Shiver, J. W.; Danishefsky, S. J. *J. Am. Chem. Soc.* **2007**, *129*, 11042–11044.

[103] Jeon, I.; Lee, D.; Krauss, I. J.; Danishefsky, S. J. *J. Am. Chem. Soc.* **2009**, *131*, 14337–14344.

17

DESIGN AND SYNTHESIS OF CARBOHYDRATES AND CARBOHYDRATE MIMETICS AS ANTI-INFLUENZA AGENTS

MAURO PASCOLUTTI AND MARK VON ITZSTEIN

Institute for Glycomics, Griffith University, Gold Coast, Queensland, Australia

17.1 INTRODUCTION

The name influenza comes from an Italian word "*influentia*" meaning influence, and when first used (possibly in the mid-eighteenth century), it related to the appearance of an epidemic attributed to an influence of the stars [1]. Influenza, the disease commonly known as flu, is a viral infection of the respiratory tract. Its common symptoms are fever, myalgia, headache, and pharyngitis [2]. In more serious cases, influenza can sometimes lead to pneumonia and be fatal particularly in young and elderly people [2]. The infection reaches peak prevalence in winter and is transmitted human to human through the air via respiratory tract secretions. Influenza has been recognized for several centuries as one of the biggest and the most uncontrollable threats to human health, causing sickness and death to millions of people. It is estimated that the annual influenza epidemics result in between three and five million cases of severe illness and between 250,000 and 500,000 deaths around the world [3, 4].

Glycochemical Synthesis: Strategies and Applications, First Edition.
Edited by Shang-Cheng Hung and Medel Manuel L. Zulueta.

17.2 INFLUENZA VIRUSES

The influenza viruses belong to the Orthomyxoviridae family and can be classified based on their internal proteins [5] and surface glycoproteins as influenza A, B, or C [2, 6, 7]. Influenza A viruses are known to be the most infectious of the three and can cause severe and acute respiratory disease, which can lead to annual epidemics [7]. Influenza A viruses infect a wide variety of mammals (e.g., humans, horses, pigs, and ferrets) and birds. Pigs and birds are the biggest reservoirs, generating diverse viruses (genetically and antigenically), which can get transferred back into the human population by close contact between humans and animals [7]. Influenza B viruses are generally not as virulent as influenza A types. They infect mostly humans and do not have an animal reservoir [2]; therefore, they do have the same possibility for wide genetic reassortment [8]. In addition, influenza B viruses mutate more slowly than the A viruses [2, 9]. Influenza C viruses infect only humans but rarely cause disease like influenza A and B viruses [10]. They are genetically and morphologically distinct from A and B types [11] and usually seem to cause only mild disease in children [10].

17.2.1 Virus Morphology

The influenza A, B, and C virus particles are highly pleomorphic; they are mainly spherical or ovoid in shape (80–120 nm diameter), but many have been found as long filamentous particles [7, 12]. The virus has a single-stranded RNA genome that is closely linked with a helical nucleoprotein and is present in eight (A and B) or seven (C) separate segments of ribonucleoprotein, each of which is necessary for replication. Each segment is surrounded by three viral polymerase polypeptides PB1, PB2, and PA, forming the ribonucleoprotein complex [6, 7, 13].

The genome in influenza A and B types is enclosed within an outer lipoprotein envelope (Fig. 17.1). The M1 protein lines the inside of the envelope and is chemically bound to the ribonucleoprotein [14]. The M1 protein plays an important role in the mediation of nuclear export of viral ribonucleoproteins and also in virus assembly and budding during the infectious cycle [5, 14, 15]. An antigenic protein M2, which functions as a proton-selective ion channel, is present in the viral membrane of influenza A viruses [6, 13]. In influenza B, the ion channel activity to aid virus uncoating in the endosome is carried out by the similar integral membrane protein BM2 [11].

The viral surface consists of the lipoprotein envelope in which the M2 protein and two types of glycoprotein spikes are embedded [13]. These glycoproteins are hemagglutinin (HA) and neuraminidase (NA) or sialidase. Influenza A viruses are further classified according to their HA and NA antigens. There are currently 17 known HA (H1–H17) serotypes [16] and 9 known NA (N1–N9) serotypes [6, 7, 17]. HA has a carbohydrate-recognizing lectin site and can interact with the host cell through sialic acid-terminated oligosaccharides on the cell surface, causing the first step in host cell infection [18, 19]. It is also involved in viral fusion with the host cell [17, 18]. Sialidase is involved at the budding stage of the infection cycle when the virus is ready to be released. Another important role is associated with the mobility of the virus particles through the respiratory tract mucus [20, 21]. HA and NA are described in more detail in later sections.

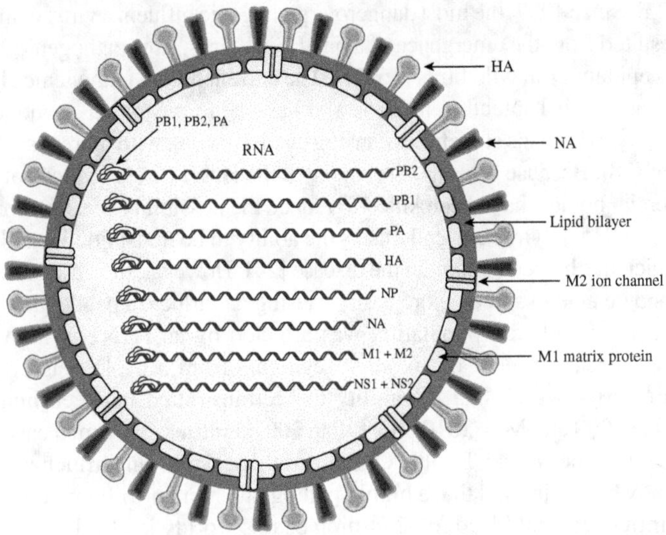

FIGURE 17.1 Representation of influenza A virus.

17.2.2 Influenza A Virus Epidemiology

The influenza A and B viruses constantly change over time due to faulty replication of the viral RNA by RNA polymerase [2]. This plays an important role in the epidemiology of influenza virus, especially with the antigenic HA and NA proteins. There are two different kinds of mutation of influenza A viruses. The first, *antigenic drift*, occurs over time due to a gradual change in the HA and NA antigens [2, 13]. The second one, *antigenic shift*, is a genetic recombination or a reassortment of viral RNA from different influenza strains infecting one organism. It is generally through antigenic shift that a pandemic virus arises [7]. Type A viruses incorporate mutations through both antigenic drift and shift, whereas influenza B type viruses change only by the more gradual process of antigenic drift and, as a result, do not cause pandemics [21].

Constant changes in the viral proteins enable the virus to escape detection by the immune system of the host. When a host is infected with influenza virus, it develops a precise antibody against that virus, in particular, against the HA and NA proteins. After mutation, the original antibody will not recognize the mutated virus, and the virus can then reinfect the original host. This scenario makes the development of a universal influenza virus vaccine difficult [13, 21].

In the last 100 years, five major influenza A pandemics have occurred. Spanish Flu (1918, H1N1), Asian Flu (1957, H2N2), Hong Kong Flu (1968, H3N2), Russian Flu (1977, H1N1), and Swine Flu (2009, H1N1) occurred due to antigenic shift involving only a restricted combination of HA (H1, H2, H3) and NA (N1, N2). These pandemics caused millions of deaths around the world [5, 7, 22]. Investigations into the mutations of these viruses have shown that when a new pandemic strain of influenza occurs, the virus has a new HA and, even sometimes, a new NA [23].

In the past 15 years, four new types of influenza A viruses, H5N1, H7N3, H7N7, and H9N2, have been discovered to cause disease and sometimes death in humans

[5, 21, 24, 25]. Since 1997, the most dangerous pathogenic influenza virus with pandemic potential resulted from the emergence of avian H5N1 virus. Low pathogenic avian influenza causes epidemics in wild birds [26, 27]. The current highly pathogenic H5N1 avian influenza causes lethal infection in birds and has been transmitted zoonotically from infected poultry to humans sometimes causing death (>50%), with more than 400 human deaths so far [28]. Because birds are the source of the H5N1 virus, millions of wild birds and commercial poultry have been killed to reduce the possibility of spread of the virus. Fortunately, the H5N1 virus currently lacks the ability to be transmitted readily between humans, which has helped to contain the disease [24]. However, the virus may mutate in the future and be able to spread more easily causing an influenza pandemic.

In 2009, the worldwide population was affected by an H1N1 virus, which was derived from a reassortment of six gene segments from a triple reassortment of a swine-origin virus, which had the ability to be transmitted between humans [29]. Between June 2009 and May 2010, more than 200 countries confirmed cases of H1N1 (swine flu), and the World Health Organization declared an influenza pandemic. A recent study has estimated that a highly pathogenic human-to-human transmissible pandemic influenza could lead to 62 million deaths worldwide [30].

17.2.3 Influenza A Virus Life Cycle

The first step in viral replication and infection (Fig. 17.2) [31] is the interaction between the viral surface HA and sialic acid-containing glycoconjugates on the surface of respiratory tract cells. After binding, the virus is endocytosed, and the low pH of the endosome triggers HA-dependent fusion. The M2 ion channel functions to

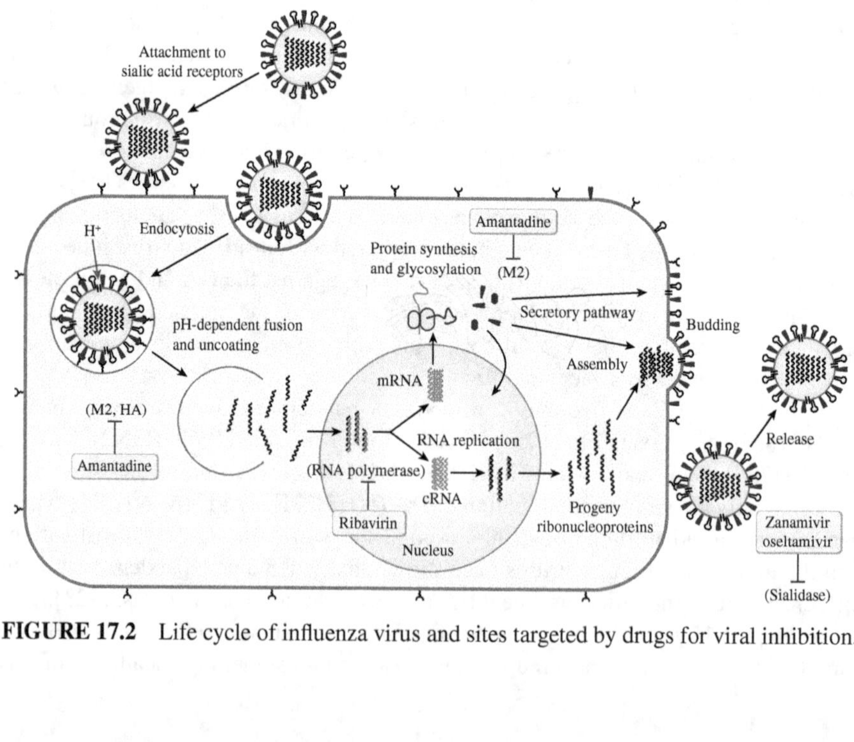

FIGURE 17.2 Life cycle of influenza virus and sites targeted by drugs for viral inhibition.

increase proton concentration in the virus, which is uncoated in the endosome, ultimately leading to the release of the viral genes into the cytoplasm [5, 7]. The viral genome is then replicated and used for the production of additional viral proteins important for progeny viruses. Once the viral proteins and genes are assembled, the budding process from the host cell begins. During this process, NA plays an important role by cleaving the terminal *N*-acetylneuraminic acid (Neu5Ac) residues from the surface of the host cell and the progeny viruses, facilitating the release of the new virus progeny that will then be ready to infect new host cells [6, 13, 32].

17.3 DEVELOPMENT OF ANTI-INFLUENZA THERAPEUTICS

Various stages of the virus life cycle have been targeted to develop drugs against influenza virus [13, 33–35]. Treatment, until recently, was achieved by adamantane-based M2 ion channel inhibitors (Fig. 17.3a) [36, 37]. Rimantadine (**1**) and amantadine (**2**) were the two most successful specific drugs against influenza A viral infection [35, 38, 39]. The adamantanes showed no inhibitory activity against the corresponding influenza B BM2 ion channel protein [35, 40, 41]. However, because of the rapid emergence of drug resistance [42, 43] and significant side effects [39, 40, 44], adamantanes were widely discontinued as therapeutics.

A second class of antiviral compounds has been developed to inhibit viral replication by acting on the RNA polymerase (Fig. 17.3b). T-705 (favipiravir, **3**) undergoes ribosylation and then phosphorylation and is believed to function similarly to a nucleobase in viral RNA replication [35, 45]. T-705 has been demonstrated to inhibit viral replication of influenza A, B, and C viruses, and recent studies have shown encouraging results toward it becoming a potential drug candidate [45, 46]. Compared to the most familiar polymerase inhibitor ribavirin (**4**), T-705 has a more favorable therapeutic index in preclinical test of toxicity [34, 35].

The currently most widely used class of antiviral inhibitors, however, has been developed by targeting the influenza virus sialidase, which is involved in the last stage of the life cycle of influenza A and B viruses [35]. These inhibitors are discussed in detail in Section 17.7.

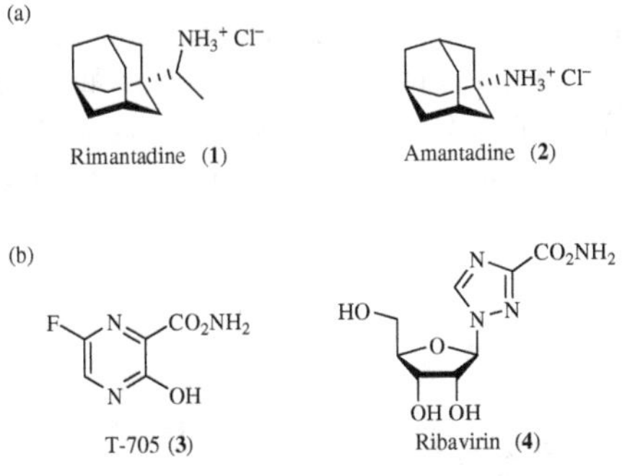

FIGURE 17.3 Anti-influenza compounds that act as (a) M2 ion channel inhibitors and (b) RNA polymerase inhibitors.

17.4 SIALIC ACID: THE VIRAL CELL-SURFACE RECEPTOR LIGAND

Sialic acids are a family of nine-carbon acidic monosaccharides based on 3-deoxy nonulosonic acid (Fig. 17.4, also see Chapter 5). In 1952, the biochemist Gunnar Blix introduced the term sialic acid from the Greek *σαλια* (*salia*) or "saliva" after isolating this molecule from the salivary gland mucin. Sialic acid is now the generic term for the family that includes derivatives with a C5-amino substituent called neuraminic acid. Neu5Ac (**5**) is the type of sialic acid found in humans [47].

Sialic acids are found widely distributed in animal tissues, fungi, bacteria and, to a lesser extent, plants [47]. Aside from Neu5Ac, the most common members of this family are *N*-glycolylneuraminic acid (Neu5Gc, **6**) and the 5-hydroxy derivative KDN (**7**). The sialic acid family is now known to contain 50 or more derivatives of sialic acid due to modifications at C4, C7, C8, and C9 [47]. Sialic acids have been found linked exclusively in the α configuration in sialoglycoconjugates. It is most often located on the terminal part of oligosaccharides, and this is most important in biological phenomena as it participates in various cell-surface interactions by protein–carbohydrate recognition [47–49].

17.5 HEMAGGLUTININ

As mentioned in Section 17.2.1, HA is a spike glycoprotein anchored to the virus lipid membrane [50]. This glycoprotein functions as a receptor-binding protein and is responsible for the first step of viral infection when it binds to sialic acid residues of receptor glycoproteins on host cells [18]. When the virus is endocytosed into the cell, the low pH (5–6) changes the structure of HA, and this new fusion-active state triggers the fusion of the viral membrane and the endosome membrane, ultimately allowing entry of the viral nucleocapsid into the cytosol of the host cell [18].

HA (Fig. 17.5 [51]) is a trimer and measures approximately 135 Å from insertion in the envelope membrane to its tip [19]. The three sites where sialic acid should bind are approximately 40 Å apart from each other [19, 52].

HA is a lectin, and it is well known that lectins do not bind very strongly with carbohydrates [33, 53–55]. In 1989, it was demonstrated that HA has low-binding affinity with monomeric sialic acid receptors: Neu5Ac α-methyl glycoside binds with a K_d of 2.8×10^{-3} M [56]. A number of research groups have tried to synthesize compounds (e.g., functionalized polymers with sialic acid-terminated chains) for drug discovery against HA, but none of these have progressed to drug development [55, 57–59].

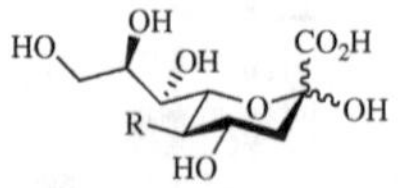

Neu5Ac (**5**): R = $NHC(=O)CH_3$
Neu5Gc (**6**): R = $NHC(=O)CH_2OH$
KDN (**7**): R = OH

FIGURE 17.4 Representation of the three major sialic acid subfamilies.

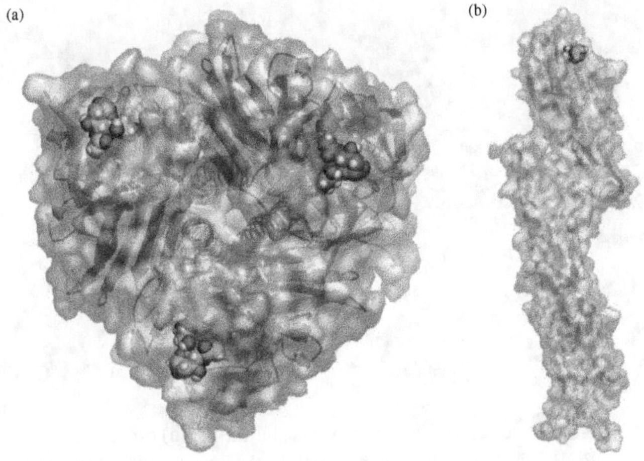

FIGURE 17.5 X-ray structures of influenza A virus HA with sialyllactose bound (PDB 1hgg) [51]. (a) Trimer of HA showing Neu5Ac (as spheres) bound in the binding site. (b) Monomer of HA showing the receptor-binding site in complex with Neu5Ac (as spheres) on the top.

17.6 SIALIDASE

Sialidase or NA is the second glycoprotein on the virion surface. One of the most important functions of sialidase is associated with the release of virus progeny from infected host cells to repeat the cycle of infection [6, 50]. The enzyme catalyzes the hydrolysis of sialic acid residues from glycoconjugates on the cell surface with either Neu5Ac(α2→3)Gal or Neu5Ac(α2→6)Gal linkages (Gal = galactose) [60, 61].

Sialidase is a homotetramer with C_4 symmetry composed of identical disulfide-linked subunits (Fig. 17.6a [62]). Each monomer is a glycosylated polypeptide with six β-sheets assuming the appearance of six blades of a propeller [6, 64] with a right-handed twist. The catalytic site is observed to be at the center of the sixfold pseudosymmetry axis, which passes through the center of each monomer and relates the six β-sheets to each other (Fig. 17.6b [63]) [64, 65]. The active site contains a large number of conserved amino acid residues, which are involved in binding to sialic acid in the substrate sialoglycoconjugate [64].

17.6.1 Influenza Virus Sialidase Active Site

X-ray crystal structure studies of influenza A/N2, A/N9, and B sialidases bound with α-Neu5Ac [66] show that the active site contains 18 invariant amino acid residues that either interact with the bound α-Neu5Ac or support these residues. These residues are conserved in all strains of influenza A and B viruses, suggesting their involvement in the enzymatic activity [65, 66]. The residues helped define the topology of the active site [66, 67]. Of those conserved amino acids interacting with the substrate, many are polar, but there are also a number of nonpolar residues

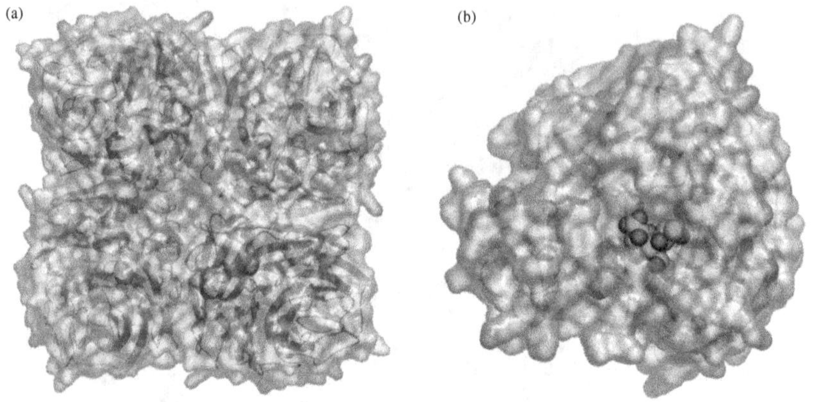

FIGURE 17.6 X-ray structures of influenza virus sialidases. (a) Structure of influenza B virus sialidase tetramer (PDB 1a4g) viewed from the top [62]. (b) Structure of influenza A virus sialidase N9 monomer (PDB 1mwe) with α-Neu5Ac (shown as spheres) bound in the active site [63].

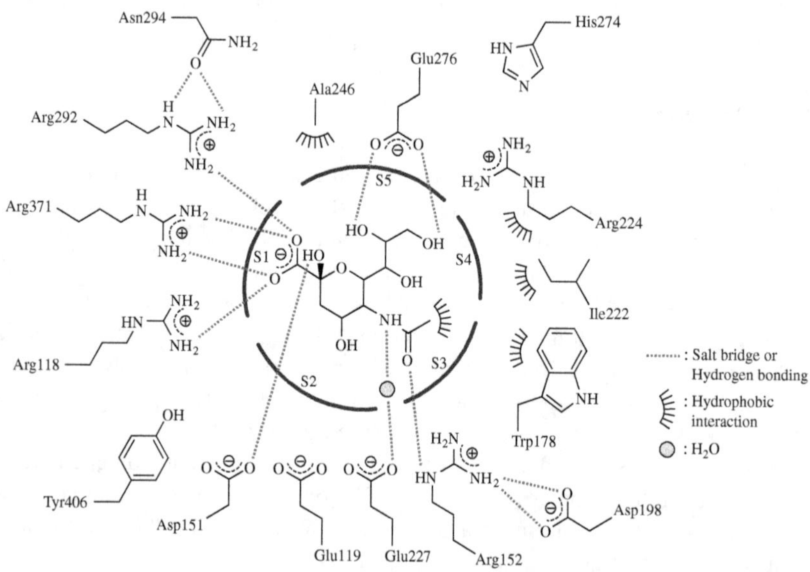

FIGURE 17.7 Some important direct interactions between conserved amino acid residues of the active site of influenza A virus sialidase and α-Neu5Ac. Ala, alanine; Arg, arginine; Asn, asparagine; Asp, aspartic acid; Glu, glutamic acid; His, histidine; Ile, isoleucine; Typ, tryptophan; Tyr, tyrosine.

[65]. Figure 17.7 represents some important interactions of α-Neu5Ac with the amino acid residues surrounding the active site [68]. The main interactions involved are charge–charge and hydrogen-bonding interactions [64, 66].

To facilitate the description of the binding between α-Neu5Ac and NA, the influenza virus sialidase active site can be divided into five regions (Fig. 17.7) [31, 69]. Subsite 1

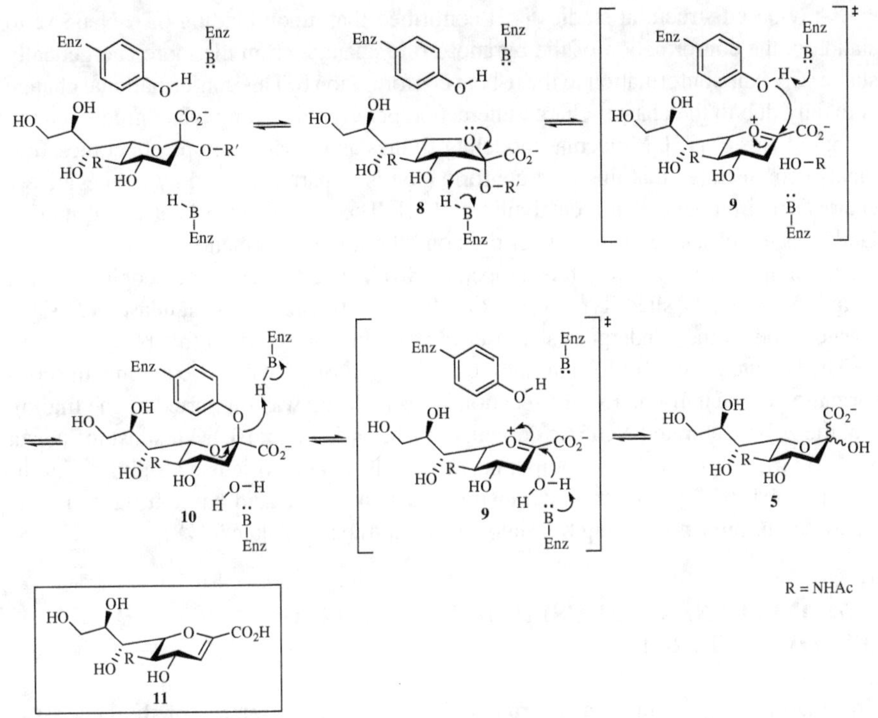

SCHEME 17.1 Proposed enzymatic mechanism for the cleavage of an α-sialoside by the influenza virus sialidase.

(S1) is a positively charged region in which the anionic carboxylate of α-Neu5Ac hydrogen bonds with three arginines called the triarginyl cluster (Arg118, Arg292, and Arg371). Subsite 2 (S2) is a negatively charged region due to the presence of Glu119 and Asp151. Subsite 3 (S3) is a predominantly hydrophobic pocket that accommodates the methyl moiety of the acetamido group of α-Neu5Ac, whereas the carbonyl group is hydrogen bonded to the positive Arg152 residue. Subsite 4 (S4) is a hydrophobic region derived from the side chain of Ile222 and Ala246 and the hydrophobic face of Arg224. The last subsite (S5) is normally a negatively charged region in which the carboxylate of Glu276 interacts with the glycerol side chain of α-Neu5Ac. However, it has been shown that replacement of the glycerol side chain of α-Neu5Ac with an aliphatic chain can lead to a reorientation of Glu276 toward Arg224 forming a salt bridge, and consequently, the region assumes a hydrophobic character [70].

17.6.2 Catalytic Mechanism of Influenza Virus Sialidase

The mechanism by which influenza virus sialidases cleaves the Neu5Ac(α2→3)Gal or Neu5Ac(α2→6)Gal linkage has been a topic of much interest for many years (e.g., see [71, 72]). Recently, it has been shown by structural analysis [73] that it involves a covalent enzyme–substrate intermediate as has been reported for other sialidases. A proposed mechanism is depicted in Scheme 17.1.

X-ray-based structural studies [72] confirmed that, upon binding of α-Neu5Ac to sialidase, the conformation of the pyranose ring changes from the more energetically stable $^{2}C_{5}$ chair conformation to the α-boat conformation **8**. This conformational change is mainly due to the charge–charge interaction between the carboxylate group and the triarginyl cluster [72]. Molecular modeling studies and kinetic isotope effect measurements demonstrated that this boat conformation is in part stabilized by the negatively charged environment of the catalytic site [74]. The intermediate boat conformation facilitates the cleavage of the glycosidic bond through the formation of the oxocarbenium ion intermediate **9**, which is trapped as covalent adduct **10** by nucleophilic attack of an appropriately sited Tyr residue (Tyr406 in influenza virus sialidase N2). This covalent adduct then undergoes stereoselective hydrolysis to afford free Neu5Ac.

The change in ring conformation of α-Neu5Ac during catalysis to assume the conformation of the half-chair sialosyl cation intermediate **9** was supported by the finding that the 2,3-unsaturated Neu5Ac derivative **11** has a K_i of 4×10^{-6} M against influenza A (N2) virus sialidase [75]. Compound **11**, which was identified as a sialidase inhibitor in 1969 [76, 77], was the first transition state-like sialic acid-based influenza virus sialidase inhibitor mimicking the sialosyl cation transition state **9** [78].

17.7 INFLUENZA VIRUS SIALIDASE AS A DRUG DISCOVERY TARGET

Considering that the influenza virus sialidase allows the release of the new virus particles from infected host cells, sialidase became a target for inhibitor development studies [60]. A considerable number of approaches to the discovery of new influenza virus sialidase inhibitors were undertaken. Random screening programs [79] and development of hydrolysis-resistant N- or S-glycosides of sialic acid [80, 81] identified sialidase inhibitors with limited success ($K_i = 10^{-3}$ to 10^{-6} M). Further development led to the synthesis of substrate mimetics, such as 3-fluoro-Neu5Ac ($K_i = 2.4\times10^{-6}$ M) [82] and Neu5Ac phosphonate analogues [83], which showed low micromolar levels of inhibition ($K_i = 6\times10^{-5}$ M at best). With already micromolar inhibitory activity, **11** was initially modified at C5. Replacement of the *N*-acetyl group with an *N*-trifluoroacetyl moiety led to an increase in the *in vitro* inhibition of influenza A (N1) virus sialidase ($K_i = 8\times10^{-7}$ M) [77]. Unfortunately, this compound did not exhibit inhibitory activity *in vivo* [84].

X-ray crystal structures of α-Neu5Ac and **11** in complex with influenza virus sialidases in the 1980s and early 1990s led to the opportunity for structure-based design and development of influenza virus sialidase inhibitors [66, 85].

17.7.1 Structure-Based Sialidase Inhibitor Design on a Sialic Acid Scaffold: Development of Zanamivir

Knowing that the natural ligand Neu5Ac is an inhibitor of sialidase in the millimolar range ($K_i = 5\times10^{-3}$ M), the development of new potent inhibitors started from Neu5Ac and, in particular, its unsaturated analogue **11** [75]. Derivatization and

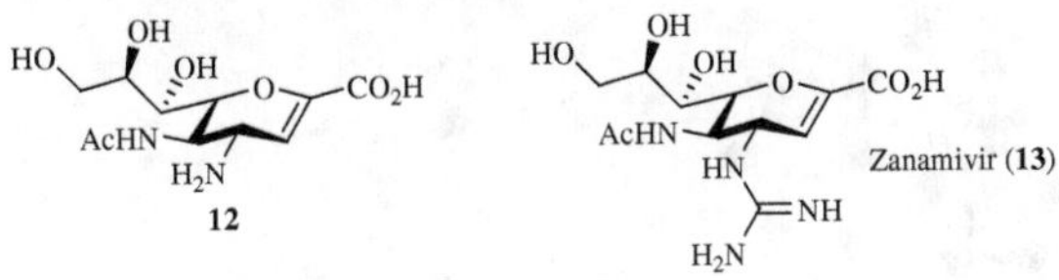

FIGURE 17.8 Developed sialidase inhibitors following structural modifications of compound **11**.

modification of **11**, based on the binding of α-Neu5Ac and **11** to sialidase as seen in the A/N2 X-ray crystal structure [66, 86], opened a new avenue to potent inhibition of the influenza virus sialidase [64, 66, 87]. Initially, the molecular modeling program GRID [88] was used to determine the regions of the catalytic site that would show, theoretically, higher affinity for different functional group probes (e.g., carboxyl, amino, or hydrophobic probes) [87]. From this, a series of modifications at different positions of **11** were studied. These studies brought the synthesis of new inhibitors, **12** and **13** (Fig. 17.8), which had improved enzyme inhibition compared to **11** [75, 89].

Analysis of the active site and GRID studies showed that near C4 of **11** was a pocket (S2) containing acidic residues (Glu119 and Glu227) that could be targeted by the introduction of a basic functional group [87, 89]. When an amino group was introduced at C4 (i.e., compound **12**), an increase in binding affinity over **11** was observed due to salt-bridge interaction of the C4-amino group with Glu119 in S2 (**12**: $K_i = 4 \times 10^{-8}$ M) [75, 87, 89, 90].

Further analysis predicted that it could be possible to fill S2 with a more basic group like the guanidino moiety [89]. Binding studies between the guanidino-functionalized **13** and NA [89, 91] revealed that the primary guanidinyl nitrogens form a hydrogen bond with the carboxylate of Glu227, oxygen of Trp178, and a water-mediated hydrogen bond (Fig. 17.9 [92, 93]). The secondary guanidinyl nitrogen interacts with Asp151 and Glu119, which is slightly moved from the original position through van der Waals and electrostatic interactions [91]. The guanidino moiety of **13**, when bound in the active site of influenza A sialidase, was found to displace a water molecule from S2, gaining entropy, resulting in slow-binding kinetics [89, 91]. The 4-guanidino derivative **13** ($K_i = 10^{-11}$ M) improved the sialidase inhibition by 1000-fold over the 4-amino derivative **12** [75, 89]. This potent inhibition was seen against a wide range of influenza A and B virus strains [94] and against all NA subtypes known at the time [31, 89, 95–97].

Derivative **13**, with its extremely potent sialidase inhibition *in vitro* and *in vivo* [31, 89], was subsequently developed as zanamivir (Relenza®) by Glaxo for treatment of influenza A and B viral infection. Zanamivir was revealed to be rapidly excreted and to have poor oral bioavailability [98, 99] because of its highly polar nature. It is therefore administered through inhalation; however, intravenous administration has also been explored and proven effective [100].

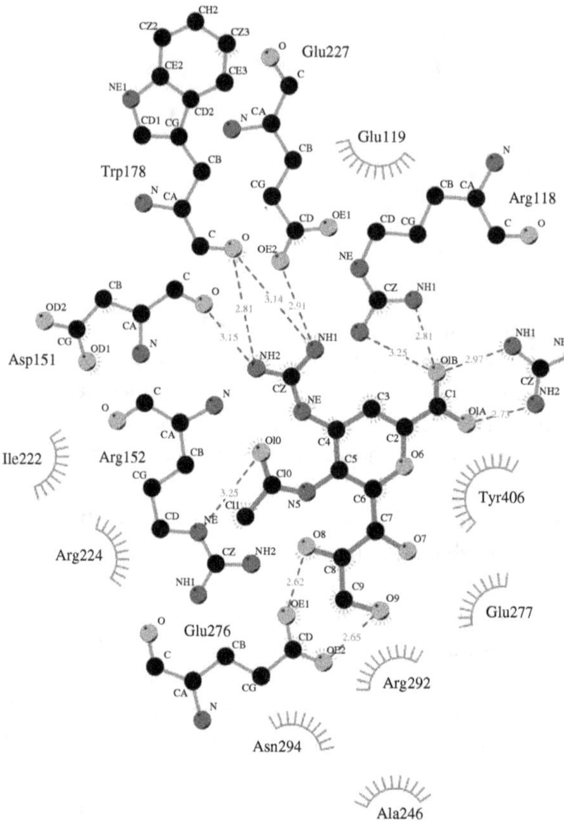

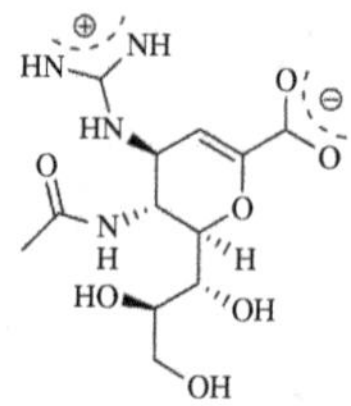

13

FIGURE 17.9 Key interactions of zanamivir (**13**) with the active site of influenza A virus sialidase (Figure generated from crystal structure data (PDB 1nmc) using LIGPLOT [92, 93]). To the right is **13** shown in the same orientation.

17.7.2 Second-Generation Zanamivir

In the challenge to develop potent influenza virus sialidase inhibitors, a large amount of research has been dedicated to the manipulation of every position on **11** except C3. Structure–activity relationship (SAR) studies carried out on compounds derived from **11** before and during the development of zanamivir (reviewed in [101–103]) revealed structural requirements to conserve the main interactions between the substrate inhibitor and the active site of NA, particularly with regards to the carboxylate, C4-guanidino, and C5-acetamido moieties.

To overcome the limits of oral bioavailability of zanamivir, the development of second-generation zanamivir was pointed toward improving its pharmacokinetic properties. In this context, the most encouraging results were obtained with the phosphonate analogue of zanamivir, with modification of the C7-hydroxy group or with a replacement of the glycerol side chain that also improved oral bioavailability compared to the original inhibitor **13** (Fig. 17.10).

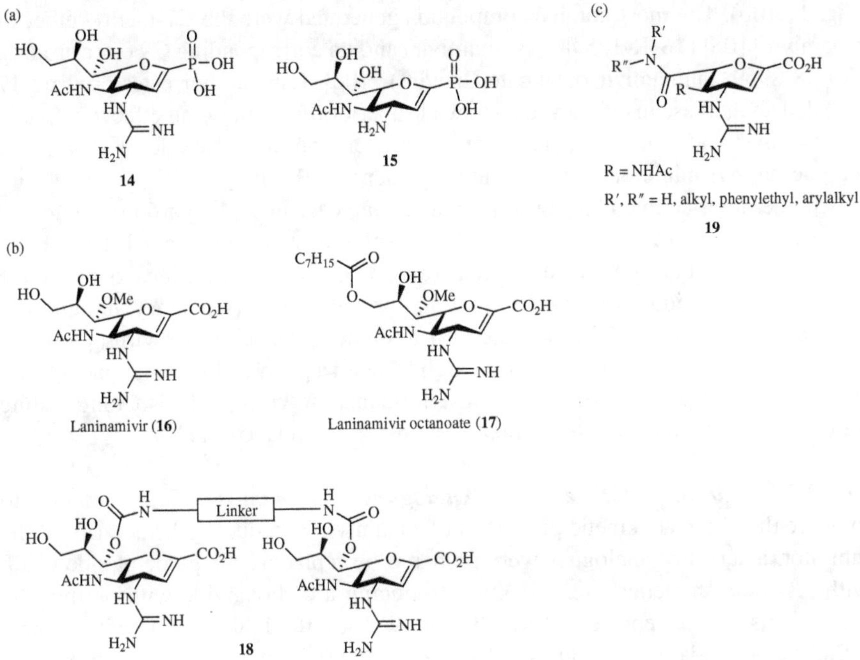

FIGURE 17.10 Analogues of zanamivir with improved bioavailability: (a) phosphonate analogues, (b) C7-functionalized derivatives, and (c) hydrophobic side-chain analogues.

17.7.2.1 Phosphonate Analogues of Zanamivir The main interactions by which α-Neu5Ac binds into the sialidase active site are with S1 and S3 (Fig. 17.7) [72, 93]. It has been calculated that the interaction between S1 (the triarginyl cluster) and the carboxylate of α-Neu5Ac is around 50% of the overall binding energy [72]. It is not surprising that, just recently, Chi-Huey Wong and coworkers [104] demonstrated that the phosphonate zanamivir analogue **14** (Fig. 17.10a) could possibly be a more potent inhibitor of influenza virus sialidases than the carboxylate derivative **13**. The replacement of the carboxylic acid group with a phosphonate would be expected to increase the charge–charge interaction with the triarginyl cluster.

Both the 4-guanidino and 4-amino derivatives **14** and **15**, respectively, were revealed to be very effective NA inhibitors with IC_{50} values in the low nanomolar range [104]. Because of its less polar group at C4 (leading to possibly higher bioavailability), the 4-amino compound **15** could be considered as a starting point for drug development studies.

17.7.2.2 Functionalization of the C7 Position of Zanamivir X-ray studies of influenza virus sialidase in complex with α-Neu5Ac [66] and zanamivir [91] showed that the C7-OH is not involved in any direct interaction in the active site of influenza virus NA, and only the C8- and C9-hydroxy groups are involved in the binding interactions. Based on this knowledge, different manipulations have been attempted at C7, for example, by a simple substitution of the C7-hydroxy group or alkylation

(Fig. 17.10b). The most notable compounds generated were the C7-methyl ether of zanamivir [105] (**16**, R-125489, laninamivir) and its corresponding C9 ester prodrug **17** (CS-8959, laninamivir octanoate, Inavir®) [106]. Laninamivir ester prodrug **17** revealed an increase in efficacy and showed superior life-prolonging effect compared to zanamivir and laninamivir [106] and has been approved for sale in Japan as a long-acting NA inhibitor for treatment of influenza [107, 108].

Another long-acting inhibitor series with an increase in binding affinity compared to zanamivir was a result of a multivalent display of zanamivir linked through C7 [105, 109, 110]. For dimeric compounds (e.g., **18**), a desirable distance between the two zanamivir residues of 20–23 Å gave an optimal inhibitory potency and allowed for a cross-linkage of sialidase either on a single virus or across virus particles. Connecting two zanamivir residues through C7 with a poly-L-glutamine spacer [105] showed an increase of efficacy against influenza A virus and also long-lasting protective effects in mice after intranasal administration [105, 111].

17.7.2.3 ***Hydrophobic Side-Chain Analogues of Zanamivir*** In an attempt to improve the pharmacokinetic properties of zanamivir by reducing the polarity of the inhibitor, a series of analogues were prepared by replacing the glycerol side chain with less polar moieties [62, 70]. By introducing a carboxamide with optimal N-substituents at C6 (represented by **19**, Fig. 17.10c), the inhibitory activity against influenza A sialidase was found similar to zanamivir's inhibition. These compounds, however, showed weaker inhibition against influenza B sialidases. A rearrangement of Glu276 to form a salt bridge with neighboring Arg224 due to the presence of the hydrophobic alkyl substituent on the carboxamide side chain led S5 to assume a lipophilic character. However, this conformational rearrangement of Glu276 in influenza B virus appeared to be less energetically favored [70].

17.7.3 Sialidase Inhibitors Based on a Cyclohexene Scaffold: Development of Oseltamivir

The worldwide interest in fighting influenza virus brought a novel approach to synthesize new potent sialidase inhibitors on noncarbohydrate scaffolds. Using cyclohexene as a scaffold with the double bond in a position (C1–C2) to mimic the double bond in the transition state **9** of the sialidase enzymatic mechanism (see Scheme 17.1), a new sialidase inhibitor **20** (Fig. 17.11), showing nanomolar levels of influenza sialidase inhibition, was developed [112]. Not surprisingly, when unsaturation was placed as in compound **11**, a loss in inhibitory activity against influenza A virus sialidase was observed, confirming the importance of the double bond position [112, 113].

The hydrophobic backbone of the glycerol side chain of **11** and zanamivir had been shown to interact with the NA active site. This observation suggested that a complete replacement of the glycerol side chain with an alkyl moiety may lead to a new potent sialidase inhibitor [112]. The glycerol side chain was replaced with a series of alkyl ether side chains, and the main groups of **11** (carboxylate, C4-amino, and C5-acetamido groups) on the cyclic ring were maintained to allow the important interactions in the active site of NA [112]. Considerable SAR studies were

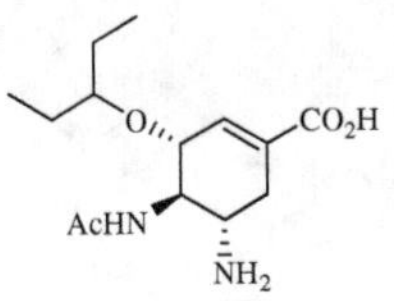

Oseltamivir carboxylate (**20**)

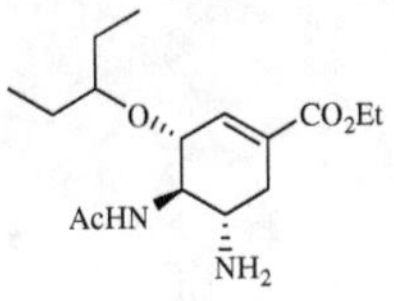

Oseltamivir (**21**)

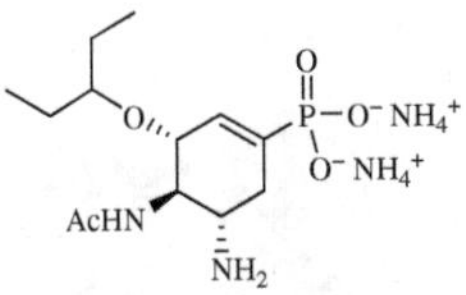

22

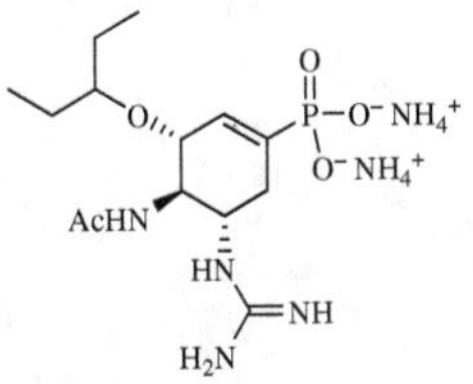

23

FIGURE 17.11 Oseltamivir and its analogues.

undertaken, leading to the development of the carbocyclic analogue oseltamivir carboxylate (**20**) [112]. Oseltamivir carboxylate was found to be a potent inhibitor of both influenza A and B viral sialidases ($K_i = 10^{-9}$ M).

An X-ray crystal structure of **20** in complex with influenza A virus N1 sialidase indicated that the cyclohexene ring was oriented in the active site, maintaining all of the interactions as observed for **11** [112]. The 3-pentyl side chain is accommodated in a hydrophobic pocket, with subsequent rearrangement of the amino acid residues in S4 and S5, especially with a reorientation of Glu276 to form a salt bridge with Arg224 (Fig. 17.12) as seen for the zanamivir carboxamide derivatives **19** [62]. The replacement of the C4-amino group of **20** with a guanidino group did not significantly improve the level of inhibition [113]. SAR studies revealed that the overall binding seem to be dominated by the lipophilic interactions of the hydrophobic side chain [113, 114].

Drug development on inhibitor **20** led to the introduction of the ethyl ester prodrug form **21** that revealed good oral bioavailability [112]. This effect was attributed to a combination of both the hydrophilic 3-pentyl tail that replaced the polar glycerol side chain of zanamivir and the use of the less basic amino group in place of the guanidino group. Oseltamivir (**21**) is marketed as Tamiflu® by Roche and has been stockpiled by governments worldwide as part of their preparedness plans for pandemic influenza.

Previously reported biological data on a Neu5Ac phosphonate analogue [115] and molecular modeling studies [116] showed the possibility of replacing the carboxylate group with a phosphonate to gain strong interactions with the Arg residues within S1 of the active site. In comparison with the carboxylate–guanidinium ion pair, it was believed that interactions between the phosphonate and guanidino ions of the Arg residues would have much stronger electrostatic character [116]. Phosphonate analogues **22** and **23** (Fig. 17.11) showed significant low nanomolar inhibition against influenza A virus sialidase [116]. Importantly, the guanidino derivative **23** also showed nanomolar levels of inhibition of oseltamivir carboxylate-resistant strains.

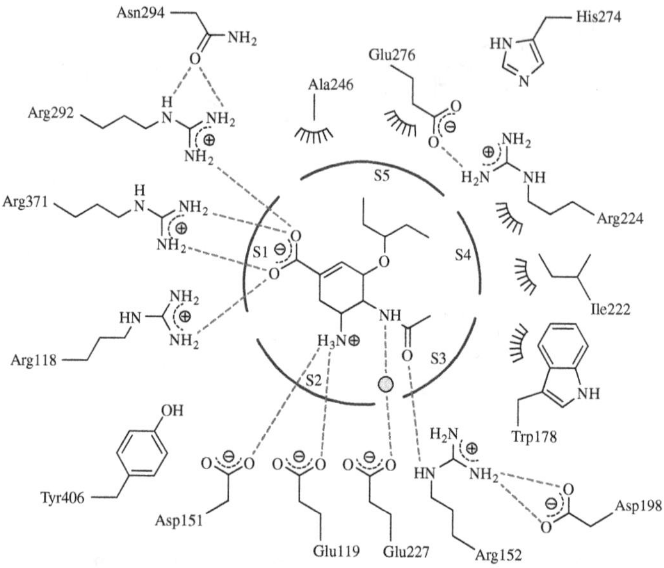

FIGURE 17.12 Interaction of oseltamivir carboxylate (**20**) with the influenza virus sialidase active site.

17.7.4 Sialidase Inhibitors Based on Five-Membered Ring Scaffolds

A number of researchers have reported and demonstrated that maintaining the appropriate position of the substituents on a cyclic scaffold to interact with the established conserved amino acid residues involved in substrate binding can lead to development of new classes of influenza virus sialidase inhibitors [117]. Two drugs based on five-membered ring scaffolds have been developed as potent sialidase inhibitors. Cyclopentane derivative **24** (BCX-1812, peramivir) [117, 118] and pyrrolidine derivative **25** (ABT-675) [119] show nanomolar levels of inhibition of both influenza A and B viral sialidases (Fig. 17.13).

An X-ray crystal structure of peramivir (**24**) in complex with influenza A N9 showed interactions of the 3-pentyl group with S4 and S5 [117]. Due to the different stereochemistry of the guanidino moiety of **24** compared to zanamivir, a water molecule was displaced from S2 when **24** was bound. The different binding mode of the guanidine moiety within the active site provided the reason by which **24** showed inhibitory activity also for zanamivir-resistant influenza virus sialidase strains [117, 120]. Compound **24**, which showed comparable or better efficacy *in vivo* than zanamivir and oseltamivir [121, 122], successfully completed animal studies and is in phase III clinical trials.

Another remarkable influenza virus sialidase inhibitor, derivative **25** [119], was developed based on a pyrrolidine core by Abbott Laboratories [119, 123]. X-ray crystal studies surprisingly observed hydrophobic interactions of the *cis*-propenyl group within S2, which is normally involved in charge–charge interaction as with the guanidino moiety in zanamivir [69]. The side-chain methyl and methoxy groups of

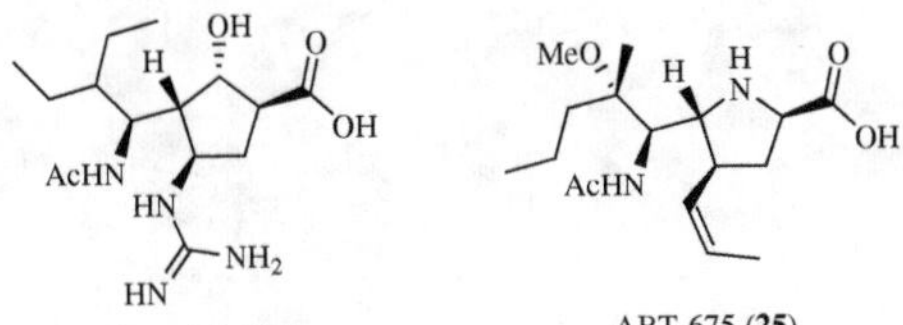

Peramivir (**24**)

ABT-675 (**25**)

FIGURE 17.13 Sialidase inhibitors with five-membered rings.

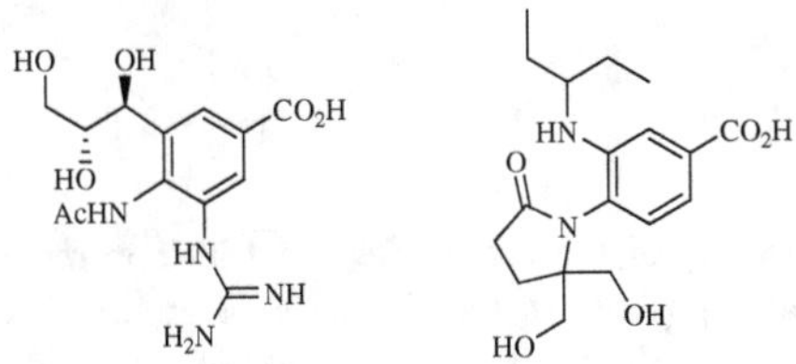

26 **27**

FIGURE 17.14 Sialidase inhibitors with core aromatic rings.

25 interact with S5, whereas the *n*-propyl chain interacts within S4. Compound **25** exhibits similar *in vitro* activity to zanamivir and oseltamivir against both influenza A and B sialidases [119]. Because of its zwitterionic nature and, therefore, limited oral bioavailability, **25** was further developed as the corresponding ethyl ester prodrug, which showed oral bioavailability of over 80% [119]. Interestingly, a lack of cross-resistance toward oseltamivir-resistant influenza sialidase N1 and N2 subtypes was observed [124].

17.7.5 Sialidase Inhibitors Based on an Aromatic Ring Scaffold

In order to have an easy access to the synthesis of a new class of sialidase inhibitors, a benzene ring was considered as a starting point for the scaffold. The aromatic analogue (**26**) of zanamivir was synthesized (Fig. 17.14); however it only had low inhibitory activity against influenza A virus N2 sialidases ($K_i > 1 \times 10^{-4}$ M) [125]. This weak inhibitory activity was believed to be due to the lack of interactions of the ring substituents, which clearly assumed different orientations within the active site to those seen with zanamivir [125]. Further development based on the benzoic acid scaffold [126] led to the design of derivative **27** incorporating a hydrophobic 3-pentyl side chain, which improved the activity against influenza A virus N9 sialidase ($IC_{50} = 50$ nM) [127, 128].

17.8 STRUCTURAL DIFFERENCES RECENTLY IDENTIFIED IN INFLUENZA A VIRUS SIALIDASE SUBTYPES

Until recently, the X-ray structures solved for sialidases from a limited number of influenza A (N2 and N9) and B subtypes had shown very similar structures, both with and without bound ligand or inhibitor, which was not unexpected as the amino

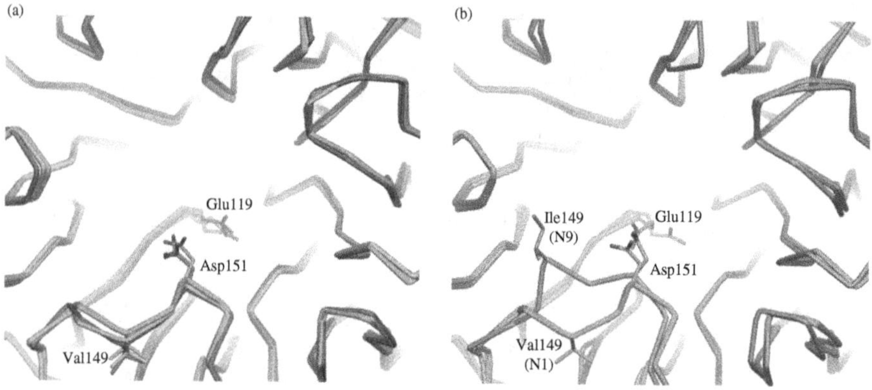

FIGURE 17.15 Comparison of the active site and 150-loop regions of *apo* group 1 and group 2 sialidases. (a) Superposition of the sialidases of group 1 (N1, N4, N8) showing an "open" conformation of the 150-loop. (b) Superposition of the sialidases N1 and N9 showing "open" (N1) and "closed" (N9) conformations of the 150-loop. Val, valine.

acid residues in the active site were conserved in these strains. However, in 2006, work carried out by John Skehel and coworkers [129] suggested two distinct families of influenza A virus sialidases, which are structurally and phylogenetically distinct: group 1 containing N1, N4, N5, and N8 subtypes and group 2 containing N2, N3, N6, N7, and N9 subtypes. Structural studies of the sialidases of group 1, including N1 from the avian flu H5N1, revealed that the enzyme active sites in the *apo* form (Fig. 17.15a) have a considerably different three-dimensional structure from the group 2 sialidases (Fig. 17.15b) [129].

In the *apo* structures of the N1, N4, and N8 sialidases close to the S2 domain of the catalytic site, there is a cavity due to the opening of a flexible protein loop—the 150-loop (residues 147–152). This is different to what was seen in group 2 sialidase when the 150-loop is closed (Fig. 17.15b) [129]. However, X-ray crystallography of group 1 sialidase N1 complexed with oseltamivir carboxylate (Fig. 17.16) revealed that, after 30 min of incubation, the conformation of the 150-loop slowly changes, ending in a similar conformation to the group 2 NA complexes [129].

When the flexible 150-loop is in the open form in group 1 sialidases, the cavity is accessible because the side-chain position of Asp151 is changed with the carboxylate group pointing in a different direction to that of Asp151 of group 2 sialidases. The side chain of Glu119 also had adopted a different orientation. The 150-cavity in group 1 sialidases is about 10 Å long and 5 Å wide [129].

The discovery of the large 150-cavity adjacent to the active site in X-ray structures of group 1 sialidases suggested new opportunities for drug design. Development of new influenza virus NA inhibitors is of major relevance because of the threat of pandemic influenza (particularly from strains such as H5N1 and H1N1, containing the group 1 sialidase N1) and the potential loss of effectiveness of the more widely used anti-influenza drug Tamiflu. Reports have shown resistance to Tamiflu in adults (2%) and in children (18%), whereas for Relenza, which is structurally close to the natural substrate, only a few cases of drug resistance have been reported [35,

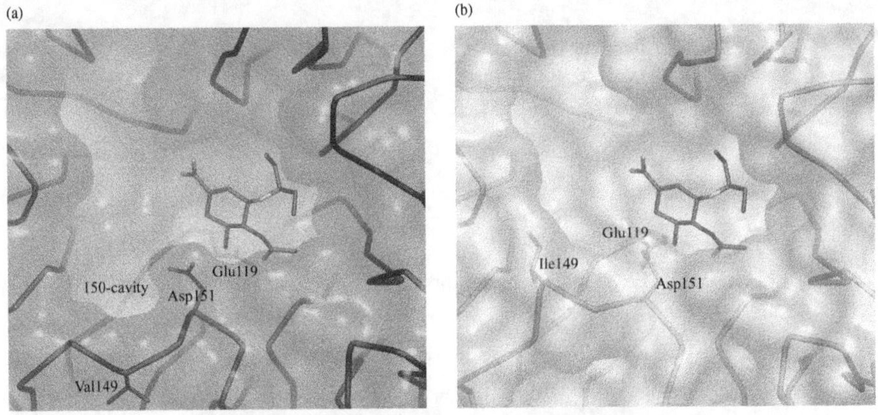

FIGURE 17.16 Comparison of group 1 and group 2 sialidase active sites. (a) Oseltamivir carboxylate binds into the open catalytic site of group 1 sialidase N1. (b) Binding between oseltamivir carboxylate and group 2 sialidase N9 with the closed 150-loop.

130–133]. More alarming was the rapid appearance of Tamiflu resistance in circulating seasonal H1N1 up to virtually 100% resistance in 2007–2009 [134].

17.9 NEW INFLUENZA VIRUS SIALIDASE INHIBITORS TARGETING THE 150-CAVITY

It is believed that engaging interactions within the 150-cavity could lead to a new class of influenza virus sialidase inhibitors. Recently, a few research groups explored the potential to bind the 150-cavity, for example, by introducing a series of modifications on the guanidino group of zanamivir (e.g., **28**) [135] as well as the manipulation of the amino group of oseltamivir (e.g., **29**) [136] to try to reach the 150-cavity (Fig. 17.17). Investigation of the binding of **29** to influenza A virus N1 by saturation transfer difference NMR (STD-NMR) spectroscopy suggested an interaction of the substituted triazole ring with the 150-cavity. This activity provided a potential reason for the better inhibition of **29** against a group 1 sialidase (N1, $K_i = 0.07\,\mu M$) than against a group 2 sialidase (N2, $K_i = 2.6\,\mu M$) [136].

Modeling studies carried out by Mark von Itzstein's group [137] on the 150-loop open structures of group 1 sialidase N1 revealed the possibility to access the predominantly hydrophobic 150-cavity by an appropriate substitution at C3 of **11**. As shown in Figure 17.18, an opportune substitution at C3 of **11** would be accommodated in the 150-cavity in the more open form of influenza virus sialidase of group 1 and could also potentially bind to the active site with minimal distortion of the normal binding mode of **11**.

The aforementioned work thus provided the novel C3 C-alkylated derivatives **30** and **31**, which selectivity inhibited a series of group 1 N1 sialidases compared to a group 2 N2 sialidase (Table 17.1) and bound group 1 N8 with the 150-loop in the open conformation [137]. In particular, **31** gives a micromolar level of inhibition against pandemic 2009 N1 (09 N1) sialidase.

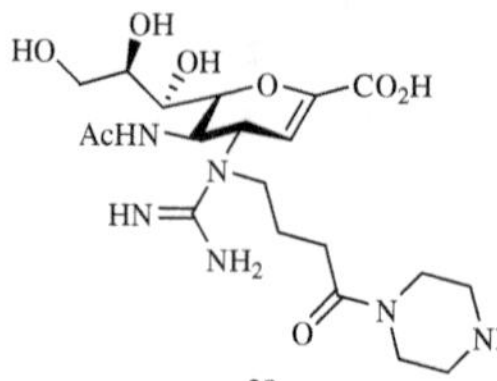

28

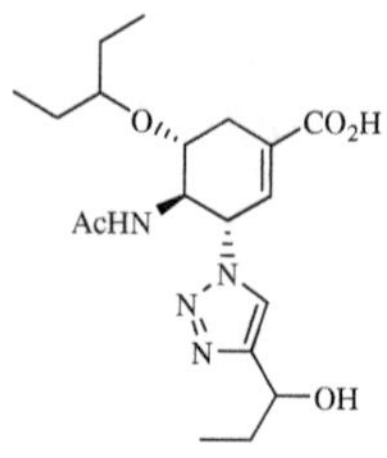

29

FIGURE 17.17 Sialidase inhibitors designed to target the 150-cavity.

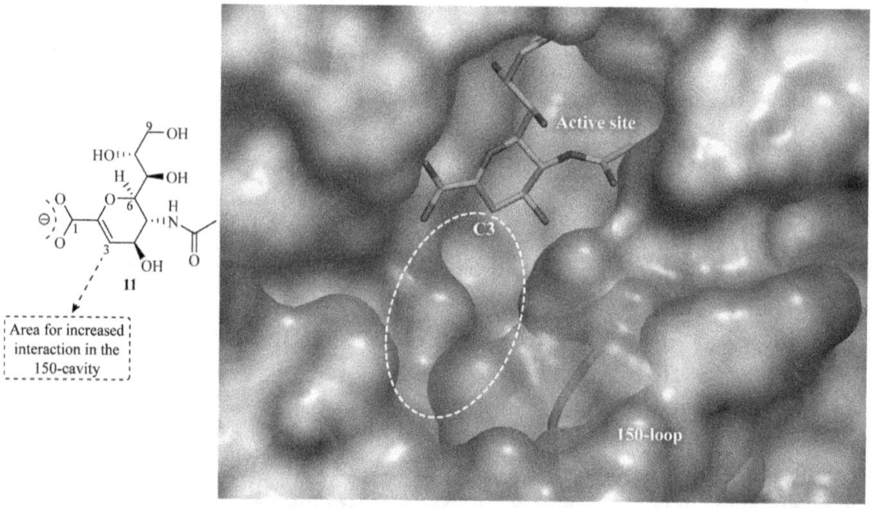

FIGURE 17.18 Compound **11** positioned in N1 active site. 150-loop open (electrostatic surface), closed (magenta surface), dotted lines indicate approximate area of the 150-cavity.

TABLE 17.1 Inhibition of Influenza A Virus Sialidases With C3 C-Alkylated Derivatives 30 and 31 [137]

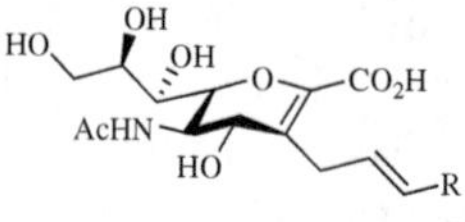

30: R = H
31: R = 4-MePh

Inhibitor	Sialidase Inhibition (K_i, μM)		
	A/N1[a]	A/N1[b]	A/N2[c]
30	222 ± 17	153 ± 18	3629 ± 2130
31	7.3 ± 0.8	1.7 ± 0.2	219 ± 30

[a] H5N1.
[b] H1N1 2009 pandemic.
[c] H3N2.

X-ray crystallography studies with group 1 sialidase N8 revealed the ability of the C3-substituent of derivative **31** to occupy the area of the 150-cavity and maintain the 150-loop in an open form (Fig. 17.19a, b) [137]. However, when **11** was bound to the active site of N8, a closed conformation of the 150-loop was observed (Fig. 17.19c).

In contrast to previous observations with (H5)N1 sialidase [129], a recent X-ray structural investigation of 09 N1 suggested the lack of a 150-cavity in this group 1 sialidase [138]. This suggestion is in contrast to recent biological assay data obtained with inhibitor **31** (Table 17.1) and has raised an interesting question: *Could the 150-loop in 09 N1 be trapped in the open form to accommodate the C3 substituent of* ***31*** *and, therefore, adopt the "usual" open conformation of group 1 NAs?* Molecular dynamics simulations of the flexibility of the 150-loop in group 1 (including 09 N1) and group 2 sialidases carried out by Rommie Amaro et al. [139] as well as by von Itzstein's group (unpublished work) suggest that the flexibility in solution may be different between group 1 NAs, and, importantly, it is not restricted only to group 1 NAs but can also be observed to some extent in group 2 NAs.

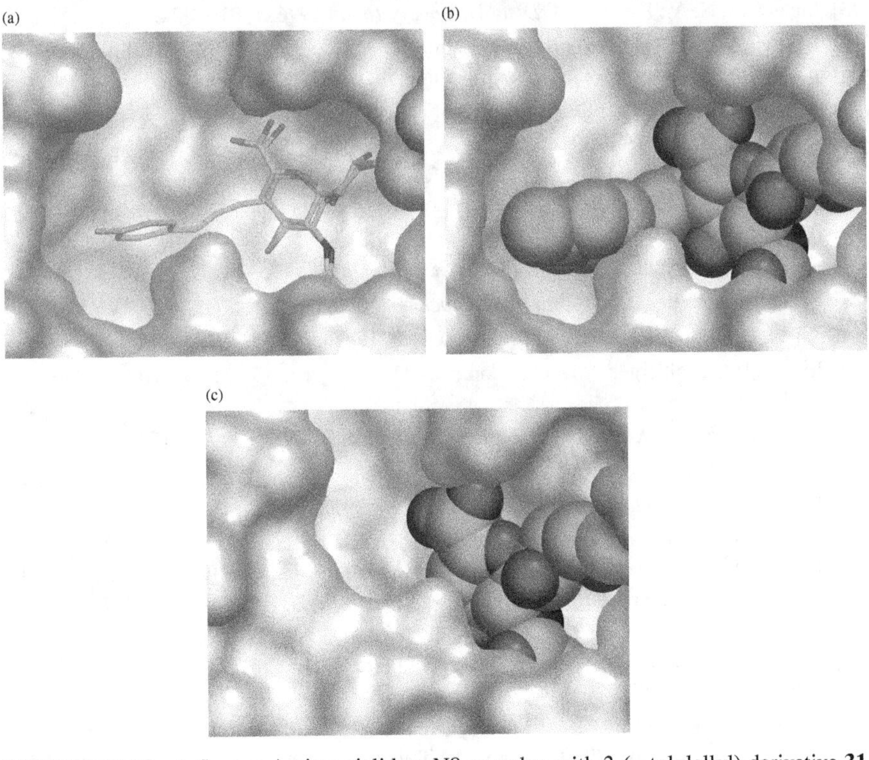

FIGURE 17.19 Influenza A virus sialidase N8 complex with 3-(*p*-tolylallyl) derivative **31**. (a) Superimposition of N8–inhibitor complexes of **31** (PDB 3O9k) and **11** (PDB 2htr). (b) N8–**31** complex with an open 150-loop. (c) N8–**11** complex with a closed 150-loop.

REFERENCES

[1] Simpsons, J.; Weiner, E. *The Oxford English Dictionary*, 2nd ed.; Oxford University: London, **1989**; Vol. *1*.

[2] Zambon, M. C. *Rev. Med. Virol.* **2001**, *11*, 227–241.

[3] World Health Organization. Influenza (seasonal). http://www.who.int/mediacentre/factsheets/fs211/en/ (accessed July 20, **2015**).

[4] Kuszewski, K.; Brydak, L. *Biomed. Pharmacother.* **2000**, *54*, 188–195.

[5] Horimoto, T.; Kawaoka, Y. *Nat. Rev. Microbiol.* **2005**, *3*, 591–600.

[6] Klenk, H. D.; Rott, R. *Adv. Virus Res.* **1988**, *34*, 247–281.

[7] Baigent, S. J.; McCauley, J. W. *BioEssays* **2003**, *25*, 657–671.

[8] Pica, N.; Chou, Y.-Y.; Bouvier, N. M.; Palese, P. *J. Virol.* **2012**, *86*, 4279–4287.

[9] Yamashita, M.; Krystal, M.; Fitch, W. M.; Palese, P. *Virology* **1988**, *163*, 112–122.

[10] Matsuzaki, Y.; Katsushima, N.; Nagai, Y.; Shoji, M.; Itagaki, T.; Sakamoto, M.; Kitaoka, S.; Mizuta, K.; Nishimura, H. *J. Infect. Dis.* **2006**, *193*, 1229–1235.

[11] Mould, J. A.; Paterson, R. G.; Takeda, M.; Ohigashi, Y.; Venkataraman, P.; Lamb, R. A.; Pinto, L. H. *Dev. Cell* **2003**, *5*, 175–184.

[12] Lamb, R. A.; Choppin, P. W. *Annu. Rev. Biochem.* **1983**, *52*, 467–506.

[13] Meanwell, N. A.; Krystal, M. *Drug Discovery Today* **1996**, *1*, 316–324.

[14] Bui, M.; Wills, E. G.; Helenius, A.; Whittaker, G. R. *J. Virol.* **2000**, *74*, 1781–1786.

[15] Reinhardt, J.; Wolff, T. *Vet. Microbiol.* **2000**, *74*, 87–100.

[16] Tong, S.; Li, Y.; Rivailler, P.; Conrardy, C.; Castillo, D. A.; Chen, L. M.; Recuenco, S.; Ellison, J. A.; Davis, C. T.; York, I. A.; Turmelle, A. S.; Moran, D.; Rogers, S.; Shi, M.; Tao, Y.; Weil, M. R.; Tang, K.; Rowe, L. A.; Sammons, S.; Xu, X.; Frace, M.; Lindblade, K. A.; Cox, N. J.; Anderson, L. J.; Rupprecht, C. E.; Donis, R. O. *Proc. Natl. Acad. Sci. U. S. A.* **2012**, *109*, 4269–4274.

[17] Gamblin, S. J.; Skehel, J. J. *J. Biol. Chem.* **2010**, *285*, 28403–28409.

[18] Skehel, J. J.; Wiley, D. C. *Annu. Rev. Biochem.* **2000**, *69*, 531–569.

[19] Wiley, D. C.; Skehel, J. J. *Annu. Rev. Biochem.* **1987**, *56*, 365–394.

[20] Liu, C.; Eichelberger, M. C.; Compans, R. W.; Air, G. M. *J. Virol.* **1995**, *69*, 1099–1106.

[21] Nicholson, K. G.; Wood, J. M.; Zambon, M. *Lancet* **2003**, *362*, 1733–1745.

[22] Suzuki, Y. *Biol. Pharm. Bull.* **2005**, *28*, 399–408.

[23] Potter, C. W. *J. Appl. Microbiol.* **2001**, *91*, 572–579.

[24] de Jong, M. D.; Simmons, C. P.; Thanh, T. T.; Hien, V. M.; Smith, G. J. D.; Chau, T. N. B.; Hoang, D. M.; Nguyen, V. V. C.; Khanh, T. H.; Dong, V. C.; Qui, P. T.; Van Cam, B.; Ha, D. Q.; Guan, Y.; Peiris, J. S. M.; Chinh, N. T.; Hien, T. T.; Farrar, J. *Nat. Med.* **2006**, *12*, 1203–1207.

[25] Sorrell, E. M.; Wan, H.; Araya, Y.; Song, H.; Perez, D. R. *Proc. Natl. Acad. Sci. U. S. A.* **2009**, *106*, 7565–7570.

[26] Beigel, J. H.; Farrar, J.; Han, A. M.; Hayden, F. G.; Hyer, R.; de Jong, M. D.; Lochindarat, S.; Tien, N. T. K.; Hien, T. T.; Nicoll, A.; Touch, S.; Yuen, K.-T. *N. Engl. J. Med.* **2005**, *353*, 1374–1385.

[27] Peiris, J. S. M.; de Jong, M. D.; Guan, Y. *Clin. Microbiol. Rev.* **2007**, *20*, 243–267.

[28] World Health Organization. Cumulative number of confirmed human cases of avian influenza A(H5N1) reported to WHO. http://www.who.int/influenza/human_animal_interface/H5N1_cumulative_table_archives/en/index.html (accessed July 20, **2015**).

[29] Sullivan, S. J.; Jacobson, R. M.; Dowdle, W. R.; Poland, G. A. *Mayo Clin. Proc.* **2010**, *85*, 64–76.

[30] Murray, C. J. L.; Lopez, A. D.; Chin, B.; Feehan, D.; Hill, K. H. *Lancet* **2006**, *368*, 2211–2218.

[31] von Itzstein, M. *Nat. Rev. Drug Discovery* **2007**, *6*, 967–974.

[32] Matrosovich, M. N.; Matrosovich, T. Y.; Gray, T.; Roberts, N. A.; Klenk, H. D. *J. Virol.* **2004**, *78*, 12665–12667.

[33] Lagoja, I. M.; De Clercq, E. *Med. Res. Rev.* **2008**, *28*, 1–38.

[34] Meanwell, N. A.; Krystal, M. *Drug Discovery Today* **1996**, *1*, 388–397.

[35] Ison, M. G. *Curr. Opin. Virol.* **2011**, *1*, 563–573.

[36] Rosenberg, M. R.; Casarotto, M. G. *Proc. Natl. Acad. Sci. U. S. A.* **2010**, *107*, 13866–13871.

[37] Hay, A. J. *Semin. Virol.* **1992**, *3*, 21–30.

[38] Douglas, R. G. J. *N. Engl. J. Med.* **1990**, *322*, 443–450.

[39] Wintermeyer, S. M.; Nahata, M. C. *Ann. Pharmacother.* **1995**, *29*, 299–310.

[40] Pinto, L. H.; Holsinger, L. J.; Lamb, R. A. *Cell* **1992**, *69*, 517–528.

[41] Pinto, L. H.; Lamb, R. A. *J. Biol. Chem.* **2006**, *281*, 8997–9000.

[42] Hayden, F. G. *Clin. Infect. Dis.* **2009**, *48*, S3–S13.

[43] Hay, A. J.; Thompson, C. A.; Geraghty, A.; Hayhurst, S.; Grambas, S.; Bennett, M. S. In *Options for the Control of Influenza Virus*; Hannoun, C., Ed.; Excerpta Medica: Amsterdam, **1993**; Vol. *II*, pp 281–288.

[44] Hayden, F. G.; Belshe, R. B.; Clover, R. D.; Hay, A. J.; Oakes, M. G.; Soo, W. *N. Eng. J. Med.* **1989**, *321*, 1696–1702.

[45] Furuta, Y.; Takahashi, K.; Shiraki, K.; Sakamoto, K.; Smee, D. F.; Barnard, D. L.; Gowen, B. B.; Julander, J. G.; Morrey, J. D. *Antivir. Res.* **2009**, *82*, 95–102.

[46] Kiso, M.; Takahashi, K.; Sakai-Tagawa, Y.; Shinya, K.; Sakabe, S.; Le, Q. M.; Ozawa, M.; Furuta, Y.; Kawaoka, Y. *Proc. Natl. Acad. Sci. U. S. A.* **2010**, *107*, 882–887.

[47] Angata, T.; Varki, A. *Chem. Rev.* **2002**, *102*, 439–470.

[48] Varki, A.; Gagneux, P. *Ann. N. Y. Acad. Sci.* **2012**, *1253*, 16–36.

[49] Schauer, R. *Adv. Carbohydr. Chem. Biochem.* **1982**, *40*, 131–234.

[50] Wagner, R.; Matrosovich, M.; Klenk, H. D. *Rev. Med. Virol.* **2002**, *12*, 159–166.

[51] Sauter, N. K.; Hanson, J. E.; Glick, G. D.; Brown, J. H.; Crowther, R. L.; Park, S. J.; Skehel, J. J.; Wiley, D. C. *Biochemistry* **1992**, *31*, 9609–9621.

[52] Wilson, I. A.; Skehel, J. J.; Wiley, D. C. *Nature* **1981**, *289*, 366–373.

[53] Lis, H.; Sharon, N. *Chem. Rev.* **1998**, *98*, 637–674.

[54] Bovin, N. V.; Tuzikov, A. B.; Chinarev, A. A.; Gambaryan, A. S. *Glycoconj. J.* **2004**, *21*, 471–478.

[55] Mammen, M.; Dahmann, G.; Whitesides, G. M. *J. Med. Chem.* **1995**, *38*, 4179–4190.

[56] Sauter, N. K.; Bednarski, M. D.; Wurzburg, B. A.; Hanson, J. E.; Whitesides, G. M.; Skehel, J. J.; Wiley, D. C. *Biochemistry* **1989**, *28*, 8388–8396.

[57] Mammen, M.; Chio, S.-K.; Whitesides, G. M. *Angew. Chem., Int. Ed.* **1998**, *37*, 2754–2794.

[58] Sun, X.-L. *Curr. Med. Chem.* **2007**, *14*, 2304–2313.

[59] Matrosovich, M.; Klenk, H.-D. *Rev. Med. Virol.* **2003**, *13*, 85–97.

[60] Moscona, A. *N. Engl. J. Med.* **2005**, *353*, 1363–1373.

[61] De Clercq, E. *Nat. Rev. Microbiol.* **2004**, *2*, 704–720.

[62] Taylor, N. R.; Cleasby, A.; Singh, O.; Skarzynski, T.; Wonacott, A. J.; Smith, P. W.; Sollis, S. L.; Howes, P. D.; Cherry, P. C.; Bethell, R.; Colman, P.; Varghese, J. *J. Med. Chem.* **1998**, *41*, 798–807.

[63] Varghese, J. N.; Colman, P. M.; van Donkelaar, A.; Blick, T. J.; Sahasrabudhe, A.; McKimm-Breschkin, J. L. *Proc. Natl. Acad. Sci. U. S. A.* **1997**, *94*, 11808–11812.

[64] Varghese, J. N.; Laver, W. G.; Colman, P. M. *Nature* **1983**, *303*, 35–40.

[65] Colman, P. M.; Varghese, J. N.; Laver, W. G. *Nature* **1983**, *303*, 41–44.

[66] Varghese, J. N.; McKimm-Breschkin, J. L.; Caldwell, J. B.; Kortt, A. A.; Colman, P. M. *Proteins* **1992**, *14*, 327–332.

[67] Colman, P. M.; Hoyne, P. A.; Lawrence, M. C. *J. Virol.* **1993**, *67*, 2972–2980.

[68] Thomson, R.; von Itzstein, M. In *Influenza Virus Sialidase—A Drug Discovery Target*; von Itzstein, M., Ed.; Springer: Basel, **2012**, pp 77–104.

[69] Stoll, V.; Stewart, K. D.; Maring, C. J.; Muchmore, S.; Giranda, V.; Gu, Y.-G. Y.; Wang, G.; Chen, Y.; Sun, M.; Zhao, C.; Kennedy, A. L.; Madigan, D. L.; Xu, Y.; Saldivar, A.; Kati, W.; Laver, G.; Sowin, T.; Sham, H. L.; Greer, J.; Kempf, D. *Biochemistry* **2003**, *42*, 718–727.

[70] Smith, P. W.; Sollis, S. L.; Howes, P. D.; Cherry, P. C.; Starkey, I. D.; Cobley, K. N.; Weston, H.; Scicinski, J.; Merritt, A.; Whittington, A.; Wyatt, P.; Taylor, N.; Green, D.; Bethell, R.; Madar, S.; Fenton, R. J.; Morley, P. J.; Pateman, T.; Beresford, A. *J. Med. Chem.* **1998**, *41*, 787–797.

[71] Miller, C. A.; Wang, P.; Flashner, M. *Biochem. Biophys. Res. Commun.* **1978**, *83*, 1479–1487.

[72] Taylor, N. R.; von Itzstein, M. *J. Med. Chem.* **1994**, *37*, 616–624.

[73] Vavricka, C. J.; Liu, Y.; Kiyota, H.; Sriwilaijaroen, N.; Qi, J.; Tanaka, K.; Wu, Y.; Li, Q.; Li, Y.; Yan, J.; Suzuki, Y.; Gao, G. F. *Nat. Commun.* **2013**, *4*, 1491.

[74] Chong, A. K.; Pegg, M. S.; Taylor, N. R.; von Itzstein, M. *Eur. J. Biochem.* **1992**, *207*, 335–343.

[75] Holzer, C. T.; von Itzstein, M.; Jin, B.; Pegg, M. S.; Stewart, W. P.; Wu, W. Y. *Glycoconj. J.* **1993**, *10*, 40–44.

[76] Meindl, P.; Tuppy, H. *Hoppe-Seylers Z. Physiol. Chem.* **1969**, *350*, 1088–1092.

[77] Meindl, P.; Bodo, G.; Palese, P.; Schulman, J.; Tuppy, H. *Virology* **1974**, *58*, 457–463.

[78] Wilson, J. C.; von Itzstein, M. *Curr. Drug Targets* **2003**, *4*, 389–408.

[79] Edmond, J. D.; Johnston, R. G.; Kidd, D.; Rylance, H. J.; Sommerville, R. G. *Br. J. Pharmacol. Chemother.* **1966**, *27*, 415–426.

[80] Khorlin, A. Y.; Privalova, I. M.; Zakstelskaya, L. Y.; Molibog, E. V.; Evstigneeva, N. A. *FEBS Lett.* **1970**, *8*, 17–19.

[81] Suzuki, Y.; Sato, K.; Kiso, M.; Hasegawa, A. *Glycoconj. J.* **1990**, *7*, 349–356.

[82] Hagiwara, T.; Kijima-Suda, I.; Ido, T.; Ohrui, H.; Tomita, K. *Carbohydr. Res.* **1994**, *263*, 167–172.

[83] Chan, T.-H.; Xin, Y.-C.; von Itzstein, M. *J. Org. Chem.* **1997**, *62*, 3500–3504.

[84] Palese, P.; Schulman, J. L. In *Chemoprophylaxis and Viral Infection of the Upper Respiratory Tract*; Oxford, J. S., Ed.; CRC Press: Cleveland, OH, **1977**; Vol. *1*, pp 189–202.

[85] Bossart-Whitaker, P.; Carson, M.; Babu, Y. S.; Smith, C. D.; Laver, W. G.; Air, G. M. *J. Mol. Biol.* **1993**, *232*, 1069–1083.

[86] Varghese, J. N. *Drug Dev. Res.* **1999**, *46*, 176–196.

[87] von Itzstein, M.; Dyason, J. C.; Oliver, S. W.; White, H. F.; Wu, W.-Y.; Kok, G. B.; Pegg, M. S. *J. Med. Chem.* **1996**, *39*, 388–391.

[88] Goodford, P. J. *J. Med. Chem.* **1985**, *28*, 849–857.

[89] von Itzstein, M.; Wu, W. Y.; Kok, G. B.; Pegg, M. S.; Dyason, J. C.; Jin, B.; Van Phan, T.; Smythe, M. L.; White, H. F.; Oliver, S. W.; Colman, P. M.; Varghese, J. N.; Ryan, D. M.; Woods, J. M.; Bethell, R. C.; Hotham, V. J.; Cameron, J. M.; Penn, C. R. *Nature* **1993**, *363*, 418–423.

[90] Smith, B. J.; Colman, P. M.; von Itzstein, M.; Danylec, B.; Varghese, J. N. *Protein Sci.* **2001**, *10*, 689–696.

[91] Varghese, J. N.; Epa, V. C.; Colman, P. M. *Protein Sci.* **1995**, *4*, 1081–1087.

[92] Wallace, A. C.; Laskowski, R. A.; Thornton, J. M. *Protein Eng.* **1995**, *8*, 127–134.

[93] von Itzstein, M.; Thomson, R. In *Antiviral Strategies*; Kräusslich, H.-G., Bartenschlager, R., Eds.; Handbook of Experimental Pharmacology 189; Springer: Berlin, **2009**, pp 111–154.

[94] Woods, J. M.; Bethell, R. C.; Coates, J. A.; Healy, N.; Hiscox, S. A.; Pearson, B. A.; Ryan, D. M.; Ticehurst, J.; Tilling, J.; Walcott, S. M. *Antimicrob. Agents Chemother.* **1993**, *37*, 1473–1479.

[95] Elliott, M. *Philos Trans. Roy Soc. B* **2001**, *356*, 1885–1893.

[96] Wutzler, P.; Vogel, G. *Infection* **2000**, *28*, 261–266.

[97] Govorkova, E. A.; Leneva, I. A.; Goloubeva, O. G.; Bush, K.; Webster, R. G. *Antimicrob. Agents Chemother.* **2001**, *45*, 2723–2732.

[98] Ryan, D. M.; Ticehurst, J.; Dempsey, M. H.; Penn, C. R. *Antimicrob. Agents Chemother.* **1994**, *38*, 2270–2275.

[99] Ryan, D. M.; Ticehurst, J.; Dempsey, M. H. *Antimicrob. Agents Chemother.* **1995**, *39*, 2583–2584.

[100] Calfee, D. P.; Peng, A. W.; Cass, L. M.; Lobo, M.; Hayden, F. G. *Antimicrob. Agents Chemother.* **1999**, *43*, 1616–1620.

[101] Bamford, M. J. *J. Enzyme Inhib.* **1995**, *10*, 1–16.

[102] Rich, J. R.; Gehle, D.; von Itzstein, M. In *Comprehensive Glycoscience*; Kamerling, J. P., Boons, G.-J., Lee, Y., Suzuki, A., Tanigichi, N., Voragen, A. G. J., Eds.; Elsevier: Oxford, **2007**; Vol. *1*, pp 885–922.

[103] Thomson, R. J.; von Itzstein, M. In *Carbohydrate-Based Drug Discovery*; Wong, C.-H., Ed.; Wiley-VCH: Weinheim, **2003**; Vol. *2*, pp 831–862.

[104] Shie, J.-J.; Fang, J.-M.; Lai, P.-T.; Wen, W.-H.; Wang, S.-Y.; Cheng, Y.-S. E.; Tsai, K.-C.; Yang, A.-S.; Wong, C.-H. *J. Am. Chem. Soc.* **2011**, *133*, 17959–17965.

[105] Masuda, T.; Yoshida, S.; Arai, M.; Kaneko, S.; Yamashita, M.; Honda, T. *Chem. Pharm. Bull.* **2003**, *51*, 1386–1398.

[106] Yamashita, M.; Tomozawa, T.; Kakuta, M.; Tokumitsu, A.; Nasu, H.; Kubo, S. *Antimicrob. Agents Chemother.* **2009**, *53*, 186–192.

[107] Kubo, S.; Tomozawa, T.; Kakuta, M.; Tokumitsu, A.; Yamashita, M. *Antimicrob. Agents Chemother.* **2010**, *54*, 1256–1264.

[108] Yamashita, M. *Antiviral Chem. Chemother.* **2010**, *21*, 71–84.

[109] Macdonald, S. J. F.; Cameron, R.; Demaine, D. A.; Fenton, R. J.; Foster, G.; Gower, D.; Hamblin, J. N.; Hamilton, S.; Hart, G. J.; Hill, A. P.; Inglis, G. G. A.; Jin, B.; Jones, H. T.; McConnell, D. B.; McKimm-Breschkin, J.; Mills, G.; Nguyen, V.; Owens, I. J.; Parry, N.; Shanahan, S. E.; Smith, D.; Watson, K. G.; Wu, W.-Y.; Tucker, S. P. *J. Med. Chem.* **2005**, *48*, 2964–2971.

[110] Watson, K. G.; Cameron, R.; Fenton, R. J.; Gower, D.; Hamilton, S.; Jin, B.; Krippner, G. Y.; Luttick, A.; McConnell, D.; MacDonald, S. J.; Mason, A. M.; Nguyen, V.; Tucker, S. P.; Wu, W. Y. *Bioorg. Med. Chem. Lett.* **2004**, *14*, 1589–1592.

[111] Weight, A. K.; Haldar, J.; Álvarez de Cienfuegos, L.; Gubareva, L. V.; Tumpey, T. M.; Chen, J.; Klibanov, A. M. *J. Pharm. Sci.* **2011**, *100*, 831–835.

[112] Kim, C. U.; Lew, W.; Williams, M. A.; Liu, H.; Zhang, L.; Swaminathan, S.; Bischofberger, N.; Chen, M. S.; Mendel, D. B.; Tai, C. Y.; Laver, W. G.; Stevens, R. C. *J. Am. Chem. Soc.* **1997**, *119*, 681–690.

[113] Lew, W.; Chen, X.; Kim, C. U. *Curr. Med. Chem.* **2000**, *7*, 663–672.

[114] Wyatt, P. G.; Coomber, B. A.; Evans, D. N.; Jack, T. I.; Fulton, H. E.; Wonacott, A. J.; Colman, P.; Varghese, J. *Bioorg. Med. Chem. Lett.* **2001**, *11*, 669–673.

[115] White, C. L.; Janakiraman, M. N.; Laver, W. G.; Philippon, C.; Vasella, A.; Air, G. M.; Luo, M. *J. Mol. Biol.* **1995**, *245*, 623–634.

[116] Shie, J.-J.; Fang, J.-M.; Wang, S.-Y.; Tsai, K.-C.; Cheng, Y.-S. E.; Yang, A.-S.; Hsiao, S.-C.; Su, C.-Y.; Wong, C.-H. *J. Am. Chem. Soc.* **2007**, *129*, 11892–11893.

[117] Babu, Y. S.; Chand, P.; Bantia, S.; Kotian, P.; Dehghani, A.; El-Kattan, Y.; Lin, T. H.; Hutchison, T. L.; Elliott, A. J.; Parker, C. D.; Ananth, S. L.; Horn, L. L.; Laver, G. W.; Montgomery, J. A. *J. Med. Chem.* **2000**, *43*, 3482–3486.

[118] Chand, P.; Kotian, P. L.; Dehghani, A.; El-Kattan, Y.; Lin, T.-H.; Hutchison, T. L.; Babu, Y. S.; Bantia, S.; Elliott, A. J.; Montgomery, J. A. *J. Med. Chem.* **2001**, *44*, 4379–4392.

[119] Kati, W. M.; Montgomery, D.; Carrick, R.; Gubareva, L.; Maring, C.; McDaniel, K.; Steffy, K.; Molla, A.; Hayden, F.; Kempf, D.; Kohlbrenner, W. *Antimicrob. Agents Chemother.* **2002**, *46*, 1014–1021.

[120] Gubareva, L. V.; Kaiser, L.; Hayden, F. G. *Lancet* **2000**, *355*, 827–835.

[121] Bantia, S.; Parker, C. D.; Ananth, S. L.; Horn, L. L.; Andries, K.; Chand, P.; Kotian, P. L.; Dehghani, A.; El Kattan, Y.; Lin, T.; Hutchison, T. L.; Montgomery, J. A.; Kellog, D. L.; Babu, Y. S. *Antimicrob. Agents Chemother.* **2001**, *45*, 1162–1167.

[122] Chand, P.; Bantia, S.; Kotian, P. L.; El-Kattan, Y.; Lin, T.-H.; Babu, Y. S. *Bioorg. Med. Chem.* **2005**, *13*, 4071–4077.

[123] Maring, C. J.; Stoll, V. S.; Zhao, C.; Sun, M.; Krueger, A. C.; Stewart, K. D.; Madigan, D. L.; Kati, W. M.; Xu, Y.; Carrick, R. J.; Montgomery, D. A.; Kempf-Grote, A.; Marsh, K. C.; Molla, A.; Steffy, K. R.; Sham, H. L.; Laver, W. G.; Gu, Y.-G.; Kempf, D. J.; Kohlbrenner, W. E. *J. Med. Chem.* **2005**, *48*, 3980–3990.

[124] Abed, Y.; Nehmé, B.; Baz, M.; Boivin, G. *Antivir. Res.* **2008**, *77*, 163–166.

[125] Williams, M.; Bischofberger, N.; Swaminathan, S.; Kim, C. U. *Bioorg. Med. Chem.* **1995**, *5*, 2251–2254.

[126] Chand, P.; Babu, Y. S.; Bantia, S.; Chu, N.; Cole, L. B.; Kotian, P. L.; Laver, W. G.; Montgomery, J. A.; Pathak, V. P.; Petty, S. L.; Shrout, D. P.; Walsh, D. A.; Walsh, G. M. *J. Med. Chem.* **1997**, *40*, 4030–4052.

[127] Atigadda, V. R.; Brouillette, W. J.; Duarte, F.; Ali, S. M.; Babu, Y. S.; Bantia, S.; Chand, P.; Chu, N.; Montgomery, J. A.; Walsh, D. A.; Sudbeck, E. A.; Finley, J.; Luo, M.; Air, G. M.; Laver, G. W. *J. Med. Chem.* **1999**, *42*, 2332–2343.

[128] Brouillette, W. J.; Bajpai, S. N.; Ali, S. M.; Velu, S. E.; Atigadda, V. R.; Lommer, B. S.; Finley, J. B.; Luo, M.; Air, G. M. *Bioorg. Med. Chem.* **2003**, *11*, 2739–2749.

[129] Russell, R. J.; Haire, L. F.; Stevens, D. J.; Collins, P. J.; Lin, Y. P.; Blackburn, G. M.; Hay, A. J.; Gamblin, S. J.; Skehel, J. J. *Nature* **2006**, *443*, 45–49.

[130] McKimm-Breschkin, J. L. *Antivir. Res.* **2000**, *47*, 1–17.

[131] Mishin, V. P.; Hayden, F. G.; Gubareva, L. V. *Antimicrob. Agents Chemother.* **2005**, *49*, 4515–4520.

[132] Monto, A. S.; McKimm-Breschkin, J. L.; Macken, C.; Hampson, A. W.; Hay, A.; Klimov, A.; Tashiro, M.; Webster, R. G.; Aymard, M.; Hayden, F. G.; Zambon, M. *Antimicrob. Agents Chemother.* **2006**, *50*, 2395–2402.

[133] Reece, P. A. *J. Med. Virol.* **2007**, *79*, 1577–1586.

[134] Okomo-Adhiambo, M.; Nguyen, H. T.; Sleeman, K.; Sheu, T. G.; Deyde, V. M.; Garten, R. J.; Xu, X.; Shaw, M. W.; Klimov, A. I.; Gubareva, L. V. *Antivir. Res.* **2010**, *85*, 381–388.

[135] Wen, W.-H.; Wang, S.-Y.; Tsai, K.-C.; Cheng, Y.-S. E.; Yang, A.-S.; Fang, J.-M.; Wong, C.-H. *Bioorg. Med. Chem.* **2010**, *18*, 4074–4084.

[136] Mohan, S.; McAtamney, S.; Haselhorst, T.; von Itzstein, M.; Pinto, B. M. *J. Med. Chem.* **2010**, *53*, 7377–7391.

[137] Rudrawar, S.; Dyason, J. C.; Rameix-Welti, M.-A.; Rose, F. J.; Kerry, P. S.; Russell, R. J. M.; van der Werf, S.; Thomson, R. J.; Naffakh, N.; von Itzstein, M. *Nat. Commun.* **2010**, *1*, 113.

[138] Li, Q.; Qi, J.; Zhang, W.; Vavricka, C. J.; Shi, Y.; Wei, J.; Feng, E.; Shen, J.; Chen, J.; Liu, D.; He, J.; Yan, J.; Liu, H.; Jiang, H.; Teng, M.; Li, X.; Gao, G. F. *Nat. Struct. Mol. Biol.* **2010**, *17*, 1266–1268.

[139] Amaro, R. E.; Swift, R. V.; Votapka, L.; Li, W. W.; Walker, R. C.; Bush, R. M. *Nat. Commun.* **2011**, *2*, 388.

18

DESIGN AND SYNTHESIS OF LIGANDS AND ANTAGONISTS OF SIGLECS AS IMMUNE RESPONSE MODIFIERS

Hajjaj H. M. Abdu-Allah[1], Hideharu Ishida[2], and Makoto Kiso[2]

[1]*Department of Pharmaceutical Organic Chemistry, Faculty of Pharmacy, Assiut University, Assiut, Egypt*

[2]*Department of Applied Bioorganic Chemistry, Gifu University, Gifu, Japan*

18.1 INTRODUCTION

Siglecs (sialic acid-binding immunoglobulin-type lectins) are the best characterized immunoglobulin (Ig)-type lectin and are important immune receptors expressed widely in mammals. Members of the siglec family are expressed on cells of the immune system, exhibiting roles in cell signaling and cell adhesion that are modulated by interaction with their sialic acid-containing glycan ligands. The restricted expression pattern and activity as endocytic receptors made these proteins as attractive molecular targets for directed therapy for immune cell-mediated diseases. Sialosides and sialomimetics have been synthesized and tested for the discovery of ligand-based probes for the siglec family. In particular, high-affinity and selective ligands/antagonists for the siglec CD22 have been developed. Development of highly potent sialomimetics as CD22 ligands represents a new arena for modulation of immune response (universal immunization), treatment of B-cell malignancies, and autoimmunity.

Glycochemical Synthesis: Strategies and Applications, First Edition.
Edited by Shang-Cheng Hung and Medel Manuel L. Zulueta.

18.2 LECTINS

The specific recognition of carbohydrate structures by carbohydrate-binding proteins [1–4] is the basis of numerous intra- and intercellular events, ranging from the control of protein folding to cell–cell communication during development, inflammation, immune responses, and cancer metastasis [5–9]. Lectins (from *lectus*, the past participle of *legere*, meaning to select or choose) are defined as carbohydrate-binding proteins other than enzymes or antibodies [10]. They exist in most living organisms, ranging from viruses and bacteria to plants and animals. Based on sequence homologies and evolutionary relatedness, most animal lectins are classified into several families, namely, C type or selectins, S type or galectins, P type or Man-6-P lectins, Ig-type or I-type lectins, R-type lectins, and F-type lectins [10].

X-ray crystallography of lectins complexed with their ligands, site-directed mutagenesis experiments, and molecular modeling allowed the identification of the chemical groups belonging to both interacting species involved in the binding and of the types of bond formed. Studies of lectin–oligosaccharide complexes are especially interesting, providing the basis for understanding protein interaction with natural ligands [3]. The most consistent structural characteristic responsible for carbohydrate recognition by lectins is the presence of a shallow pocket containing a specific pattern of hydrogen-bonding groups that interact with sugar hydroxyls and aromatic groups or hydrophobic patches that stack or interact with the nonpolar face or with the hydrophobic substituents of the sugar [11, 12].

High-affinity lectin ligands are of great medicinal interest in the diagnosis and modulation of carbohydrate-mediated processes [11–16]. The generation of high-affinity ligands, however, is not trivial, because the interaction of individual carbohydrate epitopes with lectins is, in many cases, weak and undiscriminating [17].

18.3 SIGLECS

Interest in siglecs has grown over the recent years as it became increasingly clear that these receptors play a wide range of roles in the immune system. Siglecs form the main subpopulation of the Ig-type lectins [18–20]. They comprise 14 human and nine murine members that are primarily expressed by white blood cells and mediate innate and adaptive immune functions. Siglecs are type 1 membrane proteins, characterized by an N-terminal V-set Ig domain containing the sialic acid-binding site, followed by various numbers of C2-set Ig domains (16 in case of sialoadhesin and 6 in case of CD22). Some contribution to sialic acid binding may also be attributed to the most N-terminal C2-type Ig domain. The V-set and first C2 domain contain additional cysteine residues that contribute to an interdomain disulfide bridge as well as intra-β-strand disulfide bonds. These additional disulfide bridges are thought to contribute to sialic acid recognition by siglecs. Siglecs can be further subdivided into two subsets, that is, the evolutionary conserved but distantly related (25–30% sequence identity) group containing sialoadhesin (Sn, Siglec-1, CD169), CD22 (Siglec-2), myelin-associated glycoprotein (MAG, Siglec-4), and Siglec-15, and the rapidly evolving group of CD33-related siglecs that share high sequence similarity (50–99% sequence identity) (Fig. 18.1) [8, 21]. Sialoadhesin, CD22, and MAG have been found in all mammals so far examined. In contrast, the repertoire of CD33-related siglecs varies considerably between species, with 10 known in humans and only five in mice. With the exception of MAG and Siglec-6, siglecs are primarily expressed by cells of the hematopoietic and immune systems. While some have very restricted expression patterns, others are more widely expressed. For example, sialoadhesin is mainly expressed by macrophages, CD22 by B cells, Siglec-8 by eosinophils, and Siglec-7 by natural killer (NK) cells, whereas all Siglec-9 can be found on neutrophils, monocytes, dendritic cells, and NK cells. The diverse specificity of siglecs for their sialoside

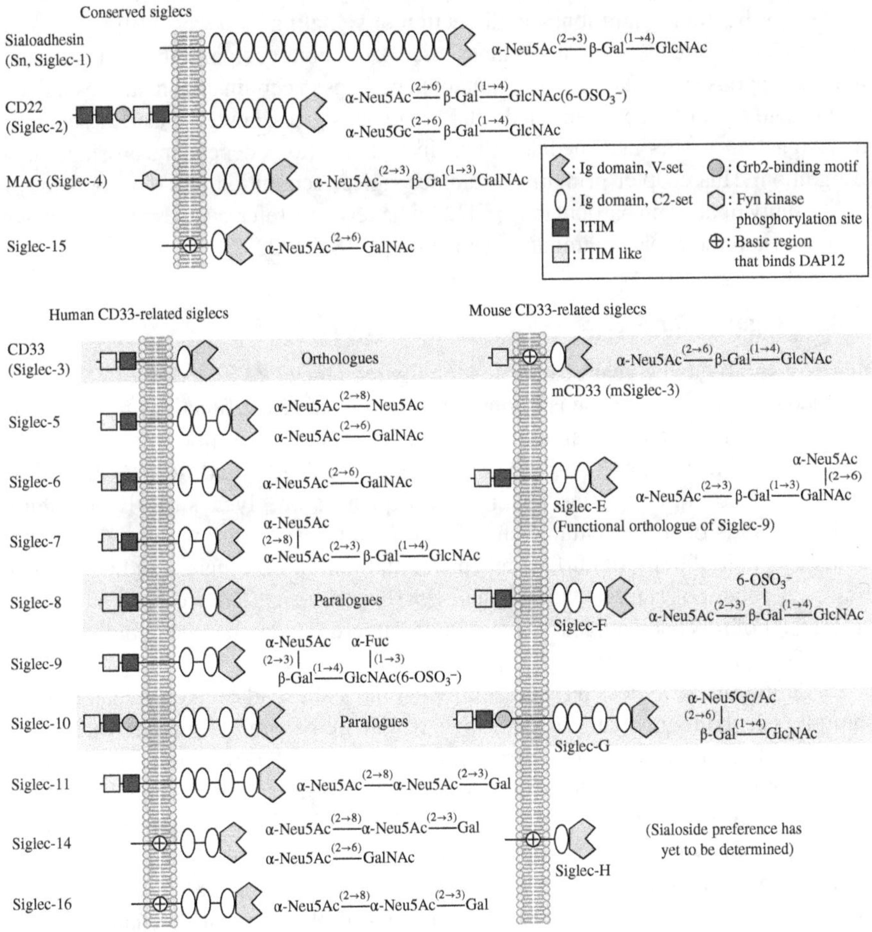

FIGURE 18.1 Representative diagram of the members of siglec family and the main sialic acid preference of the human counterpart, with the exception of Siglec-E. Fuc, L-fucose; Gal, D-galactose; GalNAc, *N*-acetyl-D-galactosamine; GlcNAc, *N*-acetyl-D-glucosamine; Neu5Ac, *N*-acetylneuraminic acid; Neu5Gc, *N*-glycolylneuraminic acid.

ligand and variable cytoplasmic regulatory elements provide them with attributes for unique roles in the cell-surface biology of each cell that expresses them. Most CD33-related siglecs and CD22 are inhibitory signaling receptors, having one or more immunoreceptor tyrosine (Tyr)-based motifs (ITIMs) in their cytoplasmic tails, whereas sialoadhesin seems to be involved in cell–cell interactions. As endocytic receptors, siglecs provide portals of entry for certain viral and bacterial pathogens, as well as therapeutic opportunities for targeting innate immune cells in diseases [21, 22].

It is clear that siglecs in the immune system have the potential to mediate cell–cell interactions and signaling functions [23]. However, defining their precise functions and determining which ligands are biologically relevant pose an important challenge. This subject is beginning to be tackled by using a combination of experimental approaches, including the production of genetically manipulated mice, biochemical analyses of ligand recognition, and dissection of signaling pathways [19].

Many siglecs are therefore important sources of inhibitory signals that can attenuate immune responses and dampen inflammation, perhaps in certain circumstances induce lymphocyte exhaustion or even cell death [24]. Other siglecs, however, lack these inhibitory signaling motifs and mediate other diverse functions dependent on sialic acid recognition. This chapter primarily examines siglec recognition and development of high-affinity ligands, in particular, for CD22. The reader is referred to excellent reviews on siglecs and their roles in immune regulation [2, 18–21, 23, 24].

18.3.1 Ligands for Siglecs

Numerous sialylated glycans can be potentially recognized as ligands by siglecs. It is assumed that this recognition is important for modulating the functions of siglecs as regulators of adhesion, cell signaling, and endocytosis [22, 23]. In general, siglecs show low affinity (K_d of 0.1–3 mM) for Neu5Ac α(2→3)- and α(2→6)-linked to Gal that are commonly found as terminal sequences on glycans of glycoproteins and glycolipids of most mammalian cells [18]. In terms of discovery, siglecs are a relatively young family of surface receptors, and our current understanding of their biological functions is mostly based on *in vitro* experiments. The present challenge is the transfer of this knowledge to *in vivo* studies in order to learn more about their role in regulating immune responses [25].

By definition, all siglecs have the ability to bind sialic acids, which are a family of naturally occurring nine-carbon acidic sugars that are generally found as α-ketosidically linked components of cell-surface glycoconjugates. The most commonly found derivatives in biological systems are those derived from Neu5Ac, Neu5Gc, and the 5-hydroxy derivative KDN [26, 27]. Natural modifications of these sialic acids generally involve acylation, phosphorylation, or sulfonation of the hydroxy groups, with over 50 structurally distinct derivatives found in nature [28]. Sialic acids are usually located at the exposed nonreducing ends of oligosaccharide chains and are transferred using α(2→3), α(2→6), or α(2→8) linkages to subterminal sugars by a family of about 20 sialyltransferases. The structural diversity of the sialic acids, together with their position within glycoconjugates, results in them being intimately involved in a number of important physiological phenomena and disease states.

Located at the terminus of numerous cell-surface oligosaccharides, sialic acids are ideally positioned to participate in biological processes including cell–biomolecule-mediated recognition phenomena, acting as receptors for viruses and bacteria, cell–cell communication, and as markers in certain diseases [9, 27, 29].

The plethora of enzymes that generate, remove, and modify these terminal sialic moieties underlies the remarkable versatility of the sialoglycome. This diversity and plasticity at the glycan termini have been exploited by the immune system over evolutionary time to provide regulatory mechanisms in a range of immune cells. Many microbial pathogens evolved the ability to synthesize sialic acids in order to evade vertebrate innate immunity either by recruiting factor H or by triggering sialic acid-binding receptors on host innate immune cells to prevent their robust activation [27, 28].

Extensive research has focused on the synthesis of structurally modified sialic acid derivatives and sialylmimetics as probes or potential inhibitors of specific sialic acid-recognizing proteins, and several comprehensive overviews have been published [26, 30–34]. An underlying theme in efforts directed toward the development of novel sialic acid derivatives as biological probes is the complexity of the chemical manipulations associated with sialic acids [30, 32]. The chemical strategies and conditions for sialylation are discussed in Chapter 5.

18.3.2 Structural Features for Siglec Recognition

The binding to sialylated glycans can be greatly influenced by the type of sialic acid, its linkage to the subterminal sugar, the structure of the underlying glycan, and other modifications such as sulfonation [18, 19]. Similar to many lectins, siglecs depend on multivalent clustering for high-avidity binding [18]. Each siglec has its own unique binding profile that varies not only in preference for linkage or sialic acid modification (Fig. 18.1) but also in selectivity. Although sialic acid is the primary binding determinant, underlying glycans can contribute to individual siglec binding [35]. For example, while sialoadhesin binds terminal α(2→3)- or α(2→8)-linked sialosides, CD22 recognizes solely α(2→6)-linked sialosides, and MAG requires α(2→3)-linked sialosides.

18.3.3 Sialic Acid Substructural Specificities for Siglecs

Contributions from several laboratories using a variety of methods led to a relatively clear picture of the interactions of siglecs with sialic acid on the atomic level. Siglecs form direct molecular contacts with the substituents of sialic acid, namely, the carboxylate group, the glycerol side chain, and the *N*-acetyl group [36]. Molecular insights into the interactions between siglecs and their ligands have been provided by siglec crystal structures, first for sialoadhesin in complex with the α(2→3)-linked sialyllactose [37] and for Siglec-7 in the absence and presence of various sialylated sugars, including the α(2→8)-disialylated GT1b oligosaccharide [35, 38]. These studies offer insights into the structural basis for siglec specificity. In both cases, sialic acid is bound to a shallow pocket in the N-terminal Ig domain of the siglec, with the negatively charged carboxyl group of sialic acid forming an important salt

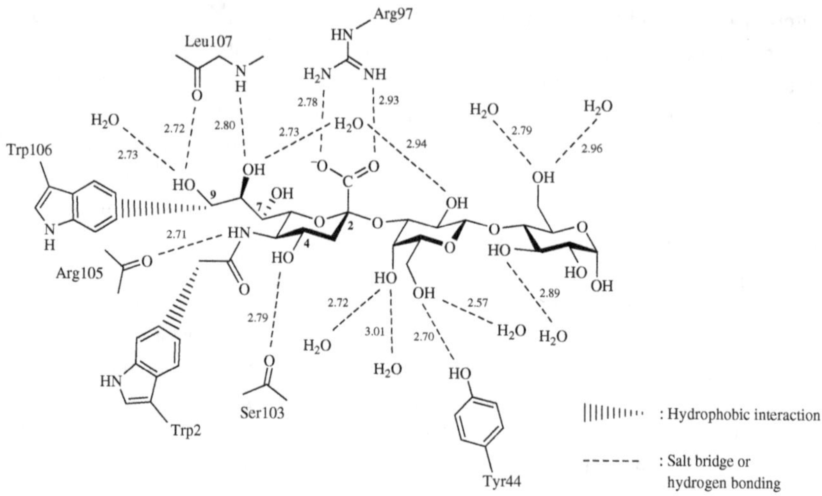

FIGURE 18.2 Interactions of the α(2→3)-linked sialyllactose in the binding site of sialoadhesin [37]. Leu, leucine; Ser, serine.

bridge with a conserved essential arginine (Arg) residue (Fig. 18.2). Both N-terminal domains are predicted to have an unusual disulfide bond that is likely to have a major influence on their orientation. However, the significance of this disulfide bond for ligand recognition is unclear because truncation–deletion studies, site-directed mutagenesis experiments, X-ray crystallography, and nuclear magnetic resonance (NMR) analysis showed that the sialic acid-binding site is contained exclusively within the N-terminal V-set domain. By solving the crystal structure of sialoadhesin in complex with α(2→3)-linked sialyllactose, it became clear that most of the molecular contacts are occurring with the sialic acid moiety rather than the attached sugar units (Fig. 18.2). An Arg residue (Arg97 in sialoadhesin), which is highly conserved in the siglec family, forms a salt bridge with the carboxylate group of the sialic acid. Two well-conserved aromatic groups [both tryptophan (Trp) for sialoadhesin] are involved in the hydrophobic interactions with the *N*-acetyl and the glycerol side chains of sialic acid.

Although the V-set domain of sialoadhesin and Siglec-7 share very similar fold and structural template for sialic acid recognition, there are also clear differences. The highly variable C—C′ loop (residues 70–75 in Siglec-7) of the V-set domain is a key determinant of siglec specificity [35]. Experiments demonstrated that the hydroxy groups at C4, C8, and C9 of sialic acid are important for binding, whereas 7-OH is not necessary [37]. Most likely, the beneficial hydroxy groups function as hydrogen bond donors to the siglec, while 7-OH appears not to be in contact with the protein.

Studies of sialoadhesin and sialoside interactions by NMR provided evidence that the molecular interactions characterized in crystals also occur in solution [39]. Although the binding between natural sialosides and siglecs is relatively weak, increased binding affinity has been seen for sialoadhesin, MAG, and CD22 using

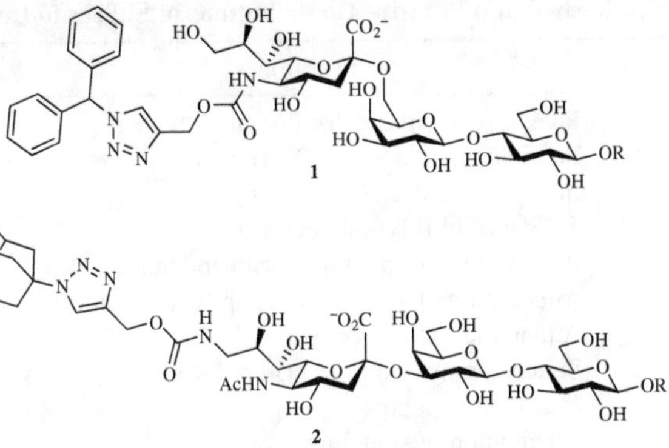

FIGURE 18.3 High-binding ligands for Siglec-9 (1) and Siglec-10 (2). Ac, acetyl.

sialic acid analogues that have aromatic substituents at the C9 position [36]. These insights have been exploited for the creation of high-affinity ligand-based probes for siglec function [36, 40].

The glycocalyx of cells is usually rich in sialic acids. Consequently, ligands for siglecs are not only presented on opposing cells (in *trans*) but also on the surface of same cell or cell type (in *cis*), the latter resulting in the masking of siglecs [41]. To detect the sugar binding activity of siglecs on cell surfaces using, for example, synthetic sialylated probes, cells usually need to be desialylated. However, it has been shown for CD22 that synthetic high-affinity sialosides can outcompete *cis* interactions [42, 43] and that CD22 is able to engage in *trans* interactions without unmasking first, demonstrating that *cis* ligands downregulate, but do not preclude, binding of ligands in a *trans* manner [40, 42, 43].

Microarray printing of a library of sialic acid analogues and screening with several members of the siglec family led to the identification of high-affinity ligands for Siglec-9 and Siglec-10 (Fig. 18.3) [44]. It was found that individual siglecs preferably bind to analogues with substituents at either C9 or C5, but not both, and the most promising were those having relatively bulky and hydrophobic substituents. Furthermore, there appeared to be little preference for the sialoside linkage (e.g., $\alpha(2\rightarrow3)$ and $\alpha(2\rightarrow6)$), suggesting that the modified sialic acid and not the underlying lactose core provides most of the binding affinity [44].

18.4 SIGLECS AND INNATE IMMUNITY

Most siglecs are expressed primarily in innate immune cells. The study of specific gene-targeted mice yielded some information on the *in vivo* function of selected murine siglecs, but many postulated aspects of siglec function in innate immune cells remain to be confirmed by *in vivo* approaches [24]. The potential functions of siglecs in innate immune cells include [24] (i) a role in the internalization of sialic acid-expressing pathogens by phagocytosis, (ii) the attenuation of inflammation in response

TABLE 18.1 Established and Putative Contributions of Siglecs to Immune Regulation

Siglec	Function	References
Sialoadhesin	Phagocytosis/internalization of pathogens	[45–48]
	Possible attenuation of DC function	[49, 50]
CD22	Inhibition of B-cell activation	[51–54]
	Tolerance in B lymphocytes	[54, 55]
Siglec-5	Phagocytosis of pathogens and apoptotic bodies	[45, 56, 57]
	Attenuation of signals by group B streptococci	[58]
Siglec-7	Attenuation of NK cells	
	Signal attenuation in immature DCs	[49, 50]
Siglec-8	Eosinophil apoptosis	[59, 60]
Siglec-9	Attenuation of signaling in activated neutrophils and macrophages	[61]
	Neutrophil apoptosis	[62]
	Attenuation of signals by group B streptococci	[63, 64]
Siglec-10	Attenuated signaling in response to DAMPs	[65, 66]
Siglec-11	Attenuating microglia-mediated inflammation	[67, 68]
Siglec-E	Attenuation of signaling in activated neutrophils and macrophages	[69]
Siglec-F	Endocytosis in eosinophils	[22]
	Eosinophil apoptosis	[70]
	Attenuating allergic inflammation	[71]
	Attenuating microglia-mediated inflammation	[72]
Siglec-G	Attenuated signaling in response to DAMPs	[65, 66]
	Attenuation of B-1a B-cell activation	[73]
	Tolerance in B lymphocytes	[74]
Siglec-H	Attenuating Toll-like receptor-linked type-I interferon secretion in plasmacytoid DCs	[75]

to both sialic acid-expressing and nonsialylated pathogens by inhibitory signaling, (iii) the attenuation of damage-associated molecular pattern (DAMP)-mediated inflammation by siglec inhibitory signaling, and (iv) the inhibition of NK cell activation by terminal sialic acid moieties on mucins and gangliosides and the potential role of siglecs in influencing the life spans of myeloid cells. Some putative and established functions of siglecs in the immune system are summarized in Table 18.1.

18.4.1 Pathogen Internalization by Innate Immune Cells: Phagocytosis and Endocytosis by Siglecs

Some siglecs are capable of phagocytosing pathogens that express cell-surface sialic acid moieties. Such interactions represent the recognition of sialic acid-containing ligands in *trans* interactions. Sialoadhesin is expressed on numerous macrophage populations and has been implicated in the interactions of macrophages with lymphocytes. It is also expressed on monocyte-derived dendritic cells (DCs). While long thought to be primarily involved in adhesion events, more recent studies indicate that sialoadhesin may participate in the phagocytosis of pathogens that synthesize or acquire terminal Neu5Ac(α2→3)Gal moieties on glycoconjugates [76]. Sialoadhesin recognizes such sialic acid-containing moieties on the lipopolysaccharide of the Gram-negative bacterium *Neisseria meningitidis* and can mediate the phagocytosis of this bacterium [45].

Comparison of wild-type macrophages and those from sialoadhesin knockout animal provides the strongest evidence for the *in vivo* role of siglec in the phagocytic process [24]. Sialoadhesin also recognizes similar sialic acid moieties on the O-linked carbohydrate termini of mucins of *Trypanosoma cruzi*, the trypanosome that causes Chagas' disease. This event facilitates macrophage–*T. cruzi* interactions and also contributes to the internalization of the trypanosome [46]. Sialoadhesin on porcine alveolar macrophages binds Neu5Ac(α2→3)Gal moieties on N-glycans of proteins of the porcine respiratory and reproductive syndrome virus and mediates virus internalization [47]. Siglec-5 can also recognize the Neu5Ac(α2→3)Gal moiety on the lipopolysaccharide of *N. meningitidis*, and *in vitro* studies employing transfected COS cells have shown Siglec-5 enabling the phagocytosis of this species [45]. Siglec-5 has also been implicated in the clearance of apoptotic bodies by macrophages [77], but there is little evidence to indicate that apoptotic cells are more avidly decorated with Neu5Ac(α2→3)Gal moieties [56, 57]. Like most siglecs with cytosolic ITIM motifs, Siglec-5 is normally bound to sialic acid-containing ligands in *cis* fashion. The endocytic activity of siglecs was also demonstrated *in vitro* for Siglec-7, Siglec-9, Siglec-F, and CD22 [22]. It is unclear whether any of these inhibitory siglecs actually participate in pathogen phagocytosis in macrophages or DCs *in vivo*. Pathogens may use some inhibitory siglecs for cellular entry. Endocytosis by CD22 has been studied in some detail, and CD22 ITIM motifs were the first siglec structural determinants linked to endocytosis [24].

18.4.2 Attenuation of Inflammatory Responses

Most siglecs appear to have evolved to attenuate innate immune responses and to control inflammation in responses to various pathogens, most of which are nonsialylated microbes. Siglecs presumably maintain an inhibitory state by binding to self-ligands in a *cis* manner. Alterations in the expression of siglecs and/or of siglec ligands may represent a means by which inhibitory signaling is upregulated in the context of infection and inflammation [24].

18.4.3 Immune Evasion by Pathogens via Siglec Ligation

Several pathogens, such as group B streptococci, *Neisseria* species, and *Campylobacter jejuni*, synthesize sialic acids, while others acquire them from the host. Although the acquisition of sialic acid by pathogens can contribute to the evasion of host complement, appropriately sialylated glycoconjugates may also trigger inhibitory siglecs on host innate immune cells and thus attenuate host protective responses. Although this type of triggering event might reflect immune evasion by pathogens, it may also represent a protective response that serves to prevent excessive inflammation in the host.

The most interesting case of sialic acid scavenging is that mediated by the unique *trans*-sialidase enzyme of *T. cruzi*. This enzyme cleaves sialic acid moieties from host glycoconjugates and transfers them in α(2→3)-linked form to *T. cruzi* mucin O-glycan termini. The more heavily sialylated *T. cruzi* strains elicit greater parasitemia, an effect attributed to the ability of sialic acid moieties to mask subterminal carbohydrate epitopes and to protect the parasite from the lytic activity of antigalactosyl antibodies, as well as to the ability of glycoproteins decorated with terminal sialic acid to inhibit complement activation [61, 78]. Immune evasion by heavily sialylated *T. cruzi* also involves siglec ligation and is considered later in the context of regulatory effects in DCs.

The capsular polysaccharide of group B streptococci displays a repeated Neu5Ac(α2→3)Gal(β1→4)GlcNAc structure that resembles that seen at the termini of many vertebrate glycoproteins. Siglec-9 on neutrophils binds to host α(2→3)-linked sialic acids in *cis* fashion, presumably to generate an inhibitory tone by Tyr phosphorylation of ITIM motifs and recruitment of the phosphatase SHP-1 or SHP-2 [24]. However, the presence of group B streptococcal polysaccharides of serotype III in *trans* induces inhibitory signaling presumably because the high valency of the α(2→3)-linked sialic acid motif on the polysaccharide displaces *cis*-acting self-glycans. Ligation of Siglec-9 attenuates oxidative burst in Siglec-5 and is, thus, subverted into contributing to a bacterial immune evasion mechanism [58].

18.4.4 Regulation of the Life Span of Myeloid Cells in the Context of Inflammation by Siglec-8, Siglec-9, and Siglec-F

One mechanism by which the half-lives of myeloid cells, such as neutrophils and eosinophils, might be regulated may involve apoptotic elimination of these cells by siglec ligation following activation. Siglec-8 in humans and Siglec-F in mice are functional convergent paralogues that are primarily expressed in eosinophils and mast cells [71, 79, 80]. Siglec-9 is expressed in neutrophils and other innate immune cells in humans [81]. Siglec-8, Siglec-F, and Siglec-9 bind in high affinity to sulfated silyl Lewis X. Eosinophils and neutrophils readily undergo apoptosis when Siglec-8 and Siglec-9 are, respectively, cross-linked with specific antibodies [59, 60, 62]. This apoptotic process occurs most readily when these granulocytes are in a proinflammatory cytokine milieu, suggesting that these siglecs might help regulate the turnover of activated granulocytes in the context of inflammation.

18.4.5 Regulation of NK Cell Function by Siglec-7

Siglec-7 is primarily expressed on human NK cells and recognizes Neu5Ac(α2→8) Neu5Ac(α2→3)Gal moieties present on certain b-series gangliosides such as GD3, GT1b, and GQ1b, but these moieties are also present on some glycoproteins. NK cells differ from other lymphocytes in that they express α(2→8)-sialyltransferase 8F, which generates disialic acid moieties on O-linked glycans of NK cells [82, 83]. NK cells express these endogenous ligands for Siglec-7, which can presumably ligate Siglec-7 and maintain these cells in an inhibited state. Certain tumors may be refractory to NK cell killing given that these tumors express GD3 and other NK cell ligands that might send inhibitory signals via Siglec-7. Siglec-7 may contribute to inhibitory signaling in NK cells that can be further enhanced as part of an immune evasion phenomenon by tumors such as melanomas and renal cell carcinomas. During HIV infection, immune evasion involves the induction of dysfunctional NK cell subsets that significantly downregulate Siglec-7 expression [84].

Sialoadhesin and Siglec-7 appear to play reciprocal roles in DCs in the context of T-cell activation and polarization. The α(2→3)-linked glycans in *C. jejuni* lipooligosaccharides that can serve as mimics for host GD1a and GM1a gangliosides interact with sialoadhesion in *trans* manner and contribute to the induction of a Th1 response, whereas α(2→8)-linked sialic acid-containing moieties mimic host GD1c, interact with Siglec-7, and can drive a Th2-like response [85]. Immune responses to sialylated *C. jejuni* lipooligosaccharides have been linked to the pathogenesis of Guillain–Barré syndrome.

Studies in murine models of *T. cruzi* infection suggest that the heightened pathogenicity of the heavily sialylated *T. cruzi* strains may also result from the modulation of the function of DCs through interactions of the parasite with Siglec-E on DCs [86].

18.4.6 Direct Role for Siglecs in T Cells

Some evidence supports the notion that siglecs are expressed in and perhaps have a functional role in T cells. Siglec-9 is expressed on about 2% of $CD4^+$ T cells and 5% of $CD8^+$ T cells [81]. Stable expression of Siglec-7 and Siglec-9 in Jurkat cells results in SHP-1 recruitment, partial colocalization with the T-cell receptor–CD3 complex, reduced phosphorylation of Tyr319 on the kinase ZAP-70, and activity diminution of the nuclear factor of activated T cells [87]. Mutation of the conserved Arg residue required for sialic acid recognition in Siglec-7 (Arg124) or Siglec-9 (Arg120) decreased the inhibitory function of transfected siglecs in an *in vitro* assay. There may be a role for sialoadhesin and sialoglycoconjugates during T-cell activation and differentiation [24]. Increased sialylation is observed in activated T cells, and the increases seen in α(2→3)-linked sialoglycoconjugates may reflect an increase in activity of the corresponding sialyltransferase. Sialoadhesin binds to glycoconjugates, especially O-linked mucins found on proteins like CD43, exhibiting α(2→3)-linked sialic acid. Sialoadhesin-deficient mice have an increased representation of $CD8^+$ T cells in spleen and lymph nodes [88]. This seems to be an effect on the homeostasis of T cells in the periphery because thymic T-cell development is normal in these mice.

18.4.7 Siglecs in B-Cell Biology and Maintenance of Immunological Tolerance

CD22 and Siglec-G are expressed at high levels on murine B lymphocytes. Siglec-E is expressed on marginal zone B cells and on a subset of $IgM^{hi}IgD^{hi}$ cells [24]. In humans, several siglecs, including Siglec-2, Siglec-5, Siglec-6, Siglec-9, and Siglec-10, are expressed in cells of the B lineage [18, 89]. Studies on the more subtle modifications of sialic acid make it very likely that the binding of CD22 to sialic acid-containing ligands is crucial *in vivo*. Moreover, the presence of sialic acid on synthetic antigens was shown to tolerize B cells in a CD22- and Siglec-G-dependent fashion, suggesting that the avid engagement of these siglecs can dampen B-cell activation [74].

18.5 DESIGN AND SYNTHESIS OF HIGH-AFFINITY LIGANDS FOR SIGLECS

This section focuses on the development of high-affinity and selective ligand for sialoadhesin and CD22 as models for the siglec family. Such ligands may provide a pathway for immunoglycotherapy strategies for autoimmune diseases and B-cell-derived non-Hodgkin's lymphoma.

18.5.1 Sialoadhesin

Sialoadhesin prefers α(2→3)-linked sialosides, which are abundant on cell surfaces and specifically recognizes the Neu5Ac moiety. [37]. The binding affinity of sialoadhesin to the simple methyl sialoside **3** (Fig. 18.4a) is in the millimolar range, and replacement of methyl group with benzyl (Bn) in the form of compound **4** improved

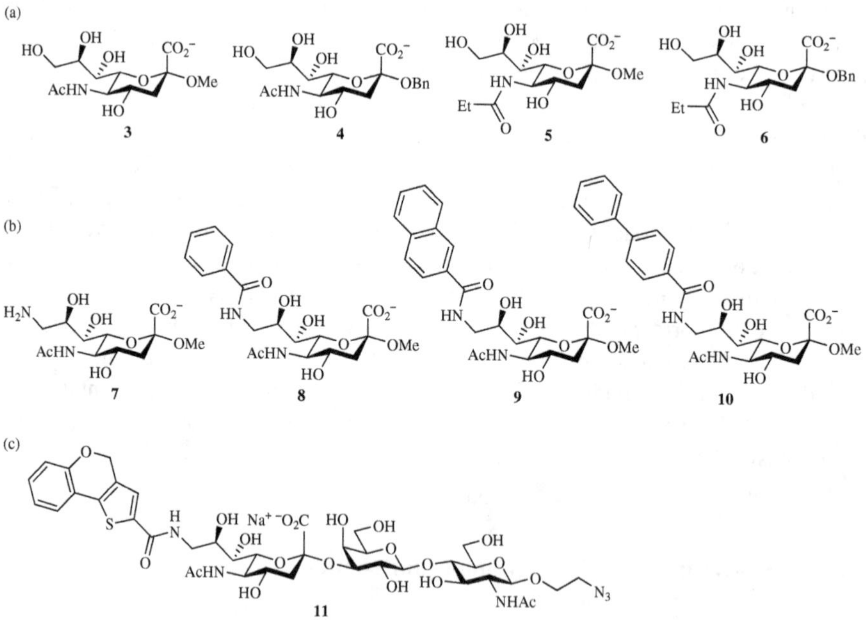

FIGURE 18.4 Modified sialosides used in determining high-affinity ligands for sialoadhesin. (a) Sialosides modified at 2- and 5-positions. (b) Methyl sialosides modified at C9. (c) A high-affinity ligand for sialoadhesin.

the affinity by eightfold [90]. Substitution of the *N*-acetyl group with propionyl (**5** and **6**) showed a slight improvement in affinity, and further modification to butanoyl did not improve the affinity [91]. The sialoadhesin structure in complex with **5** indicated that the increase in affinity could be due to additional hydrophobic interactions of the propionyl group with the Trp2 side chain of the protein. Modeling studies revealed that addition of an extra carbon chain can possibly extend away from Trp2, and hence, the butanoyl homologue could not increase the binding affinity. Replacement of the N5 acetyl with benzoyl dramatically dropped the affinity by 20-fold. This decrease in affinity suggested that addition of bulky groups could cause steric clashes with the side chain of Trp2. The combination of Bn at O2 and propionyl at N5 (**6**) exhibited lower affinity as compared to **4** [91].

Modification of the C9-hydroxy group of sialic acid by halogen substitution or by esterification greatly affected the affinity. Substitution with an amino group (**7**; Fig. 18.4b) showed a better affinity than **3**, suggesting the important role of a hydrogen donor at this position that may also be applicable to all other siglecs [90]. A series of C9-amide analogues were synthesized through a modified Mitsunobu condition [36]. Sialosides containing aromatic rings (**8**, **9**, and **10**) showed better affinity than **3**. The crystal structure complexes of **8**, **9**, and **10** with sialoadhesin suggested that aromatic groups can favorably interact with the hydrophobic cavity formed by the side chains of Tyr44, Ser45, Leu107, and valine-109. These structural data provided hints for further design of inhibitors not only against sialoadhesin but also against other family members such as CD22 and MAG [36].

Recently, *in silico*-aided design approach was employed to develop a high-affinity C9-modified sialoside ligand for sialoadhesin, which may be used for *in vivo* targeting of macrophages [92]. The most potent of these ligands, compound **11** (Fig. 18.4c), was conjugated to lipids for display on a liposomal nanoparticle. The liposomes selectively targeted cells expressing either murine or human sialoadhesin *in vitro*, and when administered to mice, they exhibited *in vivo* targeting to sialoadhesin-positive macrophages [92].

18.5.2 CD22

CD22 is a well-characterized B-cell-specific endocytic receptor. B cells depend on cues from their extracellular microenvironment for development, homeostasis, activation, proliferation, and effector function. These functions are regulated through cell-surface molecules that generate transmembrane signals, regulate intercellular communication, and direct lymphocyte localization within tissues. These events depend on signals generated by the B-cell antigen receptor (BCR). In addition to the BCR, multiple other signaling molecules provide important functional links between the cell-surface and intracellular signaling [93]. CD22 and CD19 represent two specialized cell-surface molecules [94] that also function as “response regulators” to modulate the intensity, quality, and duration of signals generated for homeostatic survival as well as antigen-induced responses [95]. Response regulators establish intrinsic signaling thresholds that provide context for other transmembrane and cytoplasmic signals.

18.5.2.1 ***Biology and Functions*** The carbohydrate ligand recognized by CD22 is the sequence Neu5Ac/Gc(α2→6)Gal(β1→4)GlcNAc found on both neighboring glycoproteins of the same cell (*cis* ligands) and on cells that interact with B cells (*trans* ligands), such as T cells. Interactions of CD22 with *cis* or *trans* ligands regulate aspects of B-cell activation, proliferation, and development. CD22 is highly conserved among mammals. Mouse CD22 is roughly 62% identical in overall amino acid sequence to human CD22 [51, 96]. CD22 is an inhibitory coreceptor that can attenuate BCR signaling. CD22 null mice possess hyperresponsive B cells that display increased Ca^{2+} flux in response to antigens [97]. Its regulation of B-cell receptor signaling is largely based on the recruitment of SHP-1 by phosphorylated Tyr residues in the CD22 cytoplasmic tail. In fact, of all phosphotyrosine-containing proteins, CD22 is the dominant binding partner for SHP-1 in B cells [98]. That CD22 contains both positive and negative signaling elements in its intracellular domain may also explain why both positive and negative signaling roles for CD22 are observed during different *in vitro* experiments [95]. Despite the complexity of B-cell signal transduction, CD22–ligand interactions are vital for normal B-cell physiology.

18.5.2.2 ***CD22 as a Therapeutic Target*** Because CD22 provides an important functional link between B-cell survival and regulation of intracellular signaling, it is not surprising that CD22-directed therapies are being developed and tested for both oncology and autoimmunity applications [99].

CD22 Therapeutic Applications in Oncology CD22 is expressed by 75–80% of B-cell lymphomas and leukemias [100]. The importance of CD22 ligand binding for normal B-cell survival suggests that blocking this interaction may influence malignant B-cell survival. Antibody binding to CD22 results in rapid internalization [101]. In light of these observations, CD22 emerged as a pertinent target for immunotherapy of B-cell malignancies. Radiolabeled CD22 monoclonal antibodies (mAbs) [102] demonstrated high sensitivities in the diagnosis and the staging of B-cell lymphomas, while therapeutic ^{131}I- or ^{90}Y-labeled CD22 mAbs generated partial to complete remissions. Because malignant B cells rapidly internalize CD22 and CD22 mAb binding leads to efficient intracellular routing of the CD22–mAb complex to intracellular compartments [103], the use of CD22 mAbs to deliver immunotoxins to non-Hodgkin's lymphoma and chronic lymphocytic leukemia cells also received considerable attention [104].

A novel approach was developed for targeting B lymphoma cells with doxorubicin-loaded liposomal nanoparticles displaying high-affinity glycan ligands of CD22 [105]. The targeted liposomes are actively bound and endocytosed by CD22 on B cells and significantly extend life in a xenograft model of human B-cell lymphoma. Moreover, they bind and kill malignant B cells from peripheral blood samples obtained from patients with hairy cell leukemia, marginal zone lymphoma, and chronic lymphocytic leukemia. The results demonstrate the potential for using a carbohydrate recognition-based approach for efficient targeting of B cells *in vivo*, which can offer improved treatment options for patients with B-cell malignancies.

CD22-Directed Therapy in Autoimmunity Because CD22 regulates both signal transduction thresholds and B-cell survival, it is not surprising that CD22 has been implicated in the development of autoimmunity, at least in mouse models of the disease. The importance of B-cell survival to autoimmune disease is well appreciated, leading to the pursuit of therapies that influence this potent biological outcome. By interfering with B-cell survival and enhancing B-cell turnover, it is envisioned that B cells or B-cell subsets can be specifically targeted without the widespread toxicity of conventional treatments including systemic corticosteroid administration. Moreover, it may be possible to affect therapeutic benefit by attenuating B-cell responses to transmembrane signals without eliminating B cells that give rise to protective humoral immunity. Accelerated B-cell apoptosis and turnover by therapeutic mAbs that block CD22 ligand engagement may therefore have considerable benefit for the treatment of autoimmunity [106].

Through its effects on both B-cell survival and signal transduction, CD22 likely plays an important role in normal B-cell selection, activation, and differentiation during immune responses. A better understanding of CD22 function and regulation may also afford mechanisms by which humoral immunity can be modulated, leading to the development of new strategies to augment antimicrobial defense and acquired immune responses, and for the development of novel approaches aimed at regulating chronic infections and inflammatory disorders [22, 23]. Because B cells contribute substantially to many human autoimmune diseases and hematologic malignancies and these diseases are often correlated with altered signal transduction or unregulated B-cell survival, new approaches targeting CD22 directly or interfering with CD22–ligand binding and function represent a valid and exciting new arena for therapeutic advancement [22, 23].

A detailed molecular understanding of the role of CD22 will continue to evolve as B-cell signaling pathways become better defined. Recent work on CD22 has provided important insights into how sialic acid recognition can modulate its signaling functions. B cells of CD22-deficient mice exhibit hyperimmune responses *in vitro* and *in vivo*, consistent with the loss of negative regulation by ITIMs of CD22 [96, 107]. Several CD22 functions, including BCR-dependent proliferation and B-cell turnover rates, depend on the ligand-binding function of CD22, as shown using mice that carry knock-in mutations of CD22 that ablate its ability to bind sialic acid [108]. In contrast to CD22-deficient mice, mice lacking the $\alpha(2\rightarrow6)$-sialyltransferase ST6GAL1 exhibit hypoimmune responses [109–111]. B cells from mice that are deficient in both CD22 and ST6GAL1 behave similarly to those from CD22-deficient mice, which indicates that the immunodeficiency of ST6GAL1-deficient mice depends on the presence of CD22. Following BCR ligation *in vitro*, the immunodeficiency caused by the absence of *cis* ligands in ST6GAL1-deficient mice is manifested by reduced B-cell proliferation and Ca^{2+} flux and increased CD22 phosphorylation and recruitment of SHP-1. Jamey Marth et al. recently showed that ablating ST6GAL1 in the lupus mouse model deficient in the Tyr kinase Lyn abrogated autoimmune responses, which indicates the therapeutic implications of inducing a CD22 ligand deficiency [109–111].

It was found that B cells deficient in CD22 undergo accelerated cell division after antigen stimulation, resulting in rapid generation of plasma cells and antibody production. This finding indicates that CD22 regulates the time course of B-cell responses and suggests that CD22 could be targeted to shorten the time required for antibody production [112].

CD22 B-cell-specific expression makes it an attractive target for immunotoxin-mediated B-cell depletion therapy for the treatment of B-cell lymphomas and autoimmune diseases. Glycan ligand-based cargo is released from CD22 and accumulates intracellularly as CD22 recycles between the cell surface and endosomal compartments. In contrast, antibodies to CD22 do not accumulate but remain bound to CD22 and recycle to the cell surface. The results have implications for the development of agents that target CD22 as an endocytic receptor for delivery of cytotoxic cargo to B cells [113].

The negative regulatory role of CD22 on B cells is well characterized, but whether siglecs play a role in inducing tolerance in immune cells had not been explored until recently. James Paulson et al. [74] showed that decoration of T-independent type 2 antigens with sialic acids induces poor immune responses and leads to tolerance. Both Siglec-G and CD22 have been shown to play a role in inducing tolerance, preventing plasma cell differentiation and survival. This is the first report of tolerance being induced through siglecs in addition to their established role in dysregulation of cell signaling. An elegant study by the same group revealed that liposomal nanoparticles displaying both antigen and CD22 ligands induce a tolerogenic program that selectively causes apoptosis in mouse and human B cells [114]. This pioneering strategy could be used to tolerize B cells in an antigen-specific manner and eliminate or prevent harmful B-cell-mediated immune responses.

18.5.2.3 Sialoside Recognition by CD22 The specificity of CD22 toward sialylated glycans was described in detail in a series of elegant studies [40, 90, 115–117]. CD22 binds specifically to the α(2→6)-linked sialyllactosamines that are commonly found on N-linked glycans of many cell-surface glycoproteins [114, 115]. At a minimum, sialic acid needs to be α(2→6)-linked to Gal or glucose (Glc) or to their corresponding acetamido forms (GalNAc or GlcNAc) to be recognized by CD22 [116, 117]. A conservative substitution of the critical Arg in the V-set domain of CD22 with a lysine results in loss of binding to sialic acid [40, 90], confirming the critical role of the axial carboxyl group of the α-configured sialic acid. The glycerol side chain of sialic acid is also important for binding. 9-O-Acetylation of sialic acid considerably reduces the binding by CD22, and mild periodate treatment abrogates the binding. Sørge Kelm's study of the affinity of a series of methyl sialosides for some siglecs (sialoadhesin, CD22, and MAG) demonstrated the requirement of 9-OH for binding [90]. Replacement of 9-OH with hydrogen or halogens did not support binding, while the 9-amino analogue showed a threefold increase in binding, suggesting that 9-OH functions as a hydrogen bond donor for one or more amino acids in the binding pocket. The introduction of a thiol group resulted in much lower affinity, probably because of the weaker hydrogen bonds with the protein and the steric constraints by the large sulfur atom. Moreover, direct functionalization of the C9-positioned amine most notably with acyl groups such as benzoyl (**8**), naphthylcarbonyl (**9**), biphenylcarbonyl (**10**), and biphenylacetamido (**12**) (Figs. 18.4 and 18.5) increased the binding dramatically. Among them, compound **10** enhanced the binding for human CD22, while the corresponding biphenylacetamide **12** showed improved potency for mouse CD22 (Table 18.2) [36, 40]. In general, the binding affinity for sialoadhesin, CD22, and MAG improved significantly when using sialic acid analogues that have aromatic substituents at C9, indeed indicating an additional hydrophobic contact at this site.

Interestingly, mouse CD22 strongly prefers Neu5Gc, in contrast to sialoadhesin and MAG. Extended investigation of the substructural specificity of the sialic acid moiety mapped the epitopes important for binding. The intact sialic acid residue is necessary for binding, while modifications at C5 and C9 can greatly enhance affinity and selectivity.

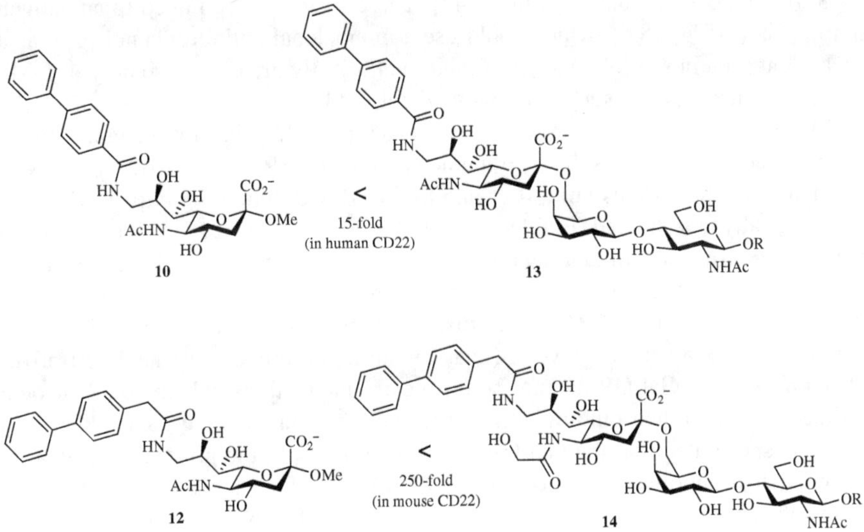

FIGURE 18.5 Some C9-modified sialosides and their comparative potencies.

TABLE 18.2 Comparison of Inhibitory Potential of Synthetic Sialosides [36, 40]

	Relative Inhibitory Potency		
Sialoside	Human CD22	Mouse CD22	Mouse Sialoadhesin
3	1	1	1
10	224	5	13
12	29	48	0.3

18.5.2.4 ***Design and Synthesis of High-Affinity Ligands/Antagonists for CD22*** High-affinity ligands suitable for *in vivo* studies are required to elucidate the role of CD22 in regulating immune responses. Makoto Kiso and coworkers designed and synthesized CD22 inhibitors using the insights about the effects of aromatic substituents at C9 of sialic acid. Computational model based on sialoadhesin/**10** crystal structure revealed that the bulky C9-biphenyl group could be sandwiched favorably between the side chains of amino acids (methionine-109 and Arg111 of human CD22) of the hydrophobic cavity as observed in sialoadhesin (Fig. 18.2). On the other hand, substitution of Arg111 (human CD22) by proline-111 in mouse CD22 is proposed as a reason for the observed affinity differences [36]. The molecular features of compounds **10** and **12** and the specificity elements preferred by the human and mouse versions of CD22 were combined into the sequences of the sialotrisaccharides **13** and **14** as high-affinity multivalent probes for B-cell targeting (Fig. 18.5) [42]. Compounds **13** (15-fold) and **14** (250-fold) showed better affinity than their parent structures for human CD22 and mouse CD22, respectively.

Conjugation of these high-affinity sialoside probes to the toxin saporin resulted in the killing of B lymphoma cell lines [42]. These ligands were linked to an antigen (nitrophenol) (Fig. 18.6), which could use a monoclonal antinitrophenol IgM, as a decavalent protein scaffold and was found to efficiently drive the assembly of IgM–CD22 complexes on the surface of native B cells [118].

To investigate the structural requirements for CD22 receptor recognition, the structure–activity relationships of *p*-methoxyphenyl (PMP) Neu5Ac(α2→6)GalNAc analogues having various amino- or amido-linked hydrophobic groups at C9 were systematically investigated (Fig. 18.7) [119]. Modifications at C9 position resulted in dramatic improvement of binding affinity for CD22 as exemplified by compounds **16** and **17**, which exhibited the highest potency for mouse CD22 and human CD22, respectively. Interestingly, **16** and **17** displayed more potent affinity for CD22 than for MAG. Docking studies revealed that C9-amido or amino sialic acid derivatives linked α(2→6) to Gal-OPMP moiety form extra interactions, which could account for the increase in their binding affinities for both human and mouse CD22 [119]. It was also speculated that the Gal-OPMP interactions with the protein are primarily driven by hydrophobic interactions within the vicinity of the active site. In support of this view, replacement of the subterminal Gal residue of **16** with Bn (**20**) or biphenylmethyl (**21**) group as aglycone led to higher potency for CD22 [120]. The hydrophobicity of the Bn or biphenylmethyl moiety could compensate for the desolvation

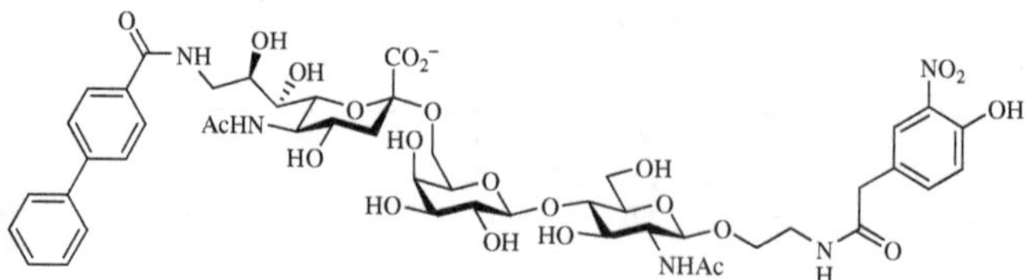

15

FIGURE 18.6 Structure of the heterobifunctional ligand that drive the assembly of IgM–CD22 complexes.

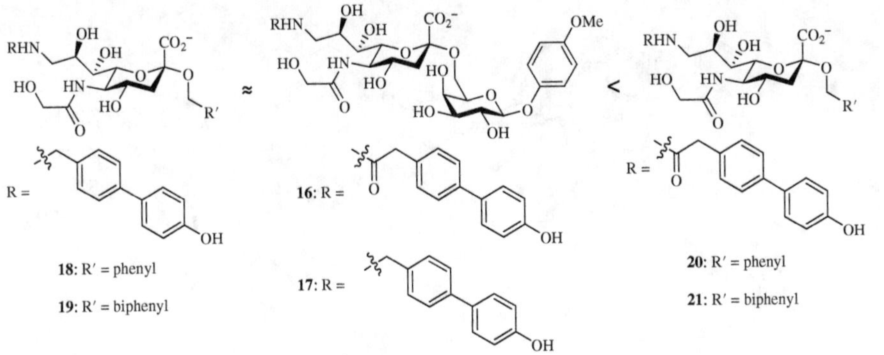

FIGURE 18.7 C9- and C2-modified sialosides as CD22 antagonists and their comparative potencies.

penalty fetched by Gal-OPMP (unfavorable desolvation enthalpy of the OH groups in Gal). Moreover, gain in entropy due to the increase in hydrophobic surface can make the binding affinity more favorable [121].

On the other hand, replacing the subterminal Gal residue of compound **17** with aromatic groups provided the equipotent compounds **18** and **19**. This may be due to the flexibility of the 9-amino substituent, which can disturb the intramolecular interactions. Both biological and computational studies suggest that the Gal-OPMP interactions (also the GlcNAc residue attached to Gal) with CD22 are not critical for the binding process and could be replaced with hydrophobic noncarbohydrate moieties with even improved potency. Compounds **20** and **21** are the most potent inhibitors for mouse CD22 (IC_{50} = 100 nM) and human CD22 (IC_{50} = 70 nM), respectively, that have been synthesized [121]. They have simpler structures, higher potency, easier synthetic access, and possibly enhanced bioavailability. Moreover, both **20** and **21** displayed significantly greater potency for human CD22 than for the closely related MAG [121]. Compound **21** was found to augment the activation of B cells, enhance B-cell tonic signaling, and promote antibody production, thereby protecting mice from influenza virus infection [122].

Recently, sialosides further modified with various hydrophobic substituents at other positions of sialic acid showed promising affinity, in particular compound **22** (Fig. 18.8) [123]. It was also found that substituents at C9 and C4 act synergistically to improve the binding affinity (**23**) [124]. Interestingly, sulfonation at the 4-position provided compound **24** with significantly improved potency [125].

18.6 CONCLUSION AND FUTURE DIRECTIONS

Siglecs are expressed abundantly on many cells of the immune system and are therefore likely to be important in both innate and acquired immune responses. The development of siglec-specific agonists and antagonists may provide new approaches for the treatment of certain autoimmune and inflammatory conditions. Moreover, the

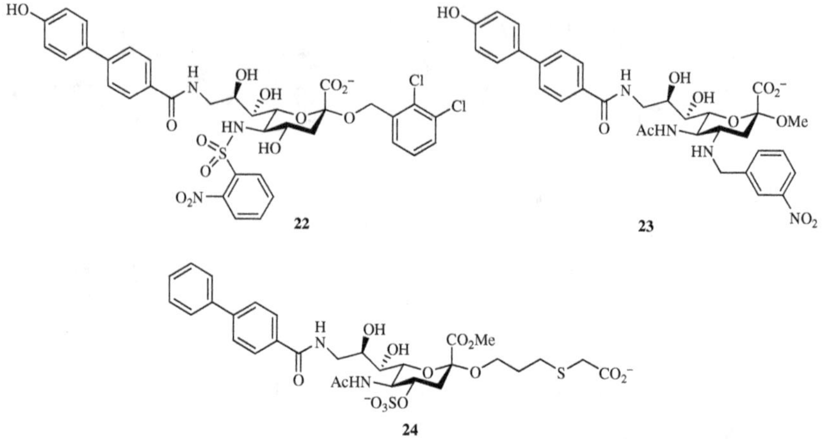

FIGURE 18.8 CD22 antagonists modified at various positions of sialic acid.

restricted expression of siglecs and the rapid progress in understanding their roles as cell signaling and endocytic receptors have made them attractive targets for cell-directed therapeutics. In particular, targeting CD22 directly or interfering with its ligand binding and function represents a valid and exciting new arena for therapeutic advancement. Several synthetic sialosides and sialomimetics have been found to be potent and selective ligands and antagonists of many members of the siglec family. However, these compounds are still at an early stage of development, and further studies are necessary to test their effects *in vivo* for concept validation purposes. In addition, it is necessary to develop drug-like mimics of the carbohydrate-based lead structures to ensure bioavailability and avoid metabolic instability.

REFERENCES

[1] Sharon, N. *J. Biol. Chem.* **2007**, *282*, 2753–2764.

[2] Gabius, H. *Biochim. Biophys. Acta* **2002**, *1572*, 163–164.

[3] Dam, T. K.; Brewer, C. F. *Chem. Rev.* **2002**, *102*, 387–429.

[4] Lis, H.; Sharon, N. *Chem. Rev.* **1998**, *98*, 637–674.

[5] Dube, D. H.; Bertozzi, C. R. *Nat. Rev. Drug Discov.* **2005**, *4*, 477–488.

[6] Gabius, H.-J.; Siebert, H.-C.; André, S.; Jiménez-Barbero, J.; Rüdiger, H. *ChemBioChem* **2004**, *5*, 740–764.

[7] Shriver, Z.; Raguram, S.; Sasisekharan, R. *Nat. Rev. Drug Discov.* **2004**, *3*, 863–873.

[8] Varki, A.; Cummings, R. D.; Esko, J. D.; Freeze, H. H.; Stanley, P.; Bertozzi, C. R.; Hart, G. W.; Etzler, M. E., Eds. *Essentials of Glycobiology*, 2nd ed.; Cold Spring Harbor Laboratory: Cold Spring Harbor, NY, **2009**.

[9] Dwek, R. A. *Chem. Rev.* **1996**, *96*, 683–720.

[10] Bianchet, M. A.; Ahmed, H.; Vasta, G. R.; Amzel, L. M. In *Animal Lectins: A Functional View*; Vasta, G. R., Ahmed, H., Eds.; CRC Press: Boca Raton, FL, **2009**, pp 13–31.

[11] Werz, D. B.; Seeberger, P. H. *Chem. Eur. J.* **2005**, *11*, 3194–3206.

[12] Nangia-Makker, P.; Conklin, J.; Hogan, V.; Raz, A. *Trends Mol. Med.* **2002**, *8*, 187–192.

[13] Rudiger, H.; Siebert, H.-C.; Solis, D.; Jimenez-Barbero, J.; Romero, A.; Lieth, C.-W.; Diaz-Maurino, T.; Gabius, H.-J. *Curr. Med. Chem.* **2000**, *7*, 389–416.

[14] Gardiner, J. M. *Expert Opin. Invest. Drugs* **1998**, *7*, 405–411.

[15] Lingwood, C. A. *Curr. Opin. Chem. Biol.* **1998**, *2*, 695–700.

[16] Lee, Y. C.; Lee, R. T. *J. Biomed. Sci.* **1996**, *3*, 221–237.

[17] Collins, B. E.; Paulson, J. C. *Curr. Opin. Chem. Biol.* **2004**, *8*, 617–625.

[18] Crocker, P. R.; Paulson, J. C.; Varki, A. *Nat. Rev. Immunol.* **2007**, *7*, 255–266.

[19] Varki, A.; Angata, T. *Glycobiology* **2005**, *16*, 1R–27R.

[20] Macauley, M. S.; Crocker, P. R.; Paulson, J. C. *Nat. Rev. Immunol.* **2014**, *14*, 653–666.

[21] O'Reilly, M. K.; Paulson, J. C. *Trends Pharmacol. Sci.* **2009**, *30*, 240–248.

[22] Tateno, H.; Li, H.; Schur, M. J.; Bovin, N.; Crocker, P. R.; Wakarchuk, W. W.; Paulson, J. C. *Mol. Cell. Biol.* **2007**, *27*, 5699–5710.

[23] von Gunten, S.; Bochner, B. S. *Ann. N. Y. Acad. Sci.* **2008**, *1143*, 61–82.

[24] Pillai, S.; Netravali, I. A.; Cariappa, A.; Mattoo, H. *Annu. Rev. Immunol.* **2012**, *30*, 357–392.

[25] Oetke, C.; Crocker, P. R. In *Animal Lectins: A Functional Overview*; Vasta, G. R., Ahmed, H., Eds.; CRC Press: Boca Raton, FL, **2009**, pp 243–253.

[26] Schauer, R. *Adv. Carbohydr. Chem. Biochem.* **1982**, 131–234.

[27] Schauer, R.; Kamerling, J. P. In *Glycoproteins II*; Montreuil, J., Vliegenthart, J. F. G., Schachter, H., Eds.; New Comprehensive Biochemistry 29, Part B; Elsevier: Amsterdam, **1997**, pp 243–402.

[28] Angata, T.; Varki, A. *Chem. Rev.* **2002**, *102*, 439–470.

[29] Varki, A. *Glycobiology* **1993**, *3*, 97–130.

[30] Kiefel, M. J.; von Itzstein, M. *Chem. Rev.* **2002**, *102*, 471–490.

[31] Brossmer, R.; Gross, H. J. In *Neoglycoconjugates, Part B: Biomedical Applications*; Lee, Y. C., Lee, R. T., Eds.; Methods in Enzymology 247; Elsevier: Amsterdam, **1994**, pp 153–176.

[32] Boons, G.-J.; Demchenko, A. V. *Chem. Rev.* **2000**, *100*, 4539–4566.

[33] Danishefsky, S. J.; Allen, J. R. *Angew. Chem. Int. Ed.* **2000**, *39*, 836–863.

[34] Roy, R. In *Carbohydrates in Drug Design*; Witczak, Z. J., Nieforth, K. A., Eds.; Marcel Dekker: New York, **1997**, pp 83–135.

[35] Attrill, H.; Imamura, A.; Sharma, R. S.; Kiso, M.; Crocker, P. R.; van Aalten, D. M. F. *J. Biol. Chem.* **2006**, *281*, 32774–32783.

[36] Zaccai, N. R.; Maenaka, K.; Maenaka, T.; Crocker, P. R.; Brossmer, R.; Kelm, S.; Jones, E. Y. *Structure* **2003**, *11*, 557–567.

[37] May, A. P.; Robinson, R. C.; Vinson, M.; Crocker, P. R.; Jones, E. Y. *Mol. Cell* **1998**, *1*, 719–728.

[38] Alphey, M. S.; Attrill, H.; Crocker, P. R.; van Aalten, D. M. F. *J. Biol. Chem.* **2002**, *278*, 3372–3377.

[39] Bhunia, A.; Jayalakshmi, V.; Benie, A. J.; Schuster, O.; Kelm, S.; Rama Krishna, N.; Peters, T. *Carbohydr. Res.* **2004**, *339*, 259–267.

[40] Kelm, S.; Gerlach, J.; Brossmer, R.; Danzer, C.-P.; Nitschke, L. *J. Exp. Med.* **2002**, *195*, 1207–1213.

[41] Razi, N.; Varki, A. *Proc. Natl. Acad. Sci. U. S. A.* **1998**, *95*, 7469–7474.

[42] Collins, B. E.; Blixt, O.; Han, S.; Duong, B.; Li, H.; Nathan, J. K.; Bovin, N.; Paulson, J. C. *J. Immunol.* **2006**, *177*, 2994–3003.

[43] Collins, B. E.; Blixt, O.; DeSieno, A. R.; Bovin, N.; Marth, J. D.; Paulson, J. C. *Proc. Natl. Acad. Sci. U. S. A.* **2004**, *101*, 6104–6109.

[44] Rillahan, C. D.; Schwartz, E.; McBride, R.; Fokin, V. V.; Paulson, J. C. *Angew. Chem. Int. Ed.* **2012**, *51*, 11014–11018.

[45] Jones, C.; Virji, M.; Crocker, P. R. *Mol. Microbiol.* **2003**, *49*, 1213–1225.

[46] Monteiro, V. G.; Lobato, C. S. S.; Silva, A. R.; Medina, D. V.; de Oliveira, M. A.; Seabra, S. H.; de Souza, W.; DaMatta, R. A. *Parasitol. Res.* **2005**, *97*, 380–385.

[47] Vanderheijden, N.; Delputte, P. L.; Favoreel, H. W.; Vandekerckhove, J.; Van Damme, J.; van Woensel, P. A.; Nauwynck, H. J. *J. Virol.* **2003**, *77*, 8207–8215.

[48] Zou, Z.; Chastain, A.; Moir, S.; Ford, J.; Trandem, K.; Martinelli, E.; Cicala, C.; Crocker, P.; Arthos, J.; Sun, P. D. *PLoS One* **2011**, *6*, e24559.

[49] Bax, M.; Garcia-Vallejo, J. J.; Jang-Lee, J.; North, S. J.; Gilmartin, T. J.; Hernandez, G.; Crocker, P. R.; Leffler, H.; Head, S. R.; Haslam, S. M.; Dell, A.; van Kooyk, Y. *J. Immunol.* **2007**, *179*, 8216–8224.

[50] Lock, K.; Zhang, J.; Lu, J.; Lee, S. H.; Crocker, P. R. *Immunobiology* **2004**, *209*, 199–207.

[51] O'Keefe, T. L.; Williams, G. T.; Davies, S. L.; Neuberger, M. S. *Science* **1996**, *274*, 798–801.

[52] Otipoby, K. L.; Andersson, K. B.; Draves, K. E.; Klaus, S. J.; Farr, A. G.; Kerner, J. D.; Perlmutter, R. M.; Law, C.-L.; Clark, E. A. *Nature* **1996**, *384*, 634–637.

[53] Nitschke, L.; Carsetti, R.; Ocker, B.; Köhler, G.; Lamers, M. C. *Curr. Biol.* **1997**, *7*, 133–143.

[54] O'Keefe, T. L.; Williams, G. T.; Batista, F. D.; Neuberger, M. S. *J. Exp. Med.* **1999**, *189*, 1307–1313.

[55] Cariappa, A.; Takematsu, H.; Liu, H.; Diaz, S.; Haider, K.; Boboila, C.; Kalloo, G.; Connole, M.; Shi, H. N.; Varki, N.; Varki, A.; Pillai, S. *J. Exp. Med.* **2008**, *206*, 125–138.

[56] Meesmann, H. M.; Fehr, E. M.; Kierschke, S.; Herrmann, M.; Bilyy, R.; Heyder, P.; Blank, N.; Krienke, S.; Lorenz, H. M.; Schiller, M. *J. Cell Sci.* **2010**, *123*, 3347–3356.

[57] Malagolini, N.; Chiricolo, M.; Marini, M.; Dall'Olio, F. *Glycobiology* **2008**, *19*, 172–181.

[58] Carlin, A. F.; Chang, Y. C.; Areschoug, T.; Lindahl, G.; Hurtado-Ziola, N.; King, C. C.; Varki, A.; Nizet, V. *J. Exp. Med.* **2009**, *206*, 1691–1699.

[59] Nutku, E. *Blood* **2003**, *101*, 5014–5020.

[60] Nutku, E.; Hudson, S. A.; Bochner, B. S. *Biochem. Biophys. Res. Commun.* **2005**, *336*, 918–924.

[61] Ando, M.; Tu, W.; Nishijima, K.; Iijima, S. *Biochem. Biophys. Res. Commun.* **2008**, *369*, 878–883.

[62] von Gunten, S. *Blood* **2005**, *106*, 1423–1431.

[63] Carlin, A. F.; Lewis, A. L.; Varki, A.; Nizet, V. *J. Bacteriol.* **2006**, *189*, 1231–1237.

[64] Carlin, A. F.; Uchiyama, S.; Chang, Y. C.; Lewis, A. L.; Nizet, V.; Varki, A. *Blood* **2009**, *113*, 3333–3336.

[65] Chen, G. Y.; Tang, J.; Zheng, P.; Liu, Y. *Science* **2009**, *323*, 1722–1725.

[66] Liu, Y.; Chen, G.-Y.; Zheng, P. *Curr. Opin. Immunol.* **2011**, *23*, 41–45.

[67] Linnartz, B.; Wang, Y.; Neumann, H. *Int. J. Alzheimer's Dis.* **2010**, 587463.

[68] Wang, Y.; Neumann, H. *J. Neurosci.* **2010**, *30*, 3482–3488.

[69] Boyd, C. R.; Orr, S. J.; Spence, S.; Burrows, J. F.; Elliott, J.; Carroll, H. P.; Brennan, K.; Gabhann, J. N.; Coulter, W. A.; Johnston, J. A.; Jefferies, C. A. *J. Immunol.* **2009**, *183*, 7703–7709.

[70] Zimmermann, N.; McBride, M. L.; Yamada, Y.; Hudson, S. A.; Jones, C.; Cromie, K. D.; Crocker, P. R.; Rothenberg, M. E.; Bochner, B. S. *Allergy* **2008**, *63*, 1156–1163.

[71] Cho, J. Y.; Song, D. J.; Pham, A.; Rosenthal, P.; Miller, M.; Dayan, S.; Doherty, T. A.; Varki, A.; Broide, D. H. *Respir. Res.* **2010**, *11*, 154.

[72] Lunnon, K.; Teeling, J. L.; Tutt, A. L.; Cragg, M. S.; Glennie, M. J.; Perry, V. H. *J. Immunol.* **2011**, *186*, 7215–7224.

[73] Ding, C.; Liu, Y.; Wang, Y.; Park, B. K.; Wang, C.-Y.; Zheng, P.; Liu, Y. *PLoS One* **2007**, *2*, e997.

[74] Duong, B. H.; Tian, H.; Ota, T.; Completo, G.; Han, S.; Vela, J. L.; Ota, M.; Kubitz, M.; Bovin, N.; Paulson, J. C.; Nemazee, D. *J. Exp. Med.* **2009**, *207*, 173–187.

[75] Blasius, A.; Vermi, W.; Krug, A.; Facchetti, F.; Cella, M.; Colonna, M. *Blood* **2004**, *103*, 4201–4206.

[76] Delputte, P. L.; Van Gorp, H.; Favoreel, H. W.; Hoebeke, I.; Delrue, I.; Dewerchin, H.; Verdonck, F.; Verhasselt, B.; Cox, E.; Nauwynck, H. J. *PLoS One* **2011**, *6*, e16827.

[77] Rapoport, E. M.; Sapot'ko, Y. B.; Pazynina, G. V.; Bojenko, V. K.; Bovin, N. V. *Biochemistry (Moscow)* **2005**, *70*, 330–338.

[78] Pereira-Chioccola, V. L.; Acosta-Serrano, A.; Correia de Almeida, I.; Ferguson, M. A.; Souto-Padron, T.; Rodrigues, M. M.; Travassos, L. R.; Schenkman, S. *J. Cell Sci.* **2000**, *113*, 1299–1307.

[79] Bochner, B. S.; Alvarez, R. A.; Mehta, P.; Bovin, N. V.; Blixt, O.; White, J. R.; Schnaar, R. L. *J. Biol. Chem.* **2004**, *280*, 4307–4312.

[80] Gao, P.-S.; Shimizu, K.; Grant, A. V.; Rafaels, N.; Zhou, L.-F.; Hudson, S. A.; Konno, S.; Zimmermann, N.; Araujo, M. I.; Ponte, E. V.; Cruz, A. A.; Nishimura, M.; Su, S.-N.; Hizawa, N.; Beaty, T. H.; Mathias, R. A.; Rothenberg, M. E.; Barnes, K. C.; Bochner, B. S. *Eur. J. Hum. Genet.* **2010**, *18*, 713–719.

[81] Zhang, J. Q.; Nicoll, G.; Jones, C.; Crocker, P. R. *J. Biol. Chem.* **2000**, *275*, 22121–22126.

[82] Avril, T.; North, S. J.; Haslam, S. M.; Willison, H. J.; Crocker, P. R. *J. Leukocyte Biol.* **2006**, *80*, 787–796.

[83] Nicoll, G.; Avril, T.; Lock, K.; Furukawa, K.; Bovin, N.; Crocker, P. R. *Eur. J. Immunol.* **2003**, *33*, 1642–1648.

[84] Brunetta, E.; Fogli, M.; Varchetta, S.; Bozzo, L.; Hudspeth, K. L.; Marcenaro, E.; Moretta, A.; Mavilio, D. *Blood* **2009**, *114*, 3822–3830.

[85] Bax, M.; Kuijf, M. L.; Heikema, A. P.; van Rijs, W.; Bruijns, S. C. M.; Garcia-Vallejo, J. J.; Crocker, P. R.; Jacobs, B. C.; van Vliet, S. J.; van Kooyk, Y. *Infect. Immun.* **2011**, *79*, 2681–2689.

[86] Erdmann, H.; Steeg, C.; Koch-Nolte, F.; Fleischer, B.; Jacobs, T. *Cell. Microbiol.* **2009**, *11*, 1600–1611.

[87] Ikehara, Y.; Ikehara, S. K.; Paulson, J. C. *J. Biol. Chem.* **2004**, *279*, 43117–43125.

[88] Oetke, C.; Vinson, M. C.; Jones, C.; Crocker, P. R. *Mol. Cell. Biol.* **2006**, *26*, 1549–1557.

[89] Cao, H.; Crocker, P. R. *Immunology* **2010**, *132*, 18–26.

[90] Kelm, S.; Brossmer, R.; Isecke, R.; Gross, H.-J.; Strenge, K.; Schauer, R. *Eur. J. Biochem.* **1998**, *255*, 663–672.

[91] Zaccai, N. R.; May, A. P.; Robinson, R. C.; Burtnick, L. D.; Crocker, P. R.; Brossmer, R.; Kelm, S.; Jones, E. Y. *J. Mol. Biol.* **2007**, *365*, 1469–1479.

[92] Nycholat, C. M.; Rademacher, C.; Kawasaki, N.; Paulson, J. C. *J. Am. Chem. Soc.* **2012**, *134*, 15696–15699.

[93] Tedder, T. F. *Semin. Immunol.* **1998**, *10*, 259–265.

[94] Nitschke, L.; Tsubata, T. *Trends Immunol.* **2004**, *25*, 543–550.

[95] Sato, S.; Tuscano, J. M.; Inaoki, M.; Tedder, T. F. *Semin. Immunol.* **1998**, *10*, 287–297.

[96] Nitschke, L. *Curr. Opin. Immunol.* **2005**, *17*, 290–297.

[97] Powell, L. D.; Varki, A. *J. Biol. Chem.* **1995**, *270*, 10628–10636.

[98] Sato, S.; Jansen, P. J.; Tedder, T. F. *Proc. Natl. Acad. Sci. U. S. A.* **1997**, *94*, 13158–13162.

[99] Smith, S. H.; Tedder, T. F. In *Animal Lectins: A Functional Overview*; Vasta, G. R., Ahmed, H., Eds.; CRC Press: Boca Raton, FL, **2009**, pp 265–274.

[100] Press, O. W.; Leonard, J. P.; Coiffier, B.; Levy, R.; Timmerman, J. *Hematology* **2001**, 221–240.

[101] Shan, D.; Press, O. W. *J. Immunol.* **1995**, *154*, 4466–4475.

[102] Stein, R.; Belisle, E.; Hansen, H. J.; Goldenberg, D. M. *Cancer Immunol. Immunother.* **1993**, *37*, 293–298.

[103] Press, O. W.; Howell-Clark, J.; Anderson, S.; Bernstein, I. *Blood* **1994**, *83*, 1390–1397.

[104] Sieber, T.; Schoeler, D.; Ringel, F.; Pascu, M.; Schriever, F. *Br. J. Haematol.* **2003**, *121*, 458–461.

[105] Chen, W. C.; Completo, G. C.; Sigal, D. S.; Crocker, P. R.; Saven, A.; Paulson, J. C. *Blood* **2010**, *115*, 4778–4786.

[106] Tuscano, J. M.; Harris, G. S.; Tedder, T. F. *Autoimmun. Rev.* **2003**, *2*, 101–108.

[107] Tedder, T. F.; Poe, J. C.; Haas, K. M. *Adv. Immunol.* **2005**, *88*, 1–50.

[108] Poe, J. C.; Fujimoto, Y.; Hasegawa, M.; Haas, K. M.; Miller, A. S.; Sanford, I. G.; Bock, C. B.; Fujimoto, M.; Tedder, T. F. *Nat. Immunol.* **2004**, *5*, 1078–1087.

[109] Collins, B. E.; Smith, B. A.; Bengtson, P.; Paulson, J. C. *Nat. Immunol.* **2006**, *7*, 199–206.

[110] Grewal, P. K.; Boton, M.; Ramirez, K.; Collins, B. E.; Saito, A.; Green, R. S.; Ohtsubo, K.; Chui, D.; Marth, J. D. *Mol. Cell. Biol.* **2006**, *26*, 4970–4981.

[111] Ghosh, S.; Bandulet, C.; Nitschke, L. *Int. Immunol.* **2006**, *18*, 603–611.

[112] Onodera, T.; Poe, J. C.; Tedder, T. F.; Tsubata, T. *J. Immunol.* **2008**, *180*, 907–913.

[113] O'Reilly, M. K.; Tian, H.; Paulson, J. C. *J. Immunol.* **2011**, *186*, 1554–1563.

[114] Macauley, M. S.; Pfrengle, F.; Rademacher, C.; Nycholat, C. M.; Gale, A. J.; von Drygalski, A.; Paulson, J. C. *J. Clin. Invest.* **2013**, *123*, 3074–3083.

[115] Kelm, S.; Pelz, A.; Schauer, R.; Filbin, M. T.; Tang, S.; de Bellard, M.-E.; Schnaar, R. L.; Mahoney, J. A.; Hartnell, A.; Bradfield, P.; Crocker, P. R. *Curr. Biol.* **1994**, *4*, 965–972.

[116] Hanasaki, K.; Powell, L. D.; Varki, A. *J. Biol. Chem.* **1995**, *270*, 7543–7550.

[117] Powell, L. D.; Jain, R. K.; Matta, K. L.; Sabesan, S.; Varki, A. *J. Biol. Chem.* **1995**, *270*, 7523–7532.

[118] O'Reilly, M. K.; Collins, B. E.; Han, S.; Liao, L.; Rillahan, C.; Kitov, P. I.; Bundle, D. R.; Paulson, J. C. *J. Am. Chem. Soc.* **2008**, *130*, 7736–7745.

[119] Abdu-Allah, H. H. M.; Tamanaka, T.; Yu, J.; Zhuoyuan, L.; Sadagopan, M.; Adachi, T.; Tsubata, T.; Kelm, S.; Ishida, H.; Kiso, M. *J. Med. Chem.* **2008**, *51*, 6665–6681.

[120] Abdu-Allah, H. H. M.; Watanabe, K.; Hayashizaki, K.; Takaku, C.; Tamanaka, T.; Takematsu, H.; Kozutsumi, Y.; Tsubata, T.; Ishida, H.; Kiso, M. *Bioorg. Med. Chem. Lett.* **2009**, *19*, 5573–5575.

[121] Abdu-Allah, H. H. M.; Watanabe, K.; Completo, G. C.; Sadagopan, M.; Hayashizaki, K.; Takaku, C.; Tamanaka, T.; Takematsu, H.; Kozutsumi, Y.; Paulson, J. C.; Tsubata, T.; Ando, H.; Ishida, H.; Kiso, M. *Bioorg. Med. Chem.* **2011**, *19*, 1966–1971.

[122] Tsubata, T.; Kiso, M.; Ishida, H.; Abdu-Allah, H. H. M. Compound Having High Affinity for CD22 and Capable of Enhancing Proliferation of B Cells. WO/2011/013398; PCT/JP2010/054406, February 3, **2011**.

[123] Mesch, S.; Lemme, K.; Wittwer, M.; Koliwer-Brandl, H.; Schwardt, O.; Kelm, S.; Ernst, B. *ChemMedChem* **2011**, *7*, 134–143.

[124] Kelm, S.; Madge, P.; Islam, T.; Bennett, R.; Koliwer-Brandl, H.; Waespy, M.; von Itzstein, M.; Haselhorst, T. *Angew. Chem. Int. Ed.* **2013**, *52*, 3616–3620.

[125] Prescher, H.; Schweizer, A.; Kuhfeldt, E.; Nitschke, L.; Brossmer, R. *ACS Chem. Biol.* **2014**, *9*, 1444–1450.

19

SUGAR–PROTEIN HYBRIDS FOR BIOMEDICAL APPLICATIONS

Macarena Sánchez-Navarro[1,2] and Benjamin G. Davis[2]

[1] *Institute for Research in Biomedicine, The Barcelona Institute of Science and Technology, Barcelona, Spain*

[2] *Department of Chemistry, Chemistry Research Laboratory, University of Oxford, Oxford, UK*

19.1 INTRODUCTION

The aim of this chapter is to give an overview of the preparation and applications of biomedically relevant (and potentially relevant) synthetic glycoproteins. The crucial role of natural sugar–protein conjugates in nature makes synthetic variants an interesting instrument to better understand the complex processes in which they are involved [1–3]. Glycoproteins participate in communication events, like those resulting in inflammation, fertilization, or host immune response [4–9]. In addition, carbohydrates are able to stabilize the tertiary structure of proteins by inducing more compact structures [10]. Glycosylation is one of the most multifaceted posttranslational modifications of proteins. Indeed, 40% of the approved therapeutic proteins are glycosylated [11].

This chapter first examines the different methods for the preparation of relevant synthetic glycoproteins largely where the linkage between the proteic part and the saccharidic moiety is unnatural. The later discussion focuses on the application of

Glycochemical Synthesis: Strategies and Applications, First Edition.
Edited by Shang-Cheng Hung and Medel Manuel L. Zulueta.

these synthetic hybrids in the field of biomedicine through their use as vaccine candidates, anti-infective agents, or drug delivery systems, among others.

19.2 CHALLENGES IN THE DEVELOPMENT OF GLYCOPROTEIN-BASED THERAPEUTICS

The main drawback of glycoprotein isolation and extraction from natural sources is the complexity of purification arising from their expression as mixtures of different glycoforms [6]. These protein isoforms differ only with respect to the number, type of attached glycans, or glycosylation site yet may result in many hundreds of variants. Homogeneous glycoforms cannot be readily accessed from natural sources because glycosylation, as with other posttranslational modifications, is not directly genetically driven.

To elucidate the different processes in which glycoforms are involved, methodologies to prepare glycoproteins in a selective and well-defined manner are needed. In addition, the determination of exact structure–activity relationships can only be established with pure and well-defined glycoforms.

Due to their exceptional roles in many biological processes (and now presence in many drugs), glycoproteins are gaining attention in the biopharmaceutical sector [11, 12], making the control of glycosylation profile an imperative priority to guarantee reproducible clinical performance. The need for high batch-to-batch consistency [13] has been evaluated [14] in the analysis of three commercial glycoprotein drugs (darbepoetin alfa, rituximab, and etanercept), where significant variations were detected that could result in an increasing risk of changes in their clinical effectiveness.

19.3 WHY UNNATURAL?

When designing a therapeutic product, some essential characteristics have to be met. The product has to have good selectivity toward the target, high affinity and/or efficacy, and stability under physiological conditions. Both preparative control and enhanced stability can be achieved by using unnatural linkages. This appealing method allows for the preparation of different protein–carbohydrate conjugates to study their implication in diverse biological processes. Different methodologies have been developed to prepare glycoproteins with unnatural linkages. Many groups attempt to exploit chemoselective strategies to create more homogeneous glycoconjugates. Several considerations have to be taken before deciding which assembly strategy to use [15]. Some unnatural linkages may better mimic natural structures and hence, more representatively participate in binding or recognition processes, resulting in better interaction with receptors. For instance, triazole moieties are among the most developed unnatural linkers (see Section 19.5.8) [16]. Yet despite the advantages inherent to their formation, water solubility, and others, their more crucial feature may be that triazoles can prove to be effective amide surrogates, retaining some hydrogen bonding and dipolar interactions [17]. In addition, the triazole moiety has been suggested to be immunosilent and hence, potentially of great, largely untapped value in vaccine development [18]. It is such focus on functional, rather than structural goals that should drive the development of glycoconjugates.

19.4 RETROSYNTHETIC ANALYSIS

In order to tackle the preparation of precisely modified glycoproteins, three assembly strategies can be envisaged (Scheme 19.1). Although this classification is subjective, it includes terms used by glycoscientists that are helpful to fully understand the different approaches reported. The first approach involves the assembly (b) of a previously prepared (a) glycopeptide with other peptides or proteins by using methodologies such as thioester-mediated native chemical ligation, solid-phase peptide synthesis, or enzymatic ligation, among others. The second tactic (c) consists of the convergent attachment of the glycan directly to the protein. The last strategy (d) of glycan elaboration includes the combination of trimming of the existing glycan structures (usually by enzymes) followed by elaboration to alternative ones.

A general aspect of the chemical modification of proteins is the need for chemoselective reactions that only modify the desired positions and do not detrimentally interfere in the activity of the proposed glycoconjugate. Practically, two approaches are often taken: (i) introduction of the carbohydrates at frequent natural residues of the protein such as lysine (Lys) or aspartic/glutamic acids (Asp/Glu) (indiscriminate convergent glycosylation) or (ii) introduction of a controlled number of sugars (usually one) at a preselected amino acid (natural or unnatural) that can be endogenous or introduced by mutagenesis (site-specific or site-directed glycosylation). The latter method can be described conceptually in the context of a "tag-and-modify" approach [19]. Using the first method, highly glycosylated proteins can be obtained rapidly, albeit often indiscriminately. Although the resulting products are a mixture

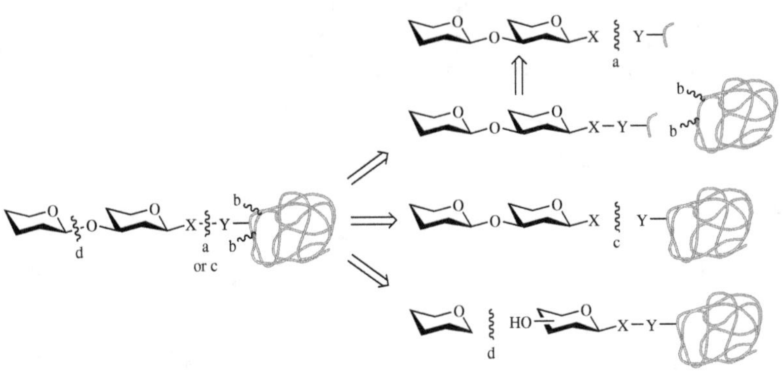

SCHEME 19.1 Glycoprotein retrosynthetic analysis. Three strategies utilize four major disconnection pathways: (a) amino acid/peptide glycosylation, (b) linear assembly of glycoamino acids/glycopeptides, (c) convergent protein glycosylation, and (d) glycoprotein remodeling.

of differently glycosylated proteins, the methodology can be highly reproducible, giving the same distribution of mixed compounds.

19.5 LINKAGES

This section highlights the currently most common and important unnatural linkages used to prepare glycoproteins (Fig. 19.1). The increasing demand for new therapeutics will lead to new members on this list.

19.5.1 Thioether

Known S-glycosidic linkages in nature are limited to small glycopeptides [20]. The high stability of thioether linkages and the absence of charge make them a very interesting option for the preparation of unnatural glycoproteins. Thioethers are commonly introduced into proteins via modification of cysteine (Cys) residues by taking advantage of its privileged features, that is, low natural abundance

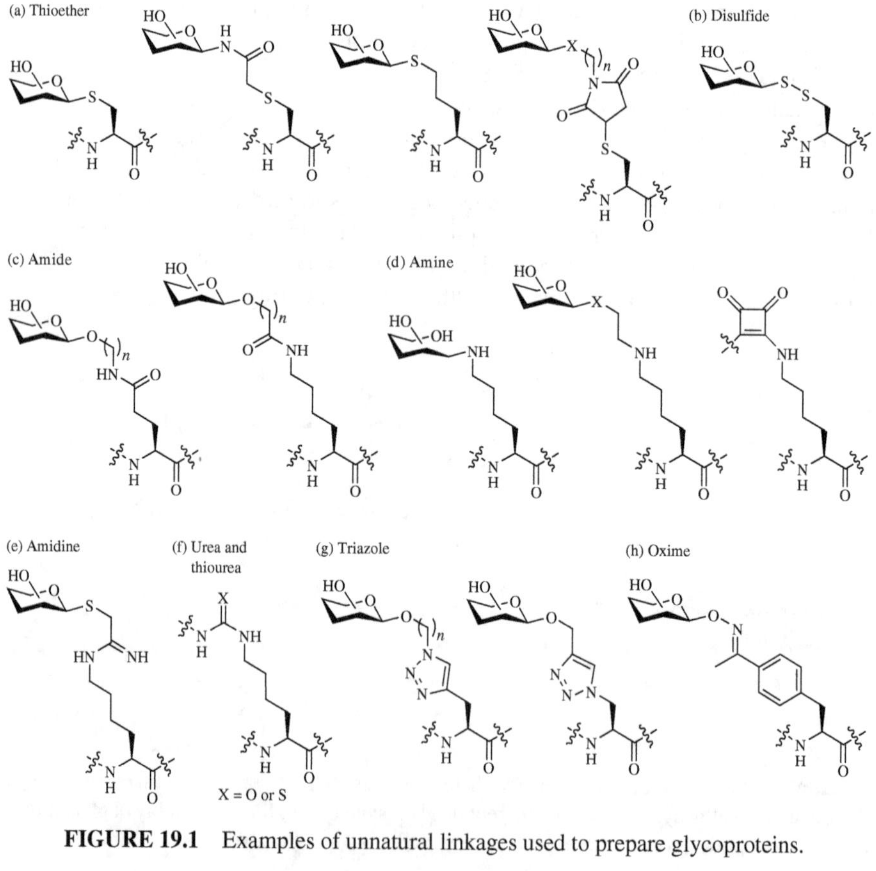

FIGURE 19.1 Examples of unnatural linkages used to prepare glycoproteins.

(<1.6%) [21], lower pK_a, and strong nucleophilic properties, which can be affected by the chemical environment within the protein [22].

Alkylation is one of the traditional methods to modify Cys because the reaction can at least be partially selective in the presence of Lys, histidines, and disulfide bonds by controlling the pH [23]. Different electrophiles can be used, with α-halo-carbonyls being one of the most frequent, especially α-iodoacetamide. First used to introduce different oligosaccharides to bovine serum albumin (BSA) [24], alkylation with α-iodoacetamide (Scheme 19.2) has been employed to prepare homogeneously glycosylated erythropoietin constructs [25] and to study enzymatic extension [26] and effects upon thermal stability [27, 28]. A more recent example of this methodology developed glycosylated human carbonic anhydrase II variants for *in vivo* fluorescence imaging [29].

Nucleophilic addition of Cys to Michael-type acceptors such as maleimides (Scheme 19.2) [30], vinyl sulfones, or α,β-unsaturated systems is another useful way of selectively modifying Cys with carbohydrates [31]. One of the first examples that illustrated the high efficiency of the maleimide–thiol coupling involved the modification of hemoglobin with a maleimido-lactose derivative, leading to an increase of the oxygen affinity of the construct [32]. In addition, this method has allowed the preparation of synthetic cancer [33, 34] or malaria [35] vaccine constructs and the conjugates of the ganglioside GM1 to keyhole limpet hemocyanin as tools to study Guillain–Barré syndrome in disease models [36].

Recently, some methods were developed for the generation of thioether linkages without direct reliance on nucleophilic Cys reactivity. The modification of dehydroalanine (Dha) with reactive thiols in essence reverses the roles of sugar and protein as nucleophile and electrophile (Scheme 19.3). Dha is a selective chemical handle that can be introduced by oxidative elimination of alkylcysteine, selenocysteine [37, 38], or phenylselenocysteine [39]; the reaction of Cys to give Dha can be accomplished with *O*-mesitylenesulfonylhydroxylamine [40] or by using α,α′-dibromo-adipyl(bis)amide [41]. Applications of this technique toward the modification of a wide range of proteins, including the glycosylation of an antibody [41] and the modification of histones with a broad spectrum of posttranslational modifications, including glycosylations, have been studied [42].

Other methods such as disulfide contraction [43] or photochemical addition of alkenes [44] or alkynes [45] have also been used for the preparation of thioether-linked

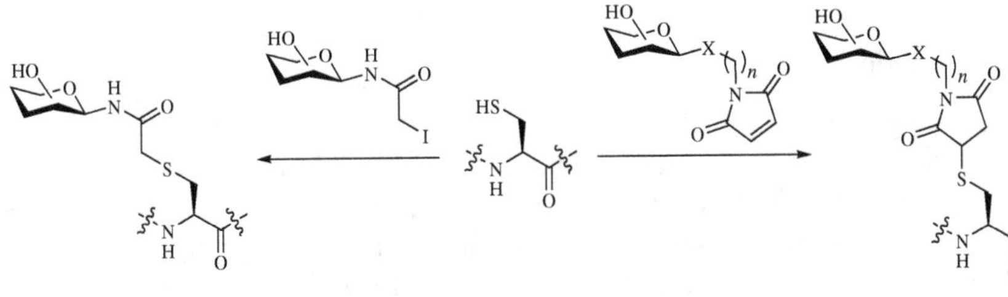

SCHEME 19.2 Examples of Cys alkylation to prepare glycoconjugates.

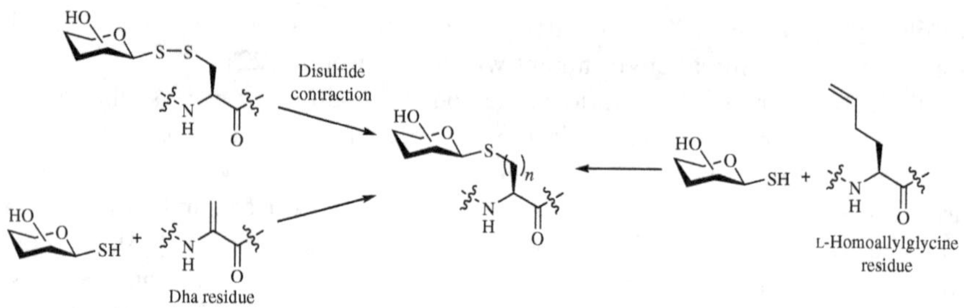

SCHEME 19.3 Formation of thioethers by nonnucleophilic methods.

homogeneous glycoprotein constructs (Scheme 19.3). The installation of alkenes into proteins by means of genetic methods [46, 47] was established and has allowed precise glycosylation of a broad range of proteins by using free radical addition of glycosyl 1-thiols under photochemical conditions. These methods were also successfully applied to the preparation of synthetic vaccine candidates by conjugation of complex oligosaccharides (up to 5 rhamnose units) to different proteins including the well-defined bacteriophage capsid Qβ [48].

19.5.2 Disulfides

The formation of disulfides is a versatile method to introduce selective modification on proteins at Cys residues [23, 49]. The most straightforward method to introduce disulfides at Cys is air oxidation [50, 51]. However, this process is typically determined by thermodynamic product control, longer reaction times are needed, and the possibility of dimer formation as well as side products instead of mixed disulfide has to be taken into account [23].

The formation of disulfides on proteins can be achieved by disulfide exchange. Glycosyl-activated electrophiles, such as 5-nitropyridine-2-sulfenyl, have been used for the formation of a range of glycoprotein conjugates [52]. One of the most developed methods to generate disulfides is the use of activated thiol-selective electrophiles such as methanethiosulfonate derivatives [53], phenylthiosulfonates [54], or phenylselenenylsulfides [55]. This method has been used, for example, to prepare glycodendriproteins—a new kind of glycoconjugate ideal for the study of multivalent binding [56, 57]—and has even allowed the preparation of a trisulfide-linked glycoprotein variant [58].

19.5.3 Amide

Amide linkages can, in principle, be introduced by modification of Asp or Glu side chains with amines or by conjugation of Lys with activated esters, as well as through Staudinger ligation [59]. One method, applied preferentially to functionalize carboxylate groups in a protein, is the use of carbodiimides as coupling reagent, especially the water-soluble 1-ethyl-3-(3-dimethylaminopropyl)carbodiimide (EDC) [60]. For instance, this methodology was used to study how glycosylation of BSA with

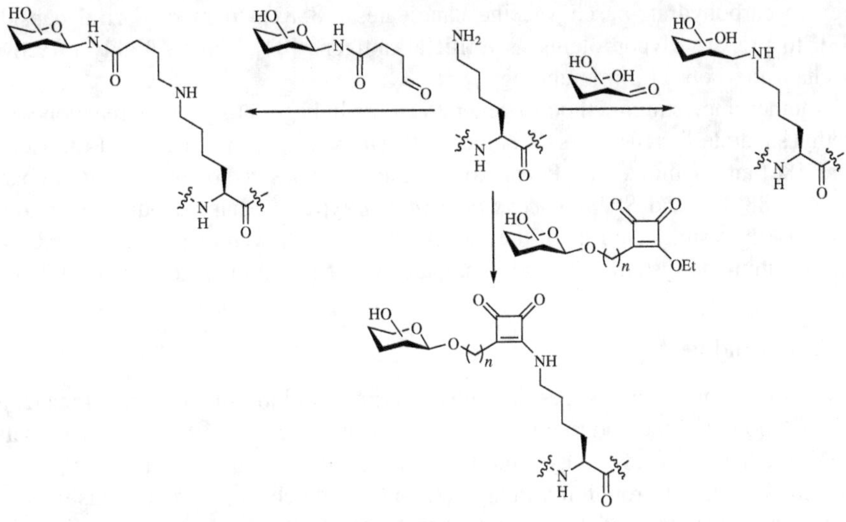

SCHEME 19.4 Different strategies to form amine linkages.

L-fucose, D-mannose (Man), and L-galactose affects macrophage migration [61], to prepare vaccine candidates [62], to study the uptake of fucosylated human serum albumin (HSA) by macrophages [63], and in targeted drug delivery strategies [64].

More popular is the amidation of Lys by reaction with activated esters such as *N*-hydroxysuccinimide, disuccinimidyl glutarate, disuccinimidyl suberate, or diethyl squarate linkers [65]. This approach leads to the preparation of different glycoproteins with variable numbers of attached ligands [66, 67]. Despite being straightforward, however, some have highlighted limitations such as the low loading efficiency and the immunogenicity of the resulting linkages [68, 69]. In addition, it should be noted that the pK_a of the Lys group is modified, resulting in a possible change of the protein charge and hence properties.

Staudinger ligation and its traceless version have been explored in the preparation of glycoconjugates [70, 71]. It was used to label cell-surface carbohydrates [72, 73] and, more recently, to prepare glycoproteins by modification of L-azidohomoalanine-containing proteins [74, 75].

19.5.4 Amine

Secondary amine linkages arise from reductive alkylation of Lys (Scheme 19.4). First reported by Gary Gray [76], this methodology was once used frequently to glycosylate proteins, although it has low efficiency. The amino group of Lys is condensed with an aldehyde group to yield an imine that can be reduced with $NaBH_3CN$ to give a secondary amine. When using carbohydrates, Lys reacts with the reducing end of the sugar, altering the constitution of the glycan. To overcome this problem, aldehyde-containing linkers can be used. Addition of Na_2SO_4 is reported to improve the efficiency of conjugation [77]. Reductive amination has been used to prepare carbohydrate-based vaccine candidates [78–82], to modify viral capsids [83], to prepare glycoproteins as artificial antigens [84], and in highly effective mechanistic probes for immunology [85].

Another interesting method to generate amine linkages at Lys is by reaction with diethyl squarate-linked sugars (Scheme 19.4). This strategy, introduced by Lutz Tietze et al. [86] and optimized by Pavol Kováč et al. [87], has been employed to prepare cholera [88, 89] and *Streptococcus pneumoniae* type 3 vaccine candidates, among others [90]; to study the role of the fucosylated LacdiNAc antigen in the interaction with the immune system [91]; and to prepare kojidextrin–protein conjugates [92].

19.5.5 Amidine

Installation of amidine linkages through the use of 2-imino-2-methoxyethyl reagents [93] is a powerful method to create glycoconjugates (Scheme 19.5). The selectivity of this methodology toward Lys and the conservation of the net charge of the protein after modification (through minimal alteration of side-chain pK_a) make this method attractive. It was applied to glycoengineer a rhamnosidase as part of a bipartite system (see Section 19.6.3) with demonstrated utility for the resulting glycoprotein in targeted drug delivery [94] and even to viral glycosylation [95].

19.5.6 Urea and Thiourea

Urea and thiourea linkages can be introduced into proteins by reaction of the side chain of Lys with cyanates and isothiocyanates, respectively [59, 65]. Usually, the cyanate or isothiocyanate functionalities are introduced by reaction of phosgene or thiophosgene with an amine in basic media (Scheme 19.6). Subsequently, this moiety reacts with the nucleophilic side chain of Lys [96]. This method is one of the most commonly used in glycoconjugate vaccine generation and has been employed to prepare KO–BSA conjugates (KO = D-*glycero*-D-*talo*-2-octulosonic acid) [97], BSA conjugates bearing a Lewis X analogous trisaccharide [98], neoglycoproteins containing epitopes of the *Pseudomonas aeruginosa* RNA group I lipopolysaccharide [99], and KDO(2→8)[KDO(2→4)]KDO(2→4)–BSA derivatives (KDO = 3-deoxy-D-*manno*-2-octulosonic acid) [100], among others. One more recent application of this technique is the varied preparations of different conjugates to investigate possible HIV vaccine candidates [101, 102].

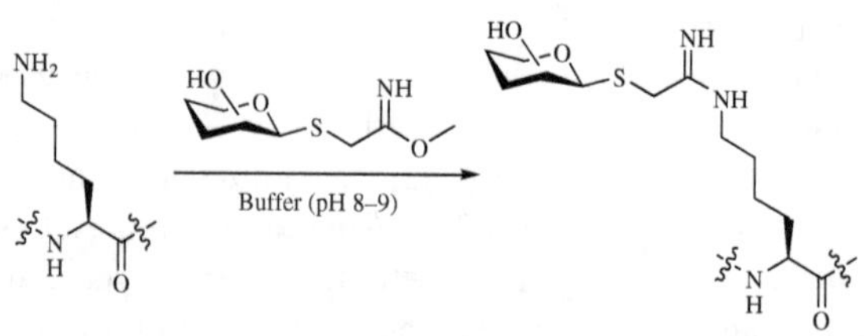

SCHEME 19.5 Conjugation of carbohydrates by amidine formation with Lys side chains.

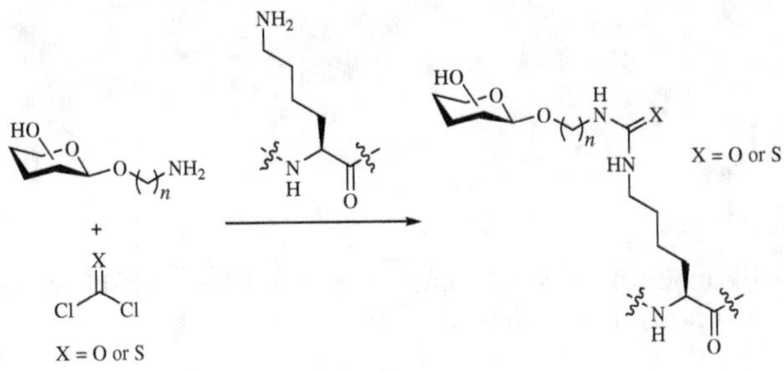

SCHEME 19.6 Modification of the Lys side chain to form urea or thiourea linkages.

19.5.7 Diazonium

First reported for glycoconjugates in 1929 [103], modification of tyrosine (Tyr) residues with diazonium salts to form the ortho-modified diazotyrosine derivatives has been used to prepare different glycoproteins, such as glycosylated BSA [104] to study multivalent binding to human α-defensin 5 [105]. One of the main drawbacks of this methodology is the poor control achieved in the presence of histidines. Nonetheless, this proved to be effective as the first source of conjugates for the pioneering work of Oswald Avery and Walther Goebel in the 1930s [106].

19.5.8 Triazole

The reaction between azides and alkynes to form triazoles is a reliable method to prepare different glycoconjugates [107–109]. First described by Arthur Michael [110] and Otto Dimroth [111] and then studied by Rolf Huisgen [112], triazole formation was found to be accelerated using Cu(I) catalysis in the early part of the last decade when using terminal alkynes, thereby regioselectively yielding 1,4-disubstituted 1,2,3-triazoles [113, 114]. Since then, this reaction has been used in many areas [115, 116], including chemical modification of proteins [117] and synthetic glycoprotein formation [118, 119]. A suggested drawback is the cytotoxicity of Cu(I). To overcome this problem, different laboratories have introduced the use of chelators to increase the biocompatibility of the catalyst, leading to copper complexes compatible with certain *in vivo* reactions [120–122]. Ring strain-promoted [3+2] cycloaddition has been suggested as an alternative to the use of Cu(I). The acetylene bond within cyclooctyne is extremely constrained (163°, 18 kcal/mol) [123, 124], reacting more easily with azides to yield triazole rings. This reactivity has allowed for the detection of azides in labeling experiments [125]. However, although initial experiments with β-mercaptoethanol suggested the stability of cyclooctyne in the presence of thiols [125], secondary reactions were observed when employing more reactive cyclooctynes [45, 126, 127].

The introduction of the chemical handle needed (azide or alkyne) can be achieved by means of indiscriminate functionalization of natural amino acids (via Lys [128–131] or Cys [129]) or by genetic incorporation. The latter strategy permits the preparation of more precisely defined glycoconjugates. One comparison of the two methods was

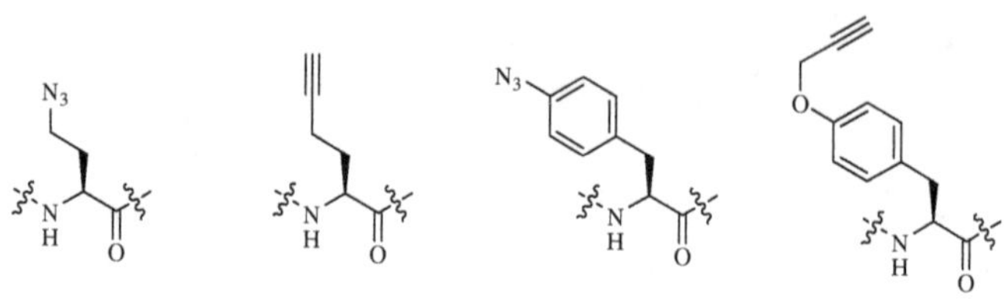

FIGURE 19.2 Examples of unnatural amino acid residues that can be biologically introduced and used to prepare triazole-linked proteins.

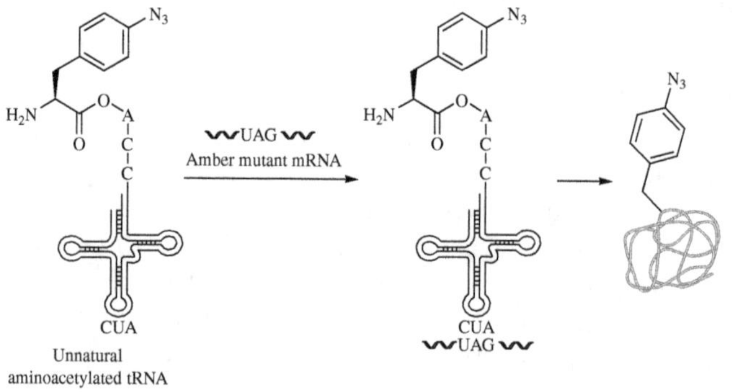

SCHEME 19.7 Amber codon suppression strategy for unnatural azido amino acid incorporation in proteins.

performed by M. G. Finn, Dennis Burton, and coworkers who modified icosahedral virus capsid scaffolds with different Man-based oligosaccharides to create HIV vaccine candidates [131]. To introduce unnatural amino acids by means of genetic incorporation, two approaches can be used. The first comprises the introduction of methionine (Met) surrogates by reassignment of the Met (ATG) codon. This approach has been used to introduce azidohomoalanine [132] and homopropargylglycine [47] (Fig. 19.2). The second consists of the incorporation of the selected unnatural amino acid in response to the nonsense codon TAG (amber codon). The introduction of azide/alkyne amino acids is then achieved by the use of orthogonal (e.g., tyrosyl) tRNA synthetase/$tRNA_{CUA}$ pairs (Scheme 19.7) [133].

19.5.9 Oxime

The formation of oximes and hydrazones can be used to modify proteins due to some attractive properties, that is, some stability in aqueous media and compatibility with many functionalities of proteins [134]. Their synthesis can be accomplished through the nucleophilic addition–elimination (condensation) of hydrazines and alkoxyamines with aldehydes and ketones, respectively. Typically, it has been used to modify the N-terminus [135–137] of peptides and proteins. For example, oxime formation has been used to modify BSA and diphtheria toxin with *Shigella sonnei* O-specific oligosaccharides [138].

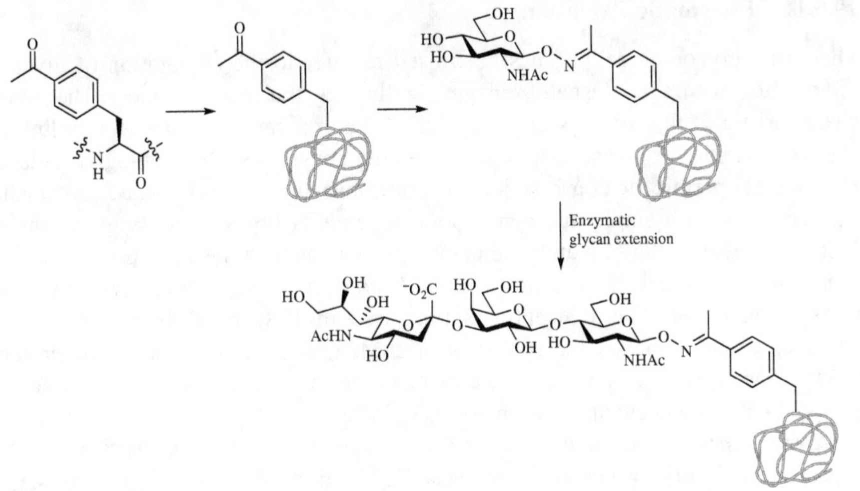

SCHEME 19.8 Site-specific modification of aldehydes. Ac, acetyl.

Among the methods used for introducing oxime linkages into proteins, the genetic codification of the keto and the aldehyde functionality in response to the amber stop codon [139–141] (Scheme 19.8) or the use of natural sequences recognized by enzymes capable of generating the aldehyde moiety (e.g., formylglycine-generating enzyme) [142] has proven usefully flexible. These carbonyl "tags" have been used to introduce fluorophores, biotin derivatives [139, 143], peptides [142], and carbohydrates [140]. In the latter case, enzymatic glycan extension was performed after glycosylation via the oxime linkage, thereby demonstrating the functional processing by enzymes despite the unnatural linkage.

19.5.10 CC Bond Formation

The formation of CC bonds to prepare glycoconjugates has also gained some attention. For example, Diels–Alder cycloaddition has allowed the preparation of synthetic glycoproteins [144] and the construction of a synthetic vaccine candidate against *Neisseria meningitidis* A [145]. These methods relied on a "pretagging" step to create suitable protein reactants.

One interesting example is olefin metathesis—an efficient and chemoselective way of forming C=C linkages [146–148]. Its application in biological systems was previously restricted due to the lack of suitable aqueous metathesis conditions, but this problem was recently solved through the development of privileged protein tags, allowing its application in protein modification [149] and chemical biology [150]. This strategy has been used to prepare different protein glycoconjugates [151].

Similarly, Pd-mediated sp^2–sp^2 C–C (Suzuki–Miyaura) coupling has been enabled in biology through the development of an aqueous biologically compatible Suzuki catalyst system [152]. This has now allowed the synthetic glycosylation of whole cell surfaces to create bacterial strains with a "synthetic glycocalyx" [153].

19.5.11 Enzymatic Extension

The term “glycoremodeling” has been used to refer to the elaboration of glycans often after trimming (usually enzymatically) of the natural ones. The main characteristic of this strategy is the use of the natural carbohydrate–protein link as chemical handle for glycan extension. Trimming enzymes such as endoglycosidase-H (Endo-H) permit the conversion of a natural mixture of glycosylated proteins to side chains with a single and hence homogeneous N-linked *N*-acetylglucosamine (GlcNAc) residue. Subsequently, other endoglycosidase enzymes like Endo-M (from *Mucor hiemalis*) and Endo-A (from *Arthrobacter protophormiae*) allow enzymatic transglycosylation. This concept was first used in 1986 by Robert Trimble and coworkers who achieved the synthesis of a $Man_6GlcNAc_2$-modified glycoprotein [154]. This work has now led to the elegant use of several Endo glycosidases to remodel naturally occurring glycoproteins [155].

Apart from natural linkages, these endoglycosidases have also been used to remodel glycoproteins carrying unnatural linkages [156]. As an extension of the “tag-and-modify” strategy [19], Benjamin Davis and coworkers showed the ability of Endo-A glycosidase to recognize and transglycosylate unnaturally linked glycoproteins (Scheme 19.9). In addition, Endo-A not only elongated GlcNAc-protein hybrids but also a range of different proteins with varied sugar acceptors. Notably, triazole-linked glycoproteins proved to be useful substrates.

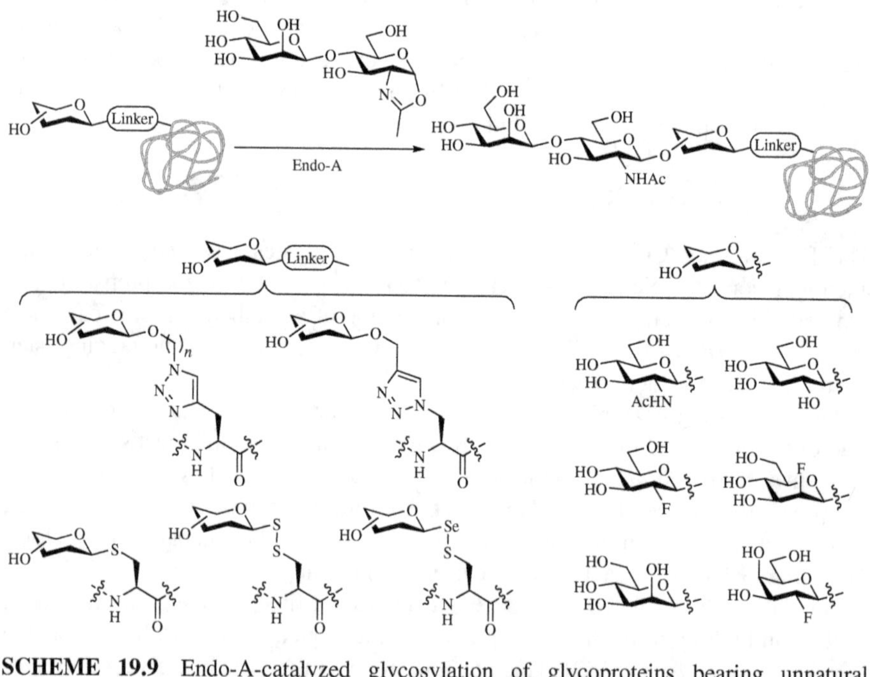

SCHEME 19.9 Endo-A-catalyzed glycosylation of glycoproteins bearing unnatural linkages.

19.6 GLYCOPROTEIN-BASED THERAPEUTICS

Carbohydrate–protein interactions govern countless biological processes even though the corresponding lectin–carbohydrate association constants are often very low ($\sim 10^6 M^{-1}$). This weak affinity can be compensated in nature by multivalent presentation of both the ligand and the receptor [157]. The combination of multiple interactions between two or more carbohydrate units (of the correct type and in the right orientation) and the multimeric receptor is sometimes referred to as glycoside cluster effect [158–160] (also see Chapter 16). Besides increasing the affinity of the binding process, multivalency often results in augmentation of the specificity of the interaction and therefore of the molecular recognition process. Thus, targeting a particular receptor by modifying a selected cargo with carbohydrates seems an appealing strategy. Synthetic glycoproteins are one of the most attractive examples of active carrier-mediated targeting agents [161], and although these methods are not the sole use of synthetic glycoproteins, they do provide a recurrent theme in this section. Many of the examples chosen here explore constructs where carbohydrates play a direct mechanistic role. However, they may also play vital indirect physicochemical roles in modulating solubility, *in vivo* lifetime, and others.

19.6.1 Carbohydrate–Protein-Based (Glycoconjugate) Vaccines

Carbohydrate-based vaccines are usually formed by a weak antigen (hapten) linked to a protein (carrier) to create an immunogen. This concept, first introduced in a clearly synthetic form in 1929 [106], is particularly attractive because the glycosylation patterns of, for example, pathogens or diseased cells are often quite unique. In addition, the use of carbohydrate-only immunogens (e.g., capsular polysaccharide vaccines) has some limitations because the immune response generated by carbohydrates does not typically induce memory; carbohydrates alone are typically T-cell-independent antigens [162]. Instead, they directly interact with polysaccharide-specific B cells, which differentiate afterward into plasma cells to produce antibodies (Fig. 19.3). However, when conjugated to a suitable carrier protein, memory is induced. This effect is traditionally attributed to the processing of the protein by polysaccharide-specific B cells and the subsequent presentation of the consequential peptides/glycopeptides to protein–peptide-specific T cells in association with major histocompatibility complex class II receptors, thereby inducing polysaccharide-specific memory B cells. However, other models are also gaining favor with the discovery of T cells that recognize polysaccharide alone and independent of context [85].

Several reviews [163–165] analyze different approaches to designing synthetic glycoconjugate vaccines, but all molecular models for vaccine efficacy draw heavily on the need for better control of conjugation. A shining example of the relevance of this field is that of a synthetic oligosaccharide–protein conjugate vaccine against *Haemophilus influenzae* type b, which reached the clinic [166] after a sublime program of design, synthesis, and translation [167].

Some key aspects have to be considered when designing such synthetic vaccines because the immunogenic activity of the conjugate depends on the nature of the hapten, the loading onto the protein, and the characteristics of the carrier. In

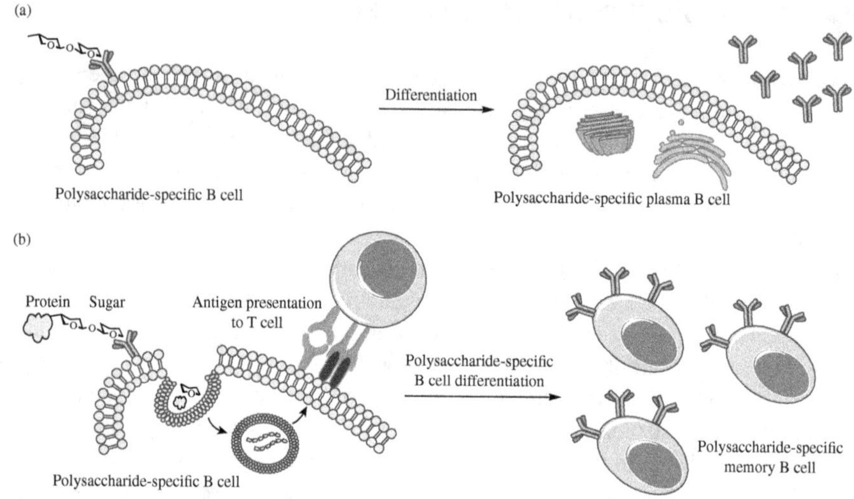

FIGURE 19.3 Traditionally proposed mechanism for immune response of (a) polysaccharide-based vaccines and (b) polysaccharide–protein conjugate-based vaccines.

carbohydrate–protein vaccines, the sugar and the proteic part are linked by a covalent bond, and importantly, the nature of the carbohydrate unit has to remain (at least partially) intact to preserve the activity [168]. The emerging and changing mechanistic picture [85] for this effect continues to provide striking opportunities for synthetic glycoconjugates.

19.6.2 Synthetic Glycoproteins as Pathogen-Process Inhibitors

The increase in viral and bacterial diseases highlights the urgent need for novel approaches [169]. Synthetic glycoproteins as inhibitors are attractive because they share essential structural features with natural glycoproteins of pathogens, thereby potentially mimicking natural interactions. Specificity in blocking can therefore be achieved, in principle, by tuning the characteristics of the hybrid, choosing an adequate protein with specific activity, and selecting the sugar to target the receptor of interest.

The potential of synthetic glycoproteins as inhibitors of bacterial infection was shown in a binding assay that demonstrates how differently glycosylated (with different linkers) HSA and BSA proteins inhibited the adhesion of type 1 fimbriated *Escherichia coli* to highly mannosylated ligands [170]. Inspired by the need for new treatments to neutralize bacterial infections, Davis et al. developed the synthesis of a new class of glycoconjugate—the so-called glycodendriproteins [56]—formed by a glycodendron covalently attached to a protein. The aptitude of these synthetic glycoproteins as inhibitors of bacterial infection was tested using an infection model through coaggregation of *Actinomyces naeslundii* (responsible for oral infections) [171] with its copathogen *Streptococcus oralis*. The initial interaction of the highly galactosylated glycoproteins of *S. oralis* with the adhesins present on the *A. naeslundii* bacteria triggers infection of hosts. In order to create a synthetic adhesin-binding

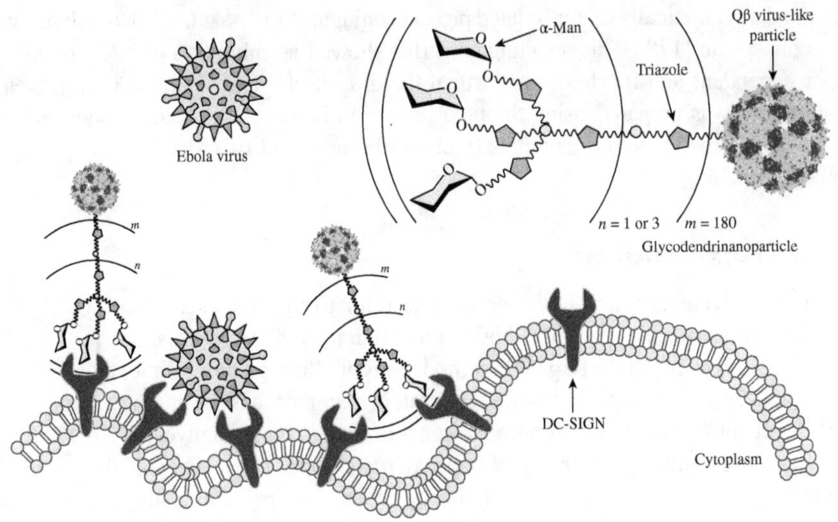

FIGURE 19.4 Inhibition strategy in which the synthetic glycodendrinanoparticle competes with Ebola virus for binding to DC-SIGN-displaying cells.

protein, a galactose-based glycodendron was conjugated with a mutant protease using disulfide chemistry [57]. The synthesized glycodendriprotein was able to inhibit the aggregation of pathogenic bacteria at a nanomolar level. This potency was attributed to both the initial binding to adhesins coupled with bacterial membrane disruption as a consequence of the proteolytic activity of the glycoconjugate.

Human corneal epithelial cells are infected by the adenovirus serotype Ad37 after its attachment to cellular glycoproteins bearing terminal sialic acid. Niklas Arnberg and coworkers demonstrated that sialic acid–HSA conjugates were able to inhibit the infection of human corneal epithelial cells by Ad37 [172]. Subsequent structural analysis of the sialic acid-binding domain of the Ad37 fiber protein in complex with sialyllactose allowed the design of a library of *N*-acyl-modified sialic acids to increase binding. These designed carbohydrates were conjugated to HSA by squarate ester chemistry. Although the inhibitory activity of the conjugates was inferior to that displayed by multivalent sialic acid, all of the synthetic glycoproteins were more effective than the corresponding monomeric sialic acid derivatives [173].

The development of antiviral glycoprotein-based inhibitors exploits the structural analogy with the sometimes highly glycosylated envelope proteins of viruses such as the HIV gp120 that is involved in the first stages of HIV infection. A more recent antiviral example [174] describes the use of the triazole linkage to prepare homogeneous glycodendriprotein multimer assemblies displaying up to 1620 sugar moieties in a single construct. The synthesis was performed using Cu(I)-mediated modification of an alkyne-bearing viruslike particle with selected azidoglycodendrons to create approximately 20–30 nm-sized protein-based "glycodendrinanoparticles." These synthetic glycodendrinanoparticles showed promising inhibition of dendritic cell infection in a pseudotyped Ebola virus model (Fig. 19.4). One intriguing anti-HIV approach exploited a synthetically mannosylated protein conjugated to 3′-azido-3′-deoxythymidine monophosphate [175]. This resulting construct showed an inhibition of infection that is sugar dependent and fourfold greater than the drug itself. The synthetic glycoprotein–drug hybrid was prepared using thiourea formation between monosaccharides and the side chain of Lys of serum albumins (rabbit or human) and EDC-mediated coupling to conjugate the drug.

19.6.3 Targeted Delivery

The term “glycotargeting” has been used to refer to the proposed ability of glycan-modified carriers to be recognized by mammalian lectins [161]. This idea for targeted drug delivery, as for other targeted methods, might allow greater efficiency and lower toxicity [176]. The hepatic asialoglycoprotein receptor has been one of the most studied receptors under this premise for gene- or drug-targeted delivery [177–181]. For instance, discriminatory delivery of naproxen to different cell types of the liver was demonstrated by attaching the selected drug to neoglycoproteins with an appropriate type and number of sugar groups [182].

Lectin-directed enzyme-activated prodrug therapy (LEAPT) (Scheme 19.10) is an interesting extension of this idea [94]. Based on a two-component strategy (a prodrug capped with a nonmammalian sugar and a glycosylated enzyme), LEAPT allows selective drug release in a selected organ (liver in this case) through targeting of the release enzyme. Glycotargeting was performed by enzymatic deglycosylation and subsequent chemical glycosylation of Lys using amidine-linkage chemistry. The method has been applied to the use of two different prodrugs (of doxorubicin and 5-fluorouracil, both capped by the nonmammalian L-rhamnosyl sugar unit) [183].

Targeted delivery has also been used to investigate the structure of carbohydrates needed to obtain discrimination of cellular uptake between liver and tumor cells [184]. A study was performed by doubly modifying BSA with both a fluorophore (via Lys isothiocyanate derivatization) and different fucose derivatives (via Tyr diazotization). Similar studies have used EDC-mediated coupling to also attach therapeutic agents for delivery to human embryonal carcinoma cells [185].

Targeted gene delivery can take advantage of adenovirus as a privileged carrier [95]. Adenovirus participates in a Lys-dependent interaction with the coxsackie adenovirus receptor (CAR) of some host cells [186], and this has allowed their use as gene delivery systems [187]. Taking advantage of the low natural glycosylation of adenovirus [188], Davis et al. explored the use of adenovirus with an appropriate synthetic glycosylation pattern to introduce a new selectivity to selected target cells. The Lys side chains were modified using amidine-linkage glycosylation, resulting in the chemical masking of the Lys residues involved in the recognition by CAR and in the redirection of adenovirus toward new routes of viral infection. Pleasingly, while the unmodified viruses were infected via CAR, mannosylated glycoviruses were able to infect macrophages, most likely via Man-binding protein interaction [95]. This work highlights great potential for the modification of natural residues to introduce unnatural linkages and hence

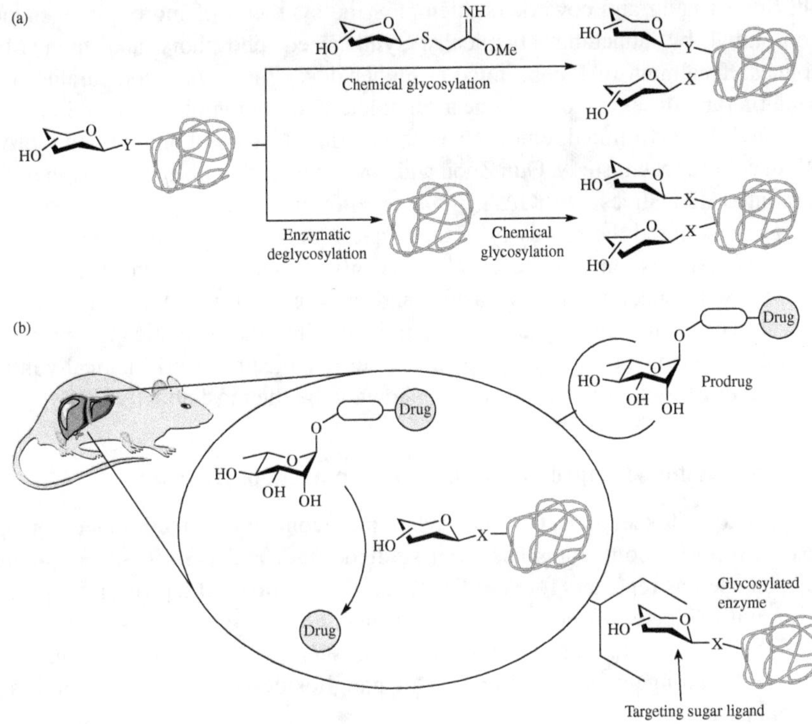

SCHEME 19.10 The LEAPT strategy. (a) Synthetic scheme used to prepare LEAPT glycoconjugates for enzyme-mediated drug release. (b) Schematic representation of LEAPT.

modulate the activity of synthetic glycoproteins. It will be interesting to see the potential of this system in more relevant test systems and with therapeutically relevant genes.

19.6.4 Enzyme-Replacement Therapy

As another interesting example of glycotargeting, enzyme-replacement therapy has received attention as a strategy to treat lysosomal storage diseases [189]. One of the most tractable examples is the type II glycogen storage disease also known as Pompe disease. It has been shown that treatment with recombinant human α-glucosidase (rhGAA) reduces the abnormal level of glycogen in heart and skeletal muscles [190]. The uptake in skeletal muscle is lower probably due to the low level of the cation-independent Man 6-phosphate receptor responsible for the endocytosis of the enzyme by the target cell and because of the low affinity of this receptor toward the protein of interest. The attachment of phosphorylated Man-containing oligosaccharides to rhGAA resulted in a higher uptake of the enzyme by the targeted cells and, importantly, in a significant reduction of the glycogen level [191]. In this case, introduction of the phosphorylated Man units was performed by modification of the endogenous carbohydrates present on rhGAA through reductive amination. Encouraged by this result, Seng Cheng and coworkers attempted the synthesis of more homogeneous glycoproteins by attaching chemically synthesized phosphorylated high Man analogues. Treatment of Pompe mice at higher doses (40 mg/kg, comparable with enzyme therapy doses) resulted in near complete elimination of glycogen in all the muscles, while only limited removal was observed in the case of unmodified rhGAA [192]. In a more recent study, Qun Zhou and coworkers rationalized the design of the glycosylated derivatives of rhGAA by using different functionalization methodologies. They explored the reactivity of the single Cys residue present in the protein, the 15 Lys residues, and the seven N-linked glycosylation sites and used glycan derivatives with succinimide, hydrazide, and aminoxy linkers. After studying the conjugation efficiency, aggregation, and activity of all of the synthetic glycoproteins, the conjugations via oxime linkages at aldehydes chemically or enzymatically introduced at endogenous carbohydrates were shown to be the most effective [193].

19.6.5 Synthetic Glycoproteins as *In Vivo* Reporters of Disease

Due to the exquisite specificity that glycosylation can confer on synthetic glycoproteins, its use in reporter conjugates has been studied. BSA and HSA glycoconjugates have been used as reporters for serodiagnosis, for example, of leprosy after glycosylation with selected disaccharides of *Mycobacterium leprae* [194, 195]. Early examples showed value in receptor elucidation, such as the use of the ganglioside GT1b–BSA conjugate in the discovery of ganglioside-specific brain membrane receptors [196].

Synthetic glycoproteins allow varied approaches to be tested. Davis et al. described the synthesis of glycoprotein probes able to detect mammalian brain inflammation and disease by using a "dual tag" approach [118]. They established two different orthogonal modifications in a selected protein (Scheme 19.11) [118, 119]. These modifications were introduced by disulfide formation at a single Cys position and by

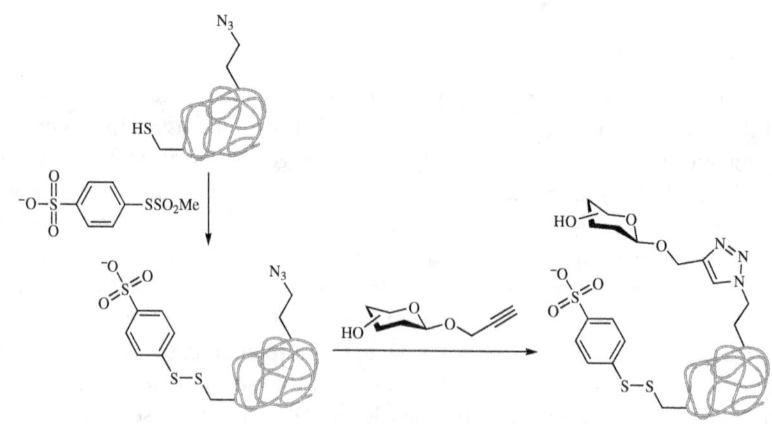

SCHEME 19.11 Double-differential modification of a protein by triazole and disulfide conjugations allowed the creation of a synthetic glycoprotein reporter.

means of Cu(I)-catalyzed triazole formation at the unnatural amino acids azidohomoalanine or homopropargylglycine introduced by reassignment of the Met codon. In that way, two different posttranslational modification mimics were precisely introduced, highlighting the potential of synthetic strategies to create modified proteins to study their roles in different processes. The resulting doubly modified protein proved to be a successful *in vivo* probe.

Other *in vivo* glycoprotein probes will likely utilize strategies for dual glycosylation and radiolabeling in order to allow exploitation by currently popular radioactivity-based imaging methods such as single-photon emission computed tomography and positron emission tomography [197, 198].

19.7 CONCLUSION

Currently, 40% of protein-based therapeutics are glycosylated and are considered as high-value products [11]. During the last few years, efforts have focused on the preparation of potential synthetic glycoprotein therapeutics with a variety of uses. Highlighted in this chapter are some synthetic methods to synthesize such glycoproteins, together with illustrative applications as therapeutic tools. The use of unnatural linkages is a flexible option to introduce enhanced properties such as higher stability or control of glycosylation site. Although impressive progress has been made in biological glycoengineering, high yielding strategies to prepare well-defined glycoproteins are still desirable and may be more readily addressed chemically in many cases. The rapid evolution of the field will continue to be translated into the approval of sugar–protein hybrids as biopharmaceuticals [11].

REFERENCES

[1] Wong, S. Y. C. *Curr. Opin. Struct. Biol.* **1995**, *5*, 599–604.

[2] Yarema, K. J.; Bertozzi, C. R. *Curr. Opin. Chem. Biol.* **1998**, *2*, 49–61.

[3] Davis, B. G. *J. Chem. Soc. Perkin Trans. 1* **1999**, 3215–3237.

[4] Bertozzi, C. R.; Kiessling, L. L. *Science* **2001**, *291*, 2357–2364.

[5] Dwek, R. A. *Chem. Rev.* **1996**, *96*, 683–720.

[6] Rademacher, T. W.; Parekh, R. B.; Dwek, R. A. *Annu. Rev. Biochem.* **1988**, *57*, 785–838.

[7] Rudd, P. M.; Elliott, T.; Cresswell, P.; Wilson, I. A.; Dwek, R. A. *Science* **2001**, *291*, 2370–2376.

[8] Talbot, P.; Shur, B. D.; Myles, D. G. *Biol. Reprod.* **2003**, *68*, 1–9.

[9] Varki, A. *Glycobiology* **1993**, *3*, 97–130.

[10] Helenius, A.; Aebi, M. *Science* **2001**, *291*, 2364–2369.

[11] Walsh, G. *Nat. Biotechnol.* **2010**, *28*, 917–924.

[12] Walsh, G., Ed. *Post-Translational Modification of Protein Biopharmaceuticals*; Wiley-VCH: Weinheim, **2009**.

[13] Schiestl, M.; Stangler, T.; Torella, C.; Cepeljnik, T.; Toll, H.; Grau, R. *Nat. Biotechnol.* **2011**, *29*, 310–312.

[14] Chirino, A. J.; Mire-Sluis, A. *Nat. Biotechnol.* **2004**, *22*, 1383–1391.

[15] Stephanopoulos, N.; Francis, M. B. *Nat. Chem. Biol.* **2011**, *7*, 876–884.

[16] Kolb, H. C.; Sharpless, K. B. *Drug Discov. Today* **2003**, *8*, 1128–1137.

[17] Brik, A.; Alexandratos, J.; Lin, Y.-C.; Elder, J. H.; Olson, A. J.; Wlodawer, A.; Goodsell, D. S.; Wong, C.-H. *ChemBioChem* **2005**, *6*, 1167–1169.

[18] Poolman, J.; Frasch, C.; Nurkka, A.; Käyhty, H.; Biemans, R.; Schuerman, L. *Clin. Vaccine Immunol.* **2011**, *18*, 327–336.

[19] Chalker, J. M.; Bernardes, G. J. L.; Davis, B. G. *Acc. Chem. Res.* **2011**, *44*, 730–741.

[20] Lafite, P.; Daniellou, R. *Nat. Prod. Rep.* **2012**, *29*, 729–738.

[21] Fodje, M. N.; Al-Karadaghi, S. *Protein Eng.* **2002**, *15*, 353–358.

[22] Nagahara, N.; Matsumura, T.; Okamoto, R.; Kajihara, Y. *Curr. Med. Chem.* **2009**, *16*, 4490–4501.

[23] Chalker, J. M.; Bernardes, G. J. L.; Lin, Y. A.; Davis, B. G. *Chem. Asian J.* **2009**, *4*, 630–640.

[24] Davis, N. J.; Flitsch, S. L. *Tetrahedron Lett.* **1991**, *32*, 6793–6796.

[25] Macmillan, D.; Bill, R. M.; Sage, K. A.; Fern, D.; Flitsch, S. L. *Chem. Biol.* **2001**, *8*, 133–145.

[26] Müllegger, J.; Chen, H. M.; Warren, R. A. J.; Withers, S. G. *Angew. Chem. Int. Ed.* **2006**, *45*, 2585–2588.

[27] Tey, L.-H.; Loveridge, E. J.; Swanwick, R. S.; Flitsch, S. L.; Allemann, R. K. *FEBS J.* **2010**, *277*, 2171–2179.

[28] Swanwick, R. S.; Daines, A. M.; Tey, L.-H.; Flitsch, S. L.; Allemann, R. K. *ChemBioChem* **2005**, *6*, 1338–1340.

[29] Liu, Y.; Chan, Y. M.; Wu, J.; Chen, C.; Benesi, A.; Hu, J.; Wang, Y.; Chen, G. *ChemBioChem* **2011**, *12*, 685–690.

[30] Shin, I.; Jung, H.-J.; Lee, M.-R. *Tetrahedron Lett.* **2001**, *42*, 1325–1328.

[31] Ni, J.; Singh, S.; Wang, L.-X. *Bioconjug. Chem.* **2003**, *14*, 232–238.

[32] Zhang, Y.; Bhatt, V. S.; Sun, G.; Wang, P. G.; Palmer, A. F. *Bioconjug. Chem.* **2008**, *19*, 2221–2230.

[33] Betting, D. J.; Kafi, K.; Abdollahi-Fard, A.; Hurvitz, S. A.; Timmerman, J. M. *J. Immunol.* **2008**, *181*, 4131–4140.

[34] Phalipon, A.; Tanguy, M.; Grandjean, C.; Guerreiro, C.; Bélot, F.; Cohen, D.; Sansonetti, P. J.; Mulard, L. A. *J. Immunol.* **2009**, *182*, 2241–2247.

[35] Schofield, L.; Hewitt, M. C.; Evans, K.; Siomos, M.-A.; Seeberger, P. H. *Nature* **2002**, *418*, 785–789.

[36] Yoshikawa, T.; Kato, Y.; Yuki, N.; Yabe, T.; Ishida, H.; Kiso, M. *Glycoconj. J.* **2008**, *25*, 545–553.

[37] Rich, D. H.; Tam, J.; Mathiaparanam, P.; Grant, J. A.; Mabuni, C. *J. Chem. Soc. Chem. Commun.* **1974**, 897–898.

[38] Seebeck, F. P.; Szostak, J. W. *J. Am. Chem. Soc.* **2006**, *128*, 7150–7151.

[39] Wang, J.; Schiller, S. M.; Schultz, P. G. *Angew. Chem. Int. Ed.* **2007**, *46*, 6849–6851.

[40] Bernardes, G. J. L.; Chalker, J. M.; Errey, J. C.; Davis, B. G. *J. Am. Chem. Soc.* **2008**, *130*, 5052–5053.

[41] Chalker, J. M.; Gunnoo, S. B.; Boutureira, O.; Gerstberger, S. C.; Fernandez-Gonzalez, M.; Bernardes, G. J. L.; Griffin, L.; Hailu, H.; Schofield, C. J.; Davis, B. G. *Chem. Sci.* **2011**, *2*, 1666–1676.

[42] Chalker, J. M.; Lercher, L.; Rose, N. R.; Schofield, C. J.; Davis, B. G. *Angew. Chem. Int. Ed.* **2012**, *51*, 1835–1839.

[43] Bernardes, G. J. L.; Grayson, E. J.; Thompson, S.; Chalker, J. M.; Errey, J. C.; El Oualid, F.; Claridge, T. D. W.; Davis, B. G. *Angew. Chem. Int. Ed.* **2008**, *47*, 2244–2247.

[44] Dondoni, A.; Massi, A.; Nanni, P.; Roda, A. *Chem. Eur. J.* **2009**, *15*, 11444–11449.

[45] Lo Conte, M.; Staderini, S.; Marra, A.; Sanchez-Navarro, M.; Davis, B. G.; Dondoni, A. *Chem. Commun.* **2011**, *47*, 11086–11088.

[46] van Hest, J. C. M.; Tirrell, D. A. *FEBS Lett.* **1998**, *428*, 68–70.

[47] van Hest, J. C. M.; Kiick, K. L.; Tirrell, D. A. *J. Am. Chem. Soc.* **2000**, *122*, 1282–1288.

[48] Grayson, E. J.; Bernardes, G. J. L.; Chalker, J. M.; Boutureira, O.; Koeppe, J. R.; Davis, B. G. *Angew. Chem. Int. Ed.* **2011**, *50*, 4127–4132.

[49] Kenyon, G. L.; Bruice, T. W. In *Enzyme Structure Part E*; Hirs, C. H. W., Timasheff, S. N., Eds.; Methods in Enzymology 47; Academic: New York, **1977**, pp 407–430.

[50] Watt, G. M.; Lund, J.; Levens, M.; Kolli, V. S. K.; Jefferis, R.; Boons, G.-J. *Chem. Biol.* **2003**, *10*, 807–814.

[51] Watt, G. M.; Boons, G.-J. *Carbohydr. Res.* **2004**, *339*, 181–193.

[52] Macindoe, W. M.; van Oijen, A. H.; Boons, G.-J. *Chem. Commun.* **1998**, 847–848.

[53] Davis, B. G.; Maughan, M. A. T.; Green, M. P.; Ullman, A.; Jones, J. B. *Tetrahedron: Asymmetry* **2000**, *11*, 245–262.

[54] Gamblin, D. P.; Garnier, P.; Ward, S. J.; Oldham, N. J.; Fairbanks, A. J.; Davis, B. G. *Org. Biomol. Chem.* **2003**, *1*, 3642–3644.

[55] Gamblin, D. P.; Garnier, P.; van Kasteren, S.; Oldham, N. J.; Fairbanks, A. J.; Davis, B. G. *Angew. Chem. Int. Ed.* **2004**, *43*, 828–833.

[56] Davis, B. G. *Chem. Commun.* **2001**, 351–352.

[57] Rendle, P. M.; Seger, A.; Rodrigues, J.; Oldham, N. J.; Bott, R. R.; Jones, J. B.; Cowan, M. M.; Davis, B. G. *J. Am. Chem. Soc.* **2004**, *126*, 4750–4751.

[58] Bernardes, G. J. L.; Marston, J. P.; Batsanov, A. S.; Howard, J. A. K.; Davis, B. G. *Chem. Commun.* **2007**, 3145–3147.

[59] Hermanson, G. T., Ed. *Bioconjugate Techniques*, 2nd ed.; Academic: New York, **2008**.

[60] Rao, M. J.; Acharya, A. S. In *Hemoglobins Part B: Biochemical and Analytical Methods*; Everse, J., Vandegriff, K. D., Winslow, R. M., Eds.; Methods in Enzymology 231; Academic: New York, **1994**, pp 246–267.

[61] Takata, I.; Chida, K.; Gordon, M. R.; Myrvik, Q. N.; Ricardo, M. J.; Kucera, L. S. *J. Leukocyte Biol.* **1987**, *41*, 248–256.

[62] Pozsgay, V. *Angew. Chem. Int. Ed.* **1998**, *37*, 138–142.

[63] Sarkar, K.; Sekhar Sarkar, H.; Kole, L.; Das, P. K. *Mol. Cell. Biochem.* **1996**, *156*, 109–116.

[64] Fiume, L.; Bassi, B.; Busi, C.; Mattioli, A.; Spinosa, G. *Biochem. Pharmacol.* **1986**, *35*, 967–972.

[65] Izumi, M.; Okumura, S.; Yuasa, H.; Hashimoto, H. *J. Carbohydr. Chem.* **2003**, *22*, 317–329.

[66] Johansson, S. M. C.; Arnberg, N.; Elofsson, M.; Wadell, G.; Kihlberg, J. *ChemBioChem* **2005**, *6*, 358–364.

[67] Bundle, D. R.; Rich, J. R.; Jacques, S.; Yu, H. N.; Nitz, M.; Ling, C.-C. *Angew. Chem. Int. Ed.* **2005**, *44*, 7725–7729.

[68] Ingale, S.; Wolfert, M. A.; Gaekwad, J.; Buskas, T.; Boons, G.-J. *Nat. Chem. Biol.* **2007**, *3*, 663–667.

[69] Buskas, T.; Li, Y.; Boons, G.-J. *Chem. Eur. J.* **2004**, *10*, 3517–3524.

[70] Koehn, M.; Breinbauer, R. *Angew. Chem. Int. Ed.* **2004**, *43*, 3106–3116.

[71] van Berkel, S. S.; van Eldijk, M. B.; van Hest, J. C. M. *Angew. Chem. Int. Ed.* **2011**, *50*, 8806–8827.

[72] Prescher, J. A.; Dube, D. H.; Bertozzi, C. R. *Nature* **2004**, *430*, 873–877.

[73] Saxon, E.; Bertozzi, C. R. *Science* **2000**, *287*, 2007–2010.

[74] Loka, R. S.; Sadek, C. M.; Romaniuk, N. A.; Cairo, C. W. *Bioconjug. Chem.* **2010**, *21*, 1842–1849.

[75] Bernardes, G. J. L.; Linderoth, L.; Doores, K. J.; Boutureira, O.; Davis, B. G. *ChemBioChem* **2011**, *123*, 1383–1386.

[76] Gray, G. R. *Arch. Biochem. Biophys.* **1974**, *163*, 426–428.

[77] Gildersleeve, J. C.; Oyelaran, O.; Simpson, J. T.; Allred, B. *Bioconjug. Chem.* **2008**, *19*, 1485–1490.

[78] Pozsgay, V.; Coxon, B.; Glaudemans, C. P. J.; Schneerson, R.; Robbins, J. B. *Synlett* **2003**, 743–767.

[79] Chefalo, P.; Pan, Y.; Nagy, N.; Harding, C.; Guo, Z. *Glycoconj. J.* **2004**, *20*, 407–414.

[80] Pan, Y.; Chefalo, P.; Nagy, N.; Harding, C.; Guo, Z. *J. Med. Chem.* **2005**, *48*, 875–883.

[81] Ariosa-Alvarez, A.; Arencibia-Mohar, A.; Madrazo-Alonso, O.; Garcia-Imia, L.; Sierra-Gonzalez, G.; Verez-Bencomo, V. *J. Org. Chem.* **1998**, *17*, 1307–1320.

[82] Sabbatini, P. J.; Kudryashov, V.; Ragupathi, G.; Danishefsky, S. J.; Livingston, P. O.; Bornmann, W.; Spassova, M.; Zatorski, A.; Spriggs, D.; Aghajanian, C.; Soignet, S.; Peyton, M.; O'Flaherty, C.; Curtin, J.; Lloyd, K. O. *Int. J. Cancer* **2000**, *87*, 79–85.

[83] Espenlaub, S.; Wortmann, A.; Engler, T.; Corjon, S.; Kochanek, S.; Kreppel, F. *J. Gene Med.* **2008**, *10*, 1303–1314.

[84] Mahajan, R.; Dixit, S.; Khare, N. K.; Khare, A. *J. Carbohydr. Chem.* **1994**, *13*, 63–73.

[85] Avci, F. Y.; Li, X.; Tsuji, M.; Kasper, D. L. *Nat. Med.* **2011**, *17*, 1602–1610.

[86] Tietze, L. F.; Schroeter, C.; Gabius, S.; Brinck, U.; Goerlach-Graw, A.; Gabius, H. J. *Bioconjug. Chem.* **1991**, *2*, 148–153.

[87] Chernyak, A.; Karavanov, A.; Ogawa, Y.; Kováč, P. *Carbohydr. Res.* **2001**, *330*, 479–486.

[88] Rollenhagen, J. E.; Kalsy, A.; Saksena, R.; Sheikh, A.; Alam, M. M.; Qadri, F.; Calderwood, S. B.; Kováč, P.; Ryan, E. T. *Vaccine* **2009**, *27*, 4917–4922.

[89] Xu, P.; Alam, M. M.; Kalsy, A.; Charles, R. C.; Calderwood, S. B.; Qadri, F.; Ryan, E. T.; Kováč, P. *Bioconjug. Chem.* **2011**, *22*, 2179–2185.

[90] Paulovičová, L.; Paulovičová, E.; Karelin, A. A.; Tsvetkov, Y. E.; Nifantiev, N. E.; Bystrický, S. *Int. Immunopharmacol.* **2012**, *14*, 179–187.

[91] Tefsen, B.; van Stijn, C. M. W.; van den Broek, M.; Kalay, H.; Knol, J. C.; Jimenez, C. R.; van Die, I. *Carbohydr. Res.* **2009**, *344*, 1501–1507.

[92] Pozsgay, V.; Dubois, E. P.; Pannell, L. *J. Org. Chem.* **1997**, *62*, 2832–2846.

[93] Lee, Y. C.; Stowell, C. P.; Krantz, M. J. *Biochemistry* **1976**, *15*, 3956–3963.

[94] Robinson, M. A.; Charlton, S. T.; Garnier, P.; Wang, X.-T.; Davis, S. S.; Perkins, A. C.; Frier, M.; Duncan, R.; Savage, T. J.; Wyatt, D. A.; Watson, S. A.; Davis, B. G. *Proc. Natl. Acad. Sci. U. S. A.* **2004**, *101*, 14527–14532.

[95] Pearce, O. M. T.; Fisher, K. D.; Humphries, J.; Seymour, L. W.; Smith, A.; Davis, B. G. *Angew. Chem. Int. Ed.* **2005**, *44*, 1057–1061.

[96] Buss, D. H.; Goldstein, I. J. *J. Chem. Soc. C* **1968**, 1457–1461.

[97] Wimmer, N.; Brade, H.; Kosma, P. *Carbohydr. Res.* **2000**, *329*, 549–560.

[98] Bartek, J.; Müller, R.; Kosma, P. *Carbohydr. Res.* **1998**, *308*, 259–273.

[99] Reiter, A.; Brade, L.; Sanchez-Carballo, P.; Brade, H.; Kosma, P. *J. Endotoxin Res.* **2001**, *7*, 125–131.

[100] Müller-Loennies, S.; Gronow, S.; Brade, L.; MacKenzie, R.; Kosma, P.; Brade, H. *Glycobiology* **2006**, *16*, 184–196.

[101] Astronomo, R. D.; Lee, H.-K.; Scanlan, C. N.; Pantophlet, R.; Huang, C.-Y.; Wilson, I. A.; Blixt, O.; Dwek, R. A.; Wong, C.-H.; Burton, D. R. *J. Virol.* **2008**, *82*, 6359–6368.

[102] Doores, K. J.; Fulton, Z.; Hong, V.; Patel, M. K.; Scanlan, C. N.; Wormald, M. R.; Finn, M. G.; Burton, D. R.; Wilson, I. A.; Davis, B. G. *Proc. Natl. Acad. Sci. U. S. A.* **2010**, *107*, 17107–17112.

[103] Goebel, W. F.; Avery, O. T. *J. Exp. Med.* **1929**, *50*, 521–531.

[104] McBroom, C. R.; Samanen, C. H.; Goldstein, I. J.; Victor, G. In *Complex Carbohydrates Part B*; Ginsburg, V., Ed.; Methods in Enzymology 28; Academic: New York, **1972**, pp 212–219.

[105] Lehrer, R. I.; Jung, G.; Ruchala, P.; Andre, S.; Gabius, H. J.; Lu, W. *J. Immunol.* **2009**, *183*, 480–490.

[106] Avery, O. T.; Goebel, W. F. *J. Exp. Med.* **1929**, *50*, 533–550.

[107] Miller, N.; Williams, G. M.; Brimble, M. A. *Int. J. Pept. Res. Ther.* **2010**, *16*, 125–132.

[108] Dondoni, A. *Chem. Asian J.* **2007**, *2*, 700–708.

[109] Aragao-Leoneti, V.; Campo, V. L.; Gomes, A. S.; Field, R. A.; Carvalho, I. *Tetrahedron* **2010**, *66*, 9475–9492.

[110] Michael, A. *J. Prakt. Chem.* **1893**, *48*, 94–95.

[111] Dimroth, O. *Ber. Dtsch. Chem. Ges.* **1903**, *36*, 909–913.

[112] Huisgen, R. *Angew. Chem. Int. Ed.* **1963**, *2*, 565–598.

[113] Rostovtsev, V. V.; Green, L. G.; Fokin, V. V.; Sharpless, K. B. *Angew. Chem. Int. Ed.* **2002**, *41*, 2596–2599.

[114] Tornøe, C. W.; Christensen, C.; Meldal, M. *J. Org. Chem.* **2002**, *67*, 3057–3064.

[115] Wu, P.; Fokin, V. V. *Aldrichimica Acta* **2007**, *40*, 7–17.

[116] Meldal, M.; Tornøe, C. W. *Chem. Rev.* **2008**, *108*, 2952–3015.

[117] Gamblin, D. P.; Scanlan, E. M.; Davis, B. G. *Chem. Rev.* **2008**, *109*, 131–163.

[118] van Kasteren, S. I.; Kramer, H. B.; Jensen, H. H.; Campbell, S. J.; Kirkpatrick, J.; Oldham, N. J.; Anthony, D. C.; Davis, B. G. *Nature* **2007**, *446*, 1105–1109.

[119] van Kasteren, S. I.; Kramer, H. B.; Gamblin, D. P.; Davis, B. G. *Nat. Protoc.* **2007**, *2*, 3185–3194.

[120] Hong, V.; Steinmetz, N. F.; Manchester, M.; Finn, M. G. *Bioconjug. Chem.* **2010**, *21*, 1912–1916.

[121] Soriano-del-Amo, D.; Wang, W.; Jiang, H.; Besanceney, C.; Yan, A. C.; Levy, M.; Liu, Y.; Marlow, F. L.; Wu, P. *J. Am. Chem. Soc.* **2010**, *132*, 16893–16899.

[122] Besanceney-Webler, C.; Jiang, H.; Zheng, T.; Feng, L.; Soriano-del-Amo, D.; Wang, W.; Klivansky, L. M.; Marlow, F. L.; Liu, Y.; Wu, P. *Angew. Chem. Int. Ed.* **2011**, *50*, 8051–8056.

[123] Meier, H.; Petersen, H.; Kolshorn, H. *Chem. Ber.* **1980**, *113*, 2398–2409.

[124] Turner, R. B.; Jarrett, A. D.; Goebel, P.; Mallon, B. J. *J. Am. Chem. Soc.* **1973**, *95*, 790–792.

[125] Agard, N. J.; Prescher, J. A.; Bertozzi, C. R. *J. Am. Chem. Soc.* **2004**, *126*, 15046–15047.

[126] Chang, P. V.; Prescher, J. A.; Sletten, E. M.; Baskin, J. M.; Miller, I. A.; Agard, N. J.; Lo, A.; Bertozzi, C. R. *Proc. Natl. Acad. Sci. U. S. A.* **2010**, *107*, 1821–1826.

[127] Fairbanks, B. D.; Sims, E. A.; Anseth, K. S.; Bowman, C. N. *Macromolecules* **2010**, *43*, 4113–4119.

[128] Sen, G. S.; Kuzelka, J.; Singh, P.; Lewis, W. G.; Manchester, M.; Finn, M. G. *Bioconjug. Chem.* **2005**, *16*, 1572–1579.

[129] Wang, Q.; Chan, T. R.; Hilgraf, R.; Fokin, V. V.; Sharpless, K. B.; Finn, M. G. *J. Am. Chem. Soc.* **2003**, *125*, 3192–3193.

[130] Bundle, D. R.; Nycholat, C.; Costello, C.; Rennie, R.; Lipinski, T. *ACS Chem. Biol.* **2012**, *7*, 1754–1763.

[131] Astronomo, R. D.; Kaltgrad, E.; Udit, A. K.; Wang, S.-K.; Doores, K. J.; Huang, C.-Y.; Pantophlet, R.; Paulson, J. C.; Wong, C.-H.; Finn, M. G.; Burton, D. R. *Chem. Biol.* **2010**, *17*, 357–370.

[132] Kiick, K. L.; Saxon, E.; Tirrell, D. A.; Bertozzi, C. R. *Proc. Natl. Acad. Sci. U. S. A.* **2002**, *99*, 19–24.

[133] Deiters, A.; Cropp, T. A.; Mukherji, M.; Chin, J. W.; Anderson, J. C.; Schultz, P. G. *J. Am. Chem. Soc.* **2003**, *125*, 11782–11783.

[134] Calzadilla, M.; Malpica, A.; Cordova, T. *J. Phys. Org. Chem.* **1999**, *12*, 708–712.

[135] Geoghegan, K. F.; Stroh, J. G. *Bioconjug. Chem.* **1992**, *3*, 138–146.

[136] Li, X.; Zhang, L.; Hall, S. E.; Tam, J. P. *Tetrahedron Lett.* **2000**, *41*, 4069–4073.

[137] Gilmore, J. M.; Scheck, R. A.; Esser-Kahn, A. P.; Joshi, N. S.; Francis, M. B. *Angew. Chem. Int. Ed.* **2006**, *45*, 5307–5311.

[138] Robbins, J. B.; Kubler-Kielb, J.; Vinogradov, E.; Mocca, C.; Pozsgay, V.; Shiloach, J.; Schneerson, R. *Proc. Natl. Acad. Sci. U. S. A.* **2009**, *106*, 7974–7978.

[139] Cornish, V. W.; Hahn, K. M.; Schultz, P. G. *J. Am. Chem. Soc.* **1996**, *118*, 8150–8151.

[140] Liu, H.; Wang, L.; Brock, A.; Wong, C.-H.; Schultz, P. G. *J. Am. Chem. Soc.* **2003**, *125*, 1702–1703.

[141] Wu, P.; Shui, W.; Carlson, B. L.; Hu, N.; Rabuka, D.; Lee, J.; Bertozzi, C. R. *Proc. Natl. Acad. Sci. U. S. A.* **2009**, *106*, 3000–3005.

[142] Carrico, I. S.; Carlson, B. L.; Bertozzi, C. R. *Nat. Chem. Biol.* **2007**, *3*, 321–322.

[143] Wang, L.; Zhang, Z.; Brock, A.; Schultz, P. G. *Proc. Natl. Acad. Sci. U. S. A.* **2003**, *100*, 56–61.

[144] Pozsgay, V.; Vieira, N. E.; Yergey, A. *Org. Lett.* **2002**, *4*, 3191–3194.

[145] Berkin, A.; Coxon, B.; Pozsgay, V. *Chem. Eur. J.* **2002**, *8*, 4424–4433.

[146] Connon, S. J.; Blechert, S. *Angew. Chem. Int. Ed.* **2003**, *42*, 1900–1923.

[147] Fuerstner, A. *Angew. Chem. Int. Ed.* **2000**, *39*, 3012–3043.

[148] Hoveyda, A. H.; Zhugralin, A. R. *Nature* **2007**, *450*, 243–251.

[149] Lin, Y. A.; Chalker, J. M.; Davis, B. G. *ChemBioChem* **2009**, *10*, 959–969.

[150] Binder, J. B.; Raines, R. T. *Curr. Opin. Chem. Biol.* **2008**, *12*, 767–773.

[151] Lin, Y. A.; Chalker, J. M.; Floyd, N.; Bernardes, G. J. L.; Davis, B. G. *J. Am. Chem. Soc.* **2008**, *130*, 9642–9643.

[152] Chalker, J. M.; Wood, C. S. C.; Davis, B. G. *J. Am. Chem. Soc.* **2009**, *131*, 16346–16347.

[153] Spicer, C. D.; Davis, B. G. *Chem. Commun.* **2013**, *49*, 2747–2749.

[154] Trimble, R. B.; Atkinson, P. H.; Tarentino, A. L.; Plummer, T. H., Jr.; Maley, F.; Tomer, K. B. *J. Biol. Chem.* **1986**, *261*, 12000–12005.

[155] Wang, L.-X. *Carbohydr. Res.* **2008**, *343*, 1509–1522.

[156] Fernandez-Gonzalez, M.; Boutureira, O.; Bernardes, G. J. L.; Chalker, J. M.; Young, M. A.; Errey, J. C.; Davis, B. G. *Chem. Sci.* **2010**, *1*, 709–715.

[157] Lundquist, J. J.; Toone, E. J. *Chem. Rev.* **2002**, *102*, 555–578.

[158] Lee, Y. C.; Lee, R. T. *Acc. Chem. Res.* **1995**, *28*, 321–327.

[159] Mammen, M.; Chio, S.-K.; Whitesides, G. M. *Angew. Chem. Int. Ed.* **1998**, *37*, 2755–2794.

[160] Krishnamurthy, V. M.; Estroff, L. A.; Whitesides, G. M. In *Fragment-Based Approaches in Drug Discovery*; Jahnke, W., Erlanson, D. A., Eds.; Methods and Principles in Medicinal Chemistry 34; Wiley-VCH: Weinheim, **2006**, pp 11–53.

[161] Wadhwa, M. S.; Rice, K. G. *J. Drug Target.* **1995**, *3*, 111–127.

[162] Coutinho, A.; Moller, G. *Eur. J. Immunol.* **1973**, *3*, 608–613.

[163] Huang, Y.-L.; Wu, C.-Y. *Expert Rev. Vaccines* **2010**, *9*, 1257–1274.

[164] Costantino, P.; Rappuoli, R.; Berti, F. *Expert Opin. Drug Discov.* **2011**, *6*, 1045–1066.

[165] Liu, C.-C.; Ye, X.-S. *Glycoconj. J.* **2012**, *29*, 259–271.

[166] Toraño, G.; Toledo, M. E.; Baly, A.; Fernandez-Santana, V.; Rodriguez, F.; Alvarez, Y.; Serrano, T.; Musachio, A.; Hernandez, I.; Hardy, E.; Rodríguez, A.; Hernandez, H.; Aguilar, A.; Sánchez, R.; Diaz, M.; Muzio, V.; Dfana, J.; Rodríguez, M. C.; Heynngnezz, L.; Verez-Bencomo, V. *Clin. Vaccine Immunol.* **2006**, *13*, 1052–1056.

[167] Verez-Bencomo, V.; Fernández-Santana, V.; Hardy, E.; Toledo, M. E.; Rodrıguez, M. C.; Heynngnezz, L.; Rodriguez, A.; Baly, A.; Herrera, L.; Izquierdo, M.; Villar, A.; Valdes, Y.; Cosme, K.; Deler, M. L.; Montane, M.; Garcia, E.; Ramos, A.; Aguilar, A.; Medina, E.; Torano, G.; Sosa, I.; Hernandez, I.; Martinez, R.; Muzachio, A.; Carmenates, A.; Costa, L.; Cardoso, F.; Campa, C.; Diaz, M.; Roy, R. *Science* **2004**, *305*, 522–525.

[168] Seppala, I.; Makela, O. *J. Immunol.* **1989**, *143*, 1259–1264.

[169] Jones, P. S. *Antiviral Chem. Chemother.* **1998**, *9*, 283–302.

[170] Nagahori, N.; Lee, R. T.; Nishimura, S.-I.; Pagé, D.; Roy, R.; Lee, Y. C. *ChemBioChem* **2002**, *3*, 836–844.

[171] Ruhl, S.; Cisar, J. O.; Sandberg, A. L. *Infect. Immun.* **2000**, *68*, 6346–6354.

[172] Johansson, S. M. C.; Nilsson, E. C.; Elofsson, M.; Ahlskog, N.; Kihlberg, J.; Arnberg, N. *Antiviral Res.* **2007**, *73*, 92–100.

[173] Johansson, S.; Nilsson, E.; Qian, W.; Guilligay, D.; Crepin, T.; Cusack, S.; Arnberg, N.; Elofsson, M. *J. Med. Chem.* **2009**, *52*, 3666–3678.

[174] Ribeiro-Viana, R.; Sánchez-Navarro, M.; Luczkowiak, J.; Koeppe, J. R.; Delgado, R.; Rojo, J.; Davis, B. G. *Nat. Commun.* **2012**, *3*, 1303.

[175] Molema, G.; Jansen, R. W.; Visser, J.; Herdewijn, P.; Moolenaar, F.; Meijer, D. K. F. *J. Med. Chem.* **2001**, *34*, 1137–1141.

[176] Langer, R. *Science* **2001**, *293*, 58–59.

[177] Rensen, P. C. N.; van Dijk, M. C. M.; Havenaar, E. C.; Bijsterbosch, M. K.; Kruijt, J. K.; van Berkel, T. J. C. *Nat. Med.* **1995**, *1*, 221–225.

[178] Meijer, D. K. F.; Molema, G.; Moolenaar, F.; de, Z. D.; Swart, P. J. *J. Control. Release* **1996**, *39*, 163–172.

[179] Beljaars, L.; Molema, G.; Weert, B.; Bonnema, H.; Olinga, P.; Groothuis, G. M. M.; Meijer, D. K. F.; Poelstra, K. *Hepatology* **1999**, *29*, 1486–1493.

[180] Beljaars, L.; Olinga, P.; Molema, G.; De Bleser, P.; Geerts, A.; Groothuis, G. M. M.; Meijer, D. K. F.; Poelstra, K. *Liver* **2001**, *21*, 320–328.

[181] Fiume, L.; Di Stefano, G. *Eur. J. Pharm. Sci.* **2010**, *40*, 253–262.

[182] Franssen, E. J. F.; Jansen, R. W.; Vaalburg, M.; Meijer, D. K. F. *Biochem. Pharmacol.* **1993**, *45*, 1215–1226.

[183] Garnier, P.; Wang, X.-T.; Robinson, M. A.; van Kasteren, S.; Perkins, A. C.; Frier, M.; Fairbanks, A. J.; Davis, B. G. *J. Drug Target.* **2010**, *18*, 794–802.

[184] Lerchen, H.-G.; Baumgarten, J.; Piel, N.; Kolb-Bachofen, V. *Angew. Chem. Int. Ed.* **1999**, *38*, 3680–3683.

[185] Gabius, H. J.; Bokemeyer, C.; Hellmann, T.; Schmoll, H. J. *J. Cancer Res. Clin. Oncol.* **1987**, *113*, 126–130.

[186] Meier, O.; Greber, U. F. *J. Gene Med.* **2004**, *6*, S152–S163.

[187] Volpers, C.; Kochanek, S. *J. Gene Med.* **2004**, *6*, S164–S171.

[188] Cauet, G.; Strub, J.-M.; Leize, E.; Wagner, E.; Dorsselaer, A. V.; Lusky, M. *Biochemistry* **2005**, *44*, 5453–5460.

[189] Desnick, R. J.; Schuchman, E. H. *Annu. Rev. Genomics Hum. Genet.* **2012**, *13*, 307–335.

[190] Amalfitano, A.; Bengur, A. R.; Morse, R. P.; Majure, J. M.; Case, L. E.; Veerling, D. L.; Mackey, J.; Kishnani, P.; Smith, W.; McVie-Wylie, A.; Sullivan, J. A.; Hoganson, G. E.; Phillips, J. A., III; Schaefer, G. B.; Charrow, J.; Ware, R. E.; Bossen, E. H.; Chen, Y.-T. *Genet. Med.* **2001**, *3*, 132–138.

[191] Zhu, Y.; Li, X.; Kyazike, J.; Zhou, Q.; Thurberg, B. L.; Raben, N.; Mattaliano, R. J.; Cheng, S. H. *J. Biol. Chem.* **2004**, *279*, 50336–50341.

[192] Zhu, Y.; Li, X.; McVie-Wylie, A.; Jiang, C.; Thurberg, B. L.; Raben, N.; Mattaliano, R. J.; Cheng, S. H. *Biochem. J.* **2005**, *389*, 619–628.

[193] Zhou, Q.; Stefano, J. E.; Harrahy, J.; Finn, P.; Avila, L.; Kyazike, J.; Wei, R.; Van Patten, S. M.; Gotschall, R.; Zheng, X.; Zhu, Y.; Edmunds, T.; Pan, C. Q. *Bioconjug. Chem.* **2011**, *22*, 741–751.

[194] Chatterjee, D.; Douglas, J.; Cho, S.-N.; Rea, T.; Gelber, R.; Aspinall, G.; Brennan, P. *Glycoconj. J.* **1985**, *2*, 187–208.

[195] Zhang, J.; Chatterjee, D.; Brennan, P. J.; Spencer, J. S.; Liav, A. *Bioorg. Med. Chem. Lett.* **2010**, *20*, 3250–3253.

[196] Tiemeyer, M.; Yasuda, Y.; Schnaar, R. L. *J. Biol. Chem.* **1989**, *264*, 1671–1681.

[197] Tanaka, K.; Masuyama, T.; Hasegawa, K.; Tahara, T.; Mizuma, H.; Wada, Y.; Watanabe, Y.; Fukase, K. *Angew. Chem. Int. Ed.* **2008**, *47*, 102–105.

[198] Boutureira, O.; Bernardes, G. J. L.; D'Hooge, F.; Davis, B. G. *Chem. Commun.* **2011**, *47*, 10010–10012.

INDEX

Glycochemical Synthesis: Strategies and Applications, First Edition.
Edited by Shang-Cheng Hung and Medel Manuel L. Zulueta.

www.ingramcontent.com/pod-product-compliance
Lightning Source LLC
LaVergne TN
LVHW011857200926
848484LV00002B/466